Microscale Heat Transfer

Fundamentals and Applications

NATO Science Series

A Series presenting the results of scientific meetings supported under the NATO Science Programme.

The Series is published by IOS Press, Amsterdam, and Springer (formerly Kluwer Academic Publishers) in conjunction with the NATO Public Diplomacy Division.

Sub-Series

I. Life and Behavioural Sciences	IOS Press
II. Mathematics, Physics and Chemistry	Springer (formerly Kluwer Academic Publishers)
III. Computer and Systems Science	IOS Press
IV. Earth and Environmental Sciences	Springer (formerly Kluwer Academic Publishers)

The NATO Science Series continues the series of books published formerly as the NATO ASI Series.

The NATO Science Programme offers support for collaboration in civil science between scientists of countries of the Euro-Atlantic Partnership Council. The types of scientific meeting generally supported are "Advanced Study Institutes" and "Advanced Research Workshops", and the NATO Science Series collects together the results of these meetings. The meetings are co-organized by scientists from NATO countries and scientists from NATO's Partner countries — countries of the CIS and Central and Eastern Europe.

Advanced Study Institutes are high-level tutorial courses offering in-depth study of latest advances in a field.
Advanced Research Workshops are expert meetings aimed at critical assessment of a field, and identification of directions for future action.

As a consequence of the restructuring of the NATO Science Programme in 1999, the NATO Science Series was re-organized to the four sub-series noted above. Please consult the following web sites for information on previous volumes published in the Series.

http://www.nato.int/science
http://www.springeronline.com
http://www.iospress.nl

Series II: Mathematics, Physics and Chemistry – Vol. 193

Microscale Heat Transfer
Fundamentals and Applications

edited by

S. Kakaç
University of Miami,
Coral Gables, FL, U.S.A.

L.L. Vasiliev
Luikov Heat and Mass Transfer Institute,
Minsk, Belarus

Y. Bayazitoğlu
Rice University,
Houston, TX, U.S.A.

and

Y. Yener
Northeastern University,
Boston, MA, U.S.A.

Springer

Published in cooperation with NATO Public Diplomacy Division

Proceedings of the NATO Advanced Study Institute on
Microscale Heat Transfer – Fundamentals and Applications in Biological and Microelectromechanical Systems
Cesme-Izmir, Turkey
18–30 July 2004

A C.I.P. Catalogue record for this book is available from the Library of Congress.

ISBN 1-4020-3359-1 (HB)
ISBN 1-4020-3361-3 (e-book)

Published by Springer,
P.O. Box 17, 3300 AA Dordrecht, The Netherlands.

Sold and distributed in North, Central and South America
by Springer,
101 Philip Drive, Norwell, MA 02061, U.S.A.

In all other countries, sold and distributed
by Springer,
P.O. Box 322, 3300 AH Dordrecht, The Netherlands.

Printed on acid-free paper

Printed in the Netherlands.

Table of Contents

Preface

This volume contains an archival record of the NATO Advanced Institute on Microscale Heat Transfer – Fundamental and Applications in Biological and Microelectromechanical Systems held in Çesme – Izmir, Turkey, July 18–30, 2004. The ASIs are intended to be high-level teaching activity in scientific and technical areas of current concern. In this volume, the reader may find interesting chapters and various Microscale Heat Transfer Fundamental and Applications.

The growing use of electronics, in both military and civilian applications has led to the widespread recognition for need of thermal packaging and management. The use of higher densities and frequencies in microelectronic circuits for computers are increasing day by day. They require effective cooling due to heat generated that is to be dissipated from a relatively low surface area. Hence, the development of efficient cooling techniques for integrated circuit chips is one of the important contemporary applications of Microscale Heat Transfer which has received much attention for cooling of high power electronics and applications in biomechanical and aerospace industries. Microelectromechanical systems are subject of increasing active research in a widening field of discipline. These topics and others are the main theme of this Institute.

The scientific program starts with an introduction and the state-of-the-art review of single-phase forced convection in microchannels. The effects of Brinkman number and Knudsen numbers on heat transfer coefficient is discussed together with flow regimes in microchannel single-phase gaseous fluid flow and flow regimes based on the Knudsen number. In some applications, transient forced convection in microchannels is important.

Steady, periodic and transient-state convection heat transfer are analytically solved for laminar slip flow inside micro-channels formed by parallel-plates, making use of the generalized integral transform technique, Laplace transforms and the exact analytical solution of the corresponding eigenvalue problem in terms of the confluent hypergeometric functions. A mixed symbolic-numerical algorithm is developed under the *Mathematica* platform, allowing for the immediate reproduction of the results and comprehension of the symbolic and computational rules developed.

Analytical solutions for flow transients in microchannels are obtained, by making use of the integral transform approach, and mixed symbolic-numerical algorithm is constructed employing the *Mathematica* platform. The proposed model involves the transient fully developed flow equation for laminar regime and incomprehensible flow with slip at the walls, in either circular tubes or parallel plate channels. The solution is constructed so as to account for any general functional form of the time variation of the pressure gradient along the duct.

In several lectures discuss the measurements of single-phase pressure drop and heat transfer coefficient in micro and mini-channels. Experimental results of pressure drop and heat transfer coefficient of flow boiling are presented in mini-channels. Many correlations for flow boiling heat transfer coefficient in mini-channels have been established.

The nature of boiling heat transfer in a channel with the gap less than the capillary is also studied and presented. The condensation flow mechanisms, pressure drop and heat transfer in microchannels, role of microscale heat transfer in augmentation of nucleate boiling and flow boiling heat transfer, binary-fluid heat and mass transfers in microchannel geometries for miniaturized thermally activated absorption heat pumps, evaporation heat transfer on porous cylindrical tube disposed in a narrow channel, from macro to micro scale boiling are presented in several lectures.

In the applications, industrial heat exchanges are mini-and-microscale heat transfers, miniature and micro heat pipes and heat transfer issues in cryogenic catheters are presented. Nanotechnology and heat transfer including heat transfer characteristics of silicon film irradiated by pico to femtosecond lasers are also introduced and discussed.

During the ten working days of the Institute, the invited lecturers covered fundamentals and applications of Microscale Heat Transfer. The sponsorship of the NATO Scientific Affairs Division is gratefully acknowledged; in person we are very thankful to Dr. Fausto Pedrazzini director of ASI programs who continuously supported and encouraged us at every phase of our organization of this Institute. Our special gratitude goes to Drs. Nilufer Egrican, Hafit Yuncu, Sepnem Tavman and Ismail Tavman for coordinating sessions and we are very thankful to the Executive and Scientific Secretary Tuba Okutucu and to the Assistant Secretary Melda Koksal for their invaluable efforts in making the Institute a success. A word of appreciation is also due to the members of the session chairmen for their efforts in expediting the technical sessions. We are very grateful to Annelies Kersbergen of Kluwer Academic Publishers for her close collaboration in preparing this archival record of the Institute, to F. Arinc, Secretary General of ICHMT, to the General Scientific Coordinator of this NATO ASI Dr. Mila Avelino, Barbaros Cetin, Ozgur Bayer, Burak Yazicioglu, Cenk Kukrer and to Dr. Wei Sun and Mr. Christian Quintanilla-Aurich for their guidance and help during the entire process of the organization of the Institute. Finally our heartfelt thanks to all lecturers and authors, who provided the substance of the Institute, and to the participants for their attendance, questions and comments.

S. Kakaç
L. L. Vasiliev
Y. Bayazitoğlu,
Y. Yener

SINGLE-PHASE FORCED CONVECTION IN MICROCHANNELS
A State-of-the-Art Review

Y. YENER[1], S. KAKAÇ[2], M. AVELINO[2,3] and T. OKUTUCU[1]
[1] *Northeastern University, Boston, MA, 02115-5000, USA*
[2] *University of Miami, Coral Gables, FL, 33124-0611, USA*

1. Introduction

With the recent advances in microfabrication, various devices having dimensions of the order of microns such as, among others, micro-heat sinks, micro-biochips, micro-reactors for modification and separation of biological cells, micro-motors, micro-valves and micro-fuel cells have been developed. These found their applications in microelectronics, microscale sensing and measurement, spacecraft thermal control, biotechnology, microelectromechanical systems (MEMS), as well as in scientific investigations. The trend of miniaturization, especially in computer technology, has significantly increased the problems associated with overheating of integrated circuits (ICs). With existing heat flux levels exceeding $100\,\mathrm{W/cm^2}$, new thermal packaging systems incorporating effective thermal control techniques have become mandatory for such applications. The recent developments in thermal packaging have been discussed by Bar-Cohen [5], and experimental, as well as analytical methods have been reported by a number of researchers in Cooling of Electronic Systems, edited by Kakaç et al. [16].

The need for the development of efficient and effective cooling techniques for microchips has initiated extensive research interest in microchannel heat transfer. Microchannel heat sinks have been recommended to be the ultimate solution for removing high rates of heat in microscale systems. A microchannel heat sink is a structure with many microscale channels machined on the electrically inactive face of the microchip. The main advantage of microchannel heat sinks is their extremely high heat transfer area per unit volume. Since microchannels of noncircular cross sections are usually integrated in silicon-base microchannel heat sinks, it is important to know the fluid flow and heat transfer characteristics in these channels for better design of the systems. Moreover, the key design parameters like the pumping pressure for the coolant fluid, fluid flow rate, fluid and channel wall temperatures, channel hydraulic diameter and the number of channels in the sink have further to be optimized to make the system efficient and economical.

2. Motivations

The use of convective heat transfer in microchannels to cool microchips has been proposed over the last two decades. Many analytical and experimental studies, involving both liquids and gases, have been carried out to gain a better understanding of fluid flow and heat transfer phenomena at the micro level.

Experimental studies have demonstrated that many microchannel fluid flow and heat transfer phenomena cannot be explained by the conventional theories of transport theory, which are based on the continuum hypotheses. For friction factors and Nusselt numbers,

[3] *Mechanical Engineering Department – State University of Rio de Janeiro, 20560-013, Rio de Janeiro, RJ, BRAZIL* – **mila@uerj.br**

S. Kakaç et al. (eds.), Microscale Heat Transfer, 1–24.

there are a great deal of discrepancies between the classical values and the experimental data. For instance, the transition from laminar to turbulent flow starts much earlier than the classical limit (e.g. from Re=300); the correlations between the friction factor and the Reynolds number are very different from those predicted by the conventional theories of fluid mechanics; and the apparent viscosity and the friction factor of a liquid flowing through a microchannel may be several times higher than those in the conventional theories. Experimental data also appear to be inconsistent with one another. Such deviations are thought to be the results of the rarefaction and compressibility effects mainly due to the tiny dimensions of microchannels, the interfacial electrokinetic effects near the solid-fluid interface and various surface conditions, which cannot be neglected in microsystems because of the large surface-to-volume ratio in these systems. These effects significantly affect both the fluid flow and the convective heat transfer.

Typically, in macrochannels, fluid velocity and temperature are taken to be equal to the corresponding wall values. On the other hand, these conditions do not hold for rarefied gas flow in microchannels. For gas flow in microchannels, not only does the fluid slip along the channel wall with a finite tangential velocity, but there is also a jump between the wall and fluid temperatures. Several gaseous flow studies have been carried out for the slip flow conditions where, although the continuum assumption is not valid due to the rarefaction effects, Navier-Stokes equations were applied with some modifications in the boundary conditions. On the other hand, there does not seem to be a general consensus among the researchers regarding the boundary conditions for liquid flows. It is not clear if discontinuities of velocity and temperature exist on the channel walls.

Therefore, there is still a need for further research for a fundamental understanding of fluid flow and heat transfer phenomena in microchannels in order to explore and control the phenomena in a length scale regime in which we have very little experience.

3. Fluid Flow and Heat Transfer Modeling

There are basically two ways of modeling a flow field; the fluid is either treated as a collection of molecules or is considered to be continuous and indefinitely divisible - continuum modeling. The former approach can be of deterministic or probabilistic modeling, while in the latter approach the velocity, density, pressure, etc. are all defined at every point in space and time, and the conservation of mass, momentum and energy lead to a set of nonlinear partial differential equations (Navier-Stokes). Fluid modeling classification is depicted schematically in Fig. 1.

Navier-Stokes-based fluid dynamics solvers are often inaccurate when applied to MEMS.

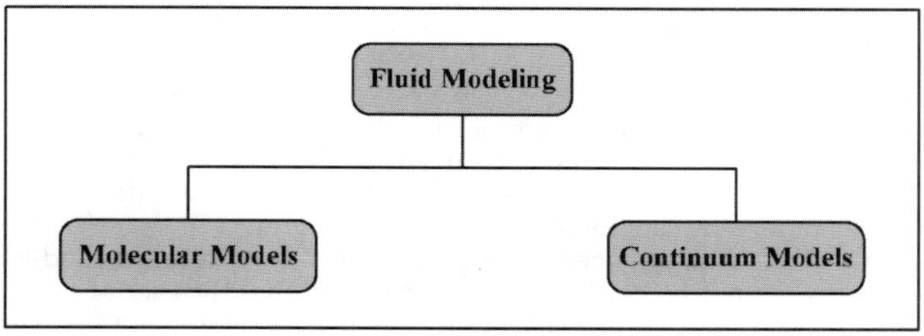

Figure 1. Classification of fluid modeling.

This inaccuracy stems from their calculation of molecular transport effects, such as viscous dissipation and thermal conduction, from bulk flow quantities, such as mean flow velocity and temperature. This approximation of microscale phenomena with macroscale information fails as the characteristic length of the (gaseous) flow gradients approaches the average distance travelled by molecules between collisions - the mean path. The ratio of these quantities is referred to as Knudsen number.

3.1. KNUDSEN NUMBER

The Knudsen number is defined as

$$\mathrm{Kn} = \frac{\lambda}{L}, \tag{1}$$

where L is a characteristic flow dimension (i.e. channel hydraulic diameter D_h) and λ is the mean free molecular path, which is given, for an ideal gas model as a rigid sphere, by

$$\lambda = \frac{\bar{k}T}{\sqrt{2}\,\pi P \sigma^2}. \tag{2}$$

Generally, the traditional continuum approach is valid, albeit with modified boundary conditions, as long as Kn< 0.1.

The Navier-Stokes equations are valid when λ is much smaller than the characteristic flow dimension L. When this condition is violated, the flow is no longer near equilibrium and the linear relations between stress and rate of strain and the no-slip velocity condition are no longer valid. Similarly, the linear relation between heat flux and temperature gradient and the no-jump temperature condition at a solid-fluid interface are no longer accurate when λ is not much smaller than L. The different Knudsen number regimes are delineated in Fig. 2.

For the small values (Kn$\leq 10^{-3}$), the flow is considered to be a continuum flow, while for large values (Kn≥ 10), the flow is considered to be a free-molecular flow. The range $10^{-3} <$Kn$<10^{-1}$ is the near continuum region.

The local value of Knudsen number determines the degree of rarefaction and the degree of validity of the continuum model in a particular flow. The different Knudsen number regimes depicted in Fig. 2 have been determined empirically and are therefore only approximate for a particular flow geometry. The pioneering experiments in rarefied gas dynamics were conducted by Knudsen in 1909 [24].

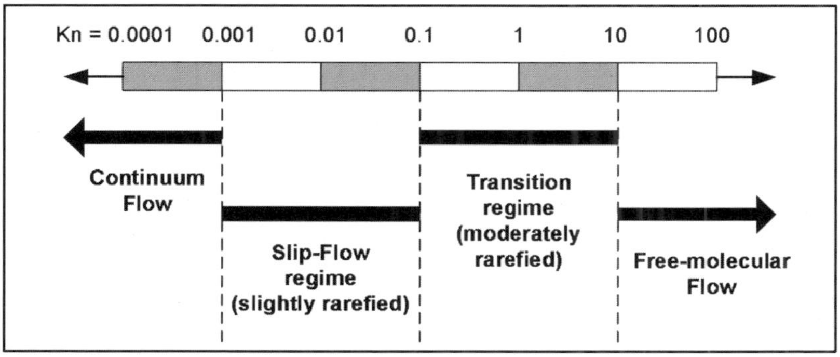

Figure 2. Knudsen number regimes.

Knudsen number can also be expressed as [12],

$$\mathrm{Kn} = \sqrt{\frac{\pi\gamma}{2}}\frac{\mathrm{Ma}}{\mathrm{Re}}, \tag{3}$$

where

$$\mathrm{Re} = \frac{V_0 L}{\nu} \tag{4}$$

is the Reynolds number and the Mach number is the ratio of the "characteristic" flow velocity to the speed of sound a_0,

$$\mathrm{Ma} = \frac{V_0}{a_0}. \tag{5}$$

The Mach number is a dynamic measure of fluid compressibility and may be considered as the ratio of the inertial forces to the elastic ones.

From the kinetic theory of gases, the mean molecular free path is related to the viscosity as follows

$$\nu = \frac{\mu}{\rho} = \frac{1}{2}\lambda \bar{V}_m, \tag{6}$$

where μ is the dynamic viscosity, and $\bar{V}_m$ is the mean molecular speed which is somewhat higher than the sound speed a_0,

$$\bar{V}_m = \sqrt{\frac{8}{\pi\gamma}}\, a_0. \tag{7}$$

4. Analysis

In this section we present the governing equations for the analysis of microchannel heat transfer in two-dimensional fluid flow. For steady two-dimensional and incompressible flow with constant thermophysical properties, the continuity, momentum and energy equations can be written in *Cartesian coordinates* as [17, 18]:

Continuity equation:

$$\frac{\partial u}{\partial x} + \frac{\partial v}{\partial y} = 0. \tag{8}$$

Momentum equations:

x-component:
$$u\frac{\partial u}{\partial x} + v\frac{\partial u}{\partial y} = -\frac{1}{\rho}\frac{\partial p}{\partial x} + \nu\left(\frac{\partial^2 u}{\partial x^2} + \frac{\partial^2 u}{\partial y^2}\right), \tag{9}$$

y-component:
$$u\frac{\partial v}{\partial x} + v\frac{\partial v}{\partial y} = -\frac{1}{\rho}\frac{\partial p}{\partial y} + \nu\left(\frac{\partial^2 v}{\partial x^2} + \frac{\partial^2 v}{\partial y^2}\right). \tag{10}$$

Energy equation:

$$u\frac{\partial T}{\partial x} + v\frac{\partial T}{\partial y} = \alpha\left(\frac{\partial^2 T}{\partial x^2} + \frac{\partial^2 T}{\partial y^2}\right) + \frac{1}{\rho c}\phi, \tag{11}$$

where ϕ is the viscous dissipation given by

$$\phi = 2\mu\left[\left(\frac{\partial u}{\partial x}\right)^2 + \left(\frac{\partial v}{\partial y}\right)^2 + \frac{1}{2}\left(\frac{\partial v}{\partial x} + \frac{\partial u}{\partial y}\right)^2 - \frac{1}{3}\left(\frac{\partial u}{\partial x} + \frac{\partial v}{\partial y}\right)^2\right] . \tag{12}$$

In *cylindrical coordinates* the governing equations, under the same conditions, are:

Momentum equations:

x-component:
$$u\frac{\partial u}{\partial x} + v\frac{\partial u}{\partial r} = -\frac{1}{\rho}\frac{\partial p}{\partial x} + \nu\left\{\frac{1}{r}\frac{\partial}{\partial r}\left(r\frac{\partial u}{\partial r}\right) + \frac{\partial^2 u}{\partial x^2}\right\} , \tag{13}$$

r-component:
$$u\frac{\partial v}{\partial x} + v\frac{\partial v}{\partial r} = -\frac{1}{\rho}\frac{\partial p}{\partial r} + \nu\left\{\frac{\partial}{\partial r}\left[\frac{1}{r}\frac{\partial}{\partial r}(rv)\right] + \frac{\partial^2 v}{\partial x^2}\right\} . \tag{14}$$

Energy equation:

$$u\frac{\partial T}{\partial x} + v\frac{\partial T}{\partial r} = \frac{\alpha}{r}\frac{\partial}{\partial r}\left(r\frac{\partial T}{\partial r}\right) + \alpha\frac{\partial^2 T}{\partial x^2} + \phi , \tag{15}$$

where

$$\phi = 2\mu\left\{\left(\frac{\partial v}{\partial r}\right)^2 + \left(\frac{v}{r}\right)^2 + \left(\frac{\partial u}{\partial x}\right)^2\right\} . \tag{16}$$

4.1. SPECIAL CASE

For steady and fully developed incompressible laminar flow with constant thermophysical properties through a parallel-plate microchannel, the continuity equation is automatically satisfied and the Navier-Stokes equations reduce to:

$$-\frac{1}{\rho}\frac{dp}{dx} + \nu\frac{d^2u}{dy^2} = 0 , \tag{17}$$

$$-\frac{1}{\rho}\frac{dp}{dy} = 0 . \tag{18}$$

The pressure must be constant across any cross-section perpendicular to the flow, thus

$$\frac{d^2u}{dy^2} = \frac{1}{\mu}\frac{dp}{dx} , \tag{19}$$

which gives the velocity profile between two parallel channel as

$$u = \frac{3}{2}u_m\left[1 - \left(\frac{y}{d}\right)^2\right] , \tag{20}$$

where $2d$ is the distance between the parallel plates.

The Energy equation, on the other hand, reduces to

$$u\frac{\partial T}{\partial x} = \alpha\frac{\partial^2 T}{\partial y^2} + \frac{\nu}{c_p}\left(\frac{du}{dy}\right)^2 , \tag{21a}$$

with the inlet and boundary conditions:

$$\text{at } x = 0: \qquad T = T_i\,, \tag{21b}$$

$$\text{at } y = d \quad \rightarrow \quad T = T_S\,, \tag{21c}$$

$$\text{at } y = 0 \quad \rightarrow \quad \frac{\partial T}{\partial y} = 0\,, \tag{21d}$$

where T_S is the slip temperature of the fluid, which is different from the wall temperature.

For steady and fully developed incompressible laminar flow with constant thermophysical properties through a microtube, the momentum energy equation reduce to:

$$\frac{1}{r}\frac{d}{dr}\left(r\frac{du}{dr}\right) = \frac{1}{\mu}\frac{dp}{dx}\,, \tag{22a}$$

$$\text{at } r = R \quad \rightarrow \quad u = u_S\,, \tag{22b}$$

$$\text{at } r = 0 \quad \rightarrow \quad u = \text{finite}, \tag{22c}$$

where u_S is the slip velocity.

The energy equation and the inlet and boundary conditions are given by

$$u\frac{\partial T}{\partial x} = \frac{\alpha}{r}\frac{\partial}{\partial r}\left(r\frac{\partial T}{\partial r}\right) + \frac{v}{c_p}\left(\frac{du}{dr}\right)^2, \tag{23a}$$

$$\text{at } x = 0: \qquad T = T_i\,, \tag{23b}$$

$$\text{at } r = R \quad \rightarrow \quad T = T_S\,, \tag{23c}$$

$$\text{at } r = 0 \quad \rightarrow \quad \frac{\partial T}{\partial r} = 0\,. \tag{23d}$$

4.2. BOUNDARY CONDITIONS

In microchannel heat transfer studies, the *no-slip* condition at a fluid-solid interface is enforced in the momentum equation, and an analogous *no-temperature-jump* condition is applied in the energy equation. The notion underlying the no-slip/no-jump condition is that within the fluid there cannot be any finite discontinuities of velocity/temperature. The interaction between a fluid particle and a wall is similar to that between neighboring fluid particles, and therefore no discontinuities are allowed at the fluid-solid interface either. In other words, the fluid velocity must be zero relative to the surface and the fluid temperature must be equal to that of the surface. But strictly speaking those two boundary conditions are valid only if the fluid flow adjacent to the surface is in thermodynamic equilibrium. This requires an infinitely high frequency of collisions between the fluid and the solid surface. In practice, the no-slip/no-jump condition leads to fairly accurate predictions for gases as long as Kn<0.001. Beyond that, the collision frequency is simply not high enough to ensure equilibrium and a certain degree of tangential velocity slip and temperature jump must be allowed.

4.2.1. Slip Velocity

In microchannels, the molecular mean free path, λ, becomes comparable with flow dimensions and the interactions between the fluid and the wall become more significant than intermolecular collisions in microchannels. When the gas molecules hit the surface, the molecules can be reflected either specularly or diffusely. In the case of specular reflection, the molecules will have the same tangential momentum. In the case of diffuse reflection, the tangential momentum balance at the wall yields the slip velocity as [21, 48]:

$$u_s = 2\frac{2-F_m}{F_m}\frac{\mu}{\rho\, u_m}\left(\frac{du}{dy}\right)_w , \tag{24}$$

where the viscosity is given by

$$\mu \cong \frac{1}{2}\rho\, u_m \lambda . \tag{25}$$

The slip velocity then becomes

$$u_s = \frac{2-F_m}{F_m}\lambda\left(\frac{du}{dy}\right)_w , \tag{26}$$

where F_m is the so-called tangential momentum accommodation coefficient which represents the fraction of the tangential momentum of the molecules given to the surface. In the case of an ideally perfect smooth surface at the molecular level, molecules will be reflected specularly, which means that the incident angle exactly equals the reflected angle. The molecules then conserve their tangential momentum exerting no shear on the wall, and thus $F_m = 0$. For diffuse reflection $F_m = 1$, which means that the tangential momentum is lost at the wall [6].

For real surfaces, some molecules reflect diffusively and some reflect specularly. In other words, a portion of the momentum of the incident molecules is lost to the wall and a typically smaller portion is retained by the reflected molecules. This coefficient depends on the fluid, the solid and the surface finish, and has been determined experimentally to be in the range 0.2-0.8. The lower limit is for exceptionally smooth surfaces, while the upper limit is typically for most practical surfaces.

4.2.2. Temperature Jump

In case of rarefied gas flow, there is a finite temperature difference between the wall temperature and the fluid temperature at the wall. A temperature jump coefficient has been proposed as:

$$c_j = \frac{T_s - T_w}{\left(\frac{\partial T}{\partial y}\right)_w} . \tag{27}$$

The thermal accommodation coefficient is defined as

$$F_T = \frac{E_a - E_l}{E_a - E_w} , \tag{28}$$

where E_a is the energy of the incoming stream, E_l is the energy carried away by the molecules leaving the surface, and E_w is the energy of the molecules leaving the surface at the wall temperature. Thus, $(E_a - E_l)$ is the net energy carried to the surface.

For a perfect gas, the temperature jump coefficient is obtained as [50, 51]:

$$c_j = \frac{2 - F_T}{F_T} \frac{1}{(\gamma + 1)} \frac{k\sqrt{2\pi RT}}{c_v P}, \tag{29}$$

or

$$c_j = \frac{2 - F_T}{F_T} \frac{2\gamma}{(\gamma + 1)} \frac{\lambda}{\mathrm{Pr}}, \tag{30}$$

The temperature jump can then be obtained as

$$T_S - T_W = \frac{2 - F_T}{F_T} \frac{2\gamma}{(\gamma + 1)} \frac{\lambda}{\mathrm{Pr}} \frac{\partial T}{\partial y}, \tag{31}$$

where y is measured from the wall.

For slip flow, the fully-developed velocity profile can be obtained from the momentum equations for laminar flow with constant thermophysical properties in a parallel-plate channel and a circular duct, respectively, as

$$u = \frac{3}{2} u_m \frac{\left[1 - \left(\frac{y}{d}\right)^2 + 4\mathrm{Kn}\right]}{1 + 8\mathrm{Kn}}, \tag{32}$$

and

$$u = 2u_m \frac{\left[1 - \left(\frac{r}{R}\right)^2 + 4\mathrm{Kn}\right]}{1 + 8\mathrm{Kn}}. \tag{33}$$

4.3. BRINKMAN NUMBER

The Brinkman number, Br, is defined by

$$\mathrm{Br} = \frac{\mu\, u_m^2}{k\Delta T}, \tag{34}$$

where ΔT is the wall-fluid temperature difference at a particular axial location. It measures the relative importance of viscous heating (work done against viscous shear) to heat conduction in the fluid along the microchannel. Although Br is usually neglected in low-speed and low-viscosity flows through conventionally-sized channels of short lengths, in flows through conventionally-sized long pipelines, Br may become important. For flows in microchannels, the length-to-diameter ratio can be as large as for flows through conventionally-sized long pipelines. Therefore, Br may become important in microchannels also.

Table 2 demonstrates the effects of the Knudsen and the Brinkman numbers on heat transfer in a tube flow. As it can be seen, the Nusselt number decreases with the increases in both the Brinkman number and the Knudsen number, since the increasing temperature jump decreases heat transfer. Also, under the constant wall temperature boundary conditions, the Nusselt numbers are greater than under constant heat flux boundary conditions when the Brinkman number is nonzero [51, 52].

Table 2: Nusselt Numbers for Developed Laminar Flow (q_w = Const., Pr = 0.6) [50].

Br = 0	Nu_q	Br = 0.01	Nu_q
Kn = 0.0	4.3627	Kn = 0.0	4.0353
Kn = 0.06	3.2519	Kn = 0.06	3.1833
Kn = 0.08	2.9487	Kn = 0.08	2.9052

5. Literature Survey

Following the recent developments in microfabrication, a number of major research initiatives have been launched to improve our understanding of the heat transfer and fluid flow phenomena at the micro level. A survey of the literature presented below gives a brief summary of the research carried out in single-phase forced convection in microchannels mostly in the last 15-20 years.

In early 1980s, Tuckerman and Pease [48, 49] investigated the problem of achieving compact, high-performance forced liquid cooling of planar integrated circuits. They demonstrated that the water-cooled microchannels fabricated on the circuit board on which the chips are mounted are capable of dissipating 790 W/cm^2 without a phase change and with a maximum substrate temperature rise of 71°C above the inlet water temperature. Their results also indicated that the heat transfer coefficient for laminar flow through microchannels might be higher than that for turbulent flow through conventionally sized channels. Since then there has been an unprecedented upsurge of research in convection through microchannels.

Shortly after the initial work of Tuckerman and Pease [48, 49], Wu and Little [56, 57] conducted experiments to measure the flow friction and heat transfer characteristics of gases flowing in the trapezoidal silicon/glass microchannels of widths 130 to 200 μm and depths of 30 to 60 μm, and found that convective heat transfer characteristics departed from the typical experimental results for conventionally sized channels. Their measurements, which involved both laminar and turbulent flow regimes, indicated a transition from laminar to turbulent flow at Reynolds numbers of 400-900 depending on the different test configurations. They reported that the reduction in the transition Reynolds number resulted in improved heat transfer. In addition, they found that, unlike in conventional channel flow, the surface roughness affected the values of friction factors even in the laminar flow regime and that the frictional pressure drop for laminar flow was higher than the classical prediction.

Samalam [43] modeled the convective heat transfer in water flowing through microchannels etched in the back of silicon wafers. The problem was reduced to a quasi-two dimensional non-linear differential equation under certain reasonably simplified and physically justifiable conditions, and was solved exactly. The optimum channel dimensions (width and spacing) were obtained analytically for a low thermal resistance. The calculations show that optimizing the channel dimensions for low aspect ratio channels is much more important than for large aspect ratios. However, a crucial approximation that the fluid thermophysical properties are independent of temperature was made, which could be a source of considerable error, especially in microchannels with heat transfer.

Aul and Olbricht [4] reported the results of an experimental study of low-Reynolds number, pressure-driven core-annular flow in a straight capillary tube. The annular film was thin compared to the radius of the tube, and the viscosity of the film fluid was much larger than the viscosity of the core fluid. The photographs showed that the film was unstable under all conditions investigated. It was found that the film fluid collects in axisymmetric lobes that are periodically spaced along the capillary wall. Eventually, the continued growth of the lobes results in the formation of a fluid lens that breaks the inner core.

Pfahler et al. [35] presented the results for friction factor measurements from an experimental investigation of fluid flow, N-propanol as the primary working liquid, in three extremely small channels of rectangular cross-section ranging in area from 80 to 7200 μm^2. Their objective was to determine at what length scales the continuum assumptions break down and to estimate the adequacy of the Navier-Stokes equations for predicting fluid behavior. They found that in the relatively large flow channels their observations were in rough agreement with the predictions from the Navier-Stokes equations. However, in the smallest of the channels, they observed a significant deviation from the Navier-Stokes predictions. Pfahler et al. [36] later conducted a series of experiments to measure friction factor for both liquids (isopropyl alcohol and silicone oil) and gases (nitrogen and helium) in small channels etched in silicon with depths ranging from 0.5 to 50 μm. For both liquids and gases, they obtained smaller friction factor values than those predicted by conventional, incompressible theory. Isopropyl alcohol results showed a dependency on the channel size. Silicone oil results, on the other hand, showed a Reynolds number dependency. They concluded that the small friction factor values for liquids were due to the reduction in viscosity with decreasing size, and for gases due to the rarefaction effects.

Choi et al. [9] measured the friction factors and the convective heat transfer coefficients for both the laminar and turbulent flow regimes for flow of nitrogen in microtubes of inside diameters ranging from 3 to 81 μm. The length/diameter ratio for the tubes was between 640 (81 μm tube) and 8100 (3 μm tube), so the flow was fully-developed both hydraulically and thermally. The microtubes had relative roughness values between 0.00017 and 0.0116, and absolute roughness (rms) between 10 nm and 80 nm. Their experimental results indicated significant departures from the correlations used for conventional-sized tubes. The measured friction factors in laminar flow were found to be less than those predicted from the macro tube correlation, and the friction factors in turbulent flow were also smaller than those predicted by conventional correlations. The measured heat transfer coefficients in laminar flow exhibited a Reynolds number dependence, in contrast to the conventional prediction for fully-established laminar flow, in which the Nusselt number is constant. In turbulent flow in microtubes, the measured heat transfer coefficients were larger than predicted by conventional correlations for smooth macrotubes. Neither the Colburn analogy nor the Petukov analogy between momentum and energy transport were supported by their data in microtubes. The measured Nusselt numbers in turbulent flow were as much as seven times larger than the values predicted by the Colburn analogy. They suggested the suppression of the turbulent eddy motion in the radial direction (but not in the axial direction) due to the small diameter of the channel as one reason for this result.

Weisberg et al. [54] are among other researchers who all provided additional information and considerable evidence that the behavior of fluid flow and heat transfer in microchannels or microtubes without phase change is substantially different from that which occurs in large channels and/or tubes.

Experimental measurements for pressure drop and heat transfer coefficient were made by Rahman [40]. Tests were performed on channels of different depths and using water as the working fluid. The fluid flow rate as well as the pressure and temperature of the fluid at the inlet and outlet of the device were measured. These measurements were used to calculate local and average Nusselt numbers and coefficients of friction in the device for different flow rates, channel size and configurations.

Designing small-scale fluid flow devices demands clarification of fluid dynamics on the order of 0.1-100 μm. Makihara et al. [27] described the flow of liquids in 4.5 - 50.5 μm micro-capillary tubes and developed a method of measuring it. They found that the measured values agree with the theoretical values calculated by the Navier-Stokes equations.

In an attempt to clarify some of the questions surrounding this issue, Peng et al. [30, 31], Wang and Peng [53], Peng et al. [32], and Peng and Peterson [33, 34] investigated microchannels and microchannel structures. Peng et al. [30, 31] measured both the flow friction and the heat transfer for single-phase convection of water through rectangular microchannels having hydraulic diameters of 0.133-0.367 mm and aspect ratios of H/W =0.333-1. Their measurements of both flow friction and heat transfer indicated that the laminar heat transfer ceased at a Reynolds number of 200-700, and that the fully turbulent convective heat transfer was reached at Reynolds numbers of 400-1,500. They observed that the transition Reynolds number diminished with the reduction in microchannel dimensions, and that the transition range became smaller in magnitude. For the laminar regime, the Nusselt number was found to be proportional to $Re^{0.62}$, while the turbulent heat transfer was shown to exhibit a typical relationship between Nu and Re numbers, but with a different empirical coefficient. The geometric parameters, especially the hydraulic diameter and aspect ratio, were found to be important variables having a significant effect on the flow and heat transfer. Their experiments demonstrated that the laminar convective heat transfer had a maximum value when the aspect ratio was approximately equal to 0.75, and, even at these conditions, small changes in the hydraulic diameter resulted in significant variations in the heat transfer. For turbulent conditions, the heat transfer approached an optimum value when the aspect ratio was in the range 0.5-0.75. They suggested new empirical correlations for the prediction of heat transfer based on their experimental data.

Wang and Peng [53] also experimentally studied the forced flow convection of liquids (water and methanol) in microchannels of rectangular cross-section. They found that the fully-developed turbulent convection was initiated at Reynolds numbers in the range of 1000-1500, and that the conversion from the laminar to transition region occurred in the range of 300-800. In addition, they showed that the turbulent heat transfer can be predicted by the Dittus-Boelter correlation by modifying the empirical constant coefficient from 0.023 to 0.00805. They also observed that the heat transfer behavior in the laminar and transition regions was quite unusual and complicated, and was strongly influenced by liquid temperature, velocity and microchannel size.

Peng et al. [32] experimentally analyzed the influence of liquid velocity, subcooling, property variations and microchannel geometric configuration on the heat transfer characteristics and cooling performance of methanol flowing through rectangular-shaped microchannels of different aspect ratios and a variety of center-to-center spacings. They found that for single-phase flow through the microchannels a transition region exists beyond which the heat transfer coefficient is nearly independent of the wall temperature and that the transition is a function of the heating rate or wall temperature conditions within the microchannel itself. Moreover, they noted that this transition was also a direct result of large temperature rise in the microchannels which caused significant variations in the liquid thermophysical properties and, hence, significant increases in the relevant flow parameters, such as the Reynolds number. As a result, the liquid velocity and subcooling were found to be very important parameters in determining the point or region where this transition occurs.

Peng and Peterson [33] later confirmed these experimental observations using methanol flowing through similar microchannel structures and analyzed experimentally the effects of the thermofluid and the geometric variables on heat transfer. They presented evidence to support the existence of an optimum channel size in terms of the forced convection of a single-phase liquid flowing in a rectangular microchannel.

Peng and Peterson [34] experimentally investigated the single-phase forced convective heat transfer and flow characteristics of water in microchannel plates with extremely small rectangular channels having hydraulic diameters of 0.133-0.367 mm and different geometric configurations. Their measurements indicated that the geometric configuration of the microchannel plate and individual microchannels had critical effects on the single phase convective heat transfer, and that the effects on the laminar and turbulent convection were quite different. They noted that, while the thermal conductivity of the material from which the plates were fabricated could be a factor, the microchannels were so small that the hydraulic radius was comparable to the sublayer thickness and, therefore, the resistance in the sublayer for the cases considered became much more important than for larger conventional channels. Accordingly, for channels as small as they evaluated, they concluded that the shape of the channels plays a negligible role for both the laminar and turbulent flow conditions. They found that the laminar heat transfer, however, did depend on the aspect ratio and the ratio of the hydraulic diameter to the center-to-center distance of the microchannels. It was also found that the turbulent heat transfer was further a function of a new dimensionless variable, Z, such that $Z = 0.5$ is the optimum configuration regardless of the groove aspect ratio. In addition, they suggested empirical correlations for predicting the heat transfer for both laminar and turbulent cases.

Beskok and Karniadakis [7] numerically simulated the time-dependent slip flow in complex microgeometries. The numerical scheme was based on a spectral element method they developed for flows in macrogeometries. A higher order velocity slip condition was used in the analysis. The method was verified by comparing it to the analytical solutions for simple cases. They noted the importance of the accommodation coefficient. Although the Knudsen number is small, a small value of the momentum accommodation coefficient would result in large slip velocities at the wall. Compressibility effects were also addressed especially for the cases where severe pressure drops occur. In another study, Beskok et al. [8], focused on the competing effects of compressibility and rarefaction for Knudsen numbers up to 0.3. The higher order velocity slip and temperature jump boundary conditions were modified for the numerical stability purposes. Viscous heating and thermal creep were found to be important mechanisms at the microscale. Viscous heating can result in considerable temperature gradients. They concluded that compressibility was important for pressure driven flows and rarefaction was important for shear driven flows.

Kleiner et al. [23] theoretically and experimentally investigated forced air-cooling, which employs microchannel parallel plate fin heat sinks and tubes. Optimization was performed and design trade-off was studied. Tube sizes were observed to have a significant impact on optimum heat sink design. Air-cooled heat sinks are used for micro channel heat sinks with heat loads less than 100 W/cm^2.

Yu et al. [59] experimentally investigated the flow of dry nitrogen gas and water in microtubes with diameters 19, 52, and 102 μm, for Re range 250-20,000, and for Pr range 0.7-5.0. They found a reduction in the friction factor in the turbulent regime, and that the heat transfer coefficient h was enhanced. The Reynolds analogy was found inapplicable in channels whose dimensions were of the order of the turbulent length scale, though the fluid could still be treated as a continuum. Their theoretical scaling analysis indicated the turbulent momentum and energy transport in the radial direction to be significant in the near-wall zone. They developed an analogy by considering the turbulent eddy interacting with the walls as a frequent event, thereby causing a direct mass and thermal energy transfer process between the turbulent lumps and the wall, similar to the eddies bursting phenomenon. This phenomenon significantly alters the laminar sublayer region in turbulent flows through microtubes. Since even a small eddy diffusivity in the laminar sublayer region can contribute significantly to the heat transfer rate while having a negligible effect on momentum transfer, an eddy can carry heat to a greater distance; hence, the increased h and lower friction factors in turbulent flows through microtubes.

A heat transfer analysis was performed by Gui and Scaringe [13] based on the data from Rahman and Gui's [40] experiment where they used water and refrigerants to determine the cooling capacity of a silicon chip and obtained $106\,\mathrm{W/m^2}$ heat dissipation. They found the laminar-to-turbulent transition Reynolds number as 1400 instead of 2300 for macro dimensions. They ascribed this to the surface roughness. Their analytical values were always smaller than the experimental results. They listed the reasons for more efficient heat transfer as: the reduced thermal boundary layer thickness, entrance effects-higher heat transfer at the channel inlet, pre-turbulence at the inlet and surface roughness.

Choquette et al. [10] performed analyses to obtain momentum and thermal characteristics in microchannel heat sinks. A computer code was developed to evaluate the performance capabilities, power requirements, efficiencies of heat sinks, and for heat sink optimization. Significant reductions in the total thermal resistance were found not to be achieved by designing for turbulent flows, mainly due to the significantly higher pumping power requirements realized, which offset the slight increase in the thermal performance.

Gaseous flow in microchannels was experimentally analyzed by Shih et al. [44] with helium and nitrogen as the working fluids. Mass flow rate and pressure distribution along the channels were measured. Helium results agreed well with the results of a theoretical analysis using slip flow conditions, however there were deviations between theoretical and experimental results for nitrogen.

Hydrodynamically fully-developed laminar gaseous flow in a cylindrical microchannel with constant heat flux boundary condition was considered by Ameel et al. [2]. In this work, two simplifications were adopted reducing the applicability of the results. First, the temperature jump boundary condition was actually not directly implemented in these solutions. Second, both the thermal accommodation coefficient and the momentum accommodation coefficient were assumed to be unity. This second assumption, while reasonable for most fluid-solid combinations, produces a solution limited to a specified set of fluid-solid conditions. The fluid was assumed to be incompressible with constant thermophysical properties, the flow was steady and two-dimensional, and viscous heating was not included in the analysis. They used the results from a previous study of the same problem with uniform temperature at the boundary by Barron et al. [6]. Discontinuities in both velocity and temperature at the wall were considered. The fully developed Nusselt number relation was given by

$$\mathrm{Nu} = \frac{48(2\beta-1)^2}{(24\beta^2-16\beta+3)\left[1+\dfrac{24\gamma(\beta-1)(2\beta-1)^2}{(24\beta^2-16\beta+3)(\gamma+1)\,\mathrm{Pr}}\right]} \tag{35}$$

where $\beta = 1 + 4\mathrm{Kn}$. It was noted here that, for Kn=0, in other words for the no-slip condition, the above equation gives Nu=4.364, which is the well-known Nusselt number for conventionally sized channels [18]. The Nusselt number was found to be decreasing with increasing Kn. Over the slip flow regime, Nu was reduced by about 40%. A similar decay was also observed for the gas mixed mean temperature. They determined that the entrance length increases with increasing rarefaction, which means that thermally fully developed flow is not obtained as quickly as in conventional channels. The following formula shows the relationship between the entrance length and the Knudsen number

$$x_e^* = 0.0828 + 0.14\,\mathrm{Kn}^{0.69} \tag{36}$$

Kavehpour et al. [20] solved the compressible two-dimensional fluid flow and heat transfer characteristics of a gas flowing between two parallel plates under both uniform temperature and uniform heat flux boundary conditions. They compared their results with the experimental results of Arkilic [3] for Helium in a 52.25x1.33x7500 mm channel. They observed an increase in the entrance length and a decrease in the Nusselt number as Kn takes higher values. It was found that the effects of compressibility and rarefaction is a function of Re. Compressibility is significant for high Re and rarefaction is significant for low Re.

Mala et al. [28] have investigated possible importance of the interfacial effects of the electrical double layer (EDL) at the solid-liquid interface (which is formed due to the electrostatic charges on the solid surface) on convective heat transfer and liquid flow in microchannels. They have solved the momentum and energy equations numerically for a steady hydrodynamically-developed and thermally-developing flow, considering the electrical body force resulting from the double layer field. They found that the EDL modifies the velocity profile and reduces the average velocity, thereby increasing the pressure drop and reducing the heat transfer rate. They reported that the EDL thickness ranges from a few nanometers to several hundreds of nanometers, and calculated the effect on a microchannel separation distance of 25 μm. As this is an order of magnitude smaller than the channels used in the reported experimental investigations, the true influence of EDL on the convective heat transfer is uncertain. Moreover, the EDL effects do not exist if the walls of the microchannel are conducting materials, which is the case for the reported experimental observations. Hence, the EDL effects cannot explain the unusual behavior of convective heat transfer and flow transitions observed in experiments in microchannels. Mala et al. [28] found that with water as the working fluid, the difference between the measured pressure drop per channel length and the correlation from conventional correlation was small for microtube diameter more than 1.5mm. Mala et al. [28] also conducted experiments on the EDL field. They found that the EDL results in a lower velocity of flow than in conventional theory, thus affecting the temperature distribution and reducing the Reynolds number. It is seen that without the EDL , a higher heat transfer is predicted.

Randall et al. [41] studied the classical problem of thermally developing heat transfer in laminar flow through a circular tube considering the slip-flow condition. They extended the original problem to include the effect of slip-flow, which occurs in gases at low pressures or in microtubes at ordinary pressures. A special technique was developed to evaluate the eigenvalues for the problem. Eigenvalues were evaluated for Knudsen numbers ranging between 0 and 0.12. Simplified relationships were developed to describe the effect of slip-flow on the convection heat transfer coefficient.

Adams et al. [1] have experimentally investigated the single-phase turbulent forced convection of water flowing through circular microchannels with diameters of 0.76 and 1.09 mm. Their data showed that the Nusselt numbers for the microchannels are higher than those predicted by the traditional correlations for turbulent flows in the conventionally-sized channels, such as the Gnielinski correlation. Their data suggested that the extent of enhancement in the convection increased as the channel diameter decreased and the Reynolds number increased. To accommodate this enhancement, the Gnielinski correlation was modified from a least squares fit of a combination of their experimental data and the data reported for small diameter channels. This modified correlation is applicable when the diameter is in the range 0.102 - 1.09 mm, the Reynolds number is in the range 2600 - 23000, and the Prandtl number is in the range 1.53 - 6.43.

Tso and Mahulikar [46, 47] proposed the use of the Brinkman number to explain the unusual behaviors in heat transfer and flow in microchannels. A dimensional analysis was made by the Buckingham π theorem. The parameters that influence heat transfer were determined by a survey of the available experimental data in the literature as thermal conductivity, density, specific heat and viscosity of the fluid, channel dimension, flow velocity and temperature difference between the fluid and the wall. The analysis led to the Brinkman number. They also reported that viscous dissipation determines the physical limit to the channel size reduction, since it will cause an increase in fluid temperature with decreasing channel size. They explained the reduction in the Nusselt number with the increase in the Reynolds number for the laminar flow regime by investigating the effect of viscosity variation on the Brinkman number. It was also found that, the variation of viscosity with temperature is beneficial to the heat transfer since it improves the heat transfer capacity. On the other hand, viscous dissipation is less important in the transition regime since the steep velocity gradients no longer exist. In their second paper [47], they investigated the effect of the Brinkman number on determining the flow regime boundaries in microchannels, and found that Br plays a more important role in the laminar-to-transition boundary than in the transition-to-turbulent boundary.

Xu et al. [58] investigated both experimentally and analytically laminar water flow in microchannels with diameters between 50 and 300 micrometers and Reynolds numbers between 50 and 1500. They found that the results deviated from Navier-Stokes predictions for diameters less than 100 μm. They also found that this deviation was not dependent on the Reynolds number. They proposed the use of a slip boundary condition, which estimated the velocity of the fluid at the wall by the velocity gradient at the wall. In doing so, they obtained an agreement between the theoretical findings and the experimental results, although they recognized the need for more experimental data to further understand the underlying physics at this scale.

Mahulikar [26] studied the role of the Brinkman number Br in microchannel flows to correlate the forced convective heat transfer in the laminar and transition regimes and hence explained the unusual behavior of convective heat transfer in microchannels. A dimensional analysis indicated that the Nusselt number in the laminar and transition regimes in microchannels correlates with Br in addition to Re, Pr, and a dimensionless geometric parameter of the microchannel. It was noted that the effect of Br on convective heat transfer is more in the laminar regime, than in the transition regime. In the turbulent regime Br is insignificant. The incorporation of Br in the correlations for the laminar and transition regimes explained the unusual behavior of Nu receding with increasing Re in the laminar regime and the approximately constant Nu in the transition regime.

Between the slip and the transition flow regimes, where most MEMS applications can be found, direct simulation Monte Carlo (DSMC) offers an alternative. The advantage of DSMC is that it makes no continuum assumption. Instead, it models the flow as it physically exists: a collection of discrete particles, each with a position, a velocity, an internal energy, a species identity, etc. These particles are allowed to move and interact with the domain boundaries over small time steps during the calculation. Intermolecular collisions are all performed on a probabilistic basis. Macro quantities, such as flow velocity and temperature, are then obtained by sampling the microscopic state of all particles in the region of interest. It is shown that DMCS has ability to calculate microflows in any of the four Knudsen number regions without modification. This is particularly valuable in simulating flows with different regimes.

In the study of Fan, et al. [11], a numerical simulation of gaseous flows in microchannels by the DSMC was carried out. Several unique features were obvious: to maintain a constant mass flow, the mean streamwise velocity at the walls was found to increase to make up for the density drop caused by the pressure decrease in the flow direction, which is in contrast to the classical Poisueile flow. In addition, the velocities at the walls were found to be nonzero and to increase in the streamwise direction, which highlights the slip-flow effect due to rarefaction. The results of the DSMC simulations were validated by an analytical solution in the slip regime. It was observed that the two results showed remarkable agreements.

Iwai and Suzuki [15] numerically investigated the effects of rarefaction and compressibility on heat transfer for a flow over a backward-facing step in a microchannel duct. They applied the velocity slip boundary condition to all the walls and considered temperature jump at the heated wall. Skin friction was seen to reduce when the velocity slip was taken into account. It was further reduced if the accommodation coefficient takes smaller values, which results in larger slip velocities. They found that the compressibility effects are significant for microchannel flows with flow separation and reattachment, which become more important as Kn becomes larger. Compressibility increases the Nusselt number due to the increase in the temperature difference between fluid and the wall since the thermal energy is converted into the kinetic energy. They also stated that there was not a significant effect of temperature jump on the Nusselt number distribution under the simulation conditions.

Convective heat transfer analysis for the calculation of the constant-wall-heat-flux Nusselt number for fully-developed gaseous flow in two-dimensional microchannels was performed by Hadjiconstantinou [14]. A Knudsen number range of 0.06-1.1 was considered. Since in this range the flow is in the transition regime, the continuum assumption is not valid. Accordingly, the DSMC technique was implemented. The channels considered had a length/height ratio of 20 to ensure fully developed flow, and care was taken to ensure that the Brinkman number is always small. It was concluded that the slip flow prediction is valid for Knudsen numbers less than 0.1. The results showed a reduction in Nusselt number with increasing rarefaction (Knudsen number). The effects of thermal creep were also discussed.

Larrodé et al. [25] studied heat convection in gaseous flows in circular tubes in the slip-flow regime with uniform temperature boundary condition. The effects of the degree of rarefaction and the gas-surface interaction properties, as determined by corresponding accommodation coefficients were investigated. The temperature jump at the tube wall, ignored in previous investigations, was taken into account, and was found to be of essential importance in the heat transfer analysis. A spatial scaling factor ρ_s^*, which is given by

$$\rho_s^* = \frac{1}{1 + 4\beta_v \mathrm{Kn}} \tag{37}$$

was introduced to recast the problem as a classical Graetz problem with mixed boundary condition. The scaling factor ρ_s^* incorporates both rarefaction effects through its dependence on the Knudsen number and gas–surface interaction properties through β_v, which is related to the tangential momentum accommodation coefficient α_m by

$$\beta_v = \frac{2 - \alpha_m}{\alpha_m} \tag{38}$$

A novel uniform asymptotic approximation to high–order eigenfunctions was derived that allowed an efficient and accurate determination of the region close to the entrance. The effect of the temperature jump at the wall was determined to be essential in the heat transfer analysis. In addition, it was shown that heat transfer increases or decreases with increasing rarefaction depending on whether $\beta_v < 1$ or $\beta_v > 1$, respectively. On the other hand, for a given Knudsen number (fixed degree of rarefaction) heat transfer decreases with increasing β_v. It was also noted that, under slip–flow conditions, gradients at the wall are smaller than in continuum flow due to the velocity slip and the temperature jump.

Kim et al. [22] modeled microchannel heat sinks as porous structures, while studying the forced convective heat transfer through the microchannels. From the analytical solution, the Darcy number and the effective thermal conductivity ratio were identified as variables of engineering importance.

Qu et al. [37, 38] performed an experimental investigation on pressure drop and heat transfer of water in trapezoidal silicon microchannels with a hydraulic diameters ranging from 62 to 169 μm. They also carried out a numerical analysis by solving a conjugate heat transfer problem involving simultaneous determination of the temperature field in both the solid and the fluid regions. They found that the experimentally determined Nusselt number was much lower than that predicted by their numerical analysis. They attributed the measured higher pressure drops and lower Nusselt numbers to the wall roughness, and proposed a roughness-viscosity model to interpret their experimental data. According to their model, however, the increase in wall roughness caused the decrease in the Nusselt number, which is contradictory to common sense.

Tunc and Bayazitoglu [50, 51] studied, by the integral transform technique, convective heat transfer for steady–state and hydrodynamically–developed laminar flow in microtubes with both uniform temperature and uniform heat flux boundary conditions. Temperature jump condition at the tube wall and viscous heating within the medium were included in the study. The solution method was verified for the cases where viscous heating is neglected. For the uniform temperature case, with a given Brinkman number, the viscous effects on the Nusselt number were presented at specified axial lengths in the developing range, reaching the fully–developed Nusselt number. The effect of viscous heating was also investigated for the cases where the fluid was both heated and cooled. A Prandtl number analysis showed that, as the Prandtl number was increased the temperature jump effect diminished which gave a rise to the Nusselt number. Tunc and Bayazitoglu [52] also investigated convective heat transfer in a rectangular microchannel with a both thermally and hydrodynamically fully–developed laminar flow and with constant axial and peripheral heat flux boundary conditions. Since the velocity profile for a rectangular channel is not known under the slip flow conditions, the momentum equation was first solved for the velocity. The resulting velocity profile was then substituted into the energy equation. The integral transform technique was applied twice, once for the velocity and once for the temperature. The results showed a similar behavior to previous studies on circular microtubes. The Nusselt numbers were presented for varying aspect ratios.

Yu and Ameel [60] studied laminar slip-flow forced convection in rectangular microchannels analytically by applying a modified generalized integral transform technique to solve the energy equation for hydrodynamically fully–developed flow. Results were given for the fluid mixed mean temperature, and for both the local and fully–developed mean Nusselt numbers. Heat transfer was found to increase, decrease, or remain unchanged, compared to non-slip-flow conditions, depending on the two dimensionless variables that include effects of rarefaction and the fluid/wall interaction. The transition point at which the switch from heat transfer enhancement to reduction occurs was identified for different aspect ratios.

Toh et al. [45] investigated numerically three-dimensional fluid flow and heat transfer phenomena inside heated microchannels. The steady, laminar flow and heat transfer equations were solved using a finite-volume method. The numerical procedure was validated by comparing the predicted local thermal resistances with available experimental data. The friction factor was also predicted in this study. It was found that the heat input lowers the frictional losses, particularly at lower Reynolds numbers. Also, at lower Reynolds numbers the temperature of the water increases, leading to a decrease in the viscosity and hence smaller frictional losses.

Qu and Mudawar [39] analyzed numerically the three-dimensional fluid flow and heat transfer in a rectangular microchannel heat sink consisting of a 1-cm^2 silicon wafer and using water as the cooling fluid. The micro-channels had a width of 57 μm and a depth of 180 μm, and were separated by a 43 μm wall. A numerical code based on the finite–difference method and the SIMPLE algorithm was developed to solve the governing equations. The code was carefully validated by comparing the predictions with analytical solutions and available experimental data. For the microchannel heat sink investigated, it was found that the temperature rise along the flow direction in the solid and fluid regions can be approximated as linear. The highest temperature was encountered at the heated base surface of the heat sink immediately above the channel outlet. The heat flux and Nusselt number had much higher values near the channel inlet and varied around the channel periphery, approaching zero in the corners. It was also found that flow Reynolds number affects the length of the flow developing region. For a relatively high Reynolds number of 1400, fully–developed flow may not be achieved inside the heat sink. Increasing the thermal conductivity of the solid substrate reduces the temperature at the heated base surface of the heat sink, especially near the channel outlet. It was further observed that although the classical fin analysis method provides a simplified means to modeling heat transfer in microchannel heat sinks, some key assumptions introduced in the fin method deviate significantly from the real situation, which may compromise the accuracy of this method.

Maynes and Webb [29] analyzed thermally fully developed, electro-osmotically generated convective transport in a parallel–plate microchannel and circular microtube analytically under imposed constant wall heat flux and constant wall temperature boundary conditions. Such a flow is established not by an imposed pressure gradient, but by a voltage potential gradient along the length of the channel or the tube. The result is a combination of unique electro-osmotic velocity profiles and volumetric heating in the fluid due to the imposed voltage gradient. The exact solutions for the fully–developed, dimensionless temperature profile and the corresponding Nusselt number were determined for both geometries and both thermal boundary conditions. The fully-developed temperature profile and the Nusselt number were found to depend on the relative duct radius (ratio of the Debye length to duct radius or plate gap half-width) and the magnitude of the dimensionless volumetric source.

Ryu and Kim [42] developed a robust three-dimensional numerical procedure for the thermal performance of a manifold microchannel heat sink and applied it to optimize the heat-sink design. The system of fully elliptic equations, which govern the flow and thermal fields, was solved by a SIMPLE–type finite volume method, while the optimal geometric shape was traced by a steepest descent technique. For a given pumping power, the optimal design variables that minimize the thermal resistance were obtained iteratively, and the optimal state was reached within six global iterations. Comparing with the comparable traditional microchannel heat sink, the thermal resistance was reduced by more than a half, while the temperature uniformity over the heated wall was improved by tenfold. The sensitivity of the thermal performance on each design variable was also examined and presented in the paper. It was demonstrated that, among various design variables, the channel width and depth are more crucial than others to the heat-sink performance, and the optimal dimensions and the corresponding thermal resistance have a power-law dependence on the pumping power.

More recently, Wu and Cheng [55] carried out an experimental investigation on the laminar convective heat transfer and pressure drop of (deionized) water in 13 different trapezoidal silicon microchannels having different geometric parameters, surface roughness, and surface hydrophilic properties. They found that the values of the laminar Nusselt number and apparent friction constant depend greatly on different geometric parameters (i.e. the bottom-to-top width ratio, the height-to-top width ratio, and the length-to-diameter ratio). The Nusselt number and the apparent friction constant both increase with the increase in surface roughness. They also increase with the increase in surface hydrophilic property; that is, the Nusselt number and the apparent friction constant in trapezoidal microchannels having strong hydrophilic surfaces (thermal oxide surfaces) are larger than those in microchannels having weak hydrophilic surfaces (silicon surfaces). These increases in the Nusselt number and the apparent friction constant become more obvious at larger Reynolds numbers. Moreover, the fact that the Nusselt number and the apparent friction constant both increase with the increase in surface hydrophilic property suggests that heat transfer can be enhanced by increasing the surface hydrophilic capability at the expense of increasing pressure drop. The experimental results also showed that the Nusselt number increases almost linearly with the Reynolds number at low Reynolds numbers (Re $<$ 100), but increases slowly at Reynolds numbers greater than 100. Based on 168 experimental data points, Wu and Cheng [55] further developed dimensionless correlations for the Nusselt number and the apparent friction constant. They also presented an evaluation of heat flux per pumping power and per temperature difference for the microchannels used in the experiment. A comparison of their results shows that the geometric parameters have more significant effect on the performance of the 13 microchannels than the surface roughness and the surface hydrophilic property.

A NATO Advanced Study Institute was held, between July 18 - 30, 2004, in Çeşme–İzmir, Türkiye to discuss the fundamentals and applications of microscale heat transfer in biological and microelectromechanical systems. During the institute, the most recent state-of-the-art developments have been presented in considerable depth by eminent researchers in the field. This current volume, edited by Kakaç et al. [19] brings together the important contributions from the institute as a permanent reference for the use of researchers in the field.

6. Summary of the Conclusions from Literature Survey

A number of heat transfer and fluid transport issues at the microscale surveyed can be summarized as follows:

- Convective heat transfer in microchannels is significantly enhanced, depending on the values of the Knudsen, the Prandtl and the Brinkman numbers and the aspect ratio. Heat transfer characteristics can be significantly different from conventionally sized channels.

- Convective heat transfer in liquids flowing through microchannels has been extensively experimented to obtain the characteristics in the laminar, transitional, and turbulent regimes. The observations, however, indicate significant departures from the classical correlations for the conventionally sized tubes, which have not been explained.

- Experimental investigations on convective heat transfer in liquid flows in microchannels have been in the continuum regime. Hence, the conventional Navier-Stokes equations are applicable.

- The geometric parameters of individual rectangular microchannels, namely the hydraulic diameter and the aspect ratio, and the geometry of the microchannel plate have significant influence on the single-phase convective heat transfer characteristics.

- Significant reductions in the total thermal resistances are not achieved in turbulent flow through microchannels mainly because the significantly higher pumping power requirements offset the slight improvement in the overall thermal performance. This highlights the importance of laminar flow in microchannels design considerations.

- Velocity slip and temperature jump affect the heat transfer in opposite ways: a large slip on the wall increases the convection along the surface. On the other hand, a large temperature jump decreases the heat transfer by reducing the temperature gradient at the wall. Therefore, neglecting temperature jump will result in the overestimation of the heat transfer coefficient.

- Reduction in Nusselt number is observed as the flow deviates from the continuum behavior, or as Kn takes higher values.

- For the reported experiments, the heat transfer coefficient h is representative of the entire length of the microchannels, calculated either at the downstream end of the microchannels, or based on the bulk mean wall-fluid temperature difference over the entire length of the microchannels.

- Correlations for single-phase forced convection in the laminar regime have not been reported for the parameters obtained locally and along the flow.

- For fully-developed laminar forced convection in microchannels, Nu is proportional to $\mathrm{Re}^{0.62}$, while for the fully-developed turbulent heat transfer Nu is predicted by the Dittus-Boelter correlation by modifying only the empirical constant coefficient from 0.023 to 0.00805.

- In the laminar and transition regimes in microchannels, the behavior of convective heat transfer coefficient is very different compared with the conventionally-sized situation. In the laminar regime, Nu decreases with increasing Re, which has not been explained.

- In microchannels, the flow transition point and range are functions of the heating rate or the wall temperature conditions. The transitions are also a direct result of the large liquid temperature rise in the microchannels, which causes significant liquid thermophysical property variations and, hence, significant increases in the relevant flow parameters, such as the Reynolds number. Hence, the transition point and range are affected by the liquid temperature, velocity, and geometric parameters of the microchannel.

- The unusual behavior of Nu decreasing with increasing Re in the laminar regime in microchannels may alter the status of thermal development and hence the conventional thermal entry length, since the variation of the heat transfer coefficient along the flow is a variation of the boundary condition.

- The effect of any variation of the boundary condition on thermal entry length has not been explained.

- The Nu in the laminar and transition regimes in microchannels is correlated with Br, in addition to Re, Pr, and a geometric parameter of the microchannels. The role of Br in the laminar regime is supported by an analysis of the experimental data.

- From an analysis of the experimental data, Br is found to be another dimensionless parameter in determining the flow regime boundaries from laminar-to-transition and from transition-to-turbulent, in addition to Re. The Re, however, has a higher role relative to Br. The role of Br relative to Re in determining the laminar-to-transition boundary is higher than its relative role in determining the transition-to-turbulent boundary.

- Since the ratio of surface area to volume is large, viscous heating is an important factor in microchannels. It is especially important for laminar flow, where considerable gradients exist. The Brinkman number, Br, indicates this effect. A decrease in Nu for $\mathrm{Br} > 0$ and an increase for $\mathrm{Br} < 0$ have been observed. This is due to the fact that for different cases, Br may increase or decrease the driving mechanism for convective heat transfer, which is the difference between the wall temperature and the average fluid temperature.

- Prandtl number is important, since it directly influences the magnitude of the temperature jump. As Pr increases, the difference between the wall and the fluid temperatures at the wall decreases, resulting in greater Nu values.

NOMENCLATURE

a_0	speed of sound, m/s
Br	Brinkman number, Ec/Pr, $\mu u_m^2/k\Delta T$
c_p	specific heat at constant pressure, J/kg·K
c_v	specific heat at constant volume, J/kg·K
d	one-half channel width, m
D	tube diameter, m
D_h	hydraulic diameter, m
Ec	Eckert number, $u_m^2/c_p\Delta T$
F_m	tangential momentum accommodation coefficient
F_T	thermal accommodation coefficient
Gz	Graetz number, $\mathrm{Re}\,\mathrm{Pr}\,D_h/L$
h	heat transfer coefficient, W/m^2·K
k	thermal conductivity, W/m·K
$\bar{k}$	Boltzman constant, 1.3806×10^{-23}J/K
Kn	Knudsen number, λ/D_h
L	channel length, m
Ma	Mach number, V_0/a_0
Nu	Nusselt number, hD_h/k
P, p	pressure, Pa
Pr	Prantl Number, ν/α
r	radial coordinate, m
R	tube radius, m
Re	Reynolds number, $\rho D_h u_m/\mu$
T	fluid temperature, K
T_i	fluid inlet temperature, K
T_s	slip temperature, K
ΔT	wall-fluid temperature difference, K
u	axial velocity, m/s
u_m	mean velocity, m/s
u_s	slip velocity, m/s
v	velocity in y-direction, m/s
V_m	mean molecular speed, m/s
V_o	characteristic flow velocity, m/s
x	axial coordinate, m
y	transverse coordinate, m

Greek Symbols

α	thermal diffusivity, m^2/s
γ	specific heat ratio
λ	mean free path, m
μ	dynamic viscosity, kg/m·s
ν	kinematic viscosity, m^2/s
ρ	density, kg/m^3
σ	molecular diameter, m
ϕ	heat dissipation, W/m^3

REFERENCES

1. Adams, T.M., Abdel-Khalik, S.I., Jeter, S.M. and Qureshi, Z.H., An Experimental Investigation of Single-Phase Forced Convection in Microchannels, *Int. J. Heat Mass Transfer*, 1998, 41, 851-857.
2. Ameel, T.A., Wang, X., Barron, R.F. and Warrington, R.O., Laminar Forced Convection in a Circular Tube with Constant Heat Flux and Slip Flow, *Microscale Thermophys. Eng*, 1(4), 1997, 303-320.
3. Arkilic, E.B., Breuer, K.S. and Schmidt, M.A., Gaseous Flow in Microchannels, *Application of Microfabrication to Fluid Mechanics*, ASME FED-197, 1994, 57-66.
4. Aul, R.W. and Olbricht, W.L., Stability of a Thin Annular Film in Pressure Driven, Low-Reynolds-Number Flow Through a Capillary, *Journal of Fluid Mechanics*, 1990, 215, 585-599.

5. Bar-Cohen, A., State of the Art and Trends in the Thermal Packaging of The Electronic Equipment, *ASME Journal of Electronic Packaging*, 1992, 114, 257-270.
6. Barron, R.F, Wang, X. Ameel, T.A. and Warrington, R.O., The Graetz Problem Extended to Slip-Flow, *Int. J. Heat Mass Transfer*, 1997, 40(8), 1817-1823.
7. Beskok, A. and Karniadakis, G.E., Simulation of Heat and Momentum Transfer in Complex Micro Geometries, *J. Thermophysics and Heat Transfer*, 1994, 8(4), 647-655.
8. Beskok, A., Karniadakis, G.E. and Trimmer, W., Rarefaction, Compressibility and Thermal Creep Effects in Micro-Flows, *Proceedings of the ASME Dynamic Systems and Control Division*, 1995, DSC-57-2, 877-892.
9. Choi, S.B., Barron R.F. and Warrington R.O., Fluid Flow and Heat Transfer in Microtubes, *Micromechanical Sensors, Actuators, and Systems*, ASME DSC-32, 1991, 123-134.
10. Choquette, S.F., Faghri, M., Channchi, M. and Asako, Y., Optimum Design of Microchannel Heat Sinks, *ASME Microelectromechanical Systems*, 1996, DSC-59, 115-126.
11. Fan, Q., Xue, H. and Shu, C., DSMC Simulation of Gaseous Flows in Microchannels, *5th ASME/JSME Thermal Engineering Joint Conference*, San Diego, U.S.A, AJTE99-6519, 1999.
12. Gad-El-Hak, M., Momentum and Heat Transfer in MEMS, *Congrs francais de Thermique, SFT*, Grenoble, France, 3-6 June, Elsevier, 2003.
13. Gui, F. and Scaringe, R.P., Enhanced Heat Transfer in the Entrance Region of Microchannels, *IECEC Paper No. ES-40*, ASME, 1995, 289-294.
14. Hadjiconstantinou, N.G., Convective Heat Transfer in Micro and Nano Channels: Nusselt Number Beyond Slip Flow, *Proceedings of the ASME Heat Transfer Division*, 2000, 366-2, 13-22.
15. Iwai, H. and Suzuki, K., Effects of Velocity Slip and Temperature Jump Conditions on Backward-Facing Step Flow in a Microchannel, *Proceedings of the 5th ASME/JSME Joint Thermal Engineering Conference*, 1999, 1-8.
16. Kakaç, S., Yüncü, H. and Hijkata, K., (eds.), Cooling of Electronic Systems, *NATO ASI series E.*, 1992, 258, Kluwer, The Netherlands.
17. Kakaç, S. and Yener, Y. Heat Conduction, 3rd ed., *Taylor & Francis*, 1993.
18. Kakaç, S. and Yener, Y. Convective Heat Transfer, *CRC Press*, 2nd ed., 1995.
19. Kakaç, S., Vasiliev, L.L., Bayazıtoğlu, Y. and Yener, Y., (eds.), *Microscale Heat Transfer – Fundamentals and Applications*, 2005, Kluwer, The Netherlands.
20. Kavehpour, H.P., Faghri, M. and Asako, Y., Effects of Compressibility and Rarefaction on Gaseous Flows in Microchannels, *Numerical Heat Transfer*, 1997, Part A, 32, 677-696.
21. Kennard, E.H., Kinetic Theory of Gases, *McGraw-Hill Book Company, Inc.*, New York, 1938.
22. Kim, S.J., Kim. D. and Lee, D.Y., On the Thermal Equilibrium in Microchannel Heat Sinks, *Int. J. Heat Mass Transfer*, 2000, 43, 1735-1748.
23. Kleiner, M.B., Kuehn, S.A. and Haberger, K., High Performance Forced Air Cooling Scheme Employing Microchannel Heat Exchangers, *IEEE Transactions on Components, Packaging, and Manufacturing Technology*, Part A, 18, Issue 4, 1995, 795-804.
24. Knudsen, M., Die gesetze der molekularstrmung und der inneren reibungsstrmung der gase durch rhren, *Annalen der Physik*, 28, 1909, 75-130.
25. Larrodé, F.E., Housiadas, C. and Drossinos, Y., Slip Flow Heat Transfer in Circular Tubes, *Int. J. Heat Mass Transfer*, 2000, 43, 2669-2680.

26. Mahulikar, S.P., PhD Thesis, Heat Transfer Studies in Microchannels, School of Mechanical and Production Engineering, Nanyang Technological University, Singapore, 1999.
27. Makihara, M., Sasakura, K. and Nagayama, A., The Flow of Liquids in Micro-Capillary Tubes - Consideration to Application of the Navier-Stokes Equations, *Journal of the Japan Society of Precision Engineering*, 1993, 59(3), 399-404.
28. Mala, G.M., Li, D. and Dale, J.D., Heat Transfer and Fluid Flow in Microchannels, *Int. J. Heat Mass Transfer*, 1997, 40(13), 3079-3088.
29. Maynes, D. and Webb, B.W., Fully Developed Electro-Osmotic Heat Transfer in Microchannels, *Int. J. Heat Mass Transfer*, 2003, 46, 1359-1369.
30. Peng, X.F., Peterson, G.P. and Wang, B.X., Frictional Flow Characteristics of Water Flowing Through Micro-Channels, *Experimental Heat Transfer*, 1994, 7, 249-264.
31. Peng, X.F., Peterson, G.P., and Wang, B.X., Heat Transfer Characteristics of Water Flowing Through Micro-Channels, *Experimental Heat Transfer*,1994, 7, 265-283.
32. Peng, X.F., Wang, B.X., Peterson, G.P. and Ma, H.B., Experimental Investigation of Heat Transfer in Flat Plates with Rectangular Microchannels, *Int. J. Heat Mass Transfer*, 1995, 38(1), 127-137.
33. Peng, X.F. and Peterson, G.P., The Effect of Thermofluid and Geometrical Parameters on Convection of Liquids Through Rectangular Microchannels, *Int. J. Heat Mass Transfer*, 1995, 38(4), 755-758.
34. Peng, X.F. and Peterson, G.P., Convective Heat Transfer and Flow Friction for Water Flow in Microchannel Structures, *Int. J. Heat Mass Transfer*, 1996, 39(12), 2599-2608.
35. Pfahler, J., Harley, J., Bau, H. and Zemel, J., Liquid Transport in Micron and Submicron Channels, *Sensors and Actuators*, A21-A23, 1990, 431-434.
36. Pfahler, J., Harley, J., Bau, H. and Zemel, J., Gas and Liquid Flow in Small Channels, *Micromechanical Sensors, Actuators, and Systems*, ASME DSC-32, 1991, 49-60.
37. Qu, W., Mala, G.M. and Li, D., Pressure-driven Water Flows in Trapezoidal Silicon Microchannels, *Int. J. Heat Mass Transfer*, 2000, 43, 353-364.
38. Qu, W., Mala, G.M. and Li, D., Heat Transfer for Water Flow in Trapezoidal Silicon Microchannels, *Int. J. Heat Mass Transfer*, 2000, 43, 3925-3936.
39. Qu, W. and Mudawar, I., Analysis of Three-Dimensional Heat Transfer in Micro-Channel Heat Sinks, *Int. J. Heat Mass Transfer*, 2002, 45, 3973-3985.
40. Rahman, M.M. and Gui, F., Experimental Measurements of Fluid Flow and Heat Transfer in Microchannel Cooling Passages in A Chip Substrate, *Advances in Electronic Packaging*, ASME EEP-4-2, 1993, 685-692.
41. Randall, F.B., Wang, X. and Ameel, T.A, The Graetz Problem Extended to slip flow, *Int. J. Heat Mass Transfer*, 1997, 40, 1817-1823.
42. Ryu, J.H., Choi, D.H. and Kim, S.J., Three-Dimensional Numerical Optimization of a Manifold Microchannel Heat Sink, *Int. J. Heat Mass Transfer*, 2003, 46, 1553-1562.
43. Samalam, V.K., Convective Heat Transfer in Microchannels, *Journal of Electronic Materials*, 1989, 18(5), 611-617.
44. Shih, J.C., Ho, C., Liu, J. and Tai, Y., Monatomic and Polyatomic Gas Flow Through Uniform Microchannels, Micro Electro Mechanical Systems (MEMS), *National Heat Transfer Conference*, DSC 59, 1996, 197-203.
45. Toh, K.C., Chen, X.Y. and Chai, J.C., Numerical Computation of Fluid Flow and Heat Transfer in Microchannels, *Int. J. Heat Mass Transfer*, 2002, 45, 5133-5141.
46. Tso, C.P. and Mahulikar, S.P., The Use of the Brinkman Number for Single Phase Forced Convective Heat Transfer in Microchannels, *Int. J. Heat Mass Transfer*, 1998, 41(12), 1759-1769.

47. Tso, C.P. and Mahulikar, S.P., The Role of the Brinkman Number in Analysing Flow Transitions in Microchannels, *Int. J. Heat Mass Transfer*, 1999, 42, 1813-1833.
48. Tuckerman, D.B. and Pease, R.F.W., *IEEE Electron Device Letter*, 2(5), 1981, 126-129.
49. Tuckerman, D.B. and Pease, R.F.W., Optimized Convective Cooling Using Micromachined Structure, *J. Electrochem. Soc.*, 1982, 129(3),98C.
50. Tunc, G. and Bayazitoglu, Y., Heat Transfer for Gaseous Flow in Microtubes with Viscous Heating, *Proceedings of the ASME Heat Transfer Division*, HTD 366-2, 2000, 299-306.
51. Tunc, G. and Bayazitoglu, Y., Heat Transfer in Microtubes with Viscous Dissipation, *Int. J. Heat Mass Transfer*, 2001, 44, 2395-2403.
52. Tunc, G. and Bayazitoglu, Y., Heat Transfer in Rectangular Microchannels, *Int. J. Heat Mass Transfer*, 2002, 45, 765-773.
53. Wang, B.X. and Peng, X.F., Experimental Investigation of Liquid Forced-Convection Heat Transfer Through Microchannels, *Int. J. Heat Mass Transfer*, 1994, 37, Suppl. 1, 73-82.
54. Weisberg, A., Bau, H.H. and Zemel, J.N., Analysis of Microchannels for Integrated Cooling, *Int. J. Heat Mass Transfer*, 1992, 35(10), 2465-2474.
55. Wu, H.Y. and Cheng, P., An Experimental Study of Convective Heat Transfer in Silicon Microchannels with Different Surface Conditions, *Int. J. Heat Mass Transfer*, 2003, 46, 2547-2556.
56. Wu, P.Y. and Little, W.A., Measurement of Friction Factor For The Flow of Gases in Very Fine Channels Used For Microminiature Joule-Thompson Refrigerators, *Cryogenics*, 1983, 23(5), 273-277.
57. Wu, P.Y. and Little, W.A., Measurement of Heat Transfer Characteristics of Gas Flow in Fine Channels Heat Exchangers Used For Microminiature Refrigerators, *Cryogenics*, 1984, 24(5), 415-420.
58. Xu, B., Ooi, K.T., Wong, N.T., Liu, C.Y., and Choi, W.K., Liquid Flow in Microchannels, *Proceedings of the 5th ASME/JSME Joint Thermal Engineering Conference*, 1999, 1-7.
59. Yu, D., Warrington, R., Barron, R. and Ameel, T., An Experimental and Theoretical Investigation and Heat Transfer in Microtubes, *Proceedings of ASME/JSME Thermal Engineering Conference 1*, 1995, 523-530.
60. Yu, S. and Ameel, T.A., Slip-Flow Heat Transfer in Rectangular Microchannels, *Int. J. Heat Mass Transfer*, 2001, 44, 4225-4234.

MEASUREMENTS OF SINGLE-PHASE PRESSURE DROP AND HEAT TRANSFER COEFFICIENT IN MICRO AND MINICHANNELS

A. BONTEMPS
Université Joseph Fourier
LEGI/GRETh - CEA Grenoble, 17 avenue des martyrs, 38 054 Grenoble cedex 9, France
andre.bontemps@cea.fr

1. Introduction

The development of MEMS technology during the 80s induced a strong research effort focused on fluid and heat flow studies in microchannels. Since then, various silicon-based systems such as micro-biochips, MOEMS, etc... have contributed to reinforce this trend and a lot of experimental results were published. In parallel to these studies, very compact heat exchangers for air conditioning purposes were developed and have lead to research programs on minichannels. In the same manner, the possible use of such minichannels in other systems such as reformers, fuel cells,... has also produced considerable interest in this field.
To design the microdevices accurately, it is necessary to know the behaviour of the fluid flow and the values of the transport parameters as precisely as possible. Unfortunately, many contradictory results have been published in the literature concerning pressure drop and heat transfer coefficients in micro- and minichannels. As an example, the values of the normalized friction factors $f_{exp}/f_{classical}$ collected by Papautsky *et al.* [1] are presented on figure 1. The "exp" and "classical" indices refer to experimental and theoretical values respectively. The theoretical values are those given by the classical theories of fluid flow. It can be seen that the coefficients can be either higher or lower than the classical ones with no apparent preference. If there are some difficulties to sort out accurate values and to find out general correlations, it is possible to observe a general trend of published measurement values. The variation of the ratio $Nu_{exp}/Nu_{classical}$ and $f_{exp}/f_{classical}$ together with their uncertainties as a function of time are given in figure 2 [2]. As can be seen, it seems that these ratios tend to one, showing that it may be sufficient to apply the classical theories.
However, it is important to know (i) how the scale effects can affect the application of these theories, and (ii) if other proposed theories are effective in minichannels (characteristic dimension between 0.2 to 3 mm) and even microchannels (characteristic dimension between few μm to 200 μm).
All references of published works will not be given here since several relevant reviews have already been presented by many authors. One can cite the first concerning Microfluidics by Gravesen *et al.* in 1993 [3]. More recently Sabry studied the scale effects on fluid flow an heat transfer in micro channels [4], Gad-el-Hak [5], Obot [6], Palm [7], Celata [8] and Kandlikar [9] made syntheses of existing results in microdevices, micropipes and/or in microchannels. A discussion is also developed by Mehendale *et al.* [10] who underline the need for experiments which can reconcile diverging results.
This article presents a selected review of the main results on pressure drop and heat transfer coefficient measurements with liquid flows and underlines the discrepancies or the agreements with classical theories. The theoretical arguments set out by the authors to explain these discrepancies are also presented. To prevent ambiguity, experimental results are presented wherein the evident effects of experimental errors have been removed. From these results a tentative critical analysis is proposed to facilitate the choice of experiment interpretations and/or system designs.

S. Kakaç et al. (eds.), Microscale Heat Transfer, 25–48.

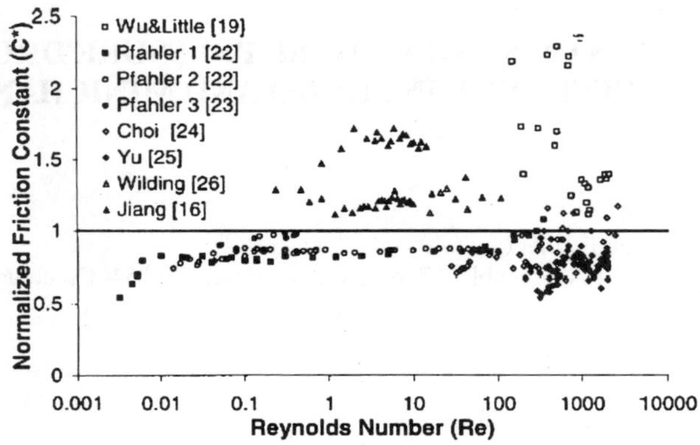

Figure 1: Normalised friction constant as a function of Reynolds number (After Papautsky *et al.* [1])

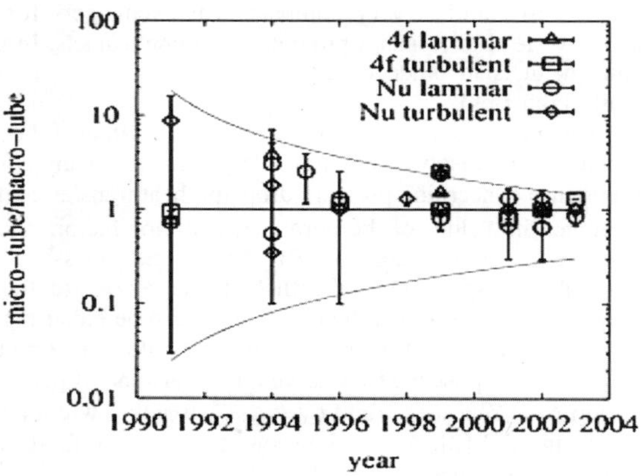

Figure 2. Convergence toward macro-tubes correlations (After Agostini *et al.* [2])

2. Classical theories

2.1 THE CONTINUUM HYPOTHESIS

In the usual macroscopic analysis of transfer phenomena, fluids are considered as continuous media and macroscopic properties are assumed to vary continuously in time and space. The physical properties (density, …) and macroscopic variables (velocity, temperature,…) are averages on a sufficient number of atoms or molecules. If $N \sim 10^4$ is a number of molecules high enough to be significant, the side length of a volume containing these N molecules is about 70 nm for a gas in standard conditions and 8 nm for a liquid. These dimensions are smallest than those of a microchannel whose characteristic dimension L_c is between 1 to 300 μm. The transport properties (heat and mass diffusion coefficients, viscosity) depend on the molecular interactions whose effects are of the order of magnitude of the mean free path L_m. These last effects can be appreciated with the Knudsen number

$$Kn = \frac{L_m}{L_c} \tag{1}$$

where L_c is a characteristic dimension of the system. For large values of Kn ($L_m \geq L_c$), The continuum model is no longer applicable. For liquids for which L_m ~ 1 – 5 nm and for systems whose L_c is of the order of few micrometers ($L_m \ll L_c$), the continuum model can be applied. On the other hand, the validity of the boundary conditions can be questioned. Figure 3 summarizes the domains of applicability of the different models and of the conventional equations [11]

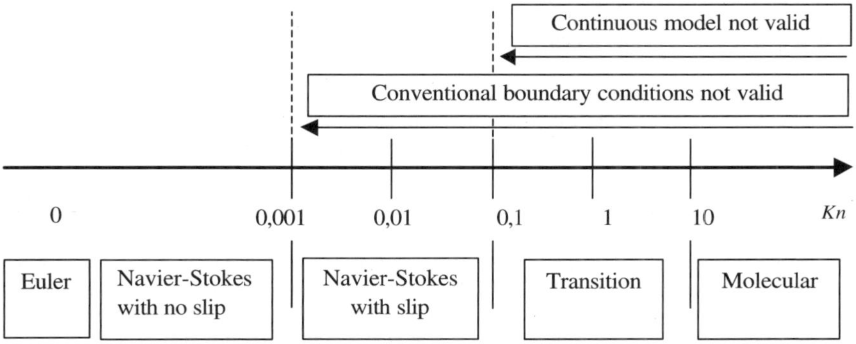

Figure 3. The different zones of applicability of conventional models according to the value of the Knudsen number.

2.2 THE CONSERVATION EQUATIONS

In a continuous medium, the classical conservation equations apply. Restricting our analysis to incompressible flows, these equations are

(i) for mass conservation

$$\vec{\nabla}.\vec{V} = 0 \tag{2}$$

where $\vec{V}$ is the flow velocity.

(ii) for momentum conservation, the Navier – Stokes equations for a Newtonian fluid of constant viscosity μ

$$\rho \frac{D\vec{V}}{Dt} = -\vec{\nabla}p + \mu \nabla^2 \vec{V} + \vec{F} \tag{3}$$

where D/Dt is the substantial or particular derivative defined by

$$\frac{D}{Dt} = \frac{\partial}{\partial t} + (\vec{V}.\vec{\nabla}) \tag{4}$$

and where ρ is the mass density, p the pressure and $\vec{F}$ a body force.

(iii) for energy conservation, the thermal energy equation can be written under the form [12]

$$\rho \frac{De}{Dt} = -p\,\vec{\nabla}.\vec{V} + \vec{\nabla}.\lambda\vec{\nabla}T + \mu\Phi + \dot{Q}_v \tag{5}$$

where e is the specific internal energy, λ is the thermal conductivity, the product $\mu \Phi$ represents the viscous dissipation and $\dot{Q}_v$ the volumetric heat source term which represents the heat power generated per unit volume of the medium. For an incompressible fluid, this equation reduces to

$$\rho c_p \frac{DT}{Dt} = \vec{\nabla}.\lambda\vec{\nabla}T + \mu \Phi + \dot{Q}_v \qquad (6)$$

2.3 BOUNDARY CONDITIONS

To solve the conservation equations, boundary conditions are needed. For the momentum equation, the flow velocity at wall is fixed. It is generally assumed that the fluid molecules near the wall are in equilibrium with those of the wall and the fluid velocity is written as:

$$V_{fluid\ at\ wall} - V_{wall} = 0 \qquad \text{(No slip condition)} \qquad (7)$$

In some cases, in particular for rarefied systems and/or low Reynolds number flows a velocity discontinuity at wall can be observed. This is the slip flow regime. The Reynolds number can be low enough for the classical boundary layer theory not to be completely valid but not low enough for the inertia terms to be neglected. The fluid velocity at the wall is not zero and

$$V_{fluid\ at\ wall} - V_{wall} = V_s \qquad \text{(Slip condition)} \qquad (8)$$

For the energy equation, the two usual boundary conditions at the wall are

(a) fixed temperature,
As for flow velocity, a temperature discontinuity at wall can be considered or not

either $T_{fluid\ at\ wall} - T_{wall} = 0$ (No temperature jump) (9)

or $T_{fluid\ at\ wall} - T_{wall} = T_j$ (Temperature jump) (10)

(b) fixed heat flux.

2.4 FLOW REGIMES

Due to the shapes of channel cross sections, pressure losses can reach values of several bars for usual lengths. This leads to small flow velocities (some mm/s or cm/s) and low Reynolds numbers. The flow is then generally laminar or transitional. For very low Reynolds numbers ($Re \ll 1$) the flow is said to be "creeping" and, neglecting the inertia term, the momentum equation becomes

$$\rho \frac{\partial \vec{V}}{\partial t} = -\vec{\nabla}p + \mu \nabla^2 \vec{V} + \vec{F} \qquad (11)$$

2.5 TYPICAL CHANNEL GEOMETRIES.

The most studied geometries are the cylinder (figure 4(a)), the rectangular channel (figure 4(c)) and the two parallel plates (figure 4(d)). Another type of channel cross section specific to silicon has received close attention. This is the trapezoid (we includes here the triangle) which corresponds to a fabrication process by chemical etching of silicon (figure 4(b)).
The example of a flow between two parallel plates will be discussed throughout the text as an Ariane thread to define the parameters of interest. This example is the limiting case of a channel of rectangular cross section whose aspect ratio ($\varepsilon = h/w$, figure 4(c)) tends to zero.

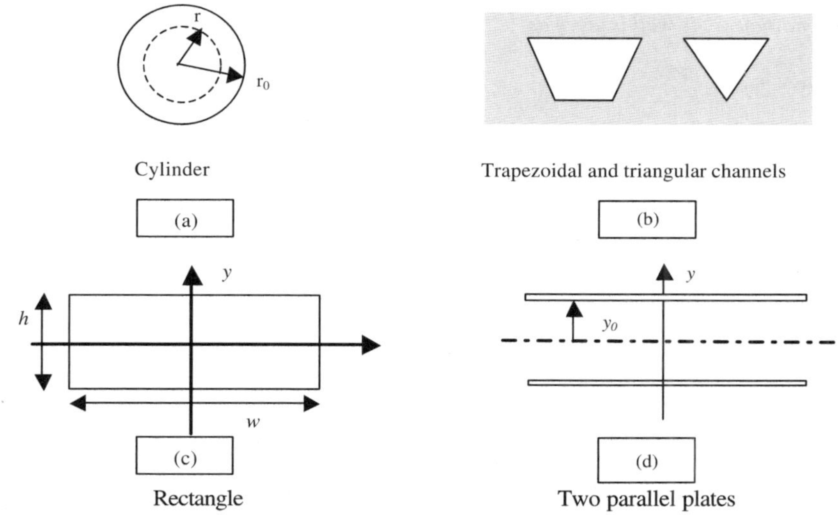

Figure 4. Usual channel geometries

2.6 PARALLEL PLATES

Limiting the analysis to a stationary fully developed one dimensional flow, the conservation equations become:

momentum conservation
$$0 = -\frac{dp}{dx} + \mu \frac{\partial^2 u}{\partial y^2} \tag{12}$$

energy conservation
$$\rho\, c_p\, u \frac{\partial T}{\partial x} = \lambda \frac{\partial^2 T}{\partial y^2} + \mu \left(\frac{\partial u}{\partial y} \right)^2 \tag{13}$$

where u is the x-component of the velocity.

- *For the no-slip condition at the wall*

$$u = 0 \text{ at } y = y_0. \tag{14}$$

A supplementary symmetry condition is

$$\frac{\partial u}{\partial y} = 0 \quad \text{at } y = 0. \tag{15}$$

Integration of the momentum equation gives the velocity profile

$$u = -\frac{y_0^2}{2\mu} \frac{dp}{dx} \left(1 - \left(\frac{y}{y_0} \right)^2 \right) = \frac{3\, V_m}{2} \left(1 - \left(\frac{y}{y_0} \right)^2 \right) \tag{16}$$

where V_m is the mean velocity

$$V_m = \frac{y_0^2}{3\mu} \left(-\frac{dp}{dx} \right). \tag{17}$$

- *A slip condition at the wall is given as*

$$u = u_s \text{ at } y = y_0 \tag{18}$$

and symmetry gives

$$\frac{\partial u}{\partial y} = 0 \text{ at } y = 0 \tag{19}$$

The wall condition is generally written as (first order condition) [13]:

$$\mu \left.\frac{\partial u}{\partial y}\right|_{y=y_0} = -\beta u_s \tag{20}$$

indicating that the fluid layer near the wall is subject to the balance of two forces, a driving force due to the shear stress and a resisting force assumed to be proportional to the slip velocity. With an elementary kinetic theory and under some conditions the β factor can also be expressed as a function of the Knudsen number remarking that the velocity gradient at wall is of the order of magnitude of u_s/L_m. The equation (20) can be written under the form:

$$u(\, y = y_0\,) = u_s = -L_m \left(\frac{\partial u}{\partial y}\right)_{y=y_0} \tag{21}$$

Solving the momentum equation gives the new velocity profile

$$u = -\frac{y_0^2}{2\mu}\frac{dp}{dx}\left(1 - \left(\frac{y}{y_0}\right)^2 + 2\frac{L_m}{y_0}\right) \tag{22}$$

and the new mean velocity

$$V_m = \frac{y_0^2}{3\mu}\left(-\frac{dp}{dx}\right) + \frac{y_0\, L_m}{\mu}\left(-\frac{dp}{dx}\right) \qquad \text{or} \tag{23}$$

$$V_m = V_{mNS} + V_{mS} \tag{24}$$

It is seen that the volumic flow rate is the sum of two components, one identical to a non slip case plus the effect of the slip velocity.

2.7 PRESSURE DROP

The friction factor f is defined by

$$f = \frac{\tau_w}{\frac{1}{2}\rho V_m^2} \tag{25}$$

where τ_w is the shear stress exerted by the fluid on the wall. If τ_w is not constant on a channel cross section, it is usual to define an average shear stress $\overline{\tau}_w$ at the x position by writing a force balance and we obtain:

$$\overline{\tau}_w = \frac{D_h}{4}\left(-\frac{dp}{dx}\right) \tag{26}$$

where D_h is the hydraulic diameter defined by

$$D_h = 4\frac{S}{P_m}$$ where S is the wetted section and P_m the wetted perimeter (27)

Then, the pressure loss can be written:

$$-\frac{dp}{dx} = \Lambda \frac{1}{D_h}\frac{1}{2}\rho V_m^2$$ where $\Lambda = 4f$ is the Darcy coefficient (28)

For the two parallel plates where $D_h = 4\ y_0$, the shear stress at wall is constant and is written as

$$\tau_w = \mu \left.\frac{\partial u}{\partial y}\right|_{y_0} = -y_0 \frac{dp}{dx} \quad (29)$$

Introducing the Reynolds number

$$Re = \frac{\rho V_m D_h}{\mu} \quad (30)$$

the Darcy cœfficient can be expressed as a function of the Reynolds number.

(i) No velocity slip at wall

From the equations (17), (19) and the definition of the friction factor (25), it is easy to see that

$$\Lambda = \frac{96}{Re} \quad (31)$$

(ii) Slip condition at wall

Taking into account the value of V_m, (Equation (23)) and the definition of the Knudsen number, the Darcy coefficient becomes:

$$\Lambda = \frac{96}{Re}\left(\frac{1}{1+6\ Kn}\right) \quad (32)$$

It can be remarked that, when $Kn \to 0$, the classical value of Λ is obtained.

2.8 THE POISEUILLE NUMBER

For fully developed laminar flows, it is obvious that the product of the friction factor and the Reynolds number is constant. This product is called the Poiseuille number.

$$Po = f\,Re \quad (33)$$

An alternative definition, which is adopted here, is

$$Po = \Lambda\, Re. \quad (34)$$

For the parallel plates with no slip condition at wall, $Po = 96$.

Scaling the lengths with D_h, $x^* = x\ /\ D_h$, it is seen that the Poiseuille number represents a non-dimensional pressure loss

$$Po = \frac{\left(dp / dx^*\right)}{\left(\frac{1}{2}\rho V_m^2\right)} \tag{35}$$

From an experimental point of view, the two measured quantities are the pressure and the flow rate. The Poiseuille number can be determined directly from these quantities as

$$Po_{exp} = -\frac{dp}{dx}\frac{2D_h^2}{\mu V_m} \tag{36}$$

and to measure the deviation from the classical theory it is usual to present the ratio

$$C_{Po} = Po / Po_{exp} \tag{37}$$

In table 1, the values of the Poiseuille number for different geometries are given

Table 1. Poiseuille numbers for fully developed laminar flows

Geometry	Po
Cylinder	64
Two parallel plates	96
Rectangular cross section	$96\,(1-1.3553\,\varepsilon + 1.9467\,\varepsilon^2 - 1.7012\,\varepsilon^3 + 0.9564\,\varepsilon^4 - 0.2537\,\varepsilon^5)$
Equilateral triangle	53

2.9 HEAT TRANSFER - THE NUSSELT NUMBER

Solving the energy equation for constant fluid properties, the temperature profile can be determined for simple cases with the two classical boundary conditions (temperature and heat flux). From this temperature profile the heat transfer coefficient is deduced from

$$\alpha = \frac{\lambda\left(\frac{\partial T}{dy}\right)_{y=y_0}}{T_w - T_m} \tag{38}$$

T_m being the bulk temperature. Introducing the expression from velocity and temperature in the preceding equation leads to an asymptotic value of the Nusselt number defined by

$$Nu = \frac{\alpha\, D_h}{\lambda} \tag{39}$$

In the fully developed laminar flow, this value is a constant. In table 2 are presented the values for a cylinder, the two parallel plates and a rectangular channel.

Table 2. Nusselt number values for fully developed laminar flow

Geometry	Fixed wall temperature	Uniform wall heat flux
Cylinder	3.657	4.364
Two parallel plates	7.541	8.235
Rectangular channel	$7.541\,(1-2.61\,\varepsilon + 4.97\,\varepsilon^2 - 5.119\,\varepsilon^3 + 2.702\,\varepsilon^4 - 0.548\,\varepsilon^5)$	$8.235\,(1-2.041\,\varepsilon + 3.0853\,\varepsilon^2 - 2.4753\,\varepsilon^3 + 1.0578\,\varepsilon^4 - 0.1861\,\varepsilon^5)$

2.10 ENTRANCE EFFECTS

At the entrance of a channel, the friction and heat flow rate are generally higher than downstream, where both the velocity and the temperature profiles are fully developed. Few experimental data exist for this region and most of studies are analytical or numerical. For cylindrical tubes a laminar flow is hydrodynamically developed (within 5 %) if :

$$x^+ = \frac{x/D}{Re} \geq 0.5 \tag{40}$$

where x is the distance from the tube inlet.
For parallel plates, Shah and London [14] propose the following law for the Poiseuille number which takes the entrance length into account:

$$Po = \frac{3.44}{\sqrt{x^+}} + \frac{24 + \frac{0.674}{4\,x^+} - \frac{3.44}{\sqrt{x^+}}}{1 + \frac{0.000029}{x^{+\,2}}} \qquad \text{with} \qquad x^+ = \frac{x/D_h}{Re} \tag{41}$$

For a cylindrical tube, a laminar flow is thermally developed if

$$\frac{x/D}{Re\,Pr} \geq 0.017\,. \tag{42}$$

For laminar flow between parallel plates, the results of Sparrow [15] are summarised in figure 5.

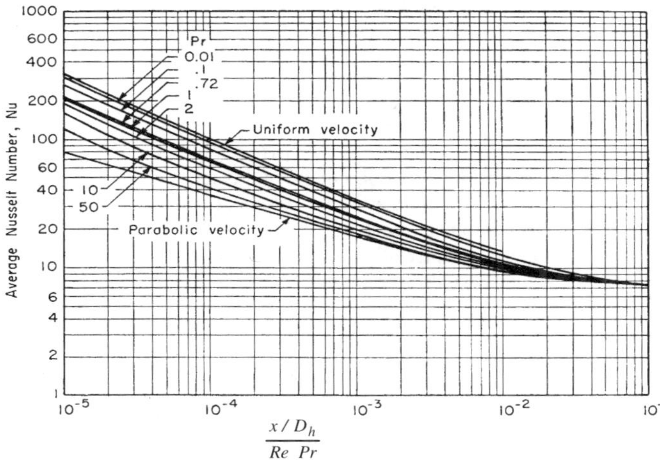

Figure 5. Entrance effect for a laminar flow between two parallel plates (After Sparrow [15])

2.11 TURBULENT AND TRANSITIONAL FLOWS

Although they are less frequently found than laminar flows it is important to recall some results concerning transitional and turbulent flows. In a conventional cylindrical channel a flow is considered to be laminar if $Re < Re_1 = 2{,}300$ and turbulent if $Re_2 > Re = 10{,}000$. Between these two values the flow is said to be transitional though the turbulence can be initiated before $Re_2 = 10{,}000$. However, for roughs tubes, as noted by Celata [8], Preger and Samoilenko cited by Idelchick [16] proposed two values of the Reynolds number depending on the roughness to determine the boundaries Re_1 and Re_2 of the transition region. This type of analysis seems to be confirmed by the work of Morini [32] in which the effect of the channel geometry is taken into account.

Concerning heat transfer,and the determination of the Nusselt number, the Gnielinski correlation valid for $Re > 2300$ is the most general [17]:

$$Nu = \frac{(\Lambda/8)\,Re\,Pr}{1{,}07 + 12{,}7\sqrt{(\Lambda/8)}\,(Pr^{2/3} - 1)}\left(1 + \frac{D_h}{L}\right)^{2/3}\left(\frac{Pr}{Pr_w}\right)^{0{,}11} \tag{43}$$

$$\text{where } \Lambda = (0{,}790 \ln(Re) - 1{,}64)^{-2} \tag{44}$$

for $2300 \leq Re \leq 5\,10^6$ and $0.5 < Pr < 2000$

The entrance effects are taken into account through the term $(1 + D_h / L)^{2/3}$. In the case of a fluid flowing in a plane wall channel, transition seems to occur at higher Reynolds numbers [18].

2.12 NUSSELT NUMBER AS A FUNCTION OF REYNOLDS NUMBER

If the Nusselt number is plotted as a function of the Reynolds number, the curve in figure 6 is obtained. One can see the constant value for laminar flow. In the turbulent regime the Gnielinski correlation is compared to a Dittus-Boelter type correlation.

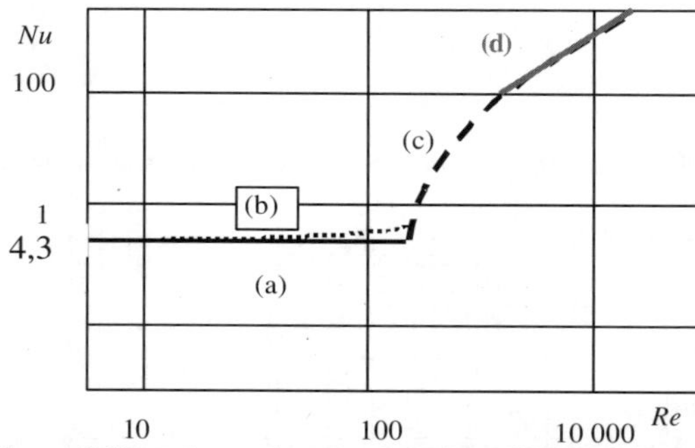

Figure 6. Nusselt number as a function of Reynolds number.

(a) $Nu = 4.36$ (Laminar flow, fixed heat flux)
(b) With entrance effects in laminar regime
(c) Gnielinski correlation
(d) Dittus-Boelter correlation

These curves will serve as a reference for the measured values.

3. Effects involved in pressure drop and heat transfer coefficient modification

3.1 EXPERIMENTAL CONDITIONS AND EXPERIMENTAL ERROR ESTIMATION

A fine review of the experimental conditions which may lead to misinterpretation of results has been carried out by Kandlikar [9]. It is instructive to recall some of those here.

3.1.1 Accuracy of channel geometry measurement

The smaller the channel dimensions, the more the errors involved in length measurements become significant. As an example, for a rectangular channel whose hydraulic diameter is

$$D_h = \frac{2\,w\,h}{w + h} \tag{45}$$

the uncertainty on the Darcy coefficient is

$$\frac{\Delta \Lambda}{\Lambda} = 2\frac{\Delta h}{h} + 2\frac{\Delta w}{w} + \frac{\Delta D_h}{D_h} \approx 7\frac{\Delta w}{w} \ . \tag{46}$$

An error of 2% on a channel dimension can lead to a 14 % error in the Darcy coefficient determination. It is essential to use an adapted instrumentation to measure the geometrical characteristics of a channel. Sometimes, the cross section may not be the same from one end of a channel to the other and, if necessary, the manufacturer's data must be verified carefully. An uncertainty analysis on the Poiseuille number determination is given by Celata [8] following the work by Holman [19]

3.1.2. Accuracy of pressure measurements

The correct position of pressure taps is essential to obtain good measurements. To measure friction losses it is best to locate them far from the inlet and the outlet of the channels to avoid entrance effects. However, if the pressure is measured by means of small holes through the wall it is important to verify that the openings do not disturb the flow streamlines. If some gas or air is to be found between the liquid and the sensor, due to the high pressure reached, dissolution of the gas in the liquid can be observed which will modify the pressure value.

3.1.3 Accuracy of temperature measurements

Several effects can play a role in the temperature measurement accuracy. Due to the small channel length, the temperature difference between the channel outlet and inlet can be as small as the sensor sensitivity. Thermocouples can have a size comparable to the channel dimensions and where is measured the temperature is questionable. Moreover, the heat flow rate through the thermocouple itself can be not negligible. The importance of these effects must be appreciated.

3.1.4. Entrance region and developing flow effects

As pointed out by Kandlikar [9] the entrance conditions can play an important role. If the pressure taps are located before and after the channel in a header with a different diameter or with elbows, the singular pressure losses can be prominent compared to the regular ones. They have been forgotten in some publications. The channels can have a short length L and the ratio L/D_h can be smaller than that in conventional channels. The developing length effects can be considerable.

3.1.5. Maldistribution condition

To obtain sufficient heat or mass flow, several channels in parallel can be used. A small defect or a different roughness in a given channel can strongly affect the pressure drop and the flow distribution. The header also can play an important role in flow distribution.

3.1.6 Longitudinal heat conduction

One-dimensional conduction, i-e between the external and internal wall only is the implicit assumption to calculate the Nusselt number from experimental data. In the case of minichannels whose wall thickness can be of the same order of magnitude as the hydraulic diameter this hypothesis may be questionable.

3.2 PHYSICAL EFFECTS

A lot of physical phenomena were advanced to explain the deviation from the classical theories. They will be briefly discussed here in order to compare the plausible mechanisms between them and between experimental results.

3.2.1 Variation of physical properties

This variation which is always taken into account for gases is often forgotten for liquids. However, very high fluxes can be obtained to or from small amounts of liquid. Reynolds numbers can be doubled between inlet and outlet of a channel, mainly due to viscosity variation [20]. Such effects could be invoked to explain the decrease in friction factors in heated channels but cannot explain results for isothermal flows.

3.2.2. Viscous dissipation

Under the effect of viscosity, the fluid itself can be heated throughout the bulk. The importance of this effect can be appreciated with the help of the Brinkman number *Br*. It is the ratio between the mechanical power degraded in heat flow and the power transferred by conduction in the fluid. It is written as

$$Br = \frac{\mu V_r}{\lambda \Delta T} \tag{47}$$

where V_r is a reference velocity.
If $Br \ll 1$, viscous heating is negligible and in most experiments, this is indeed the case. However, some authors introduced the *Br* number in correlations under the form [21]

$$Nu = A\, Re^a\, Pr^b \left(\frac{L_c}{D_h} \right)^c Br^d \tag{48}$$

where L_c is a characteristic dimension of the channel and d an exponent positive for heating and negative for cooling.

3.2.3. Micropolar theory [1]

This theory takes into account the micro-rotational effects due to rotation of molecules. This becomes important with polymers or polymeric suspensions. The physical model assigns a substructure to each continuum particle. Each material volume element contains microvolume elements which can translate, rotate, and deform independently of the motion of the microvolume. In the simplest case, these fluids are characterised by 22 viscosity coefficients and the problem is formulated in terms of a system of 19 equations with 19 unknowns. The equations for a 2-D case were solved numerically and compared to experimental results. It is concluded that the model based on the micropolar fluid theory gives a better fit than the Navier – Stokes equations. However, it seems that the difference is small.

3.2.4. Electrical double layer

Most solid surfaces bear electrostatic charges creating an electrical surface potential. Its magnitude depends on the natures of the solid and the fluid. If the liquid flowing over such a surface carries a small amount of ions the electrostatic charges on a non-conducting surface attract counter ions. Several layers of ions are created. A compact layer appears near the surface, about 0.5 nm thick, whose ions are almost immobile due to the presence of a strong electric field and a diffuse double layer due to the redistribution of ionic charges (Figure 7)..
In this diffuse layer whose thickness ranges from few nanometres to few hundred nanometres, the ions are mobile. Assuming that the surface is charged negatively there is an excess of negative ions at the centre of the channel (Figure 8) which are carried away by the flow. A "streaming" current is then created in the downstream direction. Conversely, the accumulation of negative ions downstream causes a conduction current in the opposite direction
To evaluate the influence of the EDL on fluid flow and heat transfer through microchannels the example of a flow between two parallel plates is given. From Poisson's equation Mala *et al.* [22]

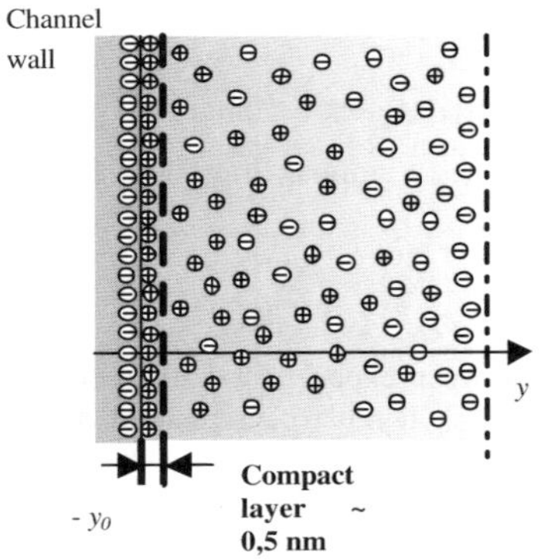

Figure 7 : Schematic representation of the electric double layer at the channel wall (After Mehendale *et al.* [9])

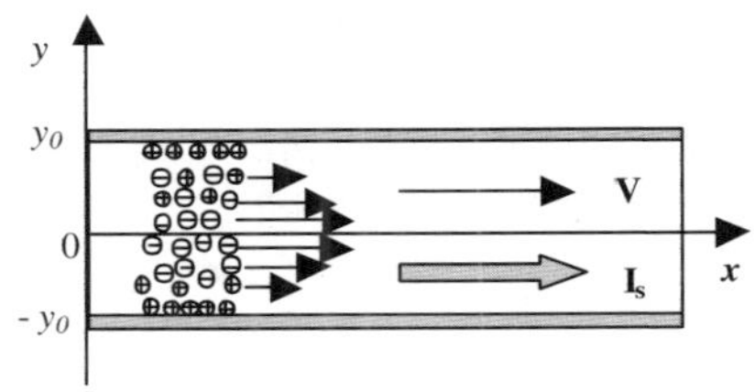

Figure 8 : The streaming current due to the electric double layer.

found an expression for the electrical potential which depends on ζ (Zeta potential), its value at the boundary between the diffusive double layer and the compact layer. They deduced the velocity profile

$$u = -\frac{y_0^2}{2\mu}\frac{dp}{dx}\left(1-\left(\frac{y}{y_0}\right)^2\right)-\frac{\varepsilon\varepsilon_0\zeta}{\mu}\left(\frac{dE_s}{dx}\right)\left(1-\left|\frac{sinh(\kappa\, y/y_0)}{sinh(\kappa)}\right|\right) \tag{49}$$

This velocity is the sum of the term without an electrostatic force and a term due to the EDL. In this formula, E_s is the streaming potential and κ is equal to:

$$\kappa = y_0\,(\,2\,n_0\,z^2e^2\,/\,\varepsilon\,\varepsilon_0\,k_b\,T)^{1/2}. \tag{50}$$

where n_0 is the average number of positive or negative ions per unit volume, z the valence of the ions, e the electron charge, ε the dielectric constant of the medium, ε_0 the permittivity of vacuum and k_b the Boltzman constant. The constant κ can be written as

$$\kappa^2 = 2\frac{y_0^2}{\lambda_D^2} \tag{51}$$

λ_D being the Debye length equal to

$$\lambda_D = \left(\frac{\varepsilon\,\varepsilon_0\,k_b\,T}{n_0\,z^2\,e^2}\right)^{1/2} \tag{52}$$

This length, which can have slightly different expressions, is characteristic of the interaction length of electric charges in a ionised medium and in our case of the EDL. For aqueous solutions at 25 °C, the ion densities of 1 mol/m^3 and 100 mol/m^3 correspond approximately to the Debye length of 10 nm and 1 nm, respectively. It must be pointed out that for high ionic concentrations the thickness of the EDL is negligible.

From the velocity expression the Poiseuille number is calculated as

$$Po = \frac{8}{V_m}\left(\frac{y_0^2}{\mu}\left(-\frac{dp}{dx}\right)+2\,\frac{y_0^2}{\mu}G\,\zeta\left(\frac{dE_s}{dx}\right)\frac{coth(\kappa)}{\kappa}\right) \tag{53}$$

where $G = \frac{n_0 z^2 e^2}{k_b T}$ (54)

It is seen that a supplementary term is added to the classical value for a laminar flow. If λ_D is small, $\kappa \to +\infty$, and this term tends to zero.

3.2.5 Surface roughness

Surface roughness is a good candidate to explain discrepancies between experimental and theoretical friction losses and its influence has been investigated by Sabry [4]. It must be remarked that for laminar flows the wall roughness should not modify either the friction factor or the Nusselt number. However, several studies give evidence of a difference between theory and experimental results. Firstly, if δ is the average roughness height of the wall, for a channel of normal size whose hydraulic diameter is D_h, it can be seen that $\delta / D_h \sim 0$. For a microchannel, with the same δ, we have $\delta / D_h \sim 0.01 - 0.05$ and the relative influence will be higher. Secondly, the wall shear stress for a 100 μm channel will be 10^4 times that of a 1 cm channel. The flow will have a strong tendency to separate over the roughness elements. This last effect should give higher friction factors. To explain the cases when these friction factors are lower than expected Sabry assumes that gases are trapped between roughness elements.

3.2.6 Trapped gas effects [4]

In a simplified approach, it is supposed that a gas blanket of thickness δ completely separates the liquid from the solid wall (Figure 9). For a liquid flowing between two plates, Sabry gave the Poiseuille number as:

$$Po = 64\left[1.5(1-\xi) + \xi/\beta\right] \tag{55}$$

where ξ is a shielding coefficient, between 1 and 0, indicating whether the gas blanket is total or not. β depends on δ / y_0. If this ratio tends to zero, $\beta \to 2/3$ and the Poiseuille number takes its classical value.

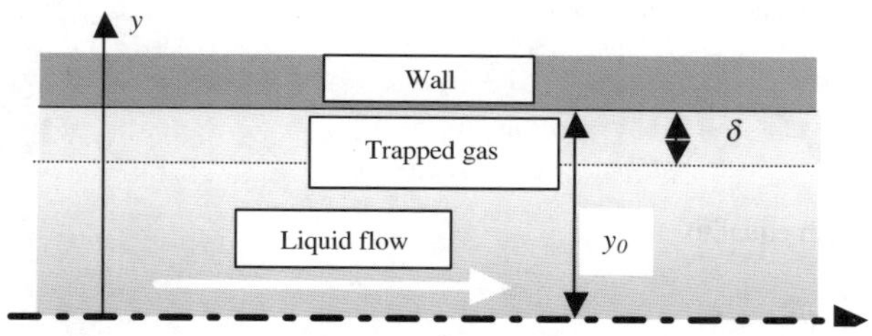

Figure 9. Simplified model of trapped gases

3.2.7. Hydrophilic or hydrophobic surfaces

The hydrophilic or hydrophobic nature of the wall surface can modify the boundary conditions and introduce a slip condition Choi *et al.*[23] used high precision microchannels treated chemically to enhance the hydrophilic and hydrophobic properties of wall surfaces.

3.2.8 Electrokinetic slip flow

The coupling of the EDL and a slip condition at the wall has been investigated theoretically [24]. This study shows the importance of the capability to control surface charge and surface hydrophobicity.

4. Pressure drop - Experimental results

In most experimental devices, the main problem is to eliminate the different sources of error. For pressure drop measurements, the pressure sensors must not be intrusive and interfere with the physical phenomenon. In most published works, the pressure sensors are added to the circuit and the fitting itself can create a singular pressure loss. Two experiments are presented. The first one has a rectangular channel whose hydraulic diameter varies from 100 μm to 1 mm with pressure sensors on either side of the test channel and includes entrance effects. The second one whose hydraulic diameter is 7.1 μm has the pressure taps far from the inlet and outlet to eliminate entrance and exit effects.

4.1 FROM MINI TO MICROCHANNELS [25] [26]

4.1.1 Experimental apparatus

The experimental apparatus consists of a closed-loop circuit which includes a pump, a filter, two flowmeters, two pressure transducers, a differential inductive pressure transducer and two K type thermocouples for the determination of the inlet and outlet temperatures (Figure 10). The test section comprises the channel between two plane bronze blocks separated by a foil whose thickness fixes the distance between the brass walls. A series of foils with several thickness enables the width to be varied (Figure 10). Details are given in [25].
The pressure losses can be measured by means of the pressure sensors. The circulating fluid is water and it can be heated by four electric cartridges inserted in the two blocks. The heat transfer coefficient is deduced from a global heat balance which takes thermal losses into account.

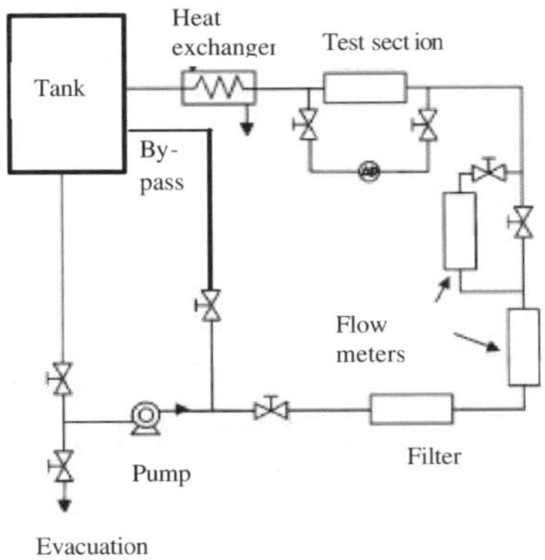

Figure 10. Test loop for the study of minichannels *(After [25])*.

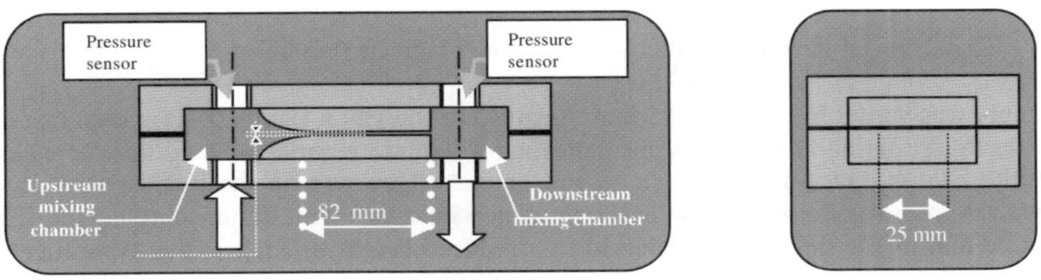

(a) section parallel to the flow direction

(b) perpendicular to the flow direction

Figure 11. Test section *(After [25])*

4.1.2 Pressure losses - Poiseuille number.

Smooth walls

The evolution of the Poiseuille number (*f Re*) as a function of the Reynolds number is shown on figure 12. It is observed that the classical value for the laminar regime is obtained if the Reynolds number is less than 2000. The laminar turbulent transition occurs for the conventional value. The authors [22] investigated the entrance effects. They conclude that the friction factor is insensitive to the channel height and that there was no sign of a faster transition to turbulence compared to conventional channel flows.

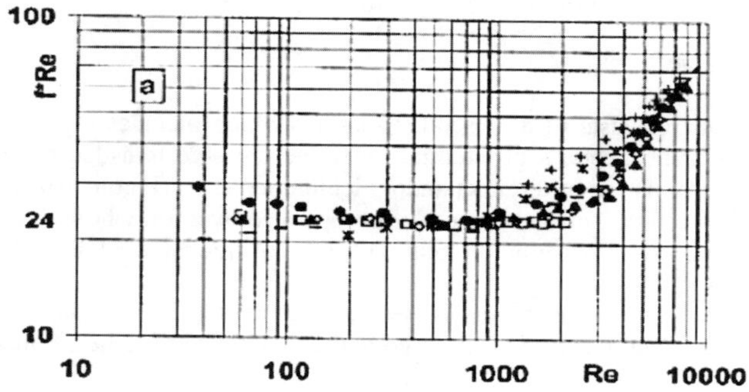

Figure 12. Influence of Reynolds number and channel height on Poiseuille number: + h = 1 mm, × h = 0.7 mm, • h = 0.5 mm, ▲ h = 0.3 mm, h = 0.1 mm, ----- Blasius law.

Influence of roughness

As pointed out by Sabry [4], roughness can play a major role in micro and minichannels, its relative importance increasing when the channel dimension decreases. There can be an infinite value of surface states and choices have to be made to control this roughness. Bavière *et al.* [26] treated the surface of the bronze blocks by anchoring SiC particles (height k between 5 and 7 μm) in a thin Ni layer deposited on the block surface (Figure 13).

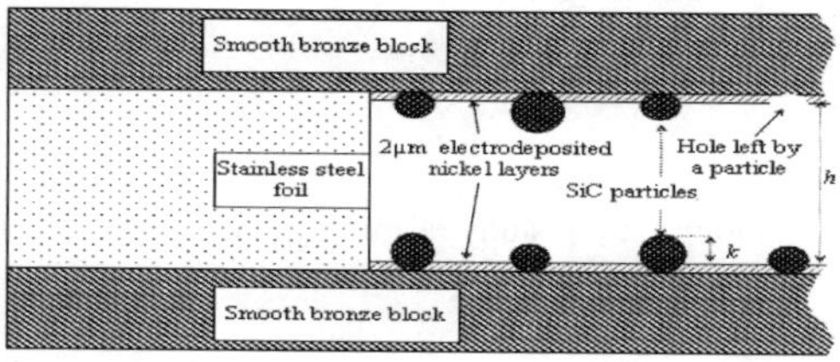

Figure 13. Transversal view of the microchannel with controlled roughness.

The channel height was varied from 0.1 to 0.3 mm. Results are shown on figure 14. It is seen that up to Re ~ 3000 the Poiseuille number is constant indicating a laminar regime. Beyond this Reynolds number value, the Blasius law for a turbulent regime applies. However, in the laminar regime, a significant deviation from the Poiseuille law is shown which increases with diminishing height. It is interesting to note that, if the the data are referred to a reduced height of 11 μm, the experimental Poiseuille number is the same as the conventional theory value ($Po = f\,Re = 24$) as seen on figure 15. interpreted by the presence of two 5.5 μm stagnant layers near the wall, this value corresponding to the k parameter. A laminar recirculating flow probably occurs behind each of the roughness elements. Such recirculating structures have been numerically calculated by Hu *et al.* [25] in 2-D microchannels with rectangular prism roughness elements. The main effect of these recirculating/stagnant zones is to reduce the effective cross-sectional area of the channel.

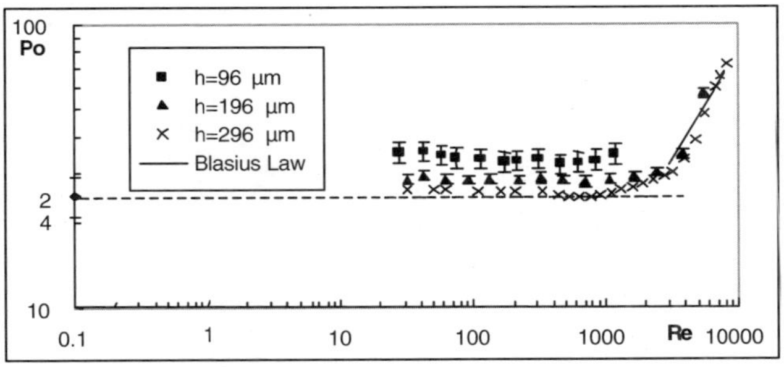

Figure 14. Influence of roughness on flow regime in microchannels (After [26])

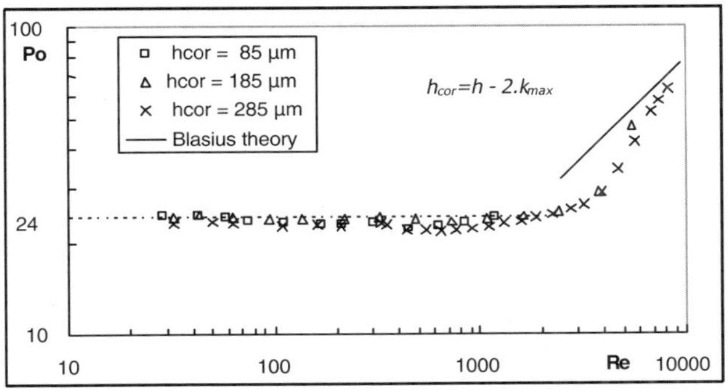

Figure 15. Channel with controlled roughness. Experimental data plotted with a modification of the hydraulic diameter (After [26])

To verify the flow regime and the laminar-turbulent transition, a bronze block was replaced by an transparent altuglas plate. Visualisation with dye revealed a very stable flow for Reynolds numbers up to 2500 for both smooth and rough channels. On the contrary, large eddies were visualised for Reynolds numbers over 3800. Between these two values a stable flow region following turbulent structures were observed.

4.2 MICROCHANNELS [26] [28]

4.2.1 Experimental apparatus

The aim is to eliminate entrance effects as much as possible and any influence on the flow of the pressure tap holes into the channels. This was achieved by integrating on the same silicon chip the microchannel, the pressure taps and the pressure sensors. The fabrication process and the operating mode are described in [28]. The pressure sensors are constituted of a membrane which is deformed under the fluid pressure and on which is deposited a thin film strain gauge. This strain gauge forms a Wheatstone bridge whose the membrane deformation modifies the electrical resistances.
The channel tested is 3 mm wide and 7.5 µm (± 0.1 µm) high. The pressure taps are longitudinally spaced out in the central zone of the channel where the flow is supposed to be established. This last point is confirmed by remarking that, for the Reynolds number range encountered, the ratio $x / (Re\ D_h)$ is greater than 65, x being the longitudinal position of the first pressure tap. In figure 16, the microchannel, the pressure taps and cavities with the membranes at the bottom can be distinguished.
According to the manufacturing process, different wall roughness can be produced.

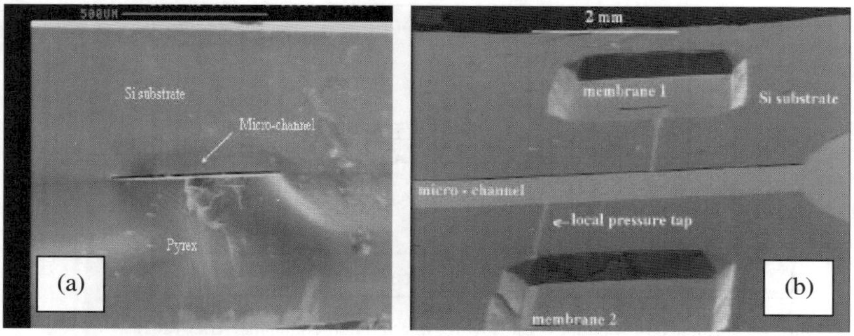

Figure 16 : Scanning Electron Microscope view of a microchannel engraved in silicon (After [26])
(a) cross section
(b) “aerial” view showing the two pressure taps.

4.2.2 Results and discussion

Experimental data were obtained for water flows at room temperature. Poiseuille number is plotted versus the Reynolds number in figure 17. It can be observed that a classical value for a laminar flow is found, as expected. The slight underestimation observed is probably due to the experimental imprecision on the estimation of the channel height

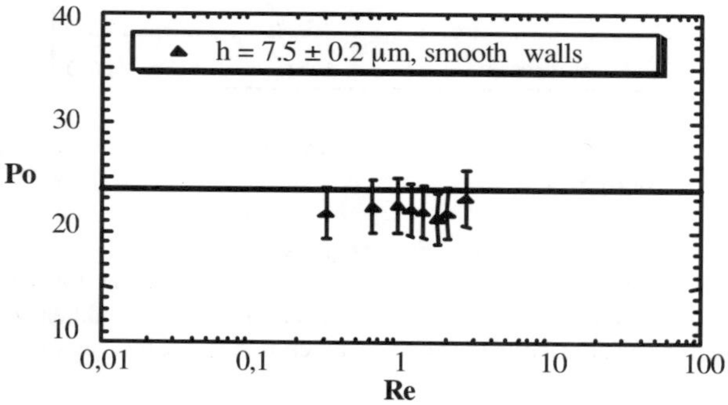

Figure 17 : Poiseuille number as a function of the Reynolds number for water flowing in a microchannel with smooth walls

After eliminating the effects of the pressure measurement instrumentation, electrokinetic and roughness effects were studied.

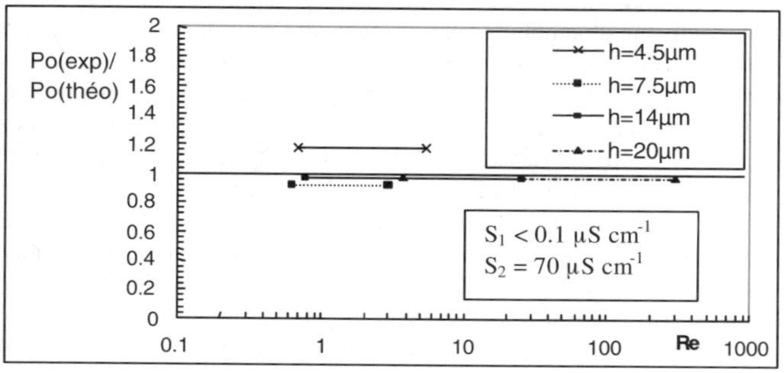

Figure 18 : Effect of water conductivity on the Poiseuille number.

Waters of different electrical conductivity were employed to detect a possible effect on pressure losses. Results are presented on figure 18. It can be seen that no evidence of an electro-viscous effect was observed . As explained in the theoretical section, EDL effects were *a priori* negligible.
Using a rough wall channel (ion etched with SF_6/O_2 plasma), it can be observed on figure 19(a) that the Poiseuille number is found to be close to the classical value. However, the error bars are particularly important due to the difficulty in measuring the channel height exactly. In this case, the nature of the wall roughness is totally different to that described in the previous section (figure 19 (b)) and further studies are in progress.

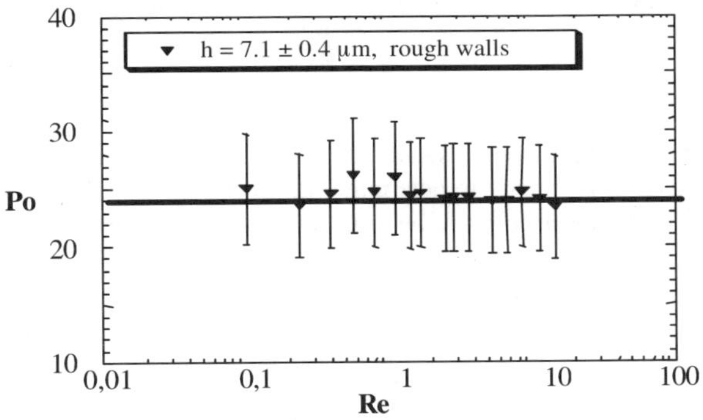

Figure 19(a) : Poiseuille number as a function of Reynolds number for a rough channel

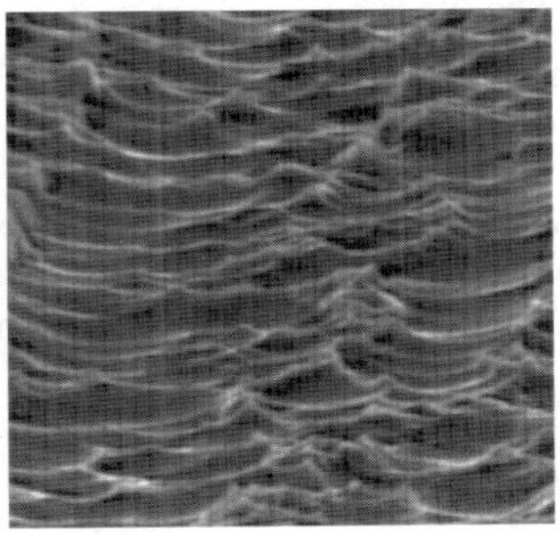

Figure 19(b) : The roughness of a silicon wall

5. Heat transfer [29], [30]

In this section are presented results for minichannels whose hydraulic diameters vary from 0.77 mm to 2.01 mm.

5.1 EXPERIMENTAL APPARATUS

The test loop used in this experiment is made of two distinct circuits: the main circuit with a refrigerant R134a flow where the test section is inserted and a secondary cooling circuit with a glycol-water mixture to cool the fluid heated in the test section (Figure 20).

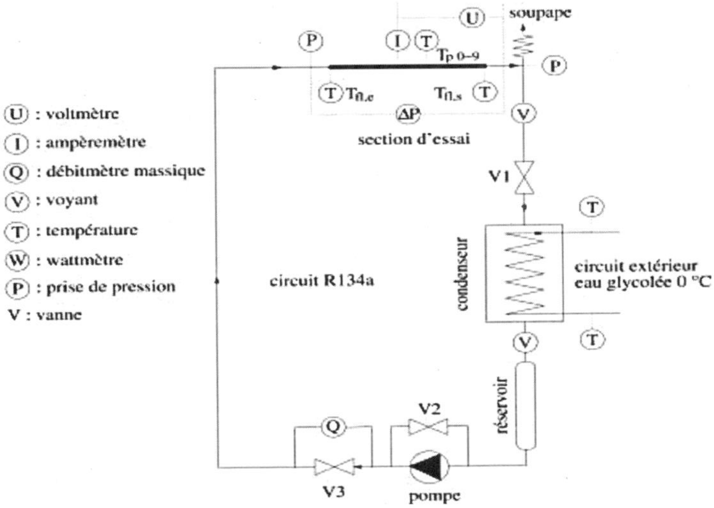

Figure 20 : The test loop

Figure 20 shows the test section and its instrumentation. Both ends are equipped with 90° manifolds for the fluid distribution. The tube diameter used for these manifolds is ten times that of the minichannels in order to suppress fluid distribution problems. The test section is made of two functional parts: an adiabatic section for the hydrodynamic entry length and a heating zone placed between two pairs of electrodes brazed on the tube to produce a Joule effect heating.
For wall temperature measurements 10 thermocouples are fixed on the heated part of the tube. Entrance and exit manifolds have pressure taps and thermocouples to measure the fluid pressure and temperature. A differential pressure sensor is also placed between the test section inlet and outlet.
Heating the test section is performed by means of a low voltage U (0 - 2 V), high intensity I (100 - 1800 A) power supply.

5.2 DATA REDUCTION

Despite the fact that the test section was thermally well insulated, a power balance taking into account heat transfer with the surroundings was necessary to determine the local heat flux $\dot{q}(x)$ and the local fluid temperature $T_{fl}(x)$, where x is the distance from the test section inlet. The heat flux to the fluid is

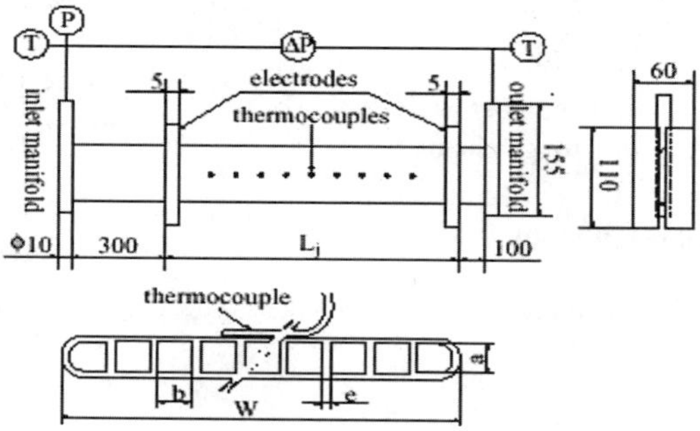

Figure 21 : The test section and the minichannels (units in mm).

$$\dot{q}(x)=\frac{U\,I}{S}-\alpha_{atm}\left(T_w(x)-T_{atm}\right) \tag{56}$$

where α_{atm} is a global heat transfer coefficient taking into account natural convection, radiation and insulation thickness and S the cross section of the test section wall. Then, the mean fluid temperature is calculated by:

$$T_{fl}(x)=T_{fl}(0)+\int_0^x \frac{\dot{q}(x).S}{\dot{M}.c_p(x)L}dx \tag{57}$$

L being the test section length and $\dot{M}$ the mass flow rate.
The global Nusselt number is calculated as follows:

$$Nu_G=\frac{\alpha_G D_h}{\lambda_{fl}}, \qquad \alpha_G=\frac{\bar{\dot{q}}}{\Delta T_{lm}} \quad \text{and} \quad \lambda_{fl}=\frac{\lambda_{fl}(0)+\lambda_{fl}(L)}{2} \tag{58}$$

ΔT_{lm} being defined as

$$\Delta T_{lm} = \frac{\left(T_w(0) - T_{fl}(0)\right) - \left(T_w(L) - T_{fl}(L)\right)}{ln\left(\frac{T_w(0) - T_{fl}(0)}{T_w(L) - T_{fl}(L)}\right)} \tag{59}$$

5.3 RESULTS AND DISCUSSION

Global Nusselt number as a function of the Reynolds number for rectangular channels with D_h = 2.01 mm, is presented on figure 22. Laminar and turbulent regimes are clearly identified with a transitional region between them. For comparison are given the correlations of Dittus-Boelter for the turbulent regime, Gnielinski for the transition and turbulent regimes and the Shah correlation corrected for the entrance effects for the laminar regime. It is observed that all the correlations are in agreement with the experimental data for Reynolds numbers greater than 500. Below this value, the deviation from the theoretical value increases as the Reynolds number decreases.

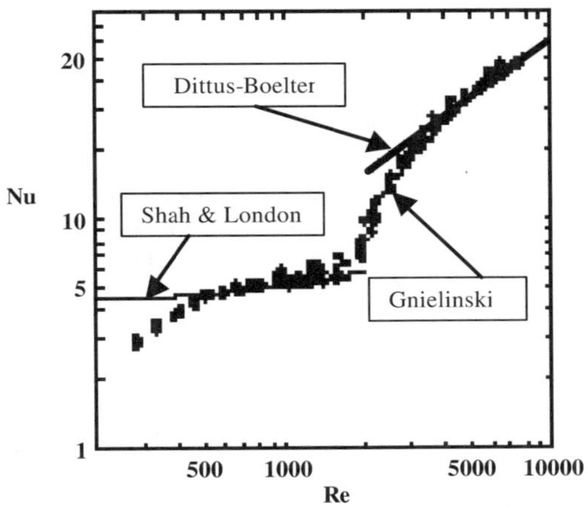

Figure 22: Nusselt number as a function of Reynolds number for R134a flowing in a minichannel (D_h = 2.01 mm)

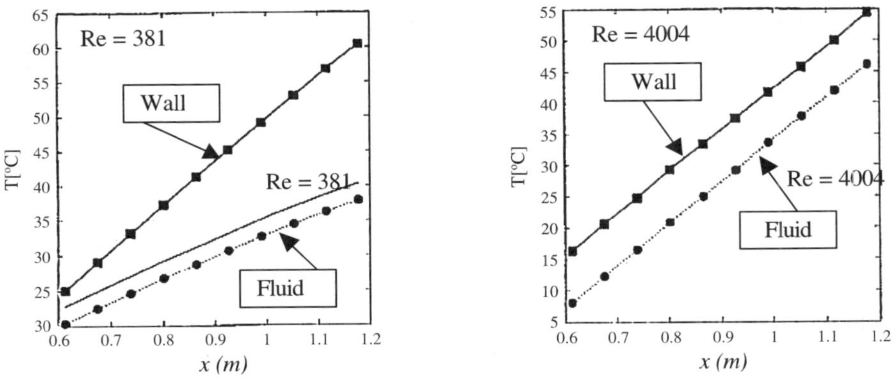

Figure 23: Wall and fluid temperatures along an electrically heated minichannel

This behaviour can be explained if we consider the wall and the fluid temperatures as a function of the channel length (Figure 23). The longitudinal profiles are presented for two Reynolds numbers. For the first, with a Reynolds number much higher than 500 (*Re* = 4004), it is seen that the two temperature profiles are parallel as expected for uniform heat flux boundary conditions. For the second Reynolds number, smaller than 500 (*Re* = 381), the two profiles are no longer parallel.
This indicates that one part of the heat does not flow directly from wall to fluid. A longitudinal heat flow exists and Agostini [30] and Commenge [31] give a rule to estimate whether or not the conditions required for a purely transversal heat flow are fulfilled. They define a Biot number which allows us to compare the convective heat flow and the conductive longitudinal heat flow. The former gives the definition

$$Bi_L = \frac{\alpha L}{\lambda_{wall}} \tag{60}$$

and the latter writes

$$Bi_{Lc} = \frac{\alpha L_c}{\lambda_{wall}} \quad \text{where } L_c \text{ is characteristic length given by } L_c = \frac{P_m L^2}{S_w} \tag{61}$$

where S_w is the wall cross section;

Agostini shows that for $Bi_L > 3$ the convective effects are prominent and for $Bi_L < 0.3$ the longitudinal heat flux produces an effect on the temperature profiles. The definition given by Commenge was calculated for counter-current heat exchangers and leads to different values. Evaluating these numbers would be useful in ensuring the heat flux is purely transversal.

6. Conclusion

The main features of the various theories have been recalled in order to facilitate the understanding of the presented results. The theories invoked to explain the discrepancy between experimental results and conventional theories were listed. To extract the proper interpretation of the different experiments, new experimental work was carried out to eliminate parasitic effects.
Concerning the friction factor, the experiments aim to eliminate (i) the entrance effects (ii) the effects of the pressure tap positioning (iii) the effect of the ion concentration of the fluid.
It was shown that, down to the characteristic dimension of 7 μm and for the fluids used, the hydrodynamics obey the conventional theories deduced from the Navier - Stokes equations. The effect of roughness on the flow behaviour needs complementary work.
Concerning the heat transfer, the experimental difficulties must be underlined. When dimensions become smaller the heat flow does not go directly through the walls. For very small dimensions and temperature difference, the heat transfer coefficients are subject to large uncertainties. The use of a longitudinal Biot number can be of help to estimate the heat flow which may not be used to calculate the heat transfer coefficient.

Acknowledgments. The author thanks B. Agostini, F. Ayela, R. Bavière, S. Le Person, M. Favre-Marinet for their results.

REFERENCES

1. Papautsky, I., Brazzle, J., Ameel, T., Bruno Frazier, A., *Laminar fluid behavior in microchannels using micropolar fluid theory*, Sensors an Actuators, Vol. 75, pp. 101-108, (1999).

2. Agostini, B. Watel, B., Bontemps, A. and Thonon. B., *Effect of geometrical and thermophysical parameters on heat transfer measurements in small diameter channels.* GRETh Grenoble. Internal report 2003. (Unpublished).

3. Gravesen, P., Branebjerg,J., Jense, O.S., *Microfluidics – A review*, J. Micromech. Microeng., Vol. 3, pp. 168 – 182, (1993).

4. Sabry, M.-N., *Scale effect on fluid flow and heat transfer in microchannels*, IEEE Transactions on components and packaging technologies, Vol. 23, N°. 3, pp. 562 – 567, (2000).

5. Gad-el-Hak, M., *The fluid mechanics of microdevices*, J. Fluid Engineering, Vol. 121, pp. 5 – 33, (1999).

6. Obot, N.T., *Towards a better understanding of friction and heat/mass transfer in microchannels – A literature review*., Proc. Int. Conf. On Heat Transfer and Transport Phenomena in Microscale, Banff, Canada, October 15-20, (2000).

7. Palm, B., *Heat transfer in microchannels*, Microscale Thermophys. Engineering, Vol. 5, pp. 155 – 175, (2001).

8. Celata, G.P., *Single-phase heat transfer and fluid flow in micropipes.* 1st Int. Conf. on Microchannels and Minichannels, Rochester, N.Y., April 24-25, (2003).

9. Kandlikar, S.G., *Microchannels and Minichannels – History, Terminology, classification and current research needs.* 1st Int. Conf. on Microchannels and Minichannels, Rochester, N.Y., April 24-25, (2003).

10. Mehendale, S.S., Jacobi, A.M., Shah, R.K., *Fluid flow and heat transfer at micro- and meso-scales with application to heat exchanger design*, Appl. Mech. Rev., Vol. 53, pp. 175 – 193, (2000).

11. Anduze, M., *Etude expérimentale et numérique de microécoulements liquides dans les microsystèmes fluidiques*, Ph. D. Thesis, Institut National des Sciences Appliquées, Toulouse, (2000).

12. Mills A.F., Heat transfer, Irwing, Boston, USA, (1992).

13. Deissler, R.G., *An analysis of second-order slip flow and temperature-jump boundary conditions for rarefied gases,* Int. J. Heat Mass Transfer, Vol. 7, pp. 681-694, (1964).

14. Shah, R.K., London, A.L., *Laminar forced convection in ducts*, Advanced heat transfer, Academic Press, New York, (1978).

15. Sparrow, E.M., N.A.C.A. note N° 3331, (1955).

16. Idelchick, I.E., *Handbook of hydraulic resistance*, Hemisphere Publishing Corporation, 2d edition, (1986).

17. Gnielinski, V., *New equations for heat and mass transfer in turbulent pipe and channel flow*, Int. Chemical Engineering, Vol. 16, N° 2, pp. 359 – 368, (1976).

18 Carlson, D.R., Widnall, S.E., Peeters, M.F., *A flow visualisation study of transition in plane Poiseuille flow*, J. Fluid Mech. Vol. 121, pp. 487 – 505, (1982).

19. Holman, J.P., Experimental methods for engineers, Mc Graw Hill, (1978).

20. Wang, B.X., Peng, X.F., *Experimental investigation on liquid forced convection heat transfer through microchannels,* Int. J. Heat Mass Transfer, Vol. 37, pp. 73 –82, (1994).

21. Tso, C.P., Mahulikar, S.P., *The role of the Brinkman number in analysing flow transitions in microchannels*, Int. J. Heat Mass Transfer, Vol. 42 pp. 1813 – 1833, (1999).

22. Mala, G.M., L.I, D., Werner, C., Jacobasch, H.J., Ning, Y.B., *Flow characteristics of water through a microchannel between two parallel plates with electrokinetic effects*, Int. J. Heat Fluid Flow, Vol. 18, pp. 489 – 496, (1997).

23. Choi, C.-H., Westin, K.J.A., Breuer, K.S., *Apparent slip flows in hydrophilic and hydrophobic microchannels,* Physics of fluids, Vol. 15, N° 10, pp. 2897 – 2902, (2003).

24. Yang, J., Kwok, D.Y., *Electrokinetic slip flow in microfluidic-based heat exchangers with rectangular microchannels*, Int. J. Heat Exchangers, Vol. 5, pp. 201 – 220, (2004).

25. Gao, P., Le Person, S., Favre-Marinet, M., *Scale effects on hydrodynamics and heat transfer in two-dimensional mini and microchannels*. Int. J. Thermal Sciences, Vol. 41, pp. 1017 – 1027, (2002).

26. Bavière, R., Ayela, F., Le Person, S. and Favre-Marinet, M., *An experimental study of water flow in smooth and rectangular micro-channels.* To be published.

27. Agostini, B., Watel, B., Bontemps, A. and Thonon, B., (2004), *Liquid flow friction factor and heat transfer coefficient in small channels: an experimental investigation.* Experimental Thermal and Fluid Science, Vol. 28, pp. 97-103

28. Hu, Y., Werner, C., Li, D., *Influence of three-dimensional roughness on pressure-driven flow through microchannels*, J. Fluids Engineering, Vol. 125, pp. 871 – 879, (2003).

29 Bavière, R., Ayela, F., *Micromachined strain gauges for the determination of liquid flow friction coefficients in microchannels*, Measurements science and technology, Vol. 15, pp. 377 – 383, (2004).

30. Agostini, B., *Etude expérimentale de l'ébullition de fluide réfrigérant en convection forcée dans les mini-canaux*, Ph. D., Thesis, Université Joseph Fourier, Grenoble, (2002).

31. Commenge, J.M., *Réacteurs microstructurés : hydrodynamique, thermique, transfert de matière et applications aux procédés*, Ph.D. Thesis, INP Lorraine, Nancy, (2001).

32. Morini, G. L., *Laminar-to-turbulent flow transition in microchannels*, Microscale Thermophysical Engineering, Vol. 8, pp. 15-30, (2004).

STEADY STATE AND PERIODIC HEAT TRANSFER IN MICRO CONDUITS

M. D. MIKHAILOV, R. M. COTTA, S. KAKAÇ[+]
Mechanical Engineering Department, Universidade Federal do Rio de Janeiro
Rio de Janeiro, Brasil
[+] Department of Mechanical Engineering - University of Miami
Coral Gables, Florida, USA

1. Introduction

The modern microstructure applications led to increased interest in convection heat transfer in micro conduits. Fluid transport in micro channels has found applications in a number of technologies such as biomedical diagnostic techniques, thermal control of electronic devices, chemical separation processes, etc.

Experimental results have been published for micro tubes [1], micro channels [2], and micro heat pipe [3]. The micro scale experimental results differ from the prediction of conventional models. Some neglected phenomena must be taken into account in micro scale convection. One of them is the Knudsen number defined as the ratio of the molecular mean free path to characteristic length of the micro conduit.

In the paper by Barron et al. [4], a technique developed by Graetz in 1885 [5] is used to evaluate the eigenvalues for the Graetz problem extended to slip-flow. The first 4 eigenvalues were found with precision of about 4 digits. The method used appears to be unstable after the fifth root, so that only the first 4 eigenvalues are reliable. The authors of the paper [4] concluded that an improved method with enhanced calculation speed would be of future interest. In reality the extended to slip-flow egenproblem has exact solution in terms of hypergeometric function and more efficient numerical methods for its solution are also available [6, 7, 8].

As demonstrated by Mikhailov and Cotta [9] the eigenvalues could be computed with specified working precision by using *Mathematica* software system [10], but the *Mathematica* rule given in [9] needs a small modification to by applicable for high-order eigenvalues.

Heat transfer by forced convection inside micro tube, generally referred as the Graetz problem, has been extended by Barron et al. [11] and Larrode and al. [12] to include the velocity slip described by Maxwell in 1890 [13] and the temperature jump [14] on tube surface, which are important in micro scale at ordinary pressure and in rarefied gases at low-pressure.

The paper by Barron et al. [11] use the first 4 eigenvalues from the above mentioned communication by Barron et al. [4] to analyze the heat convection in a micro tube. The temperature jump, although explicitly mentioned in the text, is ignored in the calculation of the eigenvalues. Therefore the temperature distribution didn't take into account the temperature jump.

The correct solution of heat convection in circular tubes for slip flow, taking into account both - the velocity

S. Kakaç et al. (eds.), Microscale Heat Transfer, 49–74.

slip and the temperature jump, is given by Larrode, Housiadas, and Drossinos [12]. These authors introduce a scaling factor that incorporate both rarefaction effect and gas-surface interaction parameters and develop uniform asymptotic approximation to high-order eigenvalues and eigenfunctions.

Heat transfer in microtubes with viscous dissipation is investigated by Tung and Bayazitoglu [15]. The temperature jump, is ignored in the calculation of the temperature distribution, but taken into account in determination of the Nusselt number.

Conventional pressure driven flow requires costly micro pumps giving significant pressures [16]. A micro scale electro-osmotic flow is a viable alternative to pressure-driven flow, with better flow control and no moving part [17]. Liquid is moved relative to a micro channel do to an externally applied electric field. This phenomena is first reported by Reuss in 1809 [18].

The fully developed velocity distribution in micro parallel plate channel and micro tube are well known [19]. Using a fully developed velocity one could investigate thermally developing heat transfer and its limiting case - thermally developed heat transfer. The corresponding solutions for electro-osmotic flow in micro parallel plate channel and micro tube are special cases from the general results given in the book by [20].

Thermally fully developed heat transfer do to electro-osmotic fluid transport in micro parallel plate channel and micro tube has been recently investigated by [21]. The dimensionless temperature profile and corresponding Nusselt number have been determined for imposed constant wall heat flux and constant temperature. The complement paper [22] study the effect of viscous dissipation. These two papers gives important physical details and references. The analyses of both papers is based on the classical simplifying assumptions that are avoided in the book by Mikhailov and Ozisik [20].

The conventional laminar forced convection in conduits at periodic inlet temperature is investigated mainly by Kakaç and coworkers [23, 24, 25, 26, 27]. The periodic heat transfer in micro conduits, to the our knowledge, is not investigated.

The solutions given here, are special cases from the general results for temperature distribution, average temperature and Nusselt numbers presented in the book [20]. Nevertheless all formulae have been derived again by using *Mathematica* software system [10].

Mathematica package is developed that computes the eigenvalues, the eigenfunctions, the eigenintegrals, the dimensionless temperature, the average dimensionless temperature, and the Nusselt number for steady state and periodic heat transfer in micro parallel plate channel and micro tube taking into account the velocity slip and the temperature jump. Some results in form of tables and plots are given bellow.

For electro-osmotic flow only the limiting Nusselt numbers for thermally fully-developed flow in parallel plate channel and circular tube are obtained as a special case from the solution for thermally developing flow.

2. Slip Flow Velocity in Parallel Plate Micro-Channel

Consider a fully developed steady flow of an incompressible constant property fluid inside a micro-channel. Let `z (0≤z<∞)` be the axial coordinate and `y (-y1≤y≤y1)` the normal coordinate. The velocity distribution `u[y]` is described by the momentum equation:

$$\mu\ u''[y] = dP / dz \qquad (1)$$

where μ is viscosity, **dP/dz** is the constant pressure gradient along the channel, and **u[y]** is the velocity profile.

Since the velocity profile is symmetric we consider only the region $0 \le y \le y1$. The boundary condition at $y=0$ is:

$$u'[0] = 0 \tag{2}$$

In conventional parallel plate channel the intermolecular collisions dominate, because the characteristic length **2*y1** is much larger than the molecular mean free path. The velocity of the fluid at the surface is zero **u[y1]=0**.

In micro parallel plate channel the interactions between the fluid and the wall become significant, because the molecular mean free path is comparable to **2*y1**. The gas slip along the wall with a finite velocity in the axial direction as described by Maxwell in 1890 [13]. The kinetic theory of gases gives the following boundary condition at the surface of the channel [28]:

$$u[y1] - \beta v \lambda u'[y1] = 0 \tag{3}$$

where

βv is $(2-\alpha_m)/\alpha_m$.

α_m is the momentum accommodation coefficient.

λ is the molecular mean free path.

To simplify the problem we define the dimensionless velocity:

$$U[Y] = -\mu u[y] / (y1^2 \, dP / dz) \tag{4}$$

The dimensionless coordinate **Y** and Knudsen number **Kn** are defined as:

$$Y = y / y1, \qquad Kn = \lambda / (2\, y1) \tag{5}$$

According to reference [1] four flow regimes for gases exist: continuum flow ($0 \le Kn < 0.001$), slip flow ($0.001 \le Kn < 0.1$), transition flow ($0.1 \le Kn < 10$), and free molecular flow ($10 \le Kn$). Continuum equations are valid for **Kn->0**, while kinetic theory is applicable for **Kn>8**. Slip flow occurs when gases are at low pressure or in micro conduits. The gas slip at the surface, while in continuum flow at the surface it is immobilized.

The equations (1), (2), and (3) in dimensionless form becomes:

$$U''[Y] = -1, \qquad U'[0] = 0, \qquad U[1] + 2\, Kn\, \beta v\, U'[1] = 0 \tag{6}$$

The solution of the problem (6) gives the velocity distribution (7), where the parameters **Kn** and βv are replaced by one parameter **Kn**β**v**:

$$U[Y] = (1 + 4\, Kn\beta v - Y^2) / 2 \tag{7}$$

The dimensionless average velocity **Uav** is defined as:

$$Uav = \int_0^1 U[Y]\, dY \tag{8}$$

Introducing eq. (7) into eq. (8) we obtain

$$Uav = (1 + 6\, Kn\beta v) / 3 \tag{9}$$

The ratio **u[y]/uav** is the same as **W[Y]=U[Y]/ Uav**. This ratio is used as dimensionless velocity in heat transfer analysis:

$$W[Y] = C0\ (1 + 4\ Kn\beta v - Y^2), \quad \text{where} \quad C0 = 1 / (2/3 + 4\ Kn\beta v) \tag{10}$$

When **Knβv=0** eq. (10) gives the classical Hagen-Poiseuille flow obtained in 1839 [29] and 1841 [30]. The velocity distribution (10) is used to plot **W[Y]** for different values of the parameter **Knβv**.

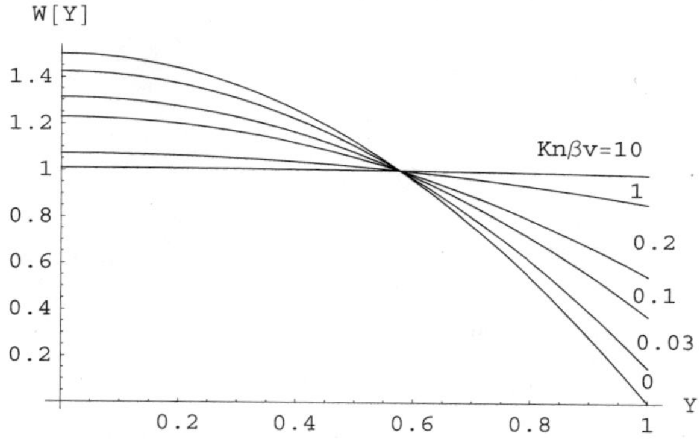

Fig. 1 Velocity distribution in parallel plate micro channel.

The figure 1 shows that even for small values of **Knβv** the considerable slip velocity appears at **Y=1**.

3. Slip Flow Velocity in Circular Micro-Tube

Consider a fully developed steady flow of an incompressible constant property fluid inside a micro tube. Let **z** ($0 \le z < \infty$) be the axial coordinate and **r** ($0 \le r \le r1$) the radial coordinate. The velocity distribution **u[r]** is described by the momentum equation:

$$\mu\ (u''[r] + u'[r] / r) = dP / dz \tag{11}$$

where μ is viscosity, **dP/dz** is the pressure gradient along the tube, and **u[r]** is the velocity.

The boundary condition at **r=0** is commonly written as **u'[0] = 0**. For this condition *Mathematica* software system is not able to find the velocity distribution. The correct condition at **r=0** is the limit of $-\mu\ \partial_r\ u[r]$ multiplied by the surface $2\ \pi\ r\ 1$ when **r->0** to be zero:

$$(r\ u'[r])_{r\to 0} = 0 \tag{12}$$

In conventional flow the velocity of the fluid at the surface is zero $u[r1]=0$ since the diameter $2*r1$ of the tube is much larger than the molecular mean free path and the intermolecular collisions dominate.

In micro tube flow the molecular mean free path is comparable to the diameter $2*r1$ and the interactions between the fluid and the wall become significant. The gas slip along the wall with a finite velocity in the axial direction [13]. The kinetic theory of gases gives the following boundary condition at the surface of the tube [28]:

$$u[r1] = -\beta v\, \lambda\, u'[r1] \tag{13}$$

where

βv is $(2-\alpha_m)/\alpha_m$

α_m is the momentum accommodation coefficient.

λ is the molecular mean free path.

To simplify the problem we define the dimensionless velocity:

$$U[R] = -\mu\, u[r] \,/\, (r1^2\, dP/dz) \tag{14}$$

The dimensionless coordinate R and Knudsen number Kn are defined as:

$$R = r/r1, \qquad Kn = \lambda/(2\, r1) \tag{15}$$

The equations (11), (12), and (13) in dimensionless form becomes:

$$U''[R] + \frac{1}{R} U'[R] + 1 = 0, \quad (R\, U'[R])_{R\to 0} = 0, \quad U[1] + 2\, Kn\, \beta v\, U'[1] = 0 \tag{16}$$

The solution of eq. (16 a) is:

$$U[R] = C_1 \operatorname{Log}[R] + C_2 - R^2/4 \tag{17}$$

Introducing eq.(17) into the eqs. (16 b, 16 c) we find the constants C_1 and C_2:

$$C_1 = 0, \qquad C_2 = 1/4 + Kn\, \beta v \tag{18}$$

The parameters Kn and βv in C_2 could be replaced by one parameter $Kn\beta v$. The velocity distribution (17) after using the constants (18) becomes:

$$U[R] = (1 + 4\, Kn\beta v - R^2)/4 \tag{19}$$

The dimensionless average velocity Uav is defined as:

$$Uav = 2 \int_0^1 R\, U[R]\, dR \tag{20}$$

Introducing eq. (19) into eq. (20) we obtain

$$Uav = (1/2 + 4\, Kn\beta v)/4 \tag{21}$$

The ratio $u[r]/uav$ is the same as $W[R]=U[R]/Uav$. This ratio is used as dimensionless velocity in heat transfer analysis:

$$W[R] = C0\,(1 + 4\,Kn\beta v - R^2), \quad \text{where} \quad C0 = 1/(1/2 + 4\,Kn\beta v) \tag{22}$$

When $Kn\beta v=0$ the eq. (22) gives the classical Hagen-Poiseuille flow obtained in 1839 [29] and 1841 [30]. The velocity distribution (22) is used to plot $W[R]$ for different value of the parameter $Kn\beta v$.

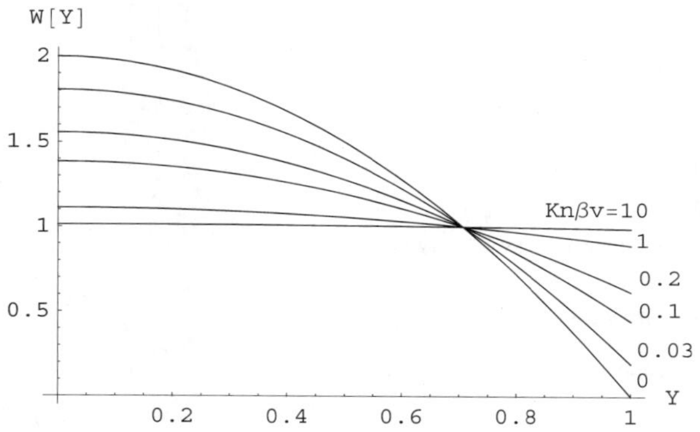

Fig. 2 Velocity distribution in micro tube.

The figure 2 shows that even for small values of $Kn\beta v$ the considerable slip velocity appears at $R=1$.

4. Electro Osmotic Velocity in Micro Channel

For fully developed electro osmotic flow in parallel plate micro channel the streamwise momentum equation [19] and boundary conditions reflecting no slip at the wall and no shear stress at the center are:

$$\mu\,u''[y] + \epsilon\,\frac{d\Phi}{dx}\,\psi''[y] = 0, \qquad u[0] = 0, \qquad u'[y1] = 0 \tag{23}$$

where $0<y<y1$, μ is the fluid viscosity, ϵ is the dielectric constant, Φ is the applied potential field, and $\psi[r]$ is the excess charge distribution. For low wall potentials the Debye-Hückel linearization holds and $\psi[r]$ becomes:

$$\psi[y] = e^{-\frac{y}{\lambda}}\,\zeta \tag{24}$$

where ζ is the zeta potential, λ is the Debye length [19]. Substitution of eq. (24) in eq. (23 a) gives:

$$e^{-\frac{y}{\lambda}}\,\epsilon\,\zeta\,\frac{d\Phi}{dx} + \mu\,\lambda^2\,u''[y] = 0, \qquad u[0] = 0, \qquad u'[y1] = 0 \tag{25}$$

The term $(\epsilon\,\zeta/\mu)\,d\Phi/dx$ represents the maximum possible electro-osmotic velocity um for a given applied potential field.

$$um = \frac{\epsilon\,\zeta}{\mu}\;\frac{d\Phi}{dx} \tag{26}$$

The dimensionless velocity is defined as:

$$U[Y] = u[y] / um \tag{27}$$

where Y is the dimensionless coordinate

$$Y = y / y1 \tag{28}$$

The ratio of plate half-width $y1$ to Debye length λ is:

$$Z = y1 / \lambda \tag{29}$$

The dimensionless form of the velocity problem (25) is:

$$U''[Y] = -e^{-Y\,Z}\,Z^2, \qquad U[0] = 0, \qquad U'[1] = 0 \tag{30}$$

The solution of eqs. (30) gives the fully-developed dimensionless electro-osmotic velocity distributions $U[Y]$:

$$U[Y] = 1 - e^{-Y\,Z} - e^{-Z}\,Y\,Z \tag{31}$$

Integration over the channel cross-sectional area yields the average velocities:

$$Uav = \int_0^1 U[Y]\,dY \tag{32}$$

Introducing eq. (31) into eq. (32) we obtain:

$$Uav = 1 + (e^{-Z} - 1) / Z - e^{-Z}\,Z / 2 \tag{33}$$

The limit of velocity (31) and average velocity (33) for $Z->0$ is zero i. e. without electric field there is no osmotic movement.

For curiosity let us find the normalized velocity usually used in conventional heat transfer analyses.

$$W[Y] = U[Y] / Uav \tag{34}$$

Introducing eq. (31) and eq. (33) into eq. (34) we obtain:

$$W[Y] == \frac{1 - e^{-Y\,Z} - e^{-Z}\,Y\,Z}{1 + (e^{-Z} - 1) / Z - e^{-Z}\,Z / 2} \tag{35}$$

For $Z=\infty$ the normalized velocity profile (35) correspond to slug flow $W[Y]=1$. For $Z=0$ the limit gives $W[1-Y]=3/2(1-Y^2)$. At $Z=0$ the osmotic movement is zero, but if the average velocity exist it is the Poiseuille parabola.

Formula (31) is used to plot velocity distribution for different values of the parameter Z.

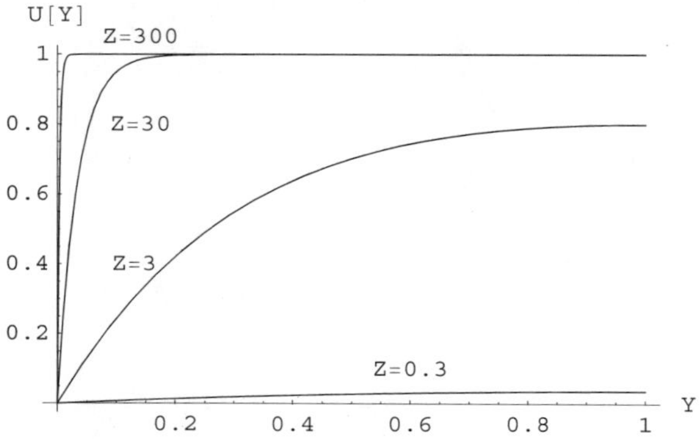

Fig. 3 Electro-osmotic velocity distribution in parallel plate micro channel.

The figure 3 shows that for large values of **Z** high velocity gradient appears near the wall surface **Y=0**.

5. Electro-Osmotic Flow in Circular Micro-Tube

For fully developed electro osmotic flow in micro tube the streamwise momentum equation [19] and boundary conditions are:

$$\mu\,(u''[r] + u'[r]/r) + \epsilon\,\frac{d\Phi}{dx}\,(\psi''[r] + \psi'[r]/r) = 0,$$
$$(r\,u'[r])_{r\to 0} = 0, \qquad u[r1] = 0 \tag{36}$$

where $r<r1$, μ is the fluid viscosity, ϵ is the dielectric constant, Φ is the applied potential field, and $\psi[r]$ is the excess charge distribution.

For low wall potentials the Debye-Hückel linearization holds and $\psi[r]$ becomes:

$$\psi[r] = \zeta\, I_0[r/\lambda]\,/\,I_0[Z] \tag{37}$$

where ζ is the zeta potential, λ is the Debye length [19], $I_0[Z]$ is the modified Bessel function of the first kind.

Substitution of eq. (37) in eq. (36) gives:

$$\mu\,(u''[r] + u'[r]/r) + \epsilon\,\zeta\,\frac{d\Phi}{dx}\,\frac{I_0[r/\lambda]}{\lambda^2\, I_0[Z]} == 0,$$
$$(r\,u'[r])_{r\to 0} == 0, \qquad u[r1] == 0 \tag{38}$$

The term $(\epsilon\,\zeta/\mu)\,d\Phi/dx$ represents the maximum possible electro-osmotic velocity **um** for a given applied potential field.

$$um = \frac{\epsilon\,\zeta}{\mu}\,\frac{d\Phi}{dx} \tag{39}$$

The dimensionless velocity is defined as:

$$U[R] = u[r] / um \quad (40)$$

where **R** is the dimensionless coordinate.

$$R = r / r1 \quad (41)$$

The ratio of tube radius to Debye length is:

$$Z = r1 / \lambda \quad (42)$$

The dimensionless form of the velocity problem (38) is:

$$U''[R] + U'[R] / R + Z^2 I_0[R Z] / I_0[Z] = 0,$$
$$(U'[R])_{R->0} = 0, \quad U[1] = 0 \quad (43)$$

The solution of eqs. (43 a) is:

$$U[R] = C_1 \text{Log}[R] + C_2 + (1 - I_0[R Z]) / I_0[Z] \quad (44)$$

After introducing eq. (44) into eqs. (43 b, 43 c) we find the constants C_1 and C_2:

$$C_1 == 0, \quad C_2 == (I_0[Z] - 1) / I_0[Z] \quad (45)$$

Than the velocity distribution (44) gives the fully-developed dimensionless electro-osmotic velocity distributions **U[R]**:

$$U[R] == 1 - I_0[R Z] / I_0[Z] \quad (46)$$

Formula (46) is used to plot velocity distribution for different values of the parameter **Z**.

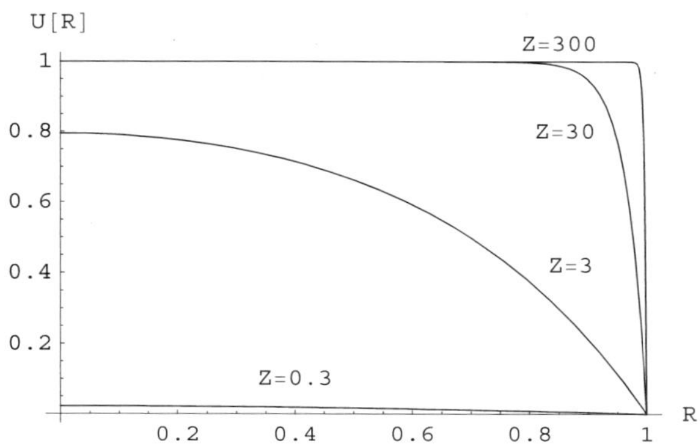

Fig. 4 Electro-osmotic velocity distribution in micro tube.

The figure 4 shows that for large values of **Z** high velocity gradient appears near the wall surface **R=1**. Integration over the channel cross-sectional area yields the average velocities:

$$Uav = 2\int_0^1 R\,U[R]\,dR \qquad (47)$$

Introducing eq. (46) into eq. (47) we obtain:

$$Uav = 1 - 2\,I_1[Z]\,/\,(Z\,I_0[Z]) \qquad (48)$$

The limit of velocity (46) and average velocity (48) for $Z\to 0$ is zero i. e. without electric field there is no osmotic movement.

For curiosity let us find normalized velocity usually used in conventional heat transfer analyses.

$$W[R] = U[R]\,/\,Uav \qquad (49)$$

Introducing eq. (46) and eq. (48) into eq. (49) we obtain:

$$W[R] = (1 - I_0[R\,Z]\,/\,I_0[Z])\,/\,(1 - 2\,I_1[Z]\,/\,(Z\,I_0[Z])) \qquad (50)$$

For $Z=\infty$ the normalized velocity profile correspond to slug flow $W[R]=1$. For $Z=0$ the limit gives $W[R] == 2\,(1-R^2)$. At $Z=0$ the osmotic movement is zero, but if the average velocity exist it is the Poiseuille parabola.

6. Steady State Heat Transfer in Micro Conduits

Consider steady-state heat transfer in thermally developing, hydrodynamically developed forced laminar flow inside a micro conduits (parallel plate micro channel or micro tube) under following assumptions:

- The fluid is incompressible with constant physical properties.
- The free heat convection is negligible.
- The energy generation is negligible.
- The entrance temperature is uniform.
- The surface temperature is uniform.

The temperature $T[r,z]$ of a fluid with velocity profile $u[r]$, diffusivity α along the channel $0\le z\le\infty$ in the region $0\le r\le r1$ is described by the following problem [20]:

$$u[r]\,\frac{\partial T[r,z]}{\partial z} = \alpha\left(\frac{\partial^2 T[r,z]}{\partial r^2} + \frac{n}{r}\,\frac{\partial T[r,z]}{\partial r}\right) \qquad (51)$$

where $n=0$ for parallel plate micro channel and $n=1$ for micro tube.

The boundary conditions at the center of the micro conduits is:

$$\frac{\partial T[0,z]}{\partial r} = 0 \quad \text{for} \quad n = 0, \qquad \left(r\,\frac{\partial T[r,z]}{\partial r}\right)_{r\to 0} = 0 \quad \text{for} \quad n = 1 \qquad (52)$$

The boundary condition (52 a) is commonly used for both - parallel plate channel and tube. The correct boundary condition for cylindrical geometry is given by eq. (52 b) [31].

The surface temperature of the micro conduits is **Ts**. As result of the temperature jump on the surface the boundary condition at **r1** becomes:

$$T[r1, z] = Ts - 2\,Kn\,r1\,\beta t\,\frac{\partial T[r1, z]}{\partial r} \tag{53}$$

where

βt is $((2-\alpha_t)/\alpha_t)(2\,\gamma/(\gamma+1))/Pr$

α_t is the thermal accommodation coefficient.

λ is the molecular mean free path.

γ is the ratio of specific heat at constant pressure c_p and specific heat at constant volume c_v.

Kn is the Knudsen number.

The entrance temperature is:

$$T[r, 0] = Ti \tag{54}$$

To simplify eqs. (51) to (54) we define the dimensionless velocity **W[R]** and dimensionless temperature $\theta[Y,Z]$ as:

$$W[R] = u[r]/uav, \qquad \theta[R, Z] = (T[r, z] - Ts)/(Ti - Ts) \tag{55}$$

where **R** is the transverse coordinate, **Z** is the axial coordinate:

$$R = r/r1, \qquad Z = z\,\alpha/(C0\,r1^2\,uav) \tag{56}$$

The eqs. (51) in dimensionless form becomes:

$$\frac{W[R]}{C0}\,\frac{\partial \theta[R, Z]}{\partial Z} = \frac{\partial^2 \theta[R, Z]}{\partial R^2} + \frac{n}{R}\,\frac{\partial \theta[R, Z]}{\partial R} \tag{57}$$

The dimensionless velocity for parallel plate micro channel, eqs. (10), and micro tube eqs. (22) could be unified as:

$$W[R] = C0\,(1 + 4\,Kn\beta v - R^2), \quad \text{where} \quad C0 = 1/(2/(n + 3) + 4\,Kn\beta v) \tag{58}$$

where **n=0** for parallel plate micro channel, and **n=1** for micro tube. After introducing the velocity (58) into eq.(57) we obtain:

$$(1 + 4\,Kn\beta v - R^2)\,\frac{\partial \theta[R, Z]}{\partial Z} = \frac{\partial^2 \theta[R, Z]}{\partial R^2} + \frac{n}{R}\,\frac{\partial \theta[R, Z]}{\partial R} \tag{59}$$

The eqs. (52) to eq.(54) in dimensionless form become:

$$\left(R^n\,\frac{\partial \theta[R, Z]}{\partial R}\right)_{R\to 0} = 0,$$
$$\theta[1, Z] + 2 * Kn\beta v * \beta\,\frac{\partial \theta[1, Z]}{\partial R} = 0, \qquad \theta[R, 0] = 1 \tag{60}$$

In eq.(60 b) the term **Kn*βt** is replaced by **Knβv*β**, where **β=βt/βv**.

The problem given by eq.(59) subject to the conditions (60) is referred to as extended Graetz problem in honor of the pioneering work [5]. To solve this problem we need the eigenvalues **m** and the eigenfunctions **y[R]** of the eigenproblem:

$$y''[R] + \frac{n}{R} y'[R] + (1 + 4\,Kn\beta v - R^2)\, m^2\, y[R] = 0,$$
$$(R^n\, y'[R])_{R\to 0} = 0, \qquad y[1] + 2 * Kn\beta v * \beta\, y'[1] = 0 \tag{61}$$

The solution of eq.(61 a) that satisfies the boundary condition (61 b) is

$$y[R] = \mathrm{Exp}[-m\,R^2 / 2]\ 1\,F1[(n + 1 - (1 + 4\,Kn\beta v)\,m) / 4;\ (n + 1) / 2;\ m\,R^2] \tag{62}$$

where **1F1[a;b;c]** is the Kummer confluent hypergeometric function.

Introducing the eq. (62) in the boundary condition (61 c) we obtain the eigencondition:

$$\begin{aligned}&(n + 1)\ (1 - 2 * Kn\beta v * \beta * m) *\\&\quad 1\,F1[(n + 1 - (1 + 4\,Kn\beta v)\,m) / 4;\ (n + 1) / 2;\ m] +\\&\quad 2 * Kn\beta v * \beta * m\ (n + 1 - (1 + 4\,Kn\beta v)\,m) *\\&\quad 1\,F1[(n + 5 - (1 + 4\,Kn\beta v)\,m) / 4;\ (n + 3) / 2;\ m] = 0\end{aligned} \tag{63}$$

The roots of (63) gives the desired eigenvalues. The **FindRoot** function of *Mathematica* software system calculates these roots starting from the values given by the asymptotic formula on p.113 of the book [20]. Fig. 5 shows the seconds per eigenvalue spend on 3 Gz computer to find 100 roots of a slightly modified eq.(63). The first 50 roots are computed much faster than the last 50 roots.

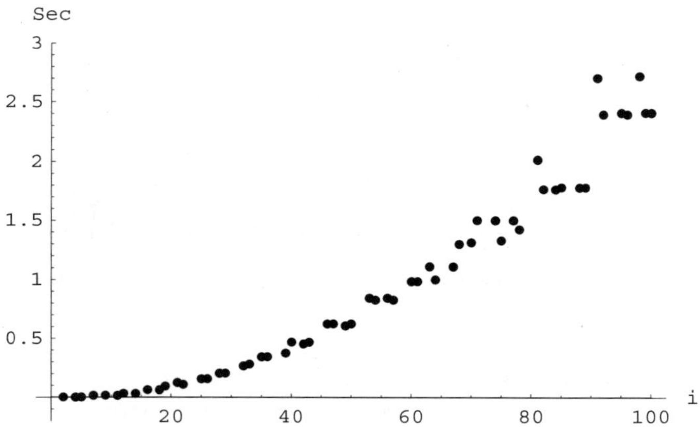

Fig. 5 CPU time in seconds per root of eq. (62) on 3 Gz PC for n=1, Knβv=0.1 and β=10.

The solution of the extended Graetz problem, eqns. (59, 60), is a special case from the solution given by Mikhailov and Ozisik in the book [20]:

$$\theta[R, Z] = \sum_{i=1}^{n} A[i]\, y[i][R]\, Exp[-Z\, m[i]^2] \tag{64}$$

The dimensionless axial coordinate defined by eq. (56 b), after taking into account eq.(58 b) could be rewritten as:

$$Z = 4\,(2/(3+n) + 4\,Kn\beta v)\,X \tag{65}$$

where **X** is the axial distance expressed through Pecklet number $Pe = uav*d/\alpha$ with characteristic length $d=2\,r1$.

$$X = \frac{z/d}{Pe} \tag{66}$$

Than the dimensionless temperature given by eq.(64) could be rewritten as:

$$\theta[R, X] = \sum_{i=1}^{n} A[i]\, y[i][R]\, Exp[-4\,(2/(3+n) + 4\,Kn\beta v)\,X\, m[i]^2] \tag{67}$$

The constants **A[i]** in the solution (67) are given by:

$$A[i] = \frac{\int_0^1 R^n\,(1 + 4\,Kn\beta v - R^2)\, y[i][R]\, dR}{\int_0^1 R^n\,(1 + 4\,Kn\beta v - R^2)\, y[i][R]^2\, dY} \tag{68}$$

The dimensionless average temperature **θav[Z]** is defined as:

$$\theta av[X] = \frac{\int_0^1 R^n\,(1 + 4\,Kn\beta v - R^2)\,\theta[R, X]\, dR}{\int_0^1 R^n\,(1 + 4\,Kn\beta v - R^2)\, dR} \tag{69}$$

Introducing **θ[R,X]** from eq. (67) into eq.(69) we obtain:

$$\theta av[X] = \sum_{i=1}^{n} Aav[i]\; Exp[-4\,(2/(3+n) + 4\,Kn\beta v)\,X\, m[i]^2] \tag{70}$$

where

$$Aav[i] = \frac{\int_0^1 R^n\,(1 + 4\,Kn\beta v - R^2)\, y[i][R]\, dR}{\int_0^1 R^n\,(1 + 4\,Kn\beta v - R^2)\, dR}\, A[i] \tag{71}$$

The heat transfer coefficient **h[z]** is determined from the balance equation:

$$h[z]\,(Tav[z] - Ts) = -k\,\frac{\partial T[r1, z]}{\partial r} \tag{72}$$

The Nusselt number **Nu[X]=h[z]*(2r1)/k** is given by:

$$Nu[X] = -\frac{2}{\theta av[X]}\frac{\partial\theta[1, X]}{\partial R} \tag{73}$$

Introducing eqs. (67) and (70) into eq. (73) we obtain the Nusselt number:

$$Nu[X] = -2\,\frac{\sum_{i=1}^{n} A[i]\, y[i]'[1]\ Exp[-4\,(2/(3+n)+4\,Kn\beta v)\,X\,m[i]^{\wedge}2]}{\sum_{i=1}^{n} Aav[i]\ Exp[-4\,(2/(3+n)+4\,Kn\beta v)\,X\,m[i]^{\wedge}2]} \tag{74}$$

For large **Z** only the first terms of both sums in eq. (74) has to be taken into account. Than we obtain:

$$Nu[\infty] = -2\,\frac{\int_0^1 R^n\,(1+4\,Kn\beta v - R^2)\,dR}{\int_0^1 R^n\,(1+4\,Kn\beta v - R^2)\,y[1][R]\,dR}\ y[1]'[1] \tag{75}$$

The integrals in eq. (75) have exact solutions and the limiting Nusselt number becomes:

$$Nu[\infty] = 2\left(\frac{1+4\,Kn\beta v}{1+n} - \frac{1}{3+n}\right) m[1]^2 \tag{76}$$

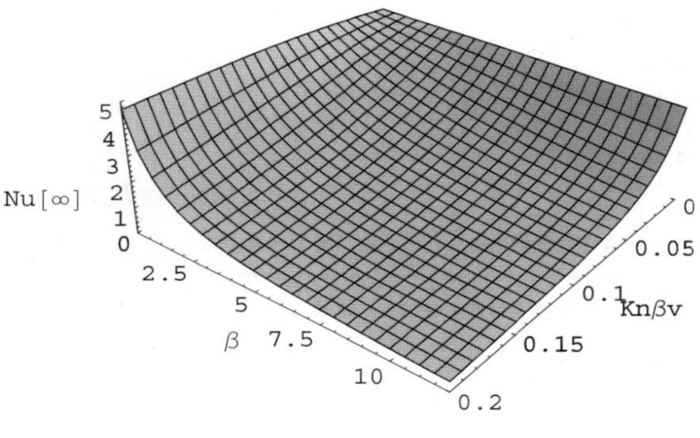

Fig. 6 The limiting Nusselt number as function of **Knβv** and **β**.

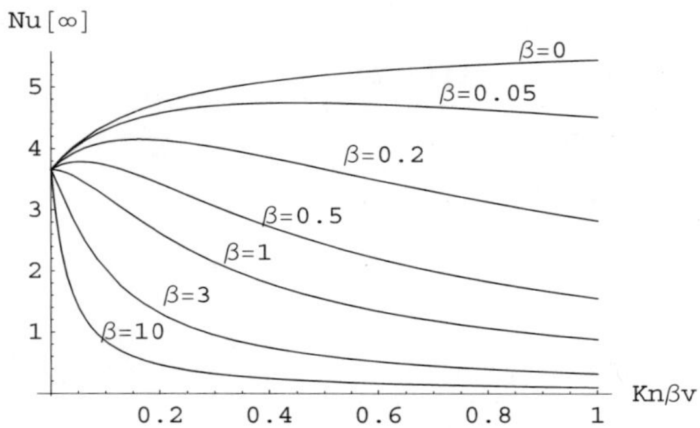

Fig. 7 The limiting Nusselt number as function of **Knβv** and parameter **β** .

The limiting Nusselt number is of great practical interest. For **n=0** (parallel plate micro channel) and **n=1** (micro tube) the limiting Nusselt number depend on 2 parameters: **Knβv** and **β**. The **Knβv** control mainly the velocity slip and have influence on the temperature jump. The parameter **β** control only the temperature jump. The limiting Nusselt number is shown on Fig 6.

In the paper [12] is discovered that for **β=0** Nusselt number increases with increasing of **Knβv**. At large **β=10** this behavior is reversed. To understand this phenomena we plot on the Fig. 7 the **Nu[∞]** versus **Knβv** from **0** to **1**. We see that the curve pass through a maximum. In the interval of practical interest, **0<Knβv<0.1**, the Nusselt number decrease with increasing of **Knβv** when **β>1**.

The dimensionless temperature given by eq.(66) are plotted on the Fig. 8 a, b. Our summation function, in contrast to the built-in *Mathematica* **Sum** function, take as many term as necessary. To avoid using extremely large number of terms we intentionally start plotting from **X=0.01**, because the missing part of the plot is not necessary for our conclusion.

The figure on the left is similar to the one (not shown) for the case **Knβv=0**, **β=0**, **n=1**. The figure on the right demonstrate the temperature jump at **R=1**. We see that the temperature jump change dramatically the temperature distribution. This observation agree with the conclusion of [12].

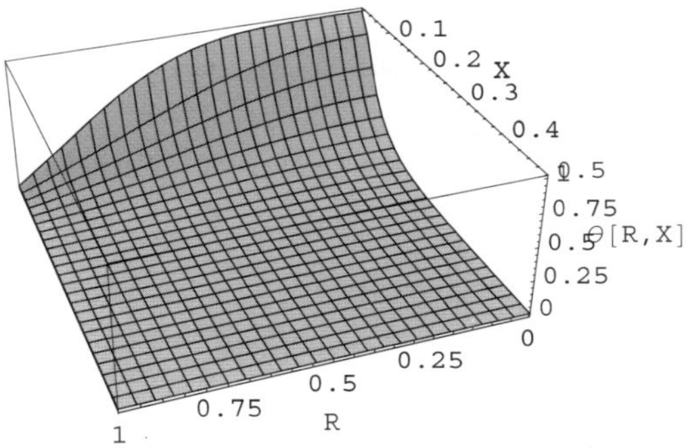

Fig 8a **θ[R,X]** for **Knβv=0.1**, **β=0**, **n=1**

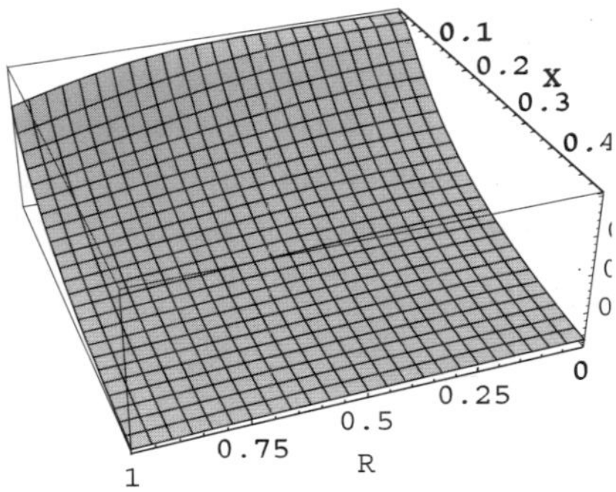

Fig 8b **θ[R,X]** for **Knβv=0.1**, **β=5**, **n=1**

7. Periodic Heat Transfer in Micro Conduits

Consider periodic heat transfer in thermally developing, hydrodynamically developed forced laminar flow inside a parallel plate micro channel or micro tube under following assumptions:

- The fluid is incompressible with constant physical properties.
- The free heat convection is negligible.
- The energy generation is negligible.

- The entrance temperature is harmonic function of time.
- The surface temperature is uniform.
- Only periodic oscillation of temperature in conduits exist.

The temperature $T[r,z,t]$ of a fluid with velocity profile $u[r]$, diffusivity α along the channel $0\le z\le\infty$ in the region $0\le r\le r1$ is described by the following problem:

$$\frac{\partial T[r, z, t]}{\partial t} + u[r] \frac{\partial T[r, z, t]}{\partial z} = \alpha \left(\frac{\partial^2 T[r, z, t]}{\partial r^2} + \frac{n}{r} \frac{\partial T[r, z, t]}{\partial r} \right) \tag{77}$$

where $n=0$ for parallel plate micro channel and $n=1$ for micro tube.

The boundary conditions at the center of the micro conduits is:

$$\frac{\partial T[0, z, t]}{\partial r} = 0 \quad \text{for} \quad n = 0,$$
$$\left(r \frac{\partial T[r, z, t]}{\partial r} \right)_{r->0} = 0 \quad \text{for} \quad n = 1 \tag{78}$$

The boundary condition (78 a) is commonly used for both - parallel plate channel and tube. The correct boundary condition for cylindrical geometry is given by eq. (78 b) [31].

The surface temperature of the micro conduits is Ts. As result of the temperature jump on the surface the boundary condition at $r1$ becomes:

$$T[r1, z, t] = Ts - 2\,Kn\,r1\,\beta t \frac{\partial T[r1, z, t]}{\partial r} \tag{79}$$

where

βt is $((2-\alpha_t)/\alpha_t)(2\gamma/(\gamma+1))/Pr$

α_t is the thermal accommodation coefficient.

λ is the molecular mean free path.

γ is the ratio of specific heat at constant pressure c_p and specific heat at constant volume c_v.

Kn is the Knudsen number.

The entrance temperature oscillate in time with amplitude Ta and frequency ω:

$$T[r,0,t] = Ts + Ta\,Exp[i\,\omega\,t] \tag{80}$$

where $i=\sqrt{-1}$.

It is not necessary to define the initial temperature $T[r,z,0]$ since only periodic temperature oscillations are considered.

To simplify eqs. (77) to (80) we define the dimensionless velocity $W[R]$ and dimensionless temperature $\theta[Y,Z,\tau]$ as:

$$W[R] = u[r] / uav, \quad \theta[R, Z, \tau] = (T[r, z, t] - Ts) / Ta \tag{81}$$

where R is the transverse coordinate, Z is the axial coordinate, and τ the dimensionless time. We define also dimensionless frequency Ω.

$$R = r / r1, \quad Z = z\,\alpha / (C0\, r1^2\, uav), \quad \tau = \alpha\, t / r1^2, \quad \Omega = r1^2\, \omega / \alpha \tag{82}$$

The eqs. (77) to (80), after using eqs.(58) and $Kn * \beta t = Kn\beta v * \beta$, in dimensionless form becomes:

$$\frac{\partial \theta[R, Z, \tau]}{\partial \tau} + (1 + 4\, Kn\beta v - R^2)\, \frac{\partial \theta[R, Z, \tau]}{\partial Z} = \frac{\partial^2 \theta[R, Z, \tau]}{\partial R^2} + \frac{n}{R}\, \frac{\partial \theta[R, Z, \tau]}{\partial R},$$

$$\frac{\partial \theta[0, Z, \tau]}{\partial R} = 0 \quad \text{for} \quad n = 0, \tag{83}$$

$$\left(R\, \frac{\partial \theta[R, Z, \tau]}{\partial R}\right)_{R->0} = 0 \quad \text{for} \quad n = 1$$

$$\theta[1, Z, \tau] + 2\, Kn\, \beta t\, \frac{\partial \theta[1, Z, \tau]}{\partial R} = 0, \qquad \theta[R, 0, \tau] = \mathrm{Exp}[i\, \Omega\, \tau]$$

The periodic solution of the problem (83) could be written as:

$$\theta[R, Z, \tau] = \Phi[R, Z]\, \mathrm{Exp}[i\, \Omega\, \tau] \tag{84}$$

Introducing eq. (84) in the problem (83) we obtain

$$(1 + 4\, Kn\beta v - R^2)\, \frac{\partial \Phi[R, Z]}{\partial Z} = \frac{\partial^2 \Phi[R, Z]}{\partial R^2} + \frac{n}{R}\, \frac{\partial \Phi[R, Z]}{\partial R} - i\, \Omega\, \Phi[R, Z],$$

$$\frac{\partial \Phi[0, Z]}{\partial R} = 0 \quad \text{for} \quad n = 0, \qquad \left(R\, \frac{\partial \Phi[R, Z]}{\partial R}\right)_{R->0} = 0 \quad \text{for} \quad n = 1, \tag{85}$$

$$\Phi[1, Z] + 2\, Kn\beta v\, \beta\, \frac{\partial \Phi[1, Z]}{\partial R} = 0, \qquad \Phi[R, 0] = 1$$

The periodic problem (85) differs from the steady state problem (59) by one term, namely $i\, \Omega\, \Phi[R, Z]$. This term change the solution to give the amplitude and phase lag of temperature oscillation in a point with coordinates R and Z.

To solve this problem we need the eigenvalues m and the eigenfunctions $y[R]$ of the eigenproblem:

$$y''[R] + \frac{n}{R}\, y'[R] + (m^2\, (1 + 4\, Kn\beta v - R^2) - i\, \Omega)\, y[R] = 0,$$

$$(R^n\, y'[R])_{R\to 0} = 0, \qquad y[1] + 2 * Kn\beta v * \beta\, y'[1] = 0 \tag{86}$$

The solution of eq.(86 a) that satisfies the boundary condition (86 b) is

```
y[R] = Exp[-m R^2 / 2]
    1F1[(n + 1 - (1 + 4 Knβv) m + ⅈ Ω / m) / 4; (1 + n) / 2; m R^2]        (87)
```

where **1F1[a;b;c]** is the Kummer confluent hypergeometric function.

Introducing the eq. (87) in the boundary condition (86 c) we obtain the following eigencondition:

```
(1 + n) (1 - 2 Knβv * β * m) *
   1 F1[(n + 1 - (1 + 4 Knβv) m + ⅈ Ω / m) / 4; (n + 1) / 2; m] +
  2 Knβv * β * m (n + 1 - (1 + 4 Knβv) m + ⅈ Ω / m) *                      (88)
   1 F1[(n + 5 - (1 + 4 Knβv) m + ⅈ Ω / m) / 4; (n + 3) / 2; m] = 0
```

The roots of (87) are the desired complex eigenvalues. For **Ω=0** eq.(88) is identical to the eigencondition (63), which has real eigenvalues. The **FindRoot** function of *Mathematica* software system calculates the complex roots starting from the corresponding real eigenvalues for the case Ω=0. For large value of Ω it is possible to use intermediate values **Ω1**, **Ω2**,..up to Ω. Such necessity never appears in our numerical experiments.

For reference purposes Table1 to Table 4 gives only 18 digits of the first 25 eigenvalues, computed with working precession 25 digits. The eigenvalues in the first column are used as a starting values to compute the second column.

The first 10 eigenvalues of the first column of Table 1 and Table 2 are given by Brown [32] and reprinted an several books.

Table 1. Eigenvalues for classical parallel plate channel

	Knβv=0,β=0,Ω=0,n=0	Knβv=0,β=0,Ω=5,n=0
1	1.68159532223898601	2.15112975153277636 + 1.32642339432191060 ⅈ
2	5.66985734589507483	5.70696770616671477 + 0.62676723217555022 ⅈ
3	9.66824246251040440	9.67664338775411890 + 0.39375507706979668 ⅈ
4	13.6676614426075439	13.67074996323983839 + 0.28822522080310368 ⅈ
5	17.6673735653492768	17.6688222684739756 + 0.2279341992852130 ⅈ
6	21.6672053243247877	21.6679918875005698 + 0.1888221115725967 ⅈ
7	25.6670964863338100	25.6675673626738869 + 0.1613455544476919 ⅈ
8	29.6670210446857057	29.6673233471742018 + 0.1409584489184913 ⅈ
9	33.6669660686664992	33.6671706169177060 + 0.1252160009837339 ⅈ
10	37.6669244562645721	37.6670686456413677 + 0.1126846783824453 ⅈ
11	41.6668920062265541	41.6669970385058385 + 0.1024675152833704 ⅈ
12	45.6668660858635695	45.6669446801505820 + 0.0939741262895628 ⅈ
13	49.6668449676188278	49.6669051132172761 + 0.0867998026261179 ⅈ
14	53.6668274742981780	53.6668743848600644 + 0.0806577072379460 ⅈ
15	57.6668127780919717	57.6668499673539652 + 0.0753387959469721 ⅈ
16	61.6668002811511388	61.6668301833852027 + 0.0706870705271318 ⅈ
17	65.6667895417112679	65.6668138840722252 + 0.0665837252519872 ⅈ
18	69.6667802267949085	69.6668002605537245 + 0.0629366655404064 ⅈ
19	73.6667720810166823	73.6667887295051847 + 0.0596733865334472 ⅈ
20	77.6667649054677503	77.6667788612489066 + 0.0567360176065488 ⅈ
21	81.6667585430920774	81.6667703332819333 + 0.0540778006400972 ⅈ
22	85.6667528683506947	85.6667628994509234 + 0.0516605399003962 ⅈ

23	89.6667477797826313	89.6667563690292977 + 0.0494527242537982 i
24	93.6667431945624228	93.6667505922184313 + 0.0474281234232819 i
25	97.6667390444589878	97.6667454499106086 + 0.0455647241780451 i

Table 2. Eigenvalues for classical tube

	Knβv=0,β=0,Ω=0,n=1	Knβv=0,β=0,Ω=5,n=1
1	2.70436441988253216	2.91820775421126864 + 1.07142745215820384 i
2	6.67903144934662777	6.70248892644280731 + 0.54324530861878286 i
3	10.6733795380537356	10.67967013533356749 + 0.35985875022369014 i
4	14.6710784627362121	14.6735786863752128 + 0.2699827808484110 i
5	18.6698718644512204	18.6710978818036623 + 0.2165265811159432 i
6	22.6691433588373313	22.6698283770102043 + 0.1810043580243207 i
7	26.6686619960114615	26.6690803849557321 + 0.1556475642222704 i
8	30.6683233409175399	30.6685959531519244 + 0.1366174849233934 i
9	34.6680738224337810	34.6682603936911956 + 0.1217967130045891 i
10	38.6678833468597878	38.6680160570524244 + 0.1099203276391844 i
11	42.6677338055420268	42.6678311879571872 + 0.1001854660245345 i
12	46.6676136978161583	46.6676870121963735 + 0.0920576607219898 i
13	50.6675153950972418	50.6675717881230725 + 0.0851671465649795 i
14	54.6674336526801355	54.6674778290284757 + 0.0792498451586722 i
15	58.6673647548702367	58.6673999088958320 + 0.0741120567713564 i
16	62.6673060011075842	62.6673343600230422 + 0.0696084372482675 i
17	66.6672553849654530	66.6672785378387396 + 0.0656277675830670 i
18	70.6672113869109676	70.6672304906190268 + 0.0620834756663163 i
19	74.6671728366289700	74.6671887485548698 + 0.0589071547704354 i
20	78.6671388192384304	78.6671521849680363 + 0.0560440277441423 i
21	82.6671086099779761	82.6671199225788567 + 0.0534497073426125 i
22	86.6670816278131802	86.6670912687207947 + 0.0510878398409221 i
23	90.6670574018969905	90.6670656696313252 + 0.0489283629282203 i
24	94.6670355469334429	94.6670426776020887 + 0.0469461986675914 i
25	98.6670157448184896	98.6670219269754647 + 0.0451202597114564 i

Table 3. Eigenvalues for parallel plate micro channel

	Knβv=1/10,β=10,Ω=0,n=0	Knβv=1/10,β=10,Ω=5,n=0
1	0.620606815630125819	1.58598905474639946 + 1.41394993934941135 i
2	3.37436334790540790	3.45857505411605252 + 0.78649462782569145 i
3	6.35778966236857578	6.36848071152635013 + 0.41245541760304200 i
4	9.38641708284046428	9.38948723919208168 + 0.27401589698266492 i
5	12.4333810175627762	12.43466961021081840 + 0.20436000099882007 i
6	15.4898417157025838	15.4905054091628506 + 0.1627619784520016 i
7	18.5519010489352100	18.5522886237567213 + 0.1351909609576084 i
8	21.6175462516132315	21.6177925622941532 + 0.1155986609976410 i
9	24.6856269913764859	24.6857933639582085 + 0.1009666699287344 i
10	27.7554367241828673	27.7555544058833817 + 0.0896251874337072 i
11	30.8265172454364277	30.8266035541280602 + 0.0805770697532097 i
12	33.8985584189896865	33.8986235948668532 + 0.0731906611063120 i
13	36.9713429188450894	36.9713933371207887 + 0.0670466055081759 i
14	40.0447140051985834	40.0447538052013288 + 0.0618556511142244 i
15	43.1185558477186627	43.1185878127196546 + 0.0574118600759709 i
16	46.1927810408854071	46.1928070985777579 + 0.0535646404841526 i

17	49.2673224208613408	49.2673439407172849 + 0.0502013291781771 ⅈ
18	52.3421275496712028	52.3421455262882217 + 0.0472359595267210 ⅈ
19	55.4171549055938807	55.4171700754178138 + 0.0446017963337620 ⅈ
20	58.4923711949841840	58.4923841127210044 + 0.0422462425009147 ⅈ
21	61.5677494189886526	61.5677605088711432 + 0.0401272828075749 ⅈ
22	64.6432674593202832	64.6432770502628308 + 0.0382109496507700 ⅈ
23	67.7189070277778361	67.7189153779578151 + 0.0364694838291102 ⅈ
24	70.7946528750683003	70.7946601894185888 + 0.0348799777209645 ⅈ
25	73.8704921873673931	73.8704986302236735 + 0.0334233594349905 ⅈ

Table 4. Eigenvalues for micro tube

	Knβv=1/10,β=10,Ω=0,n=1	Knβv=1/10,β=10,Ω=5,n=1
1	0.968387880253570957	1.79839046940073879 + 1.46363741323785155 ⅈ
2	3.97870600234012755	4.02401338098537652 + 0.66143305096659160 ⅈ
3	7.02508601797462023	7.03248414339845701 + 0.36775743918339512 ⅈ
4	10.0824159024304246	10.08482686389381933 + 0.25253378365548096 ⅈ
5	13.1457582602832749	13.14683842382091169 + 0.19194700198320030 ⅈ
6	16.2127594212002179	16.2133371263822015 + 0.1547333253457390 ⅈ
7	19.2821426413087179	19.2824882117760954 + 0.1295917801270077 ⅈ
8	22.3531573391840442	22.3533806315093538 + 0.1114785941072268 ⅈ
9	25.4253348187299712	25.4254874898170053 + 0.0978111713444075 ⅈ
10	28.4983681402710388	28.4984771428454714 + 0.0871324741825593 ⅈ
11	31.5720483712104101	31.5721289090141028 + 0.0785588731928756 ⅈ
12	34.6462286524613444	34.6462898389948998 + 0.0715236202739219 ⅈ
13	37.7208029190772781	37.7208504900928734 + 0.0656465956614858 ⅈ
14	40.7956927681576954	40.7957304810119563 + 0.0606633722880627 ⅈ
15	43.8708390654859577	43.8708694654053980 + 0.0563843287079726 ⅈ
16	46.9461964164292119	46.9462212773609240 + 0.0526699490084832 ⅈ
17	50.0217294265521819	50.0217500155878232 + 0.0494152999116561 ⅈ
18	53.0974101134128790	53.0974273551473377 + 0.0465399406490800 ⅈ
19	56.1732160779841471	56.1732306598794658 + 0.0439811703020174 ⅈ
20	59.2491291888353920	59.2491416306149612 + 0.0416893923036444 ⅈ
21	62.3251346195394530	62.3251453199951583 + 0.0396248602688932 ⅈ
22	65.4012201338982902	65.4012294031396097 + 0.0377553477039294 ⅈ
23	68.4773755479401798	68.4773836300395809 + 0.0360544494284022 ⅈ
24	71.5535923199254428	71.5535994090948893 + 0.0345003235595018 ⅈ
25	74.6298632343358741	74.6298694866632083 + 0.0330747462655671 ⅈ

The solution of the problem (85) is a special case from the general case considered in the book [20].

$$\Phi[R, X] = \sum_{i=1}^{n} A[i]\, y[i][R]\, \mathrm{Exp}[-4\,(2/(3+n) + 4\,Kn\beta v)\, X\, m[i]^{\wedge}2] \tag{89}$$

where **y[i] [R]** is given by eq.(87) and the constants **A[i]** is defined by eq.(68).

The solution (89) is used to plot Fig. 9 and Fig. 10, where the vertical distances to the surface present the amplitudes, while the color of the surface present the phase angle. As the angle moves around the circle, the color of the surface will go from red to blue, green, yellow, and back to red again. Fig. 9 shows the temperature oscillations in tube without velocity slip and temperature jump. Fig. 10 shows temperature oscillations in tube with large velocity slip and temperature jump. The comparison of Fig 9 and 10 shows that the temperature jump change dramatically the temperature oscillations.

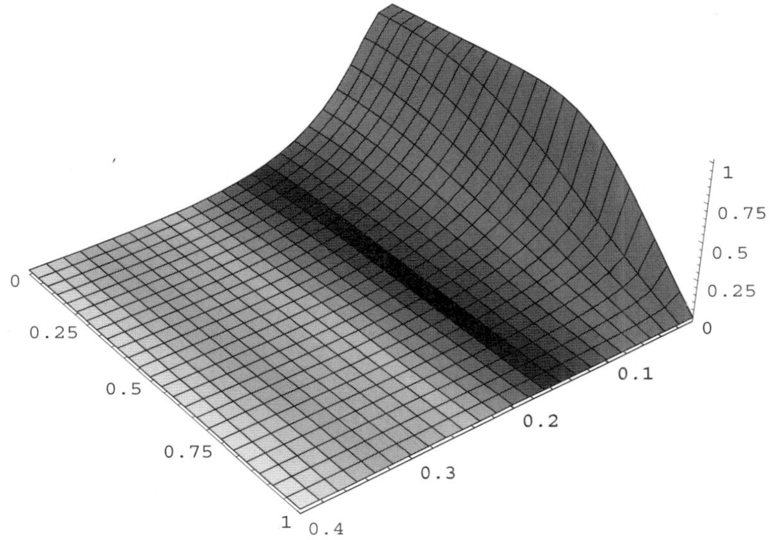

Fig. 9 The temperature oscillations in tube: **Knβv=0**, **β=0**, **Ω=5**, **n=1**

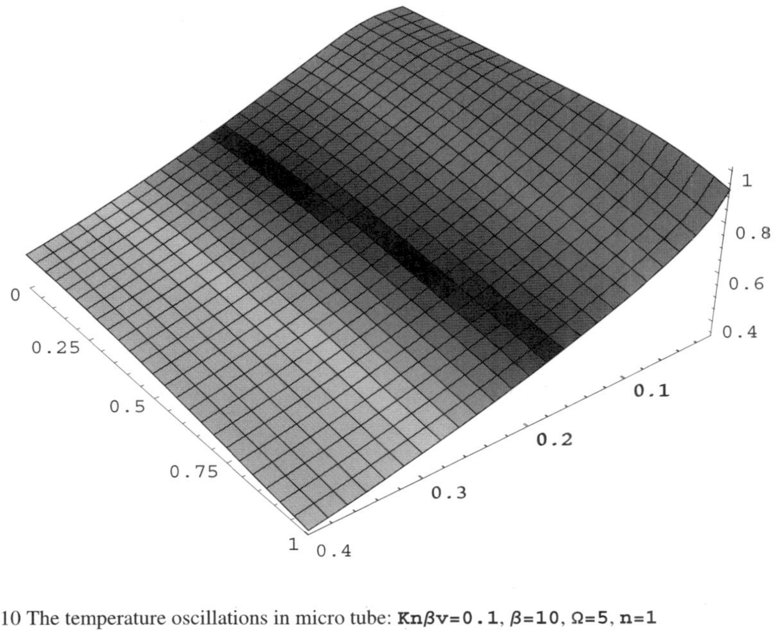

Fig. 10 The temperature oscillations in micro tube: **Knβv=0.1**, **β=10**, **Ω=5**, **n=1**

8. Electro-Osmotic Heat Transfer in Micro Conduits at Constant Wall Flux

Consider the heat transfer in thermally and hydrodynamically developed electro-osmotic flow inside a micro conduits under following assumptions:

- The fluid is a liquid with constant thermophysical properties.
- The velocity profile u[r] is fully developed.
- The free convection of heat is negligible.
- The temperature profile is stabilized.
- The heat flux at the tube wall is a constant.

The temperature **T[r,x]** in conduit (**0≤r≤r1**, **0≤x<∞**) of a fluid with velocity **u[r]**, density ρ, specific heat **c**, thermal conductivity **k**, thermal energy generation **g[r]**, constant surface flux qw, and initial temperature **Ti** is described by:

$$c\,\rho\,u[r]\,\frac{\partial T[r,x]}{\partial x} = k\left(\frac{\partial^2 T[r,x]}{\partial r^2} + \frac{n}{r}\,\frac{\partial T[r,x]}{\partial r}\right) + g[r],$$

$$\left(r^n\,\frac{\partial T[r,x]}{\partial r}\right)_{r\to 0} = 0, \qquad k\,\frac{\partial T[r1,x]}{\partial r} = qw, \qquad T[r,0] = Ti \tag{90}$$

where **n=0** for parallel plate micro channel and **n=1** for micro tube.

The dimensionless temperature **θ[R,X]**, transverse coordinate **R**, axial coordinate **X**, velocity **U[R]** and thermal energy generation **G[R]** are defined as:

$$\theta[R,X] = (T[r,x] - Ti)\,/\,(qw\,r1\,/\,k),$$

$$R = r\,/\,r1, \qquad X = (k\,x)\,/\,(c\,\rho\,um\,r1^2), \tag{91}$$

$$U[R] = u[r]\,/\,um, \qquad G[R] = r1\,g[r]\,/\,qw$$

The dimensionless form of the problem (90) is:

$$U[R]\,\frac{\partial\theta[R,X]}{\partial X} = \frac{\partial^2\theta[R,X]}{\partial R^2} + \frac{n}{R}\,\frac{\partial\theta[R,X]}{\partial R} + G[R],$$

$$\left(R^n\,\frac{\partial\theta[R,X]}{\partial R}\right)_{R\to 0} = 0, \qquad \frac{\partial\theta[1,X]}{\partial R} = 1, \qquad \theta[R,0] = 0 \tag{92}$$

Since both boundary conditions, (92 b, 92 c), are of the second kind, the average temperature is obtained below directly as described by Mikhailov and Ozisik in the book [20]:

$$\theta av[X] = X\,\frac{1 + \int_0^1 R^n\,G[R]\,dR}{\int_0^1 R^n\,U[R]\,dR} \tag{93}$$

The Nusselt number **Nu[X]=h[x]*2*r1/k** is defined as:

$$Nu[X] = \frac{2}{\theta[1, X] - \theta av[X]} \tag{94}$$

The problem (91) is splitted into a steady state term $\theta s[R]$ and transient term $\Phi[R, \tau]$. Since both boundary conditions (92 b, 92 c) are of the second kind, the average temperature has to be included into the splitting formula [20]:

$$\theta[R, X] = \theta av[X] + \theta s[R] + \Phi[R, X] \tag{95}$$

For large X the transient solution gives $\Phi[R,X]=0$ and the Nusselt number (94) becomes:

$$Nu[X] = 2 / \theta s[1] \tag{96}$$

The splitting procedure gives the following problem for the stabilized temperature profile $\theta s[R]$:

$$\theta s''[R] + n\,\theta s'[R] / R + G[R] = Cav\,U[R],$$

$$(R^n\,\theta s'[R])_{R->0} = 0, \quad \theta s'[1] = 1, \quad \int_0^1 R^n\,U[R]\,\theta s[R]\,dR = 0 \tag{97}$$

The constant Cav in eqn. (97 a) and its components Gav and Uav are defined as:

$$Cav = (n + 1 + Gav) / Uav,$$

$$Gav = (1 + n) \int_0^1 R^n\,G[R]\,dR, \quad Uav = (1 + n) \int_0^1 R^n\,U[R]\,dR \tag{98}$$

The solution of the problem (97) gives the stabilized temperature profile:

$$\theta s[R] = - \int_0^R \eta^{-n} \int_0^\eta \xi^n\,G[\xi]\,d\xi\,d\eta + Cav \int_0^R \eta^{-n} \int_0^\eta \xi^n\,U[\xi]\,d\xi\,d\eta + \frac{(1 + n)}{Uav} * \left(\int_0^1 \rho^n \left(\int_0^\rho \eta^{-n} \int_0^\eta \xi^n\,G[\xi]\,d\xi\,d\eta\right) U[\rho]\,d\rho - Cav \int_0^1 \rho^n \left(\int_0^\rho \eta^{-n} \int_0^\eta \xi^n\,U[\xi]\,d\xi\,d\eta\right) U[\rho]\,d\rho\right) \tag{99}$$

Using (99) in eq. (96) we obtain the desired Nusselt number.

As an example we consider the case of micro tube studied by Maynes and Webb [21], where the dimensionless heat generation is $G[R]=S$. The velocity profile $U[R]=1-I_0[R\,Z]/I_0[Z]$ is given above as eq. (46). For these $G[R]$ and $U[R]$ the eqs (99) gives the following stabilized temperature profile:

$$\theta s[R] = \frac{1}{4\,Z^2\,(Z\,I_0[Z] - 2\,I_1[Z])^2} (Z^2\,(-16 - 4\,S - Z^2 + 2\,R^2\,Z^2)\,I_0[Z]^2 + Z\,I_0[Z] (-4\,(2 + S)\,Z\,I_0[R\,Z] + (32 + (4 + 2\,R^2\,(-2 + S) - S)\,Z^2)\,I_1[Z]) + 4\,I_1[Z]\,(2\,(2 + S)\,Z\,I_0[R\,Z] + (2\,Z^2 + S\,(4 - (-2 + R^2)\,Z^2))\,I_1[Z])) \tag{100}$$

Using eq (100) in eq. (96) we obtain the limiting Nusselt number:

$$Nu[\infty] == (8\,Z^2\,(Z\,I_0[Z] - 2\,I_1[Z])^2) \,/\, (Z^2\,(-24 - 8\,S + Z^2)\,I_0[Z]^2 + Z\,(48 + S\,(8 + Z^2))\,I_0[Z]\,I_1[Z] + 4\,(2\,Z^2 + S\,(4 + Z^2))\,I_1[Z]^2) \tag{101}$$

It is interesting to note that (100) and (101) are different but equivalent to the formulas reported by Maynes and Webb [21].

Fig. 11 shows the plot of Nusselt number, eq.(100), as function of heat generation due to resistance **S** and the ratio of tube radius to Debye length **Z**.

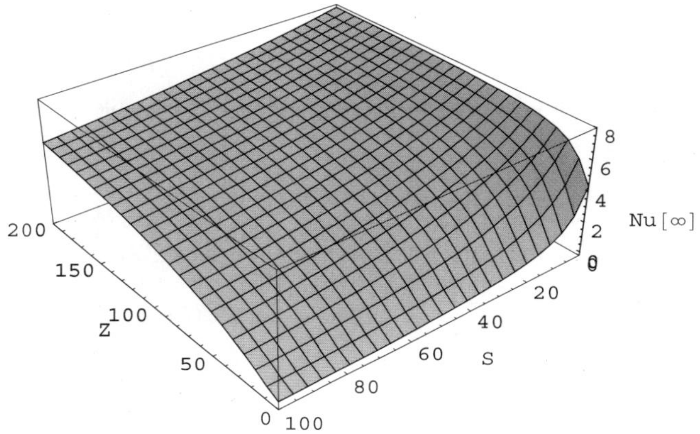

Fig. 11 Limiting Nusselt number as function of heat generation due to resistance **S** and ratio of tube radius to Debye length **Z**.

Acknowledgements

The financial support provided by FAPERJ and CNPq, Brazil is gratefully acknowledged.

References

1. Choi S. B., R. F. Barron, and R. O. Warrington, 1991, Fluid Flow And Heat Transfer In Microtubes, Micromechanical Sensors, Actuators, And Systems ASME DSC 32, 123-134.

2. Pfahler J., J. Harley, H. H. Bau, and J. Zemel, 1991, Gas and Liquid Flow In Small Channels, Micromechanical Sensors, Actuators, and Systems, ASME DSC 32, 49-60.

3. Petersen G. P., A. B. Duncan and M. H. Weichold, 1993, Experimental investigation of micro heat pipes fabricated in silicon wafer, ASME J. Heat Transfer, vol. 115, pp.751-756.

4. Barron R. F, X. Wang, R. O. Warrington, and T. Ameel, 1996, Evaluation of the eigenvalues for the Graetz problem in slip-flow, Int. Comm. Heat Mass Transfer 23 (4), 1817-1823.

5. Graetz L, 1885, Uber die Warmeleitungsfahigkeit von Flussigkeiten, Annalen der Physik und Chemie, part 1, vol. 18, pp. 79-94, (1883); part 2, vol. 25, pp. 337-357.

6. Bailey P. B., M. K. Gordon, and L. F. Shampine, 1978, Automatic Soilution of the Sturm-Liouville Problem", ACM Transactions on Mathematical Software, Vol. 4, No 3.

7. Mikhailov M. D. and N. L. Vulchanov, 1983, A computational procedure for Sturm-Liouville problems, J. Comp. Phys, 50, 323-336.

8. Mikhailov M. D. and R. M. Cotta, 1994, Integral Transform Method for Eigenvalue Problems, Comm. Num. Meth. Eng. , 10, 827-835.

9. Mikhailov M. D. and R. M. Cotta, 1997, Eigenvalues for the Graetz Problem in Slip-Flow, Int. Comm. Heat & Mass Transfer, V.24, no.3, pp.449-451.

10. *Mathematica* software system, www.wolfram.com

11. Barron R. F, X. Wang, T. A. Ameel, and R. O. Warrington, 1997, The Graetz Problem Extended To Slip-flow, Int. J. Heat Mass Transfer 40 (8), 563-574.

12. Larrode, F. E., C. Housiadas, and Y. Drossinos, 2000, Slip Flow Heat Transfer in Circular Tubes, International Journal of Heat and Mass Transfer, 43 (2000) 2669-2680.

13. Maxwell J. C., 1890, On condition to be satisfied by a gas at the surface of a solid body, In The Scientific Papers of James Clerk Maxwell, Vol. 2. Cambridge University Press, London, pp. 704.

14. Eckert E. R. G., R. M. Drake, 1972, Amalysis of the Heat and Mass Transfer, McGraw-Hill.

15. Tung G., Y. Bayazitoglu, Heat transfer in microtubes with viscous dissipation, International Journal of Heat Mass Transfer, 44 (2001), 2395-2403.

16. Gravensen P., J. Branebjerg, O. S. Jensen, 1993, Microfluids - a review, J. Micromech. Microeng. 3, 168-182.

17. Polson N. A., M. A. Hayes, 2000, Electro-osmotic flow control of fluids on a capillary electrophoresis microdevice using an applied external voltage, Anal. Chem. 72, 1088-1092.

18. Reuss F. F., 1809, Charge-induced flow, Proc. Imp. Soc. Natural, Moscow, 3, 327-344.

19. Probstein R. F., 1994, Physicochemical Hydrodynamics, second ed., Wiley.

20. Mikhailov M. D. and M. N. Ozisik, 1984 and 1994, Unified Analysis and Solutions of Heat and Mass Diffusion, John Wiley and Dover.

21. Maynes D., B. W. Webb, 2003, Fully-developed electro-osmotic heat transfer in microchannels, Int. J. Heat and Mass Transfer, 46, 1359-1369.

22. Maynes D., B. W. Webb, 2004, The effect of viscous dissipation in thermally fully-developed electro-osmotic heat transfer in microchannels, Int. J. Heat and Mass Transfer, 47, 987-999.

23. Kakaç S., Y. Yener, 1973, Exact solution of the transient forced convection energy equation for time wise variation of inlet temperature, Int. J. Heat Mass Transfer 16, 2205-2214.

24. Kakaç S., 1975, A general solution to the equation of transient forced convection with fully developed flow, Int. J. Heat Mass Transfer 18, 449-1453.

25. Kakaç S., Y. Yener, 1983, Transient Laminar Forced Convection in Ducts, in S. Kakaç, R. K. Shah, A. E. Bergles (eds), Low Reynolds Number Flow Heat Exchangers, pp. 205-227, Hemisphere, New Yourk.

26. Kim W. S., R. M. Cotta, M. N. Ozisik, 1990, Laminar internal forced convection with periodically varying, arbitrary shaped inlet temperature
Proceedings of the 9. International Heat Transfer Conference, Israel, pp. 383-388.

27. Unsal M., 1990, A solution for the complex eigenvalues and eigenfunctions of periodic Graetz problem, International Communications in Heat Mass Transfer, 25, 4, 585-592.

28. Kennard E. H., 1938, Kinetic Theory of Gases, McGraw-Hill Book Company, Inc., New York.

29. Hagen G., 1839, Uber die Bewegung des Wassers in Ergen Zylindrischen, Rohren. Pogg.Ann., 46, 423-442, 1839.

30. Poiseulle J., 1841, C. R. Acad Sci., 11, 961-967, 1041-1048 (1840); 12, 112-115.

31. Mikhailov M. D. and R. M. Cotta, 2001, Integral Transform Method with *Mathematica,* http://lttc.com.ufrj.br

32. Brown G. M., 1960, Heat or mass transfer in a fluid in Laminar flow in a circular or flat conduit, AIChE J., 6, 179-183.

FLOW REGIMES IN MICROCHANNEL SINGLE-PHASE GASEOUS FLUID FLOW

Y. BAYAZITOGLU
Department of Mechanical Engineering and Materials Science – Rice University
Houston, Texas, USA

S. KAKAÇ
Department of Mechanical Engineering – University of Miami
Coral Gables, Florida, USA

1. Introduction

As the market induces electronic chips to undergo a size reduction while increasing functionality, the use of convective heat transfer in microchannels is believed to be one of the most efficient ways to provide the necessary cooling. Many analytical and experimental investigations have been performed to provide a better understanding of liquid and gaseous flow and heat transfer at microscale, which is very important in microdevice development and design. However, these studies have yet to lead to a general conclusion. Controversial results in the literature about the boundary conditions for liquid flows show that further investigations are still needed.

The current standing on the study of gaseous slip flow in microchannels stipulates that although the continuum assumption is no longer valid within the slip region, Navier-Stokes equations are still applicable with some boundary modifications. Typically, macrochannel boundary conditions, gas velocity and temperature, that are applied to fluid flow and heat transfer equations are equivalent to the corresponding wall values. On the other hand, these conditions do not hold for rarefied gas flow in microchannels. Not only does the fluid slip along the wall with a finite tangential velocity, but there is also a jump between wall and fluid temperatures. A molecular approach using the Boltzmann equation will be employed for flow in the high Knudsen number regime that are the transition flow and the free-molecular flow regimes.

This paper will help explain to readers about the gaseous flow behavior as the molecular mean free path becomes more and more comparable to the channel size. The explanation is based on several experimental, analytical, and numerical results of the authors and many other researchers.

2. Flow Regimes

The validity of continuum flow assumption is unquestionable in solving many macroscopic heat transfer computations. However, when the flow is passed through microchannels, the continuum flow

S. Kakaç et al. (eds.), Microscale Heat Transfer, 75–92.

may no longer be valid because the ratio of the molecular mean free path to the characteristic length becomes relatively significant. This ratio is known as the Knudsen number (Kn), which is an important parameter to explain the surface effects in gaseous flows in microchannels. The mean free path is defined as the average distance traveled by a molecule before colliding with another molecule. The mean free path of some gases at atmospheric conditions is presented in Table 1.

Table 1: Mean Free Path of Some Atmospheric Gases at Room Temperature

Air	Nitrogen	Hydrogen	Helium
68 nm	66 nm	125 nm	194 nm

The continuum flow assumption will only be valid when $Kn < 10^{-3}$. As Kn increases, the flow enters the slip flow regime ($10^{-3} < Kn < 10^{-1}$), transition flow regime ($10^{-1} < Kn < 10$), and eventually the free-molecular flow regime ($Kn > 10$). These four regimes are illustrated in Figure 1.

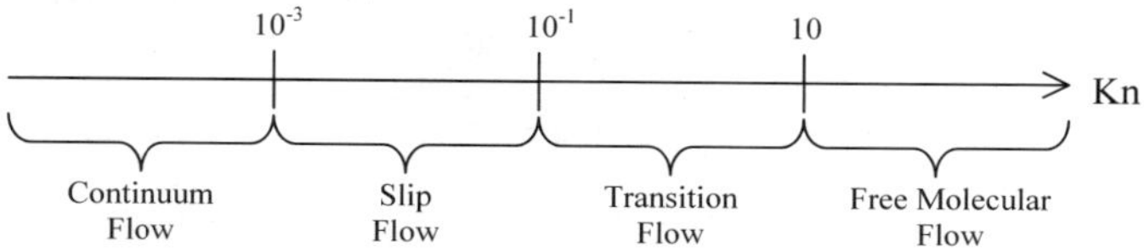

Figure 1. Flow regime classifications.

2.1 CONTINUUM FLOW REGIME

Gaseous flow in microchannels is considered continuum when the Knudsen number is less than 0.001. In this regime, the continuum assumption is valid because the molecular mean free path is much smaller compared to the channel size. This assumption is widely used for macroscopic heat transfer problems. Solutions of fluid flow in this regime are obtained using the Navier-Stokes equations. The fluid velocity and temperature in this regime are equivalent to the corresponding wall conditions. The viscous heating effect is usually neglected at moderate velocity and may only be considered at high velocity. These boundary layer approximations are known to apply [1]:

$$u >>> v \tag{2.1}$$

$$\frac{\partial u}{\partial y} >>> \frac{\partial u}{\partial x}, \frac{\partial v}{\partial x}, \frac{\partial v}{\partial y} \tag{2.2}$$

$$\frac{\partial T}{\partial y} >>> \frac{\partial T}{\partial x} \tag{2.3}$$

Equations (2.1) and (2.2) are the velocity boundary layer approximations and (2.3) is the thermal boundary layer approximation. Thus, for the steady, two-dimensional flow of an incompressible fluid with constant properties, flow continuity can be expressed as:

$$\frac{\partial u}{\partial x} = 0 \tag{2.4}$$

Then, the following equations (2.5) and (2.6) are the equations for the flow momentum and the energy in the axial direction and can be expressed as:

$$\rho\left(u\frac{\partial u}{\partial x} + v\frac{\partial u}{\partial y}\right) = -\frac{\partial P}{\partial x} + \mu\frac{\partial^2 u}{\partial y^2} \tag{2.5}$$

$$\rho c_p\left(u\frac{\partial T}{\partial x} + v\frac{\partial T}{\partial y}\right) = k\frac{\partial^2 T}{\partial y^2} + \mu\left(\frac{\partial u}{\partial y}\right)^2 \tag{2.6}$$

The flow is considered laminar for Reynolds numbers up to 2300. The values for the Poiseuille number for laminar, incompressible flow through cylindrical and parallel plate channels are 64 and 96, respectively. More Poiseuille number values are presented in Table 2.

Table 2: Poiseuille number for different geometries

	Aspect Ratio	*f*Re
Circular	-	64
Rectangular	1	57
	0.7	59
	0.5	62
	0.33	69
	0.25	73
	0.125	82
Parallel Plate	∞	96
Triangular	-	53

The Nusselt number (Nu) for laminar, fully developed flow is constant and independent of Re, Pr, and the axial location. Under these flow conditions Nu values for a cylindrical channel with uniform wall heat flux and uniform wall temperature are 4.36 and 3.66, respectively. More Nusselt number values for rectangular channels are available in Table 3.

The aforementioned continuum flow solutions are very reliable and widely acknowledged by worldwide researchers. Continuum flow solutions should be used to compare any analytical slip flow results.

2.2 SLIP FLOW REGIME

In the slip flow regime, the Navier-Stokes equations are applicable except in the layer next to the surface, the Knudsen layer as illustrated in Figure 2. To use the Navier Strokes equations throughout the entire domain, a fictitious boundary condition is derived for the velocity and temperature of the fluid next to the wall to account for the discontinuities.

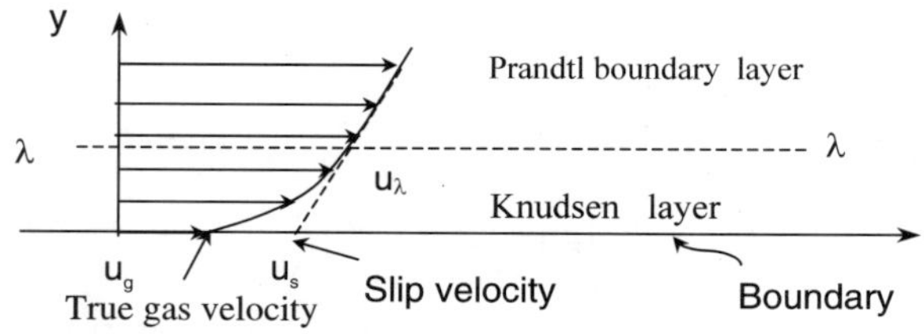

Figure 2. Schematic figure used to obtain the slip velocity.

For flows in conventional channels, the flow dimensions are much larger than the molecular mean free path. Therefore, fluid properties are determined primarily by intermolecular collisions. As the channel size is reduced, the molecular mean free path becomes comparable to channel size. Intermolecular collisions lose their importance and the interactions between the fluid and the wall become significant. The derivations of the slip flow boundary conditions using the kinetic theory of gases will be shown based on the derivations of [2] and [3] and are explained in the following manuscript in this book [4]. Briefly, the first order velocity slip is given by:

$$u_s = \frac{2-F_m}{F_m}\lambda\left(\frac{du}{dy}\right)_0 \tag{2.7}$$

where F_m is the tangential momentum accommodation coefficient, while the first order temperature jump is given by [2-9]:

$$T_s - T_w = \frac{2-F_T}{F_T}\frac{2\gamma}{\gamma+1}\frac{\lambda}{Pr}\left(\frac{\partial T}{\partial y}\right)_0 \tag{2.8}$$

where F_T is the thermal accommodation coefficient [4].

If we want to consider the higher order terms in slip flow, we refer to the development of [3] in equations (2.9) and (2.10) and [5] in equation (2.11).

$$u_s = \frac{2-F_m}{F_m}\frac{Kn}{1-b\,Kn}\left(\frac{du}{d\eta}\right)_0 + res \tag{2.9}$$

where the residual, "res", is given by:

$$res = \frac{2-F_m}{F_m}\left[\begin{array}{l} \dfrac{Kn^3}{2}\dfrac{d^3u}{d\eta^3}\Big|_0 + \dfrac{Kn^4}{6}\dfrac{d^4u}{d\eta^4}\Big|_0 + \dfrac{Kn^5}{24}\dfrac{d^5u}{d\eta^5}\Big|_0 + \cdots \\ -Kn^3\dfrac{\left(d^2u/d\eta^2\right)_0^2}{\left(du/d\eta\right)_0} - Kn^4\dfrac{\left(d^2u/d\eta^2\right)_0^3}{\left(du/d\eta\right)_0} - Kn^5\dfrac{\left(d^2u/d\eta^2\right)_0^4}{\left(du/d\eta\right)_0} - \cdots \end{array}\right] \tag{2.10}$$

$$u_s = \frac{2-F_m}{F_m}\left(Kn\frac{\partial u}{\partial \eta}\bigg|_s + \frac{Kn^2}{2}\frac{\partial^2 u}{\partial \eta^2}\bigg|_s + \frac{Kn^3}{6}\frac{\partial^3 u}{\partial \eta^3}\bigg|_s + \cdots \right) \tag{2.11}$$

In this slip flow expression, b is defined as $\left.\frac{d^2u}{d\eta^2}\right|_0 \Big/ \left.\frac{du}{d\eta}\right|_0$. In the case of gaseous flow between two parallel plates, the values of $d^2u/d\eta^2$ and $du/d\eta$ are -2 and 1, respectively.

Modified Navier-Stokes equations are also used by [10] in the range of $0.01 < Kn < 30$ and the results are compared to Direct Simulation Monte Carlo (DSMC) and linearized Boltzman solutions. They obtained good results for the centerline velocity, assuming b = -1, but deviations for the slip velocity for $0.1 < Kn < 5$.

In this relation, F_m , the tangential momentum accommodation coefficient, is a function of the interaction between gas molecules and the surface. If the surface is smooth and reflects the molecules specularly, F_m will be zero. For diffuse reflections $F_m=1$. This means that all the tangential momentum is lost at the wall. Diffuse reflection results from the penetration of the molecules into interstices in the surface where multiple impacts occur before the molecules depart.

Accommodation coefficients may be significantly different from unity for light atoms and closer to unity for heavy atoms. As shown experimentally in [11], F_m values for slip flow of argon, nitrogen, and carbon dioxide fell between 0.75 and 0.85. The results also showed that F_m is independent of pressure. Their channels are not isolated from contamination to obtain realistic values. Experimental mass flow rate values agree well with the analytical predictions using the slip boundary condition and experimentally determined momentum accommodation coefficients.

Many contributions to experimental and analytical results of gaseous flows in microchannels have been proposed the last decade. Some of them are cited in chronological order in references [12-33].

Experiments were conducted to measure flow and heat transfer characteristics of gaseous flows in microchannels in [12]. Their experimental result of the Poiseuille number is 118 for laminar flow, which is higher than the expected value. They also reported that the flow transition from laminar to turbulent occurs at Reynolds numbers around 400 to 900, which is lower than the conventional value of 2300.

The friction factors for liquids and gases were measured in [13] and [14]. Nitrogen gas and alcohol were used in channels with depths of 0.5 to 50 micrometers. They observed a lower friction factor than macro chanels which increased with Re for small Re and became independent of Re for large Re. In another analysis, they used nitrogen, helium, isopropyl, and silicone oil to determine the flow characteristics in channels with hydraulic diameters varying from 0.5 to 50 micrometers. For both gases and liquids, lower friction factor values than the macro chanel values are obtained. Isopropyl results showed a dependency on the channel size. Silicone oil results, on the other hand showed a Re dependency. They concluded that the small friction value for liquids is due to the reduction of viscosity with decreasing size, and for gases due to the rarefaction effects. It was also reported in [15] after a study of liquid flow in microchannels that there is a critical dimension below which the Navier-Stokes equations cannot be used to obtain the characteristic flow properties.

Dry nitrogen gas was used to obtain the heat transfer coefficient in both laminar and turbulent flow regimes in tubes with inner diameters between 3 and 81 micrometers [16]. Entrance effects were avoided by using long channels. Heat transfer was found to be a function of Re in the laminar regime as opposed to a constant Nusselt number for a thermally fully developed flow in a conventionally sized channel. Experimental values for turbulent flow heat transfer coefficients are as much as seven times larger than those obtained by using well-known relations for turbulent flow in macrochannels. The ratio of micro to macro turbulent Nu values are obtained as a function of Re as follows:

$$\frac{Nu_{micro}}{Nu_{macro}} = 0.000166\,Re^{1.16} \tag{2.12}$$

Another study, [17], used helium as their working fluid and carried out the experiments in 51.25 x 1.33 micrometer microchannels. They showed that, as long as the Knudsen number is in the slip flow range, the Navier-Stokes equations are still applicable and the discontinuities at the boundaries need to be represented by the appropriate boundary conditions. They obtained the following formula for the mass flow rate including the slip effects

$$\dot{m}_{slip} = \frac{H^3 W P_o^2}{24\mu LRT}\left[\left(\frac{P_i}{P_o}\right)^2 - 1 + 12\frac{2-F_m}{F_m}Kn_o\left(\frac{P_i}{P_o}-1\right)\right] \tag{2.13}$$

$$\frac{\dot{m}_{slip}}{\dot{m}_{noslip}} = 1 + 12\frac{2-F_m}{F_m}\frac{Kn_o}{1+\frac{P_i}{P_o}} \tag{2.14}$$

where H, W, and L are the height, width, and length of the microchannel.

The use of slip boundary conditions in the slip flow regime has been verified experimentally. We will show two of these studies that investigated the gaseous flow in microchannels both experimentally and analytically. The first one, [18], measured friction factor values for nitrogen, helium and argon in microchannels with 100 x (0.5-20) micrometer cross-sections. The Knudsen number at the channel outlet was in the range of 0.001-0.4, which covers slip flow and early transition regimes. The experimental data was in good agreement with the theoretical predictions assuming the slip flow boundary condition. They proposed the following expression for the friction factor in micro flows between two parallel plates:

$$\frac{(f\,Re)_{noslip}}{(f\,Re)_{slip}} = \frac{1}{1+6Kn} \tag{2.15}$$

Gaseous flow in microchannels was experimentally analyzed in [19] with helium and nitrogen as the working fluids. The mass flow rate and pressure distribution along the channels were measured. The helium results agreed well with the result of a theoretical analysis using slip flow conditions, however there were deviations between theoretical and experimental results for nitrogen.

The laminar gaseous flow heat convection problem was solved in a cylindrical microchannel with uniform heat flux boundary conditions in [20]. The fluid was assumed to be incompressible with constant properties, the flow was assumed to be steady and two-dimensional, and viscous heating was neglected. They used the results from a previous study, [21], of the same problem with uniform temperature at the boundary. Discontinuities of both velocity and temperature at the wall were considered. The fully developed velocity profile was derived as:

$$u = u_m \frac{2\left(1-(r/R)^2+4Kn\right)}{1+8Kn} \tag{2.16}$$

where the Knudsen number is given by $\mathrm{Kn} = \frac{\lambda}{\mathrm{D}}$. The fully developed Nusselt number was obtained from:

$$Nu_\infty = \frac{48(2\beta-1)^2}{\left(24\beta^2-16\beta+3\right)\left[1+\frac{24\gamma(\beta-1)(2\beta-1)^2}{\left(24\beta^2-16\beta+3\right)(\gamma+1)\mathrm{Pr}}\right]} \tag{2.17}$$

where $\beta = 1 + 4\mathrm{Kn}$. It was noted that for $\mathrm{Kn} = 0$, in other words the no-slip condition, the above equation gives Nu = 4.364, which is the well-known Nusselt number for conventionally sized channels. The Nusselt number was found to decrease with increasing Kn. Over the slip flow regime, Nu was reduced about 40%. A similar decay was also observed for the gas mixed mean temperature. Another observation they made was that the maximum temperature decreases as a result of increasing rarefaction, which also causes the temperature profile to be flat. They determined that the entrance length increases with increasing rarefaction, which means that thermally fully developed flow is not obtained as quickly as in conventional channels. The following formula shows the relationship between the entrance length and the Knudsen number

$$x_e^* = 0.0828 + 0.141 Kn^{0.69} \tag{2.18}$$

Compressible two-dimensional fluid flow and heat transfer characteristics of a gas flowing between two parallel plates with both uniform temperature and uniform heat flux boundary conditions were solved in [22]. They compared their results with the experimental results of [17]. The slip flow model agreed well with these experiments. They observed an increase in the entrance length and a decrease in the Nusselt number as Kn takes higher values. It was found that the effect of compressibility and rarefaction is a function of Re. Compressibility is significant for high Re and rarefaction is significant for low Re.

Heat convection for gaseous flow in a circular tube in the slip flow regime with uniform temperature boundary condition was solved in [23]. The effects of the rarefaction and surface accommodation coefficients were considered. They defined a fictitious extrapolated boundary where the fluid velocity does not slip by scaling the velocity profile with a new variable, the slip radius, $\rho_s^2 = 1/(1+4\beta_v \mathrm{Kn})$, where β_v is a function of the momentum accommodation coefficient, $\mathrm{F_m}$ and defined as $\beta_v = (2-\mathrm{F_m})/\mathrm{F_m}$. Therefore, the velocity profile is converted to the one used for the continuum flow, $u = 1 - r^2$. They also defined a coefficient representing the relative importance of velocity slip and temperature jump as $\beta = \beta_T/\beta_v$, where $\beta_T = \frac{2-\mathrm{F_T}}{\mathrm{F_T}}\frac{2\gamma}{\gamma+1}\frac{1}{\mathrm{Pr}}$, γ is the gas constant and $\mathrm{F_T}$ is the thermal accommodation coefficient. The analysis yields the following fully-developed Nusselt number expression

$$Nu_{\infty} = \frac{\rho_s^2\left(2-\rho_s^2\right)}{2}\lambda_1^2(\rho_s,\beta) \qquad (2.19)$$

They developed a new uniform asymptotic approximation to the eigenfunctions of the Graetz problem since higher order eigenvalues are required for the solution of the entrance region. This relation tells us heat transfer decreases with increasing rarefaction in the presence of the temperature jump due to the smaller temperature gradient at the wall. However, they noted that this was not necessarily true since the eigenvalues are also dependent on the surface-fluid interaction. Depending on the values of the accommodation coefficients, Nu may also increase or stay constant with increasing Kn. They found that for $\beta < 1$, Nu increases with increasing Kn since $\beta_v > \beta_T$ suggest increased convection at the surface. However, for $\beta > 1$, Nu decreases with increasing Kn due to the more effective temperature jump and thus reduced temperature gradient on the surface.

Convective heat transfer analysis for a gaseous flow in microchannels was performed in [24]. A Knudsen range of 0.06-1.1 was considered. In this range, flow is called transition flow. Since the continuum assumption is not valid, DSMC technique was applied. Reference [24] considered the uniform heat flux boundary condition for two-dimensional flow, where the channel height varied between 0.03125 and 1 micrometer. It was concluded that the slip flow approximation is valid for Knudsen numbers less than 0.1. The results showed a reduction in Nusselt number with increasing rarefaction in both slip and transition regimes.

Gaseous flows for parallel-plate microchannels were studied in [25]. They reported that both the Darcy friction factor and Fanning friction factor were functions of the Mach number (Ma). Their numerical results for both friction factors converged with the suggested friction factor-Mach number correlations to within a 2% discrepancy. The stagnation pressure and temperature showed negligible effects on the friction factor computations as their numerical results fell within 2% of suggested friction factor-Mach number correlations. The experimental results of the pressure distribution along the length of the channel coincided nicely with their numerical results. The friction factor computation results showed an increasing trend from the conventional incompressible value of 96 along with the increase of Ma.

More recently, [26] has confirmed the need to include the second order slip condition at higher Kn number values. Their work was both theoretical and experimental using nitrogen and helium in a silicon channels. They used the second order slip approximation to obtain the equation for the volumetric flow rate and related it to the ratio of inlet to outlet pressure. It was shown that when using the Navier-Stokes equation, the boundary conditions must be modified to include second order slip terms as the Knudsen number increases. They also studied in depth the accommodation coefficient F_v and verified the need for further study. It was shown that as the Knudsen number increases, the momentum accommodation value deviates further and further from unity; for instance Kn ~0.5 yields $F_v \sim 0.8$ for helium. The values found for nitrogen were quite similar. The measurements agreed with past studies such as [11] for lower Kn.

Effect of both compressibility and rarefaction were included in the experimental and analytical study of two-dimensional compressible gaseous flow in a parallel-plate microchannel by [27]. Nitrogen gas was utilized as the working fluid in this study. The pressure distribution along the length of the channel underwent a non-linear drop due to compressibility. This was believed to be the cause of the small discrepancies between analytical and experimental results, even though compressibility would not have been a factor in large channels due to the low Mach number flow (Ma <0.3). The pressure drop in the study was also found to be relatively small compared to that in the conventional channel.

Also, the tangential momentum accommodation coefficients obtained experimentally varied in range from 0.3 to 0.7, quite different from the conventional value of 1. This value range agreed with the earlier analytical work [11].

A three-dimensional numerical procedure to solve steady, compressible, laminar flow in long microchannels was proposed in [28]. The proposed numerical procedure was capable of solving the reduced compressible Navier-Stokes equations accurately except in the entrance region. However, the entrance region is said to be very small compared to the channel length so the error in that region was neglected. Using Nitrogen as the working fluid, the numerical analysis for the mass flow rate agreed well with analytical and experimental results of [29]. The normalized friction factor coefficient, defined as the ratio of the compressible friction factor to the incompressible friction factor, was in good agreement with the numerical results of [22] and experimental results of [15]. It was confirmed in this paper that for a fixed channel area, smaller aspect ratios contributed to larger slip effects that would produce higher f·Re. In addition, using Helium, this numerical analysis accurately predicted the increasing compressibility effects along with the increasing pressure ratio, defined as the ratio of inlet to outlet pressure along the channel.

A relationship between experimental uncertainty, Kn, and compressibility was shown in [30]. For pressure differences greater than 5% of the initial static pressure and Ma >0.3 the effect of compressibility must be considered. It was emphasized that previous works such as [12] did not consider the compressibility effect and it is this error that could have contributed to higher friction factors than conventional values.

Local fully developed Nusselt numbers for parallel plates were reported by [31]. Two experimental cases were done under different boundary conditions; two walls heated and one wall heated, the other insulated. Recovery factors as functions of dimensionless axial length, X^*, for both boundary conditions were introduced. Employing the recovery factors and plotted against the dimensionless axial length, Nusselt numbers, were found to be 8.235 and 5.385 for the boundary conditions of the two heated walls and the one heated wall the other insulated, respectively. It is noted that these values are the same as those of conventional chanels.

A summary of the contributions that explains the size effect on microscale single-phase flow and heat transfer was given by [32]. Their summary confirmed that in most cases the assumption of flow continuum might still be valid since the flow characteristic lengths in MEMS are on the order of tens to hundreds of micrometers (Kn ~ 0.001 to 0.0001). The large surface to volume ratio in microchannels tend to enhance several factors that were neglected in macroscale flow and heat transfer, such as surface friction induced flow compressibility, surface roughness, viscous force in natural convection, channel surface geometry, surface electrostatic charges, axial wall heat conduction , and measurement errors.

The experimental studies over the last two decades on the flow in microchannels was presented by [33]. The main results on the experimental friction factor of microchannels, with respect to the conventional macrochannels, highlight (1) the friction factor for laminar fully developed flow is ower than the conventional value, (2) the friction factor for turbulent fully developed flow is higher than the conventional value, (3) the dependence of the friction factor on the Reynolds number for laminar fully developed flow, (4) the decreasing friction factor for gaseous laminar fully developed flow with Knudsen number, (5) the dependence of the friction factor on the material of the microchannel walls, showing the importance of electro-osmotic phenomena at microscales, and (6) the dependence of the friction factor on the relative roughness of the walls of the microchannels. He also presented the main results on the laminar-to-turbulent transition: (1) an earlier laminar-to-turbulent transition with respect to the predictions of conventional theory, (2) the transition is characterized by a critical Reynolds number larger than the conventional value, (3) the dependence of the critical Reynolds number on the wall roughness, and (4) the decreasing critical Reynolds numbers with the microchannel hydraulic diameter. Several different Nusselt number correlations are redrawn following Ref [33] in Figure 4.

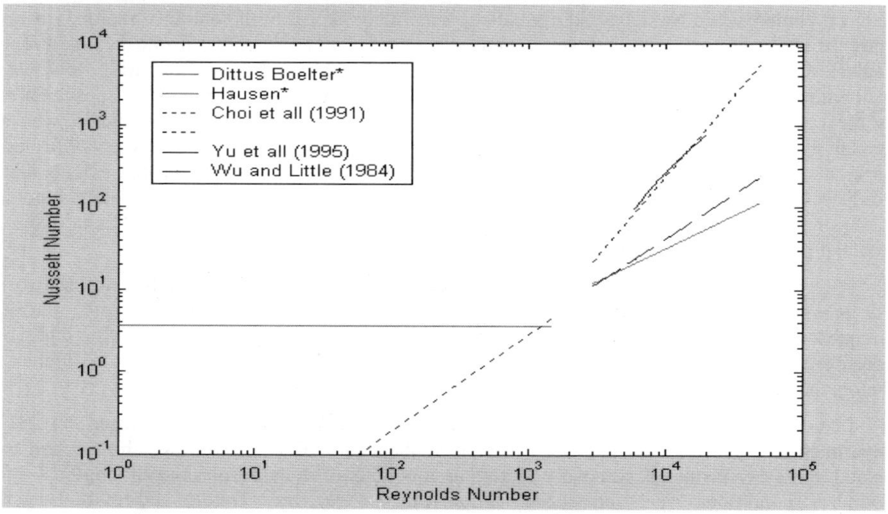

Figure 4: Comparison of experimental Nusselt number from several contributors [33].

2.2.1 SLIP FLOW NUSSELT NUMBER FOR DIFFERENT GEOMETRIES

Analytical slip-flow Nusselt numbers for different geometries are shown by Bayazitoglu et al. [34-38]. They analytically solved the continuum version of the energy equation by the integral transform technique with the appropriate jump boundary conditions. The integral transform technique has widely been used for the solution of heat transfer problems in many different applications. It is a three-step method. In the first step, the appropriate integral inversion and transform pair of the transform technique is developed. Then, partial derivatives with respect to the space variables are removed from the equation, which reduces it to an ordinary differential equation (ODE). Finally, the resulting ODE is solved subjected to the transformed inlet condition. They solved the steady state heat convection between two parallel plates [37] and in circular [35], rectangular [38] and annular [37] channels with uniform heat flux and uniform temperature boundary conditions including the viscous heat generation for thermally developing and fully-developed conditions. They also solved the transient heat convection in a circular tube including rarefaction effects and heat transfer in a double-pipe heat exchanger assuming slip conditions for both fluids and including conduction across the inner wall.

The velocity profile was assumed to be fully-developed. The velocity distribution in a circular microchannel including the slip boundary condition was taken from the literature. However, for the other geometries, they derived the fully-developed velocity profiles from the momentum equation. It is straightforward for flow between parallel plates and flow in an annulus. They applied the integral transform technique to obtain the velocity in a rectangular channel. The problem was simplified by assuming the same amount of slip at all the boundaries.

After the temperature distribution was obtained, following definitions were used to calculate the Nusselt number. Non-dimensionalizing the temperature by the fluid temperature at the wall instead of the wall temperature makes the boundary condition for the eigenvalue problem easier to handle for the uniform temperature boundary condition. Then they derived the Nusselt number equations from the energy balance at the wall so that temperature jump could be implemented. The details of this derivation can be found in the references.

The Nusselt number for flow in a tube can be expressed as:

Specified temperature
$$Nu_{x,T} = -\frac{2\left.\frac{\partial\theta}{\partial\eta}\right|_{\eta=1}}{\left(\theta_b - \frac{4\gamma}{\gamma+1}\frac{Kn}{Pr}\left.\frac{\partial\theta}{\partial\eta}\right|_{\eta=1}\right)} \tag{2.20}$$

Specified heat flux
$$Nu_{x,q} = \frac{2}{\theta_s + \frac{4\gamma}{\gamma+1}\frac{Kn}{Pr} - \theta_b} \tag{2.21}$$

These definitions are also used for flow between two parallel plates, replacing 4γ in the denominator by 2γ. Next, the Nusselt number for flow in a rectangular channel is as follows:

Specified heat flux
$$Nu_q = \frac{1}{\theta_s + \frac{2\gamma}{1+\gamma}\left(\frac{a+b}{2b}\right)\frac{Kn}{Pr} - \theta_b} \tag{2.22}$$

where a and b are the sides of the channel. Finally, the Nusselt number for flow in an annular channel can be expressed as:

$$Nu_1 = 2(1-\gamma)\frac{\left.\frac{\partial\theta}{\partial\eta}\right|_{\eta=\gamma}}{\theta_{ave}} \qquad Nu_2 = -2(1-\gamma)\frac{\left.\frac{\partial\theta}{\partial\eta}\right|_{\eta=1}}{\theta_{ave}} \qquad Nu_{ave} = \frac{\gamma Nu_1 + Nu_2}{1+\gamma} \tag{2.23}$$

Following the same integral transform method, [39] solved for the Nusselt number for flow in a rectangular microchannel subject to the uniform temperature boundary condition and included slip flow. Their results for the non-slip flow case agreed with [40] who also used the integral transform technique to solve for the Nusselt number for flow through a macrosized rectangular channel. They did not include any viscous dissipation in the study. Similar to Tunc and Bayazitoglu [38], they concluded that the Knudsen number, Prandlt number, aspect ratio, velocity jump and temperature jump can all cause the Nusselt number to deviate from the conventional value. The results for the fully developed Nusselt number in different geometries subject to either uniform heat flux or uniform temperature with both velocity and temperature jump can be seen in Table 3.

Table 3: Nusselt Number for different Geometries Subject to Slip-Flow ($\beta_T = 1.66$) [34-39]

Br = 0.0		Kn = 0.00		Kn = 0.04		Kn=0.08		Kn=0.12	
		Nu_T	Nu_q	Nu_T	Nu_q	Nu_T	Nu_q	Nu_T	Nu_q
Cylindrical		3.67	4.36	3.18	3.75	2.73	3.16	2.37	2.68
Rectangular	γ=1	2.98	3.10	2.71	2.85	2.44	2.53	2.17	2.24
	γ=0.84	3.00	3.09	2.73	2.82	2.46	2.48	2.19	2.17
Aspect	γ=0.75	3.05	3.08	2.77	2.81	2.49	2.44	2.22	2.12
Ratio	γ=0.5	3.39	3.03	2.92	2.71	2.55	2.26	2.24	2.18
γ=a/b	γ=0.25	4.44	2.93	3.55	2.42	2.89	1.81	2.44	1.68
	γ=0.125	5.59	2.85	4.30	1.92	3.47	1.25	2.8	1.12
Two Parallel Plates		7.54	8.23	6.26	6.82	5.29	5.72	4.56	4.89

Figure 5 shows the Nusselt number variations as a function of the Knudsen number and Prandtl number under both the uniform heat flux and uniform temperature conditions in microtubes. By the definition of equation (2.8), increasing the Prandtl number tends to decrease the temperature jump. As reported in [35], under the uniform temperature boundary condition, there is a 40% decrease in the Nusselt number for a Prandtl number of 0.6 and a 24.9% decrease in the Nusselt number for a Prandtl number of 1. For the uniform heat flux boundary condition, the Nusselt number decreases 43.6% and 26.8% for Prandtl numbers of 0.6 and 1, respectively.

Viscous heat generation is a result of the friction between layers of the fluid. It is the term that is generally neglected in continuum flow. However, due to the large surface area to volume ratio, viscous heating is an important factor for fluid flow in microchannels, especially for laminar flow where considerable gradients exist. The representation of the viscous heating effect is the Brinkman number (Br). For the uniform heat flux boundary condition, a positive Brinkman number represents a larger fluid temperature than wall temperature, while a negative Brinkman number represents a smaller fluid temperature than wall temperature. On the other hand, it is the opposite for the uniform temperature boundary condition.

As shown in Figure 6, the viscous dissipation cause the thermally developed Nusselt number to increase. The Nusselt number decreases as the Knudsen number increases due to the increasing temperature jump. It is noted that the decrease is larger when viscous dissipation is considered.

It is concluded for the slip flow analysis that heat transfer of fluid flow in microchannels can be significantly different from conventional channels depending on the Knudsen number, Prandtl number, Brinkman number, and the aspect ratio. Velocity slip and temperature jump effect the heat transfer in opposite ways, i.e. a large slip on the wall increases the convective heat transfer along the surface while a large temperature jump decreases the heat transfer by reducing the temperature gradient at the wall. Thus, neglecting the temperature jump will result in the overestimation of the heat transfer coefficient. The Prandtl number is an important parameter for temperature jump. An increase in the Prandtl number will cause a decrease in the temperature gradient between the fluid temperature at the wall and the wall temperature. A reduction in the Nusselt number will be obtained along with a rise in the Knudsen number. In rectangular channels, an increase in the Nusselt number is obtained along with an increase of the aspect ratio. However, the Nusselt number decreases as the Knudsen number increases regardless of the aspect ratio due to the increasing temperature jump. This decrease is more significant for a small aspect ratio.

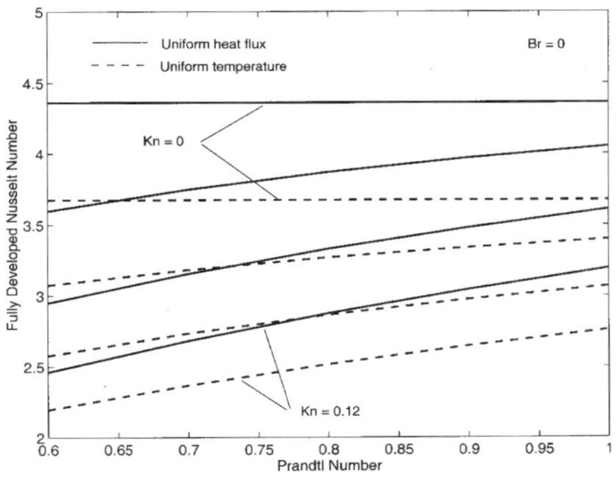

Figure 5: Nusselt Number Behavior in Microtubes [35].

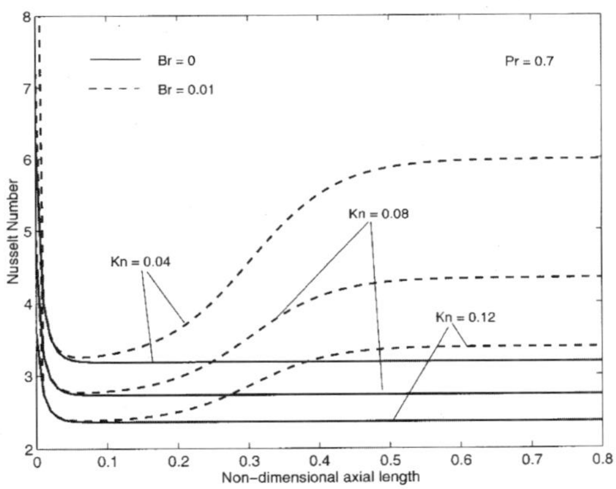

Figure 6: Brinkman Number and Knudsen Number Effects on Nusselt Number in Microtube [35].

2.3 TRANSITION FLOW AND FREE-MOLECULAR FLOW REGIME

As the flow enters the transition flow regime and continues into the free-molecular flow regime, the Knudsen number becomes significant enough that the molecular approach has to be utilized. Thus, the Boltzmann equation

$$\frac{\partial f}{\partial t} + v_i \frac{\partial f}{\partial x_i} + F_i \frac{\partial f}{\partial \xi_i} = Q(f,f) \tag{2.24}$$

should be considered to fulfill the atomic level of studies of the gaseous flows in the transition regime. The Boltzmann equation denotes v_i as the velocity, F_i as the forcing function, and f as the particle velocity distribution function in space, while Q(f,f) describes the intermolecular collisions. The density can be obtained by integrating f over time and space. The integration of the product of the density and the velocity will provide the mass velocity. The Maxwellian distribution is the simplest distribution as it is the zeroth order approximation of the Boltzmann equation.

The Boltzmann equation is solved by the particulate methods, the Molecular Dynamics (MD), the Direct Simulation Monte Carlo (DSMC) method, or by deriving higher order fluid dynamics approximations beyond Navier-Stokes, which are the Burnett Equations. The Burnett equation

$$f = f^{(0)} + Kn\, f^{(1)} + Kn\, f^{(2)} + \cdots \tag{2.25}$$

is the first three terms of the Champan-Enskog equation. The simplified Boltzmann equation can be solved using the Lattice Boltzmann Method (LBM) for the distributed function on a regular lattice. Being a deterministic approach, the MD method simulation may require a very large domain for gaseous flows while the DSMC method is a stochastic approach and is simulated more efficiently for gaseous flows. [24,41-42].

In early the transition regime, the DSMC method also requires a large number of particles, which makes it expensive in terms of computational time and memory requirements. Therefore, until recently, the advances in the gaseous flow regime in micro channels were in the slip flow regime [9].

Another mechanism that may affect the velocity profile in a microchannel is thermal creep. It is a molecular transport phenomenon that occurs when two isopressure containers at different temperatures are connected by a channel whose diameter is close to the gaseous mean free path. At this condition, gaseous molecules start to flow from the cooler container to the hotter container. Thus, a positive temperature gradient along the flow direction tends to increase the mass flow rate while a negative temperature gradient tends to reduce the mass flow rate. The inclusion of the thermal creep effect in the slip boundary condition is given in [43] by the following formula:

$$u_s = \frac{2 - F_m}{F_m} Kn \frac{\partial u}{\partial \eta} + \frac{3}{2\pi}(\gamma - 1) Kn^2\, Re \frac{\partial \theta}{\partial \zeta} \tag{2.26}$$

A study on a dilute hard-sphere gas in the transition regime using the DSMC was conducted by [44]. The simulation is for $0.02 < Kn < 2$ and unity F_m and F_T. They found a weak dependence of the Nusselt number on the Peclet number, which explains the weak dependence on the axial heat conduction. In the case of constant wall heat flux, a positive thermal creep, which occurs when the exit temperature is higher than the inlet temperature, tends to increase the Nusselt number while negative thermal creep tends to decrease the Nusselt number.

As an alternative solution, the early transition regime can simply be solved using the analytical slip flow of [3]. Instead of the velocity boundary condition, the stress boundary condition analysis provides better results to the linearized Boltzmann solution of [45]. Figure 7 on with [43], they also agree with the results of [45] in the early transition regime of $10^{-1} < Kn < 2$.

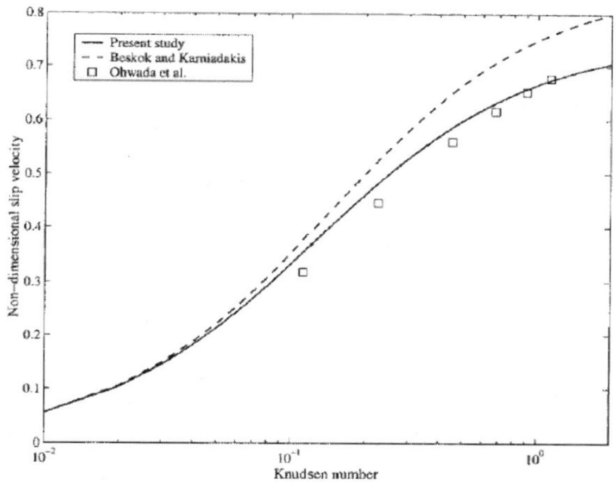

Figure 7: The analytical slip flow using stress boundary condition [3].

In the free-molecular flow regime, the molecular mean free path is of the same order as the channel characteristic length. Because Newton's 2nd Law should more or less be applied to each molecule, the analysis becomes extremely tedious and complicated. The current computational tools, the Molecular Dynamics (MD) and the Direct Simulation Monte Carlo, are still incapable of providing effective and efficient solutions.

3. Conclusion

Microscale heat transfer has attracted researchers in the last decade, particularly due to developments and current needs in the small-scale electronics, aerospace, and bioengineering industries. Although some of the fundamental differences between micro and macro heat transfer phenomenon have been identified, there still is a need for further experimental, analytical and numerical studies to clarify the points that are not yet understood, such as the effect of axial conduction, friction factors, compressibility effects, critical Reynolds number, and accommodation coefficients less then unity.

The current computational methods for analyses in transition flow and free-molecular flow regimes are ineffective and inefficient. Analyses in these two flow regime are still premature and require more extensive study.

ACKNOWLEDGEMENT: The authors acknowledge the support by the Texas State TDT program (Grant No. 003604-0039-2001).

NOMENCLATURE

a, b	Lengths of the rectangular channel
b,	Empirical parameter
Br,	Brinkman number
c_{jump},	Temperature jump coefficient
c_p,	Specific heat at constant pressure
c_v,	Specific heat at constant volume
D,	Diameter
F_M,	Tangential momentum accommodation coefficient
F_T,	Thermal accommodation coefficient
k,	Thermal conductivity
Kn,	Knudsen number
M,	Mass of the fluid
$\dot{m}$,	Mass flow rate
n,	Number of molecules per unit volume
Nu,	Nusselt number
P,	Pressure
Pr,	Prandtl number
Q,	Energy of the fluid molecules
R,	Gas constant
R,	Radius of the circular tube
Re,	Reynolds Number
T,	Temperature
U,	Internal energy of the fluid
u,	Fluid velocity, axial direction
v,	Fluid velocity, radial or y-direction
x_e*,	Entrance length
x,y,z	Cartesion coordinates
r	Cylindrical coordinate

Greek symbols

α,	Thermal diffusivity
β,	β_T/β_v,
β_v,	$(2\text{-}F_m)/F_m$.
β_T,	$((2\text{-}F_T)/F_T)*(2\gamma/(Pr(\gamma+1)))$.
γ,	Specific heat ratio, aspect ratio
λ,	Molecular mean free path
μ,	Viscosity
ρ,	Density
ρ_s,	Slip radius
υ,	Momentum diffusivity
η,	Dimen.less spatial (y/L or r/R)
θ,	Dimensionless temperature
ζ,	Dimensionless axial coordinate

Subscripts

b,	Bulk
g,	True gas condition
i,	Impinging
m,	Mean
o,	Outlet
q,	Specified heat flux
r,	Reflected
s,	Fluid properties at the wall or slip
T,	Specified temperature
w,	Wall properties

REFERENCES

1. Kakac, S. and Yener Y., (1995) Convective Heat Transfer, CRC Press, Inc.
2. Kennard, E. H., (1938) Kinetic Theory of Gases, McGraw-Hill Book Company, Inc., New York.
3. Bayazitoglu, Y., and Tunc, G., (2002) Extended Boundary Condition for Micro-Scale Heat Transfer, *AIAA/ASME Thermophysics Conference*, in St Louis, June 2002, Also published in *Proceedings of the AIAA J. of Thermophysics and Heat Transfer*, Vol.16 no. 3, pp.472-474.
4. Bayazitoglu, Y., Tunc, G., Wilson, K., and Tjahjono, I., (2005) Convective Heat Transfer for Single-Phase Gases in Microchannel Slip Flow: Analytical Solutions, *presented at NATO Advanced Study Institute, Microscale Heat Transfer – Fundamentals and Applications in Biological and Microelectromechanical Systems, July 18-30, Altin Yunus – Çeşme, Izmir, Turkey.*

5. Beskok, A., Karniadakis, G. E., (1997a) Modeling Separation in Rarefied Gas Flows, *28th AIAA Fluid Dynamics Conference, 4th AIAA Shear Flow Control Conference*, June 29-July 2.
6. Knudsen, M., (1991), *Ann. Phys. 34*, 593.
7. Smoluchowski, M. V., (1898), *Wied. Ann. 64*, 101.
8. Smoluchowski, M. V., (1910), *Ann. Physik 33*, 1559.
9. Schaaf, S., and Chambre, P., (1961) Flow of Rarefied Gases, *Princeton University Press*.
10. Beskok, A., Karniadakis, G. E., (1995) Models and Scaling Laws for Rarefied Internal Gas Flows Including Separation, *48th Annual Meeting of the American Physical Society Division of Fluid Dynamics*, Irvine, Ca. Nov 1995.
11. Arkilic, E. B., Breuer, K. S., Schmidt, M. A., (2001) Mass Flow and Tangential Momentum Accommodation in Silicon Micromachined Channels, *Journal of Fluid Mechanics*, Vol.437, pp.29-43.
12. Wu, P. Y. and Little, W. A., (1983) Measure of Friction Factor for the Flow of Gases in Very Fine Channels Used for Micr. Joule-Thompson Refrigerators, *Cryogenics*, Vol.23(5), pp.273-277.
13. Pfahler, J., Harley, J., Bau, H. H., and Zemel, J., (1990) Liquid and Gas Transport in Small Channels, *Proceedings of ASME WAM Micro Structures, Sensors, and Actuators, DSC*, Vol.19, pp.149-157.
14. Pfahler, J., Harley, J., Bau, H. H., and Zemel, J., (1990) Liquid Transport in Micron and Submicron Channels, *Sensors and Actuators*, A21-A23, pp.431-434.
15. Pfahler, J., Harley, J., Bau, H. H., and Zemel, J., (1991) Gas and Liquid Flow in Small Channels, *Micromechanical Sensors, Actuators, and Systems, ASME DSC*, Vol.32, pp.49-60.
16. Choi, S. B., Barron R. F. and Warrington R. O., (1991) Fluid Flow And Heat Transfer in Microtubes, *Micromechanical Sensors, Actuators, and Systems, ASME DSC*, Vol.32, pp.123-134.
17. Arkilic, E. B., Breuer, K. S., Schmidt, M. A., (1994) Gaseous Flow in Microchannels. *Application of Microfabrication to Fluid Mechanics, ASME FED*, Vol.197, pp.57-66.
18. Harley, J. C., Huang, Y., Bau, H. H., Zemel. J. N., (1995) Gas Flow in Micro-Channels, *Journal of Fluid Mechanics*, Vol.284, pp.257-274.
19. Shih, J. C., Ho, C., Liu, J. and Tai, Y., (1996) Monatomic and Polyatomic Gas Flow through Uniform Microchannels, *Micro Electro Mechanical Systems (MEMS), National Heat Transfer Conference, DSC*, Vol.59, pp.197-203.
20. Ameel, T. A., Barron, R. F., Wang, X., and Warrington, R. O., (1997) Laminar Forced Convection in a Circular Tube with Constant Heat Flux and Slip Flow, *Microscale Thermophys. Eng.*, Vol.4, pp. 303-320.
21. Barron, R. F, Wang, X. Ameel, T. A., and Warrington, R. O., (1997) The Graetz Problem Extended To Slip-flow, *Int. J. Heat Mass Transfer*, Vol.40(8), pp.1817-1823.
22. Kavehpour, H. P., Faghri, M., and Asako, Y., (1997) Effects of Compressibility and Rarefaction on Gaseous Flows in Microchannels, *Numerical Heat Transfer, Part A,* Vol.32, pp.677-696.
23. Larrode, F. E., Housiadas, C., and Drossinos, Y., (2000) Slip Flow Heat Transfer in Circular Tubes, *Int. J. Heat Mass Transfer*, Vol.43, pp.2669-2680.
24. Hadjiconstantinou, N. G., (2000) Heat Transfer for Gaseous Flow in Microtubes with Viscous Heating, *Proceedings of the ASME Heat Transfer Division*, Vol.366-2, pp.1-10.
25. Asako, Y., Pi, T., Turner, S. E., and Faghri, M. (2003) Effect of compressibility on gaseous flows in micro-channels, *International J. Heat Mass Transfer*, Vol.46, pp.3041-3050.

26. Maurer, J., Tabeling, P., Joseph, P., Williame, H. (2003) Second-Order Slip laws in microchannels for Helium and Nitrogen, *Physics of Fluids* Vol. 15 (9), pp. 2613-2621.
27. Hsieh, S.S., Tsai, H.H., Lin, C. Y., Huang, C. F., and Chien, C. M., (2004) Gas flow in a long microchannel, *Int. J. Heat Mass Transfer*, Vol.47, pp.3877-3887.
28. Chen, C. S., (2004) Numerical method for predicting three-dimensional steady compressible flow in long microchannels, *J. Micromech. and Microeng*., Vol.14, pp.1091-1100.
29. Arkilic, E. B., Breuer, K. S., Schmidt, M. A., (1997) Gaseous Slip Flow in Long Microchannels. *Journal of Microelectromechanical Systems*, Vol. 6, pp.167-178.
30. Morini, G. L., Lorenzini, M., and Spiga, M., (2004) A Criterion for the Experimental Validation of the Slip-Flow Models for Incompressible Rarefied Gases through Microchannels, *Proceedings of the 2nd International Conference on Microchannels and Minichannels, June 17-19, 2004 Rochester, New York, USA*, pp. 351-368.
31. Turner, S. E., Asako, Y., and Faghri, M., (2004) Effects of Compressibility on Heat Transfer in Microchannels, Proceedings of the 2nd International Conference on Microchannels and Minichannels, June 17-19, Rochester, New York, USA, pp.341-350.
32. Guo, Z. Y., and Li, Z. X., (2003) Size effect on microscale single-phase flow and heat transfer, *Int. J. Heat and Mass Transfer*, Vol.46, pp.149-159.
33. Morini, G. L., (2004) Single-Phase Convective Heat Transfer in Microchannels: A Review of Experimental Results, *International Journal of Thermal Sciences*, Vol.43, pp.631-651.
34. Tunc, G. and Bayazitoglu, Y., (2000) Heat Transfer for Gaseous Flow in Microtubes with Viscous Heating, *Proceedings of the ASME Heat Transfer Division*, HTD 366-2, pp.209-306.
35. Tunc, G., Bayazitoglu, Y., (2001) Heat Transfer in Microtubes with Viscous Dissipation, *Int. J. Heat Mass Transfer*, Vol.44(13), pp.2395-2403.
36. Tunc, G. and Bayazitoglu, Y., (2001) Nusselt Number Variation in Microchannels, *Proceedings of the 2ndInternational Conference on Computational Heat and Mass Transfer, Rio de Janeiro, Brazil,* Vol.2, pp.240-249, ISBN 85-87922-44-0.
37. Tunc, G (2002) Convective Heat Transfer in Microchannel Gaseous Slip Flow, PhD. Thesis, Rice University, Houston, TX.
38. Tunc, G., Bayazitoglu, Y., (2002) Heat Transfer in Rectangular Microchannels, *Int. J. Heat Mass Transfer*, Vol.45, pp.765-773.
39. Yu, S., and Ameel, T. (2001) Slip-flow Heat Transfer in Rectangular Microchannels, *Int. J. Heat and Mass Transfer*. Vol. 44, pp. 4225-4234.
40. Aparecido, J., and Cotta, R. (1990) Thermally Developing Laminar Flow inside Rectangular Ducts, *Int. J. Heat and Mass Transfer*. Vol. 33, pp. 341-347.
41. Beskok, A., and Karniadakis, G. E., (1994) Simulation of Heat and Momentum Transfer in Complex Micro Geometries, *J. Ther. Heat Transfer*, Vol.8(4), pp.647-653.
42. Rostami, A. A., Mujumdar, A. S., and Saniei, N. (2002) Flow and Heat Transfer for Gas Flowing in Microchannels: A Review, *Heat and Mass Transfer*, Vol.38(4-5), pp.359-367.
43. Beskok, A., Karniadakis, G. E., and Trimmer, W., (1995) Rarefaction, Compressibility, and Thermal Creep Effects in Micro-Flows, *Proceedings of the ASME Dynamic Systems and Control Division, DSC*, Vol.57-2 ,pp.877-892.
44. Hadjiconstantinou, N. G., and Simek, O., (2002) Constant-Wall-Temperature Nusselt Number in Micro and Nano-Channels, *Transaction of the ASME*, Vol.124, pp.356-364.
45. Ohwada, T., Sone, Y., and Aoki, K., (1989) Numerical Analysis of the Poiseuille and Thermal Transpiration Flows Between Two Parallel Plates on the Basis of the Boltzmann Equation for Hard-Sphere Molecules, *Physics of Fluids A*, Vol.1(12), pp.2042-2049.

MICROSCALE HEAT TRANSFER AT LOW TEMPERATURES*

RAY RADEBAUGH
Cryogenic Technologies Group
National Institute of Standards and Technology
Boulder, Colorado, USA

1. Introduction

This paper discusses the fundamentals and applications of heat transfer in small space and time domains at low temperatures. The modern trend toward miniaturization of devices requires a better understanding of heat transfer phenomena in small dimensions. In regenerative thermal systems, such as thermoacoustic, Stirling, and pulse tube refrigerators, miniaturization is often accompanied by increased operating frequencies. Thus, this paper also covers heat transfer in small time domains involved with possible frequencies up to several hundred hertz. Simple analytical techniques are discussed for the optimization of heat exchanger and regenerator geometry at all temperatures. The results show that the optimum hydraulic diameters can become much less than 100 μm at cryogenic temperatures, although slip flow is seldom a problem. The cooling of superconducting or other electronic devices in Micro-Electro-Mechanical Systems (MEMS) requires a better understanding of the heat transfer issues in very small sizes. Space applications also benefit from a reduction in the size of cryocoolers, which has brought about considerable interest in microscale heat exchangers. Some recent developments in miniature heat exchangers for Joule-Thomson and Brayton cycle cryocoolers are discussed. Both single-phase and two-phase heat transfer are covered in the paper, but the emphasis is on single-phase gas flow. Some discussion of fabrication techniques is also included.

Another application discussed here is the use of high frequency Stirling and pulse tube cryocoolers in smaller sizes and lower temperatures. This second area of microscale heat and mass transfer involves the short time scales experienced in high frequency oscillating thermodynamic systems. Models and empirical correlations for heat transfer and pressure drop obtained for steady-state flows in large systems need to be examined carefully in their use with very short time scales. Regenerative cryocoolers like the Stirling and pulse tube cryocoolers could be miniaturized much more than in current practice by utilizing high frequencies. However, the thermal penetration depths in both the gas and the regenerator matrix decrease with increasing frequency, which requires smaller hydraulic diameters for good heat transfer throughout the material in the short time available in a half cycle. These penetration depths in the helium working fluid become smaller at lower temperatures. This paper presents equations useful for the optimization of regenerator geometry that should be valid for temperatures down to about 50 K. The limitation on the maximum operating frequency and its effect on the miniaturization of regenerative cryocoolers is discussed.

S. Kakaç et al. (eds.), Microscale Heat Transfer, 93–124.

2. Temperature, Heat Transfer, and Flow Regimes

Though most heat transfer phenomena to be discussed here apply to all temperatures, we focus mostly on applications dealing with the cryogenic temperature range. Cryogenic temperatures are defined loosely as those temperatures below about 120 K. However, in reaching these low temperatures part of the refrigeration cycle will be at room temperature, and in the heat exchanger the temperature will vary from room temperature to the low temperature. Thus, we need to consider heat transfer issues at temperatures that range from cryogenic up to room temperature for any cryogenic refrigeration process.

In this paper we consider two types of microscale heat transfer issues. The first issue pertains to size. Deviations from macroscale or continuum flow heat transfer behavior in gaseous flows will occur whenever the boundary layer begins to slip at the walls, thus the term slip flow is used to describe such flow. The Knudsen number *Kn*, which is the ratio of the molecular mean free path λ to the characteristic dimension of the channel, characterizes the type of flow. Continuum flow occurs when $Kn < 10^{-3}$. As *Kn* increases, the flow enters the slip flow regime ($10^{-3} < Kn < 10^{-1}$), transition flow regime ($10^{-1} < Kn < 10$), and eventually the free-molecular flow regime ($Kn > 10$) [1]. For nitrogen at atmospheric pressure and room temperature the mean free path is about 66 nm, and for helium it is about 194 nm. Thus, slip flow will begin to occur in nitrogen with channel diameters smaller than about 66 μm and in helium with diameters less than about 194 μm for room temperature and for atmospheric pressure. In some cases characteristic dimensions of flow channels in cryocoolers may be less than these critical values where slip begins to occur. However, for *Kn* less than 10^{-2} the effects on friction factors and Nusselt numbers are less than a few percent and can usually be ignored. In this paper we calculate values of Knudsen numbers that may occur in optimized heat exchangers for cryocoolers. Of particular interest is the case where these cryocoolers are scaled to miniature sizes. We generalize the definition of the microscale region to include all channel sizes with characteristic dimensions less than about 200 μm. Fabrication of such small channels often requires techniques different than those used in larger systems. Also two-phase flow often occurs in cryogenic refrigerators and bubble size then influences the design of flow channels.

The second microscale heat transfer issue considered in this paper deals with short time scales and their influence on the dimensions required for good heat transfer. Many cryocoolers use oscillating flows and pressures with frequencies as high as about 70 Hz. Heat flow at such high frequencies can penetrate a medium only short distances, known as the thermal penetration depth δ_t. Figure 1 shows how the temperature amplitude of a thermal wave decays as it travels within a medium. The distance at which the amplitude is 1/e of that at the surface is the thermal penetration depth, which is given by

$$\delta_t = \sqrt{(2k/\omega\rho c_p)}, \quad (1)$$

where k is the thermal conductivity, ω is the angular frequency, ρ is the density, c_p is the specific heat of the medium. In a similar manner the viscous penetration depth in a fluid is given by

$$\delta_v = \sqrt{(2\mu/\omega\rho)}, \quad (2)$$

where μ is the dynamic viscosity. These penetration depths indicate how far heat and momentum can diffuse perpendicular to the surface. Higher frequencies lead to smaller penetration depths. For good

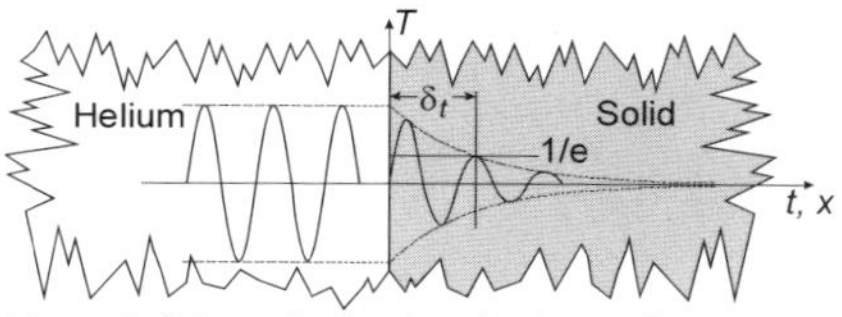

Figure 1. Schematic showing the decay of temperature amplitude inside a solid and the definition of thermal penetration depth.

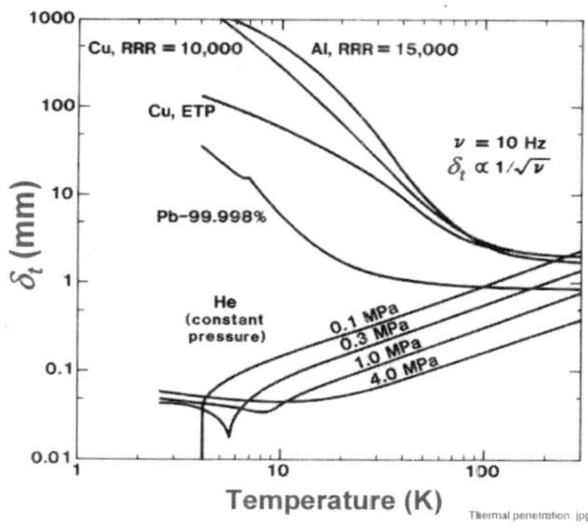

Figure 2. Thermal penetration depths at 10 Hz in helium and several pure metals.

heat transfer the lateral dimensions in the fluid or the solid must be much less than δ_t. In a fluid at distances much greater than δ_v from a wall there is no viscous contact with the wall. Figure 2 shows the temperature dependence of the thermal penetration depth in helium and several pure metals for a frequency of 10 Hz. For pure metals oscillating heat flow can penetrate large distances because of their high thermal conductivity. However, for helium gas the thermal penetration depth is quite small, especially at low temperatures. Thus, hydraulic diameters in cryogenic heat exchangers for oscillating flow should be less than about 100 μm for frequencies greater than about 10 Hz.

An additional complication that occurs with oscillating flow is the existence of several regimes of laminar and turbulent flow that are functions of frequency as well as Reynolds number, as shown in Figure 3 for the case of smooth circular tubes [2]. These flow regimes are the subject of much research [3]. They are shown as a function of the peak Reynolds number $N_{r,\,peak}$ and the ratio of channel radius R to the viscous penetration depth δ_v. This ratio is sometimes referred to as the dynamic Reynolds number and is similar to the Womersley number ($Wo = D/\sqrt{2}\delta_v$). In the weakly turbulent regime

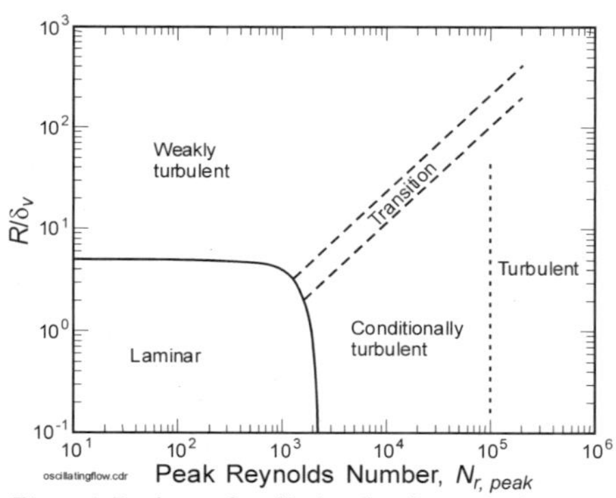

Figure 3. Regimes of oscillating flow in a smooth circular pipe as a function of peak Reynolds number and ratio of pipe radius R to viscous penetration depth δ_v [2].

turbulence occurs only in the center of the channel and not at the boundary. In the conditionally turbulent region turbulence occurs at the peak velocity and changes back to laminar or weakly turbulent when the velocity crosses through zero.

3. Cryogenic Refrigeration Techniques

3.1 CONVENTIONAL REFRIGERATION

The refrigeration techniques required to reach cryogenic temperatures are different than those of conventional vapor-compression refrigeration, which is used for most cooling applications closer to ambient temperatures. Most domestic refrigerators and air conditioners use the vapor-compression cycle. Figure 4a shows a schematic of the vapor-compression cycle, and Figure 4b shows the path of the cycle in the temperature-entropy (*T-S*) diagram. In this cycle heat is absorbed at some low temperature during the boiling of the liquid at a pressure near 0.1 MPa (1 bar). Typically the temperatures may be about 250 K (-23 °C) for most domestic refrigerators. At this temperature oil can remain dissolved in the refrigerant and not freeze. The vapor being boiled off in the evaporator then passes to the oil-lubricated compressor where it is compressed to about 2.5 MPa (25 bar). As the compressed vapor travels through the condenser it cools to ambient temperature and condenses into the liquid phase. The oil used for lubrication of the compressor is soluble in the refrigerant and a small amount of oil then completes the entire refrigerant cycle dissolved in the refrigerant. The condensed liquid then passes to the expansion capillary where the pressure is reduced to about 0.1 MPa and the temperature drops from ambient to about 250 K during this isenthalpic process between c and d in Figure 4.

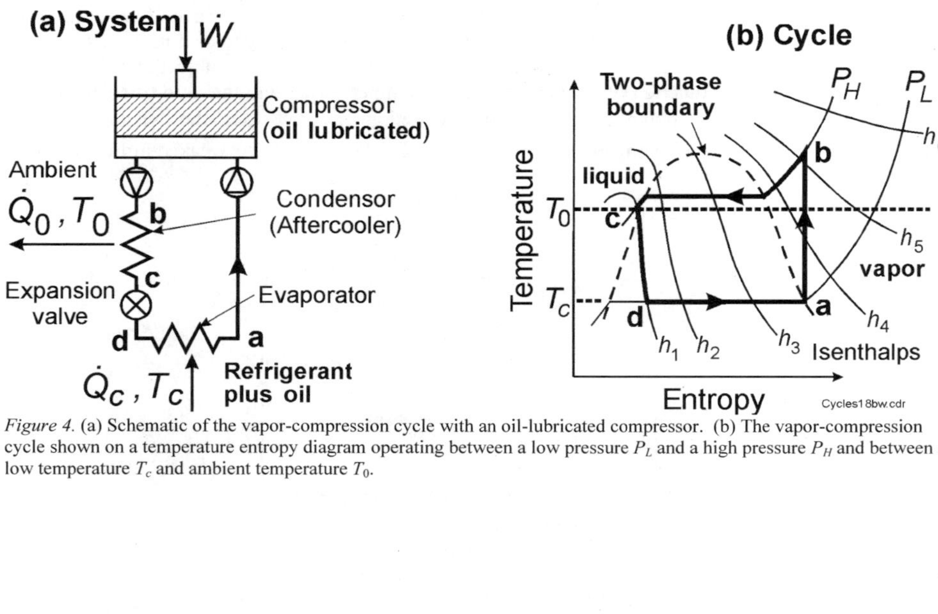

Figure 4. (a) Schematic of the vapor-compression cycle with an oil-lubricated compressor. (b) The vapor-compression cycle shown on a temperature entropy diagram operating between a low pressure P_L and a high pressure P_H and between a low temperature T_c and ambient temperature T_0.

3.2 RECUPERATIVE CYCLES

To achieve cryogenic temperatures, for example 80 K, the process shown in Figure 4 must be modified in two ways. First, the solubility of lubricating oil in the working fluid at such low temperatures is extremely small and any excess will freeze and cause plugging of the expansion channels. Thus, either the compressor must be oil free, which introduces reliability issues, or the system must have oil removable equipment utilizing complex processes (cost issues) to remove the oil before it reaches such low temperatures. Second, no fluid exists which can be expanded in an isenthalpic process (no expansion work) from room temperature to cryogenic temperatures. Even with a work-recovery process the initial pressure would need to be impractically high to achieve such low temperatures after expansion. Thus, it is necessary to precool the high-pressure gas in a heat exchanger prior to the expansion, as is shown schematically in the Joule-Thomson cryocooler in Figure 5a. The path followed on the *T*-S diagram is shown in Figure 5b. Because the heat transferred in the heat exchanger to provide sufficient precooling is much larger than the refrigeration power, the effectiveness of the heat exchanger must be very high, often higher than 95%. Small hydraulic diameters are needed in the heat exchanger to obtain such high effectiveness, especially for miniature cryocoolers. Hydraulic diameters of 50 to 100 μm may be required in some compact heat exchangers. When an expansion engine or turbine replaces the expansion orifice the cycle is called the Brayton cycle. Both it and the Joule-Thomson cryocooler are classified as recuperative types because of the use of recuperative heat exchangers throughout the cycle.

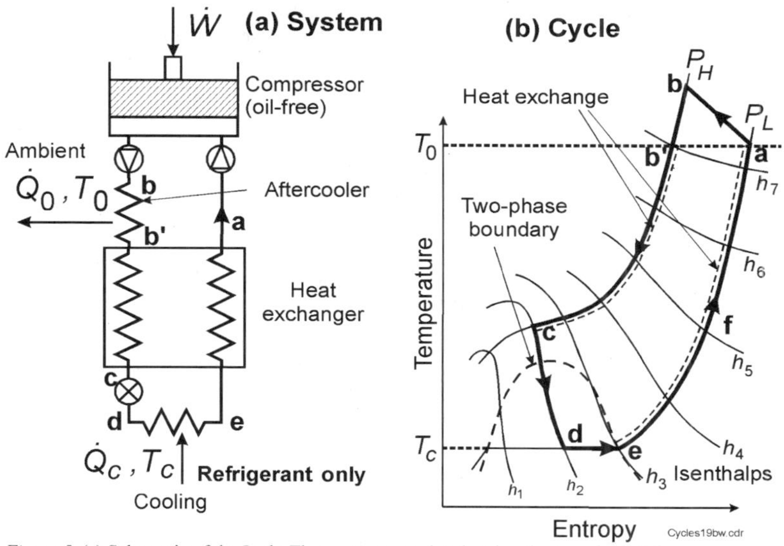

Figure 5. (a) Schematic of the Joule-Thomson cryocooler showing the use of an oil-free compressor and a high-effectiveness heat exchanger. (b) The Joule-Thomson cycle shown on a temperature-entropy diagram. Dashed lines indicate the heat exchange process in the heat exchanger.

3.3 REGENERATIVE CYCLES

In the past 20 years or so cryogenic temperatures are more commonly achieved by the use of regenerative cryocoolers. These cryocoolers, shown as schematics in Figure 6, operate with oscillating pressure and flow. They have at least one regenerative heat exchanger, or regenerator, where the hot and cold streams flow in the same channel, but at different times. Heat is stored for a half cycle in the heat capacity of the matrix. The Stirling cryocoolers and some pulse tube cryocoolers typically operate at about 20 to 70 Hz frequency and have no valves in the compressor. The Gifford-McMahon (GM) cryocoolers and some pulse tube cryocoolers operate at about 1 to 2 Hz frequency. The lack of valves in the higher frequency Stirling crycoolers and some pulse tube cryocoolers give them higher efficiencies than those of the valved systems. Thus, there is much emphasis on these higher frequency systems, particularly for space and miniature applications. For a constant power the size of the system decreases as the frequency increases. However, there are heat transfer problems at the higher frequencies that will be discussed here. More complete descriptions of the various cryocooler cycles are given by Radebaugh [4, 5].

4. Applications of Cryocoolers

We present here a few examples of cryocooler applications to show where microscale heat transfer issues at low temperatures may be of some concern. The overall size of the cryocooler usually has little bearing on whether microscale heat transfer issues are involved. It is the hydraulic diameter that is important in determining microscale effects. Small hydraulic diameters are required for very effective heat exchangers, particularly for those used in high frequency regenerative cryocoolers. For

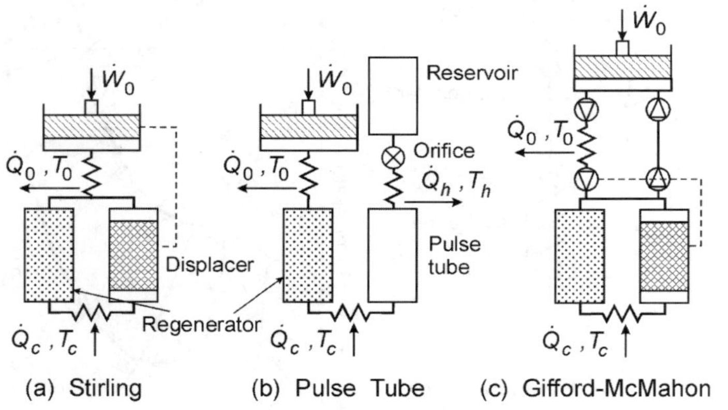

Figure 6. Schematics of the three common regenerative cryocoolers. The Stirling cryocooler (a) uses a valveless compressor or pressure oscillator and has a moving displacer operating synchronously with the piston. The pulse tube cryocooler (b) has no displacer in the cold head. The Gifford-McMahon cryocooler (c) uses a valved compressor with oil lubrication and oil removal equipment.

recuperative heat exchangers high effectiveness can still be achieved with large hydraulic diameters, but the overall size of the heat exchanger becomes large. Thus, we concentrate on applications with more compact cryocoolers. The cooling of infrared sensors to about 80 K for high-resolution night vision, primarily for the military has been one of the largest applications of cryocoolers. Figure 7 shows the cold heat exchanger and expansion orifice of a small Joule-Thomson cryocooler used to rapidly cool (few seconds) infrared sensors in the guidance system of missiles. Miniature finned tubing is used for the heat exchanger. Because there are no moving parts at the cold end of the Joule-Thomson cryocooler it can be scaled down to very small sizes. There is growing interest in developing such a cooler using MEMS technology for a cooler on a chip that might provide a few milliwatts of cooling at 80 to 100 K. We shall examine the optimization of the heat exchanger geometry for this application in the next section.

The infrared sensors on tanks, helicopters, airplanes, etc. are usually cooled to 80 K with miniature Stirling cryocoolers operating at about 50 to 60 Hz frequency. Over 140,000 such coolers have been made to date for this application [4]. Hydraulic diameters of about 50 to 60 μm are generally used in the packed-screen regenerators to obtain high effectiveness. The same hydraulic diameter and length would be used in much larger Stirling or pulse tube cryocoolers to obtain the same regenerator effectiveness. Only the cross-sectional area of the regenerator should be scaled with the refrigeration power. The smallest commercial Stirling cryocooler is shown in Figure 8. It is used primarily for commercial applications of infrared sensor cooling to 80 K, such as for process monitoring. Only 3 W of input power are required for this cooler to produce 0.15 W of cooling at 80 K. Somewhat larger Stirling cryocoolers (6 W at 77 K), shown in Figure 9, provide cooling for high-temperature superconducting microwave filters used in some cellular phone base stations for enhanced sensitivity. Over 3000 base stations now use superconducting filters (~1% of the total).

Figure 10 shows a miniature pulse tube cooler developed for cooling infrared sensors or high-T_c superconducting electronics in space applications. It provides about 0.5 W of cooling at 80 K. With this particular cooler the regenerator and the pulse tube are inline. The cold surface is in the middle between these two components. In addition to military applications the cooled infrared sensors in space are being used for studies of atmospheric phenomena, such as the ozone hole, green-house effect, and long-range weather forecasting. To gain greater sensitivity to long-wavelength infrared radiation

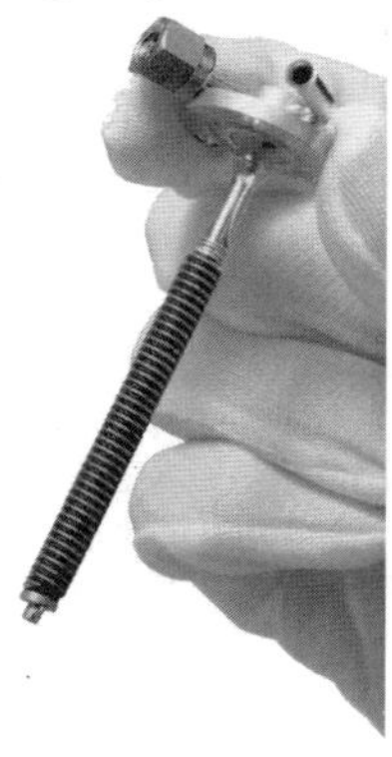

Figure 7. Joule-Thomson micro cryocooler. Courtesy APD Cryogenics.

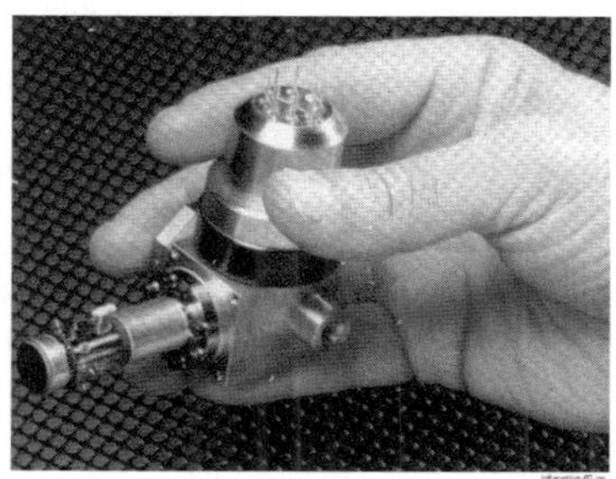

Figure 8. Stirling micro cryocooler that provides 0.15 W of cooling at 80 K with 3 W of input power. Courtesy Inframetrics/FLIR Systems.

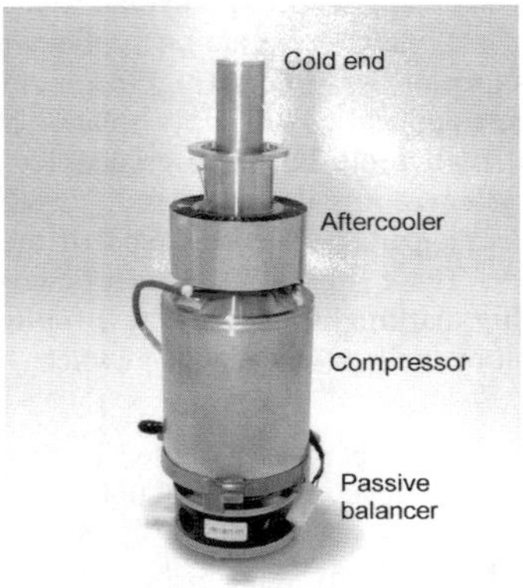

Figure 9. Stirling cryocooler used for cooling high-Tc superconducting microwave filters to 77 K in mobile phone base stations. Courtesy STI.

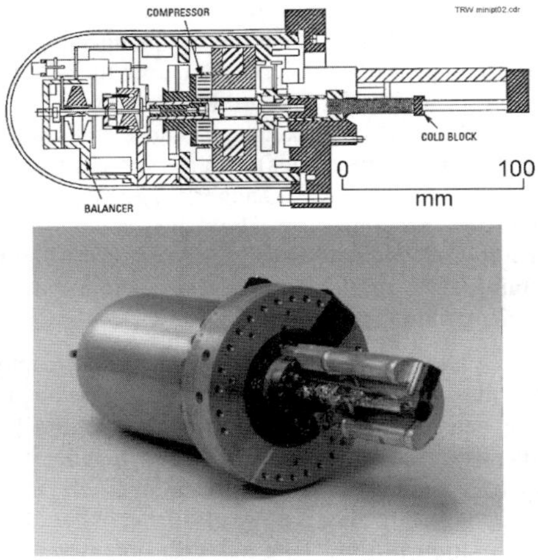

Figure 10. Mini pulse tube cryocooler used for cooling infrared sensors or superconducting devices in space. Courtesy TRW/NGST.

for astronomy missions space agencies are now pushing for cryocoolers to reach temperatures of 10 K or lower. To maintain liquid hydrogen fuel in space for long-range exploration missions there is now a growing need for efficient and compact 20 K cryocoolers. The pulse tube cryocooler is being considered for these new applications, but there is a need to better understand the heat transfer problems in regenerators at these low temperatures and at high frequencies. Frequencies of at least 20 Hz are required to keep the compressor small. As shown in Figure 2 the thermal penetration depth for this frequency in helium at a pressure of about 2 MPa and temperatures of 20 K and below is about 50 μm. The hydraulic diameter within the regenerator should then be significantly less than that value for good heat transfer. Such small hydraulic diameters are challenging to achieve in practice, particularly with parallel plates or tubes. With packed spheres the resulting void volume (38%) is higher than optimum for these low temperatures, and there is difficulty in containing such small spheres under oscillating flow conditions.

5. Optimization of Heat Exchanger Geometry

Our objective in optimizing the heat exchanger geometry is minimizing the volume of the heat exchanger. That objective is particularly important for the development of micro cryocoolers. The heat exchanger is usually the largest component of the Joule-Thomson cryocooler, except for the compressor. Of interest here is the value of the Knudsen number in optimized micro cryocoolers. In the optimization procedure we choose to fix the fractional losses associated with imperfect heat transfer, pressure drop, and axial heat conduction. These losses are normalized by the gross refrigeration power $\dot{Q}_r$ of the cryocooler. The pressure drop in a flow channel of cross-sectional area A_g, length L, and hydraulic diameter D_h is given by

$$\Delta P = \frac{2 f_r (\dot{m} / A_g)^2 L}{\rho D_h}, \tag{3}$$

where f_r is the Fanning friction factor, $\dot{m}$ is the mass flow rate in the flow channel, and ρ is the density of the gas. The density should be evaluated at the average temperature. We can relate the friction factor to the Stanton number N_{St}, a dimensionless heat transfer number, by the Reynolds analogy

$$\alpha = \frac{N_{St} N_{Pr}^{2/3}}{f_r}, \tag{4}$$

where N_{Pr} is the Prandlt number and α is a dimensionless number that is a function of the geometry. Figure 11 shows how α varies with the Reynolds number for various geometries. The behavior of the Stanton number and the friction factor discussed here are based on experiments with macrosystems in continuum flow [6]. We note that α is a rather weak function of the Reynolds number, which makes the optimization procedure being described here very simple and powerful. Radebaugh and Louie [7] describe details of this procedure for regenerators. Both f_r and N_{St} are strong functions of Reynolds number but the ratio is only a weak function. The Stanton number is defined as

$$N_{St} = \frac{h}{(\dot{m} / A_g) c_p}, \tag{5}$$

where h is the heat transfer coefficient and c_p is the specific heat of the gas at constant pressure. We can proceed with the optimization by taking α to be a specific number independent of the Reynolds

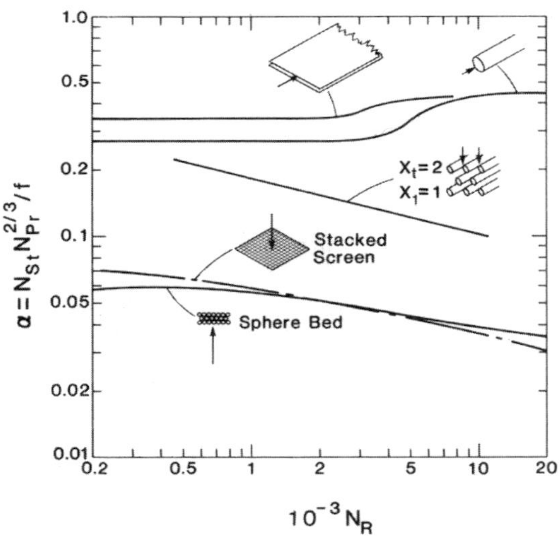

Figure 11. Reynolds analogy for several geometries as a function of the Reynolds number [6].

number, which is not known at this point in the optimization.

By substituting equation (4) into equation (3) we can express the pressure drop in terms of the heat transfer parameter N_{St} as

$$\Delta P = \frac{2N_{St}N_{Pr}^{2/3}(\dot{m}/A_g)^2 L}{\alpha\rho D_h}. \tag{6}$$

The number of heat transfer units N_{tu} on any given side of the heat exchanger is

$$N_{tu} = \frac{hA}{\dot{m}c_p}, \tag{7}$$

where A is the surface area for heat transfer on one side of the heat exchanger. The surface area A is related to the hydraulic diameter through the definition of hydraulic diameter by

$$D_h \equiv 4LA_g / A. \tag{8}$$

By combining equations (5), (7), and (8) we have

$$N_{tu} = 4N_{St}L / D_h. \tag{9}$$

By using equation (9) to find N_{St} and substituting it into equation (6) the pressure drop becomes

$$\Delta P = \frac{N_{tu}N_{Pr}^{2/3}(\dot{m}/A_g)^2}{2\alpha\rho}. \tag{10}$$

5.1 GAS CROSS-SECTIONAL AREA

After rearranging equation (10) the specific cross-sectional area is then given by

$$\frac{A_g}{\dot{m}} = \left[\frac{N_{tu}N_{Pr}^{2/3}}{2\alpha\rho\Delta P}\right]^{1/2}. \tag{11}$$

The problem with equation (11) is that we do not know *a priori* a reasonable design value for N_{tu} on each side of the heat exchanger. A more fundamental design parameter would be the heat loss associated with the imperfect heat exchanger compared with the gross refrigeration power. The heat exchanger ineffectiveness (1 - ε), where ε is the effectiveness, is defined by

$$1-\varepsilon = \frac{\dot{Q}_{hx}}{\dot{m}(h_h - h_c)_{\min}}, \tag{12}$$

where $\dot{Q}_{hx}$ is the heat flow or loss to the cold end due to imperfect heat transfer in the heat exchanger, h_h is the specific enthalpy of the gas at the hot end, and h_c is the specific enthalpy of the gas at the cold end. The subscript min refers to the stream (high or low pressure) with the minimum difference in specific enthalpy between the hot and cold ends. Figure 12 shows the specific enthalpy curves for

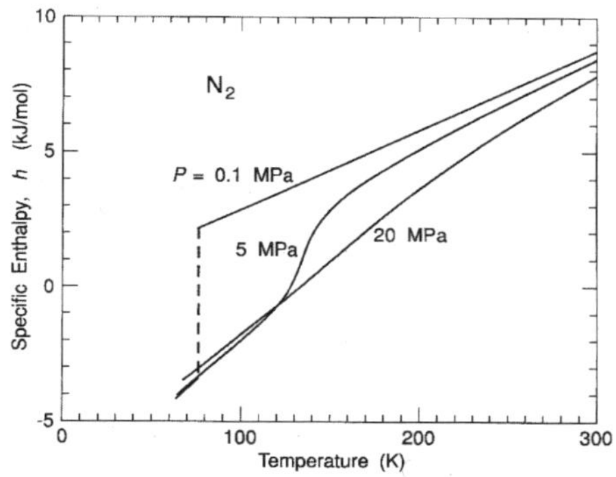

Figure 12. Specific enthalpy of nitrogen for different pressures.

nitrogen for two different pressures. As shown by these curves the minimum enthalpy difference occurs in the low-pressure stream. The gross refrigeration power provided by the Joule-Thomson cryocooler is given by

$$\dot{Q}_r = \dot{m}\Delta h_{\min}, \tag{13}$$

where $\Delta h_{\min}$ is the minimum difference in specific enthalpy between the high and low pressure. As shown in Figure 12 this minimum for nitrogen occurs at the warm end of the heat exchanger. The specific heat exchanger loss and the specific refrigeration power can be defined as

$$q_{hx} = \dot{Q}_{hx} / \dot{m} \qquad\qquad q_r = \dot{Q}_r / \dot{m} = \Delta h_{\min}. \tag{14}$$

We can combine equations (12) and (13) to obtain

$$1 - \varepsilon = (\dot{Q}_{hx} / \dot{Q}_r) q_r^*, \tag{15}$$

where q_r^* is the ratio of the refrigeration power to the heat flow in the heat exchanger, or relative refrigeration power, as given by

$$q_r^* = \frac{q_r}{(h_h - h_c)_{\min}}. \tag{16}$$

We note from equation (15) that q_r^* is the maximum allowed value of 1 - ε, that is $(\dot{Q}_{hx} / \dot{Q}_r)$ must be less than 1 to allow for any net refrigeration. If $q_r^* < 1$, a heat exchanger is required to yield any net refrigeration power. That condition usually is a distinguishing feature of cryocoolers compared with vapor-compression refrigerators.

The ineffectiveness of a heat exchanger and the associated heat flow to the cold end as given by equation (12) are normally associated with the complete heat exchanger. The number of heat transfer units for the complete heat exchanger is given by

$$\frac{1}{N_{tu0}} = \frac{1}{N_{tu1}} + \frac{1}{N_{tu2}}, \tag{17}$$

where N_{tu1} and N_{tu2} are the number of heat transfer units on each side of the heat exchanger. Published curves [6] and tables give the ineffectiveness (1 - ε) as a function of N_{tu0}. For N_{tu0} > 10 (a necessity for nearly all cryocooler heat exchangers) the heat exchanger ineffectiveness, is approximated very well by

$$1-\varepsilon = B/N_{tu0}, \tag{18}$$

where the constant B is given by B = 1 whenever the specific heats in both streams are equal. If they are not equal, equation (18) can still be used by making B < 1, but in that case B becomes somewhat a function of N_{tu0} [7]. For B independent of N_{tu0} or a weak function of it, we can expand equation (18) according to equation (17) into

$$1-\varepsilon = \frac{B}{N_{tu1}} + \frac{B}{N_{tu2}} = (1-\varepsilon)_1 + (1-\varepsilon)_2, \tag{19}$$

where each side has its own ineffectiveness and associated heat flow $\dot{Q}_{hx}$ according to equation (15).

By using equation (15) to represent each side of the heat exchanger equation (19) allows us to express N_{tu} for each side of the heat exchanger as

$$N_{tu} = \frac{B}{(\dot{Q}_{hx}/\dot{Q}_r)q_r^*}, \tag{20}$$

where $(\dot{Q}_{hx}/\dot{Q}_r)$ now is the relative heat flow from each side of the heat exchanger. We substitute this expression for N_{tu} into equation (11) to obtain the specific gas cross-sectional area for one side as

$$\frac{A_g}{\dot{m}} = \left[\frac{BN_{Pr}^{2/3}}{2\alpha\rho q_r^* \Delta P(\dot{Q}_{hx}/\dot{Q}_r)}\right]^{1/2}. \tag{21}$$

Equation (21) gives the specific area in terms of the relative heat loss, which is a better design parameter than the number of heat transfer units used in equation (11). However, for highly unbalanced heat capacity flows the actual overall heat loss will deviate from the sum of the individual losses used in these calculations. The more accurate heat loss for the complete heat exchanger is given by equation (15) when the overall ineffectiveness is found from the N_{tu0} evaluated from the individual N_{tu} by equation (17). In this expression the density ρ is a strong function of temperature.

For an ideal gas we use the relation

$$\rho = P/RT, \tag{22}$$

where R is the gas constant per unit mass and T is the absolute temperature. For an ideal gas, equation (21) can then be written as

$$\frac{A_g}{\dot{m}} = \left[\frac{BRTN_{Pr}^{2/3}}{2\alpha P_0^2 q_r^* (\Delta P/P_0)(\dot{Q}_{hx}/\dot{Q}_r)}\right]^{1/2}, \tag{23}$$

where P_0 is the average pressure within the flow channel being optimized. The temperature should be taken as the average between the hot and cold ends of the heat exchanger. Equation (23) can be converted to a molar flow basis by replacing R with MR_0, where M is the gas molecular weight and R_0 is the universal gas constant 8.314 J/(mol·K).

In order to obtain any finite refrigeration power each of the loss terms ($\Delta P/P_0$) and ($\dot{Q}_{hx}/\dot{Q}_r$) should be sufficiently small. Heat flow to the cold end by axial conduction $\dot{Q}_{cond}$ is the remaining loss that must be considered in the optimization of the heat exchanger geometry. We will use that later in the calculation of the optimum length. The pressure drop causes a loss in the refrigeration power that can be denoted as $\dot{Q}_{\Delta P}$. In order to have a finite net refrigeration power the losses must satisfy the condition

$$(\dot{Q}_{\Delta P}/\dot{Q}_r)_1 + (\dot{Q}_{hx}/\dot{Q}_r)_1 + (\dot{Q}_{cond}/\dot{Q}_r)_1 + (\dot{Q}_{\Delta P}/\dot{Q}_r)_2 + (\dot{Q}_{hx}/\dot{Q}_r)_2 + (\dot{Q}_{cond}/\dot{Q}_r)_2 < 1, \qquad (24)$$

where the subscripts 1 and 2 refer to the two sides of the heat exchanger. Typically the losses on each side must be less than about 0.5. The relationship between ($\Delta P/P_0$) and ($\dot{Q}_{\Delta P}/\dot{Q}_r$) depends on the refrigeration cycle and the operating conditions. For the Joule-Thomson cycle the ratio of the two terms is given by

$$\frac{(\dot{Q}_{\Delta P}/\dot{Q}_r)}{(\Delta P/P_0)} = \frac{P_0}{\Delta h_{\min}}\left(\frac{\partial h}{\partial P}\right)_T, \qquad (25)$$

where the partial derivative should be evaluated at the same temperature as $\Delta h_{\min}$. The ratio in equation (25) may vary from about 0.6 to 1.2 on the high pressure side of most Joule-Thomson systems. However, the low pressure side may be much more complicated. In most cases the ratio in equation (25) for the low pressure side may only be 0.3 or smaller. For pure gases the specific heats of the two streams are not equal and $\Delta h_{\min}$ occurs at the high temperature end. Thus, expanding the gas at the cold end only to the pressure where the enthalpy change is the same as $\Delta h_{\min}$ and allowing for a pressure drop through the low pressure side of the heat exchanger forces the enthalpy change in both sides of the heat exchanger to be equal. In that case there is no loss of refrigeration power, but with the balanced heat exchanger the value of B will be 1 rather than some lower value.

For a practical cooler the sum of the normalized loss terms in equation (24) should usually be less than about 0.6, which can be satisfied with each individual loss being about 0.1. In that case an approximate upper limit for the product $(\Delta P/P_0)(\dot{Q}_{hx}/\dot{Q}_r)$ in equation (23) is about 0.01 for the high pressure stream. An important aspect of equation (23) to note is that the right hand side is independent of the flow rate or size of the refrigerator. Thus, equation (23) shows that the proper scaling relationship for the gas cross-sectional area A_g is for it to be proportional to the flow rate, which is also proportional to the refrigeration power.

5.2 HEAT EXCHANGER LENGTH

Equation (23) shows that the volume of the heat exchanger could be made as small as desired by making the length small. However, in that case the conduction loss $\dot{Q}_{cond}$ becomes large. The conduction loss on each side of the heat exchanger is given by

$$\dot{Q}_{cond} = (A_s/L)\int_{T_c}^{T_h} k\,dT, \qquad (26)$$

where A_s is the cross-sectional area of the solid structure of the flow channel. We shall ignore the thermal conduction through the gas. The term A_s is related to A_g through the porosity n_g by

$$n_g = \frac{A_g}{A_s + A_g}. \tag{27}$$

The heat exchanger length for a given conduction loss on each side is expressed as

$$L = \frac{(A_g / \dot{m})(1 - n_g) \int kdT}{n_g q_r (\dot{Q}_{cond} / \dot{Q}_r)}. \tag{28}$$

As to be expected, the length can be made small by decreasing the solid fraction (1 – n_g), but the strength of the material containing the gas imposes a limit to how high the porosity can be made. For a circular tube with internal pressure P the minimum solid fraction can be taken as

$$(1 - n_g) = 2P / \sigma, \tag{29}$$

where σ is the maximum allowable tensile stress in the tube wall.

5.3 HEAT EXCHANGER VOLUME

The gas volume in the heat exchanger is given by combining equation (28) with equation (21), or with equation (23) for an ideal gas. For the general case the gas volume becomes

$$\frac{V_g}{\dot{m}} = \frac{A_g L}{\dot{m}} = \frac{B N_{Pr}^{2/3} (1 - n_g) \int kdT}{2\alpha n_g q_r^* q_r \rho \Delta P (\dot{Q}_{hx} / \dot{Q}_r)(\dot{Q}_{cond} / \dot{Q}_r)}. \tag{30}$$

The volume per unit of gross refrigeration can be expressed as

$$\frac{V_g}{\dot{Q}_r} = \frac{B N_{Pr}^{2/3} (1 - n_g) \int kdT}{2\alpha n_g q_r^* q_r^2 \rho \Delta P (\dot{Q}_{hx} / \dot{Q}_r)(\dot{Q}_{cond} / \dot{Q}_r)}. \tag{31}$$

The total volume of the heat exchanger is given by dividing V_g by n_g. We note that in these last two equations that the volume of the heat exchanger is made small by maximizing α. Values of α for several different geometries are shown in Figure 11. As this figure shows parallel plates will produce the smallest gas volume, with parallel tubes giving slightly larger volumes. Because α is nearly independent of Reynolds number, the volume determined by this optimization procedure is also independent of the Reynolds number.

5.4 HYDRAULIC DIAMETER

We now derive the hydraulic diameter of the heat exchanger that must be used to achieve the area, length, and volume given above with the specified thermal loss in the heat exchanger. By multiplying equation (9) by $N_{Pr}^{2/3}$ we obtain

$$N_{St}N_{Pr}^{2/3} = \frac{N_{tu}N_{Pr}^{2/3}D_h}{4L}. \tag{32}$$

By using equation (4) we obtain

$$\alpha f_r = \frac{N_{tu}N_{Pr}^{2/3}D_h}{4L}. \tag{33}$$

The Reynolds number is given as

$$N_r = \frac{D_h}{\mu(A_g / \dot{m})}, \tag{34}$$

where μ is the viscosity. For every geometry of interest there is some function g where

$$f_r = g(N_r). \tag{35}$$

Combining equation (33) with equation (35) yields

$$\frac{N_{tu}N_{Pr}^{2/3}D_h}{\alpha 4L} = g(N_r). \tag{36}$$

Equation (36) is nonlinear and can be solved for D_h by trial and error in the most general form. However, for laminar flow it becomes linear and is easily solved. In laminar flow the function g becomes

$$g(N_r) = b / N_r, \tag{37}$$

where b is a constant that depends only on the geometry and is sometimes called the Poiseuille number. For long parallel plates $b = 24$, for long square tubes $b = 14.25$, for long circular tubes $b = 16$, and for long triangular tubes $b = 13.33$ [6]. These numbers correspond to the use of the Fanning friction factor. Equations (36) and (34) are solved in the laminar flow regime by

$$D_h = \left[\frac{4\alpha b\mu(A_g / \dot{m})L}{N_{tu}N_{Pr}^{2/3}}\right]^{1/2}. \tag{38}$$

Each term in equation (38) has been calculated previously. The solution to D_h then allows the Reynolds number to be calculated from equation (34). With this known value of the Reynolds number a new and more precise value of α could be determined from Figure 11 and the calculations repeated if more accuracy is desired. Because α is only a weak function of the Reynolds number the second calculation is seldom necessary. Equation (38) is expressed in terms of the original variables as

$$D_h = \left[\frac{2b\mu(1-n_g)\int kdT}{n_g \rho q_r \Delta P(\dot{Q}_{cond} / \dot{Q}_r)}\right]^{1/2}. \tag{39}$$

For the case where the density is given by an ideal gas equation of state equation (39) becomes

$$D_h = \left[\frac{2b\mu RT(1-n_g)\int kdT}{n_g P_0^2 q_r (\Delta P / P_0)(\dot{Q}_{cond} / \dot{Q}_r)} \right]^{1/2}. \tag{40}$$

It is interesting to note that D_h is independent of the thermal loss in the heat exchanger $\dot{Q}_{hx}$. It is also independent of the size of the heat exchanger. For a minimum volume the hydraulic diameter is independent of the flow rate, volume, or refrigeration power of the cryocooler. According to equation (21) only the cross-sectional area varies proportional to the flow rate. For some common geometries the relation between the hydraulic diameter, defined by equation (8), and the characteristic dimension is given by

Gap thickness $\quad t_g = D_h / 2 \quad$ (41)

Tube diameter: $\quad d = D_h \quad$ (42)

Square channel side: $\quad s = D_h \quad$ (43)

Equilateral triangle side: $\quad s = \sqrt{3} D_h \quad$ (44)

Sphere diameter: $\quad d = \dfrac{3D_h(1-n_g)}{2n_g} \quad$ (packed spheres) (45)

Wire diameter: $\quad d \approx D_h(1-n_g)/n_g. \quad$ (stacked screen) (46)

The practical porosity n_g of most packed spheres is about 0.38, and that of most commercial screen is about 0.65.

Equation (40) shows that the optimum hydraulic diameter decreases at lower temperatures. Thus, microscale effects may be more important in low temperature applications, such as in heat exchangers of cryocoolers. However, temperature has little effect on the relative importance of slip flow. The mean free path of gas molecules is given by

$$\lambda = 3.62 \frac{\mu}{P} \sqrt{\frac{T}{M}}, \tag{47}$$

where M is the molecular weight and the units are: λ (m), μ (Pa·s), P (Pa), T (K), and M (kg/mol). The Knudsen number $Kn = \lambda/D_h$ for an optimized flow channel in a heat exchanger will vary with temperature only through the temperature dependence of viscosity and the thermal conduction integral.

5.5 EXAMPLES

We now consider some examples to illustrate the geometries that minimize the volume of cryocooler recuperative heat exchangers. In all cases we use a conservative calculation with the constant $B = 1$. For case A we consider an ideal refrigeration cycle using helium gas with isothermal compression at 320 K and isothermal expansion at some low temperature T_c. The low pressure P_1 is 0.3 MPa and the high pressure P_2 is 0.6 MPa. We then optimize a heat exchanger operating between T_c and some high temperature T_h where $T_h = 4T_c$. We assume there is some other perfect heat exchanger between T_h and the compressor at 320 K. The specific refrigeration power provided by the reversible isothermal expansion is given by

$$q_r = \dot{Q}_r / \dot{m} = RT_c \ln(P_2 / P_1). \qquad (48)$$

For this case where $T_h = 4T_c$ the normalized specific refrigeration power q_r^* becomes (2/15)ln(P_2/P_1). For the pressure ratio of 2 considered here we have $q_r^* = 0.0924$. This value is rather high for such a low pressure ratio, but can only be achieved with reversible isothermal expansion. Such a process can only be approximated in practice and is particularly difficult to accomplish in miniature sizes. Values of the input parameters for all the examples are summarized in Table 1. For the examples used here we consider parallel plate geometry. For case A we assume both sides have a porosity of 0.5 initially. Figure 13 gives the specific cross-sectional areas calculated from equation (21) for the two gas streams as a function of the cold-end temperature. The dashed line is for the high-pressure stream when the length and width of that stream are made the same as that of the low-pressure stream. To force such dimensions the porosity on the high-pressure side was made 0.39 and the reduced heat exchanger loss was made 0.063. Figure 14 shows the temperature dependence of the heat exchanger lengths when both sides have a porosity of 0.5. Figure 15 shows the gap thickness for the two sides, including the case where the high-pressure side is forced to have the same length and width as the low-pressure side. The Knudsen number at the average temperature is shown in Figure 16. At the high temperature end of the heat exchanger the Knudsen number will be about $\sqrt{2}$ higher. This figure shows that at the warm end of heat exchangers at the lowest temperature slip flow may just begin to occur, but because Kn is less than 10^{-2} there is very little effect on the friction factor and the heat transfer coefficient. Figure 17 shows the width of the gap for the particular case where the mechanical input power is 1 W, which provides a flow rate of 2.17 mg/s. The figure also shows the temperature variation of the resulting gross refrigeration power. If the refrigeration power at 80 K were reduced by two orders of magnitude to about 2 mW, then the gap width is reduced to 30 μm, which is less than the gap thickness of about 50 μm on the low-pressure side. The optimization procedure described here then begins to become invalid for such small sizes.

A few other examples are examined here to illustrate the range of geometries that may exist in optimized heat exchangers for cryocoolers. These examples are for Joule-Thomson cryocoolers, which are easily miniaturized. Table 1 lists the operating conditions and important input parameters for all of

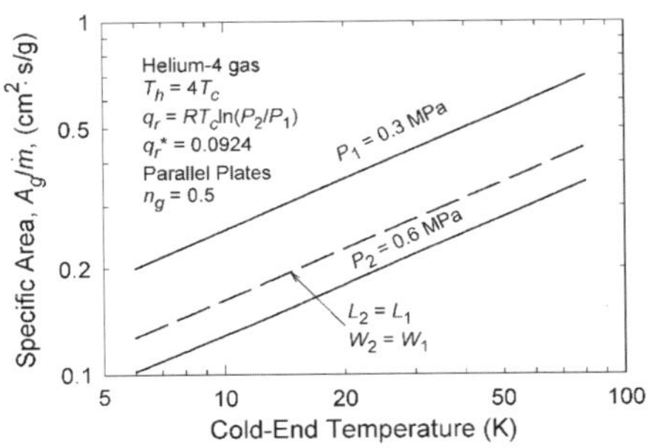

Figure 13. Gas cross-sectional area per unit mass flow rate as a function of cold-end temperature for case A.

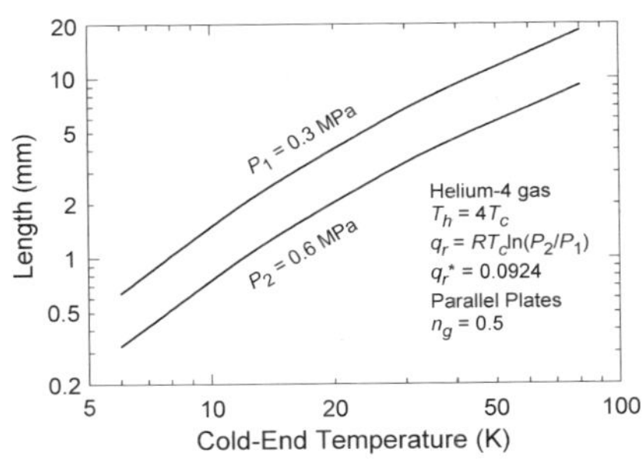

Figure 14. Calculated optimum length of heat exchanger of case A as a function of cold-end temperature.

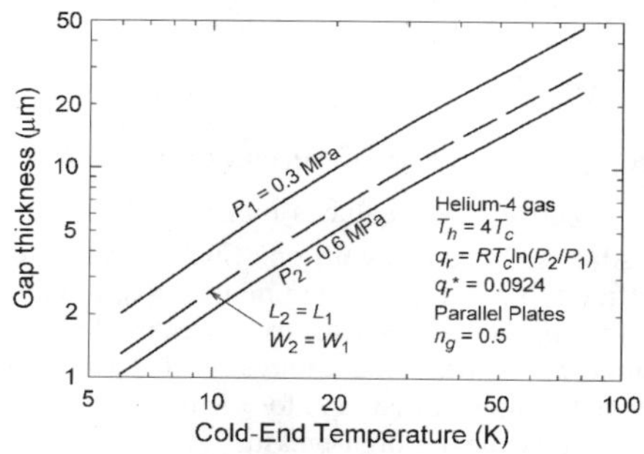

Figure 15. Optimum gap thickness between parallel plates for case A as a function of cold-end temperature.

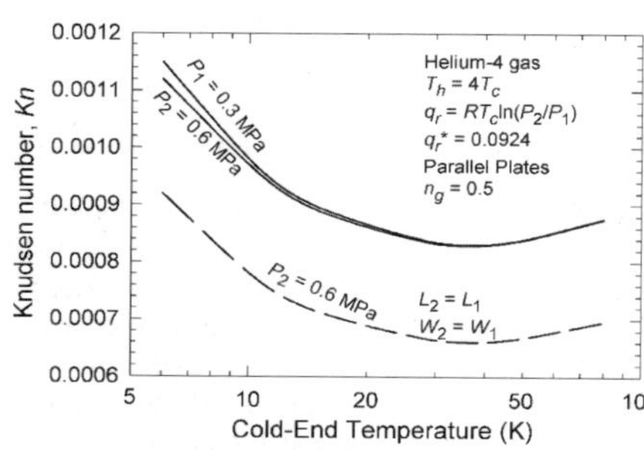

Figure 16. Knudsen number for both sides of the heat exchanger for case A.

these examples. For case B pure nitrogen is the working fluid operating between 80 K and 300 K with a low pressure of 0.1 MPa and a high pressure of 15 MPa provided by the compressor at the warm end of the heat exchanger. The pressure ratio of 150 is very high and requires a special compressor with several stages of compression. For case C the temperatures of 80 K and 300 K and the low pressure of 0.1 MPa are kept the same, but the high pressure is reduced to 2.5 MPa, which can be provided by conventional compressors used in household refrigerators. For case D a gas mixture of nitrogen and hydrocarbons operates between the 80 K and 300 K with 0.1 MPa and 2.5 MPa pressures. For case E helium gas operates between 0.3 and 0.6 MPa with temperatures of 6 and 18 K, and for case F helium has the same pressures as in case E but the temperatures are between 140 K and 300 K. In the last case the Joule-Thomson expansion will not provide any cooling because the temperature is above the inversion temperature. In this case we specify some fixed heat exchanger ineffectiveness $1 - \varepsilon$ and calculate the required external specific refrigeration power required according to equations (15) and (16) to absorb the heat flow at the cold end of the heat exchanger.

Table 2 gives the optimized geometry and other important parameters for the examples discussed above. The case designations of, for example B1, B2, B2*, correspond to (1) the low-

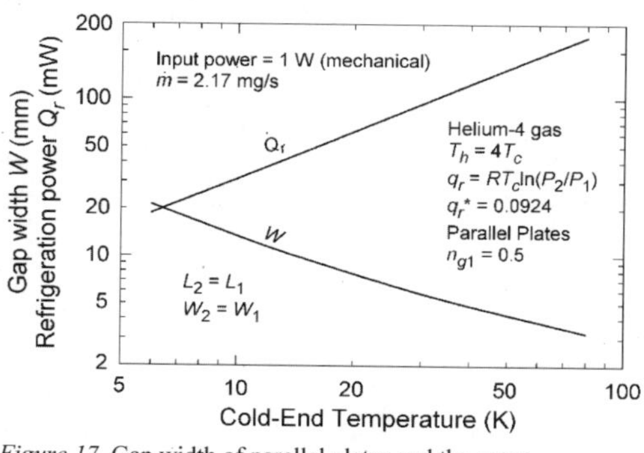

Figure 17. Gap width of parallel plates and the gross refrigeration power for case A with 1 W of mechanical input power as a function of cold end temperature.

Table 1. Input parameters used in example calculations of optimized recuperative heat exchangers.

Case	Gas	Cycle	T_c, T_h (K)	P_1, P_2 (MPa)	q_r (J/g)	q_r*	n_g	$\Delta P/P_0$	Q_{hx}/Q_r	Q_{cond}/Q_r
A	He	Isoth. exp.	$T_h = 4T_c$	0.3, 0.6	$1.440T_c$	0.0924	0.5	0.1	0.1	0.1
B	N_2	JT	80, 300	0.1, 15	26.57	0.1157	0.4	0.4, 0.1	0.1	0.1
C	N_2	JT	80, 300	0.1, 2.5	5.170	0.0224	0.4	0.4, 0.1	0.1	0.1
D	mix	JT	80, 300	0.1, 2.5	62.48	0.090	0.4	0.2, 0.1	0.1	0.1
E	He	JT	6, 18	0.3, 0.6	1.874	0.0253	0.4	0.1, 0.1	0.15	0.05
F	He	precool	140, 300	0.3, 0.6	8.309	0.01	0.4	0.05, 0.05	0.5	0.25

Table 2. Calculated optimum geometry and flow parameters for example heat exchangers.

Case	n_g	Q_{cond}/Q_r	A_g/m (cm^2s/g)	L (mm)	t_g (μm)	Kn	N_r	Q_r (mW)	m (mg/s)	W (mm)
B1	0.4	0.1	0.356	53.9	56.2	0.00066	255	10	0.376	0.238
B2	0.4	0.1	0.00496	0.75	0.925	0.00050	161	10	0.376	0.202
B2*	0.05	0.05	0.00496	47.5	7.36	0.00006	1284	10	0.376	0.025
C1	0.4	0.1	0.809	630	127.5	0.00029	254	10	1.934	1.23
C2	0.4	0.1	0.0648	50.4	16.6	0.00009	391	10	1.934	0.755
C2*	0.1	0.05	0.0648	605	35.2	0.00004	830	10	1.934	0.356
D1	0.4	0.1	0.0867	5.58	6.56	0.0395	39.9	10	0.160	0.212
D2	0.4	0.1	0.0164	1.06	3.27	0.00413	32.9	10	0.160	0.080
D2*	0.2	0.05	0.0164	5.64	7.55	0.00179	76.0	10	0.160	0.035
E1	0.4	0.05	0.309	6.88	3.56	0.00050	85.5	10	5.34	46.4
E2	0.4	0.05	0.139	3.09	2.21	0.00043	110	10	5.34	33.6
E2*	0.25	0.05	0.139	6.19	3.13	0.00031	155	10	5.34	23.7
F1	0.4	0.25	1.417	215	173	0.00026	151	-66.5	5.34	4.37
F2	0.4	0.25	0.709	108	86.7	0.00026	151	-66.5	5.34	4.36
F2*	0.25	0.25	0.709	215	123	0.00019	214	-66.5	5.34	3.09

pressure stream, (2) the high-pressure stream, and (3) the high-pressure stream with the length made the same as that of the low-pressure side. Matching the lengths is a practical requirement for most heat exchangers, but it does lead to an increase in the required volume. Matching the widths, as was done for case A, may also be a practical requirement in most cases, but that has not been done for the rest of the examples. There are several important points we wish to point out regarding the various examples. First, the gap thickness for all cases except C1, F1, and F2* are less than 100 μm. Thus, microscale heat transfer is of interest here. The optimum gap thickness decreases at lower temperatures. However, the Knudsen numbers for all cases except the mixed refrigerant case are less than 10^{-3}. Thus, slip flow is of little concern and the use of the friction factor and Stanton number correlations for continuum flow should be valid. The Reynolds numbers are much less than 2000, so the laminar flow assumption in calculating the hydraulic diameter is valid.

The need for high effectiveness heat exchangers in cryocoolers is indicated by the low values of q_r^*, which gives the maximum ineffectiveness allowed for the heat exchanger, at which point net refrigeration is eliminated. With pure gases in a Joule-Thomson (JT) cryocooler starting from temperatures much above the normal boiling point, for example nitrogen at 300 K, very low ineffectiveness (<0.0224) is required to reach 80 K when the high pressure is only 2.5 MPa. That requirement leads to relatively large heat exchangers and difficulty in practice of actually achieving the very low ineffectiveness. The use of mixed refrigerants, as in case D, increases q_r^* to 0.090 for the same pressures and permits a reduction in the size of the heat exchanger. As shown by the results in Table 2 the Knudsen numbers for the mixed refrigerant case are rather large and indicate significant slip flow. However, the calculations for the mean free path were made assuming a gas phase, whereas both gas and liquid phases exist in much of the heat exchanger, especially in the high-pressure side. Thus, two-phase heat transfer becomes important here, and heat transfer will be enhanced beyond that assumed here for a single phase.

A JT cryocooler stage using helium with a pressure ratio of 2 can provide refrigeration at 6 K when the helium is precooled to about 18 K. Such JT stages are under development by two aerospace companies for 6 K cryocoolers for space applications [8]. Still, the ineffectiveness of the heat exchanger must be quite small (<0.0253). Stirling or pulse tube cryocoolers are used to precool the helium gas to 18 K. Case F represents an example of a heat exchanger for precooling helium gas from 300 K to 140 K as part of a system that uses case E for achieving 6 K. For this example the specific enthalpy difference between 300 K and 140 K for helium at either 0.3 MPa or 0.6 MPa is 8309 J/g. The flow rate is set from case E, which then gives the total heat transferred in the heat exchanger. For this case we must select some desired ineffectiveness on each side to determine the heat $\dot{Q}_{hx}$ from each side that must be absorbed at the cold end using equation (12). The resulting N_{tu} from published graphs can then be used directly in equation (11). For this case we arbitrarily select a total ineffectiveness of 0.015. To maintain consistency with the other examples we then choose $q_r^* = 0.010$ as the maximum ineffectiveness and $(\dot{Q}_{hx} / \dot{Q}_r) = 0.5$ on each side to give a total actual ineffectiveness of 0.010. According to equation (16) the specific refrigeration power must be q_r = 8.309 J/g. With $(\dot{Q}_{cond} / \dot{Q}_r) = 0.25$ on each side the total loss from both sides of the heat exchanger becomes $1.5q_r$ = 12.46 J/g. For a gross refrigeration power of 10 mW at 6 K (case E) the flow rate is 5.34 mg/s, which then causes a heat flow of 66.5 mW at the 140 K heat exchanger that must be absorbed by some other cryocooler. The gap thickness and length of this higher temperature heat exchanger is significantly larger than its 6 K counterpart mainly because of the much lower density at these higher temperatures.

The small size of the heat exchangers described here are an advantage in many applications besides the development of micro cryocoolers. The small size reduces the radiation heat load to the cold end and the conducted heat load through mechanical supports, particularly for space applications where rigid launch support is needed. The lower mass is also important for space applications. The achievement of ineffectiveness values less than 1% is difficult in practice because of problems with nonuniform flow in parallel channels. Figure 18 shows an example of a parallel plate heat exchanger designed for conditions similar to those for case F above, except for somewhat higher flows [9]. Flow paths are formed by photoetching completely through a foil of the proper thickness rather than relying on depth etching. These foils are alternated with a barrier foil and diffusion bonded to form the complete heat exchanger. The uniform gap thickness leads to more uniform flow in each channel. The measured ineffectiveness of this heat exchanger was about 2.7%, which was higher than the design value of about 0.5%. The difference was attributed to flow imbalances. Improvements in fabrication techniques can lead to even lower ineffectiveness in similar compact heat exchangers [9].

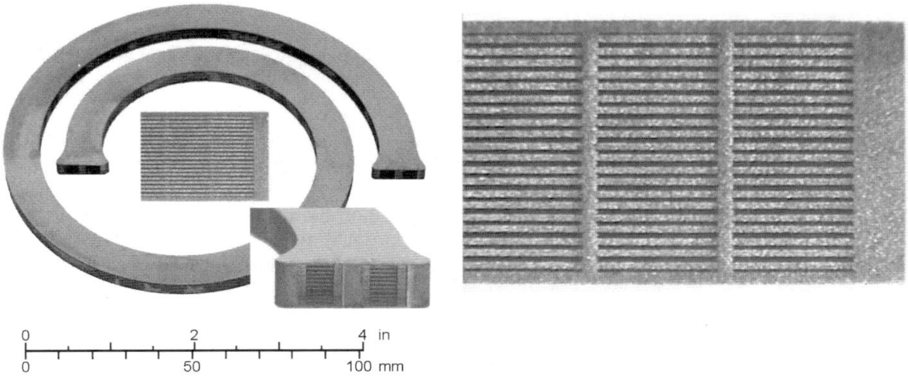

Figure 18. Photographs of a miniature heat exchanger developed for the precooling stage of a helium Joule-Thomson cryocooler. The construction used photoetched stainless steel foil diffusion bonded together [9].

6. Optimization of Regenerator Geometry and Frequency Limitations

As discussed in Section 2 the thermal penetration depth δ_t gives the effective distance oscillating heat flow can penetrate into any medium. According to equation (1) δ_t decreases with increasing frequencies. Thus, smaller hydraulic diameters are required at higher frequencies for good heat transfer within the helium working fluid. Also, in order to effectively utilize the heat capacity of the entire matrix volume for half-cycle heat storage the characteristic dimension of the matrix must also be less than δ_t for that material. As shown by figure 2 the thermal penetration depth in helium decreases at lower temperatures. For temperatures below about 20 K hydraulic diameters should be significantly less than about 50 μm for frequencies of about 10 Hz. The concept of a thermal penetration depth applies to conduction heat transfer, which is valid only in the laminar flow regime. However, we have seen in the previous section that for compact heat exchangers the Reynolds numbers are always much less than 2000. The same is true for regenerators.

6.1 ACOUSTIC OR PV POWER FLOW

The aspect of regenerative cryocoolers of interest here is their miniaturization, possibly for use in MEMS devices. The largest component of regenerative cryocoolers is the compressor, or pressure oscillator. The mechanical power, or PV power, output from the pressure oscillator is given by

$$\dot{W} = \tfrac{1}{2} P_1 \dot{V}_1 \cos\theta, \qquad (49)$$

where P_1 is the amplitude of the sinusoidal pressure oscillation, $\dot{V}_1$ is the amplitude of the volume flow at the piston, and θ is the phase angle between the volume flow and the pressure. In terms of the total swept volume V_{co} (peak to peak) of the pressure oscillator the PV power is given by

$$\dot{W} = \tfrac{1}{2}\pi f P_1 V_{co} \cos\theta = \tfrac{1}{2}\pi f \left(\frac{P_r - 1}{P_r + 1} \right) P_0 V_{co} \cos\theta, \qquad (50)$$

where f is the operating frequency and P_r is the pressure ratio (maximum divided by minimum). For a given power output the pressure oscillator size (proportional to swept volume) is reduced when increasing f or P_1 and keeping θ small. The pressure amplitude P_1 is increased when increasing the pressure ratio P_r or the average pressure P_0. Higher pressure ratios occur only by increasing the swept volume relative to the volume of the cold head, so that approach does not lead to reduced volumes. Pressure ratios in the range of 1.2 to 1.5 are typical for most high-frequency regenerative cryocoolers. We now examine the limits on the average pressure and the frequency.

As the average pressure increases the wall thickness and the solid fraction $(1 - n_g)$ must increase. As indicated by equation (29) the solid fraction becomes significant for $P/\sigma \approx 0.1$. For higher pressures the outside dimensions will no longer decrease much. Typically the allowable tensile stress may be about 70 MPa, which gives an upper limit of about 7 MPa for the average pressure for any significant reduction in system size. As shown by equation (50) higher power densities in the pressure oscillator can be achieved with higher frequencies. Typically 60 to 70 Hz is about the maximum frequency currently being used, but that is not a limit set by the pressure oscillator. An upper frequency limit for the oscillator could be at least 1 kHz or more. The internal power density of the pressure oscillator according to equation (50) for an average pressure of 7 MPa, a pressure ratio of 1.3, a frequency of 1 kHz and a phase angle of 0 is 1.43 kW/cm^3. The ratio of external volume to the swept volume is of the order 10, so the maximum external power density is of the order 100 W/cm^3. There are questions regarding the upper limits to the power density of linear motors and whether they can provide this high a power density. However, as we shall see the regenerator has the major impact on the power density in the system, so we shall not dwell on the issues in the pressure oscillator.

The purpose of the regenerator is to transmit oscillating PV power, or acoustic power, from ambient temperature to some lower temperature with a minimum of losses. Just as with the recuperative heat exchangers discussed in section 5, the losses are those due to imperfect heat transfer $\dot{Q}_{reg}$, pressure drop ΔP, and conduction $\dot{Q}_{cond}$. The gross refrigeration power $\dot{Q}_r$ available at the cold end is simply the PV power delivered to the cold end whether the system is a Stirling or pulse tube cryocooler. There also would be some additional losses associated with the expansion process in the Stirling displacer or the pulse tube that are not part of the regenerator losses. Those losses may only be about 15 to 20% of the gross refrigeration power. Most of the losses are in the regenerator. Optimization of the regenerator geometry can be carried out much like that discussed in section 5 for recuperative heat exchangers. Regenerators have oscillating mass flows with an amplitude of $\dot{m}_1$. In most cases there is no steady or DC component of mass flow even though there is a steady-flow component of PV power from the compressor to the cold end of the regenerator. The relation between the time-averaged PV power flow and the mass flow amplitude for an ideal gas is given by

$$\dot{W} = \tfrac{1}{2} RT (P_1 / P_0) \dot{m}_1 \cos\theta, \qquad (51)$$

where θ is the phase angle between the flow and pressure and the relative pressure amplitude is related to the pressure ratio by

$$\frac{P_1}{P_0} = \frac{P_r - 1}{P_r + 1}. \tag{52}$$

6.2 GEOMETRY OPTIMIZATION

The optimization procedure to be discussed here uses some approximations that are good for temperatures above about 70 K, but are not very good for lower temperatures. For lower temperatures numerical analyses are necessary for good results. The amplitude of the pressure drop ΔP_1 for flow through the regenerator is given by equation (10) when $\dot{m}_1$ is used for the flow. The specific cross sectional area of the regenerator then becomes

$$\frac{A_g}{\dot{m}_1} = \left[\frac{N_{tu} N_{Pr}^{2/3}}{2\alpha \rho_0 \Delta P_1}\right]^{1/2}, \tag{53}$$

where ρ_0 is evaluated at the average pressure and temperature. There exists a relationship between the number of heat transfer units N_{tu} during peak flow and the regenerator thermal loss, but it is more complicated than that for recuperative heat exchangers. Numerical techniques must be used for the most general case. Kays and London [6] give a simplified expression that is a good approximation for temperatures of about 80 K and above. They show that the ineffectiveness of a regenerator is a function not only of N_{tu} but also of the heat capacity ratio between the matrix and the gas that passes through the regenerator. However, for temperatures of 80 K and above the volumetric heat capacity of nearly all matrix materials is much larger than that of helium gas. In that case the relation between the ineffectiveness and N_{tu} is the same as for recuperative heat exchangers given by equation (19). Because both the hot and cold blows flow through the same flow channel in a regenerator we have

$$N_{tu1} = N_{tu2} = N_{tu} = 2N_{tu0}. \tag{54}$$

Then according to equation (18) the total ineffectiveness of the regenerator is approximated by

$$1 - \varepsilon = 2B / N_{tu}. \tag{55}$$

Here B can be made greater than 1 to better approximate the case where the heat capacity ratio is not larger than about 10. For a heat capacity ratio of 2, equation (55) is still a good approximation with $B = 2$. The relationship between the ineffectiveness and the regenerator thermal loss is given by

$$1 - \varepsilon = (\dot{Q}_{reg} / \dot{Q}_r) q_r^*, \tag{56}$$

which is the same as for the recuperative heat exchanger given by equation (15).

The specific gross refrigeration power of the regenerative cryocooler is given by

$$q_r = \dot{Q}_r / \dot{m}_1 = \dot{W}_c / \dot{m}_1 = \tfrac{1}{2} R T_c (P_1 / P_0) \cos\theta. \tag{57}$$

The relative refrigeration power for an ideal gas is then

$$q_r^* = \frac{2q_r}{h_h - h_c} = \frac{T_c(P_1/P_0)\cos\theta}{(c_p/R)(T_h - T_c)}, \tag{58}$$

where the factor of 2 is used to account for the fact that the denominator refers to heat flow between the gas and the matrix for a half cycle whereas the numerator refers to the full cycle. With the same conditions as case A for recuperative heat exchangers we have $q_r^* = 0.0444\,(\theta = 0)$ compared with 0.0924 for the ideal recuperative cycle. With the more common pressure ratio of 1.3 instead of 2.0 we have $q_r^* = 0.0174$. Thus, the ineffectiveness of regenerators must be less than that of recuperative heat exchangers for the same fractional heat loss. That is relatively easy to accomplish in practice because of the simple construction of regenerators with only one flow stream. The specific gas cross-sectional area of the regenerator is then expressed as

$$\frac{A_g}{\dot{m}_1} = \left[\frac{BN_{Pr}^{2/3}}{\alpha\rho_0 q_r^* \Delta P_1(\dot{Q}_{reg}/\dot{Q}_r)}\right]^{1/2}. \tag{59}$$

This equation is the same as equation (21) except for the factor of 2 in the denominator in equation (21). In this case $\dot{Q}_{reg}$ is for the complete regenerator whereas $\dot{Q}_{hx}$ in equation (21) is for one side of the heat exchanger. We note that the specific area is independent of frequency. Another assumption that is a part of equation (59) is that the amplitude of the mass flow rate is the same throughout the regenerator. In practice the flow rate may vary by 20 to 30% from one end to the other in an optimized design. The pressure drop in equation (59) can be made into a relative form ($\Delta P_1/P_1$) that equals the fractional loss of PV power and the gross refrigeration power. With that modification and with the ideal gas assumption equation (59) becomes

$$\frac{A_g}{\dot{m}_1} = \left[\frac{RT_0 BN_{Pr}^{2/3}}{\alpha q_r^* P_0^2 (P_1/P_0)(\Delta P_1/P_1)(\dot{Q}_{reg}/\dot{Q}_r)}\right]^{1/2}. \tag{60}$$

The calculations for the length and the hydraulic diameter of the regenerator are the same as for the recuperative heat exchangers. Thus, equation (28) gives the optimum length and equation (39) gives the optimum hydraulic diameter. They are repeated here for convenience:

$$L = \frac{(A_g/\dot{m}_1)(1-n_g)\int k_{eff}\,dT}{n_g q_r(\dot{Q}_{cond}/\dot{Q}_r)} \tag{28*}$$

$$D_h = \left[\frac{2b\mu(1-n_g)\int k_{eff}\,dT}{n_g\rho_0 q_r \Delta P_1(\dot{Q}_{cond}/\dot{Q}_r)}\right]^{1/2}, \tag{39*}$$

where k_{eff} is the effective thermal conductivity of the matrix when the effect of multiple interfaces are taken into account, such as with stacked screen or packed spheres [10]. The gas volume in the regenerator is found by combining equations (59) and (28*) to give

$$\frac{V_{rg}}{\dot{m}_1} = \frac{A_g L}{\dot{m}_1} = \frac{BN_{Pr}^{2/3}(1-n_g)\int k_{eff}\,dT}{\alpha n_g q_r^* q_r \rho_0 \Delta P_1(\dot{Q}_{reg}/\dot{Q}_r)(\dot{Q}_{cond}/\dot{Q}_r)}. \tag{61}$$

In general both $(\dot{Q}_{cond}/\dot{Q}_r)$ and $(\dot{Q}_{reg}/\dot{Q}_r)$ will be double the corresponding values for recuperative heat exchangers because they refer to the complete regenerator. With the ideal gas assumption the gas volume becomes

$$\frac{V_{rg}}{\dot{m}_1} = \frac{RT_0 BN_{Pr}^{2/3}(1-n_g)\int k_{eff}\,dT}{\omega n_g q_r^* q_r P_0^2 (P_1/P_0)(\Delta P_1/P_1)(\dot{Q}_{reg}/\dot{Q}_r)(\dot{Q}_{cond}/\dot{Q}_r)}. \tag{62}$$

This equation shows that for fixed relative losses, frequency has no effect on the volume, length or hydraulic diameter, but an increase in the average pressure for the same pressure ratio has a significant effect in decreasing the volume and a smaller effect on the hydraulic diameter. We can relate the mass flow rate to the total mass m_f (peak to peak) of gas that flows through the regenerator and out the ends by

$$\dot{m}_1 = \pi f m_f. \tag{63}$$

We can replace the mass of gas by the swept volume at either end of the regenerator. For the compressor end and with an ideal gas we have

$$\dot{m}_1 = \pi f P_0 V_{co}/RT_{co}, \tag{64}$$

where the compressor temperature T_{co} is usually equal to T_h. Substituting this expression for the flow rate into equation (62) gives us the gas volume ratio between the regenerator and the compressor as

$$\frac{V_{rg}}{V_{co}} = \frac{\pi f (T_0/T_{co}) BN_{pr}^{2/3}(1-n_g)\int k_{eff}\,dT}{\omega n_g q_r^* q_r P_0 (P_1/P_0)(\Delta P_1/P_1)(\dot{Q}_{reg}/\dot{Q}_r)(\dot{Q}_{cond}/\dot{Q}_r)}. \tag{65}$$

6.3 LIMITED MATRIX HEAT CAPACITY

In regenerators the porosity n_g of the matrix is not constrained by pressure considerations, but the overall porosity that includes the confining tube must not exceed the value given by equation (29). However, high values of the matrix porosity may lead to insufficient matrix heat capacity. The heat capacity ratio is given by

$$\frac{C_r}{C_f} = \frac{V_{rg}(1-n_g)\rho_m c_m}{n_g m_f c_p}, \tag{66}$$

where $\rho_m c_m$ is the volumetric heat capacity of the matrix, m_f is the total mass (peak to peak) of gas that flows through the regenerator, and c_p is the specific heat at constant pressure for the gas (usually helium). The total mass of fluid can be given by

$$m_f = \rho_{co} V_{co}, \tag{67}$$

where ρ_{co} is the density of the gas at the compressor temperature. Equation (66) can then be written as

$$\frac{C_r}{C_f} = \frac{(V_{rg}/V_{co})(1-n_g)(\rho_m c_m/\rho_0 c_p)(T_{co}/T_{ave})}{n_g}, \tag{68}$$

where T_{ave} is the average temperature of the regenerator. By using equation (63) for the fluid mass we can also express the heat capacity ratio as

$$\frac{C_r}{C_f} = \frac{\pi f (V_{rg} / \dot{m})(1 - n_g)\rho_m c_m}{n_g c_p}. \tag{69}$$

In order to use the relationship between ineffectiveness and N_{tu} given by equation (55) with $B = 1$, the heat capacity ratio must be at least 10. Even larger ratios are needed when the matrix specific heat drops rapidly with decreasing temperatures to account for a low ratio at the cold end of the regenerator. We now see that increasing frequency has the beneficial effect of increasing the heat capacity ratio, even though it has no effect on decreasing the gas volume in the regenerator according to equation (62). However, as we mentioned earlier, an increase in frequency does decrease the swept volume of the pressure oscillator, and in a like manner the swept volume at the cold end of the regenerator. The reduced swept volume at the cold end decreases the size the displacer in a Stirling cryocooler or the pulse tube in the pulse tube cryocooler.

6.4 FREQUENCY EFFECTS

So far in our analyses everything looks good as far as the use of increased pressures and frequencies to decrease the size of regenerative cryocoolers. An increased frequency and average pressure decreases the swept volume of the pressure oscillator and an increased average pressure decreases the size of the regenerator. We now calculate what happens when the frequency becomes too high. The frequency affects both the magnitude and the phase of the mass flow rate between the two ends of the regenerator. According to the mass conservation equation the mass flow at the hot end of the regenerator is related to the mass flow at the cold end by

$$\dot{\boldsymbol{m}}_h = \dot{\boldsymbol{m}}_c + \frac{d\boldsymbol{m}_g}{dt}, \tag{70}$$

where the bold variables are complex or phasor variables. For an ideal gas we have

$$\dot{\boldsymbol{m}}_h = \dot{\boldsymbol{m}}_c + \frac{\dot{\boldsymbol{P}} V_{rg}}{R T_r}, \tag{71}$$

where T_r is the log-mean temperature of the regenerator. For sinusoidal pressure oscillations of amplitude P_1 equation (71) becomes

$$\dot{\boldsymbol{m}}_h = \dot{\boldsymbol{m}}_c + \frac{i 2\pi f \boldsymbol{P}_1 V_{rg}}{R T_r}, \tag{72}$$

where i is the imaginary unit. The second term on the right hand side of equation (72) is referred to as the compliance component and is analogous to capacitance in electrical systems. In an optimized design of regenerators the flow at the cold end lags the pressure in time and the flow at the warm end leads the pressure in time. The time derivative of the pressure $\dot{\boldsymbol{P}}$ forms the third leg of a roughly equal lateral triangle with the three phasors as shown in Figure 19. The magnitude of the compliance component is roughly the same as the other two terms in equation (72). In that case the magnitude of the mass flow rate does not vary much throughout the regenerator. However, as the frequency

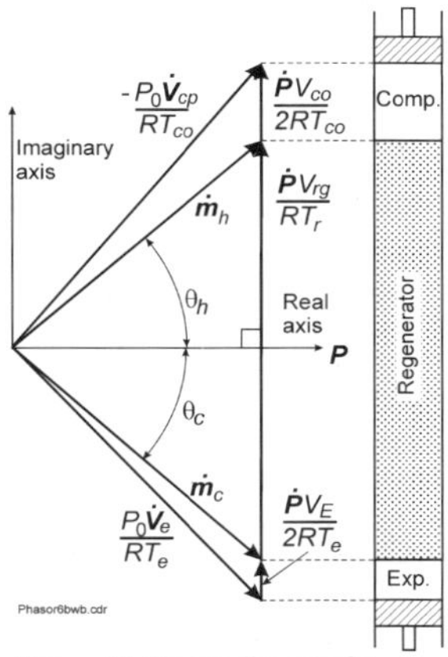

Figure 19. Phasor diagram for mass conservation in a Stirling cryocooler.

increases the vertical phasor in Figure 19 or the compliance component increases in proportion to the frequency. For $\dot{m}_c$ fixed (both amplitude and phase θ with respect to pressure) $\dot{m}_h$ increases (both amplitude and phase θ) as the frequency becomes large. The higher amplitude of the flow in the regenerator near the warm end then requires a larger regenerator area (taper) according to equation (60) and leads to an increase in the volume. We have performed extensive numerical analyses of regenerators and have shown that the system efficiency begins to decrease for phase angles at the warm end greater than about 50°. The higher flow rate and phase angle means that the swept volume of the compressor is increased for the same PV power according to equation (51). Thus, there is some upper limit to the frequency, beyond which volumes no longer decrease. To quantify this effect we divide equation (72) by the flow rate amplitude $\dot{m}_1$, which we consider to be the average amplitude throughout the regenerator. To keep the phase angle between the flow at either end and the pressure from becoming much larger than about 50° at the warm end we then require the compliance component to approximately satisfy the condition

$$\frac{2\pi f P_0 (P_1 / P_0)(V_{rg} / \dot{m}_1)}{RT_r} < \sqrt{2}. \tag{73}$$

This equation can be used to find the maximum $V_{rg} / \dot{m}_1$ for a given frequency or the maximum frequency for the calculated $V_{rg} / \dot{m}_1$. We recall that the calculated $V_{rg} / \dot{m}_1$ is independent of frequency. Substituting equation (62) into equation (73) gives the approximate upper limit to the frequency as

$$f < \frac{\alpha n_g q_r^* q_r P_0 (\Delta P_1 / P_1)(\dot{Q}_{hx} / \dot{Q}_r)(\dot{Q}_{cond} / \dot{Q}_r)}{\sqrt{2}\pi B N_{Pr}^{2/3} (1 - n_g) \int k_{eff}\, dT}. \tag{74}$$

For a given frequency the maximum $V_{rg} / \dot{m}_1$ is best expressed using the volume ratio

$$\frac{V_{rg}}{V_{co}} < \frac{1}{\sqrt{2}(P_1 / P_0)(T_{co} / T_r)}, \tag{75}$$

where equation (64) for $\dot{m}_1$ was substituted into equation (73). An increase in frequency increases the volume ratio, but the maximum frequency given in equation (74) causes the ratio to reach the maximum given in equation (75). We now substitute this upper limit on the volume ratio into equation (68) to obtain an upper limit on the porosity that is given by

$$n_g < (n_g)_{\max} = \frac{\Omega}{1+\Omega}, \qquad \Omega = \frac{(\rho_m c_m / \rho_0 c_p)}{\sqrt{2}(P_1 / P_0)(C_r / C_f)}, \tag{76}$$

where we made the assumption that $T_r \approx T_{ave}$. Figure 20 shows how the maximum porosity from equation (76) decreases with decreasing temperatures for $C_r/C_f = 100$. For these calculations the matrix volumetric heat capacity $\rho_m c_m$ was taken as a smooth curve representing the best available regenerator materials. For a regenerator with this maximum porosity the maximum frequency decreases with decreasing temperatures, as shown in Figure 21 where the conditions were taken as those of case G discussed in the next section. The various geometrical parameters, such as area, length, and hydraulic diameter can be calculated using the maximum porosity from equation (76).

6.5 EXAMPLES

The equations given above for regenerator optimization are reasonably accurate for temperatures above about 50 K. At lower temperatures real gas effects and the effect of compression and expansion of the gas within the regenerator cause additional heat transfer that makes these simple equations no longer valid. At lower temperatures numerical analyses are required to obtain reasonably accurate results[11]. These simple equations are useful for understanding the physics of the regenerative processes and for understanding the effect of various variables. The equations given here are used to examine several cases of regenerative cryocoolers. In all cases the regenerators are made with stainless steel. Some

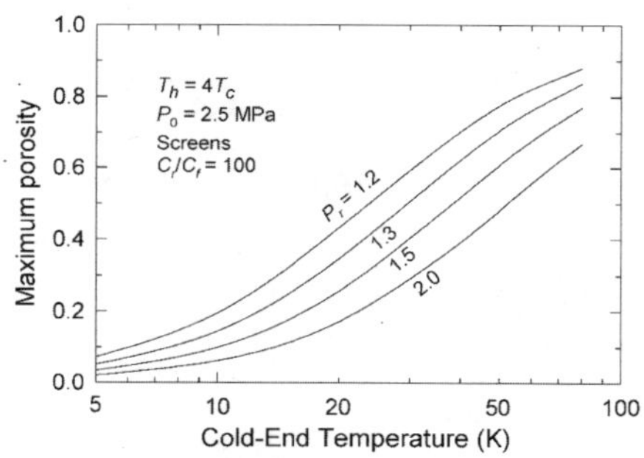

Figure 20. Calculated maximum porosity in screen regenerators as a function of cold end temperature in order to maintain adequate matrix heat capacity.

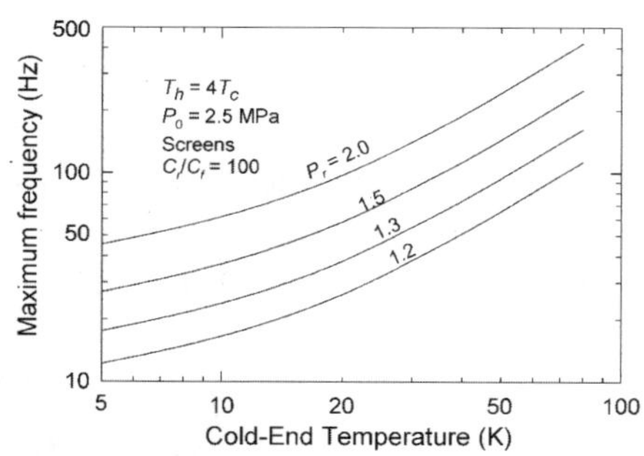

Figure 21. Calculated maximum frequency to prevent an excessive compliance in screen regenerators with the maximum porosity given in Figure 20.

cases use screen, which has a thermal conductivity degradation factor of 0.13 [10]. However we used a factor of 0.31 here to account for a combination of the screen and the tube containing the screen. Some cases use parallel plates, but we continue to use the conduction factor of 0.31 to make a clearer comparison between the two geometries.

Case G is typical of many Stirling or pulse tube refrigerators operating between 80 and 320 K. The input parameters are given in Table 3 and the results are given in Table 4. Case H is for the same conditions except the lower temperature is 40 K and the upper temperature is 160 K, which gives a temperature ratio of 4 as for case G. Some enhancement to the stainless steel heat capacity was made for this calculation to simulate the use of some improved regenerator material at this low temperature. Note that as the temperature is lowered the optimum hydraulic diameter decreases. Case I is for temperatures between 80 and 320 K, but with a lower pressure ratio. The lower pressure ratio increases the hydraulic diameter as expected from equation (39*) for the same relative pressure loss. The lower pressure ratio also decreases q_r. The relative losses given in Table 3 for case I match those found to maximize the overall efficiency of the cycle using numerical analyses [11], shown as case I* in the tables with data taken from REGEN3.2 run #2880. The geometrical parameters used in the numerical calculation are then compared with those calculated from the simple equations given here. The calculated values for the area and the length agree within about 10% with those of the numerical model, but the hydraulic diameter differs by about 25%. The 55.4 μm hydraulic diameter is achieved in practice with screen of #400 mesh with 25.4 μm wire diameter. Finer mesh screen suggested by the

Table 3. Input parameters used in example calculations for optimized regenerators.

Case	T_c, T_h (K)	P_0 (MPa)	P_r	f (Hz)	q_r (J/g)	q_r*	Geom.	n_g	$\Delta P_1/P_1$	Q_{reg}/Q_r	Q_{cond}/Q_r
G	80, 320	2.5	1.30	60	10.84	0.0166	screen	0.69	0.21	0.33	0.07
H	40, 160	2.5	1.30	60	5.42	0.0174	screen	0.69	0.21	0.33	0.07
I	80, 320	2.5	1.25	60	8.83	0.0142	screen	0.69	0.21	0.33	0.10
I*	80, 320	2.5	1.25	60	8.83	0.0142	screen	0.69	0.21	0.33	0.10
J	80, 320	8.0	1.50	400	16.6	0.0266	screen	0.69	0.21	0.33	0.07
K	80, 320	8.0	1.50	400	16.6	0.0266	plates	0.69	0.21	0.33	0.07
L	80, 320	8.0	1.50	400	16.6	0.0266	plates	0.20	0.21	0.33	0.07
M	80, 320	8.0	1.50	4000	16.6	0.0266	plates	0.69	0.21	0.33	0.07

Table 4. Calculated optimized geometry and flow parameters for example regenerators.

Case	A_g/m (cm²s/g)	L (mm)	D_h (μm)	D_h/δ_t	N_r	C_r/C_f	f_{max} (Hz)	V_{rg}/V_{co}	W_h/A_t (W/cm²)	Q_r (mW)	m_1 (mg/s)	d (mm)
G	0.801	44.8	41.0	0.29	33	177	69	2.54	37	10	0.923	0.370
H	0.566	22.5	19.5	0.24	35	40	97	1.81	26	10	1.85	0.441
I	0.962	46.3	41.2	0.29	28	219	66	3.15	25	10	1.13	0.449
I*	0.916	42.0	55.4	0.39	45	137	60	1.58	26	10	1.13	0.438
J	0.163	5.95	8.36	0.27	33	32	522	1.47	280	10	0.602	0.135
K	0.0586	2.14	6.47	0.21	71	4.1	4030	0.19	778	10	0.602	0.081
L	0.0586	18.7	19.1	0.61	211	14.4	461	1.66	227	10	0.602	0.150
M	0.0586	2.14	6.47	0.65	71	41	4030	1.90	778	10	0.602	0.081

equations given here is usually not commercially available.

Case J is an extreme case that attempts to miniaturize the overall system. In that case the average pressure is 8 MPa, the pressure ratio is 1.5, and the frequency is 300 Hz. Everything appears valid in that example including the maximum frequency of 522 Hz. The heat capacity ratio of 31.9 is somewhat low and may lead to a higher relative regenerator loss than 0.33 assumed here. Numerical analyses would be required to verify the results. Achieving the results of case J in practice will be difficult because of the problems associated with obtaining a hydraulic diameter of 8.36 μm in screen that has good flow and heat transfer characteristics. Further miniaturization is achieved by using parallel plates instead of screen, as shown in case K. Here the heat capacity ratio of 4.13 is far too low for the N_{tu} calculations to be valid. The use of lower porosity, 0.20, in case L helps to increase the heat capacity ratio to 14.4, but the size of the regenerator has increased. Case M is with the original porosity of 0.686, but the frequency is increased to 4000 Hz. The results in Table 4 show a heat capacity ratio of 133, which may be sufficient for high effectiveness, but numerical analyses would be needed to verify whether the regenerator behavior is dominated by heat transfer due to compression and expansion of the gas stored in the regenerator rather than by heat transfer due to the flow of gas through the temperature gradient. Even though the 4000 Hz frequency is very high and the time for heat transfer is very short, the classical correlations may still be valid because the 282 μm distance traveled by a parcel of gas at the cold end in case M is still large compared with the hydraulic diameter of 6.47 μm.

The PV power flux at the hot end of the regenerator $\dot{W}_h / A_t$ in Table 4 becomes very high for cases K and M. This power flow is equal to the heat that must be rejected to ambient before entering the regenerator. This is accomplished in an aftercooler whose volume must be small compared with that of the regenerator. Usually the diameter equals that of the regenerator and its length is much shorter. For a regenerator length of 2.14 mm in cases K and M the aftercooler should be no longer than about 0.5 mm. The heat flow density in this aftercooler then becomes 15.6 kW/cm^3, which is too high to be able to reject the heat without a very large temperature difference between the aftercooler and the ambient temperature heat sink. For comparison case G has a heat flow density of only about 40 W/cm^3, which is typical of present-day regenerative cryocoolers. This problem of rejecting heat to the environment becomes a serious problem in miniaturizing regenerative cryocoolers.

7. Conclusions

We have derived a set of simple equations to find the geometry of both heat exchangers and regenerators that minimizes their volume for a given refrigeration power. The equations are useful for the design of micro cryocoolers. With these equations we have shown that the optimum hydraulic diameters decrease with decreasing temperatures. For temperatures of 80 K and below the calculated hydraulic diameters are almost always less than 100 μm, and in some cases can be less than 5 μm. However, the Knudsen numbers are almost always less than 10^{-3}, which indicates slip flow does not occur and that continuum flow correlations from macrosystems can be used for analyses of these systems. The one exception may be in the use of mixed refrigerants, but that case is also complicated by the presence of two-phase flow. The use of mixed refrigerants in a Joule-Thomson cryocooler offers the potential of the smallest system for cooling to 80 K with no moving parts at the cold end. The equations developed for optimization of the heat exchangers show that for refrigeration powers less than about 10 mW at 80 K the required gap width is about comparable to the gap thickness. Thus, for lower refrigeration powers, the optimization procedure described here is no longer valid.

The use of photoetched stainless steel foil diffusion bonded together was described as one fabrication method currently under study for developing miniature heat exchangers. Very uniform gap spacing is required to maintain uniform flow distribution and high effectiveness in the heat exchanger. Measured effectiveness was lower than the calculated value, which indicates a possible problem with non-uniform flow.

In regenerative cryocoolers, such as Stirling and pulse tube cryocoolers, the use of high frequency (>10 Hz) leads to thermal penetration depths in the helium working fluid that become less than about 100 μm at 80 K and even smaller at lower temperatures. For good heat transfer the regenerative heat exchangers (regenerators) must have hydraulic diameters significantly less than the thermal penetration depth. Thus hydraulic diameters less than 50 μm are commonly used in high frequency regenerative cryocoolers. Equations similar to those for recuperative heat exchangers were developed here and are useful for minimizing the volume of regenerators and the entire cryocooler. Correlations for steady flow should be valid for most cases with regenerative crycoolers because the amplitude of gas motion is usually much greater than the hydraulic diameter. Significant miniaturization of 80 K cryocoolers according to these equations can be achieved by the use of average pressures up to about 8 MPa and frequencies of 1 kHz or more. The required hydraulic diameters become less than 10 μm and represent a challenging fabrication problem. However, an even more serious problem is the very high heat flow density in the aftercooler, which makes it difficult to reject heat to ambient without a large temperature difference. Further research in miniaturizing regenerative cryocoolers would be useful and would require the use of numerical methods in the regenerator, particularly for temperatures below about 50 K where matrix heat capacities become low and real gas effects become pronounced.

REFERENCES

1. Bayazitoglu, Y., and Kakac, S., (2005) Flow Regimes in Microchannel Single-Phase Gaseous Fluid Flow, *Microscale Heat Transfer-Fundamentals and Applications*, S. Kakac (ed.), Kluwer Academic Publishers, Dordrecht (This publication).
2. Iguchi, M., Ohmi, M., and Maegawa, K., (1982) Analysis of Free Oscillating Flow in a U-Shaped Tube, *Bull. JSME,* Vol. 25, p.1398.
3. Kurzweg, U. H., Lindgren, E. R., and Lothrop, B., (1989) Onset of Turbulence in Oscillating Flow at Low Womersley Number, *Phys. Fluids A,* Vol. 1, pp. 1972-1975 and references therein.
4. Radebaugh, R., (2003) Cryocoolers and High-T_c Devices, *Handbook of High-Temperature superconductor Electronics*, N. Khare (ed.), Marcel Dekker, New York, pp. 379-424.
5. Radebaugh, R., (2003) Pulse Tube Cryocoolers, *Low Temperature and Cryogenic Refrigeration*, S. Kakac, *et al.* (eds.) Kluwer Academic Publishers, Dordrecht, pp. 415-434.
6. Kays, W. M., and London, A. L., (1984) Compact Heat Exchangers, third edition, McGraw-Hill, New York.

7. Radebaugh, R., Louie, B., (1985) A Simple, First Step to the Optimization of Regenerator Geometry, *Proceedings of the Third Cryocooler Conference*, NBS Special Publication 698, pp. 177-198.

8. Ross, R. G., and Boyle, R. F., (2003) NASA Space Cryocooler Programs – An Overview, *Cryocoolers 12*, R. G. Ross (ed.), Kluwer Academic/Plenum Publishers, New York, pp. 1-8.

9. Marquardt, E. D., and Radebaugh, R., (2003) Compact HighEffectiveness Parallel Plate Heat Exchangers, *Cryocoolers 12*, R. G. Ross (ed.), Kluwer Academic/Plenum Publishers, New York, pp. 507-516.

10. Lewis , M. A., and Radebaugh, R., (2003) Measurement of Heat Conduction Through Bonded Regenerator Matrix Materials, *Cryocoolers 12*, R. G. Ross (ed.), Kluwer Academic/Plenum Publishers, New York, pp. 517-522.

11. Gary, J., and Radebaugh, R., (1991) An Improved Numerical Model for Calculation of Regenerator Performance (REGEN3.1), *Proceedings of the 4th Interagency Meeting on Cryocoolers*, David Taylor Research Center DTRC-91/003, pp. 165-176.

CONVECTIVE HEAT TRANSFER FOR SINGLE-PHASE GASES IN MICROCHANNEL SLIP FLOW: ANALYTICAL SOLUTIONS

Y. BAYAZITOGLU, G. TUNC, K. WILSON, and I. TJAHJONO
Department of Mechanical Engineering and Materials Science – Rice University
Houston, Texas, USA

1. Introduction

Heat transfer in microchannels has gained more interest in the last decade due to developments in the aerospace, biomedical and electronics industries. It has been a critical issue since the performance of the devices is primarily determined by temperature. As the size decreases, more efficient ways of cooling are sought due to the reduction in the heat transfer area.

Convection and conduction are the two major heat transfer mechanisms that have been investigated at microscale. Convective heat transfer in microchannels has been intensively analyzed by both experimental and analytical means. Conduction studies have focused mostly on thin films in recent years to address such questions as: How is the heat transferred? How does it differ from large-scale conduction?

As far as convective heat transfer is concerned, liquid and gaseous flows must be considered separately. Liquid flow has been investigated experimentally, whereas analytical, numerical and molecular simulation techniques have been applied to understand the characteristics of gaseous flow and heat transfer. While the Navier-Stokes equations can still be applied, due to the small size of microchannels, some deviations from the conventionally sized applications have been observed. Flow regime boundaries are significantly different, as well as flow and heat transfer characteristics.

Gaseous flow has usually been investigated by theoretical means. Some experiments were also performed to verify the theoretical results. When gases are at low pressures, or are flowing in small geometries, the interaction of the gas molecules with the wall becomes as frequent as intermolecular collisions, which makes the boundaries and the molecular structure more effective on flow. This type of flow is known as rarefied gas flow.

The Knudsen number (Kn) is used to represent the rarefaction effects. It is the ratio of the molecular mean free path to the characteristic dimension of the flow. For Knudsen numbers close to zero, flow is still assumed to be continuous. As the Knudsen number takes higher values, due to a higher molecular mean free path by reduced pressure or a smaller flow dimension, rarefaction effects become more significant and play an important role in determining the heat transfer coefficient.

The commonly used slip boundary conditions are called Maxwellian boundary conditions [1]. Since they are first order in accuracy, an extended set of boundary conditions was proposed by [2], which can be used in early transition of the slip flow regime. To do so, the velocity and temperature gradients at the boundary are written in terms of the Taylor expansion of the gradients within the layer one mean free path away from the boundary (called the Knudsen Layer).

S. Kakaç et al. (eds.), Microscale Heat Transfer, 125–148.

The laminar gaseous flow heat convection problem in the slip flow region was solved both analytically and numerically for various geometries [3-6]. The compressibility effects were included in [7],[8-11] and the results were compared with the experimental results of [12]. Thermal creep effects were studied by [13]. Exact solutions for flows in circular, rectangular, and parallel plate microchannels were given in [14-17].

2. Velocity Slip

In the Knudsen layer, the Maxwellian velocity slip boundary condition approximates the true gas velocity at the boundary by the velocity that the molecules would have if a linear velocity gradient existed as shown in figure 1 [18-19]. In other words, the magnitude of the slip is calculated from the velocity gradient evaluated at $y = \lambda$.

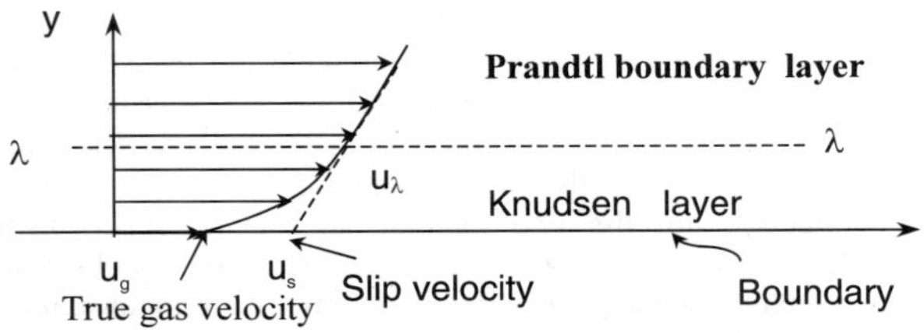

Figure 1. Schematic figure that shows the first order slip velocity approximation.

When a gas flows over a surface, the molecules leave some of their momentum and create shear stress on the wall. As shown in figure 2, specular reflections conserve the tangential momentum of molecules, and diffuse reflections results in vanishing tangential momentum. The fraction of the molecules that are diffusely reflected by the wall is defined as the tangential momentum accommodation coefficient, F_m. It is also defined as the fraction of the momentum the molecules leave on the surface.

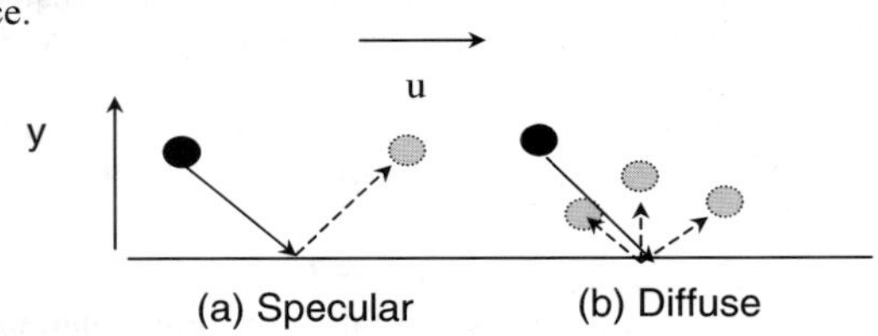

Figure 2. Specular and diffuse reflections of gas molecules at a solid boundary.

Let's assume that the molecules give fraction F_m of their tangential momentum to the surface [1]. To balance the viscous force, the molecules must be allowed to slip along the surface. We can write the tangential momentum balance at the wall as follows:

Total momentum carried by the approaching molecules:

$$\frac{1}{4} n m u_m u_s + \frac{1}{2} \mu \frac{du}{dy}, \tag{2.1}$$

where the first term is the momentum due to the slip velocity, the second term is the momentum transmitted in the gas by a molecular stream, n is the number of molecules per unit volume, m is the mass of a molecule, u_m is the mean velocity and μ is the viscosity. Then we can write the momentum given up to the surface as

$$F_m\left(\frac{1}{4}\rho u_m u_s+\frac{1}{2}\mu\frac{du}{dy}\right). \tag{2.2}$$

This will be equal to the shear stress at the wall, $\mu\frac{du}{dy}$:

$$F_m\left(\frac{1}{4}\rho u_m u_s+\frac{1}{2}\mu\frac{du}{dy}\right)=\mu\frac{du}{dy}. \tag{2.3}$$

We can then solve for the slip velocity, using the definition of viscosity as $\mu \cong \frac{1}{2}\rho u_m\lambda$, where λ is the molecular mean free path, to obtain the first order approximation to the velocity slip as:

$$u_s=\frac{2-F_m}{F_m}\lambda\frac{du}{dy}. \tag{2.4}$$

Another mechanism that may effect the velocity profile in a microchannel is thermal creep. When two containers at the same pressure but different temperatures are connected by a microchannel, mass flow starts from the cold container to the hot one. Thermal creep may increase or decrease the mass flow rate depending on the sign of the axial temperature gradient. If this gradient is negative in the flow direction, thermal creep will be in the opposite direction of the flow thus decreasing the mass flow rate. If the fluid temperature increases in the flow direction, then thermal creep will be in the same direction. Therefore, mass flow rate also increases. The slip boundary condition including the thermal creep effect is given in [13] by the following formula:

$$u_s=\frac{2-F_m}{F_m}Kn\frac{\partial u}{\partial \eta}+\frac{3}{2\pi}(\gamma-1)Kn^2\,\mathrm{Re}\frac{\partial \theta}{\partial \varsigma}. \tag{2.5}$$

3. Temperature Jump

Another characteristic of rarefied gas flow is that there is a finite difference between the fluid temperature at the wall and the wall temperature. Temperature jump is first proposed to be

$$T_s-T_w=c_{jump}\frac{\partial T}{\partial y}. \tag{3.1}$$

The thermal accommodation coefficient is defined as $F_T=\frac{Q_i-Q_r}{Q_i-Q_w}$, where Q_i is the energy of the impinging stream, Q_r is the energy carried by the reflected molecules and Q_w is the energy of the molecules leaving the surface at the wall temperature [1]. F_T can be defined as the fraction of molecules reflected by the wall that accommodated their energy to the wall temperature. Now, we will relate the accommodation coefficient to the temperature jump coefficient, c_{jump}. Let's assume that the temperature of the approaching molecules is T_s. The energy of these molecules can be written as the summation of the kinetic energy, internal energy and contribution of the incoming molecules to the conduction as

$$Q_i = M(2RT_s + U_s) + \tfrac{1}{2}k\frac{\partial T}{\partial y}, \tag{3.2}$$

where M is mass, R is the gas constant, U is the internal energy and k is the thermal conductivity of the gas. The internal energy is given by

$$U_s = c_v T_s - \tfrac{3}{2} R T_s. \tag{3.3}$$

Similarly, the energy of the outgoing molecules at the wall temperature is

$$Q_w = M(2RT_w + U_w). \tag{3.4}$$

The difference is calculated as

$$Q_i - Q_w = M\left(c_v + \tfrac{1}{2}R\right)\left(T_s - T_w\right) + \tfrac{1}{2}k\frac{\partial T}{\partial y}. \tag{3.5}$$

The following definitions are substituted into Eq. (3.5)

$$M = \frac{P}{\sqrt{2\pi RT}}, \qquad R = c_p - c_v, \qquad \gamma = \frac{c_p}{c_v},$$

then

$$Q_i - Q_w = \frac{c_v\left(T_s - T_w\right)(\gamma + 1)P}{2\sqrt{2\pi RT}} + \tfrac{1}{2}k\frac{\partial T}{\partial y}. \tag{3.6}$$

On the other hand, the net energy carried to the surface, $Q_i - Q_r$, is equal to the heat flux at the wall, which can be written as

$$Q_i - Q_r = k\frac{\partial T}{\partial y}. \tag{3.7}$$

From the definition of the accommodation coefficient, we can write the following

$$k\frac{\partial T}{\partial y} = F_T\left(\tfrac{1}{2}k\frac{\partial T}{\partial y} + \tfrac{1}{2}(\gamma+1)\frac{c_v(T_s - T_w)P}{\sqrt{2\pi RT}}\right), \tag{3.8}$$

then

$$\left(T_s - T_w\right) = \frac{2 - F_T}{F_T}\frac{k\sqrt{2\pi RT}}{(\gamma+1)c_v P}\frac{\partial T}{\partial y}. \tag{3.9}$$

Using the definition of the jump coefficient

$$c_{jump} = \frac{2 - F_T}{F_T}\frac{1}{(\gamma+1)}\frac{k\sqrt{2\pi RT}}{c_v P} \tag{3.10}$$

and after making the following substitutions,

$$\mu = \tfrac{1}{2}\rho u_m \lambda, \qquad u_m = 2\sqrt{\frac{2RT}{\pi}}, \qquad \rho = \frac{P}{RT},$$

c_{jump} is obtained as

$$c_{jump} = \frac{2 - F_T}{F_T}\frac{2}{(\gamma+1)}\frac{k}{c_v \mu}\lambda = \frac{2 - F_T}{F_T}\frac{2\gamma}{(\gamma+1)}\frac{k}{c_p \mu}\lambda,$$

$$c_{jump} = \frac{2-F_T}{F_T}\frac{2\gamma}{(\gamma+1)}\frac{k}{c_p\mu\frac{\rho}{\rho}}\lambda = \frac{2-F_T}{F_T}\frac{2\gamma}{(\gamma+1)}\frac{\alpha}{\nu}\lambda = \frac{2-F_T}{F_T}\frac{2\gamma}{(\gamma+1)}\frac{\lambda}{\Pr}.$$

Next the temperature jump is determined from Eq. (3.1) as

$$T_s - T_w = \frac{2-F_T}{F_T}\frac{2\gamma}{\gamma+1}\frac{\lambda}{\Pr}\frac{\partial T}{\partial y}. \tag{3.11}$$

4. Extended Boundary Conditions

Let's assume that the molecules give a fraction F_m of their tangential momentum to the surface. The momentum leaving the surface can then be written as [2] $M_{out}=(1-F_m)M_{in}$. The momentum balance at the wall is $M_{wall} = F_m M_{in}$. The shear stress at the wall is then given by $M_{wall}= \mu(du/dy)_0$ and the incoming momentum is given by

$$M_{in} = \frac{1}{2}\mu\left(\frac{du}{dy}\right)_0 + \frac{1}{2}\rho u_m u_o. \tag{4.1}$$

Evaluating $M_{wall} = F_m M_{in}$ at the Knudsen layer boundary, λ, we obtain

$$\mu\left(\frac{du}{dy}\right)_\lambda = F_m\left[\frac{1}{2}\mu\left(\frac{du}{dy}\right)_\lambda + \frac{1}{2}\rho u_m u_\lambda\right]. \tag{4.2}$$

Viscosity is given by $\mu = \rho u_m \lambda/2$ where λ is the molecular mean free path. This is substituted into Eq. (4.2) to obtain

$$u_\lambda = \frac{2-F_m}{F_m}\lambda\left(\frac{du}{dy}\right)_\lambda. \tag{4.3}$$

At y = λ, the velocity gradient can then be expanded around y = 0 as:

$$\left(\frac{du}{dy}\right)_\lambda = \left(\frac{du}{dy}\right)_0 + \lambda\left(\frac{d^2u}{dy^2}\right)_0 + \frac{\lambda^2}{2}\left(\frac{d^3u}{dy^3}\right)_0 + \frac{\lambda^3}{6}\left(\frac{d^4u}{dy^4}\right)_0 + ... \tag{4.4}$$

The following substitutions are made to non-dimensionalize the equation by letting $u^* = u/u_m$, $\eta = y/L$ and $Kn = \lambda/L$ where L is the characteristic length of the channel. In the non-dimensional form, after substitution of (4.4) into (4.3) the following can be obtained after rearrangement of the terms (Note that the superscript * has been dropped from the equation):

$$u_s = \frac{2-F_m}{F_m}Kn\left(\frac{du}{d\eta}\right)_0\left[1 + Kn\left(\frac{(d^2u/d\eta^2)_0}{(du/d\eta)_0}\right) + \frac{Kn^2}{2}\left(\frac{(d^3u/d\eta^3)_0}{(du/d\eta)_0}\right) + \frac{Kn^3}{6}\left(\frac{(d^4u/d\eta^4)_0}{(du/d\eta)_0}\right) + ...\right] \tag{4.5}$$

Then we use the following expansion:

$$1\Big/\left[1 - Kn\left(\frac{d^2u/d\eta^2}{du/d\eta}\right)_0\right] = 1 + Kn\left(\frac{d^2u/d\eta^2}{du/d\eta}\right)_0 + Kn^2\left(\frac{d^2u/d\eta^2}{du/d\eta}\right)_0^2 + Kn^3\left(\frac{d^2u/d\eta^2}{du/d\eta}\right)_0^3 + ... \tag{4.6}$$

The substitution of (4.6) into (4.5) yields:

$$u_s = \frac{2-F_m}{F_m} Kn \left(\frac{du}{d\eta}\right)_0 \left[\frac{1 \Big/ \left[1 - Kn\left(\frac{d^2u/d\eta^2}{du/d\eta}\right)_0\right] + \frac{Kn^2}{2}\left(\frac{d^3u/d\eta^3}{du/d\eta}\right)_0 + \frac{Kn^3}{6}\left(\frac{d^4u/d\eta^4}{du/d\eta}\right)_0 + \ldots\ldots}{-\left(Kn\frac{d^2u/d\eta^2}{du/d\eta}\right)_0^2 - \left(Kn\frac{d^2u/d\eta^2}{du/d\eta}\right)_0^3 - \left(Kn\frac{d^2u/d\eta^2}{du/d\eta}\right)_0^4 - \ldots\ldots} \right] \qquad (4.7)$$

Next Eq. (4.7) can be simplified as [2]:

$$u_s = \frac{2-F_m}{F_m} \frac{Kn}{1-bKn} \left(\frac{du}{d\eta}\right)_0 + res \qquad (4.8)$$

where b = $(d^2u/d\eta^2)/(du/d\eta)$ and "*res*" is given by:

$$res = \frac{2-F_m}{F_m} \left[\begin{array}{l} \frac{Kn^2}{2}\left(\frac{d^3u}{d\eta^3}\right)_0 + \frac{Kn^3}{6}\left(\frac{d^4u}{d\eta^4}\right)_0 + \frac{Kn^5}{24}\left(\frac{d^5u}{d\eta^5}\right)_0 \ldots\ldots\ldots \\ - Kn^3 \frac{\left(d^2u/d\eta^2\right)_0^2}{\left(du/d\eta\right)_0} - Kn^4 \frac{\left(d^2u/d\eta^2\right)_0^3}{\left(du/d\eta\right)_0^2} - Kn^5 \frac{\left(d^2u/d\eta^2\right)_0^4}{\left(du/d\eta\right)_0^3} - \ldots\ldots\ldots \end{array} \right] \qquad (4.9)$$

In this slip flow expression, b is defined as $(d^2u/d\eta^2)/(du/d\eta)$. In the case of gaseous flow between two parallel plates, the values of $d^2u/d\eta^2$ and $du/d\eta$ are -2 and 1, respectively.

Another study [13] attempted to develop a model that can be used for the whole Kn range, $0 < Kn < \infty$. They proposed a simpler second order boundary condition which does not diverge as Kn goes to infinity. This is the same as equation (4.8) without including the residual terms. They applied this new boundary condition to the Navier-Stokes equations in the range of $0.01 < Kn < 30$ and the results are compared to Direct Simulation Monte Carlo (DSMC) and linearized Boltzman solutions. They obtained good results for the centerline velocity, assuming b = -1, but deviations for the slip velocity for $0.1 < Kn < 5$. The reason for this is that for these intermediate values of Kn, both viscous and Knudsen layers exist. A parabolic velocity profile assumption in this range ignores the effect of the Knudsen layer. For Kn = 1 and b = -1 results are in 10% error. This is also verified by the DSMC results, which show deviations from a parabolic velocity profile in the transition regime due to the growth of the Knudsen layer.

Equation (3.11) gives the first-order approximation to the temperature jump if it is assumed that the temperature gradient at the wall is the same as that at $y = \lambda$. To obtain the higher-order approximation, the same approach is applied as that which was used to obtain the second-order velocity slip equation [2]. This results in (with $\theta = T/T_{reference}$):

$$\theta_s - \theta_w = \frac{2-F_T}{F_T} \frac{2\gamma}{\gamma+1} \frac{Kn}{\Pr(1-aKn)} \left(\frac{\partial\theta}{\partial\eta}\right)_0, \qquad (4.10)$$

where the variable a is given as $a = (d^2\theta/d\eta^2)_0/(d\theta/d\eta)_0$.

5. Slip Flow Nusselt Number for Different Geometries

In this section the result from the analyses of Bayazitoglu et al. [14-17] will be shown. They analytically solved the continuum version of the energy equation by the integral transform technique with the appropriate jump boundary conditions. The integral transform technique has widely been used for the solution of heat transfer problems in many different applications. It is a three-step method. In the first step, the appropriate integral transform pair is developed: the inversion and transform formulas. Then, partial derivatives with respect to the space variables are removed from the equation, which reduces it to an ordinary differential equation (ODE). Finally, the resulting ODE is solved subjected to the transformed inlet condition. They solved the steady state heat convection between two parallel plates [14] and in circular [15], rectangular [16] and annular [14] channels with uniform heat flux and uniform temperature boundary conditions including the viscous heat generation for thermally developing and fully-developed conditions. They also solved the transient heat convection problem in [17], which is the problem of a circular tube including rarefaction effects and heat transfer in a double-pipe heat exchanger assuming slip conditions for both fluids and including conduction across the inner wall.

The velocity profile was assumed to be fully-developed. The velocity distribution in a circular microchannel including the slip boundary condition was taken from the literature. However, for the other geometries, they derived the fully-developed velocity profiles from the momentum equation. It is straightforward for flow between parallel plates and flow in an annulus. They applied the integral transform technique to obtain the velocity in a rectangular channel. The problem was simplified by assuming the same amount of slip at all the boundaries.

5.1 CONVECTION IN A CIRCULAR TUBE

5.1.1 Uniform Wall Temperature

In this section the results of [15] will be discussed. First, a detailed analysis for the flow of gases through a microchannel in the slip-flow regime subject to both the constant wall temperature and constant heat flux boundary conditions is given. The results of the analysis will then be discussed.

Beginning with the two dimensional energy equation, after making the following substitutions, $\theta = \frac{T - T_s}{T_o - T_s}$, $\varsigma = \frac{x}{L}$, $\eta = \frac{r}{R}$, $u^* = \frac{u}{u_m}$, $u = \frac{2u_m\left(1-\eta^2+4Kn\right)}{(1+8Kn)}$, $Gz = \frac{\mathrm{RePr}D}{L}$, $Br = \frac{\mu u_m^2}{k\left(T_o - T_s\right)}$, the energy equation takes the form:

$$\frac{u^* Gz}{4}\frac{\partial\theta}{\partial\varsigma} = \frac{1}{\eta}\frac{\partial}{\partial\eta}\left(\eta\frac{\partial\theta}{\partial\eta}\right) + \frac{16Br}{(1+8Kn)^2}\eta^2 . \tag{5.1}$$

Subject to the boundary conditions:

$$\theta = 0 \quad \text{at} \quad \eta = 1$$
$$\theta = 1 \quad \text{at} \quad \varsigma = 0$$
$$\frac{\partial\theta}{\partial\eta} = 0 \quad \text{at} \quad \eta = 0$$

It is important to note that the last term in Eq. (5.1) which is the viscous generation term has been included in this analysis. At the microchannel level viscous generation is significantly more important. The integral transform technique is then applied. The appropriate integral transform pair is developed:

Eigenvalue Problem:

$$\frac{1}{\eta}\frac{d}{d\eta}\left(\eta\frac{d\psi}{d\eta}\right)+(1-\eta^2 4Kn)\lambda_m^2\psi=0 \tag{5.2}$$

$$\frac{d\psi}{d\eta}=0 \quad \text{at} \quad \eta=0$$

$$\psi=0 \quad \text{at} \quad \eta=1$$

The orthogonality condition is given as:

$$\int_0^1 \eta(1-\eta^2+4Kn)\psi(\lambda_m,\eta)\psi(\lambda_n,\eta)d\eta=\begin{cases}0 & m\neq n\\ N(\lambda_m) & m=n\end{cases} \tag{5.3}$$

where

$$N(\lambda_m)=\int_0^1 \eta(1-\eta^2+4Kn)\left[\psi(\lambda_m,\eta)\right]^2 d\eta. \tag{5.4}$$

Then the appropriate transform pair is given by:

Transformed Formula: $\bar{\theta}(\lambda_m,\varsigma)=\int_0^1 \eta\left(1-\eta^2+4Kn\right)\psi(\lambda_m,\eta)\theta(\eta,\varsigma)d\eta$ (5.5)

Inversion Formula: $\theta(\eta,\varsigma)=\sum_{m=1}^{\infty}\frac{1}{N(\lambda_m)}\psi(\lambda_m,\eta)\bar{\theta}(\lambda_m,\varsigma)$ (5.6)

Each side of the energy equation (5.1) is then operated on by $\int_0^1 \eta\psi(\lambda_m,\eta)d\eta$. The transformed energy equation is then:

$$\frac{d\bar{\theta}}{d\varsigma}+\frac{2(1+8Kn)}{Gz}\lambda_m^2\bar{\theta}_m=\frac{32Br}{Gz(1+8Kn)}\int_0^1 \eta^3\psi(\lambda_m,\eta)d\eta. \tag{5.7}$$

The solution to this ordinary differential equation is

$$\bar{\theta}_m=\frac{K_m}{P_m}+(\bar{1}_m-\frac{K_m}{P_m})e^{-P_m\varsigma}, \tag{5.8}$$

where

$$K_m=\frac{32}{Gz(1+8Kn)}\int_0^1 \eta^3\psi(\lambda_m,\eta)d\eta \tag{5.9}$$

$$P_m=\frac{2(1+8Kn)}{Gz}\lambda_m^2 \tag{5.10}$$

$$\bar{1}_m=\int_0^1 \eta(1-\eta^2+4Kn)\psi(\lambda_m,\eta)d\eta. \tag{5.11}$$

Finally by substitution of equation (5.8) into the inversion formula, the non-dimensional temperature is given by:

$$\theta(\eta,\varsigma)=\sum_{m=1}^{\infty}\frac{\psi(\lambda_m,\eta)}{\int_0^1\eta(1-\eta^2+4Kn)\left[\psi(\lambda_m,\eta)\right]^2 d\eta}\left(\begin{array}{l}\dfrac{16Br\int_0^1\eta^3\psi(\lambda_m,\eta)d\eta}{(1+8Kn)^2\lambda_m^2}\\ +\left(\begin{array}{l}\int_0^1\eta(1-\eta^2+4Kn)\psi(\lambda_m,\eta)d\eta\\ -\dfrac{16Br\int_0^1\eta^3\psi(\lambda_m,\eta)d\eta}{(1+8Kn)^2\lambda_m^2}\end{array}\right)e^{-P_m\varsigma}\end{array}\right) \quad (5.12)$$

The convective heat transfer coefficient is then solved and rearranged as:

$$\frac{Dh_x}{k}=-\frac{2}{\dfrac{T_b-T_s}{T_o-T_s}-\dfrac{T_w-T_s}{T_o-T_s}}\left(\frac{\partial\theta}{\partial\eta}\right)_{\eta=1}. \quad (5.13)$$

The first term in the denominator is the non-dimensional bulk temperature definition, $\theta_b=2\int_0^1\left(\frac{u}{u_m}\right)\theta(\eta,\varsigma)\eta d\eta$. The second term can be determined from the temperature jump boundary condition given in section 3 of this paper. The final form of the Nusselt number (Nu) is then:

$$\text{Isothermal: } Nu_x=\frac{h_xD}{k}=-\frac{2\left(\dfrac{\partial\theta}{\partial\eta}\right)_{\eta=1}}{\left(\theta_b-\dfrac{4\gamma}{\gamma+1}\dfrac{Kn}{\Pr}\left(\dfrac{\partial\theta}{\partial\eta}\right)_{\eta=1}\right)}. \quad (5.14)$$

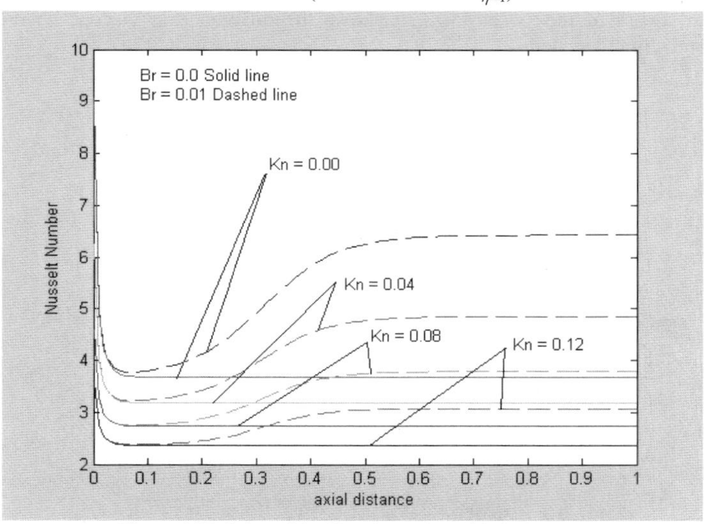

Figure 3. Variation of the Nusselt number with the Knudsen number at the entrance region for uniform temperature at the wall [15].

For the uniform temperature boundary condition in a cylindrical channel, the fully developed Nusselt number decreases as Kn increases. For the no-slip condition Nu_{∞}=3.6751, while it drops down to 2.3667 for Kn = 0.12, which is a decrease of 35.6 %. This decrease is due to the fact that the temperature jump reduces heat transfer. As Kn increases, the temperature jump also increases. Therefore, the denominator of Eq. (5.14) takes larger values. Similar results were found by [18]. They report approximately a 32 % decrease. However, [20] extended the Graetz problem to slip flow, where they find an increase in the Nusselt number for the same conditions without considering the temperature jump. We can see the same trend in the other two cases of constant wall heat flux for cylindrical and rectangular geometries.

In figure 3, we show the Nusselt number values in the thermally developing range in a cylindrical channel with a prescribed temperature at the wall. For both cases, as Kn increases, the Nusselt number decreases due to the increasing temperature jump. We note here that the decrease is greater when we consider viscous dissipation. While the fully developed Nusselt number for the no-slip condition is 6.4231 when Br = 0.01, it is 3.0729 for Kn = 0.12 (52.2 % decrease as opposed to a 35.6 % decrease for the no-viscous heating case).

To have a better understanding of the viscous heating effects, one needs to come up with a parameter to combine the effects of the Brinkman number and the Graetz number, since the viscous effects start becoming significant at a certain distance. We computed the ratio of these two non-dimensional groups as

$$\frac{Br}{Gz} = \frac{\nu u_m}{c_p (T_o - T_s)} \frac{L}{D^2}. \qquad (5.15)$$

This parameter appears to be the coefficient of the viscous term in the non-dimensionalized energy equation and determines the magnitude of the viscous heat generation. Since it is inversely proportional to the square of the system size, it shows the difference in viscous heating effects between a macro and a micro system.

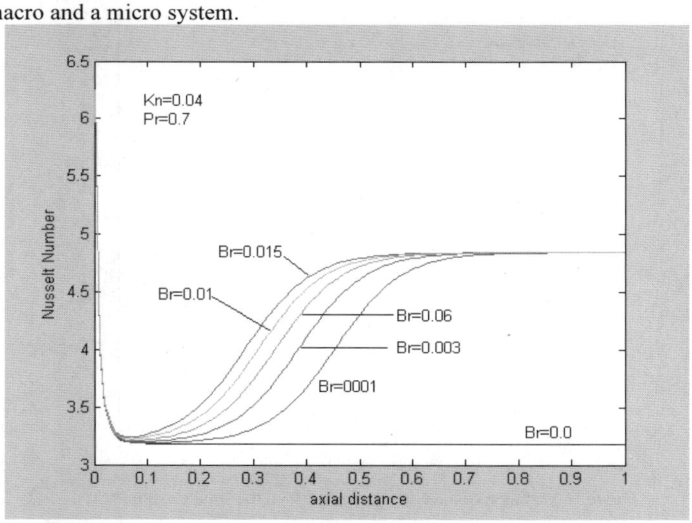

Figure 4. The effect of viscous heating on heat transfer at the channel entrance for uniform wall temperature [15].

5.1.2 Uniform Wall Heat Flux

In the case of uniform heat flux, the governing equations of convective heat transfer will be the same as the previous section. There will be a need to modify the non-dimensional numbers, the boundary conditions and the manner in which the problem is solved. The energy equation will be:

$$\frac{Gz(1-\eta^2+4Kn)}{2(1+8Kn)}\frac{\partial\theta}{\partial\varsigma}=\frac{1}{\eta}\frac{\partial}{\partial\eta}\left(\eta\frac{\partial\theta}{\partial\eta}\right)+\frac{32Br}{(1+8Kn)^2}\eta^2 \tag{5.16}$$

where the following modifications have been made: $Br=\frac{\mu u_m^2}{q''D}$, $\theta=\frac{T-T_o}{q''R/k}$, and $\left(\frac{\partial\theta}{\partial\eta}\right)_{\eta=1}=1$ from the constant temperature case. We then define $\theta(\eta,\varsigma)=\phi(\eta,\varsigma)+\theta_\infty(\eta,\varsigma)$, where $C_{Br}=\frac{32Br}{(1+8Kn)^2}$ and θ_∞ is solved from the boundary conditions [15] and given as:

$$\theta_\infty=4\varsigma+\left(1+\frac{C_{Br}}{4}\right)\frac{\left(\eta^2-\frac{\eta^4}{4}+4Kn\eta^2\right)}{(1+8Kn)} -\frac{(4+C_{Br})(7+112Kn+384Kn^2)}{96(1+8Kn)^2}-\frac{C_{Br}\eta^4}{16}+\frac{C_{Br}(1+16Kn)}{96(1+8Kn)} \tag{5.17}$$

The remaining equation and boundary conditions are given below:

$$\frac{Gz(1-\eta^2+4Kn)}{4(1+8Kn)}\frac{\partial\phi}{\partial\varsigma}=\frac{1}{\eta}\frac{\partial}{\partial\eta}\left(\eta\frac{\partial\phi}{\partial\eta}\right)+\frac{32Br}{(1+8Kn)^2}\eta^2 \tag{5.18}$$

$$\frac{\partial\phi}{\partial\eta}=0 \quad \text{at} \quad \eta=0$$

$$\frac{\partial\phi}{\partial\eta}=0 \quad \text{at} \quad \eta=1$$

$$\phi=\phi_o=-\theta_\infty \quad \text{at} \quad \varsigma=0$$

The following eigenvalue problem is solved in this case. The orthogonality condition, normalization integral and integral transform pair remain the same.

$$\frac{1}{\eta}\frac{d}{d\eta}\left(\eta\frac{d\psi}{d\eta}\right)+(1-\eta^2 4Kn)\lambda_m^2\psi=0 \tag{5.19}$$

$$\frac{d\psi}{d\eta}=0 \quad \text{at} \quad \eta=1$$

$$\frac{d\psi}{d\eta}=0 \quad \text{at} \quad \eta=0$$

The solution is then obtained in the following form by the same manner in which the solution to the uniform temperature case was obtained.

$$\phi(\eta,\varsigma)=\sum_{m=1}^{\infty}\frac{\psi(\lambda_m,\eta)}{N(\lambda_m)}\left[\frac{K_m}{P_m}-\left(\bar{\phi}_o+\frac{K_m}{P_m}\right)e^{-P_m\varsigma}\right] \tag{5.20}$$

where $K_m=\frac{64Br}{Gz(1+8Kn)}\int_0^1\eta^3\psi(\lambda_m,\eta)d\eta$, $P_m=\frac{2(1+8Kn)}{Gz}\lambda_m^2$ and

$$\bar{\phi}_o=\int_0^1\eta(1-\eta^2 4Kn)\left(\begin{array}{c}\left(1+\frac{C_{Br}}{4}\right)\frac{\left(\eta^2-\frac{\eta^4}{4}+4Kn\eta^2\right)}{(1+8Kn)}-\frac{C_{Br}\eta^4}{16}\\ -\frac{(4+C_{Br})7+112Kn+384Kn^2}{96(1+8Kn)^2}-\frac{C_{Br}(1+16Kn)}{96(1+8Kn)}\end{array}\right)\psi(\lambda_m,\eta)d\eta.$$

The Nusselt number is then determined to be:

$$Nu_x=\frac{2}{\theta_s+\frac{4\gamma}{\gamma+1}\frac{Kn}{\Pr}-\theta_b}. \tag{5.21}$$

Figure 5 shows the effect of positive or negative Br values (Br ± 0.01) on heat transfer. As we mentioned before, for this type of boundary condition, a negative Br means that the fluid is being cooled. Therefore, the Nusselt number takes higher values for Br < 0 and lower values for Br > 0.

Since the definition of the Brinkman number is different for the case of the uniform heat flux boundary condition, a positive Br means that the heat is transferred to the fluid from the wall as opposed to the uniform temperature case. Therefore, we see in figure 6 that Nu decreases as Br increases when Br > 0.

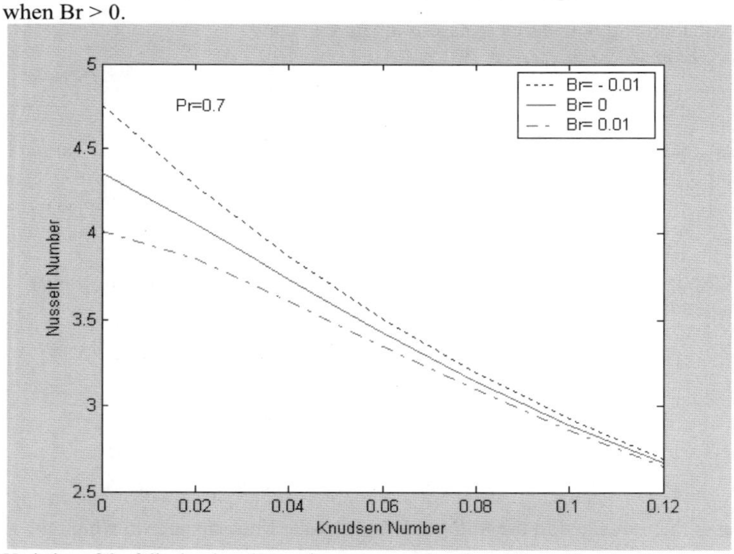

Figure 5. Variation of the fully developed Nusselt number as a function of Kn, with and without considering viscous heating for uniform heat flux at the wall [15].

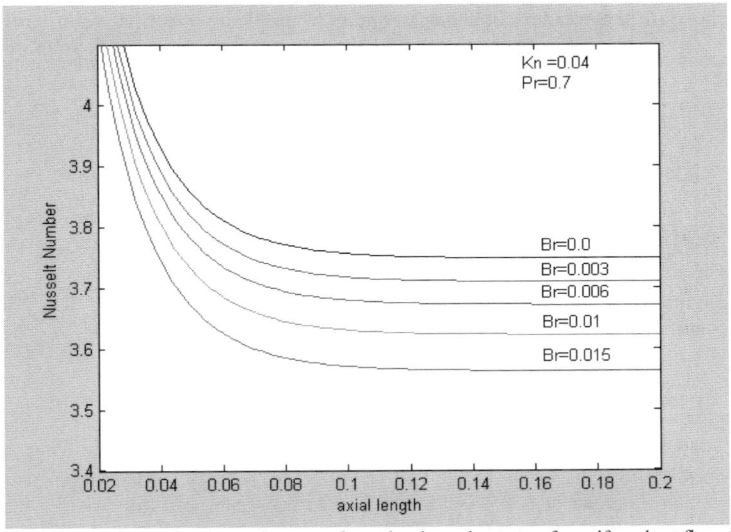

Figure 6 .The effect of viscous heating on heat transfer at the channel entrance for uniform heat flux at the wall [15].

5.2 CONVECTION BETWEEN PARALLEL PLATES

5.2.1 Velocity Profile

The axial direction momentum equation shown in Eq. (5.22) is solved in order to determine the fully developed velocity profile [14].

$$\mu \frac{d^2u}{dy^2} = \frac{dP}{dx}. \tag{5.22}$$

Integrating it twice and using the boundary conditions $u = u_s$ at $y = 0$ and $y = \ell$, Eq (5.22) yields:

$$u = \frac{6}{1+6Kn} u_m \left(Kn + \frac{y}{\ell} - \frac{y^2}{\ell^2} \right) \tag{5.23}$$

where u_m is the bulk velocity and $Kn = \lambda/\ell$.

5.2.2 Temperature Profile

After the fully developed velocity profile is known, we can determine the developing temperature profile. The steady, two dimensional, thermally developing energy equation for flow between two parallel plates is

$$u \frac{\partial T}{\partial x} = \alpha \frac{\partial^2 T}{\partial y^2} + \frac{\nu}{c_p} \left(\frac{du}{dy} \right)^2. \tag{5.24}$$

For the uniform wall temperature boundary condition problem, we first substitute the non-dimensional parameters, $\theta = \frac{T - T_s}{T_0 - T_s}, \eta = \frac{y}{\ell}, \zeta = \frac{x}{L}, u^* = \frac{u}{u_m}$, the determined velocity profile Eq. (5.23) and the following definitions $\alpha = \frac{k}{\rho c_p}$, $\mu = \rho \nu$, $Br = \frac{\mu}{k(T_0 - T_s)} u_m^{\ 2}$, $\mathrm{Re} = \frac{\rho}{\mu} u_m \ell$, $\mathrm{Pr} = \frac{\mu}{k} c_p$, $Gz = \frac{\mathrm{Re}\,\mathrm{Pr}\,\ell}{L}$ into the energy equation Eq.(5.24) yields:

$$\frac{6Gz\left(Kn + \eta - \eta^2\right)}{(1 + 6Kn)} \frac{\partial \theta}{\partial \zeta} = \frac{\partial^2 \theta}{\partial \eta^2} + \frac{36Br}{(1 + 6Kn)^2} (1 - 2\eta)^2 . \tag{5.25}$$

The boundary conditions can now be adjusted by the non-dimensional parameters. This produces new boundary conditions for Eq. (5.25).

$$\theta = 0 \quad \text{at} \quad \eta = 0$$
$$\theta = 0 \quad \text{at} \quad \eta = 1$$
$$\theta = 1 \quad \text{at} \quad \zeta = 0$$

To find the desired temperature profile, Eq. (5.25) is solved by using the following eigenvalue problem

$$\frac{d^2 \psi}{d\eta^2} + \left(Kn + \eta - \eta^2\right) \lambda_m^{\ 2} \psi = 0 \tag{5.26}$$

with its boundary conditions

$$\psi = 0 \quad \text{at} \quad \eta = 0$$
$$\psi = 0 \quad \text{at} \quad \eta = 1$$

Its eigenfunctions obey the orthogonality condition

$$\int_0^1 \left(Kn + \eta - \eta^2\right) \psi(\lambda_m, \eta) \psi(\lambda_n, \eta) d\eta = \begin{cases} 0 & \text{for} \quad \mathrm{m} \neq \mathrm{n} \\ N(\lambda_m) & \text{for} \quad \mathrm{m} = \mathrm{n} \end{cases} \tag{5.27}$$

and are normalized by

$$N(\lambda_m) = \int_0^1 \left(Kn + \eta - \eta^2\right) [\psi(\lambda_m, \eta)]^2 d\eta . \tag{5.28}$$

The integral transform pair for this particular case is

Transformed Formula: $$\bar{\theta}(\lambda_m, \zeta) = \int_0^1 \left(Kn + \eta - \eta^2\right) \psi(\lambda_m, \eta) \theta(\eta, \zeta) d\eta \tag{5.29}$$

Inversion Formula: $$\theta(\eta, \zeta) = \sum_{m=1}^{\infty} \frac{1}{N(\lambda_m)} \psi(\lambda_m, \eta) \bar{\theta}(\lambda_m, \zeta) \tag{5.30}$$

After applying the integral transform method to the equation, it becomes

$$\frac{6Gz}{1 + 6Kn} \frac{d\bar{\theta}_m}{d\zeta} = -\lambda_m^{\ 2} \bar{\theta}_m + \frac{36Br}{(1 + 6Kn)^2} \int_0^1 (1 - 2\eta)^2 \psi(\lambda_m, \eta) d\eta . \tag{5.31}$$

It can now be rewritten in the form

$$\frac{d\bar{\theta}_m}{d\zeta} + P_m \bar{\theta}_m = K_m \tag{5.32}$$

where

$$K_m = \frac{6Br}{Gz(1+6Kn)}\int_0^1 (1-2\eta)^2 \psi(\lambda_m,\eta)d\eta \quad \text{and} \quad P_m = \frac{(1+6Kn)}{6Gz}\lambda_m^{\ 2}.$$

Given that

$$\bar{\theta}_m = \frac{K_m}{P_m} + (\bar{1}_m - \frac{K_m}{P_m})e^{-P_m\varsigma} \tag{5.33}$$

and since

$$\bar{I}_m = \int_0^1 (Kn+\eta-\eta^2)\psi(\lambda_m,\eta)d\eta\ , \tag{5.34}$$

we can now solve for the developing temperature profile by using the inversion equation, Eq. (5.30), and Eq (5.33).

$$\theta = \sum_{m=1}^{\infty}\left(\frac{1}{N(\lambda_m)}\psi(\lambda_m,\eta)\left(\frac{K_m}{P_m} + \left(\int_0^1 (Kn+\eta-\eta^2)\psi(\lambda_m,\eta)d\eta - \frac{K_m}{P_m}\right)e^{-P_m\zeta}\right)\right). \tag{5.35}$$

Once the developing temperature profile has been determined, it is used to solve for the Nusselt number:

$$\text{Isothermal: } Nu_x = \frac{h_x \ell}{k} = -\frac{\left.\frac{\partial\theta}{\partial\eta}\right|_{\eta=1}}{\left(\theta_b - \frac{2\gamma}{\gamma+1}\frac{Kn}{\Pr}\left.\frac{\partial\theta}{\partial\eta}\right|_{\eta=1}\right)} \tag{5.36}$$

where the bulk temperature is defined as $\theta_b = \frac{6}{1+6Kn}\int_0^1 (Kn+\eta-\eta^2)\theta(\eta,\zeta)d\eta$.

For the case of the uniform wall heat flux boundary condition, we again apply the integral transform technique using a method similar to the cylindrical solution. The Nu number is given as:

$$\text{Isoflux: } Nu_x = \frac{2}{\theta_s + \frac{2\gamma}{\gamma+1}\frac{Kn}{\Pr} - \theta_b} \tag{5.37}$$

Compressible two-dimensional fluid flow and heat transfer characteristics of a gas flowing between two parallel plates with both uniform temperature and uniform heat flux boundary conditions were solved in [21]. They compared their results with the experimental results of [12]. The slip flow model agreed well with these experiments. They observed an increase in the entrance length and a decrease in the Nusselt number as Kn takes higher values. It was found that the effect of compressibility and rarefaction is a function of Re. Compressibility is significant for high Re and rarefaction is significant for low Re.

5.3 UNSTEADY CONVECTION

Steady flow through a microtube has been presented. In this section, convection at the entrance of a micropipe with a sudden wall temperature change will be discussed [17]. For the analytical solution the integral transform technique and the Laplace transform will both be used. The effects of velocity slip, temperature jump and viscous heating will all be included. The fully developed velocity profile will be steady and is identical to that given in section 5.1. The non-dimensional energy equation is given as:

$$\frac{\partial\theta}{\partial\tau} + u^* \frac{\partial\theta}{\partial\varsigma} = \frac{1}{\eta}\frac{\partial}{\partial\eta}\left(\eta\frac{\partial\theta}{\partial\eta}\right) + Br\left(\frac{du^*}{d\eta}\right)^2 \tag{5.38}$$

where the following variables are different than those defined in section 5.1, $\varsigma=\frac{x\alpha}{u_m R^2}$, $\tau=\frac{t\alpha}{R^2}$, $\theta=\frac{T-T_s}{T_i-T_s}$ and subject to the following boundary conditions:

$\theta=1$	at	$\tau = 0,\ \varsigma \geq 0,\ 0 \leq \eta \leq 1$
$\theta=1$	at	$\varsigma = 0,\ \tau \geq 0,\ 0 \leq \eta \leq 1$
$\frac{\partial\theta}{\partial\eta}=0$	at	$\eta = 0,\ \tau \geq 0,\ \varsigma \geq 0$
$\theta=0$	at	$\eta = 1,\ \tau > 0,\ \varsigma \geq 0$

To simplify the analysis, we write the non-dimensional temperature as the summation of two components $\theta(\tau,\eta,\varsigma)=\theta_1(\tau,\eta)+\theta_2(\tau,\eta,\varsigma)$ which may then be used to solve the following two equations:

$$\frac{\partial\theta_1}{\partial\tau}=\frac{1}{\eta}\frac{\partial}{\partial\eta}\left(\eta\frac{\partial\theta_1}{\partial\eta}\right)+Br\left(\frac{du^*}{d\eta}\right)^2 \tag{5.39}$$

with boundary conditions

$\theta_1=1$	at	$\tau = 0$
$\frac{\partial\theta_1}{\partial\eta}=0$	at	$\eta = 0$
$\theta_1=1$	at	$\eta = 1$

and:

$$\frac{\partial\theta_2}{\partial\tau}+u^*\frac{\partial\theta_2}{\partial\varsigma}=\frac{1}{\eta}\frac{\partial}{\partial\eta}\left(\eta\frac{\partial\theta_2}{\partial\eta}\right) \tag{5.40}$$

with boundary conditions

$\theta_2=0$	at	$\tau = 0$
$\frac{\partial\theta_2}{\partial\eta}=0$	at	$\eta = 0$
$\theta_2=0$	at	$\eta = 1$
$\theta_2=1-\theta_1$	at	$\varsigma = 0$

The first component to the temperature profile is solved by selecting the appropriate eigenvalue problem to the first problem given as:

$$\frac{d}{d\eta}\left(\eta\frac{dX_m}{d\eta}\right)+\gamma_m^2\eta X_m=0 \tag{5.41}$$

$\frac{\partial X_m}{\partial\eta}=0$	at	$\eta = 0$
$X_m=0$	at	$\eta = 1$

where X_m and γ_m are the eigenfunctions and eigenvalues respectively. The orthogonality condition gives $N_m = \int_0^1 \eta X_m^2 d\eta$. The transform and inversion formulas are then given as

Transform:
$$\overline{\theta}_{1m} = \int_0^1 \eta X_m \theta_1 d\eta \qquad (5.42)$$

Inversion:
$$\theta_1 = \sum_{m=1}^{\infty} \frac{X_m}{N_m} \overline{\theta}_{1m} \qquad (5.43)$$

The next step is to remove the spatial derivatives from the governing equation, reducing it to an ordinary differential equation. To do so, both sides of Eq. (5.41) is operated on by $\int_0^1 \eta X_m d\eta$. The transformed equation is then obtained as:

$$\frac{d\overline{\theta}_{1m}}{d\tau} + \gamma^2 \overline{\theta}_{1m} = K_m \qquad (5.44)$$

where $K_m = Br \int_0^1 \eta \left(\frac{du^*}{d\eta} \right)^2 X_m d\eta$ and then the first component of the temperature profile is given as:

$$\theta_1 = \sum_{m=1}^{\infty} \frac{X_m}{N_m} \left[\frac{K_m}{\gamma_m^2} + \left(\int_0^1 \eta X_m d\eta - \frac{K_m}{\gamma_m^2} \right) e^{-\gamma_m^2 \tau} \right] \qquad (5.45)$$

The second component must then be solved. The Laplace transform is first applied to Eq. (5.40) as is shown below:

$$L\left\{ \frac{\partial \theta_2}{\partial \tau} \right\} + u^* L\left\{ \frac{\partial \theta_2}{\partial \varsigma} \right\} = L\left\{ \frac{1}{\eta} \frac{\partial}{\partial \eta} \left(\eta \frac{\partial \theta_2}{\partial \eta} \right) \right\} \qquad (5.46)$$

which, using $\widetilde{\theta} = L\{\theta_2\}$ and the appropriate transformed boundary conditions, gives:

$$\widetilde{\theta}_2 = \frac{1}{s} - \frac{1}{s} \sum_{m=1}^{\infty} \frac{X_m K_m}{N_m \gamma_m^2} - \sum_{m=1}^{\infty} \frac{X_m \left(\int_0^1 \eta X_m d\eta - \frac{K_m}{\gamma_m^2} \right)}{N_m (s + \gamma_m^2)} \qquad (5.47)$$

The integral transform technique is then applied to Eq (5.46). The details of this process will not be given here [17], since the appropriate eigenvalue problem is identical to that of Eq. (5.2). The final result for the second component of the temperature profile is given as:

$$\theta_2 = \sum_{n=1}^{\infty} \frac{\psi_n}{N_n} e^{-\frac{\lambda_n^2 (1+8Kn)}{2} \varsigma} U\left(\tau - \frac{P_n (1+8Kn)}{2} \varsigma \right) \cdot \left(D_n - \sum_{m=1}^{\infty} H_{mn} - \sum_{m=1}^{\infty} J_{mn} e^{-\gamma_m^2 \left(\tau - \frac{P_n (1+8Kn)}{2} \varsigma \right)} \right) \qquad (5.48)$$

where

$$U\left(\tau - \frac{P_n (1+8Kn)}{2} \varsigma \right) = \begin{cases} 0 & for \quad \tau < \frac{P_n (1+8Kn)}{2} \varsigma, \\ 1 & for \quad \tau > \frac{P_n (1+8Kn)}{2} \varsigma \end{cases}$$

$$D_n = \int_0^1 \eta(1-\eta^2+4Kn)\psi_n d\eta\,, \qquad G_{mn} = \int_0^1 \eta(1-\eta^2+4Kn)\psi_n X_m d\eta\,,$$

$$J_{mn} = \frac{G_{mn}\left(\int_0^1 \eta X_m d\eta - \dfrac{K_m}{\gamma_m^2}\right)}{N_m}, \qquad H_{mn} = \frac{K_m G_{mn}}{N_m \gamma_m^2}, \qquad P_n = \frac{1}{N_n}\int_0^1 \eta \psi_n^2 d\eta\,.$$

The separate solutions to Eq. (5.45) and Eq. (5.48) are then substituted back into $\theta(\tau,\eta,\varsigma)=\theta_1(\tau,\eta)+\theta_2(\tau,\eta,\varsigma)$ to obtain the temperature distribution. Finally the Nusselt number can be obtained by:

$$Nu_{x,t} = -\frac{2\left(\dfrac{\partial\theta}{\partial\eta}\right)_{\eta=1}}{\left(\theta_b - \dfrac{4\gamma}{\gamma+1}\dfrac{Kn}{\Pr}\left(\dfrac{\partial\theta}{\partial\eta}\right)_{\eta=1}\right)}. \tag{5.49}$$

5.4 CONVECTION IN AN ANNULUS

The details of convection in a microannulus subject to the uniform wall temperature boundary condition will not be given in full due to the sizable resulting equations. The complete details of the derivation for convection in an annulus subject to slip-flow can be found in [14]. Here the solution to the velocity profile and constant temperature boundary condition will be given. The flow is assumed to be fully developed and therefore the momentum equation is given as:

$$\frac{dP}{dx} = \frac{\mu}{r}\frac{d}{dr}\left(r\frac{du}{dr}\right) \tag{5.50}$$

subject to the following boundary conditions:

$$\lambda\left(\frac{du}{dr}\right)_{r=a} = u_{r=a} \quad \text{and} \quad -\lambda\left(\frac{du}{dr}\right)_{r=b} = u_{r=b}\,.$$

After integrating twice and applying the boundary conditions, the resulting velocity profile is given as :

$$u^* = A_1\eta^2 + A_2\ln\eta + A_3 \tag{5.51}$$

where:

$$A_1 = \frac{\dfrac{1}{2}}{\dfrac{1-\gamma^2}{4} + \dfrac{(2Kn\gamma+\gamma(1-\gamma))\left(\ln\gamma + \dfrac{1-\gamma^2}{2}\right)}{(1-\gamma)(2\gamma\ln\gamma - 2Kn(\gamma+1))} + Kn\left(\dfrac{2Kn(\gamma+1)+(1-\gamma^2)}{2\gamma\ln\gamma - 2Kn(\gamma+1)} + \gamma\right)},$$

$$A_2 = \frac{\dfrac{2Kn\gamma(\gamma+1)+\gamma(1-\gamma^2)}{2\gamma\ln\gamma - 2Kn(\gamma+1)}}{\dfrac{1-\gamma^2}{4} + \dfrac{(2Kn\gamma+\gamma(1-\gamma))\left(\ln\gamma + \dfrac{1-\gamma^2}{2}\right)}{(1-\gamma)(2\gamma\ln\gamma - 2Kn(\gamma+1))} + Kn\left(\dfrac{2Kn(\gamma+1)+(1-\gamma^2)}{2\gamma\ln\gamma - 2Kn(\gamma+1)} + \gamma\right)}$$

$$A_3 = \frac{Kn\left(\dfrac{2Kn(\gamma+1)+(1-\gamma^2)}{2\gamma\ln\gamma - 2Kn(\gamma+1)} + \gamma\right) - \dfrac{\gamma^2}{2} - \dfrac{2Kn\gamma(\gamma+1)+\gamma(1-\gamma^2)}{2\gamma\ln\gamma - 2Kn(\gamma+1)}\ln\gamma}{\dfrac{1-\gamma^2}{4} + \dfrac{(2Kn\gamma+\gamma(1-\gamma))\left(\ln\gamma + \dfrac{1-\gamma^2}{2}\right)}{(1-\gamma)(2\gamma\ln\gamma - 2Kn(\gamma+1))} + Kn\left(\dfrac{2Kn(\gamma+1)+(1-\gamma^2)}{2\gamma\ln\gamma - 2Kn(\gamma+1)} - \gamma\right)}$$

and $\gamma=a/b$ is the aspect ratio.

To find the heat transfer coefficient, we begin with the non-dimensional energy equation given by:

$$(A_1\eta^2 + A_2\ln\eta + A_3)\frac{\partial\theta}{\partial\varsigma} = \frac{1}{\eta}\frac{\partial}{\partial\eta}\left(\eta\frac{\partial\theta}{\partial\eta}\right) + Br\left(2A_1\eta + \frac{A_2}{\eta}\right)^2 \qquad (5.52)$$

where:

$$\theta = \frac{T-T_s}{T_o-T_s} \qquad Br = \frac{\mu u_m^2}{k(T_o - T_s)}$$

The temperature jump boundary conditions and inlet condition are also used in the same form for this case. The non-homogeneous boundary conditions can be written as follows

$$\theta(\gamma) - \frac{2\gamma}{\gamma+1}\frac{Kn}{\Pr}\left(\frac{\partial\theta}{\partial\eta}\right)_{\eta=\gamma} = 0$$

$$\theta(1) - \frac{2\gamma}{\gamma+1}\frac{Kn}{\Pr}\left(\frac{\partial\theta}{\partial\eta}\right)_{\eta=1} = 0$$

The solution method again starts with the selection of the appropriate eigenvalue problem. The boundary conditions of the eigenvalue problem deserve special attention in terms of the similarity between them and the boundary conditions of the original problem.

$$\frac{1}{\eta}\frac{d}{d\eta}\left(\eta\frac{d\psi}{d\eta}\right) + \eta(A_1\eta^2 + A_2\ln\eta + A_3)\lambda^2\psi = 0 \qquad (5.53)$$

$$\psi - \frac{2\gamma}{\gamma+1}\frac{Kn}{\Pr}\frac{d\psi}{d\eta} = 0 \text{ at } \quad \eta=\gamma$$

$$\psi + \frac{2\gamma}{\gamma+1}\frac{Kn}{\Pr}\frac{d\psi}{d\eta} = 0 \text{ at } \quad \eta=1$$

The transformation of the governing equation is performed by applying the same term. The following term is obtained from the partial integration of the conduction term in the energy equation

$$\psi(1)\frac{\partial\theta}{\partial\eta}(1) - \theta(1)\frac{\partial\psi}{\partial\eta}(1) + \gamma\theta(\gamma)\frac{\partial\psi}{\partial\eta}(\gamma) - \gamma\psi(\gamma)\frac{\partial\theta}{\partial\eta}(\gamma)$$

and is identically equal to zero. This can be reached after some manipulations to the combination of the boundary conditions for both the original problem and the eigenvalue problem. Finally, the transformed version of the energy equation is obtained as follows

$$\frac{d\bar{\theta}_m}{d\varsigma} + \lambda_m^2 \bar{\theta}_m = K_m \tag{5.54}$$

where

$$K_m = \int_\gamma^1 \eta \left(2A_1\eta + \frac{A_2}{\eta} \right)^2 \psi_m d\eta$$

The transformed temperature can easily be obtained from this ODE as

$$\bar{\theta}_m = \frac{K_m}{\lambda_m^2} + \left(\bar{1}_m - \frac{K_m}{\lambda_m^2} \right) e^{-\lambda_m^2 \varsigma} \tag{5.55}$$

where the transformed inlet condition is calculated from

$$\bar{1}_m = \int_\gamma^1 \eta (A_1\eta^2 + A_2 \ln(\eta) + A_3) \psi_m d\eta$$

The non-dimensional temperature profile is then obtained from the inversion formula. Once the temperature is obtained, the inner wall, the outer wall and the average Nusselt number values are calculated from the following equations respectively.

$$Nu_1 = 2(1-\gamma)\frac{\left.\frac{\partial\theta}{\partial\eta}\right|_{\eta=\gamma}}{\theta_{ave}} \qquad Nu_2 = -2(1-\gamma)\frac{\left(\frac{\partial\theta}{\partial\eta}\right)_{\eta=1}}{\theta_{ave}} \qquad Nu_{ave} = \frac{\gamma Nu_1 + Nu_2}{1+\gamma} \tag{5.56}$$

5.5 CONVECTION IN A RECTANGULAR CHANNEL

5.5.1 Uniform Wall Heat Flux

The results of [16] will now be given. This is the case of the uniform heat flux (H_2) boundary condition for convection in a rectangular microchannel. The details will not be given since the techniques used in the integral transform technique to solve the energy and momentum equation were described in the previous sections. After first applying filtering, a technique discussed in [22], the integral transform technique is then applied. The resulting equation for the Nusselt number is:

$$Nu_q = \frac{1}{\theta_s + \frac{2\gamma}{1+\gamma}\left(\frac{a+b}{2b}\right)\frac{Kn}{\mathrm{Pr}} - \theta_b} \tag{5.57}$$

The effect of the Nusselt number was plotted against the aspect ratio for different Knudsen number values. The results compared well with those of [23], and [24].

5.5.2 Uniform Wall Temperature

Using the integral transform method, [25] solved for the Nusselt number for flow in a rectangular microchannel subject to the constant temperature and slip flow boundary conditions.

Their results for the non-slip flow case agreed with [26], who also used the integral transform technique to solve for the Nusselt number for flow through a macrosized rectangular channel. They did not include viscous dissipation in the work, but they did include variable thermal accommodation coefficients. Similar to [15], they concluded that the Knudsen number, Prandtl number, aspect ratio, velocity slip and temperature jump can all cause the Nusselt number to deviate from the conventional value.

6. Conclusion

We have shown the solution of the temperature distribution of a gas flowing in four different geometries. They are a cylindrical channel, two parallel plates, an annulus and a rectangular channel. Steady state, hydrodynamically fully developed laminar constant flow properties assumptions are made. The unsteady case was also considered. Thermally developing Nusselt numbers for cylindrical pipes, parallel-plates and rectangular channels can be obtained. A straightforward analytical solution method, the integral transform technique, is used. It is found that the heat transfer coefficient is strongly influenced by the Knudsen number as can be seen in Table 1.

Table 1:Nusselt Number for different Geometries Subject to Slip-Flow (β_T=1.66) ([14-17], and [24]).

Br = 0.0		Kn = 0.00		Kn = 0.04		Kn=0.08		Kn=0.12	
		Nu_T	Nu_q	Nu_T	Nu_q	Nu_T	Nu_q	Nu_T	Nu_q
Cylindrical		3.67	4.36	3.18	3.75	2.73	3.16	2.37	2.68
Rectangular	γ=1	2.98	3.10	2.71	2.85	2.44	2.53	2.17	2.24
	γ=0.84	3.00	3.09	2.73	2.82	2.46	2.48	2.19	2.17
Aspect	γ=0.75	3.05	3.08	2.77	2.81	2.49	2.44	2.22	2.12
Ratio	γ=0.5	3.39	3.03	2.92	2.71	2.55	2.26	2.24	2.18
γ=a/b	γ=0.25	4.44	2.93	3.55	2.42	2.89	1.81	2.44	1.68
	γ=0.125	5.59	2.85	4.30	1.92	3.47	1.25	2.8	1.12
Two Parallel Plates		7.54	8.23	6.26	6.82	5.29	5.72	4.56	4.89

The following conclusions can be made from the previous works:

- Depending on the values for the Knudsen number, the Prandtl number, the Brinkman number and the aspect ratio, heat transfer in microchannels can be significantly different from conventionally sized channels.
- Velocity slip and temperature jump effect the heat transfer in opposite ways: a large slip on the wall will increase the convection along the surface due to an increased bulk velocity. On the other hand, a large temperature jump will decrease the heat transfer by reducing the temperature gradient at the wall. Therefore, neglecting the temperature jump will result in the overestimation of the heat transfer coefficient.
- A Nusselt number reduction is observed as the flow deviates from the continuum behavior, or as Kn takes higher values.
- The Prandtl number is important, since it directly influences the magnitude of the temperature jump. Looking at the temperature jump equation, as Pr increases, the difference between wall and fluid temperature at the wall decreases. Therefore, greater Nu values for large Pr are observed.

- In rectangular channels, when Kn increases, the Nusselt number decreases regardless of the value of the aspect ratio due to the increasing temperature jump. However, the decrease in Nu is more significant for a smaller aspect ratio.
- When a fluid meets a surface, there develops a boundary layer in which each layer of fluid has a different velocity. Viscous heat generation is a result of friction between the layers. Since the ratio of surface area to volume is large for microchannels, viscous heating is an important factor. It is especially important for laminar flow, where considerable gradients exist. The Brinkman number, Br, is defined to represent this effect. Larger Nu values for the uniform temperature case with a positive Br are obtained. In this case Br > 0 meaning that the fluid is being cooled. Therefore, viscous heating increases the temperature difference between the surface and the bulk fluid. For the uniform heat flux boundary condition, the definition of Br changes such that a positive Br means that the fluid is being heated while a negative Br means the opposite [14,15]. Therefore, they observed a decrease in Nu for Br > 0 and an increase for Br < 0. This is due to the fact that for different cases, Br may increase or decrease the driving mechanism for convective heat transfer, which is the difference between wall temperature and average fluid temperature.

NOMENCLATURE

a, b Lengths of the rectangular channel
b, Empirical parameter
Br, Brinkman number
c_{jump}, Temperature jump coefficient
c_p, Specific heat at constant pressure
c_v, Specific heat at constant volume
D, Diameter
F_M, Tangential momentum accommodation coefficient
F_T, Thermal accommodation coefficient
K, Thermal conductivity
Kn, Knudsen number
M, Mass of the fluid
Ma Mach number
$\dot{m}$, Mass flow rate
n, Number of molecules per unit volume
Nu, Nusselt number
P, Pressure
Pr, Prandtl number
q'' uniform wall heat flux
Q, Energy of the fluid molecules
R, Gas constant
R, Radius of the circular tube
Re, Reynolds Number
T, Temperature
U, Internal energy of the fluid
u, Fluid velocity
x* Entrance length
x,y,z Cartesion coordinates
r Cylindrical coordinate

Greek symbols

α, Thermal diffusivity
β, β_T/β_v,
β_v, $(2-F_m)/F_m$.
β_T, $(2-F_T)/F_T$.
γ, Specific heat ratio, aspect ratio
λ, Molecular mean free path
μ, Viscosity
η, r/R
ρ, Density
ρ_s, Slip radius
υ, Momentum diffusivity
θ, Dimensionless temperature
ζ, Dimensionless axial coordinate

Subscripts

Ave *average*
b, Bulk
g, True gas condition

i, Impinging
m, Mean
o, Outlet
q, Specified heat flux
r, Reflected
s, Fluid properties at the wall or slip
T, Specified temperature
w, Wall properties
λ Properties at the Knudsen layer

Acknowledgments: The authors acknowledge the support by the Texas state TDT program (grant No. 003604-0039-2001.), and Daniel Newswander.

REFERENCES

1. Kennard, E. H., (1938) *Kinetic Theory of Gases*, McGraw-Hill Book Company, Inc., New York.
2. Bayazitoglu, Y., and Tunc, G., (2002) Extended Slip Boundary Conditions for Microscale Heat Transfer, *AIAA Journal of Thermophysics and Heat Transfer*. Vol. 16. no 3. pp. 472-475.
3. Ameel, T. A., Barron, R. F., Wang, X., and Warrington, R. O., (1997) Laminar Forced Convection In a Circular tube With Constant Heat Flux and Slip Flow, Microscale Thermophys. Eng., Vol. 4, pp. 303-320.
4. Barron, R. F, Wang, X. Ameel, T. A., and Warrington, R. O.,(1997) The Graetz Problem Extended To Slip-flow, *Int. J. Heat Mass Transfer*, Vol. 40, no. 8, pp.1817-1823.
5. Larrode, F. E., Housiadas, C., and Drossinos, Y., (2000) Slip Flow Heat Transfer in Circular Tubes, *Int. J. Heat and Mass Transfer*, Vol. 43, pp. 2669-2680.
6. Choi, S. B., Barron R. F. and Warrington R. O., (1991) Fluid Flow And Heat Transfer in Microtubes, *Micromechanical Sensors, Actuators, and Systems, ASME DSC*, Vol.32, pp.123-134.
7. Kavehpour, H. P., Faghri, M., and Asako, Y., (1997) Effects of Compressibility And Rarefaction On Gaseous Flows In Microchannels, *Numerical Heat Transfer, Part A*. Vol. 32, pp.677-696.
8. Asako, Y., Pi, T., Turner, S. E., and Faghri, M. (2003) Effect of compressibility on gaseous flows in micro-channels, *International J. Heat Mass Transfer*, Vol.46, pp.3041-3050.
9. Chen, C. S., (2004) Numerical method for predicting three-dimensional steady compressible flow in long microchannels, *J. Micromech. and Microeng.*, Vol.14, pp.1091-1100.
10. Hsieh, S.S., Tsai, H.H., Lin, C. Y., Huang, C. F., and Chien, C. M., (2004) Gas flow in a long microchannel, *Int. J. Heat Mass Transfer*, Vol.47, pp.3877-3887.
11. Morini, G. L., Lorenzini, M., and Spiga, M., (2004) A Criterion for the Experimental Validation of the Slip-Flow Models for Incompressible Rarefied Gases through Microchannels, *Proceedings of the 2nd International Conference on Microchannels and Minichannels, June 17-19, 2004 Rochester, New York, USA*, pp. 351-368.
12. Arkilic, E. B., Breuer, K. S., Schmidt, M. A., (1994) Gaseous Flow in Microchannels, *Application of Microfabrication to Fluid Mechanics, ASME FED*, Vol. 197, pp. 57-66.
13. Beskok, A., W. Trimmer, and G. Karniadakis, (1995) Rarefaction compressibility and thermal creep effects in gas microflows *Proc. ASME DSC*, Vol. 57, no.2, pp.877-892.
14. Tunc, G., (2002) Convective Heat Transfer in Microchannel Gaseous Slip Flow, PhD. Thesis, Rice University, Houston, TX.
15. Tunc, G., Bayazitoglu, Y., (2001) Heat Transfer in Microtubes with Viscous Dissipation, *Int. J. Heat Mass Transfer*, Vol. 44, pp.2395-2403.
16. Tunc, G., Bayazitoglu, Y., (2002) Heat Transfer in Rectangular Microchannels, *Int. J. Heat Mass Transfer*, Vol.45, pp.765-773.
17. Tunc, G., and Bayazitoglu, Y., (2002) Convection at the Entrance of Micropipes with Sudden Wall Temperature Change, *Proceedings of the ASME IMECE*.

18. Sparrow, E.M., and Lin, S.H., (1962) Laminar Heat Transfer in Tubes Under Slip-Flow Conditions. *J. Heat Transfer*. pp.363-369 .
19. Maxwell, J. C., (1965) *The Scientific Papers of James Clerk Maxwell*, Dover Publications, Inc., New York.
20. Barron, R. F, Wang, X. Ameel, T. A., Warrington, R. O., (1997) The Graetz Problem Extended To Slip-flow, *Int. J. Heat Mass Transfer*, Vol. 40, pp.1817-1823
21. Kavehpour, H. P., Faghri, M., and Asako, Y., (1997) Effects of Compressibility and Rarefaction on Gaseous Flows in Microchannels, *Numerical Heat Transfer, Part A,* Vol.32, pp.677-696.
22. Cotta, R. M., Mikhailov, M. D., Heat Conduction: Lumped Analysis, Integral Transforms, Symbolic Computation, (1997) John Wiley & Sons, New York.
23. Morini G.L. Spiga, M. (1998) Slip-Flow in Rectangular Microtubes, *Microscale Thermophysical Eng.*, Vol. 2., pp.273-282.
24. Spiga, M. Morini, G.L. (1996) Nusselt Numbers in Laminar Flow for H2 Boundary Conditions, *Int J. Heat Mass Transfer,* Vol. 39, pp 1165-1174.
25. Yu, S., Ameel, T. (2001) Slip-flow Heat Transfer in Rectangular Microchannels, *Int. J. Heat and Mass Transfer*. Vol. 44, pp. 4225-4234.
26. Aparecido, J., Cotta, R. (1990) Thermally Developing Laminar Flow Inside Rectangular Ducts, *Int. J. Heat and Mass Transfer*. Vol. 33, pp. 341-347

MICROSCALE HEAT TRANSFER UTILIZING MICROSCALE AND NANOSCALE PHENOMENA

A. YABE
AIST CHUGOKU, National Institute of Advanced Industrial Science and Technology
Kure, Hiroshima, 737-0197 JAPAN

1. Introduction

In this paper, "Nanotechnology" has been analyzed from the viewpoints of industrial science and technology and from the manufacturing technology. The trends and characteristics of nanotechnology have been described and the importance of nano-manufacturing has been stressed. The role of heat transfer and the transport phenomena in the microscale effects has been explained and the importance of the active control of heat & mass transfer and the transport phenomena was focused for controlling the microscale phenomena. Microscale heat transfer has been successfully researched for realizing advanced thermal engineering by utilizing microscale and nanoscale phenomena. From the viewpoint of nanotechnology, several advanced heat transfer characteristics have been realized and actual examples of microscale heat transfer for promoting energy conservation have been explained.

2. Relationship Between Nanotechnology and Industrial Technology,

The role of the manufacturing engineering among various kinds of industrial engineering is systematically shown in Fig.1. For the promotion of the highly information-oriented society, the recycling & environmental-friendly society and safe & human-friendly society, the contribution from various kinds of the industrial engineering is strongly requested. Especially, since the manufacturing engineering would have been directly relating to the actual society and to the daily lives of the human being by making many

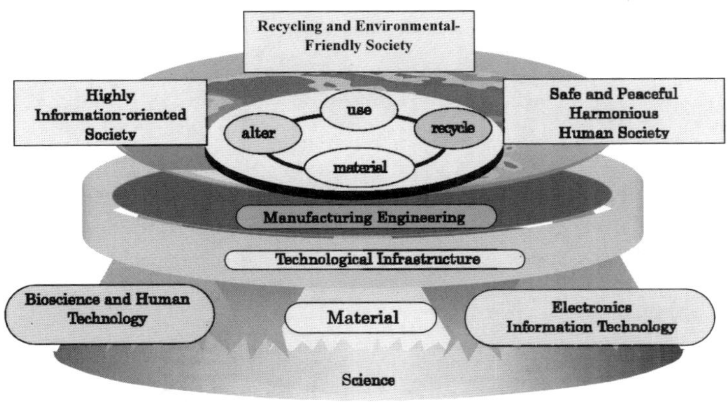

Fig.1 Role of Manufacturing in Industrial Engineering

S. Kakaç et al. (eds.), Microscale Heat Transfer, 149–156.

kinds of products and through the usage of the product, the manufacturing engineering should contribute largely. For this purpose, the manufacturing engineering should be modified and more advanced to promote the ideal society.

The future directions of the manufacturing engineering would have three important key trends. They are (1) the direction to the micro scale and nano scale, (2) the direction to the advanced self-control and (3) the direction to the environmental friendly and sustainable harmonization. Especially, the direction to the micro scale and nano-scale should be composed of three categories, which are 1) advanced manufacturing on nano-scale science and engineering, 2) micro-machine, MEMS & nano-scale machine and 3) micro-factory technology. As for the advanced manufacturing on nano-scale science and engineering, the nano-manufacturing would be developed to make the widely applicable mother machines of nano-processing. Furthermore, the promotion of the leading technology would be important for the realization of the technical seeds by analyzing the phenomena fundamentally. Concerning the present status of nanotechnology, the following features can be drawn as shown in Fig.2. Since the industrial application fields of nanotechnology to the information technology area have been large enough in the economical scale, the research and development of nanotechnology would start from the application to the information technology area and then would spread to the application areas of bio technology and environmental technology. In this process the nano-scale manufacturing technology would play the important role by providing the common manufacturing technology applicable to the fields of information technology(IT), bio and environment.

One key technology for establishing the nano-manufacturing would be the nano-scale processing by use of the laser beam. There are two kinds of nano-manufacturing technology so far for the processing of materials. One is processing by handling atoms, which is called the build-up process and the other is the machining operation such as lathe machining, which is called the break-down process. As shown in Fig.3, there would not existed the widely applicable and useful processing method for the processing unit between 1μm3 and 1nm3. One promising method would be the laser beam processing technology which has characteristics applicable to various kinds of material and atmosphere.

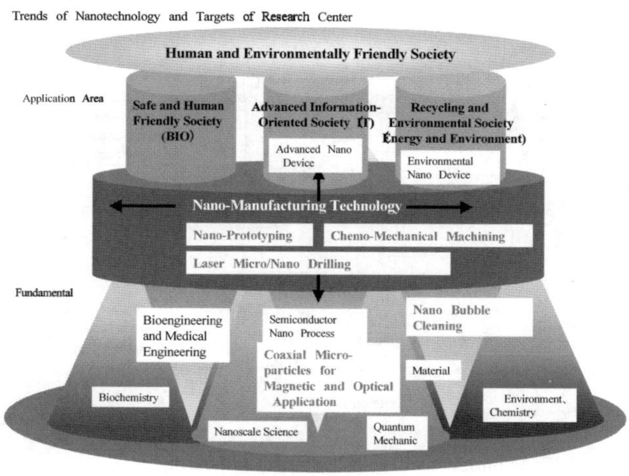

Fig.2 Systematic Trend of Nanotechnology and Key Technologies

For estimating the future trends of nano-manufacturing, the construction of the roadmap on nano-manufacturing would be very important. Fig.4 shows the example of the roadmap on micro and nano manufacturing. In this roadmap many key words are shown for the design technology, fabrication technology, assemble technology and the measurement technology. One important characteristic of this roadmap is the trend of the combination and fusion of the processing and the assembling technologies in the micro scale and the nano scale region, which forces the manufacturing technology more complicated and more advanced.

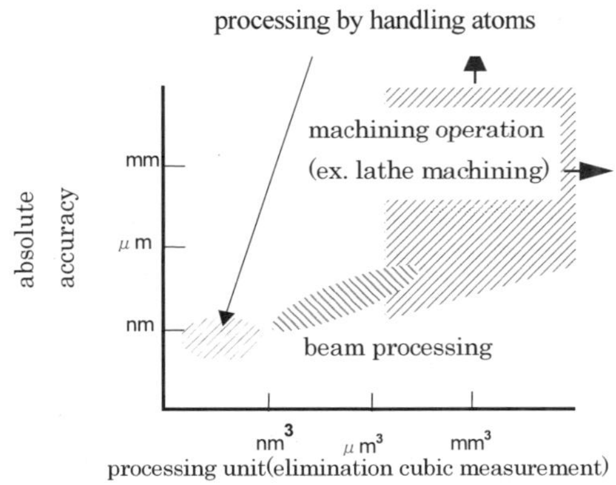

Fig.3 Trend of Nanomanufacturing Processing Technologies

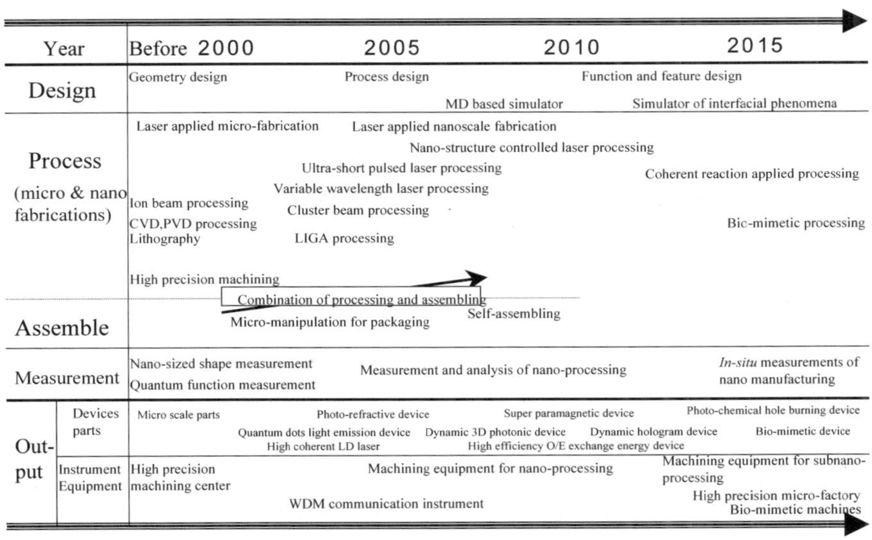

Fig.4 Roadmap of Nanomanufacturing

3. Future Trends of Microengineering and Nanoengineering

Future trends of micro-engineering and nano-engineering would be categorized into four directions. They would be (1) Reducing the size of systems while maintaining their functions (compact personal computer utilizing MEMS), (2) Integrating functions to compact size systems (DNA chips for medical application utilizing MEMS), (3) Innovating and improving system performance by adding micro- and nano-scale function to macroscale engineering applications (macroscale engineering application of microscale and nanoscale phenomena), (4) Increasing the efficiency of existing macroscale products by improving microscale factors that limit their performance. (increasing turbine efficiency & compressor efficiency) The future trends of micro-engineering and nano-engineering have been analyzed and evaluated based on the questionnaire to the leading persons of research and development of manufacturing. These four trends have several characteristics. Reducing the size would be actually effective for information technology field. Integrating functions to compact size would be useful and feasible for medical application and bio-technology. Innovating and improving system performance by utilizing microscale phenomena would have the possibility of manufacturing innovation. Furthermore, increasing the efficiency by decreasing the clearance would be extremely effective for the energy conservation.

The answers of technological leaders showed the following tendency: Reducing the size and Integrating functions would be feasible in 5 and 10 years for promoting IT and medical engineering firelds. For promoting the environmental protection and sustainable energy supply, Innovating system performance and increasing the system efficiency would be essentially important.

Consequently, for the future trends of engineering applications of nanotechnology, the reducing the size and integrating functions would be promising for the first challenge in this decade. Then, innovating and improving the system performance and improving the efficiency would be increasingly important in the next decade.

As the attempt to systematize the nanotechnology from the manufacturing technology, the viewpoint from the application area, the manufacturing technology and the fundamental phenomena has been introduced. The viewpoints from manufacturing technology would be composed of design and simulation technology, fabrication technology, assembly and accumulation technology and maintenance and reliability. Fig.5 show the keywords and typical research topics related to thermal engineering to each viewpoints. Fundamental area related to nanometer scale effects would have much research topics of nano-science with quantum mechanics, mesoscopic interactions and continuum mechanics and their complicated interactions, which would make newly functional effects.

From the viewpoints of transport phenomena, microscale order would mean the scale of various parameters. They would be the scale of ultra-short time interval, extremely small distance, nanometer-order three dimensional structure and ultra small temperature difference, which have been introduced by the advanced engineering

These thermal engineering phenomena control the microsclae and nanoscale structure such as self organization phenomena. Therefore, it becomes considerably important to utilize self-assembly phenomenon using condensation phenomenon, crystal growth phenomenon, and convective heat transfer phenomenon based on unsteadiness as well as assembly technique in micro area using light and ultrasound together. Also it becomes very important to realize the controlled organization by achieving self-assembly control, shape and property control and defect control by all the possible means such as

Nanotechnology from the Viewpoints of Thermal Engineering

	Keywords & Topics	Typical Technology & Research Topics
Application Area		
Information Technology(IT)	Optical Information & Telecommunication Device	・Memory Utilizing Nanometer- order Thermal Deformation by Heating
Bio- engineering	DNA Chip	・Simulation of Bioenergy Conversion Mechanism ・Thermal Engineering Using Balance of Hydrogen Bond, Electrostatic Power, Intermolecular Force and Thermal Effect
Environment Engineering	Energy Conversion, Environmental Sensor	・Environmental Cleaning by Micro- bubble and Nano- bubble ・Removal of Nano- particle from Wall Surface by Ultrasound
Manufacturing Area		
Design & Simulation	Quantum & Molecule Dynamics	・Heat Conductivity (Nanotube, Superlattice) Calculation of Various Nano- stuructures ・Thermal Property of Nano- structure, Thermal
Fablication & Processing	Nano- manufacturing, Lithography, Laser Manufacturing	・Engineering Study on Promoting High- efficiency in Generation Method of Carbon Nanotube ・Phase Change Phenomenon and Crystal Structure Change in Short- pulse Laser Ablation and Laser Annealing
Assembly & Accumulation	Controlled Assembly and Organization, High- precision Micro- assembly	・Drag Reduction of Turbulence by Nanoscale Micelle and Network Sturucture ・Slurrying Ice by Nanoscale Crystal Growth Control
Maintenance & Reliability	Cooling System, Micro Liquid Transfer System, Nano- measurement Technology for Complex Shape	・High- heat Flux Cooling in Micro- area and Temperature Control ・Thermal Property of Semiconductor Thin- film, Reliability ・Temperature Homogenization of Solid Laser, High- heat Flux Cooling
Fundamental Area		
Nanometer Scale Effects	Carbon Nanotube	・Heat Conductivity of Various Nano- sturucture(Nanotube, Superlattice) ・Heat Conduction of Nano-scale contact ・Thermal Property of Nano-structure ・Nano-area, Heating Phanomenon of Three- dimensional Nano-structure ・Promotion of High Efficiency in Nano-level Catalysis of Fuel Battery Electrode
Micrometer Scale Effects	MEMS Micro- machine	・Resistance Reduction of Liquid Flow by Micro- scale Concave- convex Surface ・Radiation and Absoption Property from Micro- scale Concave- convex Surface ・High- heat Flux Cooling Using MEMS ・Promotion of High Efficiency in Thermoelectric Cooling Device by Low- heat Conduction and High- electricity Conduction of Semiconductor

Fig.5 Relationship between Nanotechnology and Thermal Engineering Including Heat Transfer

electromagnetic field, temperature field, fluid field, control by light and ultrasound and also cluster addition, seed crystal addition, grain boundary utilization and phase boundary utilization.

The controlled assembly and organization technology generating micro-/nano-phenomena on dynamic implementation would include monomolecular film formation, crystal growth, structure formation inside grain, catalyst formation and functional thin-film formation. Controlled-assembly phenomenon here means structure formation process with active transport phenomenon under various controlled boundary conditions and it is a key point for thermal engineering phenomena such as heat transfer phenomenon and mass transfer phenomenon. Macroscopic equation cannot be used, because it is the phenomenon of micro-scale, however, the important point is that analysis of transport phenomena such as heat transfer phenomenon contributes to technology for the generation on dynamic implementation, which is important in nanotechnology of self-assembly and its control, as key technology giving the most important control factor.

Here I would emphasize that the role of researchers on transport phenomena such as heat transfer is important for development of nanotechnology and achievement of application technology and that it has a possibility to provide key technology toward application technology. By accomplishing the actual proof research, the above mentioned technologies, which would be the examples of the macroscopic function generation technologies based on nanoscale effects, would be proposed as the leading guideline for promoting the engineering application of nanotechnology.

5. Actual Examples of Thermal Engineering Applications of Microscopic Effects

Drag Reduction of Liquid Flow by Ultra-small Concave and Convex Surfaces

By producing the micrometer order concave and convex surfaces of several micrometer height, the liquid flow along the ultra-small concave and convex surface has the smaller flow drag by 15% for the laminar flow region of liquid water. This would due to the surface tension effects for realizing and maintaining the air trap on the conjugated solid surface. This would be one example of engineering applications of micro-scale effects, which would be useful for the reduction of the pumping power of flow system and for the reduction of the necessary pressure of the micro-scale liquid flow channel.

Ice Slurry Flow Utilizing Nano-scale Effects Originated from Anti-freeze Protein .

By utilizing the antifreeze protein, the ice slurry would be successfully realized. The length of the antifreeze protein would be the order of several nanometers and this effect has been observed by the authors firstly with the scanning tunneling microscope. The mechanism of the adhesion of anti-freeze protein on the surface of the ice crystal would be the combination of hydrophilic bond of the anti-freeze protein and the hydrogen bond of ice crystals. Therefore, the selection of the organic materials such as poly-vinyl alcohol would be possible to replace the same role of slurry formation, since they would have the periodical hydrophilic bonds and the other many hydrophobic bonds. This would be also the engineering application of the nanoscale effects, which would be useful for reducing the pumping power of the latent heat cooling system of the building.

Fabrication Technology of Size and Structure controlled Nanoparticles for Nano-Prototyping

Fabrication technique of nanoparticles with uniform size and structure has been developed for

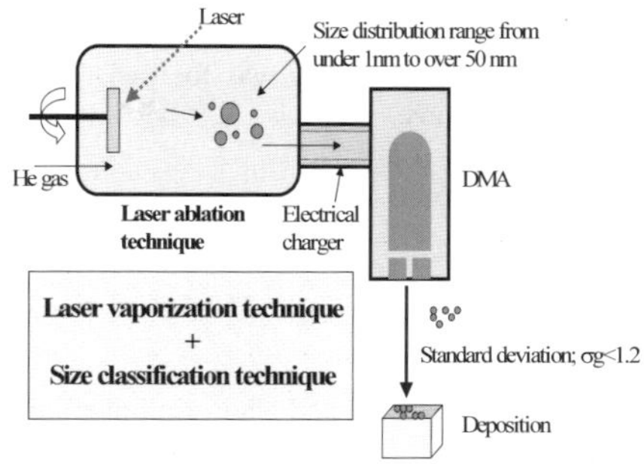

Fig.6 Schematic Draw of Nano-particle Fabrication

fabricated by the process exhibited clearly size-dependent photoluminescence at room temperature due to a confinement of the photo-excited carriers (quantum effect). We focus our research on such core / shell structured nanoparticles for optical and/or magnetic applications utilizing the noble properties originated in nanostructure.

Laser micro/nano drilling and on-site nanoscale measurement utilizing a coherence property of light and the dynamic control on wave-front

Nanoscale manufacturing tool for drilling the fairly precise columnar penetration hole which is several microns in diameter, has been researched and developed. The laser beam with tightly focused spot and long focal depth can be used to process both transparent and opaque materials. The feature of the nanoscale measurement with an interferometer is an on-site measurement, and a dynamic control of the reference wave-front.

Nanobubble Cleaning Utilizing Dispersion Effects of Fluid

Nano bubbles of which diameter would be the order of 100nm had been found for the first time by the author and their quasi-steady existence had been clarified. Since these nano bubbles would have the inner pressure of about 30atm, they would have some effects of cleaning the solid surfaces by colliding and releasing the high inner pressure. Experimental research showed the cleaning effects of nano-bubbles. Minute particles, which had the diameter of about 50nm and which were contaminated on the SiO_2 wafer, have been successfully removed from the wafer surface by impinging the liquid jet of ultra-pure water containing nano-bubbles. By impinging nano-bubble contained jet for several tens of minutes, it was revealed that 98.9% of particles were successfully removed. These would be very effective for the cleaning of wafers applicable to the information technology and MEMS systems.

6. CONCLUSION

Microscale heat engineering has been successfully researched for realizing advanced thermal engineering by utilizing microscale and nanoscale phenomena. The role of heat and mass transfer in nanotechnology has been explained from the viewpoint of manufacturing and some trends and characteristics have been analyzed. From the viewpoint of nanotechnology, several advanced heat transfer problems have been stressed and actual examples of microscale thermal engineering for promoting energy conservation have been explained. The role of thermal engineering and the transport phenomena in the microscale effects has been explained and the importance of the active control of heat and mass transfer and the transport phenomena in nano-manufacturing was focused for controlling the microscale phenomena for realizing the engineering applications.

REFERENCES

1. A. Yabe ,"Nanotechnology and Thermal Engineering":,The International Symposium on Micro-Mechanical Engineering, ISMME2003-100, Dec.2003

2. A.Yabe and et al.,"Road Map of Micro-Engineering and Nano-Engineering from Manufacturing and Mechanical Engineering Viewpoints", JSME International Journal, Series B,Vol.47, No.3,pp.534-540, July 2004

3. The following homepages would be useful for the references. http://www.nano.gov/,http://itri.loyola.edu/nano/ (U.S.A)
http://www.nedo.go.jp/informations/other/130626/pdf/gaiyou.pdf (NEDO)

MICROFLUIDICS IN LAB-ON-A-CHIP: MODELS, SIMULATIONS AND EXPERIMENTS

DONGQING LI
Department of Mechanical & Industrial Eng.
University of Toronto
Toronto, Ontario, M5S3G8, Canada
Email: dli@mie.utoronro.ca

1. Introduction

Lab-on-a-chip devices are miniaturized bio-medical or chemistry laboratories on a small glass or plastic chip. Generally, a lab-on-a-chip has a network of microchannels, electrodes, sensors and electrical circuits. Electrodes are placed at strategic locations on the chip. Applying electrical fields along microchannels controls the liquid flow and other operations in the chip. These labs on a chip can duplicate the specialized functions as their room-sized counterparts, such as clinical diagnoses, DNA scanning and electrophoretic separation. The advantages of these labs on a chip include dramatically reduced sample size, much shorter reaction and analysis time, high throughput, automation and portability.

The key microfluidic functions required in various lab-on-a-chip devices include pumping, mixing, thermal cycling, dispensing and separating. Most of these processes are electrokinetic processes. Basic understanding, modeling and controlling of these key microfluidic functions/processes are essential to systematic design and operation control of the lab-on-a-chip systems. Because all solid-liquid (aqueous solutions) interfaces carry electrostatic charge, there is an electrical double layer field in the region close to the solid-liquid interface on the liquid side. Such an electrical double layer field is responsible for at least two basic electrokinetic phenomena: electroosmosis and electrophoresis. Essentially all on-chip microfluidic processes are realized by using these two phenomena. This paper will review basics of the electrical double layer field, and three key on-chip microfluidic processes: electroosmotic flow, sample mixing and sample dispensing. A more comprehensive review of the electrokinetic based microfluidic processes for lab-on-a-chip applications can be found elsewhere [1].

2. Electrical double layer field

It is well-known that most solid surfaces obtain a surface electric charge when they are brought into contact with a polar medium (e.g., aqueous solutions). This may be due to ionization, ion adsorption or ion dissolution. If the liquid contains a certain amount of ions (for instance, an electrolyte solution or a liquid with impurities), the electrostatic charges on the solid surface will attract the counterions in the liquid. The rearrangement of the charges on the solid surface and the balancing charges in the liquid is called the electrical double layer (EDL) [2,3]. Immediately next to the solid surface, there is a layer of ions which are strongly attracted to the solid surface and are immobile. This layer is called the compact layer, normally about several Angstroms thick. Because of the electrostatic attraction, the counterions concentration near the solid surface is higher than that in the bulk liquid far away from the solid surface. The coions' concentration near the surface, however, is lower than that in the bulk liquid far away from the

S. Kakaç et al. (eds.), Microscale Heat Transfer, 157–174.

solid surface, due to the electrical repulsion. So there is a net charge in the region close to the surface. From the compact layer to the uniform bulk liquid, the net charge density gradually reduces to zero. Ions in this region are affected less by the electrostatic interaction and are mobile. This region is called the diffuse layer of the EDL. The thickness of the diffuse layer is dependent on the bulk ionic concentration and electrical properties of the liquid, usually ranging from several nanometers for high ionic concentration solutions up to several microns for pure water and pure organic liquids. The boundary between the compact layer and the diffuse layer is usually referred to as the shear plane. The electrical potential at the solid-liquid surface is difficult to measure directly. The electrical potential at the shear plane is called the zeta potential, ς, and can be measured experimentally [2,3]. In practice, the zeta potential is used as an approximation to the potential at the solid-liquid interface.

The ion and electrical potential distributions in the electrical double layer can be determined by solving the Poisson-Boltzmann equation [2,3]. According to the theory of electrostatics, the relationship between the electrical potential ψ and the local net charge density per unit volume ρ_e at any point in the solution is described by the Poisson equation:

$$\nabla^2\psi = -\frac{\rho_e}{\varepsilon} \tag{1}$$

where ε is the dielectric constant of the solution. Assuming the equilibrium Boltzmann distribution equation is applicable, which implies uniform dielectric constant, the number concentration of the type-i ion is of the form

$$n_i = n_{io}\exp\left(-\frac{z_i e\psi}{k_b T}\right) \tag{2}$$

where n_{io} and z_i are the bulk ionic concentration and the valence of type-i ions, respectively, e is the charge of a proton, κ_b is the Boltzmann constant, and T is the absolute temperature. For a symmetric electrolyte ($z_- = z_+ = z$) solution, the net volume charge density ρ_e is proportional to the concentration difference between symmetric cations and anions, via.

$$\rho_e = ze(n_+ - n_-) = -2zen_o\sinh\left(\frac{ze\psi}{k_b T}\right) \tag{3}$$

Substituting Eq.(3) into the Poisson equation leads to the well-known Poisson-Boltzmann equation.

$$\nabla^2\psi = \frac{2zen_o}{\varepsilon}\sinh\left(\frac{ze\psi}{k_b T}\right) \tag{4}$$

Solving the Poisson-Boltzmann equation with proper boundary conditions will determine the local electrical double layer potential field ψ and hence, via Eq.(3), the local net charge density distribution.

3. Electroosmotic flow in microchannels

Consider a microchannel filled with an aqueous solution. There is an electrical doubly layer field near the interface of the channel wall and the liquid. If an electric field is applied along the length of the channel, an electrical body force is exerted on the ions in the diffuse layer. In the diffuse layer of the double layer field, the net charge density, ρ_e is not zero. The net transport of ions is the excess counterions. If the solid surface is negatively charged, the counterions are the positive ions. These excess counterions will move under the influence of the applied electrical field, pulling the liquid with them and resulting in electroosmotic flow. The liquid movement is carried through to the rest of the liquid in the channel by viscous forces. This electrokinetic process is called electroosmosis and was first introduced by Reuss in 1809 [4].

Consider electroosmotic flow in a rectangular microchannel of width $2W$, height $2H$ and length L, as illustrated in Figure 1 [5]. Because of the symmetry in the potential and velocity fields, the solution domain can be reduced to a quarter section of the channel (as shown by the shaded area in Figure 1).

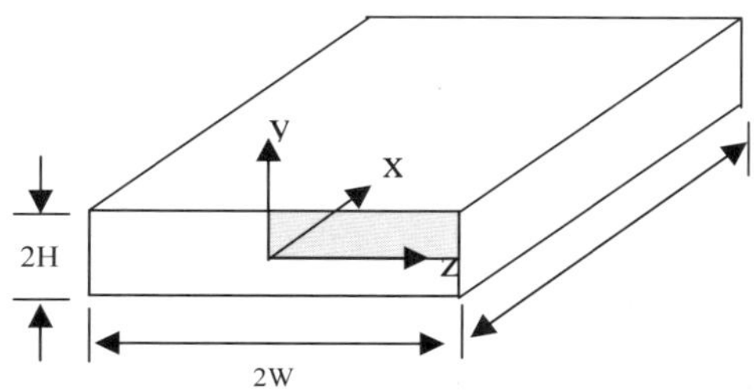

Figure 1. Geometry of microchannel. The shaded region indicates the computational domain.

The 2D EDL field can be described by the non-dimensional form of the Poisson–Boltzmann equation is given by:

$$\frac{\partial^2 \psi^*}{\partial y^{*2}} + \frac{\partial^2 \psi^*}{\partial z^{*2}} = (\kappa D_h)^2 \sinh(\psi^*) \tag{5}$$

where the dimensionless parameters are defined as: $y^* = \frac{y}{D_h}$, $z^* = \frac{z}{D_h}$, $\psi^* = \frac{ze\psi}{k_b T}$, $D_h = \left(\frac{4HW}{H+W}\right)$, κ *is* the Debye–Huckle parameter, $\kappa = \left(\frac{2n_\infty z^2 e^2}{\varepsilon\varepsilon_0 k_b T}\right)^{1/2}$. $1/\kappa$ is the characteristic thickness of the EDL. The non-dimensional parameter κD_h is a measure of the relative channel diameter, compared to the EDL thickness. κD_h is often referred to as the electrokinetic diameter. The corresponding non-dimensional boundary conditions follow:

$$\text{at} \quad y^* = 0 \quad \frac{\partial \psi^*}{\partial y^*} = 0, \quad \text{at} \quad z^* = 0 \quad \frac{\partial \psi^*}{\partial z^*} = 0$$

$$\text{at} \quad y^* = \frac{H}{D_h} \quad \psi^* = \zeta^* = \frac{ze\zeta}{k_b T} \quad \text{at} \quad z^* = \frac{W}{D_h} \quad \psi^* = \zeta^* = \frac{ze\zeta}{k_b T}$$

If we consider that the flow is steady, two-dimensional, and fully developed, and there is no pressure gradient in the microchannel, the general equation of motion is given by a balance between the viscous or shear stresses in the fluid and the externally imposed electrical field force:

$$\mu\left(\frac{\partial^2 u}{\partial y^2}+\frac{\partial^2 u}{\partial z^2}\right)=F_x=\rho_e E_x$$

where F_x is the electrical force per unit volume of the liquid, ρ_e is the local net charge density. The non-dimensionalized equation of motion can be written as:

$$\frac{\partial^2 u^*}{\partial y^{*2}}+\frac{\partial^2 u^*}{\partial z^{*2}}=ME_x^*\sinh(\psi^*) \quad (6)$$

where $u^*=\frac{u}{U}$, $E_x^*=\frac{E_x L}{\zeta}$, U is a reference velocity, L is the distance between the two electrodes, and M is a dimensionless group, a ratio of the electrical force to the frictional force per unit volume, given by: $M=\frac{2n_\infty ze\zeta D_h^2}{\mu UL}$. The corresponding non-dimensional boundary conditions are given by:

at $y^*=0$ $\frac{\partial u^*}{\partial y^*}=0$ at $z^*=0$ $\frac{\partial u^*}{\partial z^*}=0$ at $y^*=\frac{H}{D_h}$ $u^*=0$ at $z^*=\frac{W}{D_h}$ $u^*=0$

Numerically solving Eq. (5) and Eq.(6) with the boundary conditions will allow us to determine the EDL field and the electroosmotic flow field in such a rectangular microchannel. As an example, Let's consider a KCl aqueous solution. At a concentration of $1\text{x}10^{-6}$ M, $\varepsilon = 80$ and $\mu = 0.90\times10^{-3}$ kg/(m·s). An arbitrary reference velocity of U = 1 mm/s was used to non-dimensionalize the velocity. According to experimental results [6], zeta potential values changes from 100 to 200mV, corresponding to three concentrations of the KCl solution, 1×10^{-6}, 1×10^{-5} and 1×10^{-4} M. The hydraulic diameter of the channel varied from 12 to 250 μm, while the aspect ratio varied from 1:4 to 1:1. Finally, the applied voltage difference ranged from 10 V to 10 kV.

The EDL potential distribution in the diffuse double layer region is shown in Figure 2. The nondimensional EDL potential profile across a quarter section of the rectangular channel exhibits characteristic behaviour. The potential field drops off sharply very close to the wall. The region where the net charge density is not zero is limited to a small region close to the channel surface. Figure 3 shows the non-dimensional electroosmotic velocity field for an applied potential difference of 1 kV/cm. The velocity field exhibits a maximum near the wall, and then gradually drops off to a slightly lower constant velocity that is maintained through most of the channel. This unique profile can be attributed to the fact that the externally imposed electrical field is driving the flow. In the region very close to the wall, the mobile part of the EDL region, the larger electrical field force exerts a greater driving force on the fluid because of the presence of the net charge in the EDL region.

Variation of D_h affects the following nondimensional parameters: the electrokinetic diameter, and the strength of the viscous forces in the ratio of electrical to viscous forces. The volumetric flow rate increased with approximately D_h^2 as seen in Figure 4. This is expected, since the cross-sectional area of the channel also increases proportionate to D_h^2. When larger pumping flow rates are desired, larger diameter channels would seem to be a better choice.

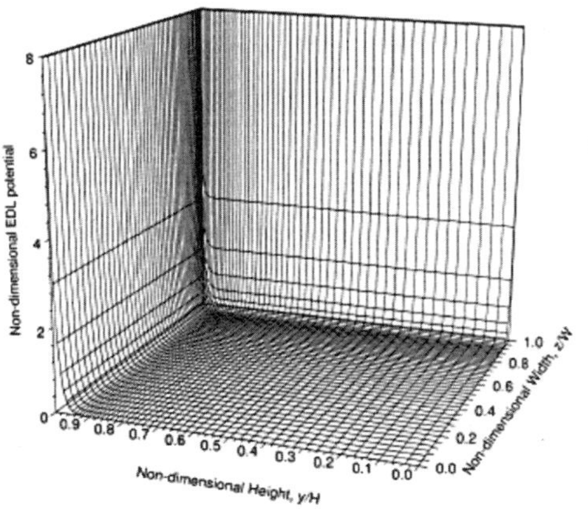

Figure 2. Non-dimensional electric double layer potential profile in a quarter section of a rectangular microchannel with $\kappa D_h = 79$, $\zeta^* = 8$ and $H{:}W = 2{:}3$.

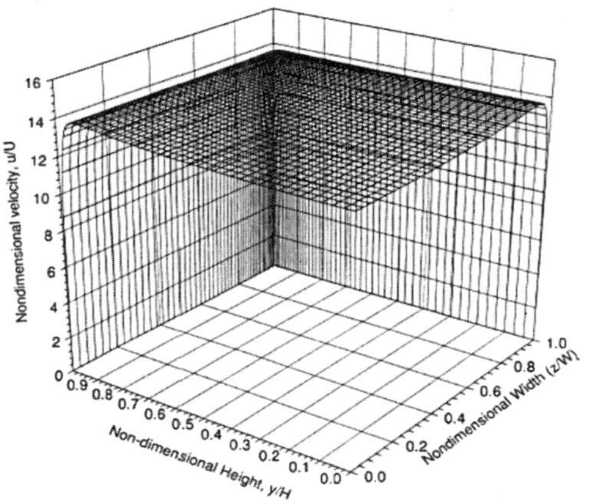

Figure 3. Non-dimensional velocity field in a quarter section of a rectangular microchannel with $\kappa D_h = 79$, $\zeta^* = 8$, $H/W = 2/3$, $E_x^* = 5000$ and $M = 2.22$.

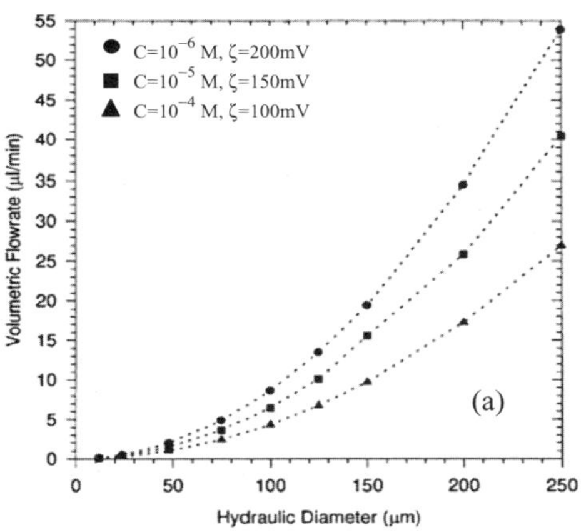

Figure 4. Variation of volumetric flow rate with hydraulic diameter for three different combinations of concentration and zeta potential, with $H/W = 2/3$, and $E_x = 1$ kV/cm.

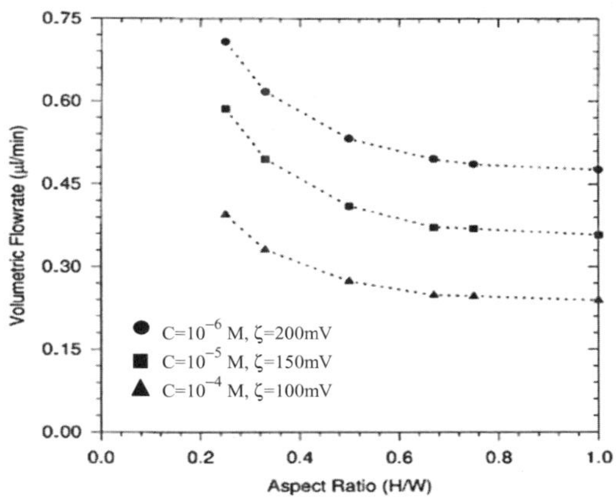

Figure 5. Variation of volumetric flow rate with aspect ratio for three different combinations of concentration and zeta potential, with $D_h = 24$ μm and $E_x = 1$ kV/cm. In this case, $z/W = 1.0$ represents the channel wall, and $z/W = 0$ represents the center of the channel.

However, there is no corresponding increase in the average velocity with increased hydraulic diameter. This is because the nature of electroosmotic flow—the flow is generated by the motion of the net charge in the electrical double layer region driven by an applied electrical field. When the double layer thickness (*1/κ*) is small, an analytical solution of the electroosmotic velocity can be derived from a one-dimensional channel system such as a cylindrical capillary with a circular cross section, given by

$$v_{av} = -\frac{E_z \varepsilon_r \varepsilon_o \zeta}{\mu} \tag{7}$$

Eq.(7) indicates that the electroosmotic flow velocity is linearly proportional to the applied electrical field strength and linearly proportional to the zeta potential. The negative sign indicates the flow direction and has to do with the sign of the ζ potential. If ζ potential is negative (i.e., a negatively charged wall surface), the excess counterions in the diffuse layer are positive, therefore the electroosmotic flow in the microchannel is towards the negative electrode.

With a rectangular microchannel not only the hydraulic diameter but also the channel shape will influence the velocity profile. This is because of the impact of the channel geometry on the EDL. Figure 5 shows the relationship between the aspect ratio (*H*/*W*) and the volumetric flow rate for a fixed hydraulic diameter. As the ratio of *H*:*W* approaches 1:1 (for a square channel), the flow rate decreases. This is because of the larger role that corner effects have on the development of the EDL and the velocity profile in square channels.

Increasing the bulk ion concentration in the liquid results in an increase in κ or a decrease in the EDL thickness *1/κ*. Correspondingly, the EDL potential field falls off to zero more rapidly with distance, i.e., the region influenced by the EDL is smaller. The ionic concentration effect on the velocity or the flow rate can be understood as follows. Since ionic concentration influences the zeta potential, as the ionic concentration is increased, the zeta potential decreases in value. As the zeta potential decreases, so does the electroosmotic flow velocity (Eq.(7)) and the volumetric flow rate.

4. Electrokinetic mixing

Let's consider a simple T-shaped microfluidic mixing system [7]. Without loss of the generality, we will consider that two electrolyte solutions of the same flow rate enter a T-junction separately from two horizontal microchannels, and then start mixing while flowing along the vertical microchannel, as illustrated in Figure 6. The flow is generated by the applied electrical field via electrodes at the upstream and the downstream positions. This simple arrangement has been used for numerous applications including the dilution of a sample in a buffer [8], the development of complex species gradients [9,10], and measurement of the diffusion coefficient [11]. Generally most microfluidic mixing systems are limited to the low Reynolds number regime and thus species mixing is strongly diffusion dominated. Consequently mixing tends to be slow and occur over relatively long distances and time. Enhanced

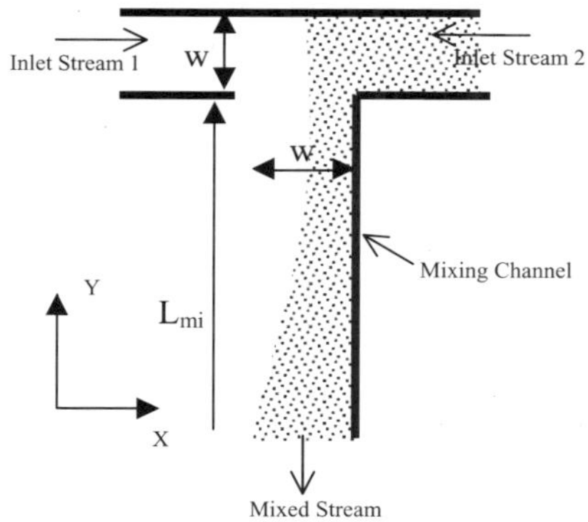

Figure 6. T-Shaped micromixer formed by the intersection of two microchannels, showing a schematic of the mixing or dilution process.

microfluidic mixing over a short flow distance is highly desirable for lab-on-a-chip applications. One possibility of doing so is to utilize the local circulation flow caused by the surface heterogeneous patches.

To model such an electroosmotic flow and mixing process, we need the following equations. The flow field is described by the Navier-Stokes Equations and the continuity Equation (given below in non-dimensional form):

$$\mathrm{Re}\left[\frac{\partial V}{\partial \tau}+\left(V\cdot\tilde{\nabla}\right)V\right]=-\tilde{\nabla}P+\tilde{\nabla}^2 V \quad (8)$$

$$\tilde{\nabla}\cdot V=0 \quad (9)$$

where V is the non-dimensional velocity ($V = v/v_{eo}$, where v_{eo} is calculated using Eq. (10) given below), P is the non-dimensional pressure, τ is the non-dimensional time and Re is the Reynolds number given by $Re = \rho v_{eo} L/\eta$ where L is a length scale taken as the channel width (w from Figure 6) in this case. The ~ symbol over the ∇ operator indicates the gradient with respect to the non-dimensional coordinates ($X = x/w$, $Y = y/w$ and $Z = z/w$). It should be noted that in order to simplify the numerical solution to the problem, we have treated the electroosmotic flow in the thin electrical double layer as a slip flow velocity boundary condition, given by:

$$v_{eo} = \mu_{eo}\nabla\phi = \left(\frac{\varepsilon_w \zeta}{\mu}\right)\nabla\phi \quad (10)$$

where $\mu_{eo} = (\varepsilon_w \zeta / \mu)$ is the electroosmotic mobility, ε_w is the electrical permittivity of the solution, μ is the viscosity, ζ is the zeta potential of the channel wall, and ϕ is the applied electric field strength. In general electroosmotic flows in microchannels has small Reynolds numbers, therefore to simplify Eq. (8) we ignore transient and convective terms.

We consider the mixing of equal portions of two buffer solutions, one of which contains a species of interest at a concentration, c_o. Species transport by electrokinetic means is accomplished by 3 mechanisms: convection, diffusion and electrophoresis, and are described by,

$$Pe\left[\frac{\partial C}{\partial \tau}+\tilde{\nabla}\cdot(C(V+V_{ep}))\right]=\tilde{\nabla}^2 C, \tag{11}$$

where C is the non-dimensional species concentration ($C = c/c_o$, where c_o is original concentration of the interested species in the buffer solution.), Pe is the Péclet number ($Pe = v_{eo}w/D$, where D is the diffusion coefficient), and V_{ep} is the non-dimensional electrophoretic velocity equal to v_{ep}/v_{eo} where v_{ep} is given by:

$$v_{ep} = \mu_{ep}\nabla\phi, \tag{12}$$

and $\mu_{ep} = (\varepsilon_w \zeta_p / \eta)$ is the electrophoretic mobility (ε_w is the electrical permittivity of the solution, μ is the viscosity, ζ_p is the zeta potential of the to-be-mixed charged molecules or particles) [12]. As we are interested in the steady state solution, the transient term in Eq. (11) can be ignored.

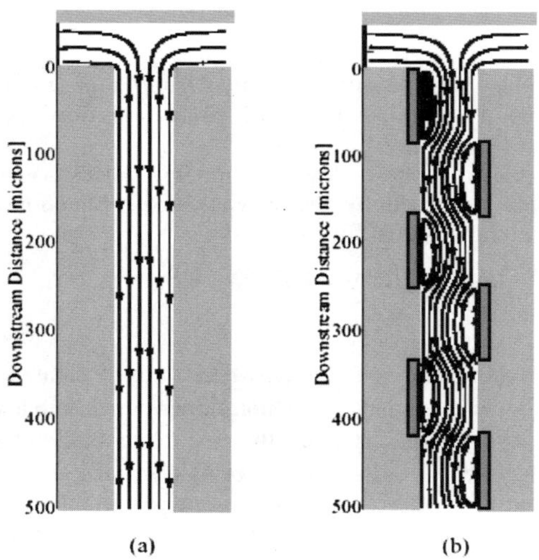

Figure 7. Electroosmotic streamlines at the midplane of a 50μm T-shaped micromixer for the a) homogeneous case with ζ = –42 mV, b) heterogeneous case with six offset patches on the left and right channel walls. All heterogeneous patches are represented by the crosshatched regions and have a ζ = + 42mV. The applied voltage is ϕ_{app} = 500 V/cm.

The above described model was solved numerically to investigate the formation of electroosmotically induced flow circulation regions near surface heterogeneities in a T-shaped micromixer and to determine the influence of these regions on the mixing effectiveness. In Figure 7 we compare the mid-plane flow field near the T-intersection of a homogeneous mixing channel with that of a mixing channel having a series of 6 asymmetrically distributed heterogeneous patches on the left and right channel walls. For clarity the heterogeneous regions are marked as the crosshatched regions in this figure. The homogeneous channel surface has a ζ potential of – 42 mV. A ζ-potential of $\zeta = +42$mV was assumed for the heterogeneous patches. Apparently, the channel with heterogeneous patches generates local flow circulations near the patches. These flow circulation zones are expected to enhance the mixing of the two streams.

Figure 8 compares the 3D concentration fields of the homogeneous and heterogeneous mixing channel shown in Figure 7. In these figures a neutral mixing species (i.e. $\mu_{ep} = 0$, thereby ignoring any electrophoretic transport) with a diffusion coefficient $D = 3\text{x}10^{-10}$ m^2/s is considered. While mixing in the homogeneous case is purely diffusive in nature, the presence of the asymmetric circulation regions, Figure 8b, enables enhanced mixing by convection.

Recently a passive electrokinetic micro-mixer based on the use of surface charge heterogeneity was developed [13]. The micro-mixer is a T-shaped microchannel structure (200µm in width and approximately 8 µm in depth) made from Polydimethylsiloxane (PDMS) and is sealed with a glass slide. Microchannels were fabricated using a rapid prototyping/soft-lithography technique. The glass surface was covered by a PDMS mask with the desired heterogeneous pattern, then treated a Polybrene solution. After removing the mask, the glass surface will have selective regions of positive surface charge while leaving the majority of the glass slide with its native negative charge [13]. Finally the PDMS plate (with the microchannel structure) will be bonded the glass slide to form the seal T-shaped microchannel with heterogeneous patches in the mixing channel surface.

A micro-mixer consisting of 6 offset staggered patches (in the mixing channel) spanning 1.8 mm downstream and offset 10 µm from the channel centerline with a width of 90 µm and a length of 300 µm, was analyzed experimentally. Mixing experiments were conducted at applied

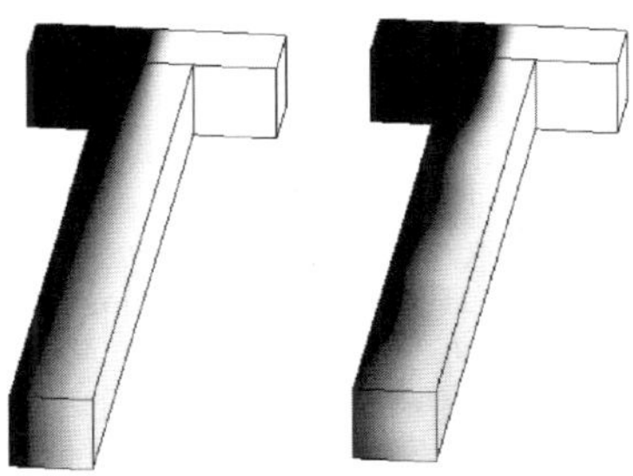

Figure 8. 3D species concentration field for a 50µm x 50µm T-shaped micromixer resulting from the flow fields shown in Figure 7. (a) homogeneous case, and (b) heterogeneous case with offset patches. Species diffusivity is $3\text{x}10^{-10}$ m^2/s and zero electrophoretic mobility are assumed.

voltage potentials ranging between 70 V/cm and 555 V/cm and the corresponding Reynolds numbers range from 0.08 to 0.7 and Péclet numbers from 190 to 1500. The liquid is a 25mM sodium carbonate/bicarbonate buffer. To visualize the mixing effects, 100 μM fluorescein was introduced through one inlet channel. As an example, Figure 9 shows the experimental images of the steady state flow for the homogenous and heterogeneous cases at 280 V/cm. The enhanced mixing effect is obvious.

This study shows that the passive electrokinetic micro-mixer with an optimized arrangement of surface charge heterogeneities can increase flow narrowing and circulation, thereby increasing the diffusive flux and introducing an advective component of the mixing. Mixing efficiencies were improved by 22-68% for voltages ranging from 70 to 555 V/cm.

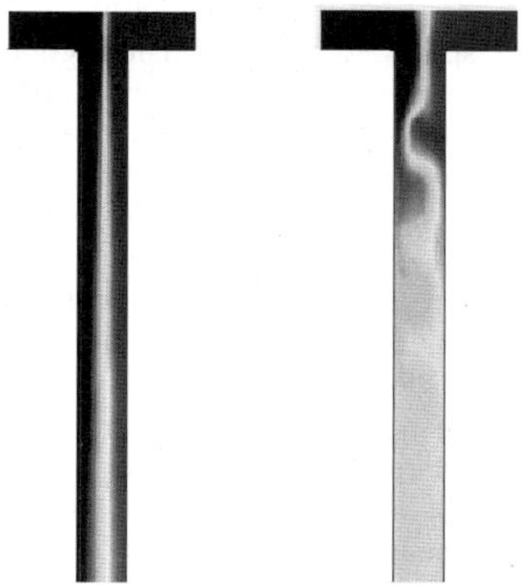

(a) (b)

Figure 9 Images of steady state species concentration fields under an applied potential of 280 V/cm for (a) the homogeneous microchannel and (b) the heterogeneous microchannel with 6 offset staggered patches.

For producing a 95% mixture, this technology can reduce the required mixing channel length of up to 88% for flows with Péclet numbers between 190 and 1500 and Reynolds numbers between 0.08 and 0.7. In terms of required channel lengths, at 280V/cm, a homogeneous microchannel would require a channel mixing length of 22mm for reaching a 95% mixture. By implementing the developed micro-mixer, an 88% reduction in required channel length to 2.6 mm was experimentally demonstrated. Practical applications of reductions in required channel lengths include improvements in portability and shorter retention times, both of which are valuable advancements applicable to many microfluidic devices.

5. Electrokinetic sample dispensing

An important component of many bio- or chemical lab-chips is the microfluidic dispenser, which employs electroosmotic flow to dispense minute quantities (e.g., 300 pico-liters) of samples for chemical and biomedical analysis. The precise control of the dispensed sample in microfluidic dispensers is key to the performance of these lab-on-a-chip devices.

Let's consider a microfluidic dispenser formed by two crossing microchannels as shown in Figure 10 [14]. The depth and the width of all the channels are chosen to be 20 μm and 50 μm, respectively. There are four reservoirs connected to the four ends of the microchannels. Electrodes are inserted into these reservoirs to set up the electrical field across the channels. Initially, a sample solution (a buffer solution with sample species) is filled in Reservoir 1, the other reservoirs and the microchannels are filled with the pure buffer solution. When the chosen electrical potentials are applied to the four reservoirs, the sample solution in Reservoir 1 will be driven to flow toward Reservoir 3 passing through the intersection of the cross channels. This is the so-called loading process. After the loading process reaches a steady state, the sample solution loaded in the intersection will be "cut" or dispensed into the dispensing channel by the dispensing solution flowing from Reservoir 2 to Reservoir 4. This can be realized by adjusting

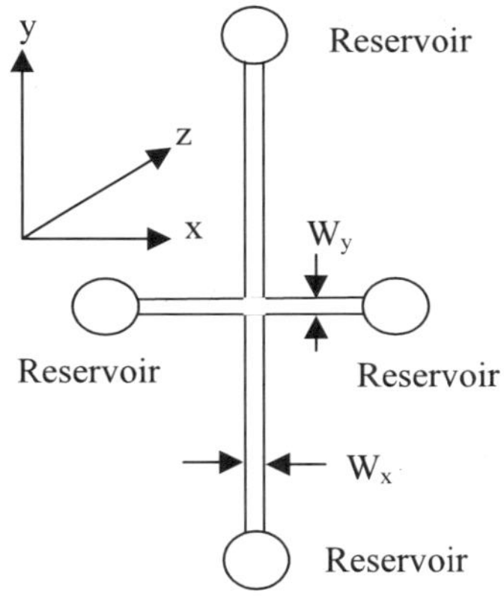

Figure 10. The schematic diagram of a crossing microchannel dispenser. W_x and W_y indicate the width of the microchannels.

the electrical potentials applied to these four reservoirs. This is the so called the dispensing process. The volume and the concentration of the dispensed sample are the key parameters of this dispensing process, and they depend on the applied electrical field, the flow field and the concentration field during the loading and the dispensing processes.

To model such a dispensing process, we must model the applied electrical field, the flow field and the concentration field. To simplify the analysis, we consider this is a 2D problem, i.e., ignoring the variation in the z-direction. The 2D applied electrical potential in the liquid can be described by

$$\frac{\partial^2 \phi^*}{\partial x^{*2}} + \frac{\partial^2 \phi^*}{\partial y^{*2}} = 0 \tag{13}$$

Here the nondimensional parameters are defined by $\phi^* = \frac{\phi}{\Phi}$, $x^* = \frac{x}{h}$, $y^* = \frac{y}{h}$, where Φ is a reference electrical potential and h is the channel width chosen as 50 μm. Boundary conditions are required to solve this equation. We impose the insulation condition to all the walls of microchannels, and the specific non-dimensional potential values to all the reservoirs. Once the electrical field in the dispenser is known, the local electric field strength can be calculated by

$$\vec{E} = -\vec{\nabla}\Phi \tag{14}$$

Because the electroosmotic flow field reaches steady state in milli-seconds, much shorter than the characteristic time scales of the sample loading and sample dispensing. Therefore, the electroosmotic flow here is approximated as steady state. Furthermore, we consider thin electrical double layer, and use the slip flow boundary condition to represent the electroosmotic flow. The liquid flow field can thus be described by the following non-dimensional momentum equation and the continuity equation.

$$u_{eo}^* \cdot \frac{\partial u_{eo}^*}{\partial x^*} + v_{eo}^* \cdot \frac{\partial u_{eo}^*}{\partial y^*} = -\frac{\partial P^*}{\partial x^*} + \frac{\partial^2 u_{eo}^*}{\partial x^{*2}} + \frac{\partial^2 u_{eo}^*}{\partial y^{*2}} \tag{15a}$$

$$u_{eo}^* \cdot \frac{\partial v_{eo}^*}{\partial x^*} + v_{eo}^* \cdot \frac{\partial v_{eo}^*}{\partial y^*} = -\frac{\partial P^*}{\partial y^*} + \frac{\partial^2 v_{eo}^*}{\partial x^{*2}} + \frac{\partial^2 v_{eo}^*}{\partial y^{*2}} \tag{15b}$$

$$\frac{\partial u_{eo}^*}{\partial x^*} + \frac{\partial v_{eo}^*}{\partial y^*} = 0 \tag{16}$$

where u_{eo}, v_{eo} are the electroosmotic velocity component in x and y direction, respectively, and non-dimensionalized as follows:

$$P^* = \frac{P - Pa}{\rho(v/h)^2}, \quad u_{eo}^* = \frac{u_{eo} \cdot h}{v}, \quad v_{eo}^* = \frac{v_{eo} \cdot h}{v}$$

The slip velocity conditions are applied to the walls of the microchannels, the fully developed velocity profile is applied to all the interfaces between the microchannels and the reservoirs, and the pressures in the four reservoirs are considered as the atmospheric pressure.

The distribution of the sample concentration can be described by the conservation law of mass, taking the following form,

$$\frac{\partial C_i^*}{\partial \tau} + \left(u_{eo}^* + u_{ep}^*\right)\frac{\partial C_i^*}{\partial x^*} + \left(v_{eo}^* + v_{ep}^*\right)\frac{\partial C_i^*}{\partial y^*} = \frac{D_i}{v}\left(\frac{\partial^2 C_i^*}{\partial x^{*2}} + \frac{\partial^2 C_i^*}{\partial y^{*2}}\right) \tag{17}$$

where C_i is the concentration of the i-th species, u_{eo} and v_{eo} are the components of the electroosmotic velocity of the i-th species, D_i is the diffusion coefficient of the i-th species, and u_{epi} and v_{epi} are the components of the electrophoretic velocity of the i-th species given by $u_{epi} = E\mu_{epi}$, where μ_{epi} is the electrophoretic mobility. The non-dimensional parameters in the above equation are defined by $C^* = C/\overline{C}$, and $\tau = t/(h^2/v)$, where $\overline{C}$ is a reference concentration.

Figure 11 shows the typical electrical field and flow field (computer simulated) for loading and dispensing process, respectively. In this figure, the non-dimensional applied electrical potentials are: For loading process: $\phi^*(1) = 1.0$, $\phi^*(2) = 1.0$, $\phi^*(3) = 0.0$, $\phi^*(4) = 1.0$ For dispensing process: $\phi^*(1) = 0.2$, $\phi^*(2) = 2.0$, $\phi^*(3) = 0.2$, $\phi^*(4) = 0.0$, where $\phi^*(i)$ represents the non-dimensional applied electrical potential to i-th reservoir. For this specific case, the electrical field and the flow field for loading process are symmetric to the middle line of the horizontal channel, and the electrical field and the flow field for the dispensing process are symmetric to the middle line of the vertical channel.

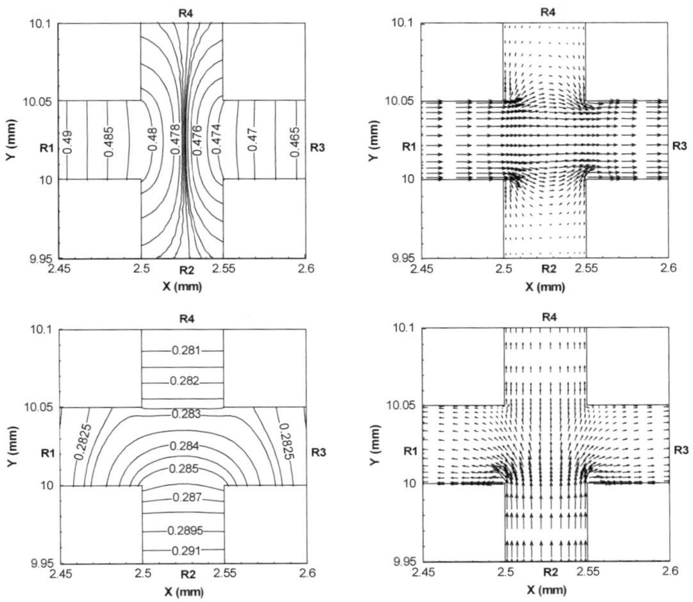

Figure 11. Examples of the applied electrical field (left) and the flow field (right) at the intersection of the microchannels in a loading process (top) and in a dispensing process (bottom).

The electrokinetic dispensing processes of fluorescent dye samples were investigated experimentally [15-17]. The measurements were conducted by using a fluorescent dye based microfluidic visualization system. Figure 12 shows a sample dispensing process and the comparison of the dispensed sample concentration profile with the numerically simulated results. Both the theoretical studies and the experimental studies have demonstrated that the loading and dispensing of sub-nanolitre samples using a microfluidic crossing microchannel chip can be controlled electrokinetically [14-17]. The ability to inject and transport large axial extent, concentration-dense samples was demonstrated. Both experimental and numerical results indicate the shape, cross-stream uniformity, and axial extent of the samples were very sensitive to changes in the electric fields applied in the loading channel. In the dispensing process, larger samples were shown to disperse less than focused samples, maintaining more solution with the original sample concentration.

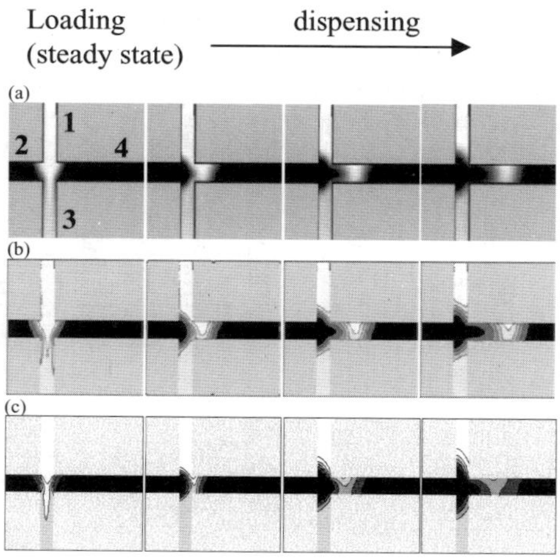

Figure 12. The loading and dispensing of a focused fluorescein sample: a.) Processed images; (b) Iso-concentration profiles at $0.1C_o$, $0.3C_o$, $0.5C_o$, $0.7C_o$, and $0.9C_o$, calculated from the images; and (c) Corresponding Iso-concentration profiles calculated through numerical simulation.

6. Experimental techniques for studying electroosmotic flow

In most electroosmotic flows in microchannels, the flow rates are very small (e.g., 0.1 μL/min.) and the size of the microchannels is very small (e.g., 10~100μm), it is extremely difficult to measure directly the flow rate or velocity of the electroosmotic flow in microchannels. To study liquid flow in microchannels, various microflow visualization methods have evolved. Micro particle image velocimetry (microPIV) is a method that was adapted from well-developed PIV techniques for flows in macro-sized systems [18-22]. In the microPIV technique, the fluid motion is inferred from the motion of sub-micron tracer particles. To eliminate the effect of Brownian motion, temporal or spatial averaging must be employed. Particle affinities for other particles, channel walls, and free surfaces must also be considered. In electrokinetic flows, the electrophoretic motion of the tracer particles (relative to the bulk flow) is an additional consideration that must be taken. These are the disadvantages of the microPIV technique.

Dye-based microflow visualization methods have also evolved from their macro-sized counterparts. However, traditional mechanical dye injection techniques are difficult to apply to the microchannel flow systems. Specialized caged fluorescent dyes have been employed to facilitate the dye injection using selective light exposure (i.e., the photo-injection of the dye). The photo-injection is accomplished by exposing an initially non-fluorescent solution seeded with caged fluorescent dye to a beam or a sheet of ultraviolet light. As a result of the ultraviolet exposure, caging groups are broken and fluorescent dye is released. Since the caged fluorescent dye method was first employed for flow tagging velocimetry in macro-sized flows in 1995 [23], this technique has since been used to study a variety of liquid flow phenomena in microstructures [24-32]. The disadvantages of this technique are that it requires expensive specialized caged dye, and extensive infrastructure to facilitate the photo injection.

Recently, Sinton and Li [30] developed a microchannel flow visualization system and complimentary analysis technique by using caged fluorescent dyes. Both pressure-driven and electrokinetically driven velocity profiles determined by this technique compare well with analytical results and those of previous experimental studies. Particularly, this method achieved a high degree of near-wall resolution. Generally, in the experiment, a caged fluorescent dye is dissolved in an aqueous solution in a capillary or microchannel. It should be noted that the caged dye cannot emit fluorescent light at this stage. Ultraviolet laser light is focused into a sheet crossing the capillary (perpendicular to the flow direction). The caged fluorescent dye molecules exposed to the UV light are uncaged and thus are able to shine. The resulting fluorescent dye is continuously excited by an argon laser and the emission light is transmitted through a laser-powered epi-illumination microscope. Full frame images of the dye transport are recorded by a progressive scan CCD camera and saved automatically on the computer. In the numerical analysis, the images are processed and cross-stream velocity profiles are calculated based on tracking the dye concentration maxima through a sequence of several consecutive images. Several sequential images are used to improve the signal to noise ratio. Points of concentration maxima make convenient velocimetry markers as they are resistant to diffusion. In many ways, the presence of clearly definable, zero-concentration-gradient markers is a luxury afforded by the photo-injection process. The details of this technique can be found elsewhere [30,31].

In an experimental study [30], the CMNB-caged fluorescein with the sodium carbonate buffer and 102μm i.d. glass capillaries were used. Images of the uncaged dye transport in four different electroosmotic flows are displayed in vertical sequence in Figure 13. The dye diffused symmetrically as shown in Figure 13(a). Image sequences given in Figures. 13(b), (c), and (d) were taken with voltages of 1000V, 1500V, and 2000V respectively (over the 14cm length of capillary). The field was applied with the positive electrode at left and the negative electrode at right. The resulting plug-like motion of the dye is characteristic of electroosmotic flow in the presence of a negatively charged surface at high ionic concentration. The cup-shape of the dye profile was observed in cases 13(b), (c), and (d) within the first 50msec following the ultraviolet light exposure. This period corresponded to the uncaging time scale in which the most significant rise in uncaged dye concentration occurs. Although the exact reason for the formation of this shape is unknown, it is likely that it was an artifact of the uncaging process in the presence of the electric field. Fortunately, however, the method is relatively insensitive to the shape of the dye concentration profile. Once formed, it is the transport of the maximum concentration profile that provides the velocity data. This also makes the method relatively insensitive to beam geometry and power intensity distribution.

Figure 14 shows velocity data for the four flows corresponding to the image sequences in Figure 13. Each velocity profile was calculated using an 8-image sequence and the numerical analysis technique described in reference [30]. The velocity profile resulting from no applied field, Figure 13(a), corresponds closely to stagnation as expected. This run also serves to illustrate that, despite significant transport of dye due to diffusion, the analysis method is able to recover the underlying stagnant flow velocity. Although the other velocity profiles resemble that of classical electroosmotic flow [22], a slight parabolic velocity deficit of approximately 4% was detected in all three flows. This was caused by a small back-pressure induced by the electroosmotic fluid motion (e.g., caused be the not-perfectly leveled capillary along the length direction).

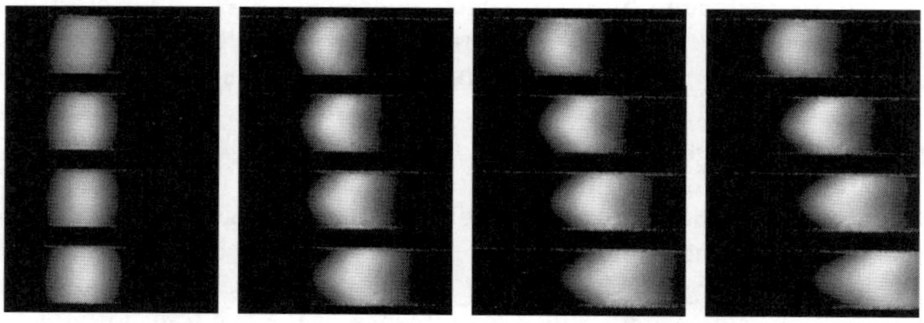

(a) (b) (c) (d)

Figure 13. Images of the uncaged dye in electroosmotic flows through a 102μm i.d. capillary at 133 msec intervals with applied electric field strength: (a) 0V/0.14m; (b) 1000V/0.14m; (c) 1500/0.14m; and (d) 2000V/0.14m.

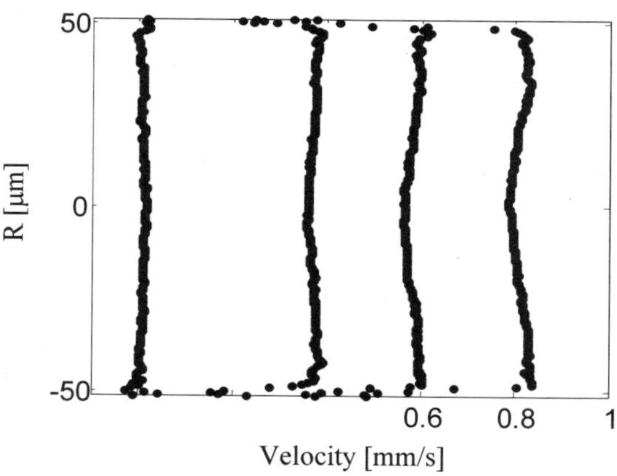

Figure 14. Plots of velocity data from four electroosmotic flow experiments through a 102μm i.d. capillary with applied electric field strengths of: 0V/0.14m; 1000V/0.14m; 1500/0.14m; and 2000V/0.14m (from left to right).

In additional to these PIV and dye-based techniques, the electroosmotic flow velocity can be estimated indirectly by monitoring the electrical current change while one solution is replaced by another similar solution during electroosmosis [29,34,35]. In this method, a capillary tube is filled with an electrolyte solution, then brought into contact with another solution of the same electrolyte but with a slightly different ionic concentration. Once the two solutions are in contact, an electrical field is applied along the capillary in such a way that the second solution is pumped into the capillary and the 1st solution flows out of the capillary from the other end. As more and more of the second solution is pumped into the capillary and the first solution flows out of the capillary, the overall liquid conductivity in the capillary is changed, and hence the electrical current through the capillary is changed. When the second solution completely replaces the first solution, the current will reach a constant value. Knowing the time required for this current change and the length of the capillary tube, the average electroosmotic flow velocity can be calculated by

$$u_{av,\exp} = \frac{L}{\Delta t} \tag{18}$$

where L is the length of the capillary and Δt is the time required for the higher (or lower)-concentration electrolyte solution to completely displace the lower (or higher)-concentration electrolyte solution in the capillary tube.

REFERENCES

1. D. Li, Electrokinetics in Microfluidics, Academic Press, London, 2004.
2. R. J. Hunter, Zeta Potential in Colloid Science: Principle and Applications. Academic Press, London, 1981, P 11--55.
3. J. Lyklema, Fundamentals of Interface and Colloid Science, Volume II, Solid-Liquid Interfaces. Academic Press, London, 1995, P 3.2--3.232.
4. F.F. Reuss, Memoires de la Societe Imperiale des Naturalistes de Moskou. 2 (1809) 327.
5. S. Arulanandam, D. Li, Colloids Surface A, 161 (2000) 89.
6. G.M. Mala, D. Li, C. Werner, H.J. Jacobasch, Y.B. Ning, Int. J. Heat Fluid Flow, 18 (1997) 489-496.
7. D. Erickson and D. Li, Langmuir, 18 (2002) 1883-1892.
8. J.D. Harrison, K. Fluri, K. Seiler, Z. Fan, C. Effenhauser, A. Manz, Science, 261 (1993) 895-897.
9. S.K.W. Dertinger, D.T. Chiu, N. L. Jeon, G. M. Whitesides, Anal. Chem. 73 (2001) 1240-1246.
10. N.L Jeon, S.K.W. Dertinger, D.T. Chiu, I.S. Choi, A.D. Stroock, G. M. Whitesides, Langmuir 16 (2000), 8311-8316.
11. A.E. Kamholz, E. A. Schiling, P.Yager, Biophys. J. 80 (2001) 1967-1972.
12. R.J. Hunter, Zeta Potential in Colloid Science; Academic Press: London, 1981.
13. E. Biddiss, D. Erickson, and D. Li, Analytical Chemistry, (in press)
14. L. Ren and D. Li, J. Colloid Interface Sci., 254 (2002) 384-395.
15. D. Sinton, L. Ren and D. Li, J Colloid Interface Sci., 260 (2003) 431 - 439.
16. D. Sinton, L. Ren and D. Li, J. Colloid Interface Sci., 266 (2003) 448-456.
17. D. Sinton, L. Ren, X. Xuan and D. Li, Lab-on-a-Chip, 3 (2003) 173 – 179.

18. J.A. Taylor, E.S. Yeung, Analytical Chemistry 65 (1993) 2928-2932.
19. J.G. Santiago, S.T. Wereley, C.D. Meinhart, D.J. Beebe, R.J. Adrian, Experments in Fluids 25 (1998) 316-319.
20. C.D. Meinhart, S.T. Wereley, J.G. Santiago, Experiments in Fluids 27 (1999) 414-419.
21. A.K. Singh, E.B. Cummings, D.J. Throckmorton, Analytical Chemistry 73 (2001) 1057-1061.
22. S.T. Wereley, C.D. Meinhart, Experiments in Fluids 31 (2001) 258-268.
23. W.R. Lempert, K. Magee, P. Ronney, K.R. Gee, R.P. Haugland, Experiments in Fluids 18 (1995) 249-257.
24. P.H. Paul, M.G. Garguilo, D.J. Rakestraw, Analytical Chemistry 70 (1998) 2459-2467.
25. A.E. Herr, J.I. Molho, J.G. Santiago, M.G. Mungal, T.W. Kenny, M.G. Garguilo, Analytical Chemistry 72 (2000) 1053-1057.
26. W.J.A Dahm, L.K. Su, K.B. Southerland, Physics of Fluids A 4(10) (1992) 2191-2206.
27. T. J. Johnson, D. Ross, M. Gaitan, and L. E. Locascio, Anal. Chem., 73 (2001), 3656.
28. J. I. Molho, A. E. Herr, and B. P. Mosier, J.G. Santiago, T. W. Kenny, R. A. Breenen, G. B. Gordon, B. Mohammadi, Analytical Chemistry, 73 (2001) 1350-1360.
29. D. Sinton, C. Escobedo, L. Ren and D. Li, J. Colloid Interface Sci., 254 (2002) 184-189.
30. D. Sinton and D. Li, International Journal of Thermal Sciences, 42 (2003) 847-855.
31. D. Sinton and D. Li, Colloids & Surfaces A., 222 (2003) 273-283.
32. D. Sinton, D. Erickson and D. Li, Experiments in Fluids, 35 (2003)178-187.
33. S. Arulanandam and D. Li, J. Colloid and Interface Sci. 225 (2000) 421-428.
34. L. Ren, C. Escobedo and D. Li, J. Colloid Interface Sci. 250 (2002) 238-242.

TRANSIENT FLOW AND THERMAL ANALYSIS IN MICROFLUIDICS

R.M. COTTA, S. KAKAÇ[*], M.D. MIKHAILOV
F.V. CASTELLÕES[**], C.R. CARDOSO
Mechanical Engineering Dept – DEM/POLI and PEM/COPPE, UFRJ
Universidade Federal do Rio de Janeiro – Brazil
[*] Department of Mechanical Engineering., University of Miami, Coral Gables, FL, USA
[**] Petrobras Research and Development Center (CENPES), Rio de Janeiro, Brazil

1. Introduction

The present lecture summarizes some of the most recent joint research results from the cooperation between the Federal University of Rio de Janeiro, Brasil, and the University of Miami, USA, on the transient analysis of both fluid flow and heat transfer within microchannels. This collaborative link is a natural extension of a long term cooperation between the two groups, in the context of fundamental work on transient forced convection, aimed at the development of hybrid numerical-analytical techniques and the experimental validation of proposed models and methodologies [1- 9]. The motivation of this new phase of the cooperation was thus to extend the previously developed hybrid tools to handle both transient flow and transient convection problems in microchannels within the slip flow regime.

The analysis of internal flows in the slip-flow regime recently gained an important role in association with the fluid mechanics of various microelectromechanical systems (MEMS) applications, as well as in the thermal control of microelectronics, as reviewed in different sources [10-16]. For steady-state incompressible fully developed flow situations and laminar regime within simple geometries such as circular microtubes and parallel-plate microchannels, explicit expressions for the velocity field in terms of the Knudsen number are readily obtainable, and have been widely employed in the heat transfer analysis of microsystems, such as in [17-23]. Only quite recently, attention has been directed to the analysis of transient flow in microchannels [24-33]. Unsteady one-dimensional models have been extended from classical works, and analytical solutions have been sought for fully developed flows in simple geometries. These recent works are also concerned with situations in which a simple and well-defined functional form for the pressure gradient time variation is prescribed or for the time dependence of the wall imposed velocity, in the case of a Couette flow application. Research findings are yet to be further pursued in the analytical and robust solution of more generalized models, which will accommodate more general conditions and parameter specifications, and thus offer a wider validation range for the automatic general purpose numerical codes.

Mikhailov and Ozisik [34] presented a unified solution for transient one-dimensional laminar flow models, with the usual no-slip boundary condition, based on the classical integral transform method. Their solution was then specialized to two situations: step change and periodically varying pressure gradient. The knowledge in regular size channels is therefore fairly well consolidated for models that use simple functional forms for the pressure gradient variation such as for the two cases cited above. One of the objectives of this paper is to illustrate the solution of a one-dimensional mathematical model for transient laminar incompressible flow in microchannels such as circular tubes and parallel-plate channels, that accounts for a source term time variation in any functional form, including electrokinetic effects for liquid flows, by making use of the Generalized Integral Transform Technique (GITT) [35-40], and thus yielding analytical expressions for the time and space dependence of the velocity fields in the fully developed region. We then demonstrate this hybrid numerical-analytical solution for transient internal slip flow, obtained employing mixed symbolic-numerical computations with the *Mathematica* platform [41]. The goal here is to improve and complement existing analytical solution implementations to study laminar fully developed flows in micro-ducts subjected to arbitrary source term disturbances in space and time.

On the other hand, the heat transfer literature of the last decade has demonstrated a vivid and growing interest in thermal analysis of flows in micro-channels, both through experimental and analytical approaches, in connection with cooling techniques of micro-electronics and with the development of micro-electromechanical sensors and actuators (MEMS), as also pointed out in recent reviews [12-16]. Since the available analytical information on heat transfer in ducts could not be directly extended to flows within microchannels with wall slip, a number of contributions have been recently directed towards the analysis of internal forced convection in the micro-scale. In the paper by Barron et al.

S. Kakaç et al. (eds.), Microscale Heat Transfer, 175–196.

[42], the original approach in the classical work of Graetz [43] is used to evaluate the eigenvalues for the Graetz problem extended to slip-flow. The method used appears to be unstable after the fifth root so that only the first four eigenvalues were then considered reliable. The authors concluded that an improved method with enhanced calculation speed would be of future interest. The problem considered in [17, 42] has also an exact solution in terms of the confluent hypergeometric function, explored in [44-45] to develop *Mathematica* [41] rules for computing the desired eigenvalues with user-specified working precision. Following the work in [42], the same technique was employed to solve the laminar flow heat convection problem in a cylindrical micro-channel with constant uniform temperature at the boundary [23], taking into account both the velocity slip and temperature jump at the tube wall. More recently [18-22], the analytical contributions were directed towards more general problem formulations, including viscous dissipation in the fluid and two-dimensional flow geometries, such as rectangular channels. For this purpose, a more flexible hybrid numerical-analytical approach was employed, based on the ideas of the same Generalized Integral Transform Technique, GITT [35-40], thus avoiding more involved analysis in relation with the eigenvalue problem inherent to the eigenfunction expansions proposed.

All such analysis are restricted to steady-state situations, and very little is apparently available on transient convective heat transfer within microchannels. Nevertheless, the understanding of unsteady phenomena in applications with MEMS devices is becoming more necessary. Then, the ability of predicting unsteady temperature fields is essential in the controlled temperature variation within the system. Only quite recently [46], an approximate analytical solution was presented for transient convection within microchannels, for a step change on wall temperature, based on a previously proposed hybrid approach that combines the Laplace and Integral transforms concepts [47]. In this context, the second goal of this paper is thus to illustrate the results obtained from a notebook also developed in the *Mathematica* platform [41] that yields hybrid numerical-analytical solutions for both the velocity and temperature distributions in a fluid flowing through parallel plate micro-channels, taking into account the velocity and temperature jumps at the surface, for the transient state. We again make use of the GITT [35-40] and the exact analytical solution of the corresponding eigenvalue problem in terms of confluent hypergeometric functions [44-45], to eliminate the transversal coordinate in the original formulation. Then, the resulting transformed partial differential system is numerically solved by the Method of Lines, implemented within the routine NDSolve of the *Mathematica* system [41]. As we wish to demonstrate in what follows this combination of solution methodologies provides a very effective eigenfunction expansion behavior, through the fast converging analytical representation in the transversal coordinate, together with a flexible and fairly reliable numerical approach for the transient and longitudinal behavior of the coupled transformed potentials. The present approach complements in scope previous developments on hybrid methods for solving fully transient forced convection problems [47-50], as recently reviewed in [51]. The present combined algorithm makes use of both the symbolic computation capabilities and novel numerical routines introduced in the latest version of the *Mathematica* system, allowing for an updated hybrid scheme for accurately handling transient convective heat transfer under any ratio of convection and diffusion effects.

2. Formal Solutions

Within the last two decades, the classical integral transform method [34] was progressively generalized under a hybrid numerical-analytical concept [35-40]. This approach now offers user-controlled accuracy and efficient computational performance for a wide variety of non-transformable problems, including the most usual nonlinear formulations in heat and fluid flow applications. Besides being an alternative computational method in itself, this hybrid approach is particularly well suited for benchmarking purposes. In light of its automatic error-control feature, it retains the same characteristics of a purely analytical solution. In addition to the straightforward error control and estimation, an outstanding aspect of this method is the direct extension to multidimensional situations, with only a moderate increase in computational effort. Again, the hybrid nature is responsible for this behavior, since the analytical part in the solution procedure is employed over all but one independent variable, and the numerical task is always reduced to the integration of an ordinary differential system over this single independent variable. More recently, however, in light of the also remarkable developments on the automatic error control of numerical solutions for partial differential equations, in particular for one-dimensional formulations[41, 52], the GITT approach has been employed in combination with well-tested algorithms for parabolic and parabolic-hyperbolic equations [49]. This possibility opened up new perspectives in the merging of numerical and analytical ideas, and in exploiting the power and flexibility of progressively more reliable and robust subroutines for partial differential equations, readily available both commercially and in public domain.

The present section reviews the concepts behind the Generalized Integral Transform Technique (GITT) [35-40] as an example of a hybrid method in convective heat transfer applications. The GITT adds to the available simulation tools, either as a companion in co-validation tasks, or as an alternative approach for analytically oriented users. We first illustrate the application of this method in the full transformation of a typical convection-diffusion problem, until an ordinary differential system is obtained for the transformed potentials. Then, the more recently introduced strategy of partial integral transformation is derived yielding a coupled system of one-dimensional partial differential equations to be numerically integrated. Finally, the different aspects in the computational implementation of each approach are critically discussed.

As an illustration of the formal integral transform procedure, a transient convection-diffusion problem of n coupled potentials (velocity, temperature or concentration) is considered. These parameters are defined in the region V with boundary surface S and including non-linear effects in the convective and source terms as follows:

$$w_k(\boldsymbol{x})\frac{\partial T_k(\boldsymbol{x},t)}{\partial t} + \boldsymbol{u}(\boldsymbol{x},t,T_\ell).\nabla T_k(\boldsymbol{x},t) + L_k T_k(\boldsymbol{x},t) = P_k(\boldsymbol{x},\mathrm{t},\mathrm{T}_\ell),$$

$$\boldsymbol{x}\in \mathrm{V},\quad \mathrm{t}>0,\quad \mathrm{k},\ell = 1,2,...,\mathrm{n} \tag{1.a}$$

with initial and boundary conditions given, respectively, by

$$T_k(\boldsymbol{x},0) = f_k(\boldsymbol{x}),\qquad \boldsymbol{x}\in \mathrm{V} \tag{1.b}$$

$$\left[\alpha_k(\boldsymbol{x}) + \beta_k(\boldsymbol{x})K_k(\boldsymbol{x})\frac{\partial}{\partial \boldsymbol{n}}\right]T_k(\boldsymbol{x},t) = \phi_k(\boldsymbol{x},t,\mathrm{T}_\ell),\qquad \boldsymbol{x}\in S,\qquad \mathrm{t}>0 \tag{1.c}$$

where the equation operator is written as

$$L_k \equiv -\nabla K_k(\boldsymbol{x})\nabla + d_k(\boldsymbol{x}) \tag{1.d}$$

and $\boldsymbol{n}$ denotes the outward-drawn normal to the surface S.

Without the convection terms and for linear source terms, i.e., $\boldsymbol{u}(x,t,T_\ell) \equiv 0$, $P \equiv P(x,t)$, and $\phi \equiv \phi(x,t)$, this example becomes a class I linear diffusion problem according to the classification in [34]. Exact analytical solutions were in this situation obtained through the classical integral transform technique. Otherwise, this problem is not *a priori* transformable, and the ideas in the generalized integral transform technique [35-40] can be utilized to develop hybrid numerical-analytical solutions to this class of problem. Following the solution path previously established for convection-diffusion and purely diffusive non-linear problems, the formal solution of the posed nonlinear problem requires the proposition of eigenfunction expansions for the associated potentials. The linear situation above commented that allows for an exact solution via the classical integral transform approach, naturally leads to the eigenvalue problems to be preferred in the analysis of the nonlinear situation as well. They appear in the direct application of separation of variables to the linear homogeneous purely diffusive version of the above problem. Thus, the recommended set of auxiliary problems is given by

$$L_k\psi_{ki}(\boldsymbol{x}) = \mu_{ki}^2 w_k(\boldsymbol{x})\psi_{ki}(\boldsymbol{x}),\qquad \boldsymbol{x}\in V \tag{2.a}$$

with boundary conditions

$$\left[\alpha_k(\boldsymbol{x}) + \beta_k(\boldsymbol{x})K_k(\boldsymbol{x})\frac{\partial}{\partial \boldsymbol{n}}\right]\psi_{ki}(\boldsymbol{x}) = 0,\qquad \boldsymbol{x}\in \mathrm{S} \tag{2.b}$$

where the eigenvalues, μ_{ki}, and related eigenfunctions, $\psi_{ki}(\boldsymbol{x})$, are assumed to be known from exact analytical expressions or application of computational methods for Sturm-Liouville type problems [35, 36]. The problem indicated by Eqs.(2.a,b) allows, through the associated orthogonality property of the eigenfunctions, definition of the following integral transform pairs:

$$\overline{T}_{k,i}(t) = \int_V \mathrm{w}_k(\boldsymbol{x})\tilde{\psi}_{ki}(\boldsymbol{x})T_k(\boldsymbol{x},\mathrm{t})\mathrm{dv},\qquad \text{transforms} \tag{3.a}$$

$$T_k(\boldsymbol{x},t) = \sum_{i=1}^{\infty}\tilde{\psi}_{ki}(\boldsymbol{x})\overline{T}_{k,i}(\mathrm{t}),\qquad \text{inverses} \tag{3.b}$$

where the symmetric kernels $\tilde{\psi}_{ki}(\boldsymbol{x})$ are given by

$$\tilde{\psi}_{ki}(\boldsymbol{x}) = \frac{\psi_{ki}(\boldsymbol{x})}{N_{ki}^{1/2}} \tag{3.c}$$

$$N_{ki} = \int_V w_k(\boldsymbol{x})\psi_{ki}^2(\boldsymbol{x})dv \tag{3.d}$$

The integral transformation of (1.a) is accomplished by applying the operator $\int_V \tilde{\psi}_{ki}(\boldsymbol{x})dv$ to yield, after using boundary conditions of Eqs. (1.c) and (2.b)

$$\frac{d\overline{T}_{k,i}(t)}{dt} + \sum_{j=1}^{\infty} a_{kij}(t,T_\ell)\overline{T}_{k,j}(t) = \overline{g}_{ki}(t,T_l), \qquad i = 1,2,..., \qquad t > 0, \quad k,\ell = 1,2,...,n \tag{4.a}$$

The initial conditions of Eq.(1.b) are also transformed through the operator $\int_V w_k(\boldsymbol{x})\tilde{\psi}_{ki}(\boldsymbol{x})dv$ to provide

$$\overline{T}_{k,i}(0) = \bar{f}_{ki} \equiv \int_V w_k(\boldsymbol{x})\tilde{\psi}_{ki}(\boldsymbol{x})f_k(\boldsymbol{x})dv \tag{4.b}$$

where,

$$\overline{g}_{ki}(t,T_l) = \int_V \tilde{\psi}_{ki}(\boldsymbol{x})P_k(\boldsymbol{x},t,T_\ell)dv + \int_S K_k(\boldsymbol{x})\left[\tilde{\psi}_{ki}(\boldsymbol{x})\frac{\partial T_k(\boldsymbol{x},t)}{\partial \boldsymbol{n}} - T_k(\boldsymbol{x},t)\frac{\partial \tilde{\psi}_{ki}(\boldsymbol{x})}{\partial \boldsymbol{n}}\right]ds \tag{4.c}$$

$$a_{kij}(t,T_\ell) = \delta_{ij}\mu_{ki}^2 + a_{kij}^*(t,T_\ell) \tag{4.d}$$

with

$$\delta_{ij} = \begin{cases} 0, & for \quad i \neq j \\ 1, & for \quad i = j \end{cases} \tag{4.e}$$

$$a_{kij}^*(t,T_\ell) = \int_V \tilde{\psi}_{ki}(\boldsymbol{x})[\boldsymbol{u}(\boldsymbol{x},t,T_\ell).\nabla\tilde{\psi}_{ki}(\boldsymbol{x})]dv \tag{4.f}$$

Equations (4) form an infinite system of coupled non-linear ordinary differential equations for the transformed potentials, $\overline{T}_{k,i}$. For computation purposes, system (4) is truncated at the Nth row and column, with N sufficiently large for the required convergence. The formal aspects behind the convergence to the infinite system solution as the truncation order N is increased have been previously investigated [35]. The non-linear initial value problem defined by eqs. (4) is likely to belong to a class of stiff ordinary-differential systems, especially for increasing values of N. Fortunately, various special numerical integrators have been developed within the last few decades, to this class of systems [41, 52]. Once the transformed potentials have been computed from numerical solution of system (4), the inversion formula Eq.(3.b) is recalled to reconstruct the original potentials $T_k(\boldsymbol{x},t)$, in explicit form.

An alternative hybrid solution strategy to the above described full integral transformation is of particular interest in the treatment of transient convection-diffusion problems with a preferential convective direction. In such cases, the partial integral transformation in all but one space coordinate, may offer an interesting combination of relative advantages between the eigenfunction expansion approach and the selected numerical method for handling the coupled system of one-dimensional partial differential equations that results from the transformation procedure. As an illustration of this partial integral transformation procedure, again a transient convection-diffusion problem of n coupled potentials (velocity, temperature or concentration) is considered, but this time separating the preferential direction that is not to be integral transformed. Thus, the vector $\boldsymbol{x}$ now includes the space coordinates that will be eliminated through integral transformation, here denoted by $\boldsymbol{x^*}$, as well as the space variable to be retained in the transformed partial differential system, z. The source term P_k includes all of the other contributions not explicitly shown in the formulation below, such as the convection terms in the $\boldsymbol{x^*}$ directions as well as diffusion in the z direction and the time dependent and non-linear components of convection terms, chosen not to be explicitly written here for conciseness:

$$w_k(\boldsymbol{x}^*)\frac{\partial T_k(\boldsymbol{x}^*,z,t)}{\partial t}+u(\boldsymbol{x}^*)\frac{\partial T_k(\boldsymbol{x}^*,z,t)}{\partial z}+L_kT_k(\boldsymbol{x}^*,z,t)=P_k^*(\boldsymbol{x}^*,z,t,T_\ell), \tag{5.a}$$

$$z_0\le z\le z_1,\ \boldsymbol{x}^*\in V^*,\ \ t>0,\ \ k,\ell=1,2,\dots,n$$

with initial and boundary conditions given, respectively, by

$$T_k(\boldsymbol{x}^*,z,0)=f_k(\boldsymbol{x}^*,z),\quad z_0\le z\le z_1,\ \boldsymbol{x}^*\in V^* \tag{5.b}$$

$$\left[\alpha_k(\boldsymbol{x}^*)+\beta_k(\boldsymbol{x}^*)K_k(\boldsymbol{x}^*)\frac{\partial}{\partial \boldsymbol{n}}\right]T_k(\boldsymbol{x}^*,z,t)=\phi_k(\boldsymbol{x}^*,z,t,T_\ell),\qquad \boldsymbol{x}^*\in S^*,\qquad t>0 \tag{5.c}$$

where the equation operator is written as

$$L_k\equiv-\nabla K_k(\boldsymbol{x}^*)\nabla+d_k(\boldsymbol{x}^*) \tag{5.d}$$

and $\boldsymbol{n}$ denotes the outward-drawn normal to the surface S^*. The boundary conditions introduced by the z variable are now explicitly provided as

$$B_{k,l}T_k(\boldsymbol{x}^*,z,t)=\varphi_{k,l}(\boldsymbol{x}^*,z,t,T_\ell),\qquad z=z_l,l=0,1,\boldsymbol{x}^*\in S^*,\qquad t>0 \tag{5.e}$$

where the boundary operator may include different combinations of first to third kind conditions at the positions z_l, $l=0,1$.

Therefore, the alternative auxiliary problem is now defined in the region V^*, with boundary S^*, formed by the space coordinates to be eliminated:

$$L_k\psi_{ki}(\boldsymbol{x}^*)=\mu_{ki}^2\,u(\boldsymbol{x}^*)\psi_{ki}(\boldsymbol{x}^*),\qquad \boldsymbol{x}^*\in V^* \tag{6.a}$$

with boundary conditions

$$\left[\alpha_k(\boldsymbol{x}^*)+\beta_k(\boldsymbol{x}^*)K_k(\boldsymbol{x}^*)\frac{\partial}{\partial \boldsymbol{n}}\right]\psi_{ki}(\boldsymbol{x}^*)=0,\qquad \boldsymbol{x}^*\in S^* \tag{6.b}$$

where the eigenvalues, μ_{ki}, and related eigenfunctions, $\psi_{ki}(\mathbf{x}^*)$, are assumed to be known.

The following integral transform pairs are now defined:

$$\overline{T}_{k,i}(z,t)=\int_{V^*}u(\boldsymbol{x}^*)\tilde{\psi}_{ki}(\boldsymbol{x}^*)T_k(\boldsymbol{x}^*,z,t)dv,\qquad \text{transforms} \tag{7.a}$$

$$T_k(\boldsymbol{x}^*,z,t)=\sum_{i=1}^{\infty}\tilde{\psi}_{ki}(\boldsymbol{x}^*)\overline{T}_{k,i}(z,t),\qquad \text{inverses} \tag{7.b}$$

where the symmetric kernels $\tilde{\psi}_{ki}(\boldsymbol{x}^*)$ are given by

$$\tilde{\psi}_{ki}(\boldsymbol{x}^*)=\frac{\psi_{ki}(\boldsymbol{x}^*)}{N_{ki}^{1/2}} \tag{7.c}$$

$$N_{ki}=\int_{V^*}u(\boldsymbol{x}^*)\psi_{ki}^2(\boldsymbol{x}^*)dv \tag{7.d}$$

The integral transformation of (5.a) is accomplished by applying the operator $\int_{V^*}\tilde{\psi}_{ki}(\boldsymbol{x}^*)dv$ to yield, after using boundary conditions of Eqs. (5.c) and (6.b)

$$\sum_{j=1}^{\infty} a_{kij}(z,t,T_\ell)\frac{\partial \overline{T}_{k,j}(z,t)}{\partial t}+\frac{\partial \overline{T}_{k,i}(z,t)}{\partial Z}=-\mu_i^2\,\overline{T}_{k,i}(x,t)+\overline{g}_{ki}(z,t,T_l),\qquad (8.a)$$

$$i=1,2,...,\ t>0,\ k,\ell=1,2,...,n$$

The initial conditions of Eq.(5.b) are also transformed through the operator $\int_{v^*} u(\boldsymbol{x}^*)\widetilde{\psi}_{ki}(\boldsymbol{x}^*)dv$ to provide

$$\overline{T}_{k,i}(z,0)=\overline{f}_{ki}(z)\equiv\int_{v^*} u(\boldsymbol{x}^*)\widetilde{\psi}_{ki}(\boldsymbol{x}^*)f_k(\boldsymbol{x}^*,z)dv \qquad (8.b)$$

where,

$$\overline{g}_{ki}(z,t,T_l)=\int_{v^*}\widetilde{\psi}_{ki}(\boldsymbol{x}^*)P_k^*(\boldsymbol{x}^*,z,t,T_\ell)dv+$$
$$\int_{S^*} K_k(\boldsymbol{x}^*)\left[\widetilde{\psi}_{ki}(\boldsymbol{x}^*)\frac{\partial T_k(\boldsymbol{x}^*,z,t)}{\partial \boldsymbol{n}}-T_k(\boldsymbol{x}^*,z,t)\frac{\partial \widetilde{\psi}_{ki}(\boldsymbol{x}^*)}{\partial \boldsymbol{n}}\right]ds \qquad (8.c)$$

$$a_{kij}=\int_{v^*}\widetilde{\psi}_{ki}(\boldsymbol{x}^*)w(\boldsymbol{x}^*)\widetilde{\psi}_{kj}(\boldsymbol{x}^*)dv \qquad (8.d)$$

with the transformed z boundary conditions

$$\int_{v^*} u(\boldsymbol{x}^*)\widetilde{\psi}_{ki}(\boldsymbol{x}^*)B_{k,l}T_k(\boldsymbol{x}^*,z,t)dv=\overline{\varphi}_{k,l,i}(z,t,T_\ell),\quad z=z_l, l=0,1,\boldsymbol{x}^*\in S^*,\quad t>0 \qquad (8.e)$$

where

$$\overline{\varphi}_{k,l,i}(z,t,T_\ell)=\int_{v^*} u(\boldsymbol{x}^*)\widetilde{\psi}_{ki}(\boldsymbol{x}^*)\varphi_{k,l}(\boldsymbol{x}^*,z,t,T_\ell)dv,\quad z=z_l, l=0,1,\boldsymbol{x}^*\in S^*,\quad t>0 \qquad (8.f)$$

Equations (8) form an infinite system of coupled non-linear partial differential equations for the transformed potentials, $\overline{T}_{k,i}$. For computation purposes, system (8) is also truncated at the Nth row and column, with N sufficiently large for the required convergence. A few automatic numerical integrators for this class of one-dimensional partial differential systems are now readily available, such as those based on the Method of Lines [41, 52]. Once the transformed potentials have been computed from numerical solution of system (8), the inversion formula Eq.(7.b) is recalled to reconstruct the original potentials $T_k(\boldsymbol{x}^*,z,t)$, in explicit form along the $\boldsymbol{x}^*$ variables.

3. Computational Procedure and Convergence Acceleration

In order to computationally solve the problem defined by eqs. (1) and (5), straightforward general algorithms can be described as follows:

- The auxiliary eigenvalue problems of eqs. (2.a, b) and (6.a, b) are solved for the eigenvalues and related normalized eigenfunctions, either in analytic explicit form when applicable or through the GITT itself [35].
- The transformed initial conditions (and z boundary conditions) are computed, either analytically or with a general-purpose procedure through adaptive numerical integration [41, 52]. Similarly, those coefficients on the transformed O.D.E. or P.D.E. system of eq. (4.a) and (8.a), respectively, which are not dependent on the transformed potentials, can be evaluated in advance.
- The truncated O.D.E. and P.D.E. systems of eqs. (4) and (8) are then numerically solved through different tools, depending on the type of problem under consideration. For the initial value problem, such as the ODE system obtained in the formal analysis, the numerical integration is performed, for instance, through subroutine NDSolve of the *Mathematica* system [41] or subroutine DIVPAG from the IMSL Library [52]. In general, these initial value problem solvers should work under the automatic selection of a stiff system situation (such as with Gear's method [41, 52]), since the resulting system is likely to become stiff, especially when increasing truncation orders. These subroutines offer interesting combination of accuracy control, simplicity in use, and reliability. For the parabolic type problem that results from the partial integral transformation, both the NDSolve function of the Mathematica

system [41] and the routine DMOLCH from IMSL [52] can be employed. These are two variations of the Method of Lines that implement a variable step and variable order discretization procedure (collocation or finite differences) in one of the independent variables.

- Since all the intermediate numerical tasks are accomplished within user-prescribed accuracy, one is left with the need of reaching convergence in the eigenfunction expansions and automatically controlling the truncation order N for the requested accuracy in the final solution. The analytic nature of the inversion formula allows for a direct testing procedure at each specified position within the medium where a solution is desired, and the truncation order N can be gradually decreased (or eventually increased), to fit the user global error requirements over the entire solution domain. The simple tolerance testing formulas employed are written as

$$\varepsilon = \max_{x \in V} \left| \frac{\sum_{i=N^*}^{N} \tilde{\psi}_{ki}(\mathbf{x}) \overline{T}_{k,i}(t)}{T_{f,k}(\mathbf{x};t) + \sum_{i=1}^{N} \tilde{\psi}_{ki}(\mathbf{x}) \overline{T}_{k,i}(t)} \right| \tag{9.a}$$

$$\varepsilon = \max_{x^* \in V^*} \left| \frac{\sum_{i=N^*}^{N} \tilde{\psi}_{ki}(\mathbf{x}^*) \overline{T}_{k,i}(z,t)}{T_{f,k}(\mathbf{x}^*;z,t) + \sum_{i=1}^{N} \tilde{\psi}_{ki}(\mathbf{x}^*) \overline{T}_{k,i}(z,t)} \right| \tag{9.b}$$

where $T_{f,k}$ is a so-called filtering solution, which may be employed for convergence improvement as later discussed. The numerator in eqs. (9) represents those terms that in principle might be abandoned in the evaluation of the inverse formula, without disturbing the final result to within the user-requested accuracy target. Therefore, this testing proceeds by reducing the value of N^* in the numerator sum until the value of ε reaches the user-requested global error at any of the selected test positions within the domain, then defining the minimum truncation order that can be adopted at that time (and z) variable value. For the next value of the time variable of interest, the system integration marches with the truncation order N changed to assume the value of this smallest N^* achieved. Thus, the accuracy testing, besides offering error estimations, in addition allows for an adaptive truncation order control along the ordinary (or partial) differential system numerical integration process.

A major aspect in the practical implementation of this methodology is the eventual need for improving the convergence behavior of the resulting eigenfunction expansions as pointed out in [35-40]. The overall simplest and most effective alternative for convergence improvement appears to be the proposition of analytical filtering solutions, which present both space and time dependence within specified ranges of the time numerical integration path. For instance, an appropriate quasi-steady filter for the above formulations could be written in general as

$$T_k(\mathbf{x},t) = \theta_k(\mathbf{x},t) + T_{f,k}(\mathbf{x};t) \tag{10.a}$$

$$T_k(\mathbf{x}^*,z,t) = \theta_k(\mathbf{x}^*,z,t) + T_{f,k}(\mathbf{x}^*;z,t) \tag{10.b}$$

where the second term in the right hand sides represents the quasi-steady filter solution which is generally sought in analytic form. The first term on the right hand side represents the filtered potentials which are obtained through integral transformation. Once the filtering problem formulation is chosen, Eqs.(10) are substituted back into Eqs.(1) or (5) to obtain the resulting formulation for the filtered potential. It is desirable that the filtering solution contains as much information on the operators of the original problem as possible. This information should include the initially posed source terms or at least their linearized versions, so as to reduce their influence on convergence of the final eigenfunction expansions. For instance, representative linearized versions of the original problem in a certain time interval, after being exactly solved through the classical integral transform approach, may partially filter the original problem source terms more effectively. These source terms are essentially those responsible for deviating the convergence behavior from the spectral exponential pattern. Then, the filter can be automatically redefined for the next time-variable range by prescribing a desirable maximum value for the system truncation order while still satisfying the user requested global accuracy target. This so-called local-instantaneous filtering (LIF) strategy has been lately preferred, as a possibly optimal scheme for enhancing convergence in eigenfunction expansions [50, 51]. Also, the LIF strategy indirectly introduces a desirable modulation effect on the transformed ODE system. While the single filter solution produces, in general, strongly stiff ODE systems which require special initial value problem solvers, the LIF solution yields, in principle, non-stiff systems, which are readily solved by standard explicit schemes at reduced computational cost [51].

In multidimensional applications, the final integral transform solution for the related potential is expressed as double or triple infinite summation for two- or three-dimensional transient problems in full integral transformation or as a double summation for a three-dimensional transient problem in the above partial integral transformation. Each of these summations is associated with the eigenfunction expansion in a corresponding spatial coordinate. Such space variables are eliminated through integral transformation from the partial differential system and are analytically recovered through these inversion formula involving multiple summations.

From a computational point of view, only a truncated version of such nested summations can be actually evaluated. However, if one just truncates each individual summation to a certain prescribed finite order, the computation becomes quite ineffective, and even a risky one. By following this path some still important information to the final result can be disregarded due to the fixed summations limits, while other terms are accounted for that have essentially no contribution to convergence of the potential in the relative accuracy required. Therefore, for an efficient computation of these expansions, the infinite multiple summations should first be converted to a single sum representation with the appropriate reordering of terms according to their individual contribution to the final numerical result. Then, it would be possible to evaluate the minimal number of eigenvalues and related derived quantities required to reach the user-prescribed accuracy target. This aspect is even more evident in the use of the GITT, when the computational costs can be markedly reduced through this reordering of terms which then represents a reduction on the number of ordinary differential equations to be solved numerically in the transformed system [36, 40]. Since the final solution is not, of course, known *a priori*, the parameter which shall govern this reordering scheme must be chosen with care. Once the ordering is completed, the remainder of the computational procedure becomes as straightforward and cost-effective as in the one-dimensional case. In fact, except for the additional effort in the numerical evaluation of double and/or triple integrals, finding a multidimensional solution may require essentially the same effort as in a plain one-dimensional situation. The most common choice of reordering strategy is based on the argument of the dominating exponential term, which offers a good compromise between the overall convergence enhancement and simplicity in use. However, individual applications may require more elaborate reordering that accounts for the influence of nonlinear source terms in the ODE system.

4. Application: Transient Flow in Microchanels

We consider fully developed incompressible laminar flow, considering slip at the walls, inside a circular micro-tube or a parallel plates micro-channel subjected to a pressure gradient *dp/dz* that varies in an arbitrary functional form with the time variable. The velocity field is represented by *u(r,t)*, which varies with the transversal coordinate, *r*, and time, *t*. The related time-dependent axial momentum equation (z-direction) is then written in dimensionless form as:

$$R^n \frac{\partial U(R,\tau)}{\partial \tau} = \frac{\partial}{\partial R}\left[R^n \frac{\partial U(R,\tau)}{\partial R}\right] + R^n P(\tau), \qquad in\, 0 < R < 1 \tag{11.a}$$

$$\left.\frac{\partial U(R,t)}{\partial R}\right|_{R=0} = 0; \quad \beta^* \left.\frac{\partial U(R,t)}{\partial R}\right|_{R=1} + U(1,t) = 0 \tag{11.b,c}$$

$$U(R,0) = U_0(R) \tag{11.d}$$

where n=0 for parallel-plates, and n=1 for circular tube, and we have considered the following dimensionless groups:

$$R = \frac{r}{r_1}; \tau = \frac{\nu t}{r_1^2}; U(R,\tau) = \frac{u(r,t)}{u_m}; P(\tau) = \frac{-(dp/dz)r_1^2}{\mu u_m};$$
$$\beta^* = 2\,Kn\,\beta_v\,; Kn = \lambda / 2r_1\,; \beta_v = (2-\alpha_m)/\alpha_m \tag{12}$$

The generalized integral transform technique (GITT) is a well-established hybrid tool in the solution of diffusion and convection-diffusion problems, reducing to the classical integral transform analysis in classes of problems that allow for an exact treatment. The generalized approach is here employed to permit a direct extension to the electrosmotic flow situation. One important aspect in this kind of eigenfunction expansion approach is the convergence enhancement achievable by introducing analytical solutions that filter the original problem source terms, which are responsible for an eventual slow convergence behavior. Thus, we start the integral transformation process by obtaining the filtering solution, based on the quasi-steady version of the present problem:

$$U(R,\tau) = U_P(R,\tau) + U_h(R,\tau) \tag{13}$$

For the present problem, the quasi-steady solution of problem (11), essentially removing the transient term in eq.(11.a) is considered

$$\frac{d}{dR}(R^n \frac{dU_P}{dR}) + R^n P(\tau) = 0 \tag{14.a}$$

$$\left.\frac{dU_P}{dR}\right|_{R=0} = 0; \quad \beta^* \left.\frac{\partial U_P(R;t)}{\partial R}\right|_{R=1} + U_P(1;t) = 0 \tag{14.b,c}$$

The above ODE is directly integrated to yield the analytical filter in terms of the dimensionless time-variable pressure gradient:

$$U_P(R;\tau) = P(\tau)\frac{(2\beta^* + 1 - R^2)}{2(n+1)} \tag{15}$$

The resulting system for the filtered potential U_h, is then given by:

$$R^n \frac{\partial U_h}{\partial \tau} = \frac{\partial}{\partial R}(R^n \frac{\partial U_h}{\partial R}) + R^n P^*(R,\tau) \tag{16.a}$$

$$\left.\frac{\partial U_h}{\partial R}\right|_{R=0} = 0; \quad \beta^* \left.\frac{\partial U_h(R,t)}{\partial R}\right|_{R=1} + U_h(1,t) = 0 \tag{16.b,c}$$

$$U_h(R,0) = U_0^*(R) = U_0(R) - U_P(R;0) \tag{16.d}$$

where the resulting source term for the filtered system becomes

$$P^*(R,\tau) = -\frac{\partial U_P}{\partial \tau} \tag{16.e}$$

The following simple eigenvalue problem is naturally selected for the integral transformation pair construction:

$$\frac{d}{dR}(R^n \frac{d\varphi(R)}{dR}) + R^n \lambda^2 \varphi(R) = 0 \tag{17.a}$$

with boundary conditions:

$$\left.\frac{\partial \varphi(R)}{\partial R}\right|_{R=0} = 0; \quad \beta^* \left.\frac{\partial \varphi(R)}{\partial R}\right|_{R=1} + \varphi(1) = 0 \tag{17.b,c}$$

The eigenfunctions $\varphi_m(R)$ are readily obtained and given by:

$$\varphi_m(R) = \cos(\lambda_m R), \text{ for n=0,} \quad \varphi_m(R) = J_0(\lambda_m R), \text{ for n=1} \tag{18.a,b}$$

and the related eigenvalues are computed from satisfaction of the boundary condition eq.(17.c), while the normalization integral is analytically computed from the definition

$$N_m = \int_0^1 R^n \varphi_m^2(R)dR \tag{19}$$

The integral transform pair is written as:

$$U(R,\tau)=\sum_{m=1}^{\infty}\frac{1}{N_m}\varphi_m(R)\overline{U}_m(\tau), \qquad \text{inverse} \tag{20.a}$$

$$\overline{U}_m(\tau)=\int_0^1 R^n\varphi_m(R)\,U(R,\tau)dR, \qquad \text{transform} \tag{20.b}$$

Operating the filtered potential equation (16.a) with $\int_0^1 \varphi_m(R)dR$ and transforming all the original potentials with the aid of the inversion formula, we obtain the following ordinary differential equations:

$$\frac{d\overline{U}_m(\tau)}{d\tau}+\lambda_m^2\overline{U}_m(\tau)=\overline{g}_m(\tau), \quad \tau>0, \quad m=1,2,3... \tag{21.a}$$

where the transformed source term is computed from

$$\overline{g}_m(\tau)=\int_0^1 R^n\varphi_m(R)P^*(R,\tau)dR \tag{21.b}$$

Similarly, the filtered initial condition (16.d) is operated on with $\int_0^1 R^n\varphi_m(R)dR$, to yield:

$$U_m(0)=\bar{f}_m, \;\; \bar{f}_m=\int_0^1 R^n\varphi_m(R)U_0^*(R)dR \tag{21.c,d}$$

Eqs.(21) are readily solved to yield the analytical expression for the transformed potential:

$$\overline{U}_m(\tau)=\bar{f}_m\exp(-\lambda_m^2\tau)+\int_0^\tau \exp[-\lambda_m^2(\tau-\tau')]\overline{g}_m(\tau')d\tau' \tag{22}$$

Once the above solution is obtained for the transformed potential, the inversion formula, equation (20.a), can be used to evaluate the filtered velocity, and then the original field from eq.(13). For computational purposes, the infinite series is evaluated to a sufficiently large finite order so as to achieve the user's requested accuracy target. The original partial differential equation presented in eqs.(11) was also solved in the *Mathematica 4.2* platform by making use of the built in function *NDSolve*, with a user prescribed relative error control. This function uses a variation of the Method of Lines [41]. A numerical analysis on these results was also performed, following the recommendations for employing this algorithm provided in [53].

5. Application: Transient Convection in Microchannels

Consider transient-state heat transfer in thermally developing, hydrodynamically developed forced laminar flow inside a microchannel under the following additional formulation choices:

- The flow is incompressible with constant physical properties.
- Free convection of heat is negligible.
- The entrance temperature distribution is uniform.
- The temperature of the channel wall is prescribed and uniform.

The temperature $T(y,z,t)$ of a fluid with developed velocity profile $u(y)$, flowing along the channel in the region $0<y<r_1$, $z>0$, is then described by the following problem in dimensionless form:

$$\frac{\partial\theta(Y,Z,\tau)}{\partial\tau}+U(Y)\frac{\partial\theta(Y,Z,\tau)}{\partial Z}=\frac{\partial^2\theta(Y,Z,\tau)}{\partial Y^2}+\frac{1}{Pe^2}\frac{\partial^2\theta(Y,Z,\tau)}{\partial Z^2}+Br\left(\frac{dU}{dY}\right)^2, \tag{23.a}$$
$$in\; 0<Y<1,\, Z>0,\, \tau>0$$

$$\left.\frac{\partial\theta(Y,Z,\tau)}{\partial Y}\right|_{Y=0}=0;\ 2Kn\,\beta\,\beta_v\left.\frac{\partial\theta(Y,Z,\tau)}{\partial Y}\right|_{Y=1}+\theta(1,Z,\tau)=0 \tag{23.b,c}$$

$$\theta(Y,0,\tau)=\theta_e(\tau)\ ;\ \left.\frac{\partial\theta(Y,Z,\tau)}{\partial Z}\right|_{Z=L}=0 \tag{23.d,e}$$

$$\theta(Y,Z,0)=0 \tag{23.f}$$

where we have considered the following dimensionless groups:

$$Y=\frac{y}{r_1};\quad \tau=\frac{\alpha\, t}{r_1^2};\quad U(Y)=\frac{u(y)}{u_m};\quad Pe=\frac{u_m\, r_1}{\alpha}\quad;\ Br=\frac{\mu\, u_m^{\ 2}}{k\,\Delta T}\quad;$$
$$\beta=\frac{\beta_t}{\beta_v}\ ;\ \theta(Y,Z,\tau)=\frac{T(y,z,t)-T_o}{\Delta T}\ ;\ Z=\frac{1}{Pe}\frac{z}{r_1}\ ; \tag{24}$$

and β_t=((2-α_t)/ α_t)(2γ/(γ+1))/Pr, α_t is the thermal accommodation coefficient, λ is the molecular mean free path, $\gamma=c_p/c_v$, while c_p is specific heat at constant pressure, c_v specific heat at constant volume, T_s is the temperature at the channel wall, and the Knudsen number is defined as $Kn=\lambda/2r_1$.

The dimensionless velocity profile is given as [45]:

$$U(Y)=\frac{6Kn\beta_v+3(1-Y^2)/2}{1+6Kn\beta_v} \tag{25}$$

The Generalized Integral Transform solution considers a Sturm-Liouville problem that includes the velocity profile, $U(Y)$, in its formulation [45]. This approach leads to an exact analytical solution in terms of confluent hypergeometric functions to eliminate the transversal coordinate, where $\psi_i(Y)$ are the eigenfunctions of the following Sturm-Liouville problem, with the corresponding normalization integral and normalized form of the eigenfunction:

$$\frac{d^2\psi_i(Y)}{dY^2}+\mu_i^2U(Y)\psi_i(Y)=0,\qquad 0<Y<1 \tag{26.a}$$

$$\left.\frac{d\psi_i(Y)}{dY}\right|_{Y=0}=0,\quad Kn\beta_v\beta\left.\frac{d\psi_i(Y)}{dY}\right|_{Y=1}=-\frac{1}{2}\psi_i(1) \tag{26.b,c}$$

$$N_i=\int_0^1 U(Y)\psi_i^2(Y)dY;\quad \tilde{\psi}_i(Y)=\frac{\psi_i(Y)}{N_i^{1/2}} \tag{27.a,b}$$

For the proposed dimensionless velocity field in micro-channels, eq.(26.a) can be rewritten in the simpler form below:

$$\frac{d^2\psi_i(Y)}{dY^2}+\nu_i^2(1+4Kn\beta_v-Y^2)\psi_i(Y)=0,\qquad 0<Y<1 \tag{28.a}$$

with the original eigenvalues to be obtained from

$$\mu_i=\sqrt{\frac{2}{3}(1+6Kn\beta_v)}\ \nu_i,\quad i=1,2,3,... \tag{28.b}$$

As discussed in [45], the solution of problem (26) is then obtained in terms of the confluent hypergeometric function, also known as Kummer function ${}_1F_1[a;b;\,z]$, readily available in the *Mathematica* system [41], as:

$$\psi_i(Y) = {}_1F_1[\frac{1-\nu_i(1+4Kn\beta_v)}{4},\frac{1}{2},\nu_i Y^2]\, e^{\nu_i \frac{Y^2}{2}} \tag{29}$$

Eq. (29) satisfies the first two eqs. (26.a,b), and the last equation (26.c) thus gives the eigencondition:

$$\begin{aligned}&\{2Kn\beta_v\beta\ {}_1F_1[\frac{5}{4}-\frac{\nu_i(1+4Kn\beta_v)}{4},\frac{3}{2},\nu_i]\nu_i(1-(1+4Kn\beta_v)\nu_i)+\\&{}_1F_1[\frac{1}{4}-\frac{\nu_i(1+4Kn\beta_v)}{4},\frac{1}{2},\nu_i](1-2Kn\beta_v\beta\nu_i)\}e^{\frac{\nu_i}{2}}=0\end{aligned} \tag{30}$$

The left hand side of eq.(30) defines a function of two parameters, $Kn\beta_v$ and β, which will be employed to provide the eigenvalues, ν_i, then allowing the computation of the original eigenvalues, μ_i. The next step is thus the definition of the transform-inverse pair, given by:

$$\bar{\theta}_i(Z,\tau)=\int_0^1 U(Y)\tilde{\psi}_i(Y)\theta(Y,Z,\tau)\,dY \quad \text{transform} \tag{31.a}$$

$$\theta(Y,Z,\tau)=\sum_{i=1}^{\infty}\tilde{\psi}_i(Y)\,\bar{\theta}_i(Z,\tau) \quad \text{inverse} \tag{31.b}$$

Here we choose to apply the GITT on equations (23) in the partial transformation strategy, resulting in the parabolic partial differential equations system below:

$$\sum_{j=1}^{N}A_{ij}\frac{\partial\bar{\theta}_j(Z,\tau)}{\partial\tau}+\frac{\partial\bar{\theta}_i(Z,\tau)}{\partial Z}=-\mu_i^2\,\bar{\theta}_i(Z,\tau)+\frac{1}{Pe^2}\sum_{j=1}^{N}A_{ij}\frac{\partial^2\bar{\theta}_j(Z,\tau)}{\partial Z^2}+\bar{g}_i \qquad i=1,2,\ldots,N \tag{32.a}$$

$$\bar{\theta}_i(Z,0)=0 \tag{32.b}$$

$$\bar{\theta}_i(0,\tau)=\bar{f}_i\,\theta_e(\tau)\ ;\quad \left.\frac{\partial\bar{\theta}_i(Z,\tau)}{\partial Z}\right|_{Z=L}=0 \tag{32.c,d}$$

where

$$\begin{aligned}&A_{ij}=\int_0^1\tilde{\psi}_i(Y)\tilde{\psi}_j(Y)\,dY\ ;\\&\bar{g}_i=Br\int_0^1\left(\tfrac{dU}{dY}\right)^2\tilde{\psi}_i(Y)\,dY\ ;\\&\bar{f}_i=\int_0^1U(Y)\tilde{\psi}_i(Y)\,dY\ ;\end{aligned} \tag{32.e,f,g}$$

The numerical Method of Lines as implemented in the routine NDSolve of the *Mathematica* system deals with system (32) by employing the default fourth order finite difference discretization in the spatial variable Z, and creating a much larger coupled system of ordinnary equations for the transformed dimensionless temperature evaluated on the knots of the created mesh. This resulting system is internally solved (still inside NDSolve routine) with Gear's method for stiff ODE systems. Once numerical results have been obtained and automatically interpolated by NDSolve, one can apply the inverse expression (31.b) to obtain the full dimensionless temperature field.

Once $\theta(Y,Z,\tau)$ is determined from (31.b), the average temperature $\theta_{av}(Z,\tau)$ can be found from:

$$\theta_{av}(Z,\tau)=\int_0^1U(Y)\theta(Y,Z,\tau)\,dY \tag{33}$$

The local Nusselt number $Nu(Z,\tau)=h(Z,\tau)D_h/k$, where $h(Z,\tau)$ is the heat transfer coefficient, can be found from:

$$Nu(Z,\tau)=-\frac{4}{\theta_{av}(Z,\tau)}\left.\frac{\partial\theta(Y,Z,\tau)}{\partial Y}\right|_{Y=1} \tag{34}$$

6. Results and Discussion

In this section we present and discuss a few numerical results for the two problems considered, transient flow and transient convection in microchannels, which were respectively handled by the full and the partial integral transformation approaches. The aim is to demonstrate the convergence behavior within each strategy and to illustrate some physical aspects on the transient phenomena at the micro-scale. Although the developed solutions are readily applicable to different physical situations of either liquid or gas flow, we here concentrate our illustration of results on typical examples of laminar gas slip flow.

For evaluation of the constructed symbolic-numerical algorithm on transient flow analysis, we considered both geometries (parallel plates and circular tube) under two different and representative transient situations: flow start up with a step change or a periodic time variation of the pressure gradient [54]. Here, due to space limitations, we present only a few of the parallel-plates case results (n=0). By assigning numerical values to the parameters, $\beta^*=0.1$, according to the chosen dimensionless formulation, we define the pressure gradient for the start-up case with a unit step change:

$$P(\tau)=\frac{3}{3\beta^*+1} \tag{35.a}$$

For the periodic case, we just change the definition of the dimensionless time variable source term, as follows for the parallel-plates geometry:

$$P(\tau)=\frac{3}{3\beta^*+1}\left(1-\frac{\sin(\Omega\tau)}{2}\right) \tag{35.b}$$

with $\Omega=\pi/15$ for the reported example.

Table 1 below illustrates the excellent convergence characteristics of the proposed eigenfunction expansion, for the case of a periodic pressure gradient in a parallel plates channel with $\beta^*=0.1$, and considering four different values of the dimensionless time. Truncation orders N=10 and 30 are explicitly shown, demonstrating that six converged significant digits at least are achieved for N as low as 10. Also presented are the numerical results obtained via the Method of Lines implemented in the built in routine NDSolve of the *Mathematica* system [41]. These results agree to within four significant digits. As was noticed along the solution procedure, the results from the integral transform solution and from the numerical built in routine are essentially coincident, since one can only observe numerical deviations in the last two significant digits. The analytical solution is also observed to be fully converged even with less than 10 terms in the expansion.

For the start up flow case, we obtain the following set of curves of the dimensionless velocity profiles evolution shown in Figure 1, where the increase in the wall velocity with time can be clearly observed. The three dimensional plot for the velocity distribution is given in Figure 2 for the periodic case, and we can observe the quasi-steady-state (periodic state) establishment, and the time variation of the dimensionless slip velocity.

Before proceeding to the analysis of transient convection with slip flow and temperature jump, we first validate the present novel strategy of combining the integral transform approach and the Method of Lines, and inspect the convergence behavior in both the partial eigenfunction expansion and the numerical procedure for the transformed partial differential system. Therefore, the test case by Gondim et al. [50, 51] for a regular parallel plates channel (Kn=0) is here analyzed for different and representative values of the Peclet number. It should be noted that gas flows in microchannels are likely to result in relatively low values of Reynolds number, in the range of incompressible flow modeling here adopted, which then produce Peclet numbers in a fairly wide range. Therefore, Figures 3.a,b, respectively for Pe=1 and Pe=10, show the excellent agreement between the present results and the full integral transformation in refs.[50, 51], where a double integral transformation in both transversal and longitudinal coordinates is employed. A truncation order of just S=15 terms was considered sufficient for convergence in the present covalidation, as we shall examine in what follows, since we are dealing with a single integral transformation, which is performed along the most diffusive direction (R) and exactly transforming the transversal convection term, as opposed to the double transformation in [50, 51] which requires significantly larger truncation orders.

Table 1: Convergence behavior of eigenfunction expansion for the dimensionless velocity and comparison with routine NDSolve [41] (parallel plates, periodic flow, β*=0.1).

U(R,τ); GITT with N=10, N=30, & NDSolve [41]					
Solution	***R***	***τ=5***	***τ=10***	***τ=15***	***τ=20***
GITT – N=10		0.827503	0.755419	1.31248	1.94167
GITT – N=30	0.0	0.827503	0.755419	1.31248	1.94167
NDSolve [41]		0.82753	0.755389	1.31251	1.9416
GITT – N=10		0.799650	0.730417	1.26918	1.87722
GITT – N=30	0.2	0.799650	0.730417	1.26918	1.87722
NDSolve [41]		0.799676	0.730387	1.26921	1.87714
GITT – N=10		0.716207	0.655336	1.13908	1.68375
GITT – N=30	0.4	0.716207	0.655336	1.13908	1.68375
NDSolve [41]		0.71623	0.655309	1.13911	1.68367
GITT – N=10		0.577521	0.529949	0.921623	1.36090
GITT – N=30	0.6	0.577521	0.529949	0.921623	1.36090
NDSolve [41]		0.577539	0.529928	0.921641	1.36084
GITT – N=10		0.384164	0.353876	0.615843	0.908121
GITT – N=30	0.8	0.384164	0.353876	0.615843	0.908121
NDSolve [41]		0.384175	0.353862	0.615853	0.908075
GITT – N=10		0.136928	0.126572	0.220405	0.324602
GITT – N=30	1.0	0.136928	0.126572	0.220405	0.324602
NDSolve [41]		0.136932	0.126567	0.220408	0.324585

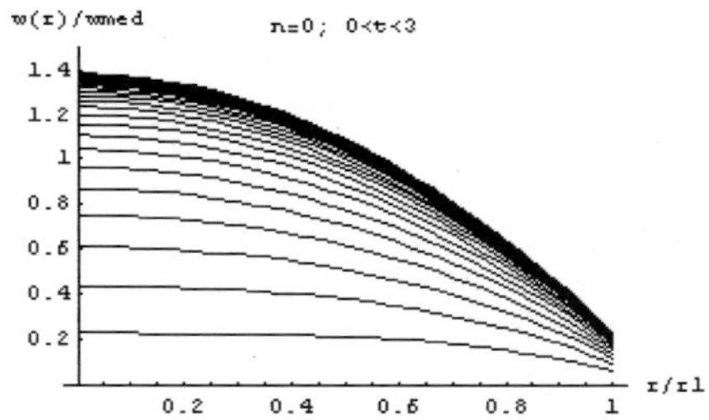

Figure 1: Transient evolution of dimensionless velocity profile for parallel-plates channel (n=0) and step change in pressure gradient, β*=0.1.

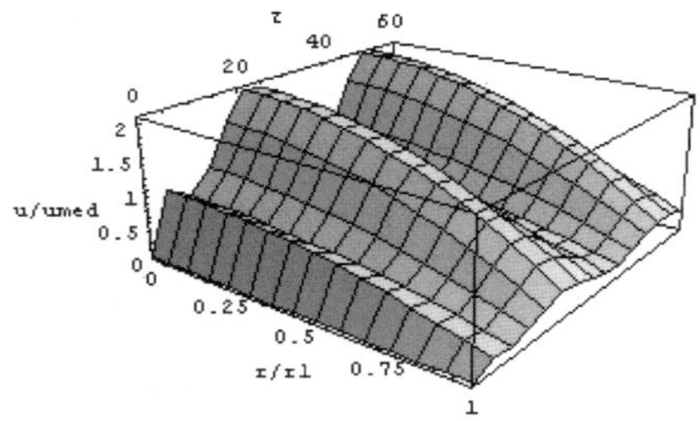

Figure 2: Transient evolution of dimensionless velocity profile for parallel-plates channel (n=0) and periodic variation in pressure gradient, $\beta^*=0.1$.

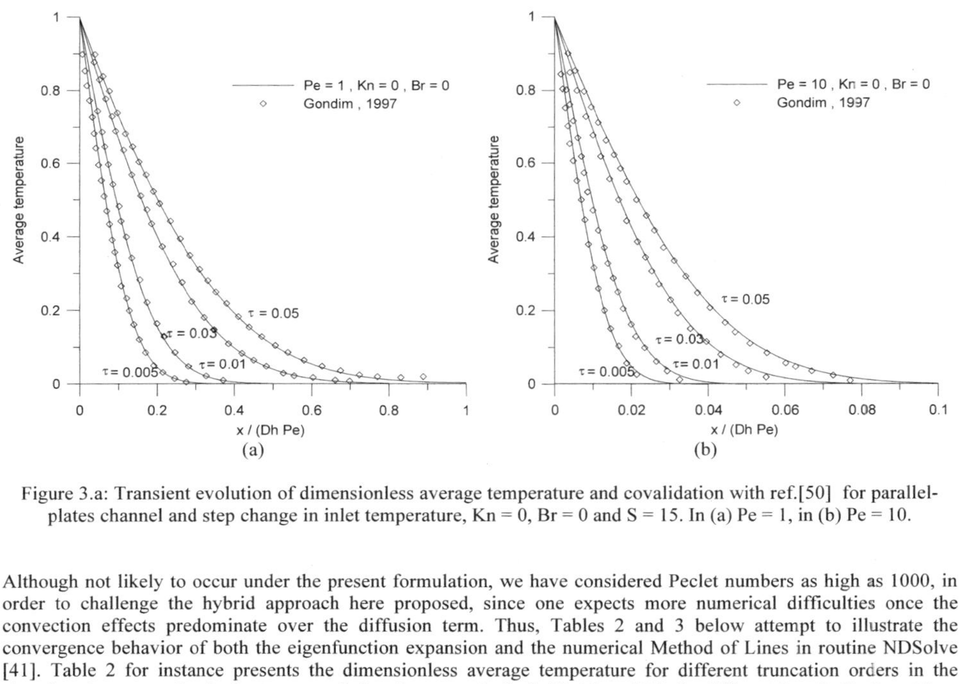

Figure 3.a: Transient evolution of dimensionless average temperature and covalidation with ref.[50] for parallel-plates channel and step change in inlet temperature, Kn = 0, Br = 0 and S = 15. In (a) Pe = 1, in (b) Pe = 10.

Although not likely to occur under the present formulation, we have considered Peclet numbers as high as 1000, in order to challenge the hybrid approach here proposed, since one expects more numerical difficulties once the convection effects predominate over the diffusion term. Thus, Tables 2 and 3 below attempt to illustrate the convergence behavior of both the eigenfunction expansion and the numerical Method of Lines in routine NDSolve [41]. Table 2 for instance presents the dimensionless average temperature for different truncation orders in the eigenfunction expansion in the transversal direction, namely S=5, 10, 15, and 20, for different longitudinal positions and time values, indicating that at least three significant digits are apparently fully converged in this range of truncation. Figure 4 essentially reconfirms this excellent convergence behavior for the dimensionless average temperature, and taking the case of Pe=10 and different dimensionless times, the distribution is practically converged to the graphical scale for truncation orders as low as S=5. This behavior naturally offers simulations of very low computational costs and still under user controllable accuracy. Table 3, on the other hand, for a fixed value of the truncation order, S=15, demonstrates the numerical error control built in the adopted routine, *NDSolve*, via a parameter named *MaxStepSize*, which controls the minimum number of nodes employed in the discretization procedure. Therefore, by decreasing the value of this parameter, we are requesting further precision to the calculation, forcing the error control to work under a more refined grid. For this example, one can observe that four significant digits are certainly unchanged by the substantial grid refinement requested.

Table 2: Convergence behavior of eigenfunction expansion for the dimensionless average temperature from partial integral transformation with routine NDSolve [41] (parallel plates, Pe = 1000, Kn = 0.0 and Br = 0, MaxStepSize = 0.0005).

	θ_m (Pe = 1000 – Kn = 0.0)				
	X / (Dh Pe)	S = 5	S = 10	S = 15	S = 20
t = 0.005	0.0000375	0.98420	0.98871	0.98987	0.99039
	0.0001500	0.94254	0.94279	0.94309	0.94324
	0.0002625	0.82085	0.82142	0.82150	0.82153
	0.0003750	0.56949	0.56959	0.56960	0.56960
	0.0004875	0.23934	0.23932	0.23932	0.23932
t = 0.01	0.0000542	0.98468	0.98926	0.99035	0.99082
	0.0002708	0.95260	0.95141	0.95162	0.95175
	0.0004875	0.85476	0.85622	0.85632	0.85634
	0.0007042	0.64431	0.64414	0.64415	0.64415
	0.0009208	0.26693	0.26694	0.26694	0.26694
t = 0.03	0.0001667	0.97637	0.97935	0.97999	0.98028
	0.0008333	0.93310	0.93119	0.93126	0.93135
	0.0015000	0.83443	0.83682	0.83695	0.83699
	0.0021667	0.62462	0.62451	0.62452	0.62452
	0.0028333	0.14691	0.14691	0.14691	0.14691
t = 0.05	0.0002292	0.97196	0.97432	0.97485	0.97509
	0.0011458	0.92169	0.92186	0.92202	0.92213
	0.0020625	0.86591	0.86513	0.86526	0.86532
	0.0029792	0.74955	0.75042	0.75049	0.75051
	0.0038958	0.52090	0.52092	0.52092	0.52092

Table 3: Convergence behavior of Method of Lines for the dimensionless average temperature from partial integral transformation with routine NDSolve [41] (parallel plates, Pe = 1000, Kn = 0.0 and Br = 0, S=15).

	θ_m (Pe = 1000 – Kn = 0.0)			
	X / (Dh Pe)	MaxStepSize 0.001	MaxStepSize 0.0005	MaxStepSize 0.00025
t = 0.005	0.0000375	0.99979	0.98987	0.98988
	0.0001500	0.93420	0.94309	0.94306
	0.0002625	0.82152	0.82150	0.82154
	0.0003750	0.57403	0.56960	0.56967
	0.0004875	0.23838	0.23932	0.23914
t = 0.01	0.0000542	0.98982	0.99035	0.99035
	0.0002708	0.95160	0.95162	0.95162
	0.0004875	0.85697	0.85632	0.85631
	0.0007042	0.64344	0.64415	0.64416
	0.0009208	0.26757	0.26694	0.26692
t = 0.03	0.0001667	0.97999	0.97999	0.97999
	0.0008333	0.93126	0.93126	0.93126
	0.0015000	0.83695	0.83695	0.83695
	0.0021667	0.62454	0.62452	0.62451
	0.0028333	0.14701	0.14691	0.14691
t = 0.05	0.0002292	0.97485	0.97485	0.97485
	0.0011458	0.92202	0.92202	0.92202
	0.0020625	0.86526	0.86526	0.86526
	0.0029792	0.75049	0.75049	0.75049
	0.0038958	0.52092	0.52092	0.52092

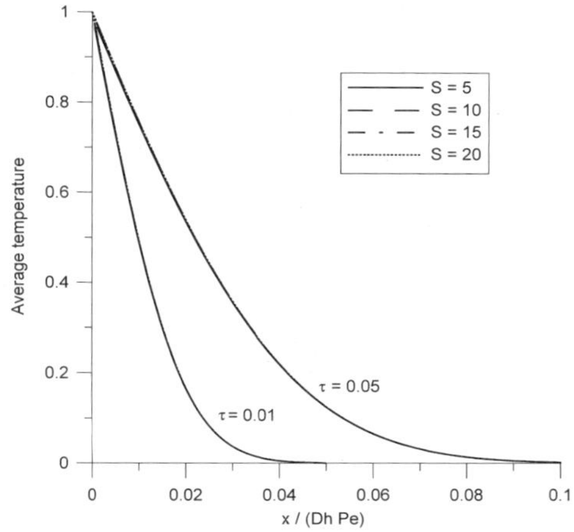

Figure 4: Convergence behavior of the dimensionless average temperature from partial integral transformation with routine NDSolve [41] (parallel plates, Pe = 10, Kn = 0.0 and Br = 0, S=5, 10, 15 & 20).

Transient heat transfer in microchannels is then studied for typical values of the accommodation factors ($\alpha_m = 1.0$ *and* $\alpha_t = 0.92$) and just for illustration considering air as the working fluid (Pr = 0.7 and $\gamma = 1.4$). Figure 5 shows the effect of Brinkman number on the transient behavior of the local Nusselt number, for the following governing parameter values, Pe = 10, Kn =0.01, and Br = 0, 0.001, 0.005, and 0.01, S=15. The effect of increasing the Nusselt number while increasing the internal heat generation via larger values of *Br*, as also evident in previous steady-state analysis, is here reproduced, while the transient solutions approach such steady configurations.

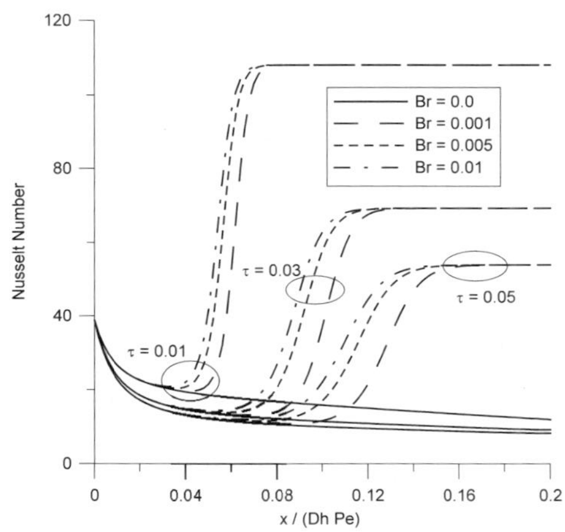

Figure 5: Influence of Brinkman number on local Nusselt number evolution (parallel plates, Pe = 10, Kn =0.01, and Br = 0, 0.001, 0.005, and 0.01, S=15).

Figure 6.a presents the deviations encountered in the average temperature within the entrance region with and without considering axial conduction in the formulation, for the steady-state, with Pe = 10, Kn = 0.01, Br = 0, and S=15. Clearly, neglecting axial diffusion along the fluid in this case, causes a much sharper average temperature drop along the channel. Figure 6.b presents the influence of axial conduction on the local Nusselt number, for two values of dimensionless time (τ), again with Pe = 10, Kn = 0.01, Br = 0, and S=15, while figure 6.c illustrates the deviations between the two formulations for the steady situation. The inclusion of axial diffusion in the model leads to higher heat transfer coefficients, with a marked difference from the formulation without axial diffusion in this case. This comparison was particularly plotted taking the dimensionless physical dimension x/D_h, removing the Peclet number from the abscissa definition.

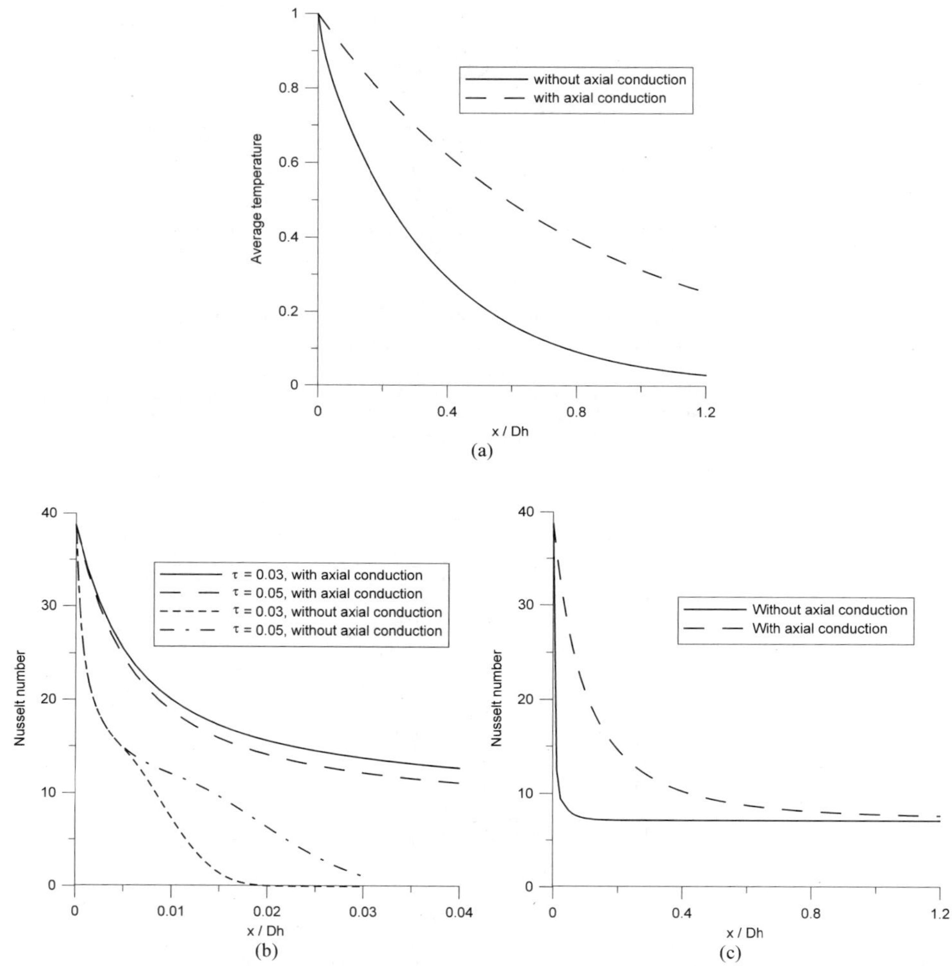

Figure 6: Influence of axial conduction on the average temperature (a) and on the local Nusselt number for (b) transient state and (c) steady-state (parallel plates, Pe = 10, Kn = 0.01 and Br = 0, S=15).

Figures 7 present the influence of Knudsen number (Kn) on the dimensionless average temperature (a) and on the local Nusselt number (b) for Pe = 10, Kn = 0, 0.001, 0.01& 0.1 and Br = 0, S =15, along the entrance region of the parallel plates channel during the transient regime. Figure 7.c presents in more detail the influence of Knudsen number on local Nusselt number within the entrance region for steady-state. It can be observed that the bulk temperature is mildly influenced by the Knudsen number variation, especially during the earlier stages of the transient regime. On the other hand, the variation of the local Nusselt number in such different levels of the microscale effect is shown in Figure 7b, for the same parameter values, where the influence is much more remarkable, with a significant increase in Nu for decreasing Kn. This set of results also allows for the inspection of the comparative transient behavior, which indicates the less pronounced transient phenomena when the Knudsen number is increased. The microscale effects practically cease for Kn=0.001, all along the transient behavior.

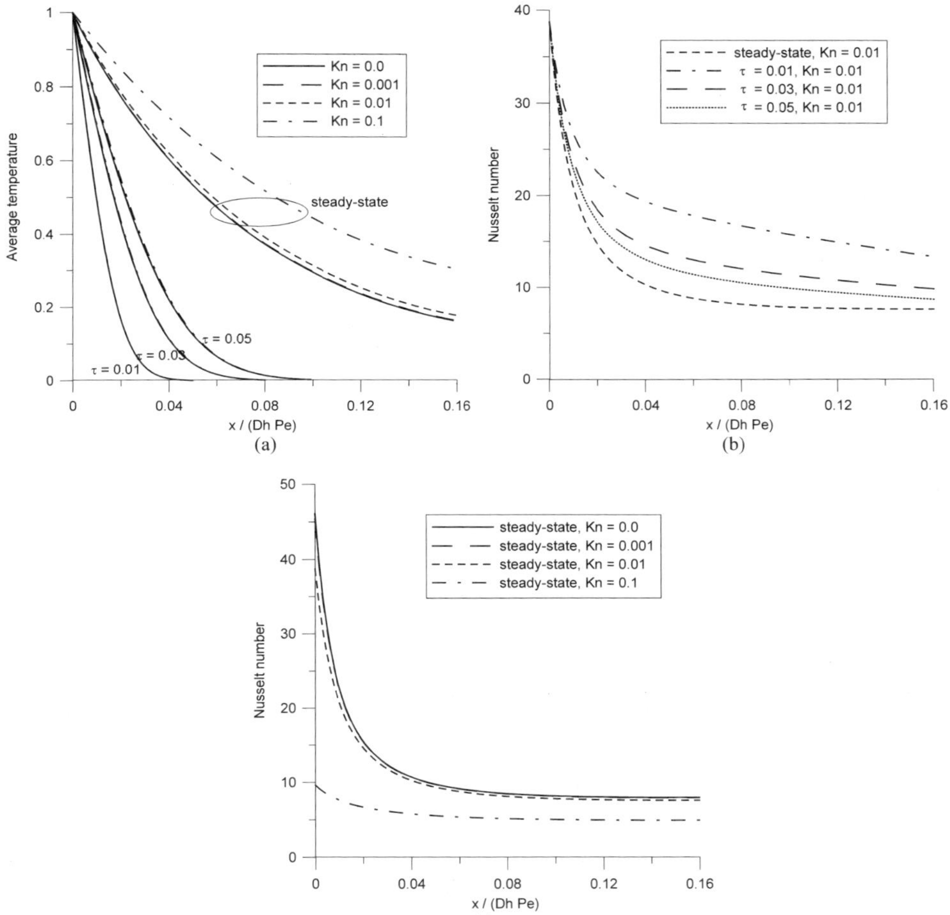

Figure 7: Influence of Knudsen number on the transient and steady behaviors of (a) - dimensionless average temperature and (b) and (c) - local Nusselt number (parallel plates, Pe = 10, Kn =0, 0.001, 0.01& 0.1 and Br = 0, S=15).

7. Conclusions

This work discusses hybrid numerical-analytical solutions and mixed symbolic-numerical algorithms for solving transient fully developed flow and transient forced convection in micro-channels, making use of the Generalized Integral Transform Technique (GITT) and the *Mathematica* system.

The first model, employed in the transient flow analysis, was described by the transient momentum equation for fully developed laminar flow of a Newtonian fluid within parallel plates and circular tubes with slip flow boundary conditions. The GITT approach proved to be very accurate and of low computational cost in solving this class of problems, due to the excellent convergence behavior provided by the time-varying filtering strategy adopted. The proposed model can be useful as a practical tool in analyzing transient flows with pressure gradient time functions fitted from experimental data, since the implementation is fully automatic for any prescribed source term input. This approach can be directly applied to the treatment of transient electrosmotic flow of liquids within microchannels as well.

The hybrid numerical-analytical solution for transient convection heat transfer within parallel-plates channels with laminar slip flow is also advanced, based on the integral transform approach and on the exact solution of the related eigenvalue problem, in terms of hypergeometric functions. A partial integral transformation strategy is employed, which results in a coupled system of one-dimensional partial differential equations for the transformed potentials, which are numerically handled by the Method of Lines implemented within the *NDSolve* routine of the *Mathematica* system. A symbolic-numerical implementation under the *Mathematica 4.2* platform is developed, for both the analytical and numerical computation of the related eigenfunction expansions and transformed PDE system. *Mathematica* rules are given and demonstrated by solving examples considered in previous papers dealing with regular channels, and in addition providing a set of new results for micro-channel configurations. The approach is also readily extendable to the analysis of transient convection in micro-channels with time-varying fluid flow, in combination with the analytical solutions obtained in the first part of this work.

8. Acknowledgements

The authors would like to acknowledge the financial support provided by FAPERJ and CNPq/Brazil. The present work is related to the PRONEX Project "Núcleo de Excelência em Turbulência", also sponsored by FAPERJ & CNPq.

References

1. Kakaç, S., Li, W. and Cotta, R.M. (1990), Unsteady Laminar Forced Convection in Ducts with Periodic Variation of Inlet Temperature, *J. Heat Transfer*, V. 112, pp. 913-920.
2. Kakaç, S., Cotta, R.M., Hatay, F.F. and Li, W. (1990), Unsteady Forced Convection in Ducts for a Sinusoidal Variation of Inlet Temperature, *9th Int. Heat Transfer Conf.,* Paper # 8-MC-01, pp. 265-270, Jerusalem, Israel, August.
3. Kakaç, S. and Cotta, R.M. (1993), Experimental and Theoretical Investigation on Transient Cooling of Electronic Systems, *Proc. of the NATO Advanced Study Institute on Cooling of Electronic Systems*, Invited Lecture, NATO ASI Series E: Applied Sciences, Vol. 258, pp. 239-275, Turkey, June/July.
4. Santos, C.A.C., Brown, D.M., Kakaç, S. and Cotta, R.M. (1995), Analysis of Unsteady Forced Convection in Turbulent Duct Flow, *J. Thermophysics & Heat Transfer*, Vol. 9, Nº 3, pp. 508-515.
5. Cotta, R.M. (1996) Integral Transforms in Transient Convection:- Benchmarks and Engineering Simulations, Invited Keynote Lecture, *ICHMT International Symposium on Transient Convective Heat Transfer*, Turkey, pp.433-453, August.
6. Santos, C.A.C., Medeiros, M.J., Cotta, R.M., and Kakaç, S. (1998), Theoretical Analysis of Transient Laminar Forced Convection in Simultaneous Developing Flow in Parallel-Plate Channel, *7th AIAA/ASME Joint Themophysics and Heat Transfer Conference*, AIAA Paper #97-2678, Albuquerque, New Mexico, June.
7. Cheroto, S., Mikhailov, M.D., Kakaç, S., and Cotta, R.M., (1999), Periodic Laminar Forced Convection:- Solution via Symbolic Computation and Integral Transforms, *Int. J. Thermal Sciences*, V.38, no.7, pp.613-621.
8. Kakaç, S., Santos, C.A.C., Avelino, M.R., and Cotta, R.M. (2001), Computational Solutions and Experimental Analysis of Transient Forced Convection in Ducts, Invited Paper, *Int. J. of Transport Phenomena*, V.3, pp.1-17.
9. Cotta, R.M., Orlande, H.R.B., Mikhailov, M.D., and Kakaç, S. (2003), Experimental and Theoretical Analysis of Transient Convective Heat and Mass Transfer:- Hybrid Approaches, Invited Keynote Lecture, *ICHMT International Symposium on Transient Convective Heat And Mass Transfer in Single and Two-Phase Flows*, Cesme, Turkey, August 17 - 22.
10. Gad-el-Hak, M. (2001) Flow Physics in MEMS, *Mec. Ind.*, Vol.2, pp.313-341.

11. Papautsky, I., Ameel, T.A., and Frazier, A.B. (2001), A Review of Laminar Single-Phase Flow in Microchannels, *ASME Int. Mechanical Eng. Congress and Exposition*, New York , USA, 1-9.
12. Bayazitoglu, Y. and Tunc, G. (2001), Convective Heat Transfer in Microchannels: Slip Flow, *2nd Int. Conf. Computational Heat & Mass Transfer, ICCHMT-2001*, Invited Lecture, Rio de Janeiro, Brasil, 1,112-121.
13. Rostami, A.A., Mujumdar, A.S. and Saniei, N. (2002) Flow and Heat Transfer for Gas Flowing in Microchannels: A Review, *Heat and Mass Transfer*, Vol.38, pp.359-367.
14. Karniadakis, G., Beskok, A. (2002) *Micro Flows*, Springer Verlag.
15. Tabeling, P. (2003) *Introduction à la Microfluidique*, Belin, Collection Échelles, Paris.
16. Yener, Y., Kakaç, S., and Avelino, M. R. (2005) Single-Phase Convective Heat Transfer in Microchannels – The State-of-the-Art Review, *NATO Advanced Study Institute on Microscale Heat Transfer*, Cesme, Turkey, July 18-30.
17. Barron, R.F., Wang, X., Ameel, T.A., and Warrington, R.O. (1997) The Graetz Problem Extended to Slip Flow, *Int. J. Heat Mass Transfer*, Vol.40, pp.1817-1823.
18. Yu, S., and Ameel, T.A. (2001) Slip Flow Heat Transfer in Rectangular Microchannels, *Int. J. Heat Mass Transfer*, Vol.44, pp.4225-4234.
19. Yu, S., and T.A. Ameel, (2002), Slip Flow Convection in Isoflux Rectangular Microchannels, *ASME J. Heat Transfer*, 124, 346-355.
20. Tunc, G., and Bayazitoglu, Y. (2001) Nusselt number variation in microchannels, *Proc. of the 2nd Int. Conf. Comp. Heat & Mass Transfer, 2nd ICCHMT*, Rio de Janeiro, Brazil, October 2001.
21. Tunc, G., and Bayazitoglu, Y. (2002) Heat Transfer in Rectangular Microchannels, *Int. J. Heat Mass Transfer*, Vol.45, pp.765-773.
22. Tunc, G. and Bayazitoglu, Y. (2001) Heat Transfer in Microtubes with Viscous Dissipation, *Int. J. Heat Mass Transfer*, Vol.44 (13), pp.2395-2403.
23. Larrodé, F.E., Housiadas, C., and Drossinos, Y. (2000) Slip Flow Heat Transfer in Circular Tubes, Int. J. Heat Mass Transfer, 43, 2669-2680
24. Bestman, A.R., Ikonwa, I. O., and Mbelegodu, I. U. (1995) Transient Slip Flow, *Int. J. Energy Research*, Vol. 19, pp.275-277.
25. Colin, S., Aubert, C., and Caen, R. (1998) Unsteady gaseous flows in rectangular microchannels: Frequency response of one or two pneumatic lines connected in series, *Eur. J. Mech., B/Fluids*, 17, no.1, 79-104.
26. Aubert, C. (1999) *Ecoulements Compressibles de Gas dans les Microcanaux: Effets de Raréfaction, Effets Instationnaires*, These de Docteur de L'Université Paul Sabatier, France.
27. DeSocio, L.M., and Marino, L. (2002) Slip flow and temperature jump on the impulsively started plate, *Int. J. Heat Mass Transfer*, 45, 2169-2175.
28. Kang, Y., Yang, C., and Huang, X. (2002) Dynamic Aspects of Electrosmotic Flow in a Cylindrical Microcapillary, *Int. J. Eng. Science*, Vol.40, pp.2203-2221.
29. Yang, C., Ng, C.B., and Chan, V. (2002) Transient Analysis of Electrosmotic Flow in a Slit Microchannel, *J. Colloid and Interface Science*, Vol. 248, pp.524-527.
30. Yang, J. and Kwok, D. Y. (2003) Time-dependent laminar electrokinetic slip flow in infinitely extended rectangular microchannels, J. Chemical Physics, V.118, no.1, pp.354-363.
31. Yang, J. and Kwok, D. Y. (2003) Analytical treatment of flow in infinitely extended circular microchannels and the effect of slippage to increase flow efficiency, *J. Micromech. Microeng.*, V.13, pp.115-123.
32. Bhattacharyya, A., Masliyah, J.H. and Yang, J. (2003) Oscillating laminar electrokinetic flow in infinitely extended circular microchannels, *J. Colloid and Interface Science*, V.261, pp.12-20.
33. Khaled, A.R.A., and Vafai, K., The effect of the slip condition on Stokes and Couette flows due to an oscillating wall: exact solutions, *Int. J. Non-Linear Mechanics*, in press.
34. Mikhailov, M.D., and Ozisik, M.N. (1984) *Unified Analysis and Solutions of Heat and Mass Diffusion*, John Wiley, New York; also, Dover Publications, 1994.
35. Cotta, R.M. (1993) *Integral Transforms in Computational Heat and Fluid Flow*, CRC Press, Boca Raton, FL.
36. Cotta, R.M., and Mikhailov, M.D. (1997) *Heat Conduction: Lumped Analysis, Integral Transforms, Symbolic Computation*, Wiley-Interscience, Chichester, UK.
37. Cotta, R.M., Ed., (1998) *The Integral Transform Method in Thermal and Fluids Sciences and Engineering*, Begell House, New York.
38. Cotta, R.M., and Mikhailov, M.D. (2001) Hybrid Approaches in Convective Heat Transfer, in: *Benchmark Results for Convective Heat Transfer in Ducts: - The Integral Transform Approach,* eds. C.A.C. Santos, J.N.N. Quaresma, and J.A. Lima, ABCM Mechanical Sciences Series, Editora E-Papers, Rio de Janeiro, Part 1, Chapter II, pp.17-38.
39. Cotta, R. M. and Orlande, H.R.B. (2003) Hybrid Approaches in Heat and Mass Transfer:- A Brazilian Experience with Applications in National Strategic Projects, *Heat Transfer Eng.*, Invited Editorial, V.24, no.4, pp.1-5.

40. Cotta, R.M., and Mikhailov, M.D. (2004) Hybrid Methods and Symbolic Computations, in: *Handbook of Numerical Heat Transfer*, 2nd edition, Chapter 16, Eds. W.J. Minkowycz, E.M. Sparrow, and J.Y. Murthy, John Wiley, New York. (in press)
41. Wolfram, S. (1991) *Mathematica - A System for Doing Mathematics by Computer*, The Advanced Book Program, Addison-Wesley Publishing Co., Redwood City.
42. Barron, R. F., Wang, X., Warrington, R. O., and Ameel, T. (1996) Evaluation of the Eigenvalues for the Graetz Problem in Slip-flow, *Int. Comm. Heat Mass Transfer,* Vol. 23 (4), pp.1817-1823.
43. Graetz, L. (1883) Uber die Warmeleitungsfahigkeit von Flussigkeiten, *Annalen der Physik und Chemie*, part 1, Vol. 18, pp. 79-94, (1883); part 2, Vol. 25, pp. 337-357, (1885).
44. Mikhailov, M. D. and Cotta, R. M. (1997) Eigenvalues for the Graetz Problem in Slip-Flow, *Int. Comm. Heat & Mass Transfer*, Vol.24, no.3, pp.449-451.
45. Mikhailov, M.D., and Cotta, R.M., Mixed Symbolic-Numerical Computation of Convective Heat Transfer with Slip Flow in Microchannels, *Int. Comm. Heat & Mass Transfer*, submitted.
46. Tunc, G., and Bayazitoglu, Y., 2002, Convection at the Entrance of Micropipes with Sudden Wall Temperature Change, *Proc. of ASME IMECE 2002*, Paper # 32438, New Orleans, November.
47. Cotta, R.M. and Ozisik, M.N. (1986) Transient Forced Convection in Laminar Channel Flow with Stepwise Variations of Wall Temperature, *Can J. Chem. Eng.*, Vol. 64, pp. 734-742.
48. Cotta, R.M., Ozisik, M.N. and McRae, D.S. (1986) Transient Heat Transfer in Channel Flow with Step Change in Inlet Temperature, *Num. Heat Transfer*, V. 9, pp. 619-630.
49. Cotta, R.M. and Gerk, J.E.V. (1994) Mixed Finite Difference/Integral Transform Approach for Parabolic-Hyperbolic Problems in Transient Forced Convection, *Numerical Heat Transfer - Part B: Fundamentals*, Vol 25, pp. 433-448.
50. Gondim, R.R (1997) *Transient Internal Forced Convection with Axial Diffusion: Solution by Integral Transforms*, D.Sc. Thesis, COPPE / UFRJ (in Portuguese).
51. Gondim, R.R., Cotta, R.M., Santos, C.A.C., and Mat, M. (2003) Internal Transient Forced Convection with Axial Diffusion: Comparison of Solutions Via Integral Transforms, *ICHMT International Symposium on Transient Convective Heat And Mass Transfer in Single and Two-Phase Flows*, Cesme, Turkey, August 17 - 22.
52. IMSL Library, MATH/LIB, Houston, TX, 1989.
53. Cotta, R.M., Alves, L.S.B. and Mikhailov, M.D. (2001) Applied Numerical Analysis with *Mathematica*, Editora E-Papers, Rio de Janeiro, Brasil.
54. Cardoso, C.R., and Cotta, R.M. (2003) Transient Slip Flow in Microchannels via Integral Transforms, *Proc. of the VI EMC-Computational Modeling Meeting*, IPRJ-UERJ, Nova Friburgo, Brazil, December 1-3.

FROM NANO TO MICRO TO MACRO SCALES IN BOILING

V. K. DHIR, H. S. ABARAJITH, AND G. R. WARRIER
Henry Samueli School of Engineering and Applied Science
University of California, Los Angeles
Los Angeles, CA 90095-1592
Ph: (310) 825-8507, Fax: (310) 206-4061
Email: vdhir@seas.ucla.edu

1. Introduction

Boiling, being the most efficient mode of heat transfer is employed in various energy conversion systems and component cooling devices. The process allows accommodation of high heat fluxes at low wall superheats. At the same time the process is very complex and its understanding imposes severe challenges. Prior to inception of boiling, at low heat fluxes, the heat transfer is controlled by convection (natural or forced). At higher heat fluxes, the heat transfer is controlled by bubble dynamics. Initially during partial nucleate boiling, discrete bubbles form on the heater surface. At moderately high heat fluxes, bubbles start merging laterally as well as vertically. The lateral and vertical merger of bubbles indicates the transition from partial nucleate boiling to fully developed nucleate boiling.

In the past, several attempts have been made to model bubble growth and bubble departure processes on a heated wall. Lee and Nydahl [1] calculated the bubble growth rate by solving the flow and temperature fields numerically. They used the formulation of Cooper and Llyod [2] for the micro layer thickness. However they assumed a hemispherical bubble and wedge shaped microlayer and thus they could not account for the shape change of the bubble during growth.

Zeng *et al.* [3] used a force balance approach to predict the bubble diameter at departure. They included the surface tension, inertial force, buoyancy and the lift force created by the wake of the previously departed bubble. But there was empiricism involved in computing the inertial and drag forces. The study assumed a power law profile for growth rate with the proportionality constant exponent determined from the experiments.

Mei *et al.* [4] studied the bubble growth and departure time assuming a wedge shaped microlayer. They also assumed that the heat transfer to the bubble was only through the microlayer, which is not totally correct for both subcooled and saturated boiling. The study did not consider the hydrodynamics of the liquid motion induced by the growing bubble and introduced empiricism through the shape of the growing bubble. Welch [5] has studied bubble growth using a finite volume method and an interface tracking method. The conduction in the solid wall was also taken in to account. However, the microlayer was not modeled explicitely.

In 1994, Sussman *et al.* [6] presented a level-set approach for computing incompressible two-phase flow. By keeping the level set as a distance function, the interface was easily captured by the zero level-set. The calculations, for air bubbles in water and falling water drops in air, yielded satisfactory results. Though the level-set method is easy to use, the numerical discretization of the level-set formulation does not satisfy mass conservation, in general. Chang *et al.* [7] introduced a volume correction step to the level-set formulation in 1996. By solving an additional Hamilton-Jacobi equation to steady state, the mass was forced to be conserved. In 1999, Son *et al.* [8] developed a model for growth of an isolated bubble on a heated surface using complete numerical simulation. The model, based on Sussman's level-set method, captures the bubble interface and offers many improvements over previously published models. It yields the spatial and temporal distribution of the wall heat flux, the microlayer contribution and the interface heat transfer. In this model a static contact angle was used both for the advancing and receding phases of the interface. However, the numerical results agreed well with data from experiments. One possible

S. Kakaç et al. (eds.), Microscale Heat Transfer, 197–216.

reason could be that the constant contact angle used in the numerical studies represents the average value of the advancing and receding contact angles (static contact angle) and the bubble is symmetrical in pool boiling case. Also, the time over which the receding contact angle prevails is much shorter than that for the advancing contact angle. In 2001, Son [9] modified Chang *et al's* formulation and included the volume correction formulation into the boiling heat transfer model.

Singh and Dhir [10] have obtained numerical results for low gravity conditions by exercising the numerical simulation model of [8], when the liquid is subcooled. The computational domain was divided in to two regions viz. micro and macro regions. The interface shape and velocity and temperature field in the liquid in the macro region were obtained by solving the conservation equations. For the micro region, lubrication theory was used, which included the disjoining pressure in the thin liquid film. The solutions of the micro region and macro region were matched at the outer edge of the micro layer.

Abarajith and Dhir [11] studied the effects of contact angle on the growth and departure of a single bubble on a horizontal heated surface during pool boiling under normal gravity conditions. The contact angle was varied by changing the Hamaker constant that defines the long-range forces. They also studied the effect of contact angle on the microlayer and macrolayer heat transfer rates.

In spite of all the advances in the computational techniques for solving boiling problems, the one variable that has still not been modeled correctly is the contact angle (static or dynamic). In all the previous studies, the contact angle is specified as an input for the simulations. The reason the contact angle has not been modeled is that it depends on the physical phenomena occurring close to or at the solid surface, which by its very nature, occurs at very small length scales (of the order of nano to micrometers). In this work, the contact angle is related to the Hamaker constant.

2. Mathematical Development of Model

The three-dimensional model discussed in this paper is an extension of the two-dimensional model developed earlier by [8]. The model is used to study both single dynamics and multiple bubble merger during subcooled and saturated pool nucleate boiling. The computational domain is divided into two regions, namely, the micro region and the macro region as shown in Fig 1. The micro region is a thin film that lies between the solid wall and bubble whereas the macro region consists of the bubble and it's surrounding. Both the regions are coupled through matching of the shape at the outer edge of the micro layer and are solved simultaneously. Microlayer modeling covers length scales from nano to micrometers, whereas the macro region includes the length scales from micrometers to millimeters and above.

2.1 MICRO REGION

A two-dimensional quasi-static model is used for the micro region and no azimuthal variations are considered. As such, the solution for the microlayer thickness is obtained in the radial direction from the center of the bubble base. This solution is assumed to be valid for all the azimuthal positions. This assumption is still applied during the multiple bubble merger process when the bubble shapes are not symmetrical, such that no cross flow occurs in the circumferential direction. Furthermore, the length of the micorlayer is assumed to remain constant (though varies with contact angle) throughout the bubble growth.

The equation of mass conservation in micro region is written as,

$$\frac{q}{h_{fg}} = -\frac{1}{r}\frac{\partial}{\partial r}\int_0^{\delta} \rho_l .rudz \,, \tag{1}$$

where q is the conductive heat flux from the wall, defined as $\frac{k_l(T_{wall}-T_{\text{int}})}{\delta}$ with δ as the thickness of the thin film. Lubrication theory has been used ([12], [13] and [14]). According to the lubrication theory, the momentum equation in the micro region is written as,

$$\frac{\partial p_l}{\partial r} = \mu \frac{\partial^2 u}{\partial z^2}, \tag{2}$$

where p_l is the pressure in the liquid. The heat conducted through the thin film must match that due to evaporation from the vapor-liquid interface. By using a modified Clausius-Clapeyron equation, the energy conservation equation for the micro region yields,

$$\frac{k_l(T_{wall}-T_{\text{int}})}{\delta} = h_{ev}\left[T_{\text{int}} - T_v + \frac{(p_l - p_v)T_v}{\rho_l h_{fg}}\right]. \tag{3}$$

The evaporative heat transfer coefficient is obtained from kinetic theory as,

$$h_{ev} = 2\left[\frac{M}{2\pi \bar{R} T_v}\right]^{1/2} \frac{\rho_v h_{fg}^2}{T_v} \text{ and } T_v = T_s(p_v). \tag{4}$$

The pressure of the vapor and liquid phases at the interface are related by,

$$p_l = p_v - \sigma K - \frac{A_0}{\delta^3} + \frac{q^2}{2\rho_v h_{fg}^2}, \tag{5}$$

where A_0 is the dispersion constant. The second term on the right-hand side of equation (5) accounts for the capillary pressure caused by the curvature of the interface, the third term is for the disjoining pressure, and the last term originates from the recoil pressure. The curvature of the interface is defined as,

$$K = \frac{1}{r}\frac{\partial}{\partial r}\left[r\frac{\partial \delta}{\partial r} / \sqrt{1+\left(\frac{\partial \delta}{\partial r}\right)^2}\right]. \tag{6}$$

Combining the mass conservation (Eq. (1)), momentum conservation equation (Eq. (2)), mass balance and energy conservation, (Eq. (3)) and pressure balance (Eq. (5)) along with Eq. (6) for the curvature for the micro-region results in a set of three nonlinear first-order ordinary differential Eqs. (7), (8) and (9), as derived in [15],

$$\frac{\partial \delta_r}{\partial r} = -\frac{\delta_r(1+\delta_r^2)}{r} + \frac{(1+\delta_r^2)^{3/2}}{\sigma}\left[\frac{\rho_l h_{fg}}{T_v}\left(T_{\text{int}} - T_v - \frac{q}{h_{ev}}\right) - \frac{A_0}{\delta^3} + \frac{q^2}{\rho_v h_{fg}^2}\right], \tag{7}$$

$$\frac{\partial T_{\text{int}}}{\partial r} = -\frac{q\delta_r}{\kappa_l + h_{ev}\delta} + \frac{3T_v h_{ev}\mu\Gamma}{(\kappa_l + h_{ev}\delta)\rho_l^2 h_{fg} r \delta^2}, \tag{8}$$

$$\frac{\partial[\Gamma]}{\partial r} = -\frac{rq}{h_{fg}}, \tag{9}$$

where the mass flow rate in the thin film, $\Gamma = -\frac{r\delta^3 \rho}{3\mu}\frac{\partial p_l}{\partial r}$.

Equations (7), (8) and (9) can be simultaneously integrated using a Runge-Kutta scheme, when boundary conditions at $r = R_0$ are given. In the present case, the interface shape obtained from micro and macro solutions is matched at the radial location R_1. As such, this is the end point for the integration of the above set of equations. The radius of the dry region beneath a bubble, R_0, is related to R_1 by the definition of the apparent contact angle, $\tan\varphi = 0.5h/(R_1 - R_0)$.

The boundary conditions for the film thickness at the end points are given as,

$$\begin{array}{ll} \delta = \delta_0, \ \ \delta_r = 0, \ \ \Gamma = 0 & \text{at} \ \ r = R_0 \\ \delta = h/2, \ \ \delta_{rr} = 0 & \text{at} \ \ r = R_1 \end{array}, \qquad (10)$$

where, δ_0 is the interline film thickness at the inner edge of micro-layer, $r = R_0$, and is calculated by combining Eqs. (3) and (4) and requiring that $T_{\text{int}} = T_{wall}$ at $r = R_0$ with h being the spacing of the three-dimensional grid for the macro-region. For a given $T_{\text{int},0}$ at $r = R_0$, a unique vapor-liquid interface is obtained.

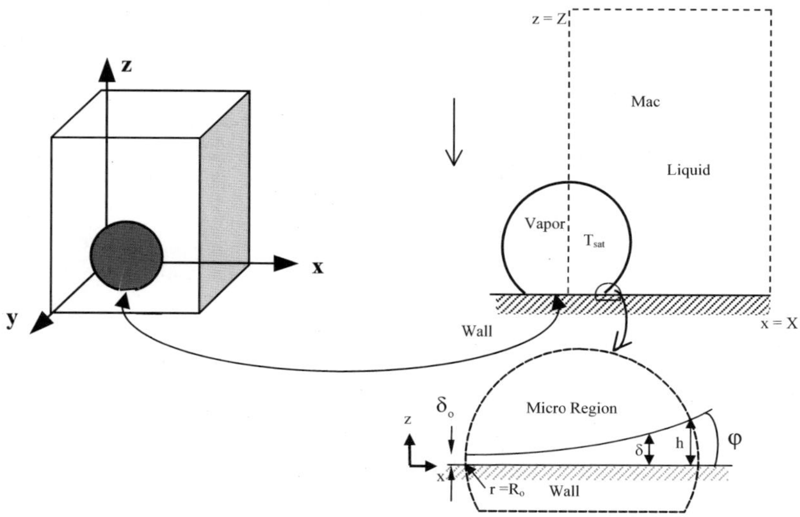

Figure 1 Computational domain for nucleate boiling showing details of micro and macro region.

2.2 MACRO REGION

For numerically analyzing the macro region, the level set formulation developed in [8] for nucleate boiling of pure liquids is used. The interface separating the two phases is captured by a distance function, ϕ, which is defined as a signed distance from the interface. The negative sign is chosen for the vapor phase and the positive sign for the liquid phase. The discontinuous pressure drop across the vapor and liquid, caused by the surface tension force, is smoothed into a numerically continuous function with a δ-function formulation (see [6] for details). The continuity, momentum, and energy conservation equations for the vapor and liquid in the macro region are written as,

$$\rho_t + \nabla\cdot(\rho\vec{u}) = 0, \tag{11}$$

$$\rho(\vec{u}_t + \vec{u}\cdot\nabla\vec{u}) = -\nabla p + \nabla\cdot\mu\nabla\vec{u} + \nabla\cdot\mu\nabla\vec{u}^T + \rho\vec{g} - \rho\beta_T(T-T_s)\vec{g}(t) - \sigma K\nabla H, \tag{12}$$

$$\rho c_p(T_t + \vec{u}\cdot\nabla T) = \nabla\cdot\kappa\nabla T \quad \text{for } H>0, \tag{13}$$

$$T = T_s(p_v) \quad \text{for } H=0. \tag{14}$$

The fluid density, viscosity and thermal conductivity of the fluid are defined in terms of the step function, H, as,

$$\rho = \rho_v + (\rho_l - \rho_v)H, \tag{15}$$

$$\mu^{-1} = \mu_v^{-1} + (\mu_l^{-1} - \mu_v^{-1})H, \tag{16}$$

$$\kappa^{-1} = \kappa_l^{-1}H, \tag{17}$$

where, H, is the Heaviside function, which is smoothed over three grid spaces as described below,

$$H = \begin{cases} 1 & \text{if } \phi \ge 1.5h \\ 0 & \text{if } \phi \le 1.5h \\ 0.5 + \dfrac{\phi}{3h} + \sin\left[\dfrac{2\pi\phi}{3h}\right]/(2\pi) & \text{if } |\phi| \le 1.5h \end{cases} \tag{18}$$

The mass conservation Eq. (11) can be rewritten as,

$$\nabla\cdot\vec{u} = -(\rho_t + \vec{u}\cdot\nabla\rho)/\rho, \tag{19}$$

The term on right hand side of Eq. (19) is the volume expansion due to liquid-vapor phase change. From the conditions of the mass continuity and energy balance at the vapor-liquid interface, the following equations are obtained,

$$\vec{m} = \rho(\vec{u}_{\text{int}} - \vec{u}) = \rho_l(\vec{u}_{\text{int}} - \vec{u}_l) = \rho_v(\vec{u}_{\text{int}} - \vec{u}_v), \tag{20}$$

$$\vec{m} = \kappa \nabla T / h_{fg} , \tag{21}$$

where $\vec{m}$ is the evaporation rate, and $\vec{u}_{\text{int}}$ is the interface velocity. If the interface is assumed to advect in the same way as the level set function, the advection equation for density at the interface can be written as,

$$\rho_t + \vec{u}_{\text{int}} \cdot \nabla \rho = 0 . \tag{22}$$

Using Eqs. (18), (20) and (21), the continuity equation (Eq. (19)) for the macro region can be rewritten as,

$$\nabla \cdot \vec{u} = \frac{\vec{m}}{\rho^2} \cdot \nabla \rho . \tag{23}$$

The vapor produced as a result of evaporation from the micro region is added to the vapor space through the cells adjacent to the heated wall, and is expressed as,

$$\left(\frac{1}{V_c} \frac{dV}{dt} \right)_{mic} = \frac{\dot{m}_{mic}}{V_c \rho_v} \delta_\varepsilon(\phi) , \tag{24}$$

where, V_c is the volume of the control volume and $\dot{m}_{mic}$ is the evaporation rate from the micro-layer which can be expressed as,

$$\dot{m}_{mic} = \int_{R_0}^{R_1} \frac{\kappa_l (T_w - T_{\text{int}})}{h_{fg} \delta} r dr . \tag{25}$$

The bubble expansion due to the vapor addition from the micro layer is smoothed at the vapor-liquid interface by the smoothed delta function as given in [6],

$$\delta_\varepsilon(\phi) = \partial H / \partial \phi . \tag{26}$$

In the level set formulation, the level set function, ϕ, is used to keep track of the vapor-liquid interface location as the set of points where $\phi = 0$, and it is advanced by the interfacial velocity while solving the following equation,

$$\phi_t = -\vec{u}_{\text{int}} \cdot \nabla \phi . \tag{27}$$

To keep the values of ϕ close to that of a signed distance function, $|\nabla \phi| = 1$, ϕ is reinitialized after every time step,

$$\frac{\partial \phi}{\partial t} = \frac{u_1 \cdot \phi_0}{\sqrt{\phi_0^2 + h^2}} (1 - |\nabla \phi|) , \tag{28}$$

where, ϕ_0 is a solution of Eq. (27) and u_1 is the characteristic interface velocity, which is set to unity.

In these numerical simulations, the independent variables are: (i) wall superheat, (ii) liquid subcooling, (iii) system pressure, (iv) thermophysical properties of test fluid, (v) contact angle, (vi) gravity level, (vii) thermophysical properties of the solid and surface quality (conjugate problem), and (viii) heater geometry.

3. Details of Computations

Figure 1 shows the computational domain used in the simulations. Details of the micro and macro regions are also shown in Fig. 1. In these simulations, the gravity vector is oriented in the –z direction (i.e., the bubble is growing on an upwards facing heated surface). The simulations are carried out on a uniform grid ($\Delta x = \Delta y = \Delta z$).

3.1 BOUNDARY CONDITIONS

The boundary conditions for the pool boiling simulations are as follows:

$$\begin{array}{llll} u=0, v=w=0, & T_x=0, & \phi_x=0, & at \quad x=0 \\ u_x=v_x=w_x=0, & T_x=0, & \phi_x=0, & at \quad x=X \\ u=v=w=0, & T=T_w, & \phi_z=-\cos\varphi, & at \quad z=0 \\ u_z=w_z=v=0, & T_z=0, & \phi_z=0, & at \quad z=Z \\ u_y=v_y=w_y=0, & T_y=0, & \phi_y=0, & at \quad y=0 \\ u_y=v_y=w_y=0, & T_y=0, & \phi_y=0, & at \quad y=Y \end{array} \qquad (29)$$

where φ is the static contact angle.

3.2 SOLUTION PROCEDURE

For the numerical calculations, the governing equations for micro and macro regions are nondimensionalized by defining the characteristic length, l_0, the characteristic velocity, u_0, and the characteristic time, t_0 as,

$$l_0 = \sqrt{\sigma/[g(\rho_l - \rho_v)]}; \; u_0 = \sqrt{g l_0}; \; t_0 = l_0 / u_0 \,. \qquad (30)$$

In performing these simulations, the wall temperature is assumed to be constant and the thermodynamic properties of the individual phases are assumed to be insensitive to the small changes in temperature and pressure. The assumption of constant property is reasonable as the computations are performed for low wall superheat range.

The simulations were performed assuming that the flow is laminar. Additionally, the contact angle is assumed to be known. The initial velocity is assumed to be zero everywhere in the domain. The initial fluid temperature profile is taken to be linear in the natural convection thermal boundary layer and the thermal boundary layer thickness, δ_T, is evaluated using the correlation for the turbulent natural convection on a horizontal plate as, $\delta_T = 7.14(\nu_l \alpha_l / g\beta_T \Delta T)^{1/3}$.

The governing equations are numerically integrated by following the procedure of Son *et al.* [8].

1) The value of A_0, the Hamaker (dispersion) constant is initially guessed for a given contact angle. This initial guess can be obtained from Molecular Dynamics simulation results (if available).
2) The macro layer equations are then solved to determine the value of R_l (radial location of the vapor-liquid interface at $\delta = h/2$.)

3) The micro layer equations are subsequently solved with the guessed value of A_0, to determine the value of R_0 (radial location of the vapor-liquid interface at $\delta = \delta_0$.)
4) The apparent contact angle is then calculated from $\tan\varphi = 0.5h/(R_1 - R_0)$ and steps 1 - 4 are repeated for a different value of A_0, if the values of the given and the calculated apparent contact angles do not match.

4. Experiments

In order to validate the results of the numerical simulations, nucleate boiling experiments need to be performed under defined conditions. The dynamics of single and multiple bubbles were experimentally studied by Qiu *et al.* ([16] and [17]) using a polished silicon wafer as the test surface. The wafer is 10 cm in diameter and 1 mm thick. Single and multiple cavities (with sizes varying from 10 to 3 μm) were etched on the wafer using standard microfabrication techniques. Thin-film strain gage heaters were attached to back surface of the wafer. By energizing the heaters individually, any number of cavities could be activated. Thermocouples attached to the bottom were used to measure the temperature in the vicinity of the cavity. The silicon wafer heater assembly was placed in the experimental apparatus shown in Fig. 2(a). Figure 2(b) shows the details of the test wafer. A CCD camera (up to 1220 frames/sec) was used to capture the boiling process.

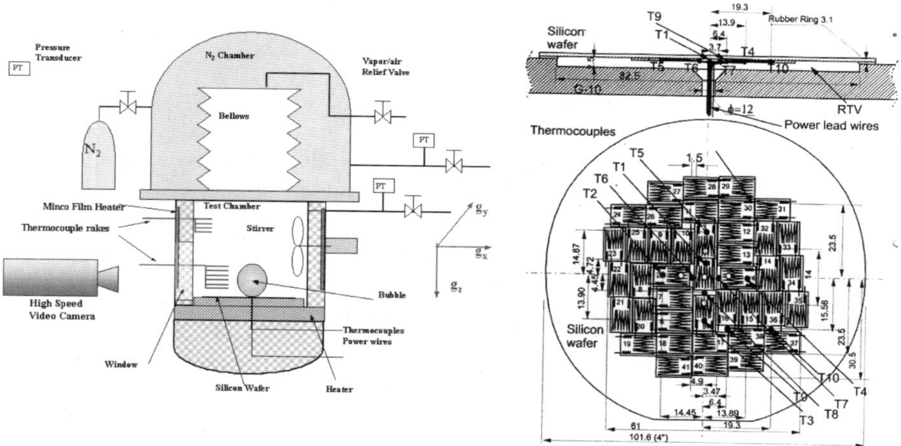

(a) (b)

Figure 2 (a) Experimental apparatus (b) details of test heater.

5. Results and Discussion

5.1 HAMAKER CONSTANT

As mentioned earlier, the value of A_0, the Hamaker constant (dispersion constant), is found by iteration so as to match the bubble shape at the outer edge of the microlayer with that of the macrolayer, for a given contact angle. Figure 3 shows the variation of the dispersion constant, A_0, with contact angle for two fluids: water and PF5060. The dispersion constant, A_0 changes from negative to positive value at around 18° indicating the change to attractive nature between the liquid and wall. The value of the dispersion constant A_0 does not vary much between water and PF5060, for the same contact angle and $\Delta T_w = 8$ °C.

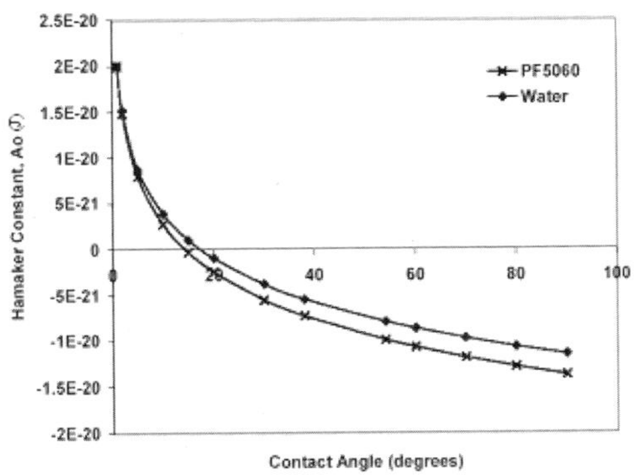

Figure 3 Variation of the dispersion constant (A_o) with contact angle (Fluids: water and PF5060, $\Delta T_w = 8$ °C, $\Delta T_{sub} = 0$ °C).

5.2 SINGLE BUBBLE

Figure 4(a) shows the variation of bubble departure diameter with wall superheat for boiling of saturated water at one atmosphere pressure. Both the bubble diameter and bubble growth period increases with wall superheat. Also shown in Fig. 4(a) is the experimental data of Qiu *et al.* [16]. Good agreement between the experimental and numerically predicted bubble departure diameters is observed, though the bubble growth time is slightly over predicted. Figure 4(b) shows the variation of bubble departure diameter and bubble growth period with liquid subcooling, for water ($\Delta T_{wall} = 8$ °C). Both the bubble departure diameter and bubble growth period increase with increasing liquid subcooling. The contribution of the various heat transfer mechanisms (microlayer, evaporation around the bubble boundary, and condensation) as a function of time are shown in Fig. 5, for boiling of subcooled water. The condensation around the bubble is zero in the initial stages of bubble growth (up to 32 ms), when the bubble is still smaller than the thermal boundary layer. Once the bubble diameter becomes larger than the thermal boundary layer, the condensation rate increases (shown in Fig.5 as a negative value).

The bubble growth history for two fluids with different contact angles (water and PF5060) is shown in Fig. 6. In general, the lower the contact angle, the smaller is the bubble departure diameter and the bubble growth time. The corresponding evaporative heat transfer rates from the micro and macro layers are shown in Fig. 7. The microlayer evaporation rate increases with increasing contact angle because the bubble base area and interfacial area increases with increasing contact angle. A corresponding increase in the evaporation rate from the macrolayer is also observed. The area-averaged Nusselt number for multiple bubble growth and departure cycles are plotted in Fig. 8. It is seen that the microlayer contributes about 20% of the total heat transfer rate. Also, it takes about 10 to 12 cycles before quasi-static conditions are achieved.

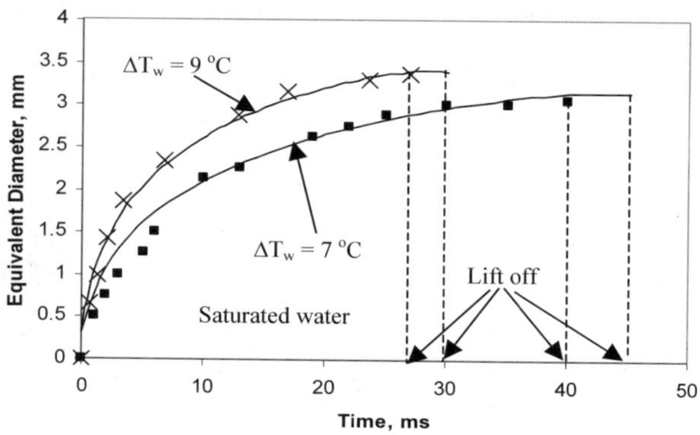

(a)

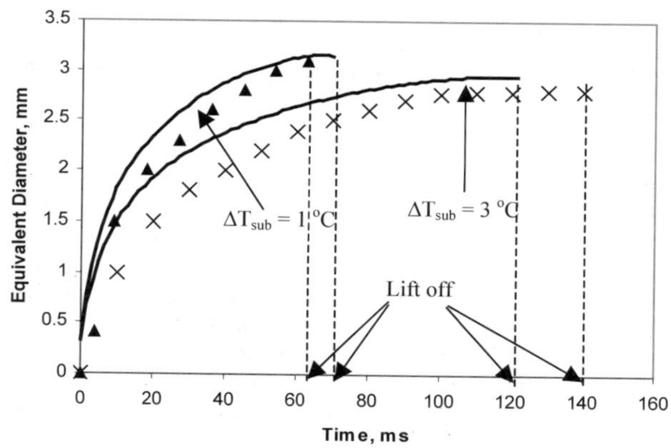

(b)

Figure 4 Comparison of numerical simulations with experimental data (a) Effect of wall superheat, (b) effect of liquid subcooling (fluid: water, $\phi = 54^{\circ}$, $g = 1.0g_e$).

The scaling of the bubble departure diameter and the bubble growth time with gravity level is shown in Fig. 9. The comparison of the experimental data of Qiu *et al.* [16] with the numerical predictions are shown in Fig. 10. It can be seen that the bubble departure diameter scales as $g^{-0.5}$, while the bubble growth time scales as $g^{-1.05}$, for water. Numerical simulations are in general agreement with the observed behavior. The data set that lie well below the single bubble curve corresponds to situations in which bubbles departed after merger. The reason for this will be clear later from the numerical results for merged bubbles.

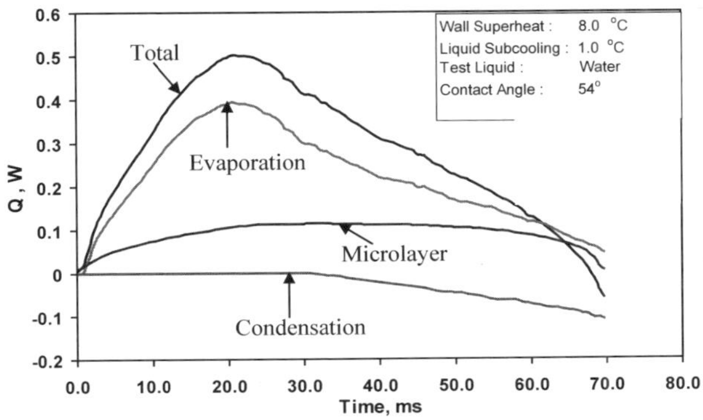

Figure 5 Contribution of the various heat transfer mechanisms during subcooled pool nucleate boiling.

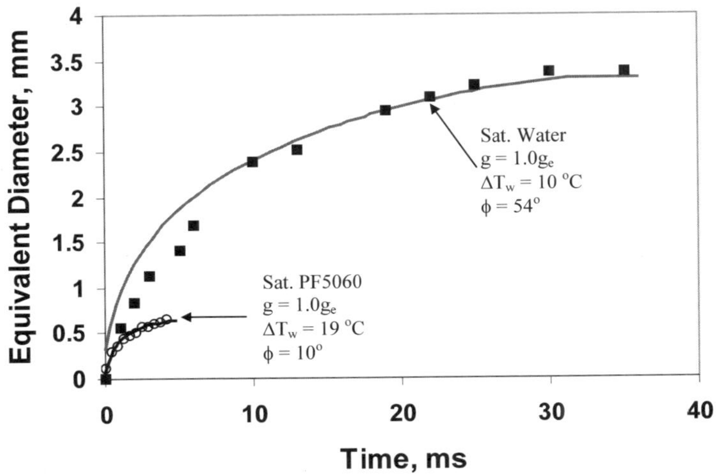

Figure 6 Comparison of bubble departure diameter and bubble growth time for water and PF5060.

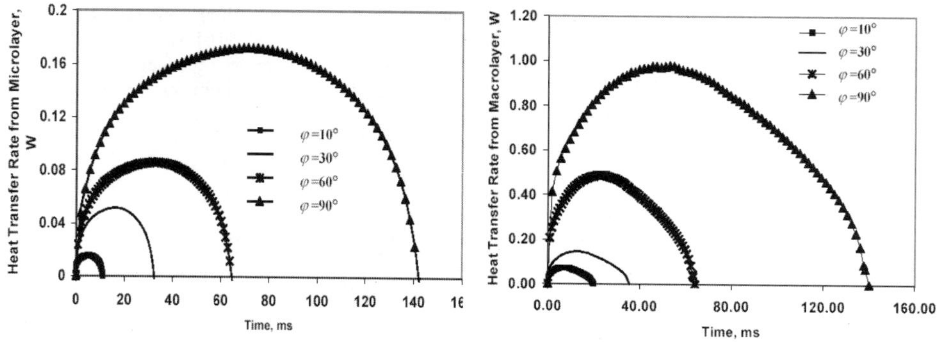

(a) (b)

Figure 7 The variation of heat transfer rates with time for various contact angles (a) from micro layer and (b) from macro region (Fluid: water, p = 1.01 bar, $\Delta T_w = 8$ °C, $\Delta T_{sub} = 0$ °C).

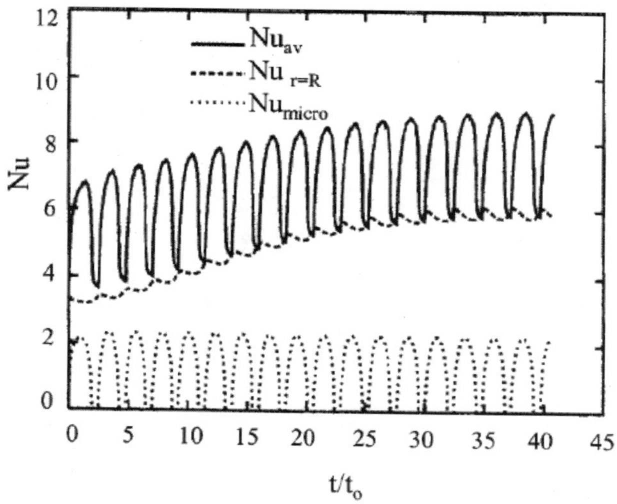

Figure 8 Variation of Nusselt number with time for various bubble growth cycles (fluid: water, ΔT_w = 6.2 °C, ΔT_{sub} = 0.0 °C, g = 1.0g_e, $\phi = 38°$).

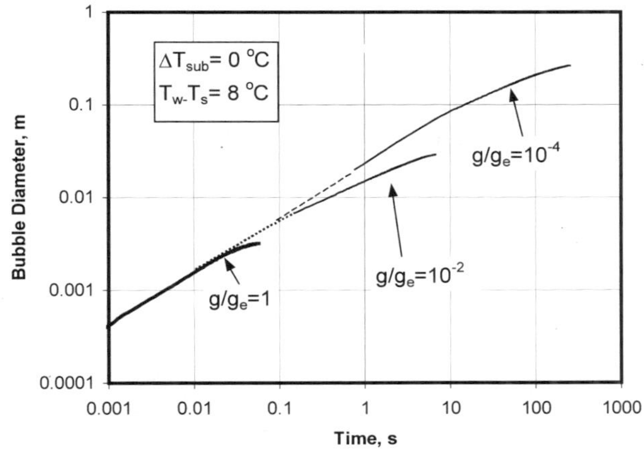

Figure 9 Scaling of equivalent bubble diameter with gravity (Fluid: water).

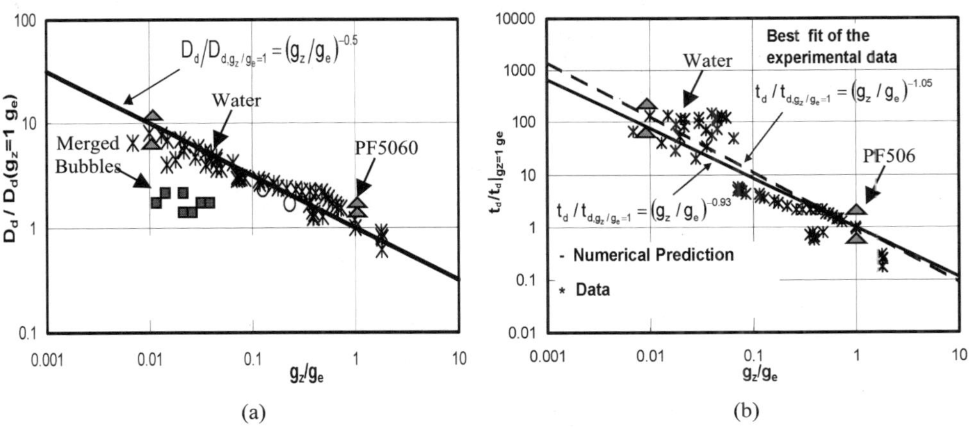

Figure 10 Comparison of experimental data with numerical prediction for various gravity levels (a) bubble departure diameter and (b) bubble growth time.

5.3 TWO BUBBLE MERGER

During nucleate boiling, increasing the wall superheat results in the increase in the bubble release frequency and in the number of nucleation sites that become active. As a result, merger of bubbles both normal and along the heater surface can occur which results in the formation of vapor columns and mushroom type bubbles. Qualitative comparison of the numerical and experimental bubble shapes during the merger of two bubbles in the vertical [17] and lateral [18] directions is shown in Figs. 11 and 12, respectively. From these figures it can be seen that there is very good agreement between the observed and predicted bubble shapes. A comparison of the bubble growth rate during lateral merger of two bubbles is shown in Fig. 13. The predicted growth rate, time of merger, time of departure, and departure diameter are in good agreement with the experimentally obtained values. Figure 14 shows a similar comparison for lateral bubble merger under low-gravity conditions.

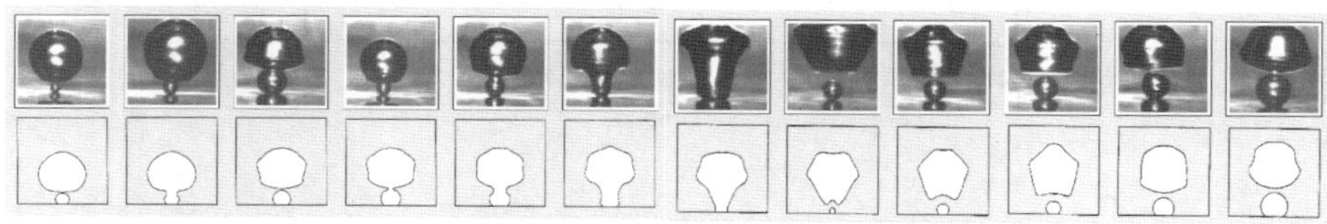

Figure 11 Comparison of numerical and experimental bubble shapes during vertical merger [17] (fluid: water, $\Delta T_w = 10$ °C, $\Delta T_{sub} = 0.0$ °C, $g = 1.0g_e$, $\phi = 38$ °).

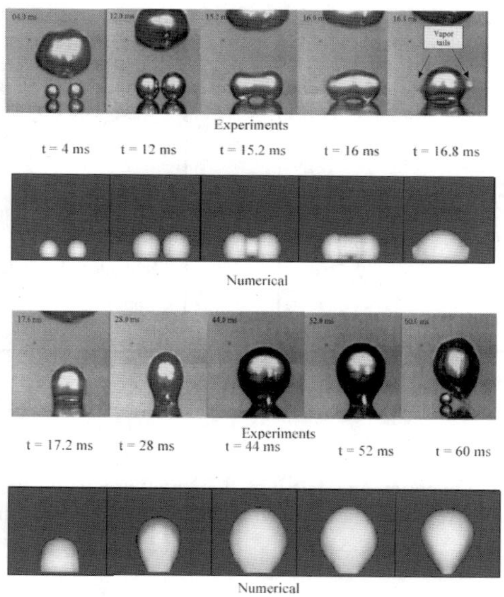

Figure 12 Comparison of the experimental data from Mukherjee and Dhir [18] and numerical bubble shapes during two bubble merger (fluid: water, $\Delta T_w = 5.0$ °C, $\Delta T_{sub} = 0.0$ °C, $g = 1.0g_e$, spacing = 1.5 mm).

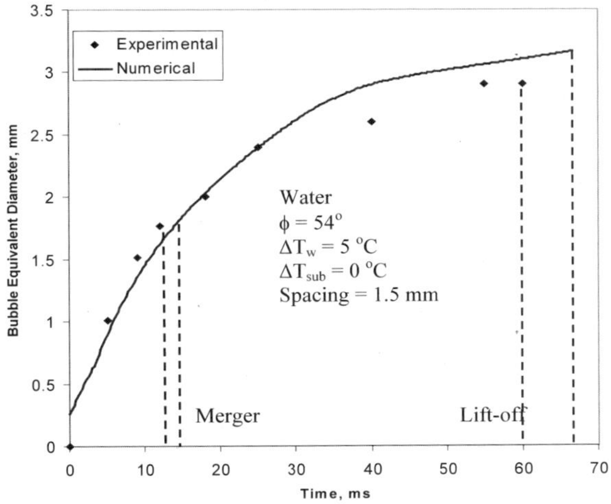

Figure 13 Comparison of numerically predicted bubble growth with experimental data of Mukherjee and Dhir [18] for saturated water at earth normal gravity.

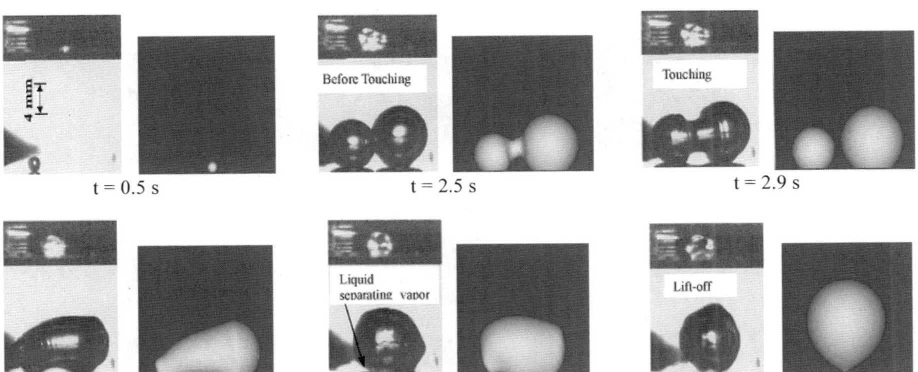

t = 0.5 s t = 2.5 s t = 2.9 s

t = 3.05 s t = 3.15 s t = 3.2 s

Figure 14 Comparison of experimental and numerical bubble shapes during the merger of two bubbles at low gravity (fluid: water, $\Delta T_w = 5$ °C, $\Delta T_{sub} = 3$ °C, $g = 0.01g_e$, $\phi = 54°$, spacing = 7 mm).

5.4 THREE BUBBLE MERGER

Figures 15 shows the bubble shapes for three bubbles located at the corners of an equilateral triangle for microgravity conditions (fluid: water, $\Delta T_w = 7\ ^oC$, $\Delta T_{sub} = 0.0\ ^oC$, $g = 0.01g_e$, $\phi = 54^o$, spacing = 6 mm). These simulations were carried out in a computational domain of 40 mm × 40 mm × 80 mm. The symmetry conditions imposed at the wall (four side walls and the to wall), given in Eq. (29), were replaced by no-slip boundary conditions ($u = v = w = 0$). All other boundary conditions remain the same.

From Fig. 15 it can be seen that the bubbles begin to merge at t = 0.5 sec. Thereafter, the merged bubble grows as a single bubble and finally lifts off at t = 4.2 sec. Figure 16(a) shows the bubble growth rate comparison of the three bubble merger process with that for a single bubble. It can be seen that the merged bubble lifts off at a much smaller diameter compared to the single bubble. The growth period for the merged bubble is also smaller than that for a single bubble. Figure 16(b) shows the net force acting on the vapor mass for the three bubble merger case and the single bubble case. The force acting downward is negative while the force acting upward is positive. It is found that during bubble merger an additional vertical force (which we call the "lift force") is induced by the fluid motion. At about 2.5 seconds when the force changes sign and the merged bubble starts to detach, the single bubble is still experiencing a negative force and continues to grow. The difference between the two at 2.5 seconds is designated as the "lift force" and this additional "lift force" causes the merged bubble to lift off earlier. The bubble merger process also increases the heat transfer rate as shown in Fig. 17. This is due to the increase in interfacial area and the fluid motion induced by bubble merger.

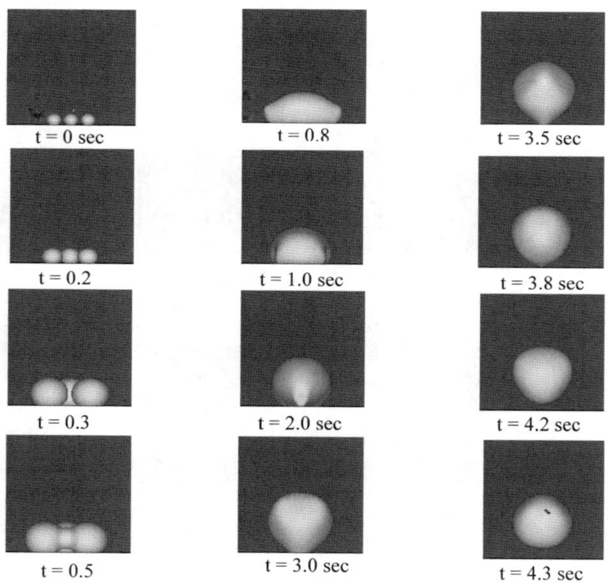

Figure 15 Growth, merger and departure of three bubble in a plane (fluid: saturated water, $g = 0.01g_e$, $\phi = 54^o$).

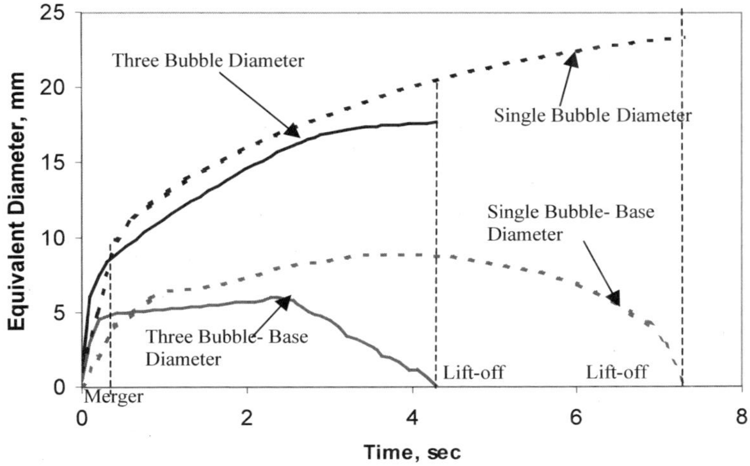

(a)

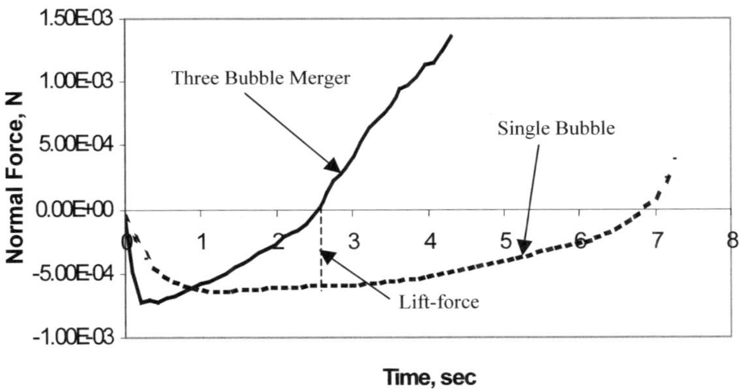

(b)

Figure 16 Comparison of (a) bubble growth history and (b) normal force for single and three bubble merger cases.

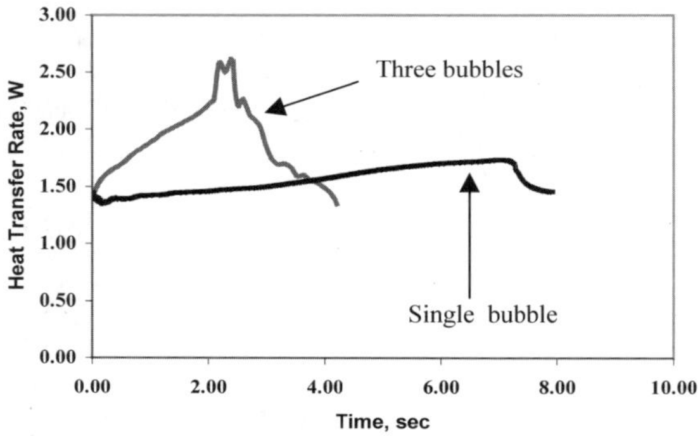

Figure 17 Comparison of heat transfer rates for single and three bubble merger cases.

6. Summary

- Numerical simulations of the bubble dynamics during pool nucleate boiling have been carried out without any approximation of the bubble shapes. The effect of microlayer evaporation is included. By focusing on the micro and macro regions, the length scales from nano to micro to macro have been connected.
- Effects of wall superheat, liquid subcooling, contact angle and level of gravity on bubble growth process, bubble diameter at departure and growth period have been quantified.
- Bubble mergers normal to the heater and along the heater leading to the formation of vapor columns and mushroom type bubbles have been studied.
- The merger process is highly nonlinear. A "lift force" leading to premature departure of bubbles from the heating surface after merger has been identified.

Acknowledgements

This work received support from NASA under the Microgravity Fluid Physics Program.

NOMENCLATURE

A_0, Hamaker constant, [J];
C_p, specific heat, [J/(kg K)];
D, diameter of the bubble, [m];
g, gravitational acceleration, [m/s^2];
h, grid spacing for the macro region;
h_{ev}, evaporative heat transfer coefficient, [W/(m^2 K)];
h_{fg}, latent heat of evaporation, [J/kg];
H, step function;
Ja, Jacob number, $(\rho_l C_p \Delta T_w / h_{fg})$;
k, thermal conductivity, [W/mK];
K, interfacial curvature, [1/m];
l_0, characteristic length scale, [m];
M, molecular weight, [g];
$\vec{m}$, evaporative mass rate vector at interface, [kg/(m^2 s)];
$\dot{m}_{micro}$, evaporative mass rate from micro layer,

	[kg/s];
p,	pressure, [bar];
q,	heat flux, [W/m^2];
r,	radial coordinate, [m];
R,	radius of computational domain, [m];
$\overline{R}$,	universal gas constant, [J/mol K];
R_0,	radius of dry region beneath a bubble, [m];
R_1,	radial location of the interface at y = h/2, [m];
t,	time, [s];
t_0,	characteristic time, [s];
T,	temperature, [^{o}C];
U,	velocity in r direction, [m/s];
$\vec{u}_{int}$,	interfacial velocity vector, [m/s];
u_0,	characteristic velocity, [m/s];
V_c,	volume of a control volume in the micro region, [m^3];
v,	velocity in y direction, [m/s];
w,	velocity in z direction, [m/s];
x,	coordinate, [m];
X,	length of computation domain in x direction, [m];
y,	coordinate, [m];
Y,	length of computation domain in y direction, [m];
z,	vertical coordinate normal to the heating wall, [m];
Z,	height of computational domain, [m];

Greek symbols

α,	thermal diffusivity, [m^2/s];
β_t,	coefficient of thermal expansion, [1/K];
δ,	liquid thin film thickness, [m];
δ_T,	thermal layer thickness, [m];
$\delta_\varepsilon(\phi)$,	smoothed delta function, [m];
φ,	apparent contact angle, [deg.];
ϕ,	level set function;
θ,	dimensionless temperature, $(T - T_{sat})/(T_w - T_{sat})$;
μ,	viscosity, [Pa.s];
ν,	kinematic viscosity, [m^2/s];
ρ,	density, [kg/m^3];
σ,	surface tension, [N/m];
Γ,	mass flow rate in the micro layer, [kg/s];

Subscripts

f,	fluid;
int,	interface;
l,	liquid;
r,	$\partial/\partial r$;
sat,	saturation;
s,	solid;
t,	$\partial/\partial t$;
v,	vapor;
w,	wall;
y,	$\partial/\partial y$;
z,	$\partial/\partial z$;

References

1. Lee, R.C and Nyadhl, J.E. (1989) Numerical calculation of bubble growth in nucleate boiling from inception to departure, *Journal of Heat Transfer*, Vol. 111, pp. 474-479.
2. Cooper, M.G. and Lloyd, A.J.P. (1969) The microlayer in nucleate boiling, *International Journal of Heat and Mass Transfer*, Vol. 12, pp. 895-913.
3. Zeng, L.Z., Klausner, J.F. and Mei, R. (1993) A unified model for the prediction of bubble detachment diameters in boiling systems-1. Pool boiling, *International Journal of Heat and Mass Transfer*, Vol. 36, pp. 2261-2270.
4. Mei, R., Chen, W. and Klausner, J. F. (1995) Vapor bubble growth in heterogeneus boiling-1. growth rate and thermal fields, *International Journal of Heat and Mass Transfer*, Vol. 38, pp. 921-934.
5. Welch, S.W.J. (1998) Direct simulation of vapor bubble growth, *International Journal of Heat and Mass Transfer*, Vol. 41, pp. 1655-1666.
6. Sussman, M., Smereka, P and Osher, S. (1994) A level set approach for computing solutions to incompressible two-phase flow, *Journal of Computational Physics*, Vol. 114, pp. 146-159.
7. Chang, Y.C., Hou, T.Y., Merriman, B., and Osher, S. (1996) A level set formulation of Eulerian interface capturing methods for incompressible fluid flows, *Journal of Computational Physics,* Vol. 124, pp. 449–464.

8. Son, G., Dhir, V.K., and Ramanujapu, N. (1999) Dynamics and heat transfer associated with a single bubble during nucleate boiling on a horizontal surface, *Journal of Heat Transfer*, Vol.121, pp.623-632.
9. Son, G. (2001) Numerical study on a sliding bubble during nucleate boiling, *KSME International Journal*, Vol. 15, pp. 931–940.
10. Singh, S. and Dhir, V.K. (2000) Effect of gravity, wall superheat and liquid subcooling on bubble dynamics during nucleate boiling, *Microgravity Fluid Physics and Heat Transfer* (editor: Dhir, V.K.), Begell House, New York, pp.106-113.
11. Abarajith, H.S. and Dhir, V.K. (2002) Effect of contact angle on the dynamics of a single bubble during pool boiling using numerical simulations, *Proceedings of IMECE2002 ASME International Mechanical Engineering Congress & Exposition*, New Orleans.
12. Stephan, P., and Hammer, J. (1994) A new model for nucleate boiling heat transfer, *Wärme- und Stoffübertragung*, Vol.30, pp. 119-125.
13. Lay, J.H., and Dhir, V.K. (1995) Numerical calculation of bubble growth in nucleate boiling of saturated liquids, *Journal of Heat Transfer*, Vol. 117, pp.394-401.
14. Wayner, P.C. (1992) Evaporation and stress in the contact line region, *Proceedings of the Engineering Fundamentals Conference on Pool and Flow Boiling*, ASME, pp. 251-256.
15. Bai, Q., and Dhir, V.K. (2001) Numerical Simulation of Bubble Dynamics in the Presence of Boron in the Liquid, *Proceedings of IMECE'01*, New York, NY.
16. Qiu, D.M., Dhir, V.K., Hasan, M.M., Chao, D., Neumann, E., Yee, G., and Witherow, J. (1999) Single Bubble Dynamics During Nucleate Boiling Under Microgravity Conditions, *Engineering Foundation Conference on Microgravity Fluid Physics and Heat Transfer*, Honolulu, HI.
17. Son, G, Ramanujapu, N, and Dhir, V.K. (2002) Numerical simulation of bubble merger process on a single nucleation site during pool nucleate boiling, *Journal of Heat Transfer*, Vol. 124, pp. 51-62.
18. Mukherjee, A. and Dhir, V.K. (2004) Numerical and experimental study of bubble dynamics associated with lateral merger of vapor bubbles during nucleate pool boiling, In Press, *Journal of Heat Transfer*.

FLOW BOILING IN MINICHANNELS

ANDRÉ BONTEMPS[(1,2)], BRUNO AGOSTINI[(3)], NADIA CANEY[(1,2)]

(1) CEA-GRETh, 17 rue des Martyrs, 38054 Grenoble, France, (2) LEGI/GRETh, Université Joseph Fourier, 17 rue des Martyrs, 38054 Grenoble, France, (3) 15 rue Denis Papin, 38000 Grenoble, France

1. Introduction

The use of mini-channel heat exchangers (hydraulic diameter about 1 mm) in compact heat exchangers improves heat transfer coefficients, and thermal efficiency while requiring a lower fluid mass. They are widely used in condensers for automobile air-conditioning and are now being used in evaporators, as well as in other applications such as domestic air-conditioning systems. However, more general use requires a better understanding of boiling heat transfer in confined spaces. Many definitions of micro and minichannel hydraulic diameter are used throughout the literature. Kandlikar and Grande (2003) proposed the following classification: conventional channels ($D_h > 3$ mm), minichannels (200 μm $< D_h < 3$ mm), micro-channels ($D_h < 200$ μm), that will be used throughout this paper.

These definitions rely upon the molecular mean free path in a single-phase flow, surface tension effects and flow patterns in two-phase flow applications. In recent studies in minichannels the hydraulic diameter ranges from 100 μm to 2–3 mm. The channel cross sections were either circular or rectangular and much of the research concerned boiling. Commonly, classical correlations have been used with or without modifications to predict flow boiling results in minichannels. However agreement was poor and the need for new correlations was evident.

It has been shown through a number of experiments that boiling is controlled by two additive components: nucleate boiling and convective boiling. Nucleate boiling is due to nucleating bubbles and their subsequent growth and removal from the heated surface. Convective boiling is due to heated fluid moving from the heated surface to the flow core. These two mechanisms cannot be separated with any precision since they are closely interconnected. Figure 1 shows a classical representation of flow boiling regimes in tubes. The successive steps, as the fluid is heated, are:

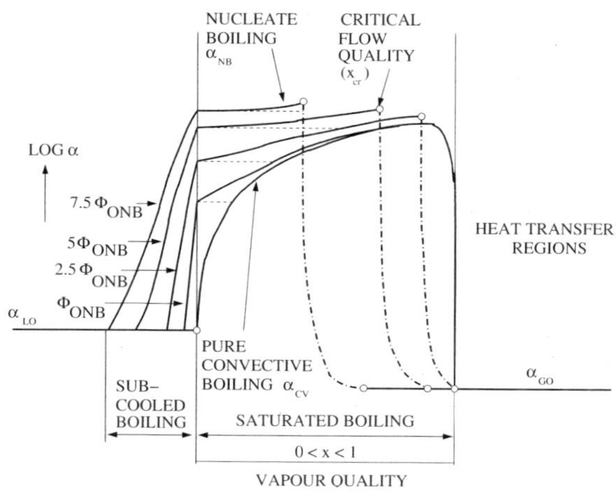

Figure 1: Boiling regimes from Collier and Thome (1994).

(i) In subcooled boiling the average fluid temperature is below the saturation temperature while the fluid at the tube wall has already reached it and therefore can boil. The heat transfer coefficient rises and depends on the heat flux, until the core of the flow, which is colder, reaches the saturation temperature. Bubbles formed at

S. Kakaç et al. (eds.), Microscale Heat Transfer, 217–230.

the wall move and condense in the flow core and increase its temperature.

(ii) In saturated boiling the flow core has reached saturation. Nucleate and convective boiling compete. In nucleate boiling the heat transfer coefficient depends on the heat flux which is the driving force of bubble generation (dashed lines). In convective boiling the heat transfer coefficient is independent of the heat flux but depends on the liquid quality and mass velocity which are the driving forces of convection. The combination of both shows almost horizontal and parallel lines at low vapour quality (nucleate boiling) which merge into a single increasing line at higher vapour quality (convective boiling). The smaller the heat flux the sooner (in terms of vapour quality) convective boiling will take over from nucleate boiling. This is further highlighted by figure 2 which represents experimental results on flow boiling regimes in tubes.

(iii) At high quality and heat flux, dry-out can occur. This is a dramatic outcome of boiling. The liquid layer wetting the wall and providing heat transportation is totally vaporised. Only gas remains which severely decreases the heat transfer from the wall. With imposed heat flux this can lead to tube meltdown.

(iv) Finally, when all the liquid is vaporised, single-phase gaseous flow governs the heat transfer with, of course, a heat transfer coefficient smaller than for a single-phase liquid flow.

The main difficulty is to establish the dependence of the heat transfer coefficient on vapour quality in relation to different mechanisms controlling flow boiling. Some correlations do not take into account the two mechanisms. Others account for convective and nucleate boiling. To the present author's knowledge, none take into account the influence of channel size. The aim here is to summarise recent work on flow boiling, to describe an experiment on the phenomenon in minichannels and to compare the results with classical correlations.

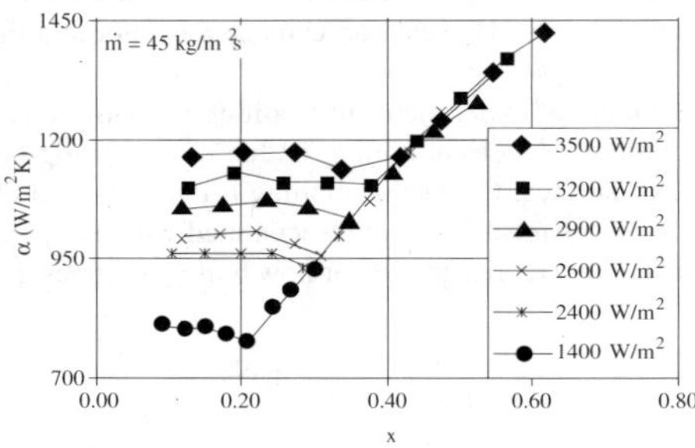

Figure 2: Boiling regimes observed Feldman (1996).

2. Review of selected flow boiling correlations for minichannels

For an extended review of experimental work on mini and microchannels, the reader is refered to the Thome (2004) and Kandlikar (2002) papers. This brief review covers a representative selection of heat transfer studies in minichannels and its aim is to illustrate the tendencies observed in the presented data. Recently Kandlikar (2004) developed a new general correlation adapted to minichannels which gives very good results for low qualities but fails to take dry-out into account, as noted by the author in question. Lately Thome *et al.* (2004) and Dupont *et al.* (2004) proposed a semi-empirical three zone model which is the only published work to predict the unique trends observed in minichannels. In this model the dominant boiling mechanism is the evaporation of the liquid film pressed under confined bubbles.

A few studies on boiling in minichannels are available in the literature. The experimental conditions are gathered in table 1. Tran *et al.* (1997), Aritomi *et al.* (1993) and Kew and Cornwell (1997) established correlations of heat transfer coefficient for various refrigerants. All noted that the local heat transfer coefficient was only dependent on the heat flux. Accordingly they concluded that the governing mechanism was nucleate boiling and no dependence on quality was considered. Recently, Huo *et al.* (2004) studied boiling of refrigerant R134a in minichannels and highlighted the prevalence of nucleate boiling and the occurrence of dry-out at low vapour quality. Kew and Cornwell (1997) defined a non dimensional confinement number, Co, and proposed that microscale boiling should prevail for Co > 0.5 while macroscale boiling would occur for Co < 0.5.

However some experimental studies in similar geometries show a dependence of the heat transfer coefficient on vapour quality. The Feldman (1996), Oh *et al.* (1998) and Kandlikar and Grande (2003) correlations illustrate a clear evolution of the heat transfer coefficient with vapour quality. These works indicate that nucleate boiling may not be the only mechanism governing boiling in minichannels and that new mechanisms may happen too.

Table 1: Summary of experimental conditions of some studies on refrigerants flow boiling in minichannels

Author	Fluid	$\dot{q}$ (kW/m^2)	$\dot{m}$ (kg/m^2 s)	x_o	D_h (mm)
Aritomi *et al.* (1993)	R113	10–100	31–620	0–0.8	1–4
Feldman (1996)	R114	1.4–3.5	20–45	0.1–0.6	1.66–2.06
Kew and Cornwell (1997)	R113,141b	9.7–90	188–212	0–0.8	1.04–3.69
Tran *et al.* (1997)	R12,113,134a	0.75-129	44–832	0.2–0.8	2.46–2.92
Oh *et al.* (1998)	R134a	10–20	240–720	0–1	1–2
Huo *et al.* (2004)	R134a	13–150	100–500	0–0.9	2.01–4.26
Present study (2004)	R134a	2.8–31.6	90–469	0–1	0.77–2.01

Another phenomenon experimentally identified in flow boiling is the oscillatory nature of the flow. Some intermittent local dry-out can occur in confined spaces. This occurrence certainly influences the evolution of the heat transfer coefficient at high vapour quality. Brutin *et al.* (2003) experimentally investigated two-phase flow instabilities in narrow channels. They observed vapour slug formation blocking the two-phase flow and pushing it back to the inlet. Kandlikar and Grande (2003) observed periodic slug flow with quick dry-out and re-wetting. This phenomenon occurs faster and lingers longer when the heat flux increases. However the exact influence on the heat transfer coefficient has not been quantified yet.

Consequently there is no clear indication that boiling phenomena in small diameter channels is either dominated by nucleate boiling or convective boiling or any new mechanism. Yet there is strong evidence that these mechanisms are not interconnected as in conventional tubes. The motivation for the present work is therefore to get a more accurate vision of boiling in minichannels, to establish a correlation for flow boiling in minichannels and consequently to identify the most adapted correlation in the literature. To this end, experimental results of ascendant boiling flow of refrigerant R134a obtained with two minichannels, whose hydraulic diameters are 2 mm and 0.77 mm, will be presented and discussed.

3. Experimental set-up

Figure 3 is a schematic of the R134a experimental facility and test section. The test loop included a liquid pump and a mixed glycol-water circuit for heat evacuation. Subcooled liquid enters the bottom inlet manifold, is then vaporised in the test section and condensed further on in the heat exchanger. The test section consisted of a vertical industrial MPE (MultiPort Extruded) aluminium tube composed of parallel rectangular channels. The whole test section was thermally insulated with wrapping foam. For heat transfer measurements, a section of the tube was heated by Joule effect with the passage of an electric current from two brased electrodes through the tube wall. Upstream of the heated region there was an adiabatic zone to ensure the flow was hydrodynamically developed. Experimental conditions are summarised in table 2.

The determination of the channel dimensions was carried out using scanning electron microscopy. The hydraulic diameter was calculated with the total flow area and wet perimeter measured from electron microscope images in order to take into account the effect of the first and last channels which are rounded. Roughness measurements were also carried-out.

Figure 3 shows the test section and instrumentation. Ten wall temperatures on the tube external surface were measured with 0.5 mm diameter calibrated type E thermocouples electrically insulated from the aluminium. Fluid inlet and outlet temperatures were measured with 1 mm diameter calibrated type K thermocouples. Calibration was carried out with a Rosemount 162-CE platinum thermometer. Due to the high thermal conductivity of the aluminium and the low thickness of the tube walls the measured temperature is very close to the wall temperature in contact with the fluid (the difference less than 0.01 K). The inlet fluid pressure was measured with a calibrated Rosemount type II absolute pressure sensor. Two calibrated differential pressure sensors measured the pressure loss through the test section. A Rosemount Micro-motion coriolis flowmeter was used to measure the mass flow rate of R134a downstream of the pump. The heating voltage and current were measured directly through a HP 3421A multiplexer. The overall system was tested with single-phase flow runs in order to check heat losses (Agostini *et al.* (2002) and Agostini (2002)). Classical turbulent single-phase flow correlations agreed with measurements to within ±10%.

The heat flux was varied for every fixed mass flow rate in order to obtain a series of outlet vapour qualities between 0.2 and 1 with a step of 0.05. Steady state values were monitored using a Hewlett Packard 3421A with a 30 minutes time lapse between each mass flow rate or heat flux change. Averaging was carried out after every 20 values and uncertainties were calculated according to the Kline and McClintock (1953) method. The total electrical power dissipated in the test section was calculated as the product of voltage and current. The variations of R134a thermophysical properties with temperature were calculated with the REFPROP 6.01 software.

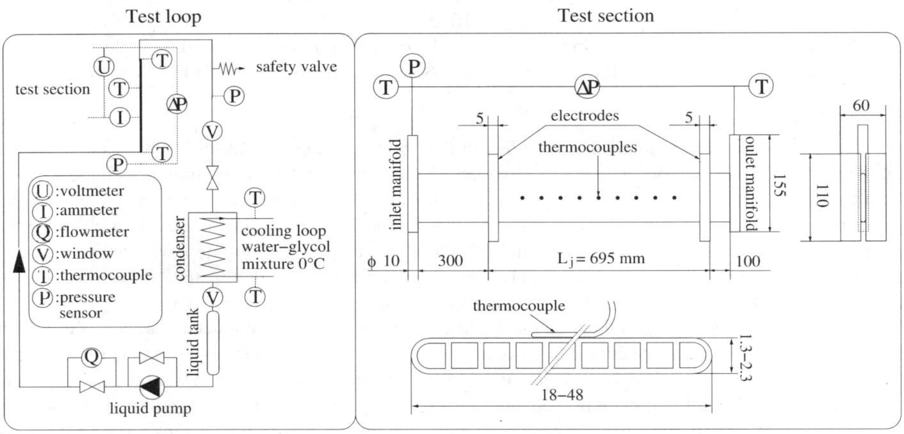

Figure 3: Test loop and test section.

Table 2: Operating conditions and uncertainties

	value	error	value	error
D_h (mm)	2.01	± 3 %	0.77	± 7 %
$\dot{m}$ (kg/m^2 s)	90–295	± 1.7–8.6 %	214–469	± 3.3–8.6 %
$\dot{q}$ (kW/m^2)	6–31.6	± 2–4.1 %	2.8–19.5	± 3 %
T_w, T_{fl} (K)	276–308	± 0.1–3 K	281–307	± 0.2–2 %
p_i (kPa)	405 & 608	± 4 %	517	± 2.7 %
Δp (kPa)	9.5–37.5	± 1.2–17 %	153–1570	± 0.3–3.7 %

4. Heat transfer results

In this section the general trends of the measurements will be presented. Electrical power, refrigerant mass flow rate, in and outward fluid temperatures and wall temperatures were measured. From these data physical parameters of interest were computed as a function of tube length: heat flux $\dot{q}(z)$, fluid temperature $T_{fl}(z)$, vapour quality $x(z)$ and heat transfer coefficient $\alpha(z)$. Table 3 presents uncertainties on some calculated parameters.

Figures 4 and 5 show the local heat transfer coefficient as a function of local quality for $D_h = 2$ mm and $D_h = 0.77$ mm. Two general trends are observed. On figure 4 a strong decrease in the heat transfer coefficient with vapour quality is visible when the vapour quality is greater than a "critical quality" whereas on figure 5 the heat transfer coefficient starts decreasing before increasing with vapour quality.

Table 3: Uncertainties on calculated parameters

	value	error	value	error
Co	0.46	± 3 %	1.14	± 7 %
$Bo \times 10^{-4}$	2.2–7.9	± 3.7–10.4 %	0.7–2.2	± 6.3–11.6 %
α (kW/m² K)	0.8–10.3	± 6–30 %	0.6–32.5	± 5–30 %
x_o	0.26–1	± 1–7 %	0.23–0.9	± 2–9 %

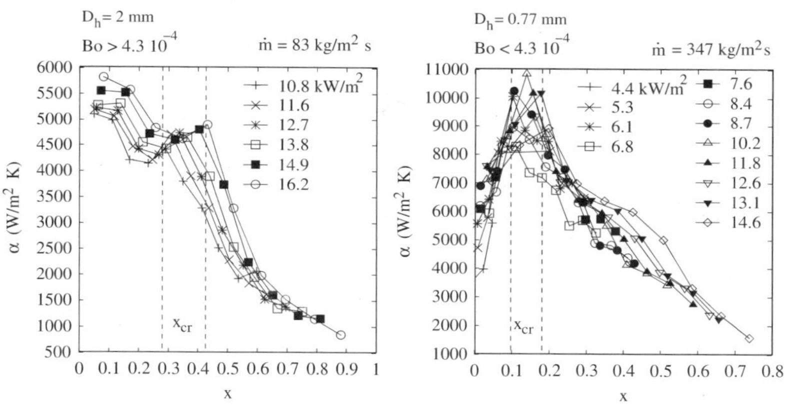

Figure 4: Local heat transfer coefficient versus local quality.

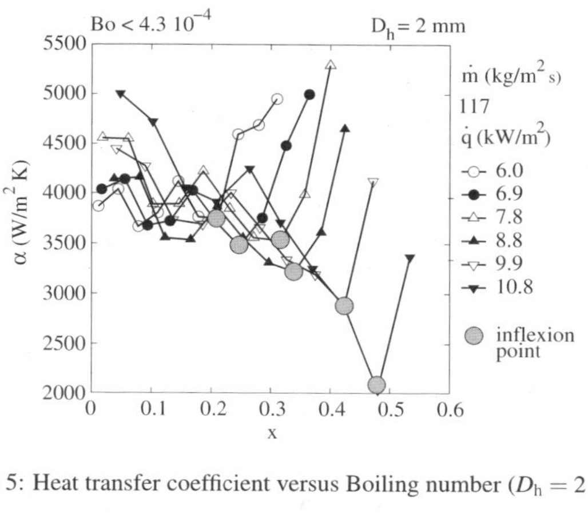

Figure 5: Heat transfer coefficient versus Boiling number ($D_h = 2$ mm).

Figure 6 shows the local heat transfer coefficient as a function of local quality, for a given heat flux and mass velocity, for $D_h = 2$ mm and $D_h = 0.77$ mm. It is clear that the heat transfer coefficient increases when the hydraulic diameter decreases. Thus the local heat transfer coefficient is increased by 74 % ± 26 % when the hydraulic diameter is decreased by 62 %.

The enhancement ratio can also be written $D_h^{-0.6\pm0.1}$ which is, given the uncertainties, close to the values proposed by Ishibashi and Nishikawa (1969) ($D_h^{-0.67}$) and Aritomi *et al.* (1993) ($D_h^{-0.75}$). It is not clear how Tran *et al.* (1997), who proposed D_h^{-1} for the confinement effect, established this expression. The range of tested hydraulic diameters, 2.4–2.92 mm, is too small to build a correlation. It is possible, from their articles, that it comes from an analogy with single-phase flow classical theory and has no experimental basis. On the contrary the $D_h^{-0.4}$ term proposed by Steiner and Taborek (1992) is based on some experimental data. Nevertheless the data concerning refrigerant fluids (R11 and R113) were performed for hydraulic diameters from 7 to 20 mm only. The lowest hydraulic diameters (from 1 to 5 mm) were tested with Helium I only so that this $D_h^{-0.4}$ factor is difficult to compare with the present results.

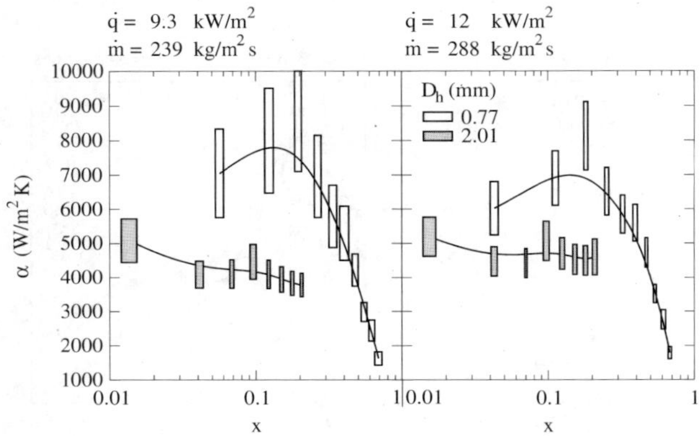

Figure 6: Influence of confinement on heat transfer coefficients.

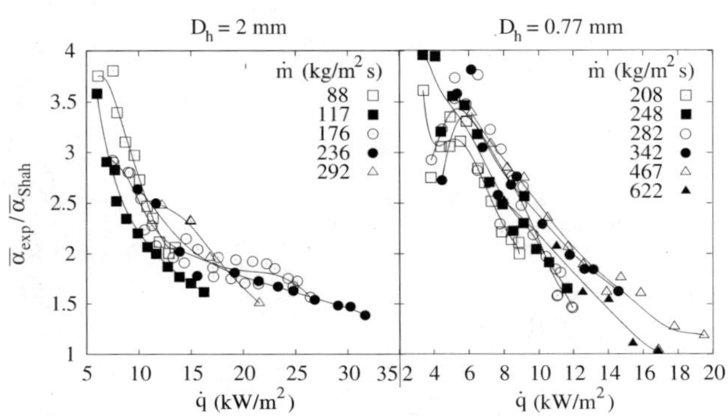

Figure 7: Average heat transfer coefficient versus heat flux.

Figure 7 represents the ratio of the measured average heat transfer coefficient to that predicted by the Shah (1976) correlation for conventional tubes, as a function of the heat flux. This figure shows the global intensification of heat transfer in MPE minichannels compared with conventional tubes. This intensification ranges from 0 to 400 % depending on the heat flux and occurs up to 35 kW/m^2 for $D_h = 2$ mm and 20 kW/m^2 for $D_h = 0.77$ mm.

5. Heat transfer analysis

An analysis and a physical interpretation of these observations will now be proposed. In order to analyse and classify the different heat transfer coefficient behaviours observed on figures 4 and 5 it is useful to represent, for a given vapour quality, the heat transfer coefficient as a function of the heat flux and the heat flux as a function of the wall-fluid temperature difference. For D_h = 2 mm, since Co < 0.5, the results were analysed in terms of macroscale boiling. This was done on figure 8, which exhibits two trends:

(i) For $T_w - T_{sat} < 3$ K and $\dot{q} < 14$ kW/m^2, $\dot{q}$ is proportional to $T_w - T_{sat}$. Thus α is independent of $\dot{q}$ and moreover decreases with $\dot{m}$. This region may correspond to a convective boiling regime and, as will be further highlighted, the decrease with $\dot{m}$ may be due to the occurrence of partial dry-out.

(ii) For $T_w - T_{sat} > 3$ K and $\dot{q} > 14$ kW/m^2, $\dot{q}$ is proportional to $(T_w - T_{sat})^3$, therefore α is proportional to $\dot{q}^{2/3}$, and the heat transfer coefficient depends only weakly on $\dot{m}$. This second region may be identified as a nucleate boiling regime.

However, as Co is very close to 0.5, these results may also be interpreted in terms of microscale boiling with the film evaporation mechanism proposed by Thome *et al.* (2004).

For D_h = 0.77 mm $\dot{q}$ is always proportional to $T_w - T_{sat}$ and α is independent of $\dot{m}$. Since Co is greater than 0.5, microscale boiling should prevail and according to the three zone model of Thome *et al.* (2004) film evaporation would be the boiling mechanism occurring in this tube.

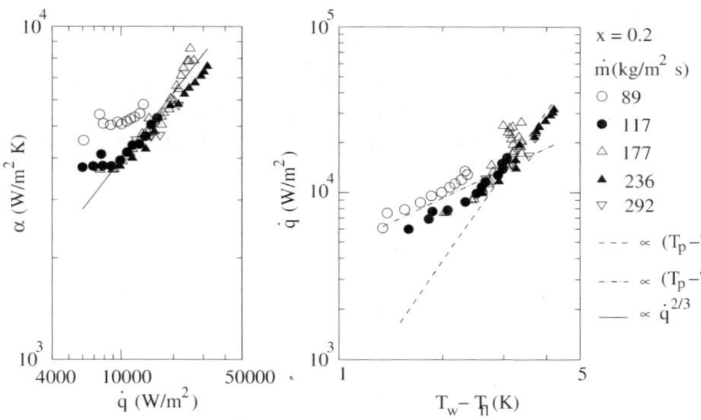

Figure 8: α versus $\dot{q}$ and $\dot{q}$ versus $T_w - T_{sat}$ ($D_h = 2$ mm).

From an analysis conducted on figures 4 and 5 with the dimensionless boiling number, the following tendencies can be outlined. For $D_h = 2$ mm:

(i) for Bo $> 4.3 \cdot 10^{-4}$ and x < 0.3–0.4, the heat transfer coefficient is weakly dependent on x and proportional to $\dot{q}^{2/3}$. Thus the nucleate boiling regime might governs this region.

(ii) for Bo $> 4.3 \cdot 10^{-4}$ and x > 0.3–0.4, the heat transfer coefficient decreases with x but is still proportional to $\dot{q}^{2/3}$. This suggests that partial dry-out occurs with nucleate boiling which is confirmed on figure 9 where the wall temperature and the statistical uncertainty on T_w suddenly rise for x > 0.3–0.4. Most of the data belong to these two regions.

(iii) for Bo $< 4.3 \cdot 10^{-4}$ the heat transfer coefficient is weakly dependent on x and proportional to $\dot{q}^{2/3}$ for low qualities. It then starts increasing with vapour quality when x is greater than a transition value. This transition value is all the greater since the heat flux is high for a given mass velocity. This behaviour may correspond to competition between a convective boiling and a dry-out regime where partial dry-out and regeneration of the liquid layer occur. Furthermore it was found that this transition occurred for a constant value of the product Bo $\cdot (1-x)$ equal to 2.2×10^{-4}.

These results are in agreement with the Huo *et al.* (2004) study which highlighted the prevalence of heat flux dependent boiling and early dry-out in a 2 mm diameter tube with refrigerant R134a. In their work the boiling number was always greater than 8×10^{-4} which is coherent with the present results.

For $D_h = 0.77$ mm, Bo is always smaller than $2.2 \cdot 10^{-4}$ and

(i) for $x < 0.1$–0.2, the heat transfer coefficient increases weakly with x and is independent of of $\dot{q}$. Film evaporation seems to dominate and the thinning of the liquid layer could explain the increase.

(ii) for $x > 0.1$–0.2, the heat transfer coefficient always decreases with x and is still independent of $\dot{q}$. This suggests that intermittent dry-out governs the boiling alongside with film evaporation.

It was found that the critical vapour quality x_{cr} did not depend on $\dot{q}$ and $\dot{m}$. This does not mean that x_{cr} does not depend on $\dot{q}$ or $\dot{m}$ but simply that such a variation is less than the uncertainty. The present study highlights a decrease in x_{cr} from 0.3–0.4 to 0.1–0.2 when D_h decreases from 2 to 0.77 mm. For comparison, Huo *et al.* (2004) found that x_{cr} decreased from 0.4–0.5 to 0.2–0.3 when D_h decreased from 4.26 to 2.01 mm. Given that the uncertainty on x_{cr} in the present study is about 30 % the results are quite close.

Examination of figure 9 strongly suggest the occurrence of early dry-out. When the liquid layer disappears from the tube wall, the heat transfer coefficient suddenly decreases because of the lesser heat transport properties of the gas. This implies a wall temperature rise since heat removal is less efficient. Moreover, it is well know that dry-out is an intermittent phenomenon and that liquid drops regularly hit the tube wall, so that the tube wall temperature is submitted to quick changes. For the authors this explains why the uncertainty on the tube wall temperature also increases tenfold when dry-out occurs. Furthermore, the occurrence of early dry-out may be explained by the thinning of the liquid layer due to bubbles confinement. This hypothesis also explains why the critical vapour quality decrease from 0.4 to 0.2 when the hydraulic diameter decreases from 2 to 0.77 mm since this increases bubble confinement. Moreover, this thinning of the liquid layer also explains why the heat transfer coefficient increases when the hydraulic diameter decreases because the heat transfer resistance due to this liquid layer also decreases as long as dry-out does not occur.

The occuring of dry-out may explain why, when combined with heat flux dependent boiling, the heat transfer coefficient decreases with the mass velocity. The greater the mass velocity, the more probable dry-out should be, because the liquid film is increasingly dragged from the wall due to shear stress. Thus dry-out should cause a decrease in the heat transfer coefficient with the mass velocity.

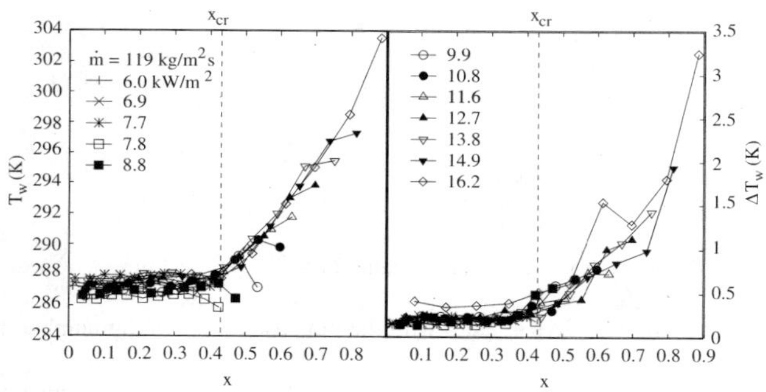

Figure 9: Wall temperature and its uncertainty versus local quality ($D_h = 2$ mm).

To explain why the boiling number seems to govern the transition between heat flux increasing α and vapour quality increasing α, the following interpretation is proposed, based on macroscale boiling mechanisms. From the Rohsenow (1952) and Kew and Cornwell (1997) analysis, an inertial characteristic time τ_{cv} for the liquid layer and a characteristic time τ_b for bubbles leaving the wall can be defined. Then, from the Kutaleladze (1981) and Rohsenow (1952) analysis it can be shown that the ratio of these two characteristic times can be written:

$$\frac{\tau_{cv}}{\tau_b} = f(\theta, g, \rho_l, \rho_g, \sigma) \cdot \mathrm{Bo} \cdot f(x). \tag{1}$$

This ratio is a comparison of convective effects in the liquid layer (causing α to increase with x) and bubble dynamics at the wall (causing α to increase with $\dot{q}$). Thus τ_{cv}/τ_b is proportional to Bo and a function of vapour quality so that the boiling number is the appropriate dimensionless number to study the transition between these two boiling regimes.

Finally, figure 10 illustrates the different boiling regimes in minichannels. This work suggests:

(i) when $\mathrm{Bo} > 4.3 \cdot 10^{-4}$ and $\mathrm{Co} < 0.5$, nucleate boiling and dry-out seems to govern boiling in minichannels. Moreover dry-out occurs at low vapour quality. Most of data for $D_\mathrm{h} = 2$ mm were in this case.

(ii) when $2.2 \cdot 10^{-4} < \mathrm{Bo} < 4.3 \cdot 10^{-4}$ and $\mathrm{Co} < 0.5$, nucleate boiling, convective boiling and dry-out seem to compete. The frontier between the nucleate boiling and convective boiling is $\mathrm{Bo} \cdot (1-x) = 2.2 \cdot 10^{-4}$. Furthermore dry-out occurs also at low vapour quality and its effects are superimposed on those of nucleate boiling and convective boiling and competes with them.

(iii) when $\mathrm{Bo} < 2.2 \cdot 10^{-4}$ and $\mathrm{Co} > 0.5$, boiling directly starts in the film evaporation regime with no heat transfer dependance since $\mathrm{Bo} \cdot (1-0) = 2.2 \cdot 10^{-4}$. The heat transfer coefficient increases with vapour quality, and does not depend upon the mass velocity and heat flux, until dry-out occurs. Then boiling is totally governed by dry-out and the heat transfer coefficient decreases sharply with vapour quality and remains independent of the mass velocity and heat flux.

This scheme illustrates the difference between the classical boiling regimes representations like figures 1 and 2, and the present results on figure 10.

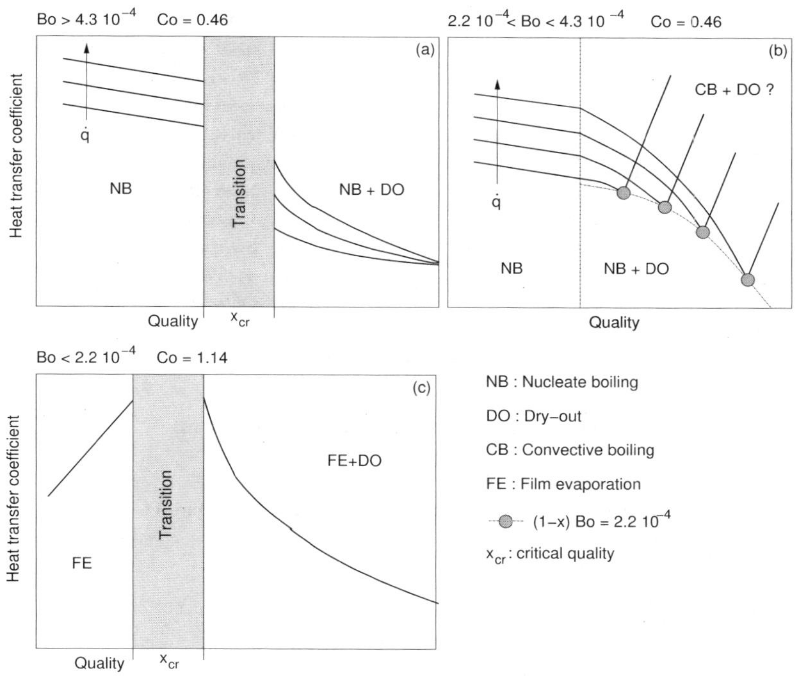

Figure 10: Summing-up of boiling regimes in minichannels observed in the present work.

6. Correlating data

Most of the present data points belong to the heat flux dependent regime so that it has been possible to correlate the heat transfer coefficient in this region with $\dot{m}$, $\dot{q}$ and x. Finally, the following expressions were obtained. For $D_\mathrm{h} = 2$ mm and $\mathrm{Bo} > 4.3 \cdot 10^{-4}$,

$$x < 0.3-0.4 : \alpha = 28 \cdot \dot{q}^{2/3} \cdot \dot{m}^{-0.26} \cdot x^{-0.10} \quad \text{and} \quad x > 0.3-0.4 : \alpha = 28 \cdot \dot{q}^{2/3} \cdot \dot{m}^{-0.64} \cdot x^{-2.08}, \tag{2}$$

for $D_\mathrm{h} = 0.77$ mm and $\mathrm{Bo} < 2.2 \cdot 10^{-4}$,

$$x < 0.1-0.2 : \alpha = 10260 \cdot x^{0.15} \quad \text{and} \quad x > 0.1-0.2 : \alpha = 10260 \cdot (1-x)^{1.57}. \tag{3}$$

Equation (2), obtained by linear least squares fitting over 723 data points, predicts 95 % of our data in the ±30% range. Equation (3), obtained by linear least squares fitting over 825 data points, predicts 85 % of our data in the ±30% range.

Figure 11 compares the ability of various correlations to predict the present data for $D_h = 2$ mm. The analytical expressions of these correlations have been reported in table 4. The Tran *et al.* (1997) and Kandlikar (2004) correlations, proposed for minichannels predicts the present data rather well in the pure heat flux dependent regime but fails as soon as dry-out occurs. The Steiner and Taborek (1992) correlation over-predicts the present data since it includes a $D_h^{-0.4}$ diameter correction term which is not well fitted for such small diameters. On the contrary the Shah (1976), Liu and Winterton (1991) correlations under-predict the present data because they do not take into account any confinement phenomenon as suggested by Cornwell and Kew (1992, 1995).

The Thome *et al.* (2004) and Dupont *et al.* (2004) three zone model was able to predict most of the trends observed, in particular for the 0.77 mm tube. It is not represented here since it implies the optimization of minimum and maximum liquid layer thickness and bubble generation frequency parameters. However the reader is invited to refer to the cited articles where this model predictions are compared with the present data.

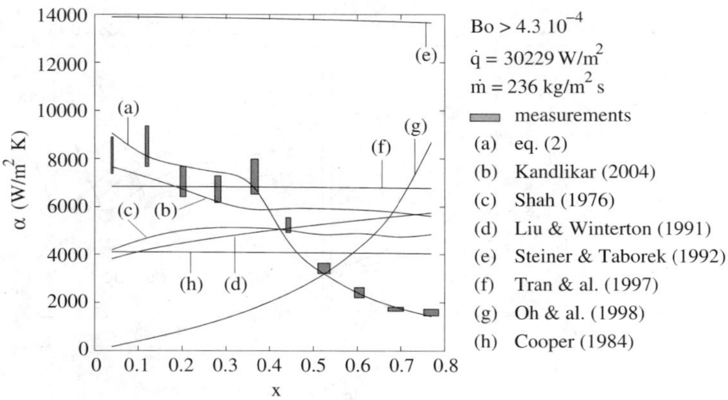

Figure 11: Comparaison with litterature correlations.

Table 4: Expressions of correlations represented on figure 11.

Author	Expression
Shah (1976)	$\alpha_{TP}/\alpha_l = f(\mathrm{Bo}, \mathrm{Cv})$
Cooper (1984)	$\alpha_{TP} = 55 \cdot p_r^{0,12} \cdot \dot{q}^{2/3}(-\log_{10} p_r)^{-0,55}\tilde{M}^{-0,5}$
Liu and Winterton (1991)	$\alpha_{TP}^2 = \alpha_{lo}^2 + \alpha_{Cooper}^2$
Steiner and Taborek (1992)	$\alpha_{TP}^3 = \alpha_{lo}^3 + \alpha_{npb}^3$
Tran *et al.* (1997)	$Nu = 770 \cdot (\mathrm{Bo} \cdot \mathrm{Re}_{lo} \cdot \mathrm{Co})^{0,62} \cdot (\rho_v/\rho_l)^{0,297}$
Oh *et al.* (1998)	$\alpha_{TP}/\alpha_l = 240/\chi_{tt} \cdot (1/\mathrm{Re}_{TP})^{0,6}$
Kandlikar (2004)	$\alpha_{TP} = \max(\alpha_{nb}, \alpha_{cb})$
Present study (2004)	eq. (2)

7. Flow regimes

Different authors have identified various flow regimes in large channels. In both vertical and horizontal configurations these include bubbly, dispersed bubbly, slug, pseudo-slug, churn, annular, annular mist and dispersed droplet flows. An important difference in minichannels is that the liquid flow is preferentially laminar. Surface tension effects have more and more influence as the hydraulic diameter is reduced. Gravity becomes negligible compared to surface tension so that the orientation is less influential.

In different studies identifying flow configurations in minichannels, fundamental configurations specific to minichannels are observed: isolated bubbles, confined bubbles and annular slug flow (Kew and Cornwell (1997)). However, some authors observed flow regimes typical of macroscale tubes: bubbly, plug, slug, wavy-annular and annular flow (Kuwahara *et al.* (2000)).

Triplett *et al.* (1999) measured pressure drop and void fraction in minichannels with air-water adiabatic flows. They observed bubbly, churn, slug, slug-annular and annular flows as in conventional tubes, but the transitions were very different. Moreover they highlighted that the homogeneous model best predicted their pressure drop measurements for every flow configuration except the annular one.

Huo *et al.* (2004) established a flow map for refrigerant R134a flowing in 2.01 mm and 4.26 mm diameter round vertical tubes. They observed six typical flow patterns, i.e. dispersed bubbles, bubbly, slug, churn, annular and mist. Significant differences were found with the existing models for normal size tubes. For example the churn flow pattern becomes a more important flow pattern compared to classical models where it shrinks to a very small area. The authors also observed that reducing the diameter shifted the transition of slug to churn and churn to annular to higher values of the gas velocity. Figure 12 presents the flow map proposed by Huo *et al.* (2004) for a 2.01 mm diameter tube. The data of the present work are reported and are all in the annular flow region. As a conclusion, the analysis in terms of flow chart does not allow us to find a clear relation with the heat transfer coefficient measurements.

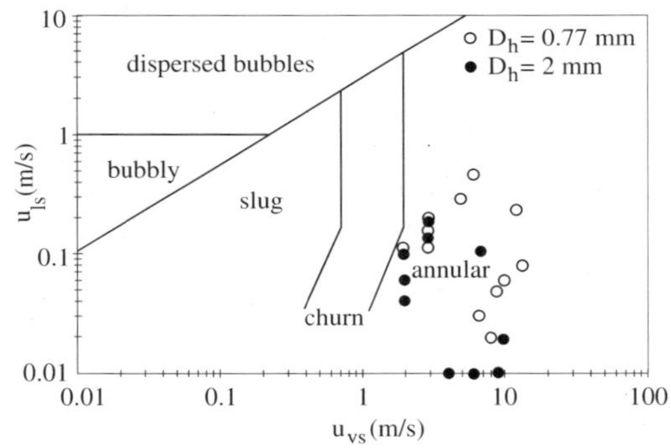

Figure 12: Flow regimes observed by Huo *et al.* (2004).

8. Pressure drop

A time averaging method was used in order to reduce wild pressure oscillations. In order to avoid non uniform distribution of coolant fluid only subcooled liquid entered the inlet manifold. Furthermore the engineering rule that the manifold diameter should be at last five times greater than the channel hydraulic diameter to equalise the fluid distribution was used. However, even if non uniform distribution occurs it will not affect the inlet and outlet measurements which are performed outside of the manifolds and it should not affect the local temperature measurements because of the averaging of wall temperatures across the *N* channels due to the very high thermal conductivity of the aluminium.

Figure 13 shows the two-phase pressure drop gradient versus the outlet quality. The solid lines represent the pressure gradient modelled with the homogeneous model. As shown in figure 13, the present measured pressure gradient is linear with x_o. This is characteristic of preponderant frictional pressure losses since integration of the homogeneous model for uniform longitudinal heating, constant thermophysical properties and friction factor shows that the frictional part of the two-phase flow pressure drop is linear with the outlet quality.

However this result and the prevalence of heat flux dependent boiling for $D_h = 2$ mm seem contradictory with the prevalence of the annular flow regime (see section 7.) which would rather suggest a separated phases model to calculate pressure losses. Yet, it is well known that the parietal heat transfer deeply disturbs the flow configuration because of the generation of bubbles at the wall. Thus the flow configuration might not be pure annular but actually slug-annular or churn flow. That would promote the mixing of liquid and vapour thus explaining the good predictions of the homogeneous model. Nevertheless this issue will be resolved only with a test section allowing flow visualisation and heat transfer at the same time.

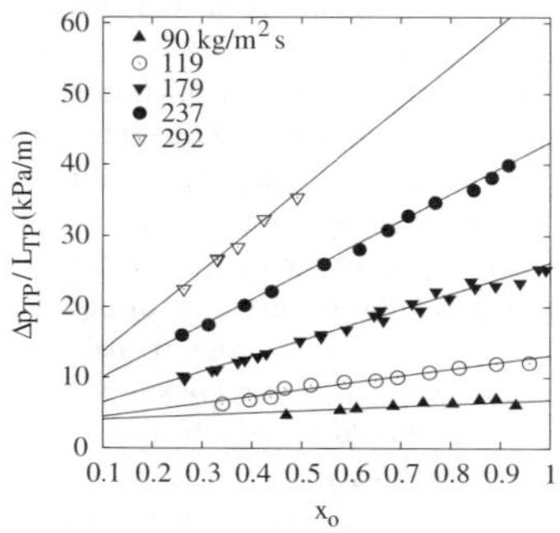

Figure 13: Pressure drop versus outlet quality ($D_h = 2$ mm).

9. Conclusions

Forced flow boiling heat transfer in minichannels in similar conditions as encountered in automobile air conditioners has been studied. Higher heat transfer coefficients than in conventional tubes are achieved but dry-out occurs at low vapour qualities thus decreasing performances. However the average heat transfer coefficient remains higher than in conventional tubes. These observations support literature studies which predict that bubble confinement leads to higher heat transfer coefficients and dry-out at low vapour quality in minichannels. The new Kandlikar (2004) general correlation for flow boiling in tubes was found to predict the present results before dry-out occurs. The Thome *et al.* (2004) and Dupont *et al.* (2004) three zone model for microscale boiling predicted most of the observed trends, including dry-out and the lack of mass velocity influence. Using it for predictions still requires testing over a consequent database.

The effect of confinement on the heat transfer coefficient before dry-out was found to be an increase of 74% when the hydraulic diameter decreased from 2 to 0.77 mm. The effect of confinement on dry-out was found to be a decrease in the critical quality from 0.3–0.4 to 0.1–0.2 for the same reduction of the hydraulic diameter. Heat flux dependent boiling prevailed in the 2 mm hydraulic diameter tube while quality dependent boiling prevailed in the 0.77 hydraulic diameter tube because of the difference in boiling and confinement numbers. The transition from one regime to another occurred for $\mathrm{Bo} \cdot (1-x) \approx 2.2 \cdot 10^{-4}$ regardless of the heat and mass velocity. Moreover it was found that dry-out could even be the dominant boiling mechanism at low qualities. The results obtained with the 2 mm hydraulic diameter tube were in total agreement with Huo *et al.* (2004)'s work. Finally frictional pressure losses seem to dominate up to mass velocities of 469 $\mathrm{kg/m^2\ s}$.

The choice of MPE tubes for the test section allowed easier measurements and results closer to industrial reality. Further studies should put the stress on the accurate influence of channel geometry and confinement on heat transfer with diabatic flow visualisation and a large variety of channels configurations.

Nomenclature

A_{fl}	total flow area	($\mathrm{m^2}$)
$\mathrm{Bo} = \dot{q}/(\dot{m} \cdot h_{\mathrm{lv}})$	Boiling number	
$\mathrm{Co} = (\sigma/(g \cdot (\rho_{\mathrm{l}} - \rho_{\mathrm{v}})))^{0.5}/D_{\mathrm{h}}$	Confinement number	
$\mathrm{Cv} = ((1-x)/x)^{0.8} \cdot (\rho_{\mathrm{v}}/\rho_{\mathrm{l}})^{0.5}$	Convection number	
$D_{\mathrm{h}} = 4A_{\mathrm{fl}}/P_{\mathrm{fl}}$	hydraulic diameter	(m)
L	tube length	(m)
$\tilde{M}$	molecular weight	g/mol
$\dot{m}$	mass velocity	($\mathrm{kg/m^2\ s}$)
N	number of channels	

P_{fl}	total wet perimeter	(m)
p	pressure	(Pa)
p_{r}	reduced pressure	(Pa)
Δp	pressure loss	(Pa)
$\dot{q}$	heat flux	$(\mathrm{W/m^2})$
$\mathrm{Re} = (\dot{m} \cdot D_{\mathrm{h}})/\mu_{\mathrm{l}}$	Reynolds number	
T	temperature	(K)
v	velocity	(m/s)
x	vapour quality	
z	z coordinate	(m)
Greek letters		
α	heat transfer coefficient	$(\mathrm{W/m^2\,K})$
χ_{tt}	Lockhart-Martinelli parameter	
ρ	mass density	$(\mathrm{kg/m^3})$
σ	surface tension	(N/m)
τ	characteristic time	(s)
θ	wetting angle	
Subscripts		
b	bubble	
cv	convective	
fl	fluid	
go	all gas flow	
i	inlet	
j	joule heated	
l	liquid	
lo	all liquid flow	
nb	nucleate boiling	
npb	nucleate pool boiling	
o	outlet	
onb	onset of nucleate boiling	
sat	saturation	
TP	two-phase	
v	vapour	
w	wall	

References

B. Agostini. Étude expérimentale de l'ébullition en convection forcée de fluide réfrigérant dans des mini-canaux. Ph.D. thesis, Université Joseph Fourier - Grenoble I (2002).

B. Agostini, B. Watel, A. Bontemps, and B. Thonon. "Friction factor and heat transfer coefficient of R134a liquid flow in mini-channels". Applied Thermal Engineering, **22**, 16, pp. 1821–1834 (2002).

M. Aritomi, T. Miyata, M. Horiguchi, and S. Sudi. "Thermohydraulics of boiling two-phase flow in high conversion light water reactors (thermohydraulics at low velocities)". International Journal of Multiphase Flow, **19**, 1, pp. 51–63 (1993).

D. Brutin, D. Topin, and L. Tadrist. "Experimental study of unsteady convective boiling in heated mini-channels". International Journal of Heat and Mass Transfer, **46**, pp. 2957–2965 (2003).

J. Collier and J. Thome. Convective boiling and condensation (Oxford University Press, 1994).

M. Cooper. "Saturation nucleate boiling. A simple correlation". 1st U.K. National Conference on Heat Transfer, **2**, pp. 785–793 (1984).

K. Cornwell and P. Kew. "Boiling in small parallel channels". In "Proceedings of the International Conference on Energy Efficiency in Process Technology", pp. 624–638 (Elsevier Applied Science, 1992).

K. Cornwell and P. A. Kew. "Evaporation in micro-channel heat exchangers". In "Proceedings of the 4th U.K. National Conference on Heat Transfer", pp. 289–294 (ImechE, 1995).

V. Dupont, J. Thome, and A. Jacobi. "Heat transfer model for evaporation in microchannels. part II: comparison with the database". International Journal of Heat and Mass Transfer, **47**, pp. 3387–3401 (2004).

A. Feldman. De l'ébullition en convection forcée dans des canaux d'échangeurs à plaques et ailettes. Ph.D. thesis, Université Henry Poincaré, Nancy I (1996).

X. Huo, L. Chen, Y. Tian, and T. Karayiannis. "Flow boiling and flow regimes in small diameter tubes". Applied Thermal Engineering, **24**, pp. 1225–1239 (2004).

E. Ishibashi and K. Nishikawa. "Saturated boiling heat transfer in narrow spaces". International Journal of Heat and Mass Transfer, **12**, pp. 863–894 (1969).

S. Kandlikar. "Fundamental issues related to flow boiling in minichannels and microchannels". Experimental Thermal Fluid Science, **26**, pp. 389–407 (2002).

S. Kandlikar. "An extension of the flow boiling correlation to transition, laminar, and deep laminar flows in minichannels and microchannels". Heat Transfer Engineering, **25**, 3, pp. 86–93 (2004).

S. Kandlikar and W. Grande. "Evolution of microchannel flow passages — thermohydraulic performance and fabrication technology". Heat Transfer Engineering, **25**, pp. 3–17 (2003).

P. A. Kew and K. Cornwell. "Correlations for the prediction of boiling heat transfer in small diameter channels". Applied Thermal Engineering, **17**, 8–10, pp. 705–715 (1997).

S. Kline and F. McClintock. "Describing uncertainties in single-sample experiments". Mechanical Engineering, pp. 3–8 (1953).

S. Kutaleladze. "Principal equations of thermodynamics of nucleate boiling". Heat Transfer Sovietic Research, **13**, 3, pp. 1–14 (1981).

K. Kuwahara, S. Koyama, J. Yu, C. Watanabe, and N. Osa. "Flow pattern of pure refrigerant HFC134a evaporating in a horizontal capillary tube". In "Proceedings of Symposium on Energy Engineering in the 21th Century", pp. 445–450 (2000).

Z. Liu and R. Winterton. "A general correlation for saturated and subcooled flow boiling in tubes and annuli, based on a nucleate pool boiling equation". International Journal of Heat and Mass Transfer, **34**, 11, pp. 2759–2766 (1991).

H. Oh, M. Katsuta, and K. Shibata. "Heat transfer characteristics of R134a in a capillary tube heat exchanger". In "Proceedings of 11th IHTC", volume 6, pp. 131–136 (1998).

W. Rohsenow. "A method of correlation heat-transfer data for surface boiling of liquids". Transactions of the ASME, **74**, pp. 969–976 (1952).

M. Shah. "A new correlation for heat transfer during boiling flow through pipes". Transactions of the ASHRAE, **82**, pp. 66–86 (1976).

D. Steiner and J. Taborek. "Flow boiling heat transfer in vertical tubes correlated by an asymptotic model". Heat Transfer Engineering, **13**, 2, pp. 43–69 (1992).

J. Thome. "Boiling in microchannels: a review of experiment and theory". International Journal of Heat and Fluid Flow, **25**, pp. 128–139 (2004).

J. Thome, V. Dupont, and A. Jacobi. "Heat transfer model for evaporation in microchannels. part I: presentation of the model". International Journal of Heat and Mass Transfer, **47**, pp. 3375–3385 (2004).

T. Tran, M. Wambsganss, M. Chyu, and D. France. "A correlation for nucleate flow boiling in small channels". In R. K. Shah, editor, "Compact Heat Exchangers for the Process Industries", pp. 353–363 (Begell House, inc., 1997).

K. Triplett, S. Ghiaasiaan, S. Abdel-Khalik, A. LeMouel, and B. McCord. "Gas liquid two-phase flow in microchannels. Part I : two-phase flow patterns". International Journal of Multiphase Flow, **25**, pp. 377–394 (1999).

HEAT REMOVAL USING NARROW CHANNELS, SPRAYS AND MICROJETS

M. FABBRI, S. JIANG, G. R. WARRIER, AND V. K. DHIR
Mechanical and Aerospace Engineering Department
Henry Samueli School of Engineering and Applied Science
University of California Los Angeles
420 Westwood Plaza, Los Angeles, CA 90095
Ph. (310) 825-9617, Fax: (310) 206-4830, Email: vdhir@seas.ucla.edu

1. Introduction

As the chip fabrication technology keeps improving, smaller and more powerful components are introduced in the market. The traditional air cooling techniques are proving to be inadequate in removing the heat fluxes generated by these new microchips and new ways are being sought to cool the components. Hence, active cooling methods such as the use of narrow channels, sprays and arrays of microjets are now being seriously considered.

One of the simplest arrangements that can be used for heat removal involves using single-phase forced convection or flow boiling in small channels. The cross-section of the channel is typically circular or rectangular. In this arrangement, the electonic device is mounted on the top and/or bottom surface of the substrate material which have the channels built into it. The heat is conducted through the substrate to the channel, where it is removed by forced convection. However, if the channel is long enough or the heat flux is high or the liquid flow rate is low, boiling can also occur in these channels. Due to the nature of the geometries (small sizes) involved, the heat transfer and the associated pressure drop can be very different from that in large tubes or channels. The coupling between the pressure drop and the heat transfer becomes stronger as the size of the channel gets smaller.

Liquid droplet spray and jet impingement cooling techniques have been widely used in the metal manufacturing industry and have been shown capable of high heat removal rates. Researchers have investigated the possibility of applying such techniques to the cooling of electronic components. The droplet sprays can have the form of a mist, and impinge on the surface with a random pattern or they can be formed by one or more streams of droplets which impinge upon the surface with a fixed pattern. If the frequency of the streams is high enough, the droplets merge forming continuous liquid jets. After hitting the surface, the liquid droplets spread and, if the spreading area is small enough a continuous thin liquid film covering the surface is formed. If the wall superheat is high, a thin vapor layer can be present underneath the droplets or the thin liquid film. The heat transfer process is transient and it involves liquid and vapor convection, thin film evaporation, and air convection. The areas not covered by the film dry out.

When continuous liquid jets are employed, the liquid film covering the surface is continuous and the heat is removed mainly by convection. Evaporation from the thin film may occur at high heat fluxes or low flow rates.

The physics governing the heat removal process by droplet sprays is very complex and still is not completely understood, and few theoretical models are available in the literature. Hence, it has turned out to be easier to investigate the various aspects of the problem by performing experimental work. Several studies have been conducted in the past on sprays, but most of them deal with the boiling regime, which was not considered in the present work. Air driven sprays obtained using atomizer nozzles are not considered in this study either because they would be impractical to use in a closed system for electronic cooling.

S. Kakaç et al. (eds.), Microscale Heat Transfer, 231–254.

The objective of this study is to experimentally compare the cooling performance of narrow channels, droplet sprays and arrays of microjets. The comparison is mostly under single-phase conditions. However, some implications of two-phase flow are also discussed. Since details regarding each of the three cooling techniques have been reported elsewhere ([1], [2] and [3]), only a brief description is presented here

2. Experimental Apparatus

2.1 NARROW CHANNELS

The schematic of the experimental apparatus is shown in Fig. 1. It consists of a supply tank, variable speed gear pump, filter, preheater, flowmeter, test section, and return tank. The supply tank is provided with a preheater (1.5 kW). The return tank is provided with cooling coils and is mounted above the supply tank and is connected to the supply tank by a valve. An inline filter and an additional inline preheater (700 W) are provided downstream of the pump. The liquid flow rate is measured using the flowmeter provided upstream of the test section.

The test channel is made of aluminum and consists of five small rectangular channels in parallel. Figure 2 shows details of the test channel. Each flow channel is rectangular in cross-section with a height of 0.5 mm and a width of 1.5 mm. Thus the hydraulic diameter (D_h) is 0.75 mm. The total length to diameter ratio (L/D) is 433.5, with the effective heated (L/D) ratio being 409.8. The distance between the centers of any two channels is 3.0 mm, with the total width of the two plates being 16 mm. Smooth inlet and outlet transition sections (127 mm long) are also provided. A total of 20 miniature thermocouples (K-type, 0.25 mm) are placed on the top and bottom of the test channel to measure the wall and fluid temperatures as shown in Fig. 2. All thermocouples were calibrated prior to their installation. All thermocouples used in the experiment had a calibration accuracy of ± 0.2 °C.

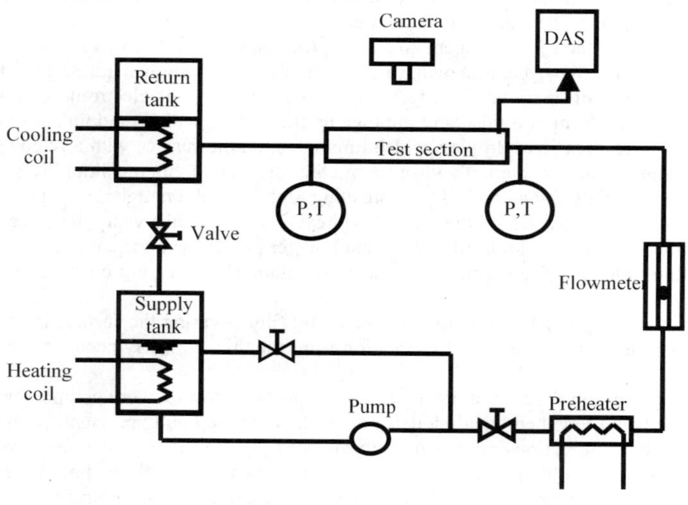

Figure 1 Schematic of narrow channels setup.

Figure 2 shows the placement of the thermocouples at axial distances of 39.0 mm, 89.0 mm, 163.0 mm, 236.0 mm, and 287.0 mm from the inlet, respectively. The thermocouples were placed into holes drilled into the test channel surface and secured in place by high conductivity thermocouple cement. The center channel was provided with thermocouples at both the top and the bottom surfaces, while the temperatures of the two end channels were only measured on one surface. The pressures were measured at locations 1 cm before the inlet and after the outlet of the test channel. The accuracy in the measurement of the inlet and outlet pressures is about ± 0.8 kPa.

The heat flux was applied using two inconel-625 heating strips placed at the top and bottom surfaces of the test channel. The heater strips were cut to the exact dimensions of the test channel (16 mm wide and 325.1 mm long) and were coated with an electrically insulating varnish on the side in contact with the test channel. Holes were cut into the strips at the thermocouple locations, so that the thermocouples could pass through these heating strips and into the holes in the test channel. The heater stip was electrically heated using a DC power supply. The temperatures, pressures, and power supply voltage and current were recorded using a data acquisition system connected to a computer.

Additionally, a few visualization experiments were performed by replacing the top aluminum plate by a clear transparent polycarbonate plate. In these experiments, movies of the flow were captured using a high speed CCD camera. The camera was capable of capturing at 1220 frames/sec.

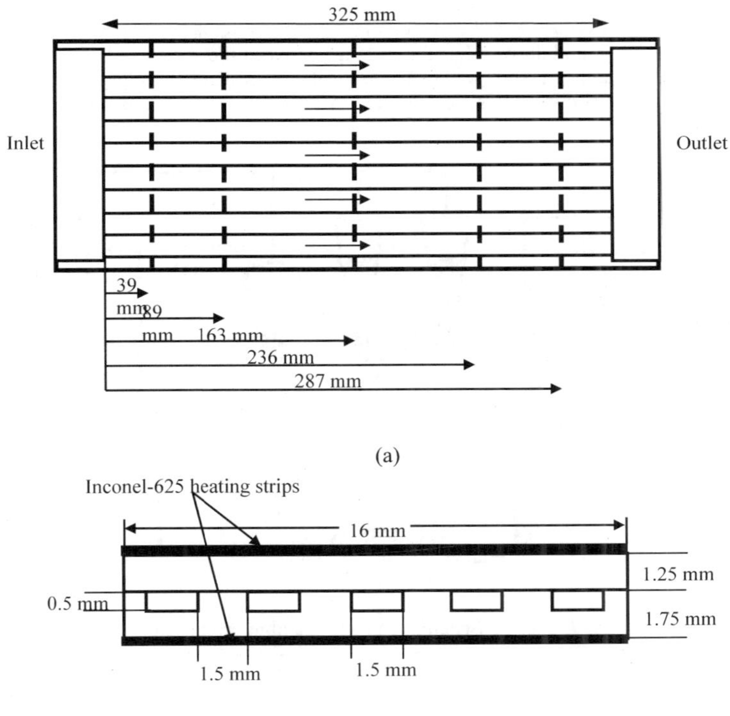

Figure 2 Details of narrow channels test section.

2.2 SPRAYS AND MICROJETS

The experiments using sprays and microjets were performed in the same experimental apparatus. Figure 3 shows a schematic of the experimental setup. The coolant is circulated with two variable speed gear pumps, installed in parallel. Two rotameters and a turbine flowmeter were installed in parallel to measure the flow rate of the liquid being sprayed. The accuracy of these flow meters is ±3%. Before entering the flowmeters, the coolant passed through a heat exchanger, where it was cooled down to the specified temperature. The sprays and microjets impinged on the top surface of a copper cylinder 19.3 mm in diameter, which simulated the backside of an electronic microchip. The copper cylinder was enclosed in a Teflon jacket that provided thermal insulation. The cylinder had a larger cylindrical base (76.2 mm in diameter and 16.5 mm in length) with six 750 W cartridge heaters embedded in it. The power to the heaters was controlled with a variac. Four K-type thermocouples, soldered at different axial locations (5 mm apart starting from 1.8 mm below the top surface) along the central axis of the cylinder were used to measure the temperature profile. The heat flux and the temperature of the heat transfer surface were calculated from the measured temperature profile.

Figure 4(a) shows details of the test chamber. The teflon jacket–copper block assembly was mounted on a stainless steel plate. The sprayed liquid was drained back to the reservoir installed below the test section. Thermocouples to measure the liquid temperature were installed both upstream of the flow meters and downstream of the orifice plate. Another thermocouple was used to measure the ambient temperature. The HAGO nozzle and the orifice plate (for microjets) were attached to the end of a pipe which could be moved up and down. Thus the distance between the sprays or jets and the heated surface could be varied. The liquid pressure was measured using a pressure transducer (0 – 2.07 MPa, 0.25% accuracy) just before the sprays or jets were formed. All the data were recorded using two IOTech 16-bit data acquisition boards.

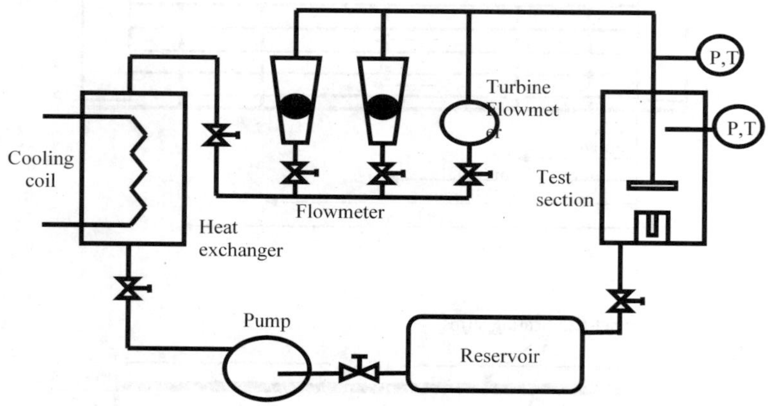

Figure 3 Schematic of spray and microjet setup.

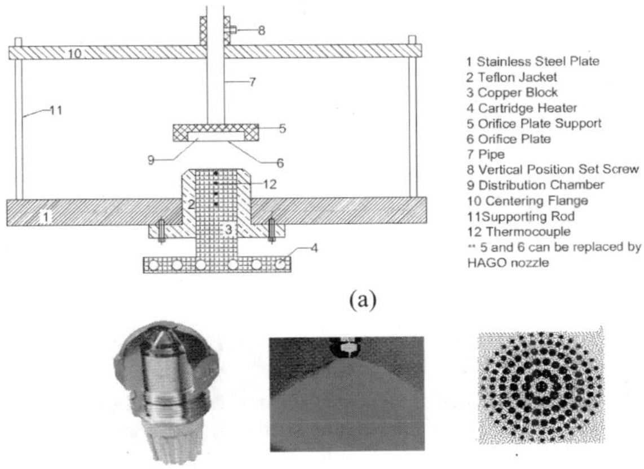

(a)

(b)

Figure 4 Details of (a) test section and (b) HAGO nozzle.

Originally designed for use with home power spray humidifiers, the HAGO nozzle produces the finest possible atomization with direct water pressure operation. The minimum operating pressure is 274.8 kPa gage but increasingly finer droplets result from higher operating pressures. Each nozzle is individually spray tested for flow rate, spray angle, and spray quality. The standard spray angle is 70°.

The extremely fine atomization achieved by the Type B nozzle has resulted in it being used extensively in applications where fine atomization and precise flow rates are required such as evaporative cooling, humidification, moisture addition and misting. The "DFN" type nozzle is designed for low flow rates (18.9 ml/min to 63 ml/min). It employs two sintered bronze filters, which give more surface filtration. The specific type of the nozzle used in this study is "DFN" B100. Figure 4(b) shows a picture of the nozzle, of the spray cone, and of the droplets pattern on the impinged surface. The mean Sauter diameter of droplets for the DFN-B100 nozzle at 274.8 kPa gage pressure is 44 μm according to the manufacturer. The orifice diameter of the spray nozzle is 356 μm.

The microjets were formed by pushing the liquid through a 0.5 mm thick stainless steel orifice plate. The holes in the plate were laser drilled and were arranged in a circular pattern giving a radial and circumferential pitch of 1 mm for 397 jets, 2 for 127 jets and 3 for 61 jets. It is important to note that the outer one or two rings of jets, depending on the pitch used, did not directly hit the copper surface but instead impinged on the surrounding Teflon surface. Thus, the actual number of jets impinging directly on the copper surface was 37, 61, and 271 for 1, 2, and 3 mm pitch, respectively.

3. Experimental Conditions and Procedure

3.1 NARROW CHANNELS

The experimental procedure adopted for all test cases is as follows: The liquid was first preheated to the required temperature in the supply tank. This liquid was then pumped from the supply tank thorough the filter, preheater, flowmeter, and then into the test section. The fluid exiting the test section was collected in the return tank. The flow rate of the liquid was measured using the flowmeter and was adjusted as required. Once the inlet subcooling and flow rate were at the desired values, the recording of all the temperatures and pressures was initiated. One set of data was always collected for a case where the input power was zero. Subsequently, the power to the test section was turned on and set to a predetermined value. The typical duration of a single heat flux experiment was approximately 20-25 minutes. All measurements were taken only when the temperatures were steady for about 10-15 minutes. At the end of each experiment, the liquid collected in the return tank was drained back into the supply tank via the valve at the bottom of the return tank. The power to the test channel was then increased and the same procedure was repeated.

The experiments using the narrow channels were performed using FC-84, deionized water and deionized water-methanol (60:40) mixture as the test fluids. The experiments were only conducted for test conditions wherein the pressure drop was less than 0.6 bar.

3.2 SPRAY AND MICROJETS

Prior to running the experiments, the copper surface was polished to obtain a smooth and shiny surface. Thereafter, in order to guarantee the same surface condition throughout the duration of the experiment, the surface was oxidized in air for five hours at 320 °C. The contact angle for a water droplet placed on the oxidized copper surface was measured to be 44°.

After the pumps were started, and the liquid flow rate set to the desired value, the cartridge heaters were energized. Once all the parameters reached steady-state, the values were recorded for 100 s at a sampling rate of 1 Hz. Thereafter, the power to the copper block was increased and a new set of data recorded. The experiment was stopped when either the surface temperature was above the boiling point or when the temperature at the base of the copper block rose above 350 °C, which could damage the Teflon jacket and the electrical wires. In these experiments, the test fluids used were deionized water and FC-84.

The experiments using sprays were performed in both open and closed loop configurations. In the closed loop configuration, the experiments could be performed with varying total system pressures and air partial pressures.

4. Data Reduction

4.1 NARROW CHANNELS

In each experiment, the net power to the channels (Q) was calculated as,

$$Q = Q_{supply} - Q_{loss} \tag{1}$$

where Q_{supply} is the total power supplied to the heaters (Q_{supply} = VI, V = voltage, I = current) and Q_{loss} is the heat loss to the environment. The heat loss was estimated by accounting for the thermal resistance between the test channel walls and the environment. The accuracy of this method was verified by comparing Q to the sensible heat gained by the liquid ($= \dot{m} c_{p,liq} \left(T_{liq,out} - T_{liq,in} \right)$). In most cases, the difference between $Q_{channel}$ and sensible heat gained was satisfied to within ± 8%. The wall heat flux is assumed to be uniform and is calculated as,

$$q_w = Q / A_s \tag{2}$$

where A_s is the total surface area of the narrow channels assembly (since heated length = 325.1 mm and width = 16 mm, A_s = 98.3 cm^2).

Since the wall temperatures measured in the different channels at the same axial location showed only small variations, they were averaged and this averaged wall temperature was used in all further calculations. In general, at any given axial location, the maximum variation between the measured wall temperatures was between ±0.1 °C and ±0.4 °C. The fluid temperature at any axial location (z) is calculated from the energy balance as,

$$T_l(z) = T_{l,in} + \frac{Qz}{L\dot{m}C_{p,liq}}, \tag{3}$$

where $T_{liq,in}$ and $\dot{m}$ are the measured inlet liquid temperature and mass flow rate, respectively. The uncertainty in the calculated liquid temperatures is estimated to be about ±9%. The local heat transfer coefficient for single-phase forced convection (h_{sp}) is calculated as,

$$h_{sp} = \frac{q_w}{T_w(z) - T_{liq}(z)}. \tag{4}$$

The corresponding Nusselt number (Nu) is calculated as Nu = $h_{sp}D_h/k_l$, where k_l is the liquid thermal conductivity. The measurement uncertainty in the single-phase heat transfer coefficients were estimated to about 15% - 25% close to the channel inlet and about 10% - 15% further downstream.

Knowing the single-phase heat transfer coefficient, the location of the onset of nucleate boiling (z_{ONB}) was taken as the axial location where (T_w - $T_{liq}(z)$) from experiment was lower than that calculated by the expression q_w/h_{sp}. As mentioned earlier, $T_{liq}(z)$ is calculated using Eq. (3). In the two-phase region (downstream of z_{ONB}), the pressure drop and heat transfer are calculated assuming that a certain fraction of the energy from the wall is utilized in producing vapor. Thus a nonequilibrium condition with vapor and subcooled liquid existing simultaneously is present before bulk boiling begins. Because of the coupling between the local nonequilibrium (or subcooled) quality and the pressure and saturation temperature, an iterative procedure was used to obtain the local heat transfer coefficient and local pressure.

In the subcooled and saturated boiling regimes, the local two-phase heat transfer coefficient was obtained as,

$$h_{tp} = \frac{q_w}{(T_w - T_{sat}(z))}, \tag{5}$$

where $T_{sat}(z)$ is the saturation temperature at the location z and it depends on the local pressure. The pumping power ($Q_{pumping}$) required was calculated as the product of the volumetric flow rate and the pressure drop across the channel.

4.2 SPRAYS AND MICROJETS

For each data set recorded, the heat flux at the heat transfer surface is given by the slope of the temperature profile obtained from the four thermocouples embedded in the copper block. The same temperature profile, which was mostly linear, allowed the temperature at the heat transfer surface to be calculated (by extrapolation).

The area-averaged heat transfer coefficient, Nusselt number, and Reynolds number were calculated as,

$$h = \frac{q}{T_w - T_l}, \quad Nu = \frac{hd_n}{k_{film}}, \quad \mathrm{Re}_{d_n} = \frac{\rho_{film} v d_n}{\mu_{film}}, \tag{6}$$

where d_n is the jet or droplet diameter and v is the velocity of the jets or droplets at the nozzle or orifice exit. All the physical properties are evaluated at the mean film temperature, T_{film} (= $(T_w + T_{liq})/2$).

Repeatability of the data was verified by randomly repeating some of the cases already tested. The minimum and maximum uncertainties of the main parameters are as follows: (i) T_w and Tliq: ±0.1 °C, (ii) q_w: ±2.7% and ±28.9%, (iii) $\overline{h}$: ±2.9% and ±29.6%, (iv) $\overline{Nu}$: ±6.7% and ±33.0%, (v) Re: ±7.2% and ±22.7%, (vi) $Q_{pumping}$: ±0.5% and ±26.0%. The uncertainty of ±28.9 % in the heat flux occurred when the heat flux was only 1.75 W/cm^2 and it was caused by the small temperature gradient in the copper block.

5. Results and Discussion

5.1 NARROW CHANNELS

Figure 5 shows the variation of the heat transfer coefficient along the length of the narrow channels. In the single-phase region h_{sp} decreases along the length of the channel due to the developing thermal boundary layer. However, with the onset of nucleate boiling (z_{ONB}), h_{tp} increases dramatically but then decreases monotonically downstream of z_{ONB}. Figure 6(a) shows the variation of Nu for single-phase forced convection as a function of the axial distance. Also, shown in Fig. 6(a) are the numerical results listed in the textbook by [4]. Similar results were obtained for all the other test cases. It can be seen from Fig. 6(a) that the measured values agree quite well with the numerical results, especially near the fully developed region. Close to the channel entrance, the experimentally measured values generally tend to be about 25% lower. It must however be kept in mind that the measurement uncertainty in Nu close to the channel entrance is about 15% - 25%, while far downstream it is about 10% - 15%.

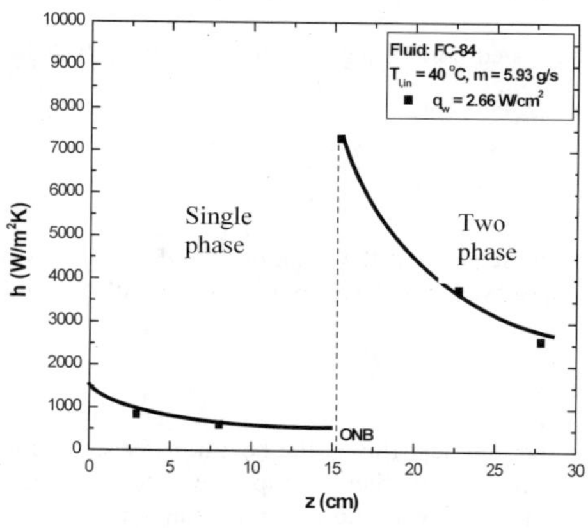

Figure 5 Variation of the heat transfer coefficient along the length of the narrow channels.

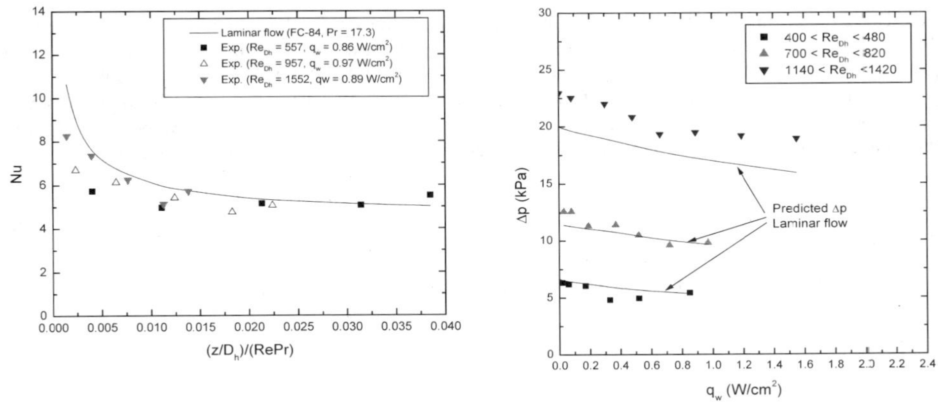

(a) (b)

Figure 6 (a) Nusselt number and (b) pressure drop for single-phase flow in narrow channels.

Figure 6(b) shows a comparison of the measured single-phase pressure drop with that predicted for laminar flow in rectangular channels. The fanning friction factor used in predicting the pressure drop was $f = 17.25/Re_{Dh}$ ([5]). From Fig. 6(b) it can be seen that there is good agreement between the experimentally measured and predicted values, especially at lower mass fluxes. For the case of G = 1552 kg/m^{2}s and $T_{liq,in} = 26$ °C, the predicted pressure drops are about 8% - 14% lower than the experimentally measured values. Similar results were obtained for other test cases. At present it is not clear why this difference exists. It is most likely due to the early transition to turbulence. Though it is not substantiated by the heat transfer data, it is possible that the improvement in heat transfer due to the transition to turbulence is not significant and lies within the scatter present in the heat transfer data.

Figure 7 shows the variation of the two-phase heat transfer coeffient (h_{tp}) as a function of the local subcooling parameter (Sc), for subcooled flow boiling. The subcooling parameter is defined as,

$$Sc(z) = \left(\frac{c_{p,l}(T_{sat} - T_l)}{h_{fg}} \right)_{ONB} - \left[\frac{Q(z - z_{ONB})}{L\dot{m}h_{fg}} - x_{sc}(z) \right], \tag{7}$$

where x_{sc} is the quality during subcooled boiling given as,

$$x_{sc}(z) = \left[\left(\frac{c_{p,l}(T_{sat} - T_l)}{h_{fg}} \right)_{ONB} - \left(\frac{c_{p,l}(T_{sat} - T_l)}{h_{fg}} \right)_z \right]^n = \left[\left(\frac{c_{p,l}(T_{sat} - T_l)}{h_{fg}} \right)_{ONB} - Sc(z) \right]^4 . \tag{8}$$

The resulting correlation for h_{tp} during subcooled flow boiling can be expressed as,

$$\frac{h_{tp}}{h_{sp\_FD}} = 1 + 6.0Bo^{1/16} + f_1(Bo)Sc^{4.15}, \quad f_1(Bo) = 290[1 - 855Bo], \tag{9}$$

where Bo is the boiling number ($Bo = q_w/Gh_{fg}$) and h_{sp_FD} is the single-phase heat transfer coefficient for fully-developed flow. Since this correlation was developed based on a limited set of data, Eq. (9) is strictly valid for the following range of parameters, $0.0 \leq Sc \leq 0.80$ and $0.00014 \leq Bo \leq 0.00089$.

Saturated flow boiling begins downstream of the location where the liquid becomes saturated (z_{sat}). In these experiments, only a few data points were obtained for saturated flow boiling. Figure 8 shows the variation of h_{tp} as a function of the local quality (x), for saturated flow boiling. Downstream of z_{sat}, x is given as,

$$x = x_{sc} + x_{eq} + x_{flash}, \tag{10}$$

where x_{eq} is the equilibrium quality given by the expression,

$$x_{eq} = \frac{Q(z - z_{sat})}{Lmh_{fg}}, \tag{11}$$

and x_{flash} is the quality due the flashing of the liquid (i.e, occurs when the local liquid temperature exceeds the local saturation tempeature brought about as a result in the rapid decrease in the pressure along the channel length) given by,

$$x_{flash} = \frac{c_{p,l}\left(T_{sat}\left(z_{sat}\right) - T_{sat}\left(z\right)\right)}{h_{fg}}. \tag{12}$$

The final correlation for h_{tp} during saturated flow boiling can be expressed as,

$$\frac{h_{tp}}{h_{sp\_FD}} = 1 + 6.0Bo^{1/16} + f_2\left(Bo\right)\left(x\right)^{0.65}, \quad f_2\left(Bo\right) = -5.3\left[1 - 855Bo\right]. \tag{13}$$

It must be noted that due to the very limited data currently available, the correlation (Eq.(13)) is only valid for the following range of parameters: $0.00027 \leq Bo \leq 0.00089$, and $0.03 \leq x \leq 0.55$.

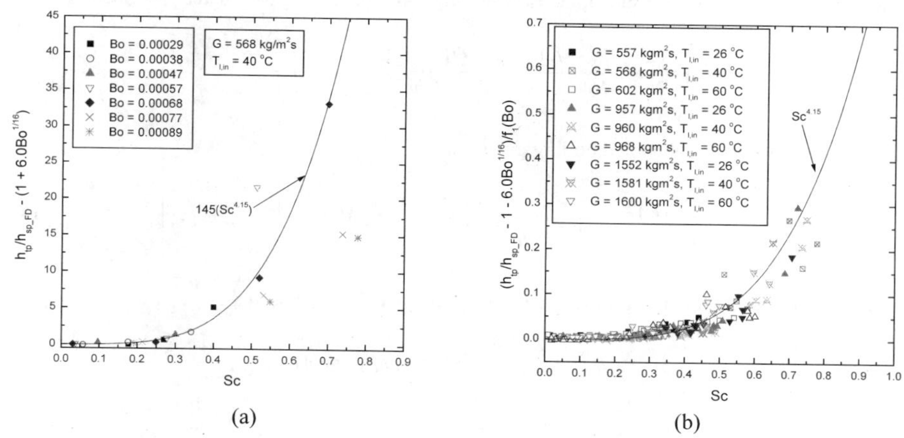

Figure 7 Subcooled boiling: h_{tp} vs. Sc for (a) $G = 568\ kg/m^2s$ and (b) $G = 557$ to $1600\ kg/m^2s$.

An expression for the local pressure at any location prior to z_{ONB} in the channel is obtained as,

$$p(z) = p_{in} - (\Delta p)_{sp}, \tag{14}$$

where $(\Delta p)_{sp}$ is the single-phase friction pressure drop. In the region beyond z_{ONB}, the local pressure is obtained from the following expression:

$$p(z) = p_{in} - (\Delta p)_{sp} - (\Delta p)_{tp}, \tag{15}$$

where $(\Delta p)_{tp}$ is the sum of the two-phase friction ($(\Delta p)_{tp,fr}$)and acceleration ($(\Delta p)_{tp,acc}$) pressure drops. Since the liquid saturation temperature depends on the local temperature, one needs to iterate between Eqs. (10) and (15) to solve for the local quality x, and the local pressure.

The two-phase pressure drop calculations were performed assuming the vapor void fraction (α) is given by the Lockhart and Martinelli correlation (as given in [6]),

$$\alpha = \left[1 + 0.28\chi^{0.71}\right]^{-1}. \tag{16}$$

In Eq. (16), χ is the Martinelli parameter ($\chi = (\Delta p/dz)_{liq}/(\Delta p/dz)_v$, where $(\Delta p/dz)_{liq}$ is the pressure gradient for liquid flowing alone and $(\Delta p/dz)_v$ is the pressure gradient for vapor flowing alone). Knowing the void fraction and vapor quality variation along the channel length, the two-phase pressure drop in the channel can be calculated using the separated flow model, i.e.,

$$\begin{aligned} \Delta p_{tp} &= \Delta p_{tp,fr} + \Delta p_{tp,acc} \\ &= \left[\int_{z_{ONB}}^{z} \frac{2 f_{lo} G^2 \phi_{lo}^2}{D_h \rho_l} dz\right] + \left[\int_{z_{ONB}}^{z} G^2 \left\{\frac{dx}{dz}\left\{\frac{2x}{\rho_v \alpha} - \frac{2(1-x)}{\rho_l(1-\alpha)}\right\} + \frac{d\alpha}{dx}\left(\frac{(1-x)^2}{\rho_l(1-\alpha)^2} - \frac{x^2}{\rho_v \alpha^2}\right)\right\} dz\right]. \end{aligned} \tag{17}$$

In Eq (17), the two-phase multiplier for only liquid flowing (ϕ_{lo}^2) (mass flux = G) is given by

$$\phi_{lo}^2 = \phi_l^2 \left(\frac{f_l}{f_{lo}}\right)(1-x)^2, \tag{18}$$

where f_l and f_{lo} are the friction factors for liquid flowing alone (mass flux = G(1-x)) and the mixture flowing as liquid (mass flux = G), respectively. The two-phase multiplier for liquid flowing alone (ϕ_l^2) (mass flux = G(1-x)) is given by the Martinelli-Nelson correlation as,

$$\phi_l^2 = 1 + \frac{C}{\chi} + \frac{1}{\chi^2}. \tag{19}$$

In the above equation, the constant C is taken to be 38, irrespective of the liquid and vapor flow regimes. Good agreement can be seen between the predicted and experimentally obtained pressure drops, as shown in Fig. 9. Further details of the pressure drop and heat transfer model can be found in [1]

One way of determining the effectiveness of an active cooling system is to compare the pumping power required ($Q_{pumping}$) to move the liquid to the actual power removed ($Q_{removed}$) by the fluid. Figure 10(a) shows $Q_{pumping}$ as a function of $Q_{removed}$, for the different test fluids. It can be seen that $Q_{pumping}$ decreases with increasing $Q_{removed}$, reaches a minimum and then increases rapidly. For a given $Q_{pumping}$ vs.

$Q_{removed}$ curve, the region to the left of the minimum represents the single-phase region; the minimum represents ONB while the region to the right of the minimum point represents the two-phase region. For water at a temperature of 25 °C at the inlet, in single-phase about 500 W (~ 8.6 W/cm^2) could be removed for $Q_{pumping}$ = 0.01 W, while in two-phase, about 900 W (13.2 W/cm^2) could be removed while consuming only about 0.03 W. Figure 10(b) shows the process efficiency (i.e., $Q_{pumping}/Q_{removed}$) as a function of the dimensionless temperature difference $(T_w - T_{liq})/(T_{sat} - T_{liq})$. Please note that here T_w is the wall temperature at the channel exit while T_{liq} is the temperature at the channel inlet. In analyzing the data plotted in Fig 10(b), it must be kept in mind that the lower the process efficiency, more efficicient is the process. Again it can be seen that water at 25 °C performs the best, while FC-84 performs the worst.

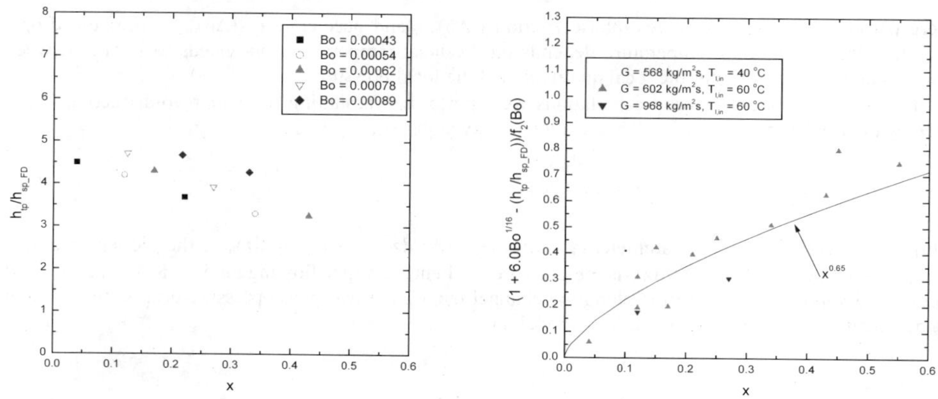

(a) (b)

Figure 8 Saturated boiling: (a) h_{tp} vs. x for (a) Bo = 0.00043 to 0.00089 kg/m^{2}s and (b) G = 568 to 968 kg/m^{2}s.

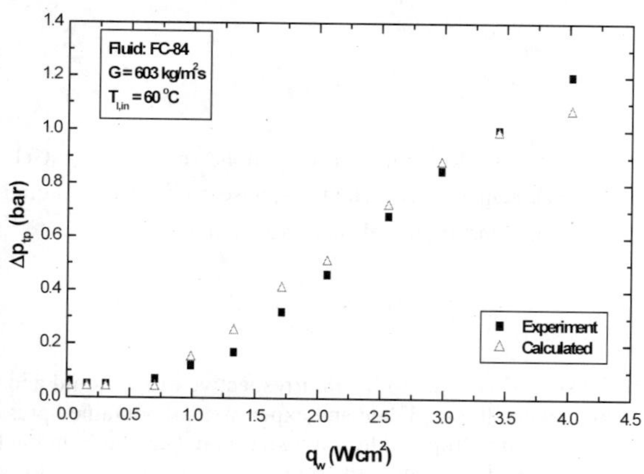

Figure 9 Comparison of measured and predicte pressure drops.

Figure 11 shows the flow patterns that were observed in the narrow channels [7]. At ONB, small bubbles can be seen nucleating and departing. Further downstream, bubbly flow can be observed. Still further downstream, these bubbles coalesce together to form large vapor slugs, which in turn merge to form still larger slugs. This merger of vapor slugs finally leads to annular flow in the narrow channels, with the liquid at the walls and the vapor at the core.

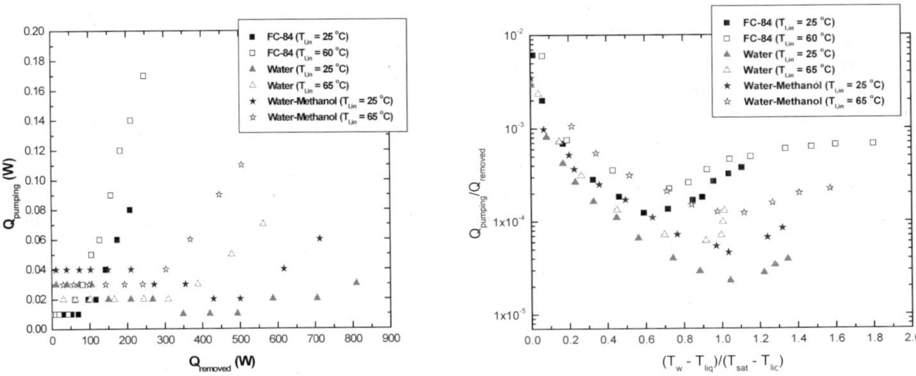

(a) (b)

Figure 10 (a) $Q_{pumping}$ as a function of $Q_{removed}$ (b) Process efficiency as a function of dimensionless temperature difference.

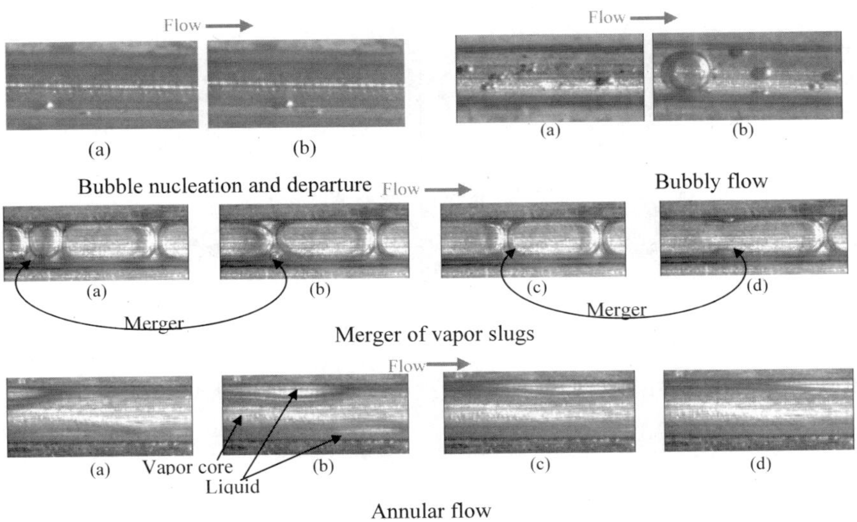

Figure 11 Flow visualization in narrow channels.

5.2 SPRAYS

Figure 12(a) shows the boiling curves obtained using the HAGO nozzle for water flow rates of 2.87 and 4.63 μl/s/mm^2, respectively. As expected, the effect of increasing the flow rate results in the boiling curves shifting to the left. Data for a flow rate of about 2.7 μl/s/mm^2, at 101 kPa total pressure and varing P_{air}, are plotted in Fig. 12(b). The last data point is the dry out point. It can be seen from Fig. 12(b) that in the single-phase regime, heat transfer coefficient is constant and independent of T_w - T_{liq}. A linear best fit is drawn on the graph for each P_{air}/P_{total}. It can be seen that the data points for each P_{air}/P_{total} fall onto a straight line. When boiling starts the slope of the heat transfer curve begins to deviate from the single-phase regime and continues to change until fully developed nucleate boiling regime, then the slope of the curve becomes large and remains almost constant until CHF is reached. It can be seen from the curves that for P_{air}/P_{total} = 98 %, the change in the slope is more discernible than for P_{air}/P_{total} = 3 %. A dotted line is drawn on the graph to distinguish the observed linear single-phase regime and the possible partial nucleate boiling regime. Figure 13 shows the variation in the single-phase heat trasfer coefficient as a function of P_{air}/P_{total}. Increasing P_{air}/P_{total} results in a decrease in Nu, with the rate of decrease being higher at low P_{air}/P_{total} and lower as P_{air}/P_{total} approaches unity.

The process efficiency ($Q_{pumping}/Q_{removed}$) for the HAGO nozzle, for P_{air}/P_{total} varying from 3% to 98% is shown in Fig. 14. The graph shows that increasing T_w results in a decrease in $Q_{pumping}/Q_{removed}$. Since the water flow rate is almost constant, $Q_{pumping}$ is almost constant, thus the decrease in $Q_{pumping}/Q_{removed}$ is due to the increase in $Q_{removed}$. Further details can be found in [2].

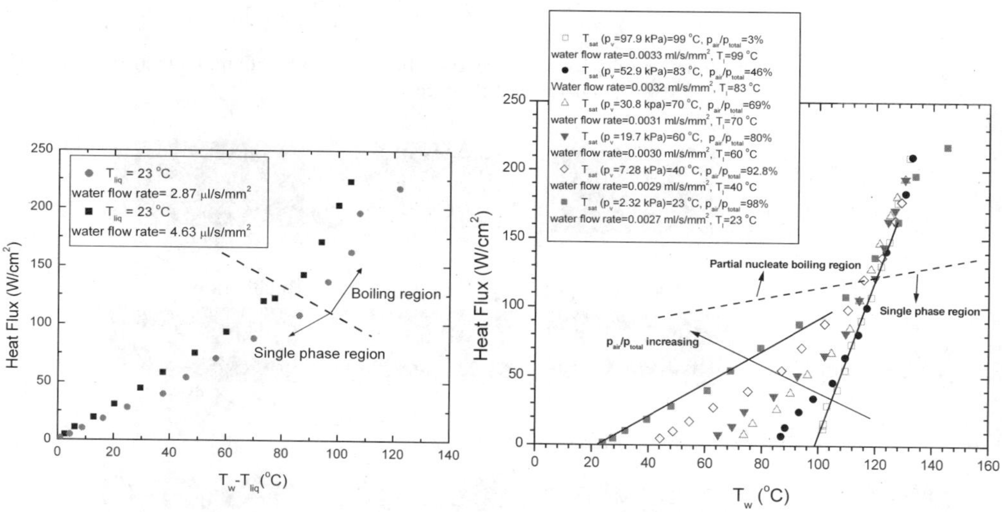

(a) (b)

Figure 12 Boiling curves for spray with HAGO nozzle (a) effect of flow rate and (b) effect of air content in the test chamber.

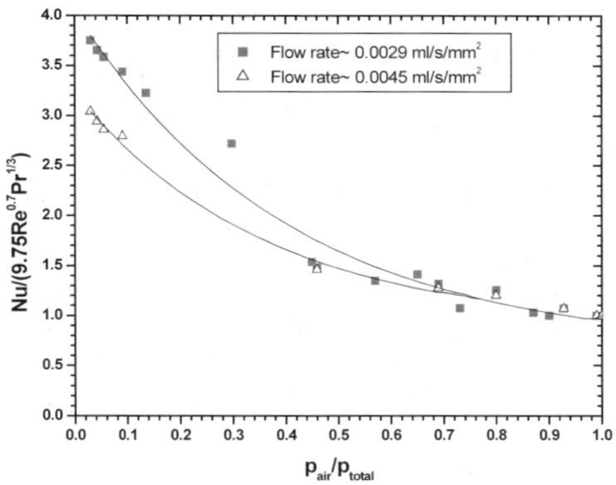

Figure 13 Variation of Nu as a function of P_{air}/P_{total}.

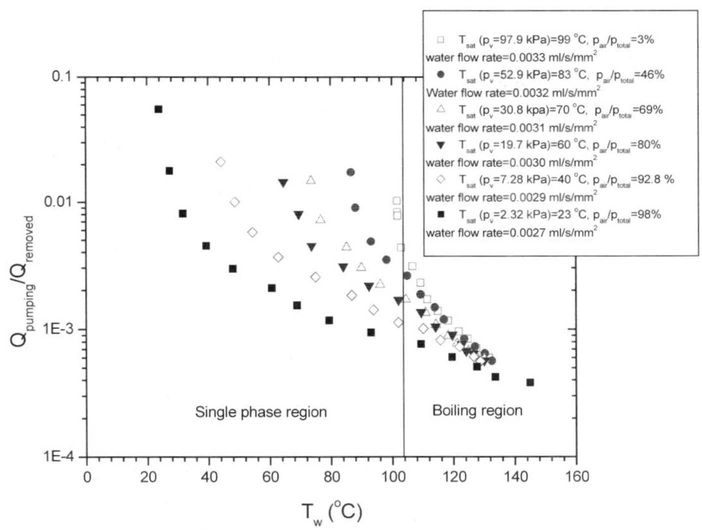

Figure 14 Process efficiency for spray cooling.

5.3 MICROJETS

For the array of microjets, the effect of varying Re_{dn}, Pr, and s/d_n on Nu are shown in Fig. 15(a) through Fig. 15(c), respectively. For a given number of jets and Pr, increasing Re_{dn} results in an increase in Nu. Similarly, for a given s/d_n and Re_{dn}, increasing Pr results in increasing Nu. Also, for given Re_{dn} and

Pr, increasing s/d_n results in an increase in Nu. Based on the experimental data obtained, a correlation for single-phase heat transfer was developed. This can be expressed as,

$$Nu = 0.143\,\mathrm{Re}_{d_n}^{0.78}\,\mathrm{Pr}^{0.48}\exp\left(-0.069\,s/d_n\right). \qquad (20)$$

The above correlation is strictly valid for thr following range of parameters: $43 \le \mathrm{Re}_{dn} \le 3813$, $2.6 \le \mathrm{Pr} \le 84$, and $4 \le s/d_n \le 26.2$.

The effect of varying the distance between the nozzle and the heater surface was found to be negligible, as can be seen from Fig. 16. From Fig. 17 it can be seen that there is good agreement between the results of the present study and those available in the literature ([8] and [9]).

For a given chip power, there exists an optimum configuration in terms of jet diameter and spacing. Details of the optimization can be found in [3]

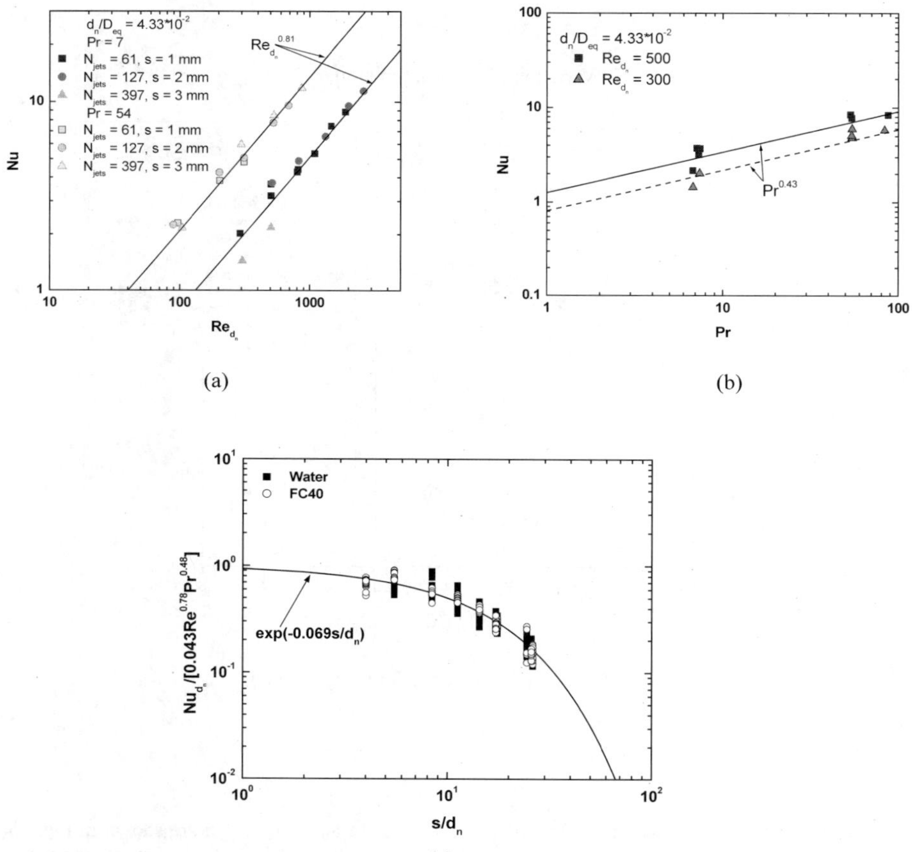

Figure 15 Effect of (a) Re_{dn} (b) Pr and (c) s/d_n on Nu for microjets.

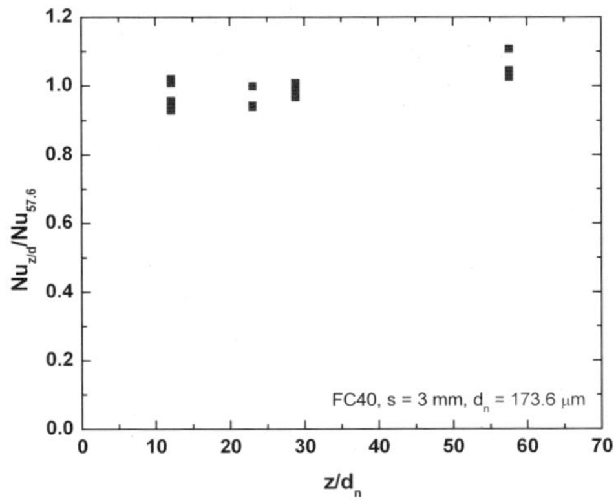

Figure 16 Effect of distance between nozzle and heater surface on Nu for microjets.

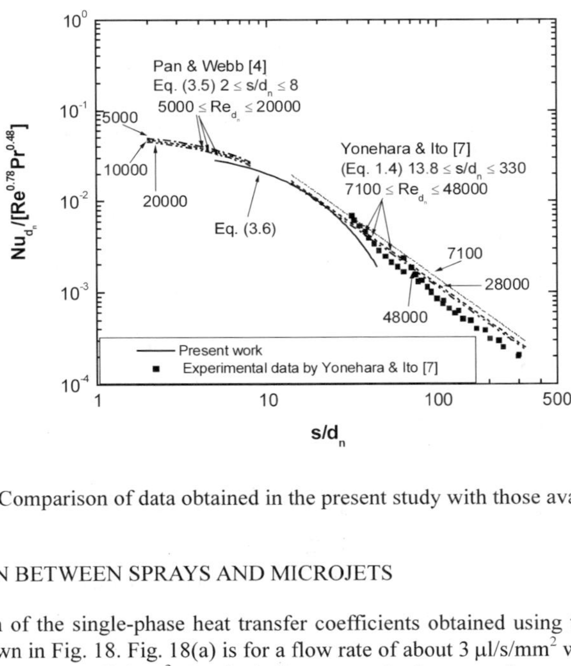

Figure 17 Comparison of data obtained in the present study with those available in the literature.

5.4 COMPARISON BETWEEN SPRAYS AND MICROJETS

Comparison of the single-phase heat transfer coefficients obtained using the HAGO nozzle and the microjets is shown in Fig. 18. Fig. 18(a) is for a flow rate of about 3 $\mu l/s/mm^2$ while Fig. 18(b) is for a higher flowrate of about 4.5 $\mu l/s/mm^2$. For both flow rates, the heat transfer coefficient for the HAGO nozzle is higher than that for the mirojets. However, if one compares the heat transfer coefficients for a given pumping power (see Fig. 19), the microjets yield higher heat transfer coefficient (by almost a factor of two). This is due the increased pressure drop that is inherent in the HAGO nozzle. Comparing the performance (process efficiency) of the HAGO nozzle and the mirojets (Fig. 20), for similar flow rates, it can be seen that in most cases the microjets outperform the HAGO nozzle.

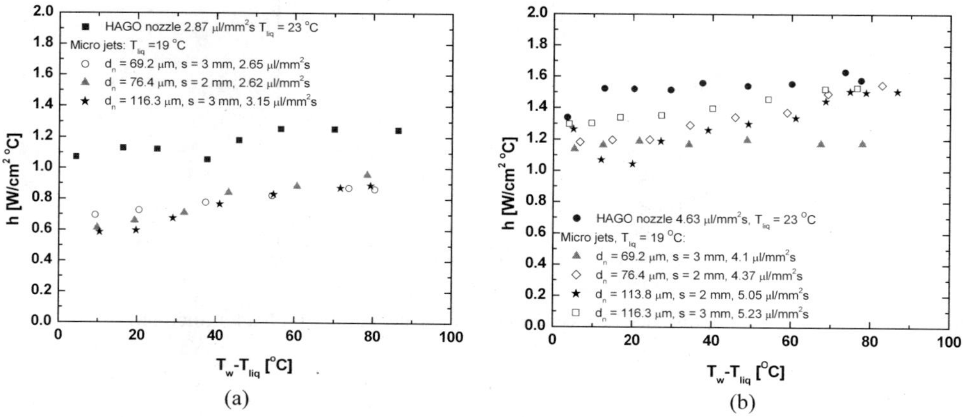

Figure 18 Comparison between sprays and microjets: heat transfer.

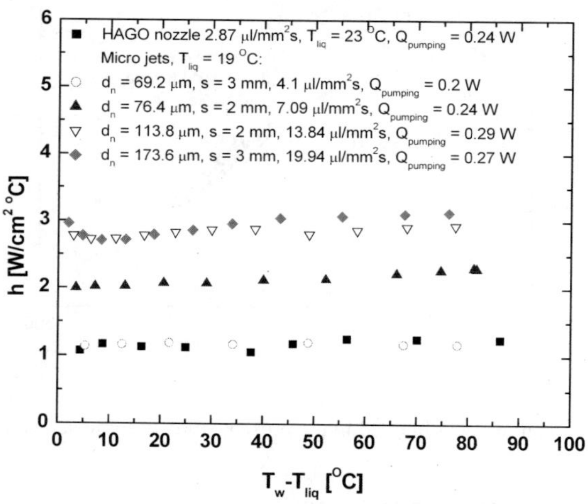

Figure 19 Comparison of the performance of sprays and microjets for the same pumping power.

5.5 COMPARISON BETWEEN NARROW CHANNELS, SPRAYS AND MICROJETS

Figure 20 shows a comparison of performance of the narrow channels with that of the sprays and microjets. The pumping power required for the narrow channels lies between those for the sprays and the microjets (see Fig. 20(a)). While $Q_{pumping}$ is constant for sprays and microjets (because $Q_{pumping}$ is independent of $Q_{removed}$), for the microchannels, in single-phase forced convection, it decreases with increasing $Q_{removed}$ (due to the decrease in viscosity with T_w). At ONB, $Q_{pumping}$ reaches a minimum and therafter $Q_{pumping}$ increases rapindly due to the rapid increase in ΔP_{tp}. Fig. 20(b) shows the comparison of $Q_{pumping}/Q_{removed}$ as a function of $(T_w - T_{liq})/(T_{sat} - T_{liq})$. It can be clearly seen that both the narrow channels and the mirojets perform far better than the sprays, with the narrow channels performing almost as well or better than the micorjets.

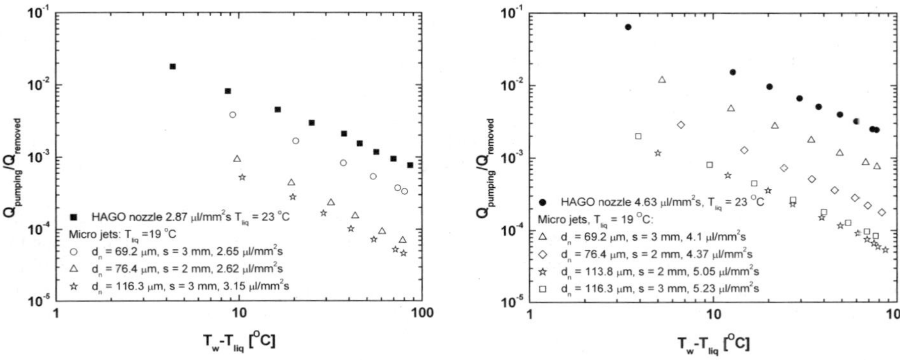

Figure 20 Comparison of the performance of sprays and microjets for similar flow rates.

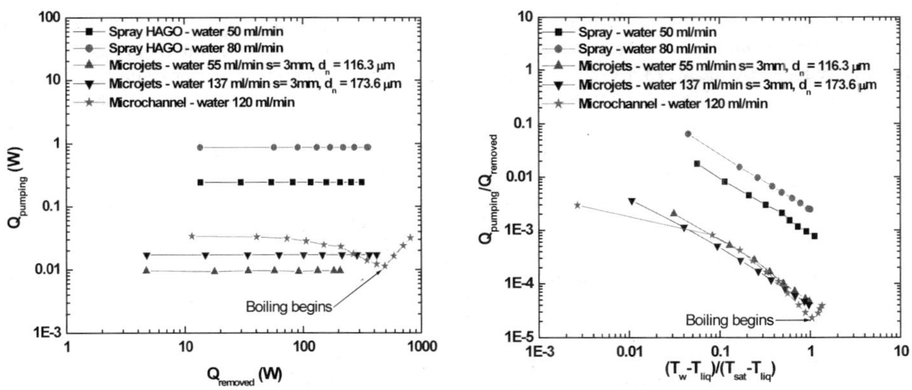

(a) (b)

Figure 21 Comparison of the performance of narrow channels, sprays and microjets (a) $Q_{pumping}$ vs. $Q_{removed}$ (b) process efficiency.

6. Implementation of a Cooling Module

Since the microjets performed better than the sprays, it was decided that the microjets would be implemented in a stand-alone cooling module for high heat flux removal. An impinging jets based cooling module requires three primary components: an orifice plate for forming jets; a containment vessel to hold the nozzle, the heat source and the cooling liquid, which also serves as a heat exchanger to the ambient; and a pump which recirculates the coolant. A fan could be used to improve the heat transfer to the ambient, and that would also allow the use of a smaller container. A brief description of a cooling module based on this idea is given below.

The cooling module, shown in Fig. 22, consists of an aluminum box with internal dimensions of 50 x 50 x 65 mm and wall thickness of 3.175 mm. At the bottom, the box is closed with a 3.175 mm thick stainless steel plate. The orifice plate was installed on a support located above the heat source and could be moved up or down. The heat source consists of a diode used in current controlled mode to avoid high voltages. The diode is mounted on a Direct Bond Copper (DBC) substrate layer, which is in turn glued on top of a G10 insulating base. The diode is 8.68 x 4.97 mm in size. The electrical connections are provided by means of two copper rods, 3.175 mm in diameter.

The orifice plate, 0.5 mm thick, had 24 holes (140 μm in diameter), distributed on a square array pattern with 2 mm spacing. All the tests were conducted keeping the jet velocity approximately constant at 4.5 m/s. Aluminum pin fins, 20 mm long and 3.175 mm in diameter, were installed on the outside of the container in a 45° staggered pattern with both pitches equal to 10.16 mm. The fin tips are inserted into holes drilled into four aluminum plates, which are welded at the corners and form an external shroud. A small DC fan is mounted at the bottom that pushes ambient air over the fins.

K-type thermocouples are used to measure the air temperature at the inlet and outlet of the fin array, the inlet water temperature, and the temperature inside the chamber. Two RTD's are used to measure the temperatures on the top of the diode and on the back of the DBC. Additionally the pressure in the chamber and the pressure drop across the fins was also monitored. The overall dimensions of the whole module, including the fan, were 100 x 100 x 130 mm.

The module was first charged with 40.8 ml of deionized water at room temperature. Subsequently the pump was started and the flow rate was set to the desired value. Thereafter, the internal pressure of the chamber was reduced by means of a vacuum pump. At steady-state chamber conditions, the vapor partial pressure was calculated, using steam tables, assuming the temperature measured in the chamber is equal to the vapor saturation temperature at the computed vapor partial pressure. Thereafter, power was supplied to the diode and the data were recorded.

Assuming that the relative humidity inside the chamber is equal to 100%, the air partial pressure can be calculated using Dalton's law, as the difference between the total pressure and the vapor partial pressure. All temperature, pressures and power supplied to the diode were recorded.

The data reduction procedure was similar to that for the sprays and microjets. The minimum and maximum uncertainties for the data are as follows: (i) Power: ±4.2% at 130 W and ±8.1% at 20 W, (ii) Heat flux: ±4.2% and ±8.1%, (iii) Temperature: ±0.1 °C, (iv) Thermal resistances: ±4.2 % and ±8.7%, (v) Heat transfer coefficient: ±4.2% and ±8.6%.

Figure 23 shows the heat flux at the diode surface as a function of the temperature difference between the top surface of the diode (T_w) and the sprayed liquid (T_{liq}). Although in this work single-phase heat transfer has been investigated, the module tests with jets have been conducted over both the single and two phase regimes in order to explore the full potential of the implementation of the concept in cooling of a device. No comparison was carried out between the performance of droplet sprays and microjet arrays in the boiling regime or when evaporation played a significant role.

In Fig. 23, the highest heat fluxes that could be achieved were limited by the high current flow through the diode, as in the 16 kPa data set, or by the fan speed reaching a maximum, as for the 104 kPa, or by critical heat flux (CHF) conditions, as in the 6.5 kPa case.

The advantageous effect of reducing the pressure in the chamber is clearly illustrated in Fig. 23. Lowering the system pressure lowers the boiling inception point. For the same T_w = 80 °C and T_{liq} = 47 °C, the heat flux increased from 130 W/cm^2, achieved with single-phase heat transfer, to 300 W/cm^2, obtained with boiling, when the pressure was reduced from 114 to 16 kPa.

Another important aspect that must be considered is the ratio of the power consumed (used for cooling the diode and operating the fan) to the power removed from the chip. The pumping power varied between 3.8 - 5.1 W, while the fan power varied from 0 to 2.2 W. From Fig. 24, it can be seen that spray cooling becomes more effective as the heat removed from the diode increases. This means that it is not convenient to employ this cooling technique if the power to be removed and the heat flux are low. For the most efficient case, the ratio of the total power spent to the power removed was around 4.4%. However this value is still a very conservative value since in several cases the power input to the chip was limited by restriction imposed by the current rather than the critical heat flux.

The values of the ratio of the power consumed to the power removed obtained from the cooling module tests are substantially higher than those shown for the orifice plates. There are several reasons for this. Because some of the jets did not impinge on the diode since the area covered by the jet array was larger than that of the diode, part of the water did not remove any heat, while still being pumped.

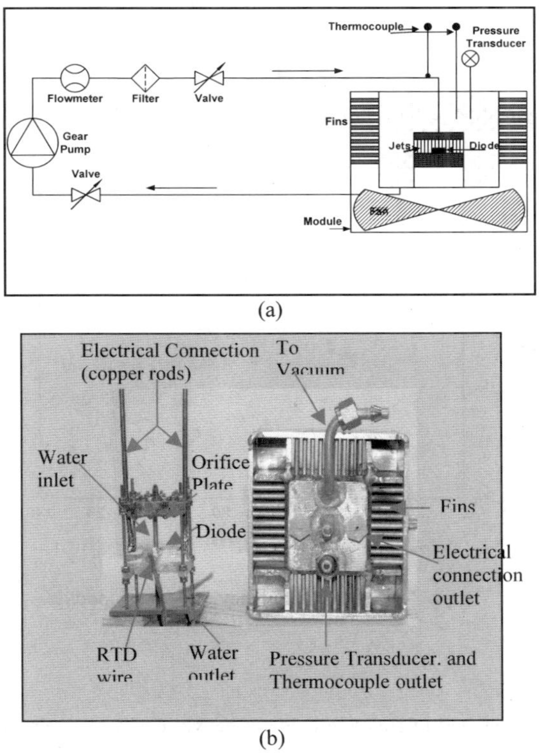

Figure 22 Implementation of microjets in a cooling module (a) schematic, (b) photograph showing details of the module.

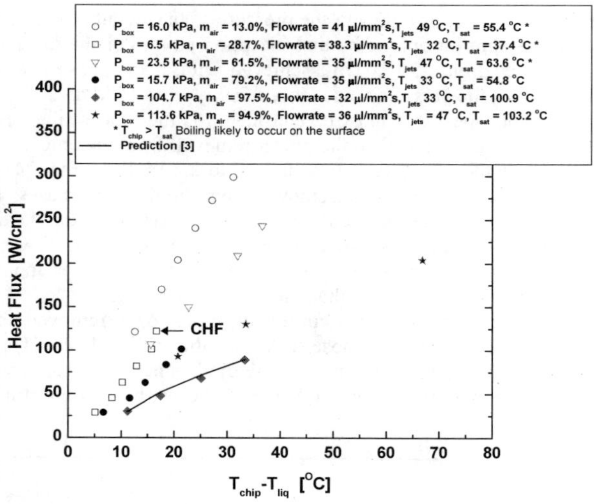

Figure 23 Cooling module results (4x6 array, d_n = 140 μm, s = 2 mm): heat flux vs. of T_{chip} - T_{liq}.

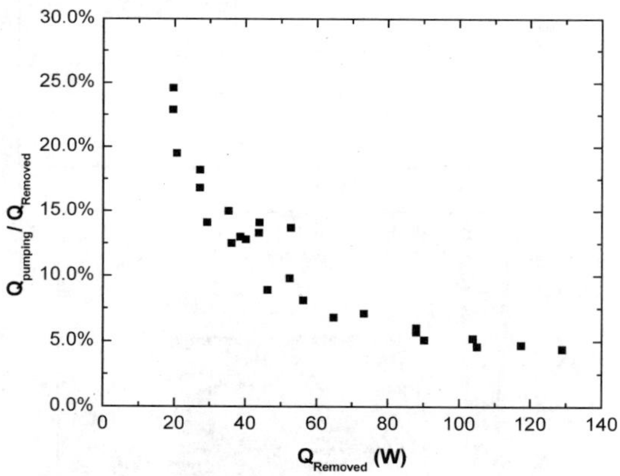

Figure 24 Process efficiency of cooling module.

7. Summary

- Narrow channels, sprays and microjets provide effective means of direct heat removal even in the absence of boiling. In general, the heat transfer coefficients increase with flow velocity.
- Sprays created with the HAGO nozzle are associated with higher pressure drops than narrow channels and microjets.

- For large surface areas to be cooled at low power densities, narrow channels are superior. At 120 ml/min (~ 2 ml/sec/mm2), the narrow channels can remove about 500 W (~ 8.6 W/cm2) with single-phase forced convection.
- Microjets and sprays are superior options when the power density is high. For equal pumping power, at $T_W - T_{liq}$ = 76 °C, the microjets can remove 240 W/cm2 while the sprays can only remove 93 W/cm2. For the same conditions, the narrow channels can remove only 8.6 W/cm2.
- For a given chip power there is an optimal array configuration in terms of jet spacing and diameter.
- The concept of using arrays of liquid microjets was successfully implemented. The module has proved capable of dissipating 129 W, with a heat flux of 300 W/cm^2 at a surface temperature of 80 °C, which is a considerable achievement at the present time.
- Reducing the system pressure had the effect of lowering the boiling inception temperature, thus allowing for higher heat removal rates at lower surface temperature.

Acknowledgements

This work received support from HRL laboratories and DARPA and was performed in the Boiling Heat Transfer Laboratory at the University of California, Los Angeles (UCLA).

Nomenclature

A, area, [m^2];
Bo, Boiling number, $Bo = q/Gh_{fg}$;
c_p, specific heat, [J/kgK];
d, jet or droplet diameter, [m];
D, diameter, [m];
ΔP, pressure drop, [Pa];
G, mass flux, [kg/m^{2}s]
h, heat transfer coefficient [W/m^{2}K], $h = q/(T_w - T_{liq})$;
h_{fg}, latent heat of vaporization, [J/kg];
k, thermal conductivity, [W/mK];
L, heater length, [m];
M, molecular weight, [kg/kmole];
N, number of jets;
Nu, Nusselt number, $Nu = hd_n/k$ or $Nu = hD_h/k$;
Pr, Prandtl number, $Pr = \mu c_p/k$;
q, heat flux, [W/m^2];
Q, power, [W];
Re, Reynolds number, $Re = \rho v D/\mu$;
s, pitch between the jets, [m];
Sc, local subcooling parameter;
T, temperature, [K];
v, velocity, [m/s];
$\dot{V}$, volumetric flowrate, [m^3/s];
x, quality;
z, nozzle to heater distance or axial distance, [m];

Greek symbols

α, thermal diffusivity, [m^2/s];
μ, dynamic viscosity, [kg/ms];
ν, kinematic viscosity, [m^2/s];
ρ, density, [kg/m^3];
σ, surface tension, [N/m];

Subscripts

air, air or noncondensable gas;
chip, surface of the chip;
d, droplet;
eq, equilibrium;
flash, flashing;
h, hydraulic;
liq, liquid;
n, at nozzle or orifice plate exit;
ONB, onset of nucleate boiling;
OSV, onset of significant void;
sat, saturation;
sp, single-phase;
tp, two-phase;
w, wall or surface;

References

1. Warrier, G.R., Dhir, V.K., and Momoda, L.A. (2002) Heat transfer and pressure drop in narrow rectangular channels, *Experimental Thermal and Fluid Science*, Vol. 26, pp. 53-64.
2. Jiang, S. (2002) Heat removal using microjet arrays and microdroplets in open and closed systems for electronic cooling, *Ph.D. Dissertation*, University of California, Los Angeles.
3. Fabbri, M. (2004) Cooling of electronic components using arrays of microjets, *Ph.D. Dissertation*, University of California, Los Angeles.
4. Kays, W.M. and Crawford, M.E. (1993) Convective Heat and Mass Transfer, McGraw-Hill, New York.
5. Mills, A.F. (1995) Heat and Mass Transfer, Richard D. Irwin, Boston.
6. Wallis, G.B. (1969) One-Dimensional Two-phase Flow, McGraw-Hill, New York.
7. Warrier, G.R. and Dhir, V.K., (2004) Visualization of flow boiling in narrow rectangular channels, *Journal of Heat Transfer*, Vol. 126, pp. 495.
8. Pan, Y. and Webb, B. W. (1995) Heat Transfer Characteristics of Arrays of Free-Surface Liquid Jets, ASME *Journal of Heat Transfer*, Vol. 117, pp. 878-883.
9. Yonehara, N. and Ito, I. (1982) Cooling Characteristics of Impinging Multiple Water Jets on a Horizontal Plane, *Technol. Rep. Kansai University,* Vol. 24, pp. 267-281.

BOILING HEAT TRANSFER IN MINICHANNELS

V.V. KUZNETSOV, O.V. VITOVSKY and A.S. SHAMIRZAEV
Institute of Thermophysics of Siberia Branch of Russian Academy of Sciences
Novosibirsk, Russia

1. Introduction

The objective of this lecture is not a comprehensive review of publication dealing with boiling heat transfer in minichannels in general, and first of all to illustrate the mechanism of flow and heat transfer in a channel with a small cross-section areas. Though there are large number of results, which describe the boiling heat exchange and two-phase flow pattern in pipes of large diameter, they can't be applied directly to a flow through minichannels. When channel gap is about capillary constant $\delta_c=[2\sigma/(\rho_l-\rho_g)g]^{1/2}$ the capillary forces predominate, and they determine flow pattern and boiling heat transfer. The interest for study of boiling heat transfer in minichannels increases at last decade due to obtained possibility to makes the channels of millimeter's and micron's size for application in wide range of industrial devices for enhancement of heat and mass transfer. They could be used in compact evaporator/condenser of cryogenic devices in form of vacuum brazed fin passage. Another area of application of microchannel architecture is micro-heat exchangers and micro-cooling assemblies referred to as micro-thermal-mechanical systems (MTMS) used as micro-cooling elements. They also could be used as micro-chemical reactors, which operate with a small residence time. The characteristic of mini/microchannel assemblies is high channel density and large surface area, which considerably increases the rate of heat transfer. When boiling/condensing occurs inside of very small and non-circular passages the capillary forces become important in determining the aspects of flow phenomena such as flow pattern, shape of interface and macroscale heat transfer. In the literature the transition between large scale and mini scale channels is not exactly defined. At paper [1], the next channel size classification on its hydraulic diameter is proposed: microchannels (size of 1 to 100μm), mesochannels (size of 100μm to 1mm), macrochannels (size of 1 to 6 mm) and conventional channels (size more than 6 mm). In contrarily in [2] proposed another channel size classification on its hydraulic diameter: microchannels (size of 50 to 600μm), minichannels (size of 600μm to 3 mm) and conventional channels (size more than 3 mm). In current lecture, we will follow classification [2] and the channels with size of 600μm to 3 mm will be under consideration.

There are many papers where the boiling heat transfer was analyzed for such channels. In contrast to boiling in conventional tubes the flow boiling heat transfer coefficients in microchannels are dependent on heat flux and pressure while only slightly dependent on flow velocity and vapor quality [3], [4], [5], [6], [7], [8]. Many experimental study have concluded that for wall superheat the nucleate boiling is dominant mechanism for evaporation in minichannels with a small convective evaporation contribution [3], [4], [5], [6], [7], [9]. Another studies concluded that for their tests with multichannel arrangement the nucleation is not important mechanism [10], [11], [12], [13]. The weak dependence of heat transfer coefficient on heat flux density was observed for boiling of low thermal conductivity liquid in vertical slot with boiling induced liquid circulation [14]. It was shown also that the heat transfer to be dependent on the existing flow pattern [8], [15]. At paper [16] suggested that the transient evaporation of thin liquid film surrounding elongated bubbles is the dominant heat transfer mechanism, not nucleate boiling. It was concluded that macroscale models are not realistic for predicting flow boiling heat transfer in mini/microchannels since they are based on the nucleate boiling and convective evaporation mechanism and not accounts the effect of surface tension [2], [6], [8], [17], [18].

S. Kakaç et al. (eds.), Microscale Heat Transfer, 255–272.

One can see that there are many disagreements in existing data, which cannot be described easily. To present our view on this problem we will focus on the description of two-phase flow peculiarities and boiling patterns in a minichannels according to our results from [8], [19], [20], [21], [22], [23], [24]. We will describe consequently what is the flow regime exists inside of minichannels, how the capillary forces which work at a small scale have effect on a macroscale heat transfer, what is the rate of heat transfer in boiling mode for small flow velocity, what is the rate of heat transfer in flow boiling mode for high flow velocity, what is more general definition could be address the threshold where macroscale model is no longer fully applicable with respect to heat transfer processes.

2. Two-Phase Flow Regimes Study in Narrow Gap

When nucleate boiling occurs in narrow channel the flow regime could be different from it's in conventional tube. In conventional tube the bubble departure size is typically less that the tube diameter, so bubble flow pattern and nucleate boiling are preferable in wide range of vapor content. In case of narrow channel the bubble departure size could be comparable with the channel size, it causes strong interaction between bubbles and change both the flow pattern and heat transfer rate.

2.1. EXPERIMENTAL APPARATUS

Figure 1 presents the test section, which was used to study the flow patterns in narrow annular gap in case of air-water co-current upflow [19], [24]. The experiments were carried out in a closed air-water loop. It consisted of a tank-separator with controlled temperature, centrifugal pump, working section and flow control system. The working section was made as two coaxial organic glass cylinders with polished internal and outside surfaces. The column length from the inlet point of the two-phase flow to the column top was 0.6 m, the annular gap between cylinders was h = 0.68 mm, the internal cylinder diameter was D = 0.15 m. The experiments were performed at gas superficial velocities $V_{G,S}$ from 0.035 up to 45 m/s and for liquid superficial velocities from $V_{L,S}$ 0.4 up to 4.0 m/s. The superficial gas velocity increases along the channel length was calculated taking into account the isothermal gas expansion. The flow rates for water and air were measured by turbine flow meter and orifice plate. The description of the measurement procedure is given in [19], [24].

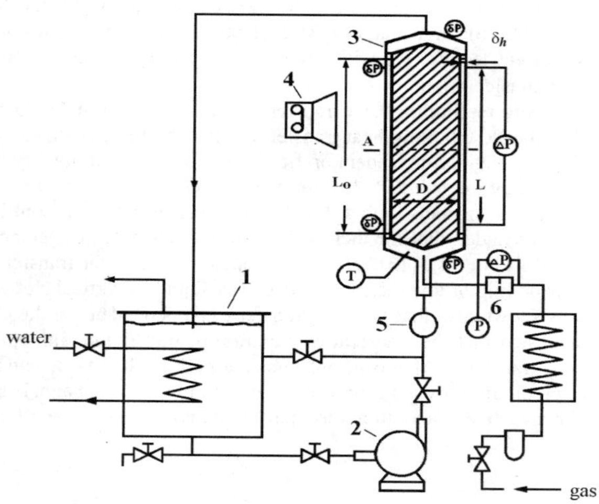

Figure 1. Schematic diagram of experimental system: 1- tank-separator, 2- pump, 3- working section.

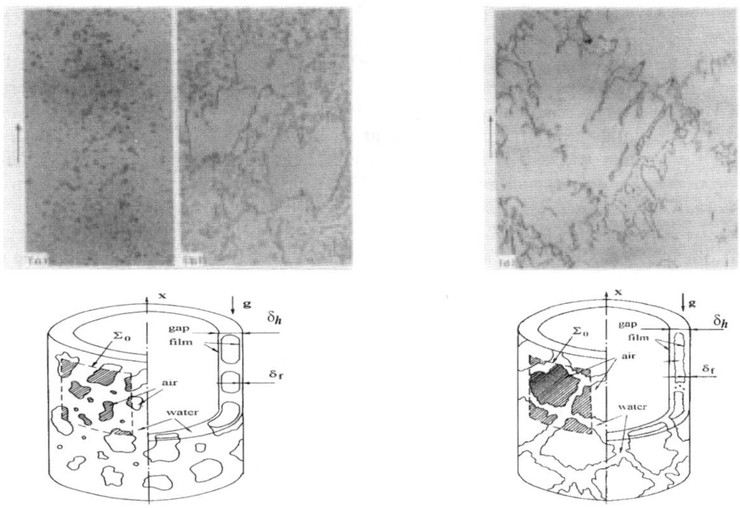

Figure 2. Typical flow patterns in the narrow gap.

2.2. UP FLOW PATTERN OBSERVATION

Figure 2 presents three typical flow patterns were observed at narrow annulus, which are the flow with small bubbles whose size is less than a channel width (see Fig. 2a), the flow with large Taylor bubbles (see Fig. 2b) and the flow with the cell structure of liquid plugs, (see Fig. 2c). The flow pattern map is presented on Fig. 3. The first type of the flow is observed at the superficial liquid velocities greater than 2 m/s when the flow becomes turbulent (point 1 and line A in Fig. 3). At such velocities the flow is turbulent and small bubbles

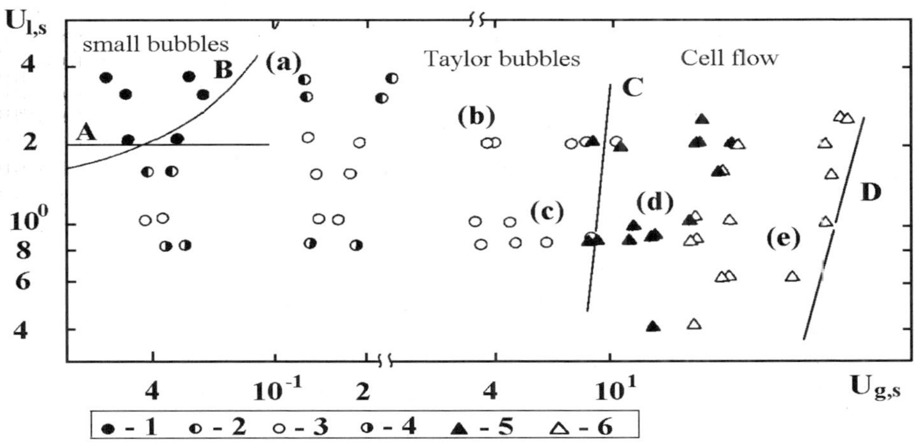

Figure 3. Air-water flow pattern map in a narrow gap, 1 - small bubbles, 2-small and Taylor bubbles, 3- Taylor bubbles, 4-branched Taylor bubbles, 5-cell flow without ripple waves, 6- cell flow with ripple waves.

are stable, large bubbles here are broken by turbulent forces [25]. The flow with Taylor bubbles exists in a wide range of gas flow rates. When the gas velocity increases, the bubble size grows due to coalescence of Taylor bubbles. At gas velocities large than 10 m/s the bubble size is so large that liquid is localized in plugs between the bubbles, and cell flow pattern arises, see points 5 and 6 on Fig. 3. The line C here separates flow pattern with Taylor bubbles and cell's structure. The Taylor bubble flow transforms to cells flow but not to an annular flow. The thing is that the cell structure with liquid accumulated mainly in plugs has smaller friction losses compared to the annular one and is more favorable energetically. Capillary forces in narrow gap prevent the plug destruction only for small gas velocities. Usually cells are unstable and both destruction of old and formation of new plugs observes.

In narrow gap the area of conventional nucleate boiling could be observed just in small bubble flow regime. The balance between capillary forces and turbulent shear stress gives the maximum stable diameter of the bubble in turbulent flow as [26]

$$d_{\max} = 0.725(\sigma/\rho_L)^{3/5}\varepsilon^{-2/5} \tag{2.1}$$

here ε is the energy dissipation on unit mass. If one set the limit diameter of the bubble as a gap size then the area of small bubbles flow could be determine as $U_{L,S} \rangle U^{cr}_{L,S}$ were critical liquid superficial velocity is [19]:

$$U^{cr}_{L,S} = \left(\frac{0.725}{\delta_h}\left(\frac{\sigma}{\rho_L}\right)^{3/5} \left(\frac{0.073}{\delta_h}\left(\frac{\nu_L}{\delta_h}\right)^{0.25}\right)^{-2/5}\right)^{11/10} \tag{2.2}$$

It was assumed here that at low gas content the rate of energy dissipation in two-phase flow closes to it value for single-phase flow. If the liquid flow rate will be less then critical value then the walls of the channel will effect on heat transfer rate.

3. Refrigerant R318C Flow Boiling in Narrow Gap

3.1. EXPERIMENTAL APPARATUS

Figure 4 shows the experimental apparatus for study of the local heat transfer for boiling in an annular channel. The experiments were performed in a test loop with refrigerant R318C. At the outlet of the test section, an additional evaporator is placed to avoid the flooding in regimes with low mass vapor content. Liquid refrigerant is cooled after pumping and the flow rate is measured by a turbine flow meter. The test

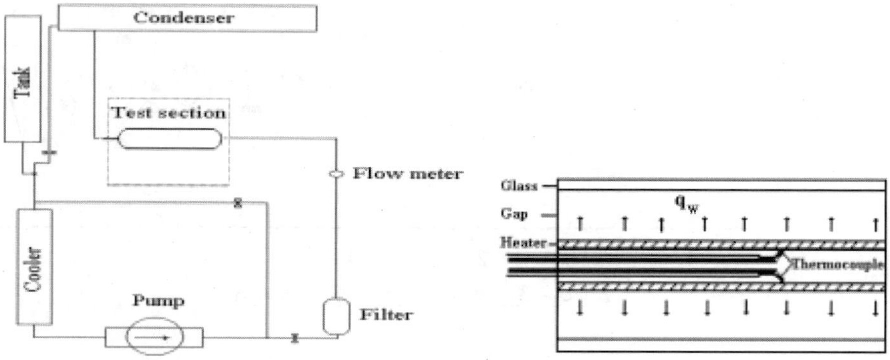

Figure 4. Scheme of experimental system and the test section.

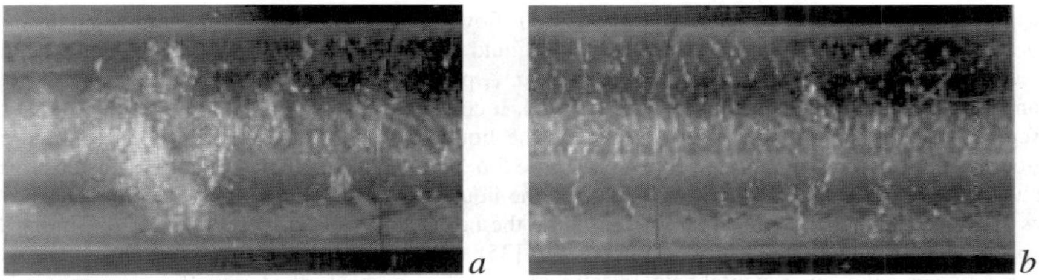

Figure 5. Flow pattern at (*a*) $x = 0.34$ and (*b*) $x = 0.51$.

section is an annular channel with a length of 0.4 m operating in the regime of horizontal flow. It was manufactured from two tubes - see Fig. 4. The outer tube is made transparent to control flow pattern and the inner one is made of stainless steel – the wall thickness is 1 mm. The inner tube diameter is 7 mm and the gap size equals to 0.95 mm. To measure the local heat transfer coefficient two insulated thermocouples were installed inside of the central tube, which was heated by an AC electric current. The thermocouples were pressed to the tube wall by a spring and the position of the thermocouple can be changed. The temperature of the heated wall of the annulus was corrected with account of the volume heat production. The refrigerant temperature is measured before and after the test section by the thermocouples. The saturation temperature was calculated through the pressure inside of the channel at the point where the wall temperature was measured. The local heat transfer coefficient is determined as a heat flux divided by the temperature difference between the inner wall of the annular and the saturation temperature. The experiments were performed for values of mass flux from 200 kg/m^2-s to 900 kg/m^2-s and heat flux density from 2 kW/m^2 to 100 kW/m^2. The description of the measurement procedure is given in [8], [24].

3.2. FLOW BOILING PATTERN AND HEAT TRANSFER RATE

Typical flow pattern for flow boiling are shown in Fig. 5. For subcooled boiling and high liquid flow rate, the observed bubble detachment size is smaller than the gap size. The treatment of the data showed that correlation [26] obtained for a large tube flow boiling can be applied to predict subcooled flow boiling heat transfer in a confined space. For saturated flow, the vapor bubbles have a tendency to merge and produce the

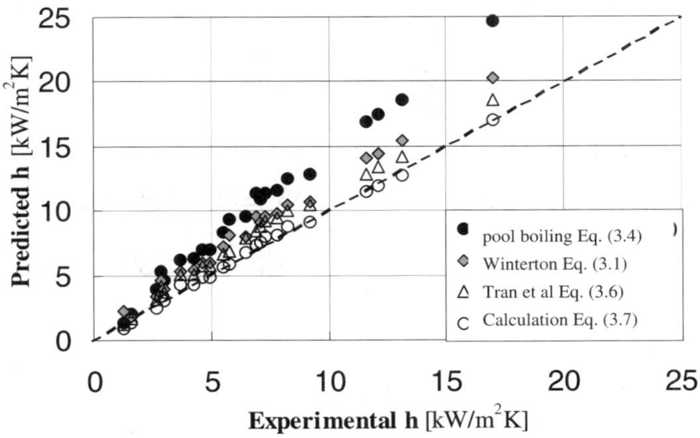

Figure 6. Predicted h vs. experimental h from refrigerant R318C saturated flow boiling with $x < 0.3$.

flow with confined bubbles (Taylor's bubbles), cell flow (annular-slug flow) and annular flow with consequent evolution from one mode to another. A liquid plug with a boiling liquid moves through the annulus with a high velocity, as it shown in Figure 5, *a*. A plug arises initially as a result of merging of the disturbance waves on the opposite walls of the channel. It collects any disturbance wave on the film surface like a scraper and the entrainment does not occur. The liquid moves either in a film or in a liquid plug. Boiling occurs both in the plug and the film, see Figure 5 *a*, and enhanced nucleate boiling was observed in liquid plug. When the vapor mass fraction increases, the liquid flow rate for plug mode decreases. The flow becomes completely annular with ripple waves where the boiling may occur, see Figure 5, *b*. Similar flow boiling patterns in an annular channel were observed at [35] also.

Figure 6 presents the experimental data on saturated flow boiling heat transfer coefficient compared with existed correlations [6], [27], [30]. For saturated flow boiling correlation [27]

$$h = \sqrt{\left(Fh_l\right)^2 + \left(Sh_{pool}\right)^2} \tag{3.1}$$

where F is force convection heat transfer enhancement factor and S is boiling suppression factor. Forced convection heat transfer for turbulent liquid flow at $\mathrm{Re}_l > 2300$ was calculated from the correlation [28]:

$$Nu_l = 0.86\left(\frac{D_i}{D_o}\right)^{-0.16} \frac{f/8\left(\mathrm{Re}_l - 1000\right)\mathrm{Pr}_l}{\left(1+12.7\sqrt{f/8}\left(\mathrm{Pr}_l^{2/3}-1\right)\right)}\left(1+\left(\frac{D_h}{L}\right)^{2/3}\right) \tag{3.2}$$

with entire mass flow rate flowing as liquid in the same channel. For $\mathrm{Re}_l < 2300$ the laminar flow correlation from [29] was used in the form:

$$Nu_l = 3.66 + 1.2\left(\frac{D_i}{D_o}\right)^{-08} + \left(1+0.14\left(\frac{D_i}{D_o}\right)^{2/3}\right)\frac{0.19\left(\mathrm{Re}_l\,\mathrm{Pr}_l\,D_h/L\right)^{0.8}}{1+0.117\left(\mathrm{Re}_l\,\mathrm{Pr}_l\,D_h/L\right)^{0.467}} \tag{3.3}$$

The pool boiling heat transfer is calculated from the correlation from [30] as

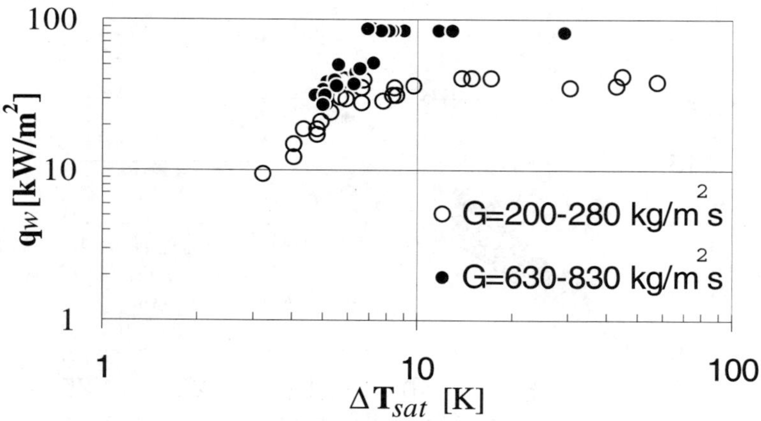

Figure 7. Heat flux density as function of the wall superheat.

$$h_{pool} = 7.42 q_W^{0.67} R_p^{-0.21 \log p/p_r} p_r^{0.12} / (-\log p_r)^{0.55} \tag{3.4}$$

Here p_r is the pressure scaled by a critical pressure, and the wall roughness Rp was 5 μm. The forced convection heat transfer enhancement factor and boiling suppression factor were used as in [27]:

$$\begin{aligned} F &= \left(1 + x\mathrm{Pr}_l\left(\rho_l/\rho_g - 1\right)\right)^{0.35} \\ S &= \left(1 + 0.055 F^{0.1}\,\mathrm{Re}_l^{0.16}\right)^{-1} \end{aligned} \tag{3.5}$$

For vapor mass quality less then 0.3 correlations (3.1), (3.4) and (3.6) over estimates the test data because of additional boiling suppression in the confined space. A better agreement with the test data has the correlation [6]

$$Nu = 770(\mathrm{BoRe}_l \mathrm{N}_{conf})^{0.62}(\rho_g/\rho_l)^{0.297} \tag{3.6}$$

which takes into account the flow boiling suppression by the confinement number introduced in [30] as $N_{conf}=1/Eo^{1/2}$. Nevertheless, this correlation over estimates test data also. We suppose that the exponent at heat flux density must depend from liquid properties as for pool boiling correlation [29] and [31]. After data processing, we find that the best correlation is

$$Nu = 680(\mathrm{BoRe}_l \mathrm{N}_{conf})^{0.67}(\rho_g/\rho_l)^{0.297} \tag{3.7}$$

where the power for complex $\mathrm{BoRe}_l\mathrm{N}_{conf}$ corresponds to pool boiling heat transfer correlation (3.6).

Figure 7 presents experimental data on the wall superheat as a function of the heat flux density for two areas of mass flow rate. There is a strong degradation of the heat transfer near limiting value of the heat flux (the latter is a function of the flow mass flow rate). The scattering in data on wall superheat is caused by variation of the real mass flow rate in present areas. The wall superheat for this regime is stable in time. The data of visualization showed that the heat transfer degradation occurs for annular flow when liquid plugs are not observed. For annular mode, the liquid film keeps evaporation on the heated wall and this decreases its thickness along the test section. For a small film thickness, the boiling will be suppressed up to complete disappearance. Test data on the heat transfer coefficient scaled by (3.1) as a function of the film thickness are shown in Fig. 8 for mass quality more and less than 0.3. The film thickness at the point of the wall temperature measurement was calculated from heat balance for liquid flow on the heated wall. A correlation for the wavy film thickness as a function of the superficial vapor velocity from [33] was used.

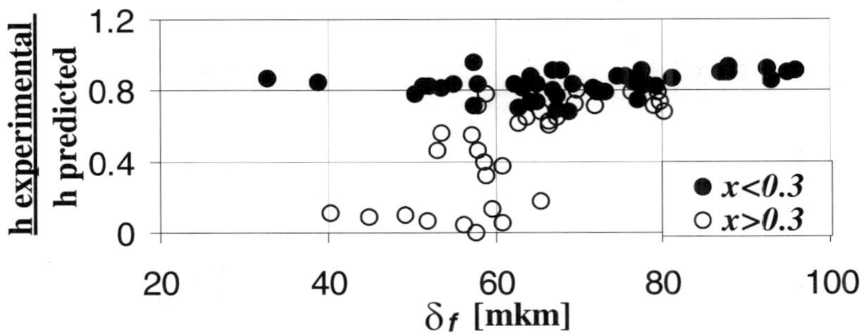

Figure 8. Heat transfer coefficient normalized by calculation through Eq.(8) as a function of film thickness.

It was supposed, that for a high vapor velocity and a thin liquid film the influence of gravity is small and the correlation for up flow was used. Total boiling suppression occurs when mass quality more than 0.3 for a film thickness less than 60 μm. That value is close to the bubble departure diameter observed for flow boiling in a film. When the film thickness is smaller than the critical one, the forced convection occurs with a small heat transfer coefficient. The crisis of the heat transfer was observed for a complete liquid evaporation on a heated wall. While the mass quality less than 0.3, we have the cell or slug flow mode, so boiling is not suppressed.

4. Two-Phase Flow Pattern in Rectangular Channel

When boiling occurs inside a very small and non-circular channel, the capillary forces become important in determining the aspects of flow phenomena such as interface shape and flow pattern. Fundamental features of interface shape and flow pattern under such regimes determine flow boiling heat transfer. In this chapter we will discuss the action of capillary forces in a passage of very small rectangular channels where they can move liquid around the perimeter of each channel because of transversal pressure gradient. At the beginning the shape of individual bubble and velocity of it's rising will be discussed and then the flow patterns of two-phase flow will be presented.

4.1. EXPERIMENTAL APPARATUS

Measurements of free rising air bubble velocity in stagnant liquid inside a vertical and tilted rectangular channels with the open top and closed bottom were carried out at setup described in [23]. The set up was made from steel with glass windows. The channel length was 0.5 m and the length of a transparent window was 0.22 m begins at 200 mm from the point of air injection at the bottom. Appointed air volume was injecting into the channel. Bubble rise velocity was measured using a digital camera. The measurements of velocity of air plugs in different glass rectangular minichannels were done.

Observations of the flow patterns inside of fin passage making the set of rectangular channels were done at apparatuses shown at Fig. 9. Subcooled liquid was pumped through electro-heating coil to provide a certain vapor quality of the flow. Then the flow was passed through adiabatic test section and later through the evaporator, for exception pulsation of flow, into the condenser. The test section can operate both in up

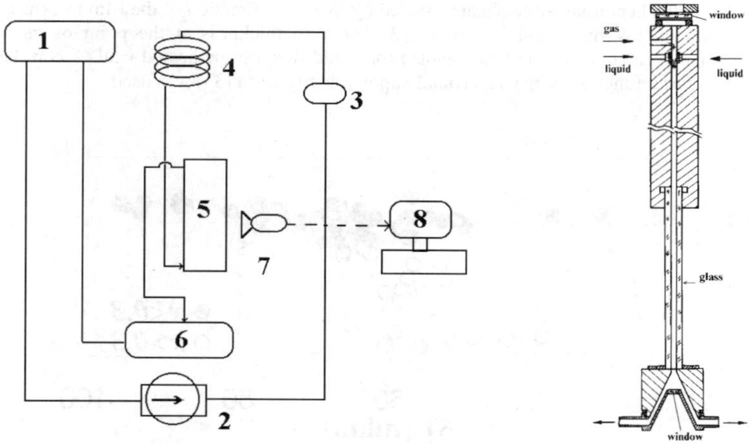

Figure 9. Experimental apparatus for flow pattern observation and test section with bottom window: 1-condenser, 2-pump, 3-flowmeter, 4-heater, 5- test sample, 6-evaporator, 7- camera, 8-PC.

upward flow and downward flow modes. For flow visualization we used a transparent test section with aluminum plain fins and transparent rectangular channel made from glass. Fin pad's length and width was 0.28 m and 0.08 m respectively. There were 25 sub-channels with a height of 7 mm, pitch of 3.2 mm (8FPI), and fin thickness of 0.5 mm. The fin pad was fixed between glass plates. To achieve the uniform flow distribution before fin pad a slot distributor was used. In case of transparent rectangular channel the bottom of the channel had a window for interface observation [20]. The schemes of test section are shown in Figure 9. The flow patterns were recorded by digital camera, placed in front of the test section or through the bottom window. The temperatures of flow before coil and in test section were measured by thermocouples. Inlet liquid flow rate was measured by turbine flow meter. The working fluid used for this study was refrigerant R21, which has a saturation temperature T_{sat}=300 K at pressure P=0.194 MPa. Flow patterns were studied approximately at this temperature in a range of mass fluxes from 20 to 90 kg/m^{2}s. Transparent rectangular channel operated with refrigerant R113 and air at room temperature. The experimental procedure in detail was described at [20], [21].

4.2. BUBBLE RISING IN VERTICAL CHANNEL

The observations showed that in rectangular minichannel, in contrarily with round tube, the bubble floats even at Etveos number less then four. It was indicated in [34] that wide side renders a define the value on the plug rise velocity and Etveos number here and hereinafter calculated on maximum width of channel b, Eo= $\rho g\ Sin(\Theta) b^2/\sigma$. Figure 10 show the influence of angle inclination of channel and channel width on non-dimensional velocity of bubble rise. The curvature of plug head is also dependent on Eo number (see Fig. 10). In rectangular channel there are liquid meniscus in the corners, and the liquid can bypasses thin film area to decrease the viscous friction in down flow area. It leads to possibility for bubble rising for small Eo number. When bubble length become more then widths of the channel, the plug forms and the rising velocity does not depend on the length of the plug.

4.3. TWO-PHASE FLOW PATTERN IN NON-CIRCULAR CHANNEL

The first results described here have been obtained using liquid refrigerant R113 with or without co-current air down flow. The liquid Reynolds number is defined as $Re_l = 4AU_{S,l}/P\nu_l$, where $U_{S,l}$ is the superficial

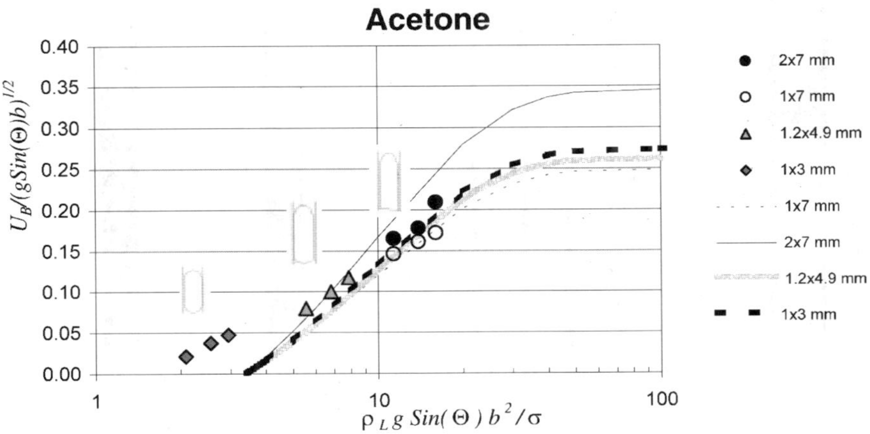

Figure 10. Experimental data on bubble rising velocity and bubble shape in dependence on Eo number. Lines are the Griffit's correlation [34].

liquid velocity, A is the channel cross sectional area, P is the channel perimeter and ν_L is the kinematical liquid viscosity. In a like manner, the gas Reynolds number is defined as $Re_g = 4AU_{S,g}/P\nu_g$. Figure 11b shows that film flow on the long walls of the rectangular channel is wavy for high liquid Reynolds numbers with a specific wave pattern. This is brought out best by photographing from the side of the apparatus under reflective light. Some typical two-phase patterns in rectangular channel observed with and without flow of air are shown in Fig 11a. All the pictures are obtained from the side of the apparatus looking straight through the wide side of the glass rectangular channel. Photographs are produced in passing of dissipative light. This technique brings out some features of the interface. Dark regions near the edges show accumulation of liquid near the corners of the rectangular channel. They appear to be dark because light is partially reflected from circular-arc meniscus near the corners where flow accumulation occurs by capillary action. On the other hand the liquid presented on the long sides passes light through easily because the film curvature is very small, and therefore does not reveal the wave structure which is shown earlier in Fig. 11b.

Liquid down flow in stagnant gas shows four types of flow regimes. At very low superficial velocities the capillary forces move a liquid quickly to the channel corners and the film on a long sides is thinned substantially by this drainage. Figure 12 shows development of liquid interface along the channel length. Channel's wall and circular-air meniscus bounded the flow area at the corners. When liquid superficial velocities increase the menisci on the short side of the channel unite together and the flow in the corners becomes unstable. Roll waves on the short side of the channel are shown in Fig. 11b at Re_l of 915. As the liquid flow rates increase the roll wave amplitude grows further, until when the superficial velocities go beyond 0.3 m/s (Re_l ~ 2500), at which point the waves on the opposite long sides of the channel wall unite together and straight arches occur. The nature of this regime is highly transient and a snapshot of this phenomenon is shown at Re_l of 3077. When the velocities are increased further the straight arches are poured into liquid plugs separated by gas slugs.

In presence of co-current gas flow the patterns do not change in principle at low gas and liquid velocities. At liquid velocities higher then 0.3 m/s the wave confluence leads to slug flow. When superficial

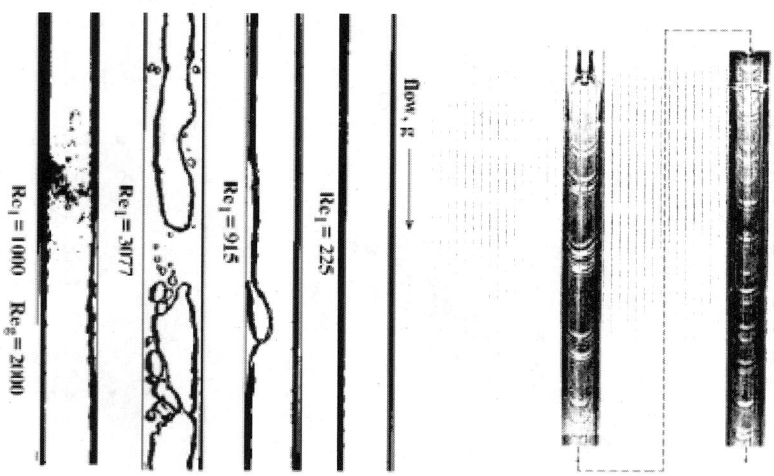

a) b)

Figure 11. Typical regimes for refrigerant R113 down flow

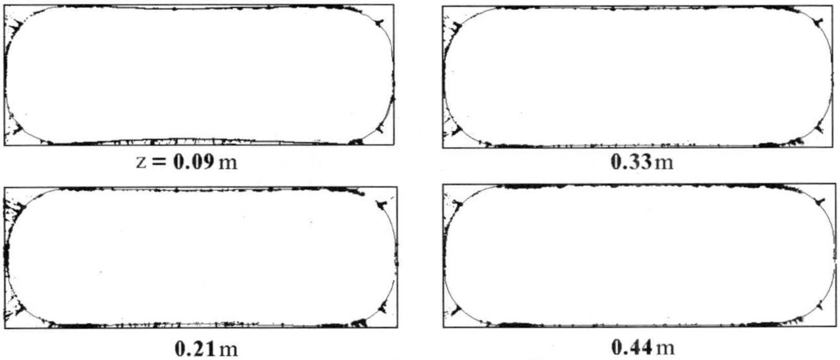

Figure 12. Liquid interface development for channel in 2x7 mm size channel at Re_L =14.

gas velocities are raised above 2.8 m/s the cell flow forms and the liquid bridge is essentially unstable, it contains many droplets also.

Figure 13 presents the pattern map for an upward vapor-liquid flow through assemblage with plain fins. Corresponding areas of modes are as follows: annular flow, cell flow, froth flow, plug and bubble flow. For an upward flow with $U_{S,g}$ > 3.3 m/s and Re_l <50, we have the annular flow mode with a smooth liquid film. For a higher liquid flow rate when 50 <Re_l <300, this film becomes wavy. When liquid superficial velocity exceeds 0.02 m/s (Re_l > 300) the large amplitude waves on interface arise and some of the liquid phase is entrained as droplets in the vapor core. For closed spaced walls the waves and droplets form bridges between walls

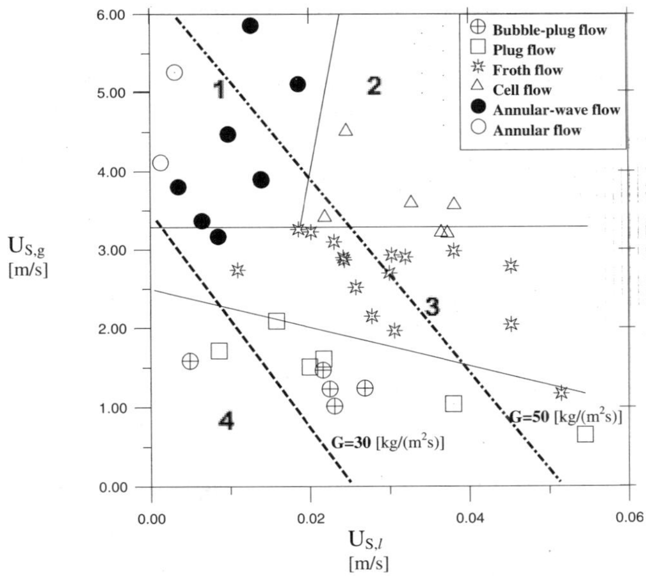

Figure 13. The pattern map for an upward vapor-liquid flow of R21 refrigerant through the assemblage with plain fins (8FPI). Here 1, 2, 3, 4 are the areas of annular flow, cell flow, froth flow, plug and bubble flows respectively. Solid lines indicate transition between flow modes and dashed lines indicate constant mass flux condition.

or concentrated droplet's waves, this regime was selected at [19] as cell flow pattern. With lower vapor flow rates, when $U_{S,g} < 3.3$ m/s, the flow regimes change from the annular and cell modes to the froth and plug modes. There is always the same steady state flow regime for all channels, except for the case of $U_{S,g} < 1$m/s. With $U_{S,g} < 1$m/s, the gas can not lift the whole liquid upwards and there is counter-current liquid flow in the channels placed near the edges of the test sample. The dash lines on Fig. 13 show the connection between vapor and liquid superficial velocity for two mass fluxes during the liquid evaporation.

5. Flow Boiling Heat Transfer in Rectangular Channel

5.1 EXPERIMENTAL APPARATUS

For investigating of heat transfer at the boiling in vertical minichannels assemblage with rectangular channels, we used the test section with 280mm length. It was mounting into refrigerant setup shown in Figure 14. Here the test section scheme is shown also. At the outlet of the test section, an additional evaporator was placed. The wall temperature was measured by 12 thermocouples, by 6 on each side. The saturation temperature was determined from pressure and temperature measurements in the center of test section. The heat was supplied to the test section from the thermoelectric module operating on Peltier's effect. A thick cooper plate was placed between the lateral sheet of the assemblage and Peltier's module to provide a uniform temperature of the lateral sheets. To keep a constant temperature difference between the cold and hot sides of Peltier's module, the internal water heat exchangers were used. We used a glue with a high thermal conductivity coefficient about ~1 W/m K to provide a good heat contact for all significant components (Peltier's elements, external water heat exchangers, tested sample). The heat transfer measurements was done for plain fin assemblage, dip brazed to aluminum plates. It had fin pad with sizes of 279x70 mm and the fin density of 20.3 FPI. Fin height and thickness were 6.7 and 0.2 mm respectively. The hydraulic diameter of channels was 1.81 mm. The working fluid used for this study was refrigerant R21. The thermocouples for measuring of inlet and outlet temperature of refrigerant were placed into the top and bottom collectors. There were 9 thermocouples for measuring of wall temperature (6 on one side and 3 on another). The hole for pressure measuring was made in the middle of a narrow side of the heat exchanger. The thermocouple for refrigerant temperature measurement was put on the opposite side.

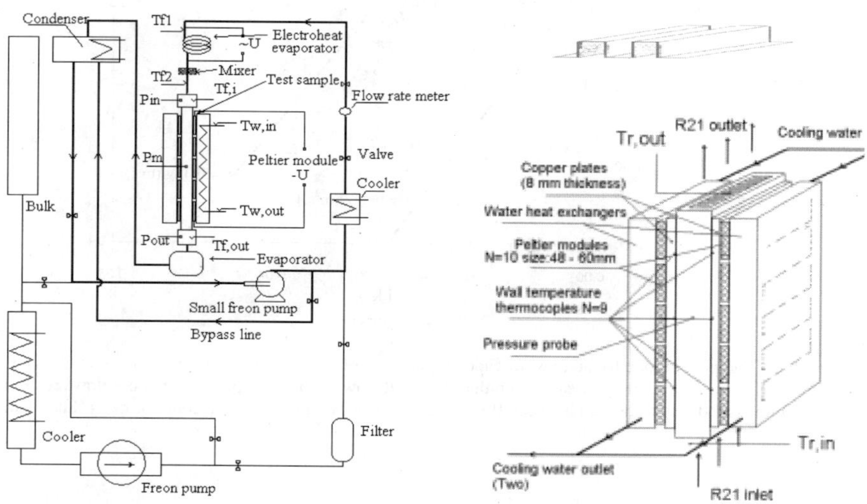

Figure 14. Experimental apparatus for boiling heat transfer study and schematic diagram of test section.

Experiments were performed at boiling induced circulation of saturated liquid and forced boiling condition. The description of measurement procedure was done at [21] and [23].

To determine the heat transfer coefficient and mass vapor quality in boiling test the following variables were measured:

- heat exchanger's wall temperature for nine points on the parting sheet;
- refrigerant temperature at the inlet/outlet of electric vapor generator;
- current and voltage in electric vapor generator;
- inlet/outlet temperature of refrigerant;
- saturation pressure in the central cross-section;
- inlet/outlet temperature of cooling water for external heat exchangers;
- mass flow rates for water and refrigerant,
- current and voltage in Peltier's elements.

The experiments were performed for mass flux of 30 kg/m^2 s and 50 kg/m^2-s and wall flow temperature difference from 1 K up to 5 K.

In boiling and condensing tests the heat rate consumed by refrigerant was calculated from measurements of water temperature drop, water flow rate, voltage and current for Peltier's elements. Wall temperature T_W was measured on an external side of parting sheet, so taking into account the parting sheet temperature drop, the heat rate consumed by refrigerant can be presented through a concept of surface efficiency $\eta_0 = 1-(A_1/A_0)(1-\eta_f)$ as:

$$Q_{refrigerant} = \frac{A_0(T_{Sat} - T_w)\eta_0 h}{1+\delta_w A_0 \eta_0 h/(kA_1)} \tag{5.1}$$

Here the fin efficiency η_f is determined through half of the fin height. An iterative solution of (5.1) gives the heat transfer coefficient for the cases of evaporation and condensing. The average refrigerant temperature was defined as a half-sum of the inlet and outlet temperature. We also calculated the heat transfer coefficient from the refrigerant temperature recalculated from the pressure in the middle of the working section. The difference between these two approaches (temperature and pressure) was less than 10%. The ultimate value of the heat transfer coefficient was defined as the averaged value.

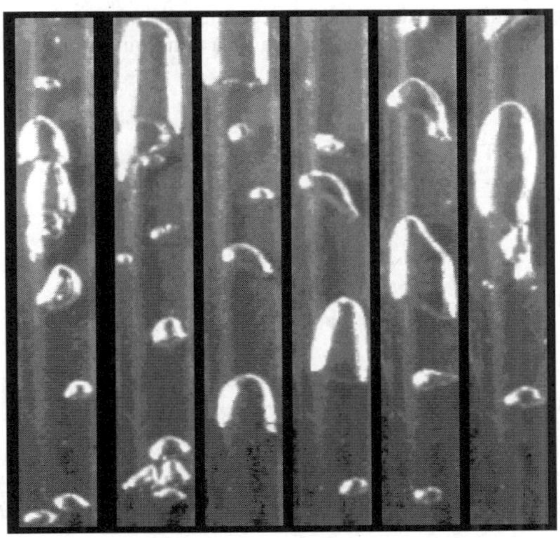

Figure 15. Evolution of vapor bubbles under saturated boiling of refrigerant R-11 at atmospheric pressure (frames consecutively). Vertical channel 2x7mm under constant heat flux about 1.5 kW/m^2. The distance is 5 cm from bottom of channel.

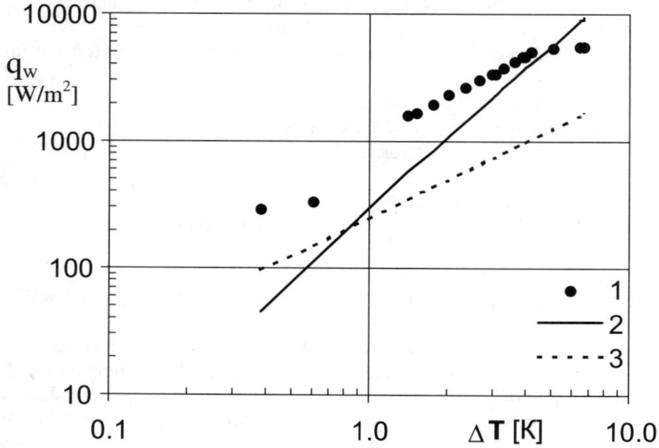

Figure 16. Heat flux density vs. wall superheat: 1-experiment, 2- boiling calculation by [13], 3-convection calculation by [14].

5.2 BOILING HEAT TRANSFER IN CONDITION OF LIQUID CIRCULATION INDUCED BY RISING BUBBLES

Figure 15 shows the visualization of boiling for saturated refrigerant R11 at pressure about 1.1 bar under constant heat flux conditions in a single vertical rectangular glass channel (size of 2x7 mm). Wall heat flux was about 1.5 kW/m^2. The digital camera with 12.5 FPS was used to obtain the consequent series of frames. It demonstrated that the heat transfer enhancement in a confined space is caused by the rapid growth of the bubbles and their merge into the plug. Vapor plugs form up a thin liquid film on the channel walls, boiling suppressed and thin film evaporation dominated.

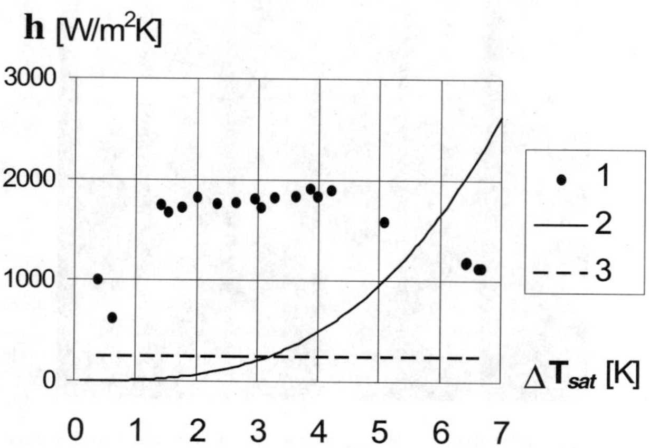

Figure 17. Heat transfer coefficient vs. wall superheat: 1-experiment, 2 boiling calculation by [13], 3-convection calculation by [14].

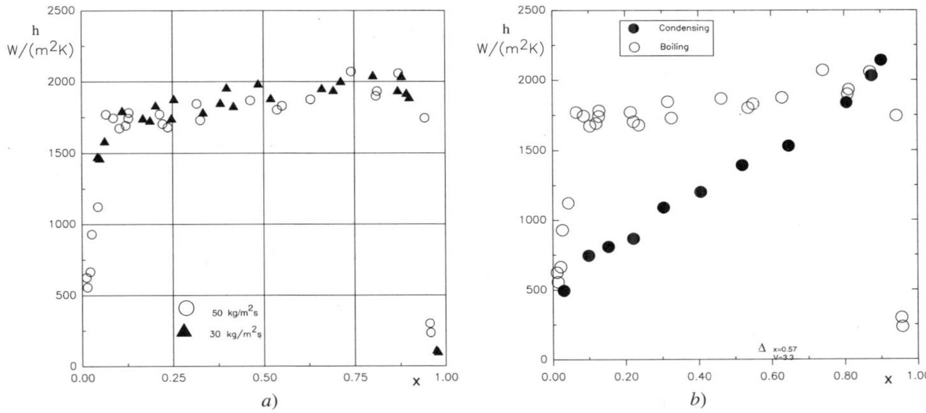

Figure 18. Heat transfer coefficients vs. mass vapor quality for upward flow R21 in heat exchanger with plain fins (20.3FPI). Wall superheat ranged from 0.9 to 1.4 K. *a*) evaporation for two mass fluxes, *b*) Comparison of heat transfer coefficients in condensation/evaporation modes at G=50 kg/m^{2}s. Mark shows the transition to the annular modes.

Figures 16, 17 present the heat transfer data in a vertical minichannels assemblage with boiling induced liquid circulation of refrigerant R21 inside. The incipience of the boiling in a confined space occurs for a wall superheat near 1K and the heat transfer coefficient slightly growth with the wall superheat until 4.5K. Then a decrease in heat transfer coefficient occurs and heat flux becomes independent from superheat. While the wall superheat less then 5.5 K, a heat transfer coefficient in a minichannels is higher than that for free boiling and for forced convection induced by boiling. The degradation of heat transfer coefficient occurs if the wall superheat keeps growing. A weak heat transfer dependence on the wall superheat indicates that boiling is suppressed.

In confined space, the heat transfer enhancement is caused by evaporation a thin liquid film in near corner area (see Fig. 12) and dry spot formation on the channel wall. The dry spot formation in this area can explain low dependence of heat transfer coefficient on wall superheat [21]. For this case heat flux in vicinity of liquid-solid-vapor contact line has higher level due to evaporation in ultra thin film area [20]. In that way high level of heat transfer in vicinity of contact line is responsible for the heat transfer enhancement during boiling in mini-channels. The possibility of dry spot formation on the wall for water boiling in narrow annular channel was observed in [35] also. At wall superheat over 4.5 K the drying-out of liquid is responsible for decrease of heat transfer when the size of dry area becomes very large.

5.3. FLOW BOILING HEAT TRANSFER IN A PASSAGE OF RECTANGULAR CHANNELS

Figure 18*a* presents heat transfer coefficients in boiling mode vs. mass vapor quality for forced upward flow through a test heat exchanger with plain fins for mass fluxes of 30 and 50 kg/m^{2}s. Heat transfer coefficient is almost constant up to mass vapor quality of x~0.9 (for the entire range of mass fluxes). If the vapor mass quality approaches 0.9, the deterioration of heat transfer takes place. Experiments were done under different wall superheat from 1 to 5 K. Maximal heat flux through parting sheet was 28 KW/m^2. Data in Fig. 18*a* are in agreement with the data from [36] for R 113 boiling with mass flux of 51.6 kg/m^{2}s at horizontal narrow channel (channel size was 1x20 mm, q_W~3-5 kW/m^2). We did not observe the dependence of heat transfer coefficient on wall superheat, which indicates small contribution of the nucleate boiling in total heat transfer. At the same time we have just slightly growth of heat transfer coefficient on vapor quality, it indicates small contribution of forced convection in total heat transfer. The possible explanation of such heat transfer coefficient behavior is again high heat flux in ultra thin liquid film area near channel corner or near vapor-liquid-solid contact line. High heat flux in these areas is responsible for high value of heat transfer in a wide range of vapor quality.

Figure 18*b* presents the comparison of heat transfer coefficient vs. mass vapor quality for upward flow condensation and upward flow boiling modes at mass flux of 50 kg/m^{2}s. At low vapor quality the boiling heat transfer coefficient is considerably higher than that for condensation. The difference in heat transfer behavior can be explained by the absence of film rupture in case of condensation. The contact line does not exist in this case and heat transfer level is much less than in case of boiling.

5. Conclusions

When a boiling liquid flows inside a very small and non-circular channel, the capillary forces become important in determining aspects of the flow phenomena such as two-phase flow pattern and heat transfer.

- Data have been presented allowing identification of different flow patterns in a narrow annular channel with a gap less than capillary constant. For large superficial velocities the flow with Taylor bubbles and cell flow regime with liquid plugs are typical.
- For horizontal annular channel with a small gap the regimes of nucleate boiling and forced convection were defined in dependence of the vapor quality. It was observed that enhanced nucleate boiling occurs at liquid plug. Flow boiling correlation was modified for prediction of boiling heat transfer which to take into account the suppression of the boiling in narrow space.
- Flow regimes in vertical rectangular minichannel were identified as a function of liquid and gas superficial velocities and Reynolds numbers. A novel laser illumination technique is described by which the interface shapes in channel cross section have been recorded using photography. The tests for bubble rising in organic liquids filling the vertical and tilted rectangular minichannel demonstrated that bubbles are free to float even if Etveos number less then four because the liquid can flow near the corners of the channel.
- Up to wall superheat 5.5K the boiling heat transfer in conditions of liquid circulation induced by rising bubbles in a vertical minichannel set are higher than that for pool boiling. The decrease in heat transfer occurs at wall superheat more than 4.5K. The heat transfer enhancement in confined space is caused by blocking of the channel area by elongated bubbles and strong heat flux both in ultra thin film area and in vicinity of vapor-liquid-solid contact line.
- Study of upward flow boiling heat transfer in a set of vertical minichannels at wall superheat in range from 0.9 K to 1.4K showed that considerable dependence of heat transfer coefficients on wall superheat was not obtained. In this tests just slightly growth of heat transfer coefficient on vapor quality was observed, it indicates small contribution of forced convection in total heat transfer for mass flux of 30 kg/m^{2}s and 50 kg/m^{2}s. The possible explanation of such heat transfer behavior is again high heat flux in ultra thin liquid film area near the channel corner or in the vicinity of liquid-vapor-solid contact line. In contrast with evaporation mode, at upward flow condensation mode the heat transfer coefficient is strongly dependent on vapor quality and for low vapor quality the boiling heat transfer coefficient is considerably higher than that for condensation.

NOMENCLATURE

A, cross-sectional area, [m^2];
A_0, total heat transfer surface area on one side of a direct transfer-type exchanger, [m^2];
A_l, fin or extended (secondary) area on one side of the exchanger, [m^2];
b, maximal wide of rectangular channel, [m];
Bo, q_w/rG boiling number;
d_{max}, maximum stable diameter of the bubble in turbulent flow [m];
D_h, hydraulic diameter, [m];
D_i, inner diameter of annuls, [m];
D_o, outer diameter of annuls, [m];
Eo, $\rho_l\, g\, Sin(\Theta)\, b^2/\sigma$, Etveos number;
f, friction factor;
F, enhancement boiling heat transfer factor,
g, acceleration due to gravity, [9.806 m/s^2];
G, mass velocity, [kg/m^2 s];
h, heat transfer coefficient, [W/m^2 K];
h_l, convective heat transfer coefficient, [W/m^2 K];
h_{pool}, pool boiling heat transfer coefficient, [W/m^2 K];
L, length of the channel, [m];
N_{conf}, $Eo^{-1/2}$ confinement number;
Nu, Nusselt number, hD_h/k_w;
P, pressure, [Pa];
P_c, critical pressure, [Pa];
Pr, $C_p\mu/k$ Prandtl number;
p_r, P/P_c, reduced pressure;
q, heat flux density, [W/m^2];

Q, heat flux, [W];
r, latent heat of evaporation, [J/kg];
Re, $\upsilon D_h \rho / \mu$ Reynolds number;
R_p, surface roughness, [m];
S, suppression boiling heat transfer factor;
T, temperature [^{o}C];
U, fluid velocity, [m/s];
x, vapor quality;

Greek symbols

δ_h, gap size of annular channel, [m];
δ_f, liquid film thickness, [m];
δ_w, wall thickness, [m];
ε, is the energy dissipation on unit mass, [J s/kg];
Π, is the perimeter of channel, [m];
Θ, angle inclination of channel;
μ, dynamic viscosity of the fluid, [Pa s];
ν, cinematic viscosity of the fluid, [m^2/s];
ρ, density, [kg/m^3];
η_0, surface efficiency, dimensionless;
η_f, fin efficiency, dimensionless;
σ, surface tension, [N/m]

Subscripts

B, bubble;
g, vapor phase;
in, inlet;
out, outlet;
l, liquid phase;
S, superficial;
sat, saturation;
w, wall;

Postscripts

cr, critical;

REFERENCES

1. Mehendale, S.S., and Jacobi, A.M., (2000), Evaporative Heat Transfer in Mesoscale Heat Exchangers, *ASHRAE Trans.*, 106(1), pp. 446-452.
2. Kandlikar, S.G., (2001), Two-phase flow patterns, pressure drop and heat transfer during boiling in minichannels and microchannels flow passages of compact evaporators, *Keynote Lecture presented at the Engineering foundation Conference on Compact Heat Exchangers, Davos, Switzerland, July 1-6.*
3. Lazarek, G.M., and Blake, S.H., (1982), Evaporative heat transfer, pressure drop and critical heat flux in small vertical tube with R-113, *International Journal Heat and Mass Transfer* 25, 7, pp 945-960.
4. Wambsganss, M.W., France, D.M., Jendrzejczyk, J.A., and Train, T.N., (1993), Boiling heat transfer in a horizontal small-diametr tube, *Journal of Heat Transfer,* 115 (November), pp 963-972.
5. Train, T.N., Wambsganss, M.W. and France, D. M., (1996), Small Circular- and a Rectangular Channel Boiling with Two Refrigerants, *International Journal Multiphase Flow*, 22, pp. 485-498.
6. Tran, T.N., Wambsganss, M.W., Chyu, M.C. and France, D.M., (1997), A Correlation for Nucleate Flow Boiling in a Small Channel, *Proc. Int. Conf. On Compact Heat Exchangers for Process Industries*, pp.291-304.
7. Peng, X. and Wang, B., (1993), Forced convection and flow boiling heat transfer in flat plates with rectangular microchannels. *International Journal of Heat and Mass Transfer* 36,14, pp. 3421-3427.
8. Kuznetsov, V.V., Shamirzaev, A.S., (1999), Two-phase flow pattern and boiling heat transfer in non-circular channel with a small gap, *Proc. of Two-Phase Flow Modeling and Experimentation*, Pisa, V.1, pp. 249-253.
9. Mertz, R., Wein, A., Groll, (1996), Experimental investigation of flow boiling heat transfer in narrow channels, *Calore Technologia* 14, 2, pp. 47-54.
10. Robertson, J.M., (1979), Boiling Heat Transfer with Liquid Nitrogen in Braised-Aluminum Plate-Fin Heat Exchangers, AIChE Symposium Series 75, Vol. 189, 151-164
11. Robertson, J.M., Lovergrove, P.C., (1983), Boiling heat transfer with Freon 11 in brazed aluminum Plate-Fin Heat Exchangers, *J. of Heat Transfer*, 105
12. Robertson, J.M., (1982), The correlation of boiling coefficients in Plate-Fin Heat Exchanger passages with a film flow model, *in 7th Int Heat Transfer Conference Munich*, pp341-345.
13. Wadekar,V.V., (1992), Flow boiling of heptane in a plate-fin heat exchanger passage, *Compact Heat Exchangers for the Process and Power Industries*, ASME HTD v201, pp. 1-6.
14. Mironov, B.M., Lobanova, L.S., Shadrin, A.O., (1978), About local heat transfer crisis at liquid boiling in vertical parallel narrow channels, *Izvestiya vuzov, Energetica*, 7, pp 85-88.
15. Cornwell, K., Kew, P.A., (1992), Boiling in small parallel channels, *Proceedings of CEC Conference on Energy Efficiency in Process Technology, Athens, October, Paper 22,* pp.624-638.

16. Jacobi, A.M., Thome, J.R., (2002), Heat transfer model for evaporation of elongated bubble flows in microchannels, *J HEAT TRANSFER*, 124, 6, pp. 1131-1136.
17. Lin Liu, H.T., (1993), Pressure Drop Type and Thermal Oscillations in Convective Boiling Systems, *Ph.D. Thesis, University of Miami, Coral Gables, FL.*
18. Kandlikar, S.G., (2002), Fundamental issues related to flow boiling in minichannels and microchannels, *Experimental Thermal and Fluid Science*, 26, pp. 389-407.
19. Nakoryakov, V.E., Kuznetsov, V.V., Vitovsky, O.V., (1992), Experimental Investigation of Upward Gas-Liqud Flow in a Vertical Narrow Annuls, *Int. Journal of Multiphase Flow*, 18, 3, pp. 313-326.
20. Kuznetsov, V.V.,Safonov, S.A.,Sunder, S.,Vitovsky, O.V., (1997), Capillary Controlred Two-Phase Flow in Rectangular Channel, *Proc. Int. Conf. on Compact Heat Exchangers for Process Industries, Utah USA June 22-27 New York*, pp.291-303.
21. Kuznetsov, V.V., Dimov, S.V., Shamirzaev, A.S., Houghton, P.A., Sunder, S., (2003), Upflow Boiling and Condensation in rectangular Minichannels, *First International Conference on Microchannels and Minichannels, April 24-25, Rochester, New York Editor S.G. Kandlikar*, pp.683-689.
22. Kuznetsov, V.V., Shamirzaev, A.S., (2003), Flow boiling heat transfer in minichannels, *Proc. of Eurotherm Seminar No 72: Thermodynamics Heat and Mass Transfer of Refrigeration Machines and Heat Pumps, Valencia, Spain, on March 31 to April 2.*
23. Kuznetsov, V.V., Shamirzaev, A.S., (1999), Two-phase flow pattern and flow boiling heat transfer in non- circular channel with a small gap, *Two-Phase Flow Modeling and Experimentation, Pisa, Italy*, v 1, pp. 249-253.
24. Kuznetsov, V.V., Shamirzaev, A.S., Ershov I.N., (2004), Flow Boiling Heat Transfer and Regimes of Upward Flow in Minichannels, *3rd Int. Symposium on Two-Phase Flow Modeling and Experimentation, 22-25 September, Pisa, Italy,* ven 03.
25. Kelessidis, V.C., Dukler, A.E., (1989), Modeling flow pattern transition for upward gas-liquid flow in vertical concentric and eccentric annuli, *Int. J. of Multiphase Flow* 15, pp. 173-191.
26. Hinze, J. O., (1955), Fundamentals of the hydrodynamic mechanism of splitting in dispersion processes, *AIChE J* 1, pp. 269-295.
27. Liu, Z., Winterton, R.H.S., (1991), A general correlation for saturated and subcooled flow boiling in tubes and annuli, based on a nucleate pool boiling equation, *International Journal of Heat and Mass Transfer*, 34, pp. 2759-2766.
28. Petuchov, B.S. and Rozen, I.I., (1964), Generalized Relationship for Heat Transfer in a Turbulent Flow of Gas in Tubes of Annular Section, *High Temp.,* 2, pp. 65-68.
29. Stephan, K., (1962), Warmeuebergang bei Turbulenter und bei Laminarer Stroemung in Ring-spalten, *Chem. Ing. Tech.,* 34, pp. 207-212.
30. Cooper, M.G., (1984), Heat Flow Rates in Saturated Nucleate Pool Boiling - A Wide Ranging Examination Using Reduced Properties, *Advances in Heat Transfer,* 16, pp. 157-239.
31. Kew, P.A., and Cornwell, K., (1994), Confined bubble flow and boiling in narrow channels, *Proc. 10th Int. Heat Transfer Conf., Brighton, UK,* pp. 473-478.
32. Danilova, G.N., (1970), Correlation of Boiling Heat Transfer Data for Freons, *HEAT TRANSFER- Soviet Research*, 2, 2, pp.-73-78.
33. Asali, J.C., Hanratty, T.J. and Andreussi, P., (1985), Interfacial Drag and Film Height for Annular Flow, *AIChE J.,* 31, pp. 886-902.
34. Wallis, G.B., (1969), One Dimensional Two-Phase Flow, *McGraw-Hill, New York.*
35. Kozelupenko, Yu. D., Smirnov, G.F., Koba, A. L., (1985), Heat transfer crisis in subcooled liquid in narrow annular channels at low velocities of motion, *Promushlennaya Teploenergetica*, 7, 1, pp 30-32.
36. Han Ju Lee, Sang Yong Lee, (2001), Heat transfer correlation for boiling flows in small rectangular horizontal channels with low aspect ratios, *Int. J. of Multiphase Flow*, 27, pp. 2043-2062.

CONDENSATION FLOW MECHANISMS, PRESSURE DROP AND HEAT TRANSFER IN MICROCHANNELS

Srinivas Garimella
George W. Woodruff School of Mechanical Engineering
Georgia Institute of Technology
Atlanta, GA 30332-0405
srinivas.garimella@me.gatech.edu

1. Introduction

Heat transfer coefficients and pressure drops for condensation inside tubes are strongly dependent on the different flow patterns that are established at different regions of the condenser as the fluid undergoes a transition from vapor to liquid. Accurate heat-transfer and pressure-drop predictions require an approach that accounts for the variation in flow patterns as the quality changes. Circular and noncircular microchannel tubes are being used in a variety of applications because of the extremely high heat transfer coefficients that these geometries offer. Coleman and Garimella [1] demonstrated in a study on two-phase flow of air-water mixtures that flow regime transitions in such geometries are different from those observed in larger diameter circular tubes. This is because of significant differences between large round tubes and the smaller noncircular tubes in the relative magnitudes of gravity, shear, and surface tension forces, which determine the flow regime established at a given combination of liquid and vapor-phase velocities. Thus, extrapolation of large round tube correlations to smaller diameters and noncircular geometries could introduce errors into pressure drop and heat transfer predictions.

Limited research has been conducted on addressing the effect of tube diameter and shape on flow regimes, pressure drop and heat transfer coefficients during condensation. Early attempts at flow regime mapping were conducted on relatively large tubes using air-water or air-oil mixtures [2-4]. Mechanistic models for transition criteria [5] have achieved limited success in predicting the available experimental data. The relatively few studies on flow regime maps for small diameter round tubes have also primarily used isothermal air-water mixtures [6], $1.0 < D < 1.6$ *mm*; [7], $4 < D < 12$ *mm*). Damianides and Westwater [8] and Fukano *et al.* [9] showed that the flow regime maps of Mandhane *et al.* [10], Taitel and Dukler [5], and Weisman *et al.* [11] cannot accurately predict transitions in small diameter tubes ($1 < D < 5$ *mm*). Most of the research on two-phase flow in small hydraulic diameter rectangular channels uses tubes of either small ($\alpha < 0.50$) or large ($\alpha > 2.0$) aspect ratios [12-15]. In a study on flow regime maps for $0.125 \leq \alpha < 0.50$ and $11.30 < D_h < 33.90$ *mm*, Richardson [16] showed that the smaller aspect ratio suppressed the stratified and wavy flow regimes and promoted the onset of elongated bubble and slug flows. Troniewski and Ulbrich [17] proposed corrections for the Baker [2] map based on the single-phase velocity profiles in rectangular channels for horizontal and vertical channels with $0.09 < \alpha < 10.10$ and $7.45 < D_h < 13.10$ *mm*. Lowry and Kawaji [15] studied rectangular geometries with $D_h < 2.0$ *mm* and $40 < \alpha < 60$ in vertical upward flows and concluded that the Taitel and Dukler [5] model was not valid for narrow channel flow. Wambsganss *et al.* [18] reported flow patterns and flow regime transitions in a single rectangular channel with aspect ratios of *6.0* and *0.167* and $D_h = 5.45$ *mm* through flow visualization and dynamic pressure measurements. Wambsganss *et al.* [19] extended this work to develop criteria for transition from slug flow based on root-mean-square pressure changes. Wang *et al.* [20] developed a two-phase flow pattern map for a *6.5 mm* round tube using refrigerant R134a.

Coleman and Garimella [1] investigated the effect of tube diameter and shape on flow patterns and flow regime transitions for air-water mixture flow in tubes with small hydraulic diameters (*1.3 mm* to *5.5 mm*). Gas

S. Kakaç et al. (eds.), Microscale Heat Transfer, 273–290.

and liquid superficial velocities were varied from *0.1 m/s* to *100 m/s*, and *0.01 m/s* to *10.0 m/s*, respectively. They showed that while pipe diameter and surface tension may have a negligible effect on flow regime transitions in tubes with diameters greater than *10 mm* [10, 11], for smaller tubes these factors play an important role. Therefore, flow regime maps such as those developed by Mandhane *et al.* [10] based upon data from larger tubes may not be applicable for a smaller tube diameter range. It was also shown that the theoretical results of Taitel and Dukler [5] and the assumptions inherent in these analyses may not be valid for small diameter tubes. It was shown that as the tube diameter decreases, the transition to a dispersed flow regime occurs at a higher value of the superficial liquid velocity. Also, the transition to annular flow occurs at a nearly constant value of the superficial gas velocity, which approaches a limiting value as the tube diameter decreases. The stratified regime is suppressed in small diameter tubes, while the size of the intermittent regime increases. Thus, this study showed that the flow patterns and the respective transitions change significantly with tube diameter and shape.

The above discussion shows that many of these studies have investigated circular channels with much larger hydraulic diameters than are of interest in microchannel applications. The few studies on non-circular geometries have investigated very large or very small aspect ratio rectangular channels that approximate flow between parallel plates, but are not pertinent to the microchannel condenser tubes being used by industry. The effect of tube diameter and shape on flow pattern transitions is not well understood, with conflicting trends reported by different investigators. Also, most of the studies on small D_h channels have used isothermal air-water mixtures to simulate two-phase flow. However, due to the adiabatic flow in these studies, the results are not directly applicable to phase-change situations.

To address the deficiencies in the literature documented above, a comprehensive study of condensation of refrigerant R134a in several different circular and noncircular channels of varying hydraulic diameters ($0.4\ mm < D_h < 4.91\ mm$) was initiated by the author's group in 1997 over the mass flux range $150 < G < 750\ kg/m^2\text{-}s$, and quality range $0 < x < 1$. The study started with the documentation of flow mechanisms in nine different tubes of round, square and rectangular cross-sections tubes with $1 < D_h < 4.91\ mm$. With the flow mechanisms thus established, pressure drop measurements were conducted on these and other circular and noncircular (square, rectangular, triangular and other shapes) channels with hydraulic diameters as small as 0.4 mm over a range of flow rates that covered each of the documented flow regimes. In addition, unique experimental techniques were developed to measure heat transfer coefficients in these same channels for the entire range of refrigerant quality and mass flux mentioned above. An overview of this research and the significant results follows.

2. Flow Regime Mapping

2.1 EXPERIMENTAL APPROACH

The test facility (Figure 1) used to conduct the flow visualization experiments was designed with the express purpose of recording flow patterns during the actual condensation process of the refrigerant. The facility and the experimental procedures are described in greater detail in [21-23]. A brief summary is provided here. Subcooled liquid refrigerant is pumped through a tube-in-tube evaporator, in which steam is used to boil and superheat the refrigerant. The superheated vapor is partially condensed to the desired quality using a set of water-cooled pre-condensers. Refrigerant exiting the pre-condensers enters the test section, which is a counterflow tube-in-tube heat exchanger, with refrigerant flowing through an inner glass tube of the cross-section of interest, and air flowing through the space between this inner tube and another transparent outer plexiglass tube, thus enabling visualization of the respective flow regimes. Heat transfer between the cold air and refrigerant causes condensation. Compressed air flowing in the annulus provided a low differential pressure for the glass microchannel, making it possible to conduct tests at saturation pressures as high as *1379-1724 kPa*. The air flow rate was varied to accommodate different condensation loads for different test conditions. The set of post-condensers downstream of the test section was used to completely condense and subcool the refrigerant.

Coriolis mass flow meters (±*0.15%* uncertainty for the refrigerant, and ±*0.2%* uncertainty for the air) were used for the measurement of refrigerant and air flow rates. Cooling water flow rates were measured using banks of precision rotameters with a maximum uncertainty of ±*2%*. Pressure transducers with uncertainties of ±*0.25%* of the span were used to measure refrigerant pressures. All temperatures were measured using Platinum RTDs. Flow regimes were recorded using a digital video camera with a shutter speed adjustable from *1/20* second to *1/10,000* second.

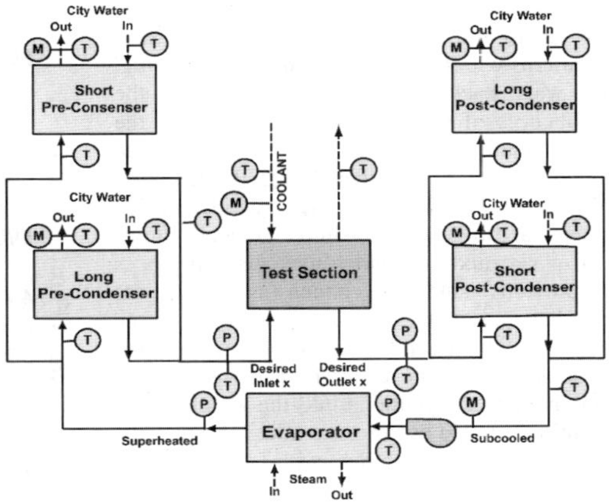

Figure 1: Schematic of Test Facility.

An energy balance on the pre-condenser yielded the refrigerant quality at the test section inlet, while a similar energy balance on the post condenser provided the refrigerant quality at the test section outlet. In addition, a redundant and independent validation of the change in refrigerant quality across the test section (Δx) was obtained from the measured inlet and outlet temperatures and the flow rate of the air in the test section. Excellent agreement was found between the measured Δx and $(x_{test,i} - x_{test,o})$ in all the experiments. The air flow rate was controlled to typically yield a refrigerant quality change Δx of about *0.05*, which ensured recording the variation of flow patterns in small increments. The average test section quality was used to represent the recorded flow regimes. This approach was used for the flow regime maps for all the different geometries, with nominally ten different refrigerant qualities ($0 < x < 1$) for five different mass fluxes ($150\text{-}750\ kg/m^2\text{-}s$) per tube; i.e., up to 50 data points per tube.

2.2 FLOW REGIME DESIGNATION

Table 1: Descriptions of Two-Phase Flow Regimes and Patterns.

	FLOW REGIMES			
	Annular	Wavy	Intermittent	Dispersed
Flow Patterns	Mist Flow	Discrete Wave (0)	Slug Flow	Bubbly Flow
	Annular Ring	Discrete Wave (1)	Slug Flow	Bubbly Flow
	Wave Ring	Discrete Wave (2)	Plug Flow	Bubbly Flow
	Wave Packet	Disperse Wave (3)	Plug Flow	
	Annular Film	Note: Numbers above denote intensity of secondary waves		

Digitized frames of the flow visualization video for each data point were used to identify four major flow ***regimes***, including annular, intermittent, wavy and dispersed flow, with the regimes further subdivided into flow ***patterns*** (Table 1.) Detailed descriptions of the flow phenomena in each of these regimes and patterns are provided elsewhere [21-23]. In the annular flow regime, the vapor flows in the core of the tube with a few entrained liquid droplets, while liquid flows along the circumference of the tube wall. The flow patterns within this regime (mist, annular ring, wave ring, wave packet and annular film) show the varying influences of gravity and shear forces as the mass flux and quality changes. The wavy flow ***regime*** used here has often been considered (e.g. Barnea *et al.* [7] as a flow ***pattern*** within the annular regime entitled the wavy-annular flow pattern. Unlike adiabatic flows (such as air-water mixtures, which have been the focus of most previous studies), condensing flows are expected to have a coating of liquid around the whole circumference for most combinations of mass flux and quality. Consolidating the different patterns for this entire combination of conditions into the annular flow regime definition would not provide adequate insights. Therefore, here, flow patterns with a significant influence of gravity (vapor flowing above the liquid, or a noticeable difference in film thickness at the top and bottom of the tube) and a wavy interface are assigned to the wavy flow regime. The waves at the liquid-vapor interface are caused by interfacial shear between the two phases moving at different velocities. Thus, this regime was subdivided into discrete waves of larger structure moving along the phase interface, and disperse waves with a large range of amplitudes and wavelengths superimposed upon one another, as shown in Table 1. The other flow regimes shown in Table 1 are well documented in the literature and in other work by the author, and are therefore not described here.

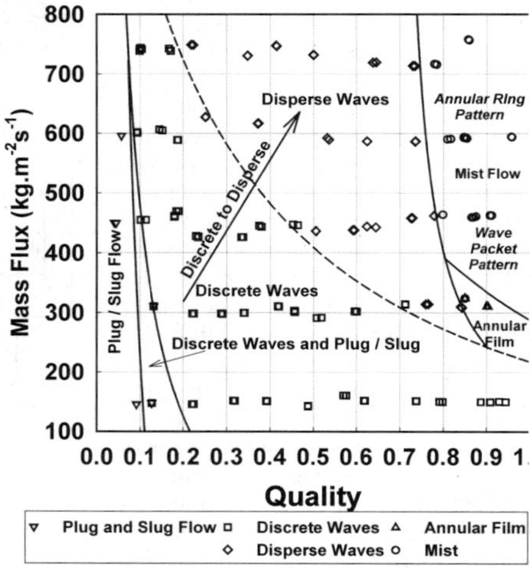

Figure 2: Typical Flow Regime Map (4.91 mm Round Tube).

The observed flow mechanisms for each data point were assigned to a particular flow pattern from among those shown in Table 1 to develop flow regime maps. In some cases, the flow mechanisms corresponded to more than one flow regime, typically indicating a transition between the respective regimes. Figure 2 shows the flow regime map for the *4.91 mm* circular tube plotted using the mass flux G, and quality x coordinates. A major portion of this map is occupied by the wavy flow regime with a small region where the plug, slug and discrete wave flow patterns coexist. The waves become increasingly disperse as the quality and mass flux is increased (shown by the arrow in Figure 2). The approximate demarcation between discrete and disperse waves is shown by the dashed line in this figure, although this transition occurs gradually. The different flow patterns within the annular flow regime are also noted in this Figure. Similar flow regime maps were also drawn for all the other tubes under consideration.

2.3 EFFECT OF HYDRAULIC DIAMETER

The effect of hydraulic diameter on the flow regime maps is shown in Figures 3 and 4. Figure 3, which depicts the transition from the intermittent regime for the 4 square tubes investigated, shows that the size of the intermittent regime increases as D_h decreases, with this effect being greater at the lower mass fluxes. The large increase in the size of the intermittent regime in the smaller hydraulic diameter tubes is because surface tension achieves a greater significance in comparison with gravitational forces at these dimensions. This also occurs because in square channels, it is easier for the liquid to be held in the sharp corners, counteracting to some extent, the effects of gravity. This facilitates plug and slug flow at higher qualities as the hydraulic diameter is decreased.

Figure 4 shows that the *4-mm* tube map is dominated by the wavy flow regime (with an absence of the annular film flow pattern). As D_h is decreased, the annular flow regime appears and occupies an increasing portion of the map. Thus, for the *4-mm* tube, the effects of gravity dominate, resulting in most of the flow regime map being covered by the wavy flow regime. As the hydraulic diameter decreases, the effects of surface tension increasingly counteract the effects of gravity, promoting and extending the size of the annular film flow pattern region instead of the more stratified wavy flow regime. Thus, as D_h decreases, the wavy flow regime is increasingly replaced by the annular flow regime, and is non existent in the D_h = *1 mm* tube. In addition to the effect on the size of the wavy flow regime, D_h also affects the flow patterns within the wavy flow regime. As D_h decreases, the waves are progressively more discrete, until the entire wavy flow regime consists of only discrete waves for the *2-mm* tube. Thus, it appears that as D_h decreases, the wavy flow regime is replaced by the annular film pattern due to liquid retention in the corners and along the entire circumference. Furthermore, surface tension stabilizes the waves, which leads to more discrete waves at small diameters.

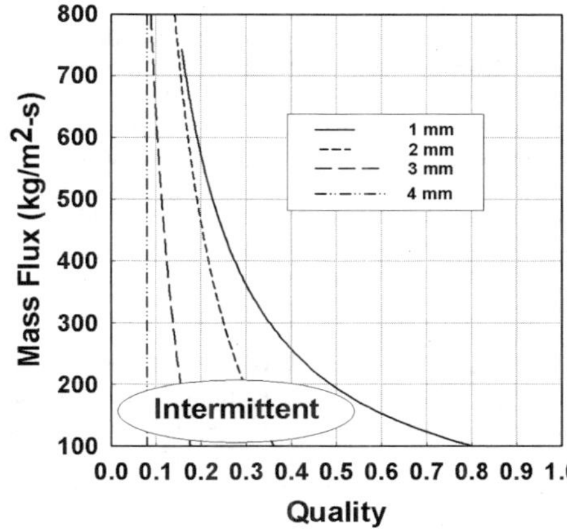

Figure 3: Effect of Hydraulic Diameter on the Intermittent Flow Regime.

2.4 EFFECT OF TUBE SHAPE

The effect of tube shape was investigated using the flow regime maps for six tubes: circular (D_h = *4.91 mm*), square *4×4 mm* (D_h = *4 mm*, α = *1*), rectangular *2×4 mm* and *4×2 mm* (D_h = *2.67 mm*, α = *0.5* and *2*), and rectangular *4×6 mm* and *6×4 mm* (D_h = *4.8 mm*, α = *0.67* and *1.5*). Here, aspect ratio (α) is the tube height divided by the tube width. Transition lines for the *4×4 mm*, *4×6 mm*, and *6×4 mm* tubes, as well as the round tube are shown together in Figure 5. (The corresponding transition lines for the *2×4* and *4×2* tubes are not shown for the sake of clarity.) The tubes in Figure 5 are of similar hydraulic diameter (*4.0 – 4.91 mm*), with the primary difference being tube shape. This figure shows that the intermittent regime is larger in the round tube than in the square tube at lower mass fluxes and approximately the same at higher mass fluxes. The extent of the intermittent regime for the rectangular tubes is in between that of the circular and square tubes. The wavy flow regime is also larger in the round tube. It appears that the square and rectangular channels help liquid retention in the corners and along the entire circumference of the tube leading to annular flow, rather than preferentially at the bottom of the tube as would be the case in the wavy flow regime. In the *4×6 mm* and *6×4 mm* tubes, the larger aspect ratio results in a slight increase in the size of the intermittent regime at the lower mass fluxes, and a small reduction in the size of this regime at the higher mass fluxes. However, these effects are small, and it can be concluded that this

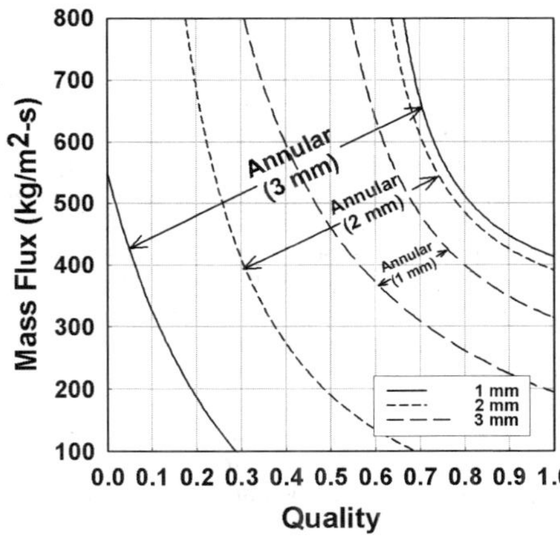

Figure 4: Effect of Hydraulic Diameter on Annular Flow Regime.

transition line is only weakly dependent on the aspect ratio. The smaller aspect ratio also results in a larger annular film flow pattern region, which is to be expected because of the reduced influence of gravity for the tubes with the smaller height. The smaller aspect ratio results in a smaller wavy flow regime, which may be viewed as a corollary to the effect on the size of the annular film flow region. Within the wavy flow regime, the transitions between discrete and disperse waves are shown with the gray lines with symbols. It appears that the discrete waves are more prevalent in the round tube compared to the square and rectangular tubes. It was also found that the smaller aspect ratio results in a smaller wavy flow regime but with a larger fraction of discrete waves. At the higher mass fluxes, the effect of the aspect ratio is negligible. In the *2×4 mm* and *4×2 mm* tubes (transition lines not shown in Figure 5), the slight increase in the size of the intermittent regime at the larger aspect ratio was found for all mass fluxes tested. The increase in the annular film flow pattern region was more pronounced in these smaller hydraulic diameter tubes, perhaps due to the greater influence of surface tension. The larger increase in the annular flow regime at the smaller aspect ratio was accompanied by a corresponding decrease in the wavy flow regime; and once again, was more pronounced in this smaller tube than in the *4×6* and *6×4 mm* tubes.

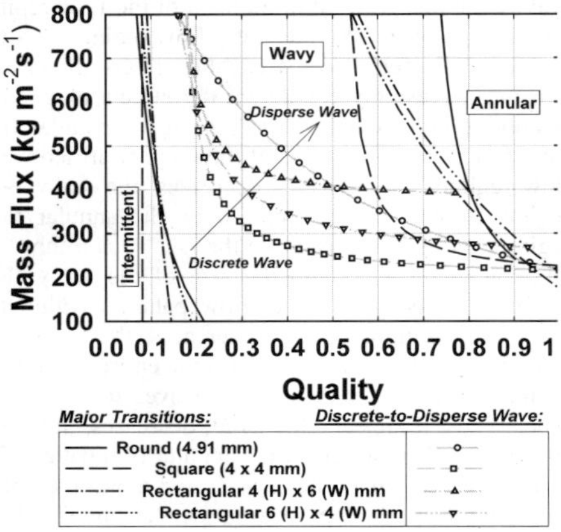

Figure 5: Effect of Tube Shape on Flow Regime Transitions.

From the above discussion, it can be seen that while tube shape (with similar hydraulic diameters) has some effect on the transitions between the various flow regimes, the influence of hydraulic diameter is far more significant. These findings about the flow patterns were used as the basis for the development of pressure drop and heat transfer models, as described in the following sections.

3. Pressure Drop Models

3.1 EXPERIMENTS

For this part of the study, the flow visualization test sections were replaced with those designed specifically for pressure drop measurements. For the two largest diameter (*4.91 mm* and *3.05 mm*) tubes, the test sections consisted of counterflow, tube-in-tube heat exchangers. For the smaller circular and noncircular channels, the test sections were fabricated as flat tubes with multiple extruded parallel channels. This method of using multiple parallel channels ensured that the refrigerant flow rates used were large enough to be adequately controlled and measured, with accurate heat balances around the test loop. Three such tubes were brazed together, as shown in Figure 6, with refrigerant flowing through the center tube, and coolant (air) flowing in counterflow through the top and bottom tubes. The cross-sections investigated include a variety of shapes as shown in Figure 6, including tubes with triangular microchannels fabricated by extrusion as well as by placing a W-shaped corrugated insert in a rectangular tube. The low thermal capacity and heat transfer coefficients of air maintained low condensation rates and small changes in quality in the test section, which in turn enabled the measurement of the pressure drop variation as a function of quality with high resolution. This small quality change across the test section minimized the likelihood of flow regime transitions within the test section for any data point. These measurements were conducted at a nominal saturation pressure of 1396 kPa which corresponded to a saturation temperature of 52.3°C; the saturation temperature was within ±3°C of this for all the data points. Pressure drops across the test section were measured using a bank of three selectable differential pressure transducers, with an accuracy of *±0.25%* of the span. Additional details of the pressure drop measurements, including validation procedures using single-phase gas and liquid pressure drop measurements, are available in Garimella *et al.* [24, 25].

3.2 ANALYSIS

The measurements described above were used to develop condensation pressure drop models for each flow regime. The measurement techniques were first verified by conducting single-phase pressure drop measurements *for each tube under consideration* over a wide range of mass fluxes for both the superheated vapor and subcooled liquid cases. The single-phase pressure drops were in excellent agreement with the values predicted by the Churchill [26] correlation, thus validating the experimental approach. The measured pressure drops included expansion and contraction losses due to the headers at both ends of the test section, and the pressure change due to deceleration caused by the changing vapor fraction as condensation takes place. The portion of the total pressure drop (change) attributable to deceleration of the fluid was estimated from void fraction and momentum change analyses; contraction/expansion losses at the inlet and outlet of the test section were estimated using two-phase "minor loss" models available in the literature. This process is described in detail by Garimella *et al.* [25]. Estimates of the deceleration effects were validated by conducting tests at the same nominal conditions with and without condensation in the test section. Similarly, estimates for the contributions of the end effects were validated using tests on tubes ranging in length from *22.5* to *508 mm*, with the same entrance and exit configuration at each length. Excellent agreement was obtained between these data and the models. The residual frictional component of the two-phase pressure drop, which generally was at least an order of magnitude larger than these minor losses, was used for developing condensation pressure drop models for the respective flow regimes.

3.3 FLOW REGIME ASSIGNMENT

The approach described above provided a sound data base of experimental values for the frictional component of the two-phase pressure drop for several tubes over a large range of conditions. The flow visualization results of Coleman and Garimella [22] described above, and the transition criteria developed by Garimella *et al.* [25] from these results were used to assign each of the pressure drop data points to the respective flow regimes. According to these criteria, transition from the intermittent to the other flow regimes occurs as follows:

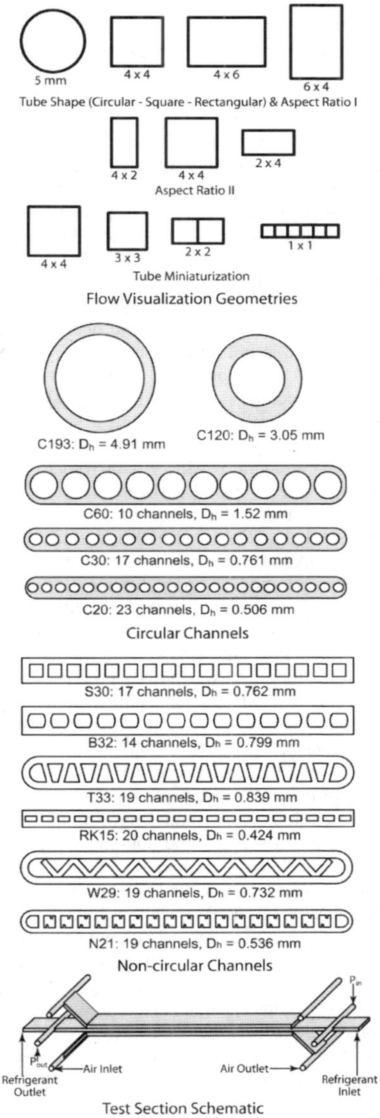

Figure 6: Microchannel Tubes and Test Section.

$$x \leq \frac{a}{G+b} \tag{1}$$

where G is the total mass flux expressed in kg/m^2-s and a and b are geometry dependant constants given by:

$$a = 69.57 + 22.60 \cdot \exp(0.259 \cdot D_h) \tag{2}$$

$$b = -59.99 + 176.8 \cdot \exp(0.383 \cdot D_h) \tag{3}$$

where D_h is the hydraulic diameter of the tubes in mm. On a mass-flux versus quality map, these transition lines appear in the lower left corner, as shown in Figure 7. For the purpose of pressure drop model development, the primary flow regimes depicted here are the intermittent regime (lower left) and the annular film/mist/disperse patterns of the annular flow regime (upper right). For clarity, the wavy flow regime that occurs between these is not labeled in Figure 7. As discussed in the previous section, the work of Coleman and Garimella [22] identified several other regimes and patterns; however, for pressure drop model development, it will be shown that this broad categorization suffices. In the absence of other valid transition criteria for phase-change flow in small hydraulic diameter circular channels, these criteria were also assumed to apply for circular channels of equivalent diameters under consideration here.

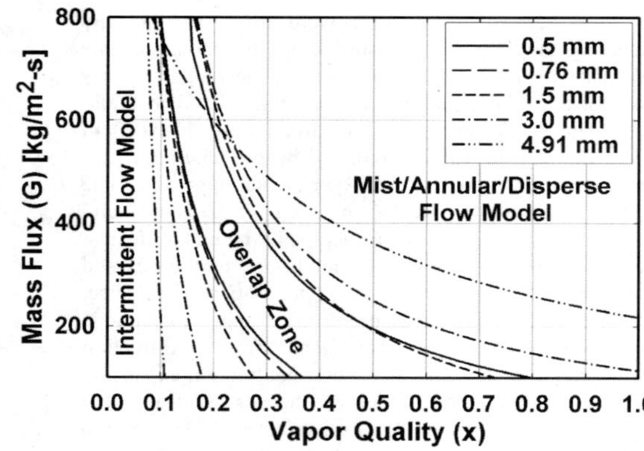

Figure 7: Flow Regime Assignment.

The wavy flow that exists at the larger D_h values consists of stratified layers of liquid and vapor flowing through the tube. As was noted in the previous section, in addition to the effect on the extent of the wavy flow regime, D_h also affects the flow patterns within the wavy flow regime. As D_h decreases, the waves are progressively more discrete, until the entire wavy flow regime consists of only discrete waves for the *2-mm* tube. Surface tension appears to stabilize the waves, which leads to more discrete waves at small diameters. As will be discussed later, these considerations allow the incorporation of the data points in the wavy regime into the annular or intermittent regimes for the purpose of developing the pressure drop models.

An intermittent flow model for circular tubes was first developed by Garimella *et al.* [25], with subsequent modifications to include the non-circular geometries shown in Figure 6 [24]. This model applies for the data points to the left of the corresponding transition lines shown in Figure 7. For annular flow, a preliminary pressure drop model was first developed by Garimella *et al.* [27], and was subsequently extended to the mist and disperse flow regions that occur at high mass fluxes and vapor qualities. It should be noted that the transitions between these various regimes do not occur abruptly at unique combinations of mass flux and quality, but rather across overlap zones, in which the flow could switch back and forth between the respective regimes, or exhibit discrete wave flow. This overlap region is also shown in Figure 7, and appropriate interpolation techniques for the pressure drops calculated from the individual models are also recommended in the subsequent sections.

For circular tubes, the experimental data set consisted of a total of 603 points. Of these, 77 points lie in the intermittent regime, 448 in the disperse/annular/mist flow regime, and the remaining 78 data points are in the overlap zone between these two regimes. Pressure drop models for these regimes are described below.

3.4 MODEL DEVELOPMENT AND RESULTS

3.4.1 Intermittent and Discrete-Wave Flow

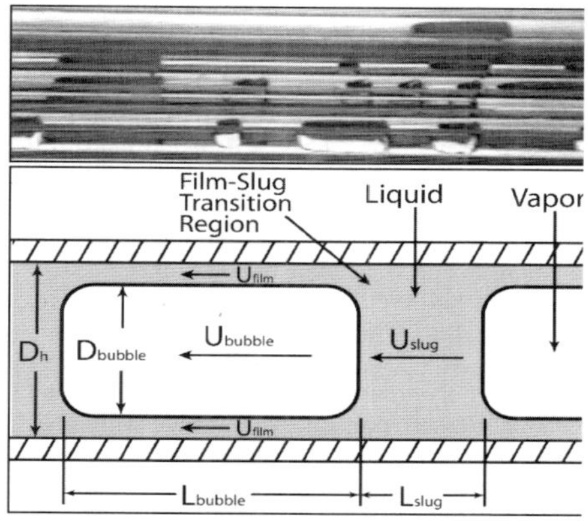

Figure 8: Modeling of Intermittent Flow.

The detailed models for intermittent flow in circular [25] and non-circular [24] channels served as one of the starting points for the development of a comprehensive pressure drop model across the overall flow regime domain, as described in Garimella *et al.* [28]. The flow visualization studies described above demonstrate that in the intermittent regime, the vapor-phase travels as long solitary bubbles surrounded by an annular liquid film and separated by liquid slugs. As the tube size decreases, surface tension forces at the bubble interface begin to dominate the gravitational forces and the bubble tends to a cylindrical shape. The corresponding "unit cell" used for the development of the model is shown in Figure 8. In general, the bubble travels faster than the liquid slug, which implies that there is a continual uptake of liquid from the film into the front of the slug. These phenomena were accounted for in the model. In addition, based on the recorded flow patterns, it was assumed that the bubble is cylindrical, and that there is no entrainment of vapor in slug, or liquid in bubble. Further, for any given condition, it was assumed that the length/frequency/speed of bubbles/slugs is constant, with no bubble coalescence, and a smooth bubble/film interface. Unlike other work in the literature, here the pressure drop in bubble/film region was not neglected. The total pressure drop for this flow pattern includes contributions from: the liquid slug, the vapor bubble, and the flow of liquid between the film and slug as follows:

$$\Delta P_{total} = \Delta P_{slug} + \Delta P_{f/b} + \Delta P_{film-slug\ transitions} \tag{4}$$

A simple control volume analysis [24, 25] similar to that performed by Suo and Griffith [6] showed that the velocity in the liquid slug can be directly calculated given the overall mass flux and quality. The results of several investigations [6, 9, 29] suggested that the bubble velocity for these conditions was 1.2 times the slug velocity. With this assumption, the diameter of the bubble, velocity within the film, and relative length of bubble and slug can all be calculated from a system of simultaneous equations including a shear balance at the bubble-film interface. Thus the Reynolds number in the liquid slug and vapor bubble (based on the relative velocity at the interface between the bubble and the surrounding film) could be directly determined. The Churchill [26] correlation was then used to calculate the friction factor and thus the pressure gradient at the respective Reynolds numbers in the liquid slug and bubble/film regions. A relationship from the literature for the pressure loss associated with the mixing that occurs in the uptake of liquid from the film to the slug was used to estimate the pressure loss due to each of these transitions. These components of the total pressure drop are shown below:

$$\frac{\Delta P}{L} = \left(\frac{dP}{dx}\right)_{\substack{film\\bubble}} \left(\frac{L_{bubble}}{L_{\substack{unit\\cell}}}\right) + \left(\frac{dP}{dx}\right)_{slug} \left(\frac{L_{slug}}{L_{\substack{unit\\cell}}}\right) + \Delta P_{\substack{one\\transition}} \left(\frac{N_{\substack{unit\\cells}}}{L}\right) \tag{5}$$

A depiction of these various contributions to the measured pressure drop for circular channels is shown in Figure 9. For the solution of the above equation, the number of unit cells per unit length is required, and can be determined from the slug frequency (which yields the unit cell length). Several models from the literature for slug frequency were considered; however, because these models were based primarily on studies of gas-liquid (instead of vapor-liquid) flows in large (> *10 mm*) diameter tubes, they did not give satisfactory results. Instead, the following correlation for slug frequency (nondimensional unit-cell length, or unit cells/length) based on slug *Re* and D_h was developed:

$$a\left(\mathrm{Re}_{slug}\right)^b = \omega\frac{D_h}{U_{bubble}} = D_h\left(\frac{N_{unit\ cells}}{L_{tube}}\right) = \left(\frac{D_h}{L_{unit\ cell}}\right) \tag{6}$$

The coefficients a and b were fit using the difference between the measured pressure drop and the pressure drop calculated as described above for the slug and bubble/film regions, i. e., the net pressure drop due to transitions. The correlation yielded $a = 2.437$, $b = -0.560$ for both circular and non-circular (except triangular, for which different coefficients were necessary, as discussed in Garimella *et al.* [24] channels. For circular channels (*0.5 – 4.91 mm*), the predicted pressure drops are on average within *±13.5%* of the measured values, with *90%* of the predicted results being within *±27%* of the measured values.

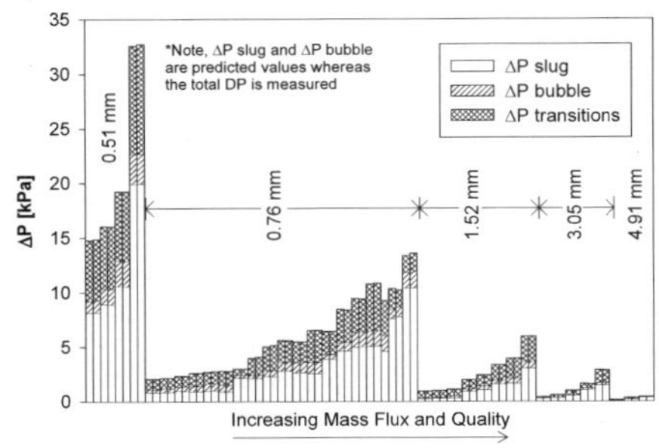

Figure 9: Contribution of Each Pressure Drop Mechanism to Total Pressure Drop.

The intermittent model was also extended to the discrete-wave flow region (Garimella *et al.* [28]). This is because as the progression from the intermittent region to discrete wave occurs, the gas bubbles start disappearing, to be replaced by stratified, well defined liquid and vapor layers. This can be treated as a phenomenon that results in a decrease in the number of unit cells per unit length. Accordingly, within the discrete wave region, traversing from the intermittent flow boundary toward the annular flow boundary, the bubbles disappear completely, with the number of unit cells per unit length approaching zero. Based on this conceptualization of the intermittent and discrete flow regions, the slug frequency model developed by Garimella *et al.* [24, 25] for intermittent flow was modified to include data from the discrete-wave flow region. Thus, 78 additional discrete-wave flow points were added to the 77 intermittent flow points in the earlier models, resulting in the following combined model for the two regions:

$$N_{UC}\left(\frac{D_h}{L_{tube}}\right) = \left(\frac{D_h}{L_{UC}}\right) = 1.573\left(\mathrm{Re}_{slug}\right)^{-0.507} \tag{7}$$

This model predicted *65%* of the intermittent and discrete wave flow data points within *±20%*. This somewhat lower prediction accuracy is due to the inclusion of the discrete-wave flow points into the data set. The above model predicted *75%* of the intermittent flow points within *±20%*, which is comparable with the results obtained with the original intermittent flow models of Garimella *et al.* [24, 25].

3.4.2 Annular/Mist/Disperse Flow

A schematic of the flow pattern used to represent annular flow is shown in Figure 10. A preliminary model for pressure drops in the annular flow regime for the circular tubes under consideration here was reported in Garimella *et al.* [27], followed by the more detailed model [28] described below. For the development of this model the following assumptions were made: steady flow, equal pressure gradients in the liquid and gas core at any cross section, uniform thickness of the liquid film and no entrainment of the liquid in the gas core. The measured pressure drops were used to compute the Darcy form of the interfacial friction factor to represent the interfacial shear stress as follows:

$$\frac{\Delta P}{L} = \frac{1}{2}\cdot f_i \rho_g V_g^2 \cdot \frac{1}{D_i} \tag{8}$$

The above equation uses the interface diameter, D_i. This same expression can be represented in terms of the more convenient tube diameter, D, through the use of a void fraction model [30] as follows:

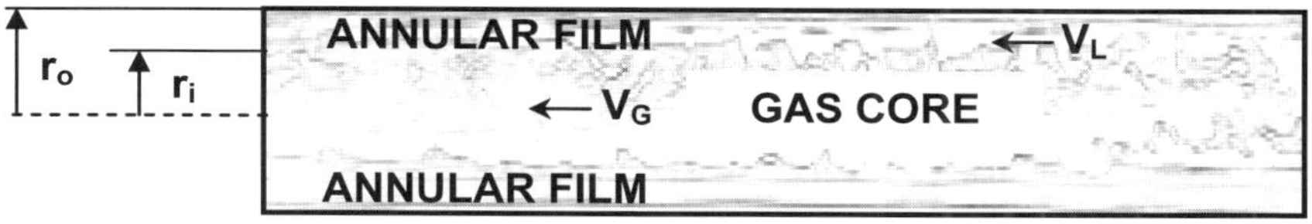

Figure 10: Annular Film Flow Pattern.

$$\frac{\Delta P}{L}=\frac{1}{2}\cdot f_i\frac{G^2\cdot x^2}{\rho_g\cdot\alpha^{2.5}}\cdot\frac{1}{D} \quad (9)$$

The ratio of this interfacial friction factor (obtained from the experimental data) to the corresponding liquid-phase Darcy friction factor was then computed and correlated as follows:

$$\frac{f_i}{f_l}=A\cdot X^a\,\mathrm{Re}_l^{\,b}\,\psi^c \quad (10)$$

Here the Martinelli parameter X is given by:

$$X=\left[\frac{(dP/dz)_l}{(dP/dz)_g}\right]^{1/2} \quad (11)$$

For this model, the liquid-phase Reynolds number required in equation (10) above and also to compute the liquid-phase pressure drop in the Martinelli parameter was defined in terms of the annular flow area occupied by the liquid phase as follows:

$$\mathrm{Re}_l=\frac{GD(1-x)}{\left(1+\sqrt{\alpha}\right)\mu_l} \quad (12)$$

Similarly, the gas-phase Reynolds number required for the calculation of the pressure drop through the gas core in the Martinelli parameter was calculated as follows:

$$\mathrm{Re}_g=\frac{GxD}{\mu_g\sqrt{\alpha}} \quad (13)$$

The friction factors required in Equation (10) above and for the individual-phase pressure drops in the Martinelli parameter (Equation 11) were computed using $f=64/\mathrm{Re}$ for $Re < 2100$ and the Blasius expression $f=0.316\cdot\mathrm{Re}^{-0.25}$ for $Re > 3400$. The effects of surface tension were accounted for by using the following non-dimensional parameter first introduced by Lee and Lee [31]:

$$\psi=\frac{j_L\mu_L}{\sigma} \quad (14)$$

where $j_L=\dfrac{G(1-x)}{\rho_l(1-\alpha)}$ is the liquid superficial velocity. Regression analysis on data grouped into two regions based on the liquid phase Reynolds number yielded the following values for the respective parameters in Equation (10):

$$\text{Laminar region } (\mathrm{Re}_l<2100):\quad A=1.308\times10^{-3};\ a=0.427;\ b=0.930;\ c=-0.121 \quad (15)$$

$$\text{Turbulent region } (\mathrm{Re}_l>3400):\quad A=25.64;\ a=0.532;\ b=-0.327;\ c=0.021 \quad (16)$$

For the transition region, Garimella *et al.* [28] recommend that the pressure drop be first independently calculated using the laminar and turbulent constants shown above at the low and high values of the mass flux and quality representing the boundaries of the laminar and turbulent regions. Linear interpolation between these bounding values should then be conducted separately based on the quality and the mass flux in consideration.

The average of these two pressure drops resulting from the interpolation based on G and x represents the two-phase pressure drop for the transition region data point. This model predicts 87% of the data within ±20%. It should be noted that these predictions include not only the annular flow region, but also the mist and disperse-wave flow data, whereas the preliminary model of Garimella *et al.* [27] applied only to the annular flow regime data. As can be seen above, the constants for the equation for the friction factor ratio are based on whether the *liquid phase* is laminar or turbulent, and do not depend on the gas-phase flow regime. This is because the gas core was consistently turbulent for all the data points under consideration in the mist, annular and disperse flow regime. Thus the model is not valid for cases with a laminar gas core; however, in such an instance, it is unlikely that the flow will be in one of these regimes.

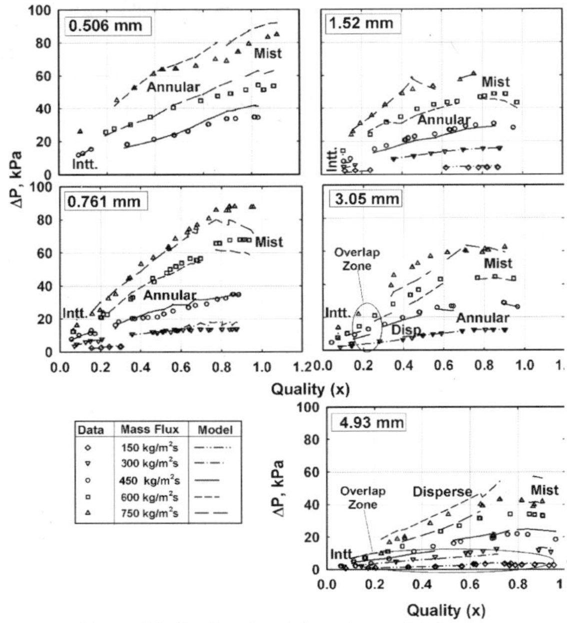

Figure 11: Predicted and Experimental ΔP vs. x.

This annular/disperse-wave/mist flow pressure drop model development above uses a physical representation where the liquid forms an annular film around a gas core; however, as noted above, the resulting correlation is also recommended for the disperse-wave and mist flow regions. This applicability may be explained as follows. The flow visualization work of Coleman and Garimella [22] considered tubes only as small as those with $D_h = 1\ mm$, whereas the pressure drop data used for this model include tubes with D_h as small as 0.5 mm. It is known from Coleman and Garimella's work that extent of the annular flow regime increases as the tube size decreases, as evidenced by the majority of the flow regime map for $D_h = 1\ mm$ being in the annular flow regime. Using this rationale, the annular flow regime is only expected to be larger for the $D_h = 0.5$ and $0.76\ mm$ tubes that are included in this paper. In the absence of actual flow visualization data for these latter tubes, and keeping the uncertainties of extrapolation of the transition criteria in mind, the $D_h = 1\ mm$ transition criteria were directly applied to the smaller tubes also in the present work. It is therefore to be expected that several of the data classified here as mist- or disperse-wave flow points using the $D_h = 1\ mm$ transition criteria would in fact be in the annular flow regime. This phenomenon explains the applicability of the model developed using an annular flow mechanism to the neighboring flow regimes also.

3.5 MODEL IMPLEMENTATION AND DISCUSSION

A comparison of the measured pressure drops and those calculated using the intermittent and annular/disperse-wave/mist flow models is shown in Figure 11 for each tube considered. In the overlap zone (Figure 7), the flow exhibits both the adjoining mechanisms (intermittent and annular/disperse-wave/mist flow). Therefore, for calculating the pressure drops in the overlap zones in Figure 11, the four-point interpolation scheme described above in connection with the transition between laminar and turbulent data was applied to the pressure drops calculated using the intermittent and annular/disperse-wave/mist flow models. This combined model for the

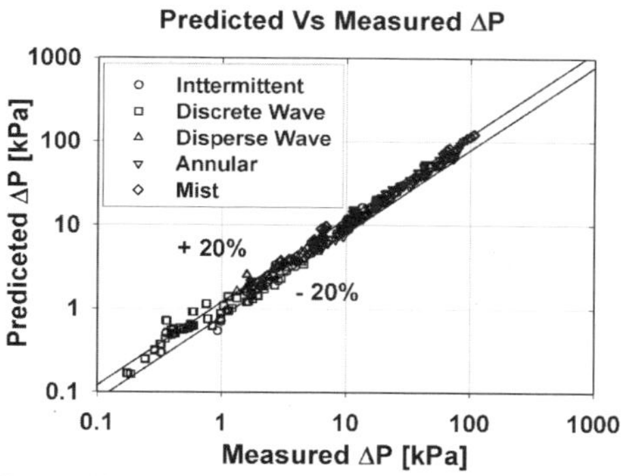

Figure 12: Comprehensive *ΔP* Model Predictions.

multiple zones predicted 82% of the data within ±20%, as shown in Figure 12. The applicable flow regimes for each data point are also shown in Figures 11 and 12, and demonstrate that the model of Garimella *et al.* [28] effectively captures the trends in the data based on the underlying physical phenomena: the variation in pressure drop with quality, mass flux, and tube diameter is well represented by the model.

The effect of hydraulic diameter on pressure drop, i.e., the demonstration of the influence of small diameter channels, is shown in Figure 13 for three representative mass fluxes. To particularly focus on this influence, all graphs have been plotted for a constant *L/D = 800*. Thus, if hydraulic diameter had no influence, the plots in each graph (at the same mass flux) would be perfectly superimposed on each other. From these graphs, it can be seen that pressure drop increases with mass flux and quality as expected, but also increases as the hydraulic diameter decreases. The model is also able to demonstrate the decrease in pressure drop toward single-phase gas flow pressure drops as the quality approaches 1, as evidenced by the maxima in the graphs around $x \cong 0.9$. The slight changes in slope at the interfaces of the respective flow regimes reflect the different flow mechanisms that occur at the respective conditions.

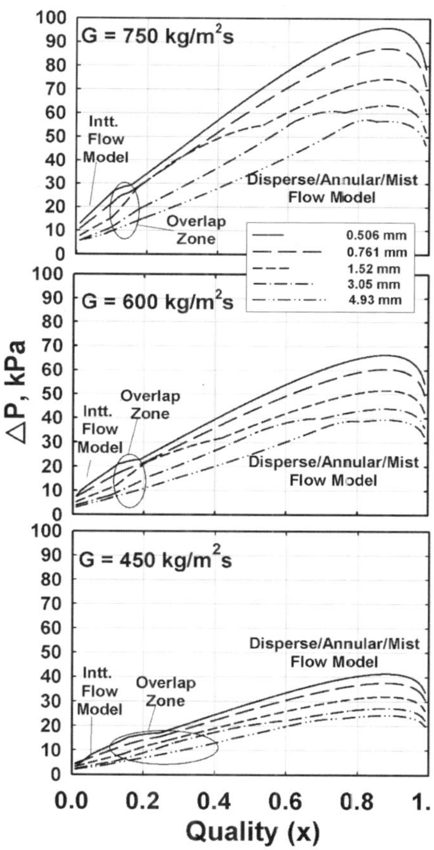

Figure 13: Effect of Hydraulic diameter on Condensation Pressure Drop.

4. Heat Transfer Coefficients

4.1 EXPERIMENTS

Garimella and Bandhauer [32] conducted heat transfer experiments using the same test sections that were used for the pressure drop experiments of Garimella *et al.* [24, 25, 27, 28] described above. The high heat transfer coefficients and low mass flow rates in microchannels necessitate modifications to the test facility and test procedures described above. For the small Δx required in the test section, the heat duties at the mass fluxes of interest are relatively small. Calculating this heat duty from the test section inlet and outlet quality measurements would result in considerable uncertainties because this would involve the difference between two similar quality values. Therefore, the heat duty must be measured on the coolant side, which must in turn be based on large ΔTs to minimize uncertainties. Ensuring large ΔTs requires low coolant flow rates. However, this makes the coolant-side thermal resistance the dominant resistance in the test section, making it difficult to deduce the refrigerant-side resistance from the measured test section *UA*.

These conflicting requirements for the accurate measurement of heat duty and the refrigerant heat transfer coefficients were resolved by developing a thermal amplification technique that decoupled these two issues. Thus, as shown in Figure 14, the test section was cooled using water flowing in a closed (primary) loop at a high flow rate to ensure that the condensation side presented the governing thermal resistance. Heat exchange between this primary loop and a secondary cooling water stream at a much lower flow rate was used to obtain a large temperature difference, which was in turn used to measure the condensation duty. The secondary coolant flow rate was adjusted as the test conditions change to maintain a reasonable ΔT and also small condensation duties in the test section.

Multiple layers of low conductivity phenolic foam insulation and small temperature differences between the primary coolant and the ambient minimized the heat loss from the primary loop to the ambient. Also, the heat addition to this loop was minimized by using a recirculation pump with an extremely low heat dissipation rate, which was calculated from the pump curves supplied by the manufacturer. With the pump heat dissipation and the ambient heat loss being small fractions of the secondary loop duty, the test section heat load was relatively insensitive to these losses and gains. Local heat transfer coefficients were therefore measured accurately in small increments for the entire saturated vapor-liquid region. Additional details of this thermal amplification technique are provided in the paper by Garimella and Bandhauer [32].

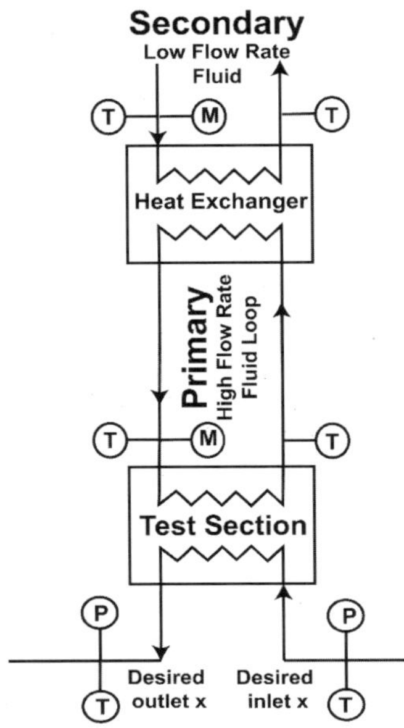

Figure 14: Thermal Amplification Technique.

4.2 ANALYSIS

The thermal amplification provided by this technique resulted in uncertainties typically as low as ±2% in the measurement of the secondary loop heat duty even at heat transfer rates less than *200 W*. Combining the errors in the secondary loop duty, the pump heat addition, and the ambient heat loss (even with a highly conservative uncertainty of ±50% assumed for these terms), the test section heat duty was typically known to within a maximum uncertainty of ±10%. With the refrigerant heat duty known, the condensation heat transfer coefficient was determined using the applicable thermal resistances. The coolant-side resistance was determined from correlations available in the *Handbook of Single-Phase Convective Heat Transfer* [33]. The large coolant flow rate and the enhancement in surface area (indirect area of about 4.7 times the direct area at an efficiency of 73.2%) provided by the coolant port walls on both sides of the microchannel tube resulted in high refrigerant-to-coolant resistance ratios (between 5 and 30). With this high resistance ratio, even an uncertainty of ±25% in the tube-side heat transfer coefficient did not appreciably affect the refrigerant-side heat transfer coefficient. For much of the data on circular and noncircular microchannels in this study, the uncertainties in condensation heat transfer coefficients were within about 20%. Representative heat transfer coefficients deduced from the measured data for a square microchannels with $D_h = 0.76$ are shown in Figure 15, along with the respective uncertainties.

4.3 MODELING

Development of flow regime-based models for heat transfer coefficients is underway. Essentially, the data are divided into the applicable flow regimes, along with regions of overlap between multiple flow mechanisms. The heat transfer coefficients are then modeled according to the applicable flow mechanism, with the associated pressure drop model providing the basis for flow-related parameters such as the interfacial shear stress, for example, in the annular flow regime. Incorporation of the pressure drop data from these very channels, rather than conventional models for larger tubes, into the heat transfer models is yielding better representation of the heat transfer data, and higher accuracies in the models. In addition, interpolations between models for adjacent flow regimes are resulting in smooth transitions in heat transfer coefficients through the overlap regions. Preliminary versions of these models are available in Bandhauer [34].

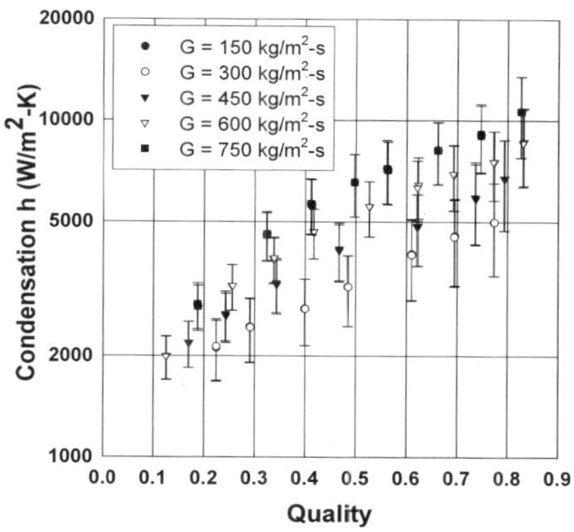

Figure 15: Condensation h as a Function of G, x (D_h = *0.76 mm*).

5. Conclusions

An overview of a comprehensive program for the investigation of condensation flow patterns, pressure drop and heat transfer in microchannels was provided in this paper. The study included circular and noncircular channels ranging in hydraulic diameter from *0.4 mm* to *4.91 mm* for mass fluxes between *150 kg/m²-s* and *750 kg/m²-s*. Unique techniques that permitted visualization of the flow patterns during the condensation process provide a thorough understanding of the flow mechanisms over this wide range of mass fluxes across the vapor-liquid dome for a variety of tube shapes and hydraulic diameters. Flow regime maps developed from these video recordings showed that transitions between the various flow regimes occur at different conditions in microchannels than what would be expected from maps for conventional geometries available in the literature. Specifically, it was found that the extent of the intermittent flow regime increases as the hydraulic diameter decreases, signifying an increasing influence of surface tension at the small diameters. Also, the wavy flow regime progressively decreases and disappears as the diameter decreases, giving way to the annular flow regime, signifying a diminishing influence of gravitational forces at the small diameters. The effect of changing tube shapes (round, square, rectangular with different aspect ratios) was also documented. Tube shape, however, was found to be less significant than hydraulic diameter in determining the applicable condensation flow pattern.

Pressure drop measurements on a multitude of circular and noncircular tubes across the vapor-liquid dome were used, in conjunction with the insights from the flow visualization studies, to develop a multiple-flow-regime, experimentally validated model for condensation pressure drop. The intermittent flow pressure drop model (also shown to apply to discrete-wave flow) treats the overall pressure drop as a combination of the contributions due to the liquid slug, the film-bubble interface region, and the transitions between the slug and the bubble. A slug frequency model was used to provide closure to the intermittent flow model. In the annular flow pressure drop model (also shown to apply to disperse-wave and mist flows), the interfacial friction factor derived from the measured pressure drops was correlated in terms of the corresponding liquid-phase Reynolds number and friction factor, the Martinelli parameter, and a surface tension-related parameter. Appropriate interpolation techniques were specified to address the regions of overlap and transition between the different regimes. The resulting model predicted 82% of the annular flow pressure drop data within ±20%. It was also shown that at the same mass flux, quality and *L/D*, the two-phase pressure drop increases as the tube diameter decreases.

Measurements of heat transfer coefficients for many circular and noncircular geometries over a wide range of conditions were also taken. A novel thermal amplification technique was developed to enable the accurate measurement of these coefficients in spite of the low heat transfer rates at the small flow rates and quality increments under consideration. The technique allowed accurate measurement of heat transfer rates while ensuring that the governing thermal resistance was on the condensation side, leading to low uncertainties in the heat transfer coefficients.

Development of additional models for pressure drop in noncircular channels, and for heat transfer coefficients and transition criteria based on nondimensional parameters is underway. This integrated approach using flow visualization, pressure drop and heat transfer measurements, and analytical modeling, is yielding a comprehensive understanding of condensation in microchannels.

Nomenclature

A, a, b, c	curve-fit constants and exponents
dP/dx	pressure gradient
D	diameter
D_h	hydraulic diameter
f	friction factor (Darcy)
G	mass flux
j	superficial velocity
L	channel length
Re	Reynolds number
U	velocity
UA	overall heat conductance
V_g	gas phase velocity
x	vapor quality
X	Martinelli parameter

Greek Symbols

α	channel aspect ratio, void fraction
ΔP	pressure drop
ΔT	temperature difference
Δx	change in quality
ρ	density
ψ	surface tension parameter
μ	viscosity
σ	surface tension

Subscripts/Superscripts

f/b	film-bubble interface
g	gas phase
i	interface
l	liquid
$slug$	slug basis
$test,i$	test section inlet
$test,o$	test section outlet

References

1. Coleman, J. W., and Garimella, S. (1999) Characterization of Two-Phase Flow Patterns in Small Diameter Round and Rectangular Tubes, *International Journal of Heat and Mass Transfer*, Vol. 42(15), pp. 2869-2881.
2. Baker, O. (1954) Simultaneous Flow of Oil and Gas, *Oil and Gas Journal*, Vol. 53, pp. 185-195.
3. Alves, G. E. (1954) Cocurrent Liquid-Gas Flow in a Pipe-Line Contactor, *Chemical Engineering Progress*, Vol. 50(9), pp. 449-456.
4. Govier, G. W., and Omer, M. M. (1962) The Horizontal Pipeline Flow of Air-Water Mixtures, *Canadian Journal of Chemical Engineering*, pp. 93-104.
5. Taitel, Y., and Dukler, A. E. (1976) A Model for Predicting Flow Regime Transitions in Horizontal and near Horizontal Gas-Liquid Flow, *AIChE Journal*, Vol. 22(1), pp. 47-55.

6. Suo, M., and Griffith, P. (1964) Two-Phase Flow in Capillary Tubes, *Journal of Basic Engineering*, Vol. 86, pp. 576-582.
7. Barnea, D., Luninski, Y., and Taitel, Y. (1983) Flow Pattern in Horizontal and Vertical Two Phase Flow in Small Diameter Tubes, *Canadian Journal of Chemical Engineering*, Vol. 61(5), pp. 617-620.
8. Damianides, C. A., and Westwater, J. W. (1988) Two-Phase Flow Patterns in a Compact Heat Exchanger and in Small Tubes, *Second UK National Conference on Heat Transfer (2 vols)*, Glasgow, Scotland, pp. 1257-1268.
9. Fukano, T., Kariyasaki, A., and Kagawa, M. (1989) Flow Patterns and Pressure Drop in Isothermal Gas-Liquid Concurrent Flow in a Horizontal Capillary Tube, *Proceedings of the 1989 ANS National Heat Transfer Conference*, Philadelphia, Pennsylvania, Vol. 4, pp. 153-161.
10. Mandhane, J. M., Gregory, G. A., and Aziz, K. (1974) A Flow Pattern Map for Gas-Liquid Flow in Horizontal Pipes, *International Journal of Multiphase Flow*, Vol. 1(4), pp. 537-553.
11. Weisman, J., Duncan, D., Gibson, J., and Crawford, T. (1979) Effects of Fluid Properties and Pipe Diameter on Two-Phase Flow Patterns in Horizontal Lines, *International Journal of Multiphase Flow*, Vol. 5(6), pp. 437-462.
12. Wilmarth, T., and Ishii, M. (1994) Two-Phase Flow Regimes in Narrow Rectangular Vertical and Horizontal Channels, *International Journal of Heat and Mass Transfer*, Vol. 37(12), pp. 1749-1758.
13. Hosler, E. R. (1968) Flow Patterns in High Pressure Two-Phase (Steam-Water) Flow with Heat Addition, *AIChE Symposium Series*, Vol. 64, pp. 54-66.
14. Jones, J., Owen C., and Zuber, N. (1975) The Interrelation between Void Fraction Fluctuations and Flow Patterns in Two-Phase Flow, *International Journal of Multiphase Flow*, Vol. 2(3), pp. 273-306.
15. Lowry, B., and Kawaji, M. (1988) Adiabatic Vertical Two-Phase Flow in Narrow Flow Channels, *Heat Transfer - Houston 1988, Papers Presented at the 25th National Heat Transfer Conference*, Houston, TX, USA, Vol. 84, Publ by AIChE, New York, NY, USA, pp. 133-139.
16. Richardson, B. L. (1959) *Some Problems in Horizontal Two-Phase Two-Component Flow*, Ph.D. Thesis, *Mechanical Engineering*, Purdue University, West Lafayette.
17. Troniewski, L., and Ulbrich, R. (1984) Two-Phase Gas-Liquid Flow in Rectangular Channels, *Chemical Engineering Science*, Vol. 39(4), pp. 751-765.
18. Wambsganss, M. W., Jendrzejczyk, J. A., and France, D. M. (1991) Two-Phase Flow Patterns and Transitions in a Small, Horizontal, Rectangular Channel, *International Journal of Multiphase Flow*, Vol. 17(3), pp. 327-342.
19. Wambsganss, M. W., Jendrzejczyk, J. A., and France, D. M. (1994) Determination and Characteristics of the Transition to Two-Phase Slug Flow in Small Horizontal Channels, *Journal of Fluids Engineering, Transactions of the ASME*, Vol. 116(1), pp. 140-146.
20. Wang, C. C., Chiang, C. S., Lin, S. P., and Lu, D. C. (1997) Two-Phase Flow Pattern for R-134a inside a 6.5-mm (0.25-in.) Smooth Tube, *Proceedings of the 1997 ASHRAE Winter Meeting, Jan 26-29 1997*, Philadelphia, PA, USA, Vol. 103, ASHRAE, Atlanta, GA, USA, pp. 803-812.
21. Coleman, J. W., and Garimella, S. (2000) Visualization of Two-Phase Refrigerant Flow During Phase Change, *Proceedings of the 34th National Heat Transfer Conference*, Pittsburgh, PA, Vol. NHTC 2000-12115, ASME.
22. Coleman, J. W., and Garimella, S. (2000) Two-Phase Flow Regime Transitions in Microchannel Tubes: The Effect of Hydraulic Diameter, *American Society of Mechanical Engineers, Heat Transfer Division*, Orlando, FL, Vol. HTD-366, American Society of Mechanical Engineers, pp. 71-83.
23. Coleman, J. W., and Garimella, S. (2003) Two-Phase Flow Regimes in Round, Square and Rectangular Tubes During Condensation of Refrigerant R134a, *International Journal of Refrigeration*, Vol. 26(1), pp. 117-128.
24. Garimella, S., Killion, J. D., and Coleman, J. W. (2003) An Experimentally Validated Model for Two-Phase Pressure Drop in the Intermittent Flow Regime for Noncircular Microchannels, *Journal of Fluids Engineering*, Vol. 125(5), pp. 887-894.

25. Garimella, S., Killion, J. D., and Coleman, J. W. (2002) An Experimentally Validated Model for Two-Phase Pressure Drop in the Intermittent Flow Regime for Circular Microchannels, *Journal of Fluids Engineering*, Vol. 124(1), pp. 205-214.
26. Churchill, S. W. (1977) Friction-Factor Equation Spans All Fluid-Flow Regimes, *Chemical Engineering Progress*, Vol. 84(24), pp. 91-92.
27. Garimella, S., Agarwal, A., and Coleman, J. W. (2003) Two-Phase Pressure Drops in the Annular Flow Regime in Circular Microchannels, *21st IIR International Congress of Refrigeration*, Washington, DC, Vol. ICR0360, International Institute of Refrigeration.
28. Garimella, S., Agarwal, A., and Killion, J. D. (2004) Condensation Pressure Drops in Circular Microchannels, *Proceedings of the Second International Conference on Microchannels and Minichannels (ICMM2004)*, Rochester, NY, United States, American Society of Mechanical Engineers, New York, NY 10016-5990, United States, pp. 649-656.
29. Dukler, A. E., and Hubbard, M. G. (1975) A Model for Gas-Liquid Slug Flow in Horizontal and near Horizontal Tubes, *Ind. Eng. Chem. Fundamentals*, Vol. 14(4), pp. 337-347.
30. Baroczy, C. J. (1965) Correlation of Liquid Fraction in Two-Phase Flow with Applications to Liquid Metals, *Chemical Engineering Progress Symposium Series*, Vol. 61(57), pp. 179-191.
31. Lee, H. J., and Lee, S. Y. (2001) Pressure Drop Correlations for Two-Phase Flow within Horizontal Rectangular Channels with Small Heights, *International Journal of Multiphase Flow*, Vol. 27(5), pp. 783-796.
32. Garimella, S., and Bandhauer, T. M. (2001) Measurement of Condensation Heat Transfer Coefficients in Microchannel Tubes, *2001 ASME International Mechanical Engineering Congress and Exposition*, New York, NY, United States, Vol. 369, American Society of Mechanical Engineers, pp. 243-249.
33. Kakaç, S., Shah, R. K., and Aung, W. (1987) *Handbook of Single-Phase Convective Heat Transfer*, Wiley, New York, NY.
34. Bandhauer, T. M. (2002) *Heat Transfer in Microchannel Geometries During Condensation of R134a*, Master of Science Thesis, *Mechanical Engineering*, Iowa State University, Ames, IA, p. 201.

HEAT TRANSFER CHARACTERISTICS OF SILICON FILM IRRADIATED BY PICO TO FEMTOSECOND LASERS

J. S. LEE* and S. PARK**
*School of Mechanical & Aerospace Engineering-Seoul National University
**Department of Mechanical & System Design Engineering-Hongik University
Seoul, Korea

1. Introduction

Modern micro-fabrication technologies are shrinking electronic components and devices into micro- and nanometer scales [1]. The technology of miniaturization has brought an unprecedented upsurge of research interests in microscale heat transfer. Microscale heat transfer phenomena are often encountered in a number of applications such as microelectronic devices, micro-fabrication, MEMS, NEMS, biotechnology and so on. The reduced devices in size have increased the importance of understanding microscale heat transfer processes over a wide temperature range confronted in processing and operation of these devices. Since the 1980s, in particular, lasers of pulse durations ranging from picoseconds to femtoseconds have been developed rapidly and commercially available [2]. Since short-pulse laser heating is capable of controlling the heating location and depth precisely, and realizing high heating/cooling rates, it has made tremendous impacts on many applications, e.g. fabrication of sophisticated microstructures [3], syntheses of advanced materials [4, 5], measurement of thin-film properties [6], laser annealing and damage [7, 8], and electronic desorption process using ultra-short pulse lasers [9~11]. These abundant applications require a better understanding of the energy transport phenomena during fast laser heating from a microscopic point of view.

Figure 1 illustrates a distinct feature between thermal and non-thermal phenomena during the drilling process of a thin steel foil by ultra-short pulse lasers. This is due to thermal non-equilibrium between energy carriers (electron-hole pairs in a semiconductor material) and lattice phonons when the pulse duration is extremely short. The energy relaxation time of carriers to phonons is typically in the range of 0.1 to 0.5 ps for silicon, necessary to transport the carrier energy to phonons. When the laser pulse duration time, however, is on the order of or shorter than the energy relaxation time, a substantial non-equilibrium occurs between the carrier and lattice temperatures, and then the carriers are extremely hot compared to the lattice because of a considerable amount of energy deposited on carriers within ultra-short time duration [11].

In order to effectively analyze the microscale heat transfer mechanisms and to accurately model the ultra-short pulse heating of materials, it is necessary to understand energy absorption, transport, and storage phenomena in detail. The primary laser–solid interaction process is the excitation of electrons from their equilibrium states to some excited states by absorption of photons. Dephasing processes take place in a very short time of about 10^{-14} s. The occupation of these primary states is rapidly changed by carrier–carrier interaction processes, and a quasi-equilibrium situation is established among the electrons on a time scale of about 10^{-13} s. The energy distribution of the carriers over the available states is described by the Fermi–Dirac distribution with an electron temperature, which is higher than the lattice temperature. The quasi-equilibrium electrons cool down on a time scale of 10^{-13} to 10^{-12} s by emission of phonons, which leads to the population of certain phonon modes. These phonons relax predominantly by inharmonic interaction with other phonon modes. The final stage of the thermalization process is the redistribution of the phonons over the entire Brillouin zone according to the Bose–Einstein distribution. At this point the temperature of the laser-excited material can be defined, and the energy distribution is characterized by the temperature. After the thermalization, the spatial distribution of the energy can be characterized by the temperature profile. Under these conditions thermal diffusion can take place on a time scale of the order of 10^{-11} s. The details depend on the thermal transport coefficients and the optical properties of the material. When a sufficient amount of energy is deposited in the material, the melting temperature is eventually reached, and a transition from the solid to the liquid state takes place. An important conclusion to be drawn from the consideration of the time scale of energy thermalization is that there is a distinct

S. Kakaç et al. (eds.), Microscale Heat Transfer, 291–302.

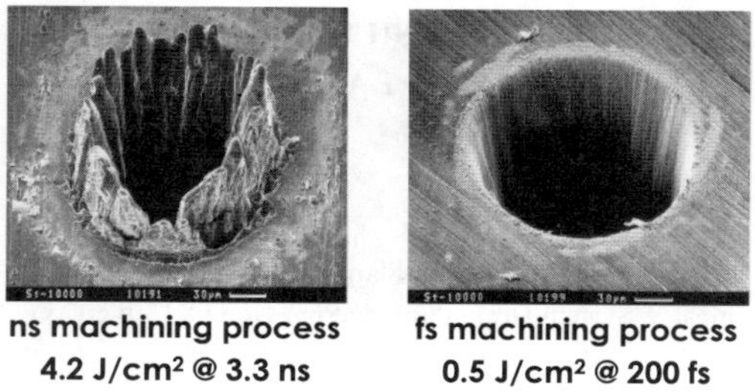

Figure 1 Drilling of a thin steel foil by ultra-fast lasers with pulse durations of nanoseconds and femtoseconds

dividing line at about 10^{-12} s separating the regimes of non-thermal and thermal processes.

The two-temperature equation is used to characterize mutual interactions among lattice temperature, and number density and temperature of carriers during pico- to femtosecond pulse laser processing [12]. In this study, a new parameter related with non- equilibrium durability is introduced and its characteristics for various laser pulses and fluences are discussed.

In the investigation of non-thermal damage to dielectrics, the Fokker-Planck equation is applied to describe the transient behaviors of electron densities, and to predict the damage threshold fluences for various laser pulse widths ranging from 10 femtoseconds to 10 picoseconds [13]. This model includes the effects of electron avalanche and multiphoton ionization on the generation of electrons.

Kang et al. [14] extended the above study to examine the spatial effect on dielectric optical breakdown by using the multivariate Fokker-Planck equation, which is modeled in terms of the electron energy distribution function. The multivariate Fokker-Planck equation is derived by adding generation and recombination source terms, and Joule heating term to kinetic equation which can describe charged particle transport in gases.

2. Transport Modeling

The spectrum of length and time scales involved in the micro/nanoscale transport is so wide that the development of a model that covers the whole range is a formidable task. In addition, the energy carriers such as electrons (holes), phonons, and photons have distinctive transport characteristics, and both quantum and classical mechanics are to be taken into account. Because of the complexity, transport models have been developed according to the confined time and length scales of interest as illustrated in Fig. 2.

In principle, the Boltzmann transport equation (BTE) can cover the regime where the length and time scales are larger than carrier mean free time τ and mean free length Λ. However, tremendous computational efforts are required in practice when the system length scale L and the process time scale t are getting larger. The BTE is, thus, usually

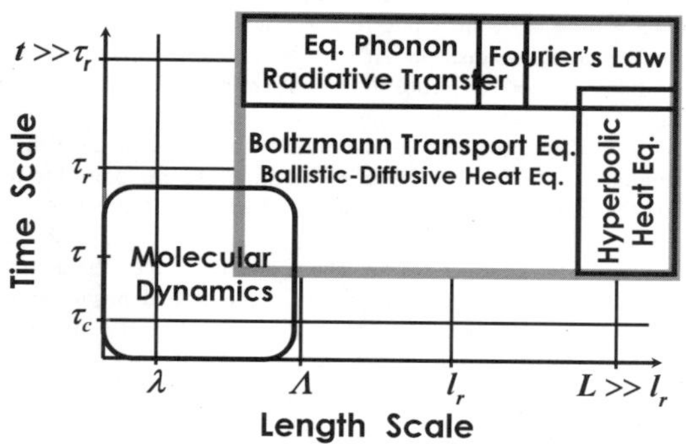

Figure 2 Regime map of transport modeling

attempted when the phenomena is strictly non-local in both space and time that the full form of statistical transport equations should be used. When the system length scale is on the same order of carrier wavelength λ, wave phenomena such as diffraction, tunneling, and interference are important. In this case, wave optics based on Maxwell's equation should be considered for the photon transport, and quantum transport laws for electrons and phonons. In the case that L is on the same order of carrier mean free path Λ, and the time scale is much larger than mean free time τ or relaxation time τ_r, transport is ballistic in nature, and thus, thermodynamic equilibrium cannot be defined. The transport is non-local in space but time-averaged statistical particle transport equations can be applied such as the equation phonon radiative transfer. On the other hand, when L is much larger than Λ or l_r, the length scale corresponding to the relaxation time, but the process takes place on a time scale of τ or τ_r, the local thermodynamic equilibrium can be assumed but the time-dependent terms cannot be averaged. In this regime, the hyperbolic type heat equation can be used.

3. Two Temperature Equation Approach

Actually, some of important parameters such as carrier number density, and lattice and carrier temperatures, which are closely connected one another during rapid or slow heating process by the ultra-short pulse lasers, are difficult to measure experimentally because the considered scales of length and time are usually smaller than the micrometers and picoseconds, respectively. Therefore, the computational modeling offers a promising alternative to obtain a detailed information on the microscale heat transfer characteristics. Qui and Tien [15, 16] have performed a numerical simulation on interactions between electrons and lattice phonons by simplifying the scattering terms in Boltzmann transport equations through the quantum analysis. They show that the excited carriers are no longer in thermal equilibrium with the other carriers, creating a non-equilibrium heating situation, when the laser pulse width is shorter than approximately five times the electron energy relaxation time.

The self-consistent theoretical models based on the Boltzmann transport theory are used to characterize the microscale heat transfer mechanism by explaining mutual interactions among lattice temperature, and number density and temperature of carriers [12]. Especially, a new parameter related with non-equilibrium durability is introduced and its characteristics for various laser pulses and fluences are discussed. This study also investigates the temporal characteristics of carrier temperature distribution, such as the one- and two-peak structures, according to laser pulses and fluences, and establishes a regime criterion between one-peak and two-peak structures for picosecond laser pulses.

As for the carrier temperature much higher than the lattice temperature owing to the ultra-short pulse laser heating, especially, a consistent theoretical model should be developed to be able to mimic the non-equilibrium between carriers and lattice. In the case that photons incident on semiconductor have energy greater than the band gap energy of the material, the main heat carrier is an electron-hole pair, whereas it is the free electron for metals [11, 17, 18]. The carrier number density of electron-hole pairs, N_C, is determined from the following conservation law given by

$$\frac{\partial N_C}{\partial t} = \frac{\alpha_1 I}{h\nu} - \gamma N_C^3 \tag{1}$$

where t is the time, γ the Auger recombination rate, N_C the carrier number density, h the Planck constant, I the laser intensity as a function of time and space, ν the photon frequency, and α_1 is the band-to-band absorption coefficient of one photon. The temporal change of the carrier number density is governed by two terms on the right-hand side of Eq. (1). One of which is the absorption source term that corresponds to direct transition that excites an electron to the conduction band and creates an electron-hole pair, and the other of which gives the loss of the carriers through Auger recombination process in which the free electrons are captured by ionized donors and lose their energy non-radiatively [19].

Since the size of the laser beam is large compared to the laser penetration depth, the short-pulse laser heating of materials can be modeled as one-dimensional. A set of consistent models for the non-equilibrium thermal system is composed of two parts, one of which is the carrier transport equation and the other the lattice one as follows [17, 20];

$$\frac{\partial U_C}{\partial t} = \frac{\partial}{\partial x}\left(k_C \frac{\partial T_C}{\partial x}\right) - \frac{3N_C k_B}{\tau_C}\left(T_C - T_L\right) + \alpha I \tag{2}$$

$$\frac{\partial U_L}{\partial t} = \frac{\partial}{\partial x}\left(k_L \frac{\partial T_L}{\partial x}\right) + \frac{3N_C k_B}{\tau_C}\left(T_C - T_L\right) \tag{3}$$

where U is the internal energy, x the spatial coordinate, k the thermal conductivity, k_B the Boltzmann constant, τ_C the carrier-to-phonon energy relaxation time, T the temperature, and α is the absorption coefficient. On the right-hand side of Eq. (2), the first term is the heat diffusion due to carriers, the second represents the energy transfer from carriers to phonons during τ_C, and the third term is a source term due to photon absorption. Note that the assumption of Fourier law is not really valid for short time scale studies, but this is an approximation that invokes closure in the present type of hydrodynamic equation. The non-equilibrium between the optical and acoustic phonons is not considered because the lattice is assumed to be a single thermodynamic system [21]. Under the above assumptions, the internal energies of carrier and lattice can be described, respectively, as

$$U_C = N_C E_g + C_C T_C \tag{4}$$

$$U_L = C_L T_L \tag{5}$$

where E_g is the band gap energy and C is the heat capacity.

Figure 3 shows the time evolution of carrier and lattice temperatures at the film surface for different laser pulses when the laser fluence $J = 3.82$ mJ/cm^2 and wavelength $\lambda = 790$ nm [12]. As expected, the carrier temperature rapidly increases compared to the lattice temperature. This non-equilibrium phenomenon is due mainly to the difference between energy relaxation times and laser pulse durations. Especially, the carrier temperature is remarkably increased in the early stage of laser exposure because the carrier heat capacity is several orders of magnitude smaller than that of the lattice. As a matter of fact, the carrier heat capacity of silicon is approximately 10^{-5} J/m^3-K at 300K [19], but its magnitude proportionally increases with the number density of carriers. As the carrier number density increases, so does the heat capacity of carriers. In the non-equilibrium state, the carriers do not lose their energy to the lattice within a certain time period, and then approach their maximum values. From Figs. 3(a) and 3(b), it should be noted that at a very early time stage, the laser pulse is mainly providing energy, whereas the number density of carriers is not significantly increasing. The carrier temperature is gradually decreasing just after reaching its maximum value, indicating that the carriers begin to transfer their energies to the lattices. As can be seen in Fig. 3(b), the number density of carriers increases with decreasing laser pulse duration. However, the maximum of the carrier number density occurs near $t/t_p = 0$, and after then the number density remains constant with time, indicating that the Auger recombination rate is small in the case when the laser fluence is relatively small. This means that the high-density carriers are present on a long time scale in the plasma for small laser fluence, as consistent with the earlier observation [17].

In Fig. 4(a) which shows carrier and lattice temperature distributions for $\lambda = 530$ nm, the carrier temperature drastically changes with laser pulse durations when the laser fluence is 50 mJ/cm^2 [12]. An interesting to note is that a two-peak structure appears in the carrier temperature variation because of heating the plasma by both the laser pulse and rapid Auger recombination as pointed out by van Driel [17]. The rapid decrease in the number density is observed in Fig. 4(b) owing to the Auger recombination in which band-to-band recombination or trapping at a band gap center occurs simultaneously with collision between two like-carriers [19]. The energy released by recombination is transferred to the surviving carrier. Subsequently, this highly energetic carrier loses its energy in small time steps through collisions with the semiconductor lattice. As can be seen in Figs. 4(a) and 4(b), the carrier temperature rises again near $t/t_p = 0$ at which the number density increases. Due to the Auger recombination, the ionization energy of the

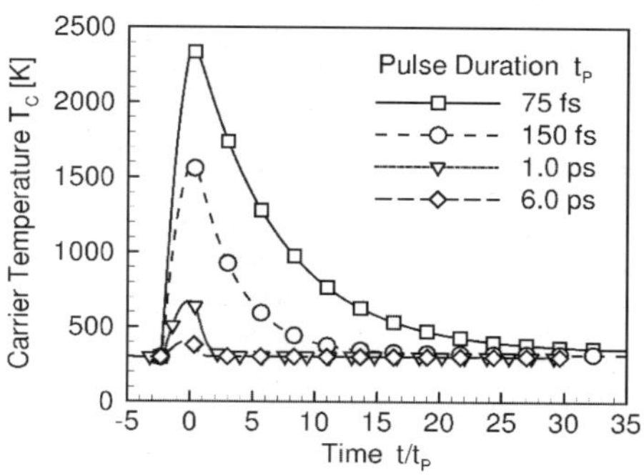

(a)

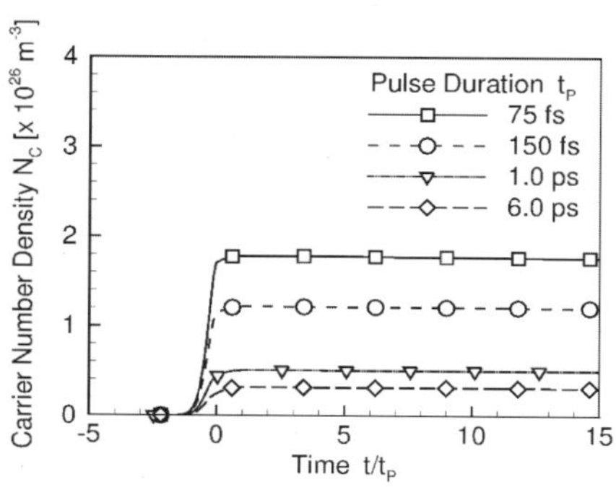

(b)

Figure 3 (a) carrier temperatures and (b) carrier number densities for different laser pulses at the silicon layer front surface when $\lambda = 790$ nm and $J = 3.82$ mJ/cm^2 [12]

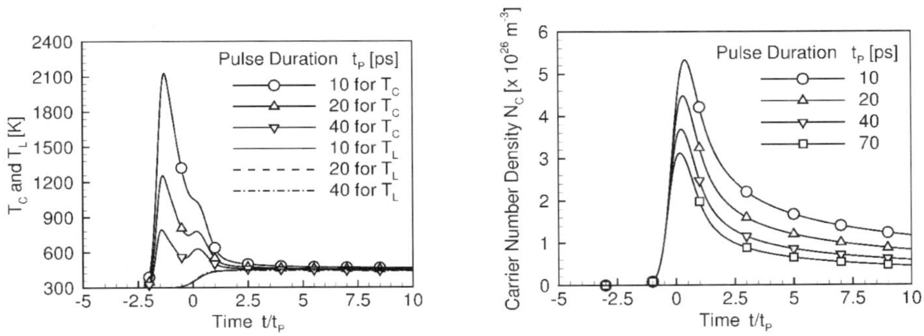

(a) (b)

Figure 4 (a) carrier and lattice temperatures and (b) carrier number densities for different laser pulses at the silicon layer front surface when λ = 530 nm and J = 50 mJ/cm^2 [12]

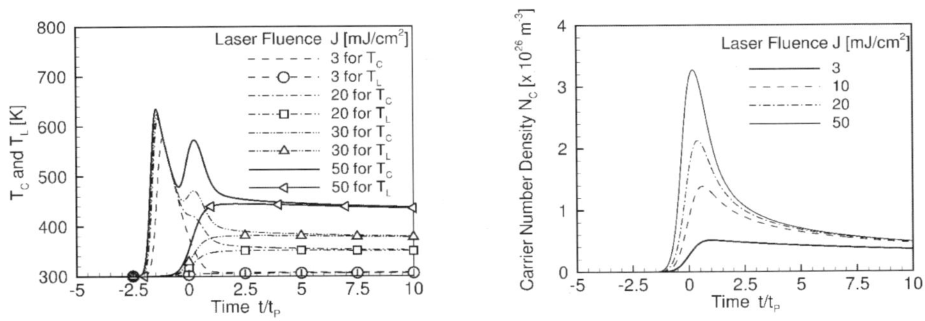

(a) (b)

Figure 5 (a) carrier and lattice temperatures and (b) carrier number densities for different laser fluences at the silicon layer front surface when λ = 530 nm and t_p = 60 ps [12]

carrier is converted into kinetic energy at a sufficiently rapid rate, which results in increase in the carrier temperature. Once the carrier number density begins to decrease, the temperature of carriers begins falling but the non-equilibrium between the carriers and lattice temperatures maintains for long times owing to the on-going Auger recombination. It seems that the occurrence of the second peak due to Auger heating depends highly on the laser pulse duration because the carrier number density affects the Auger recombination process.

Figure 5 shows time evolutions of T_C, T_L, and N_C for different laser fluences when λ = 530 nm and t_p = 60 ps [12]. In Fig. 5(a), it should be noted that the two-peak structure disappears at relatively low laser fluences, whereas it becomes prominent at higher laser fluences. Actually, extremely high carrier temperature may be realized owing to the small heat capacity of the carriers compared with the lattices, and the heat capacity of the carrier increases proportionally with the carrier number density which is strongly affected by the laser fluence. However, it is found in Figs. 5(a) and 5(b) that the first peaks of carrier temperature have already reached before the carrier number density begins to increase rapidly at the early stage of the laser incidence. In addition, the peak values of carrier temperature are hardly affected by the laser fluence values, consistent with the results of van Driel [17]. On the contrary, the lattice temperature is hardly affected by the laser pulse, whereas it gradually increases with the laser fluence as shown in Fig. 5(a). The latter indicates that the lattice temperature may be controllable by changing the laser intensity irradiated on the silicon film.

Meanwhile, it can be seen in Fig. 5(b) that the carrier number density increases drastically near t/t_p = -1.0 with increasing laser fluence, and approaches its maximum value near t/t_p = 0. At this moment the laser intensity has its maximum and the second peak of carrier temperature exhibits due to the Auger heating. Figure 5(b) shows that the laser fluence has an effect on the transient behavior of the carrier number density distribution. That is, the carrier number density increases and its decaying rate also increases owing to the Auger recombination with the laser fluence. In particular, at J = 3 mJ/cm^2 the carrier number density does not vary with time when the laser fluence is relatively small. This shows that Auger heating effect is relatively small compared to other cases and consequently the second peak of carrier temperature does not appear as seen in Fig. 5(a). According to the consideration of van Driel [17], the small variation of the carrier number density with time may be attributed to the presence of high-density carriers on a long time scale during which a significant portion of the total deposited laser energy remains in the plasma. This effect is more significant at a lower laser fluence. Thus, it is concluded that as the laser fluence becomes larger, the Auger recombination occurs faster during irradiation.

Figure 6 represents the maximum values of T_C, T_L, and N_C according to the variations of the laser fluence [12]. When the laser fluence varies from 3 mJ/cm^2 to 50 mJ/cm^2, the lattice temperature increases by about 44.5 %, whereas the carrier temperature by about 10.1 %. This indicates that the temperature difference between carriers and lattices, which can be a measure of non-equilibrium level, becomes smaller as the laser fluence increases. From this fact, it can be said that there exists a deep interrelationship between the non- equilibrium durability and the laser fluence.

An appropriate regime map for the distinction between one- and two-peak structures of the carrier temperature is established for picosecond laser pulses as shown in Fig. 7 [12]. A regime criterion is found from the predicted results under various situations where the laser fluence varies from 2 mJ/cm^2 to 100 mJ/cm^2 and the pulse duration time ranges from 10 ps to 80 ps when λ = 530 nm. The symbols represent the regime boundaries between one- and two-peak structures. A useful relationship within the regime criterion can be expressed in terms of the pulse duration time and the incident laser intensity. When t_p is 20 ps, for instance, a two-peak structure begins to appear in the carrier temperature distributions for laser fluences larger than about 40 mJ/cm^2. It is helpful in determining the threshold condition at which the two-peak structure occurs and in understanding the interrelations between the non-equilibrium duration and the change in carrier temperature over time. As referred to above, the non-equilibrium state is associated deeply with the laser fluences and pulse duration times, which have an important effect on the changes in carrier temperature such as the one- and two-peak structures.

4. Fokker-Planck Equation Approach

The laser-induced breakdowns are caused by three consequent mechanisms [22]: (i) the excitation of electrons in the conduction band by impact and multiphoton ionization (MPI), (ii) radiation-induced heating of the conduction-band electrons, and (iii) transfer of the plasma energy to the lattice. The key benefit of ultra-short femtosecond laser pulses lies in their ability to deposit energy in materials in a very short time interval. Heat diffusion is frozen during the interaction and the shock-like energy deposition leads to ablation for ultra-short pulses. This is because the pulse

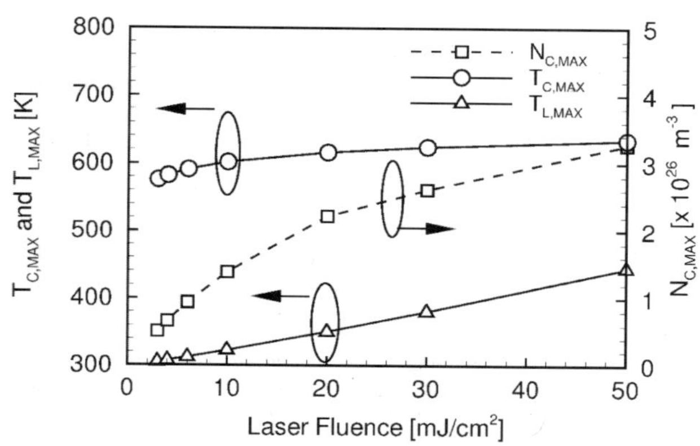

Figure 6 Maximum values of T_C, T_L, and N_C for different laser fluences when λ = 530 nm and t_p = 60 ps [12]

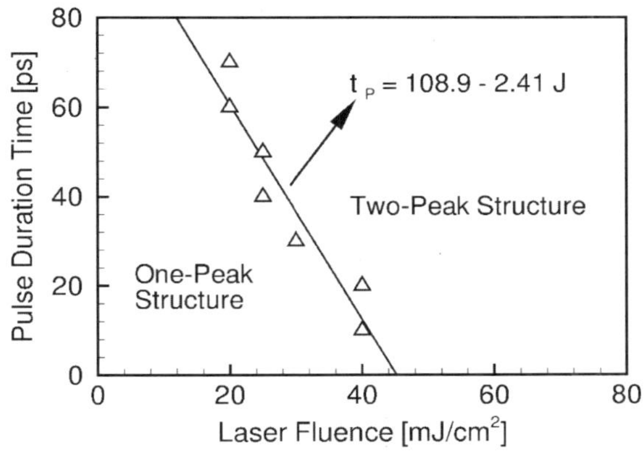

Figure 7 Regime map for the distinction between one- and two-peak structures delineated when $\lambda = 530$ nm for picosecond laser pulses [12]

deposits its energy so quickly that it does not interact at all with the plume of vaporized material, which would distort and bend the incoming beam and produce a rough-edged cut. When ultra-short pulse lasers are used, the cut surfaces become very smooth and do not require subsequent cleanup, because only a very thin layer of material is removed during each pulse irradiation of the laser. Although previous breakdown experiments [23~25] were recently extended to the sub-picosecond regime, the characteristics of the avalanche and the role of MPI have remained controversial even up to now.

Laser-induced damage can be regarded as one of the limiting factors of the transmission and deposition of laser energy on solids. The energy transfer mechanism of laser-induced damage is of great importance to the development of high-intensity lasers. A large number of theoretical and experimental studies [23~27] have been conducted on laser-induced damage in dielectrics. Stuart et al. [24] reported extensive laser-induced damage threshold measurements on dielectric materials at wavelengths of 1053 and 526 nm for pulse durations ranging from 140 fs to 1 ns. They found that a gradual transition existed from the long-pulse, thermally dominated regime to an ablative regime dominated by collisional and multiphoton ionization, and plasma formation. In an investigation of interactions between laser-induced photons and energy carriers, Holway et al. [27] conducted simulations using the Fokker-Planck (F-P) equation. They took the non-polar deformation potential effects into consideration for the electron-phonon interaction in defining the collision frequencies for momentum. Lenzner et al. [22] measured the damage threshold fluence and the ablation depth in dielectric materials for laser pulse durations ranging from 5 ps to 5 fs. They demonstrated that sub-10-fs laser pulses opened up a way to reversible nonperturbative nonlinear optics at intensities greater than 10^{14} W/cm^2 slightly below damage threshold, and to nanometer-precision laser ablation in dielectric materials.

Most recently, extensive simulations are conducted to investigate the influence of laser pulse width on the transient behavior of electrons and the damage threshold fluence in SiO_2 dielectric material, and also to survey the role of the impact ionization and the MPI in laser-induced damage [13]. In addition, the damage characteristics are investigated when both alternating current electric fields and ultra-short pulse lasers are applied simultaneously to the dielectric materials. This hybrid process is planned to confirm the possibility that the laser fluence required for breakdown would be lowered when a very high electric field is imposed on the material, compared to that without electric fields.

Radiative energy of ultra-short pulse lasers ranging from pico to femtoseconds is absorbed much faster by the newly excited electrons than it is delivered from the electrons to the lattice. These electrons gain energy from the laser field until they have sufficient energy to collisionally ionize neighboring atoms thereby producing more free electrons. When laser is irradiated on materials, thermal effects should be considered in general. In ultra-short pulse laser applications, however, there would be no need to track the energy flow into the lattice to account for thermal and mechanical stresses, which is necessary for pulses longer than typically 50 ps [24]. As a matter of fact, it is possible to describe the plasma formation quantitatively by using the time dependence of the electron energy distribution function. Since the impact ionization rate depends highly on energy, the absorption rate of laser energy requires integrating over the electron energy distribution. When the electrons are strongly driven by intense laser pulses, the energy distribution can differ substantially from the Maxwellian. For a material having a band gap energy that is much larger than the single photon energy, the heating and collisional ionization of conduction electrons can be described by the F-P equation, which is well known to be able to effectively describe the electron avalanche phenomena induced by an ultra-short pulse laser.

The electron distribution function $f(\varepsilon,t)$ can be estimated from

$$\frac{\partial f(\varepsilon,t)}{\partial t}+\frac{\partial}{\partial \varepsilon}\left[V(\varepsilon,t)f(\varepsilon,t)-D(\varepsilon,t)\frac{\partial f(\varepsilon,t)}{\partial \varepsilon}\right]=\frac{\partial f(\varepsilon,t)}{\partial t}+\frac{\partial J(\varepsilon,t)}{\partial \varepsilon}=S(\varepsilon,t), \tag{6}$$

where ε is the electron kinetic energy. The number density of electrons with a kinetic energy between ε and $\varepsilon + d\varepsilon$ at time t is given by $f(\varepsilon,t)d\varepsilon$. The square bracket in Eq. (6) represents the change in the electron distribution due to Joule heating, inelastic scattering of phonons, and electron energy diffusion $D(\varepsilon,t)$. In particular, $V(\varepsilon,t)$ consists of the Joule heating $R_J(\varepsilon,t)$ and the electron energy dissipation to phonon, $U_p\gamma(\varepsilon)$, which is given by

$$V(\varepsilon,t)=R_J(\varepsilon,t)-U_p\gamma(\varepsilon), \tag{7}$$

where U_p is the characteristic phonon energy, and $\gamma(\varepsilon)$ denotes the rate at which electron energy is transferred to the lattice. The last term in Eq. (6) denoting the total electron source can be expressed as

$$S=R_{pi}(\varepsilon,t)+R_{imp}(\varepsilon,t) \tag{8}$$

where the first term on the right hand side represents the generation of free electrons due to the MPI and second one due to the impact ionization. The detail description of $R_J(\varepsilon,t)$, $R_{pi}(\varepsilon,t)$, and $R_{imp}(\varepsilon,t)$ can be found in the literature [13].

Figure 8 represents the effects of the MPI and the avalanche (impact) ionization on the electron number density at different laser pulse widths [13]. As the pulse width decreases, the electron number density increases obviously. The maximum electron density is about 10^{21} cm^{-3} at t_p = 0.01 ps and the difference between the cases with and without considering the avalanche ionization effect is also very small. For shorter pulse widths, the MPI alone generates sufficient electrons to cause damage effectively. In strong contrast with short pulse width, as the pulse width increases, the avalanche ionization process has a strong effect on the increase of electron number density, especially after $t/t_p = 0$. For the pulse width of 1.0 ps, the significant increase in the electron number is especially observed over time when both MPI and avalanche ionizations are considered. Meanwhile, the reason why the avalanche ionization becomes dominant for longer pulse durations is that the avalanche ionization depends highly on the pulse duration. Contrary to the avalanche ionization, the MPI is mainly affected by the laser characteristics such as the laser wavelength and the pulse shape, in addition to laser intensity and duration. Thus, these basic mechanisms need to be clarified for fundamental understandings of the laser-induced damage characteristics in dielectrics.

Damage threshold fluence is defined as the critical laser energy per unit area at which the plasma is formed. Some investigators [22~25] reported that the damage threshold fluence scales approximately as $t_p^{1/2}$ in the long-pulse limit and then changes to the short-pulse limit near 20 ps. They observed a deviation from the $t_p^{1/2}$ scaling of breakdown threshold fluence and an increasingly deterministic character of breakdown for t_p < 10 to 20 ps as opposed to longer

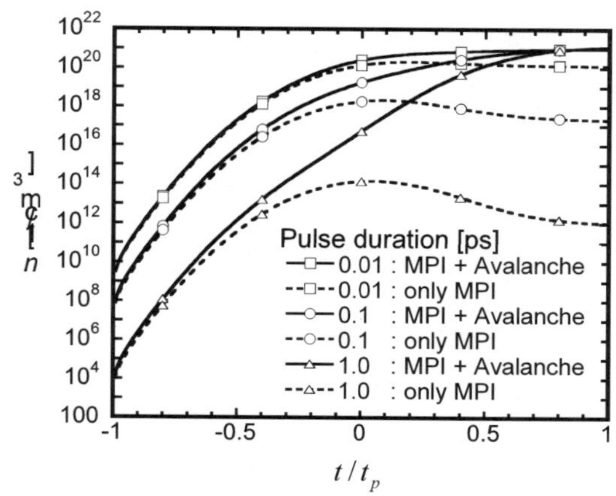

Figure 8 Influence of multiphoton and avalanche ionizations on electron number density at different pulse durations [13] for λ = 1053 nm and I_0 =10 TW/cm^2 [13]

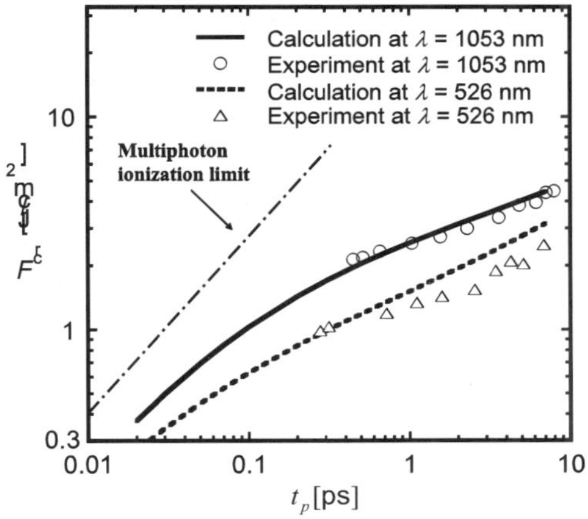

Figure 9 Comparisons of the predictions of damage threshold fluences with experimental data of Stuart et al. [24] at $\lambda = 1053$ and 526 nm for various pulse durations [13]

pulses. Figure 9 indicates that the calculated damage threshold fluences are in fairly good agreements with experimental observations of Stuart et al. [24] at λ = 1053 nm, whereas they are slightly over-predicted at λ = 526 nm [13]. For pulses shorter than 10 ps, the damage fluence no longer follows the $t_p^{1/2}$ dependence and there exists the 'frozen region' of heat diffusion or the non-thermal region due to the non-equilibrium state between electrons and phonons, unlike thermal damage in the long-pulse limit. It is also found in Fig. 9 that the damage fluence approaches to the multiphoton ionization limit as the pulse width decreases. Generally, as the damage threshold fluence becomes smaller in machining any material, the efficiency of that laser machining system does higher. It is thus evident that the damage threshold fluence would be one of the most important factors in fabricating the material with high precision and saving cost. It is also noted in Fig. 9 that the damage threshold fluence decreases with reducing laser pulse width. As previously indicated in Fig. 8, it is because of the drastic growth of electron number density with decreasing laser pulse widths. Consequently, it suggests that laser fluence needed to cause damage is smaller for shorter pulses.

Lenzner et al. [22] used 800 nm laser pulses ranging from 5 fs to 5 ps under 50 shots at 1 kHz to observe the laser damage nature in fused silica. The measurements [23, 26] have shown different dependence of damage threshold on pulse duration in the subpico-to-femto second regime. Figure 10 compares the predicted damage threshold fluences using three different MPI rates, $P(I) = (6.0 \times 10^7)\, I^6$, $(6.0 \times 10^8)\, I^6$, and $(6.0 \times 10^9)\, I^6$ which are obtained from the results of Lenzner et al. [22]. Unlike the observations in Fig. 9, the damage threshold fluence becomes no longer lower than 1 J/cm^2, even at a 20-fs pulse. This is because the MPI rate at 780 nm is smaller than that at 1053 and 526 nm. Moreover, the damage threshold fluence rather increases for pulses shorter than nearly 100 fs. This feature is surprising at first because it is thought that the enhancement of MPI or other nonlinear effects reduce the damage threshold from the scaling rule for short pulses. Similar phenomena have been observed by Du et al. [23]. However, other measurements presented earlier have shown totally different characteristics from those of Du et al. [23] and the present calculations. Nevertheless, none of the theoretical and experimental investigators obviously explain this discrepancy in measurements and predictions. As discussed by Tien et al. [26], these discrepancies are perhaps thought to arise from different experimental conditions. Another point to note is that the damage threshold fluence decreases with increasing MPI rate. As indicated previously, the MPI becomes one of dominant channels in rapid increase of free electrons for ultra-short pulses that are typically shorter than several tens of femtosecond to subpicosecond regimes. Therefore, when the pulse duration is very small, the decrease of MPI rate leads to suppress the increasing rate of free electrons and consequently much more laser fluence needs for the laser-induced damage in solids. Numerical predictions show satisfactory agreements with experimental data for both $(6.0 \times 10^7)\, I^6$ and $(6.0 \times 10^8)\, I^6$ cases. In spite of some discrepancies between predictions and measurements, it is thought that the numerical predictions are acceptable, taking into consideration of the uncertainties in experimental data.

In what follows, the laser damage characteristics are discussed when both alternating current electric fields and ultra-short pulse lasers are simultaneously applied to the dielectric materials. It is true that very high electric fields may allow the electron to reach an energy level sufficient to ionize, causing the impact ionization. As discussed earlier, the damage

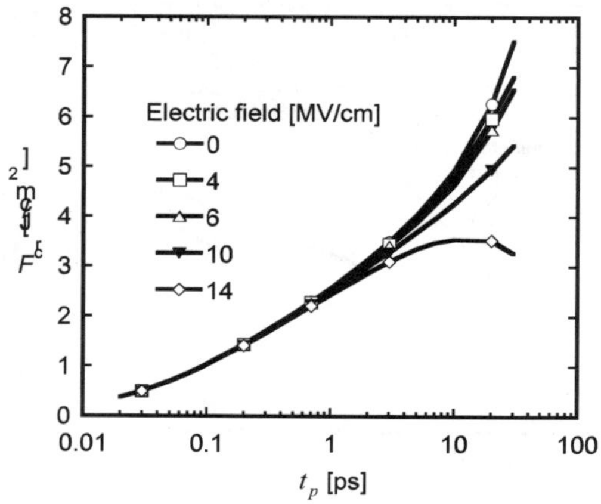

Figure 10 Effects of multi-photon ionization rates on damage threshold fluences at $\lambda = 780$ nm for various pulse durations [13]

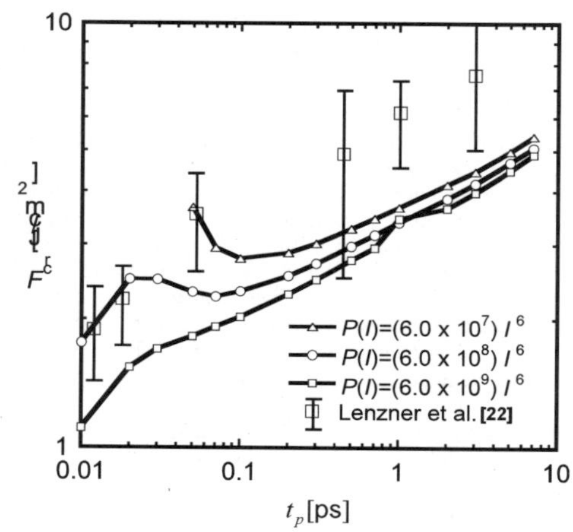

Figure 11 Influence of applied electric fields on damage threshold fluence for different laser pulse durations at $\lambda = 1053$ nm [13]

threshold fluence is closely associated with the efficiency of the laser machining system. This trial is thus planned to check whether the high electric fields can play a supplementary role in the generation of free electrons. In order to account for the electric field effects, the Joule heating term $R_J(\varepsilon,t)$ in Eq. (7) can be modified as

$$R_J(\varepsilon,t) = R_{J,\text{la}}(\varepsilon,t) + R_{J,\text{el}}(\varepsilon,t) \tag{9}$$

where the first term on the right hand side denotes the Joule heating due to pulse laser irradiation and the second one due to the applied electric field. For simplicity, the two terms in Eq. (9) are assumed to be independent of each other. Figure 11 shows the effects of applied electric fields at a fixed frequency of 1 GHz on the damage threshold fluence for different laser pulses at $\lambda = 1053$ nm [13]. For $t_p > 3$ ps, the damage threshold fluence substantially decreases with increasing electric fields. It results from the fact that the amount of Joule heating due to the external electric field is of nearly the same order as that by the laser irradiation. This indicates that the excited electrons by external electric fields partly serve to the production of seed electrons for avalanche process. On the other hand, for pulses shorter than 1 ps, the electric field does not affect the changes of the damage threshold fluence at all. It means that the laser-induced damage becomes dominant over the electric field-induced damage, because the magnitude of Joule heating by the laser irradiation is much larger than that by the electric fields. Through further calculations, it is found that the electric field required to change the damage threshold fluence is about 10^2 to 10^3 MV/cm when $t_p < 1$ ps. Since this high strength of the electric field is really difficult to achieve in practice, it seems to be feasible for only subpicosecond regimes that the electric fields can be used in accelerating the laser damage.

5. Summary

The non-equilibrium interactions between electrons (holes) and lattice phonons in ultra-fast laser processing are surveyed, including two-temperature equation and Fokker-Planck equation approaches for the simulations of thermal and non-thermal transports depending on laser pulse durations.

The influence of laser fluence and pulse duration time on microscale heat transfer mechanisms are investigated by using one-dimensional and transient equations of carrier and lattice temperatures. The scale difference between energy relaxation and laser pulse duration times results in the thermal non-equilibrium state that can be controlled by laser fluence as well as pulse duration time. In the case that a few picosecond pulse laser is irradiated over the semiconductor surface with relatively high fluence, a two-peak structure in the carrier temperature variation can be observed. As pulse duration increases, the maximum carrier temperature and the number density decrease, whereas the lattice temperature is nearly of constant values. Meanwhile, the two-peak structure due to Auger heating disappears and converts into the one-peak structure as the laser fluence decreases.

In the study of the ultra-short pulse laser-induced damage on fused silica by using the Fokker-Planck equation, the influence of laser pulses and wavelengths on laser damage threshold fluences is investigated. As the laser pulse width decreases, so does the damage threshold intensity, and the MPI effect becomes more dominant than avalanche ionization effect. On the other hand, for longer pulses of laser, the impact ionization, i.e., avalanche ionization, becomes an important factor that contributes to the damage threshold of the dielectric material.

Unlike a few hundreds of femtosecond laser pulses, when the laser pulse duration is a few tens of femtoseconds, the damage threshold fluence does not decrease rather than 1 J/cm^2 or less. This limit seems to be persisted even with decreasing pulse width. However, the damage threshold may increase as the laser pulse decreases when the MPI effect is negligible.

Additionally, a hybrid scheme applying a high electric AC field simultaneously with the intense pulse laser is investigated to reduce the damage threshold intensity. This may reduce the damage threshold fluences considerably for relatively longer pulses, whereas it becomes ineffective for laser pulses shorter than 1.0 ps. The threshold fluences are found to be independent of AC frequencies of the applied electric fields except for very high frequencies.

NOMENCLATURE

C, Heat capacity per unit volume, J/m^3 K
D, Diffusion coefficient, eV·s^{-1}
E_g, Band gap energy, eV
F_{cr} Damage threshold fluence, J·m^{-2}
f, Electron distribution function, eV^{-1}·m^{-3}
h, Planck constant (= 6.6262 × 10^{-34}), J·s
I, Laser intensity, W/m^2
I_0, Maximum laser intensity, W/m^2
J, Laser fluence, mJ/cm^2, or current, W·m^{-5}
k, Thermal conductivity, W/m·K
k_B, Boltzmann constant (= 1.38066 × 10^{-23}), J/K
L, System length scale, m
l_r, Relaxation length, m
N_C, Carrier number density, m^{-3}
n, Electron number density, m^{-3}
P, Multiphoton ionization rate, 1/cm^3·ps
R_{imp}, Electron generation due to impact ionization, eV·s^{-1}
R_J, Joule heating, eV·s^{-1}
R_{pi}, Electron generation due to MPI, eV·s^{-1}
S, Source, eV^{-1}·m^{-1}·s^{-1}
T, Temperature, K
t, Time, s
t_p, Pulse duration time, s
U, Internal energy, eV
U_p, Photon energy, eV
V, Effective electron heating, eV·s^{-1}
x, Spatial coordinate, m

Greek symbols

α, Absorption coefficient, m^{-1}
α_1, One photon band-to-band absorption coefficient, m^{-1}
γ, Auger recombination coefficient or energy loss rate
ε, Electron energy, eV
Λ, Mean free length, m
λ, Laser wavelength, m or carrier wavelength, m
ν, Photon frequency, s^{-1}
τ, Mean free time, s
τ_C, Energy relaxation time, s
τ_c, Collision time scale, s
τ_r, Relaxation time, s

Subscripts

C, Carrier
el, Electric field
L, Lattice
la, Laser

Acknowledgements: The authors gratefully acknowledge the financial support of the Micro Thermal System Research Center through the Korea Science and Engineering Foundation.

REFERENCES

1. Koshland, D.E. (1991) Engineering a Small World from Atomic Manipulation to Microfabrication, *Science*, Vol. 254, pp. 1300-1342.
2. Tien, C.L., Majumdar, A., and Gerner, F.M. (1998) *Micro-scale Energy Transport*, Taylor & Francis.
3. Chlipala, J.D., Scarforne, L.M., and Lu, C.Y. (1989) Computer-Simulated Explosion of Poly-Silicide Links in Laser-Programmable Redundancy for VLSI Memory Repair, *IEEE Trans. on Electron Devices*, Vol. 36, pp. 2433-2439.
4. Simon, R. (1991) High-T_c Thin Films and Electronic Devices, *Physics Today*, Vol. 44, pp.64-70.
5. Narayan, J., Godbole, V.P., and White, G.W. (1991) Laser Method for Synthesis and Processing of Continuous Diamond Films on Nondiamond Substrates, *Science*, Vol. 52, pp. 416-418.
6. Eesley, G.L. (1990) Picosecond Dynamics of Thermal and Acoustic Transport in Metal Films, *Int. J. Thermophysics*, Vol. 11, pp. 811-817.
7. Kuanr, A.V., Bansal S.K., and Srivastava, G.P. (1996) Laser Induced Damage in GaAs at 1.06 μm Wavelength: Surface Effects, *Optics and Laser Technology*, Vol. 28, pp. 25-34.
8. Singh, A.P., Kapoor, A., Tripathi, K.N., and Kumar, G.R. (2001) Thermal and Mechanical Damage of GaAs in Picosecond Regime, *Optics and Laser Technology*, Vol. 33, pp. 363-369.
9. Phinney L.M., and Tien, C.L. (1998) Electronic Desorption of Surface Species Using Short-Pulse Lasers, *ASME J. Heat Transfer*, Vol. 120, pp. 765-771.
10. Fushinobu, K., Phinney, L.M., Kurosaki, Y., and Tien, C.L. (1999) Optimization of Laser Parameters for Ultrashort-Pulse Laser Recovery of Stiction-Failed Microstructures, *Numerical Heat Transfer, Part A*, Vol. 36, pp. 345-357.
11. Fushinobu, K., Phinney, L.M., and Tien, N.C. (1996) Ultrashort-Pulse Laser Heating of Silicon to Reduce Microstructure Adhesion, *Int. J. Heat Mass Transfer*, Vol. 39, pp. 3181-3186.
12. Lee, S.H., Lee, J.S., Park, S., Choi, Y.K. (2003) Numerical Analysis on Heat transfer Characteristics of a Silicon Film Irradiated by Pico-to-Femtosecond Pulse Laser, *Numerical Heat Transfer, Part A*, Vol. 44, pp. 833-850.
13. Oh, Y.M., Lee, S. H., Park, S., Lee, J.S. (2004) A Numerical Study on Ultra-short Pulse Laser-induced Damage on Dielectrics Using the Fokker-Plank Equation, *Int. J. Heat Mass Transfer*, to appear.
14. Kang, K.G., Lee, S.H., Lee, J.S., Choi, Y.K., Park, S., and Ryou, H.S. (2004) Fokker-Planck Approach to Laser-Induced Damages in Dielectrics with Sub-Picosecond Pulses, *Int. Symp. on Micro/Nanoscale Energy Conversion & Transport 2004*, pp. 138-140, Seoul, Korea.
15. Qiu, T.Q., and Tien, C.L. (1994) Femtosecond Laser Heating of Multi-Layer Metals – I. Analysis, *Int. J. Heat Mass Transfer*, Vol. 37, pp. 2789-2797.
16. Qiu, T.Q., and Tien, C.L. (1993) Heat Transfer Mechanisms During Short-Pulse Laser Heating of Metals, *ASME J. Heat Transfer*, Vol. 115, pp. 835-841.
17. van Driel, H.M. (1987) Kinetics of High-Density Plasmas Generated in Si by 1.06-and 0.53 μm Picosecond Laser Pulses, *Physical Review B*, Vol. 35, pp. 8166-8176.
18. Seeger, K. (1991) *Semiconductor Physics: An Introduction*, 5th ed., Springer, New York.
19. Pierret, R.F. (1983) Advanced Semiconductor Fundamentals, *Modular Series on Solid State Device*, Vol. 6, Addison-Wesley Publishing Company.
20. Agassi, D. (1984) Phenomenological Model for Picosecond Pulse Laser Annealing of Semiconductors, *J. Applied Physics*, Vol. 55, pp. 4376-4383.
21. Majumdar, A., Fushinobu K., and Hijikata, K. (1995) Effect of Gate Voltage on Hot-Electron and Hot-Phonon Interaction and Transport in a Sub-Micron Transistor, *J. Applied Physics*, Vol. 77, pp. 6686-6694.
22. Lenzner, M., Krüger, J., Sartania, S., Cheng, Z., Spielmann, C., Mourou, G., Kautek, W., and Krausz, F. (1998) Femtosecond Optical Breakdown in Dielectrics, *Physical Review Letters*, Vol. 80, pp. 4076-4079.
23. Du, D., Liu, X., Korn, G., Squier, J., and Mourou, G. (1994) Laser-Induced Breakdown by Impact Ionization in SiO_2 with Pulse Widths from 7 ns to 150 fs, *Applied Physics Letter*, Vol. 64, pp. 3071-3073.
24. Stuart, B.C., Feit, M.D., Rubenchik, A.M., Shore, B.W., and Perry, M.D. (1996) Nanosecond-to-Femtosecond Laser-Induced Breakdown in Dielectrics, *Physical Review B*, Vol. 53, pp. 1749-1761.
25. Loesel, F.H., Niemz, M.H., Bille, J.F., and Juhasz, T. (1996) Laser-Induced Optical Breakdown on Hard and Soft Tissues and Its Dependence on the Pulse Duration: Experiment and Model, *IEEE J. Quantum Electrons*, Vol. 32, pp. 1717-1732.
26. Tien, A., Backus, S., Kapteyn, H., Murnane, M., and Mourou, G. (1999) Short-Pulse Laser Damage in Transparent Materials as a Function of Pulse Duration, *Physical Review Letters*, Vol. 82, pp. 3883-3886.
27. Holway Jr., L.H., and Fradin, D.W. (1975) Electron Avalanche Breakdown by Laser Radiation in Insulting Crystals, *J. Applied Physics*, Vol. 46, pp. 279-291.

MICROSCALE EVAPORATION HEAT TRANSFER

V.V. KUZNETSOV and S.A. SAFONOV
Institute of Thermophysics of Siberia Branch of Russian Academy of Sciences
Novosibirsk, Russia

1. Introduction

In this work, microscale evaporation heat transfer and capillary phenomena for ultra thin liquid film area are presented. The interface shapes of curved liquid film in rectangular minichannel and in vicinity of liquid-vapor-solid contact line are determined by a numerical solution of simplified models as derived from Navier-Stokes equations. The local heat transfer is analyzed in term of conduction through liquid layer. The data of numerical calculation of local heat transfer in rectangular channel and for rivulet evaporation are presented. The experimental techniques are described which were used to measure the local heat transfer coefficients in rectangular minichannel and thermal contact angle for rivulet evaporation. A satisfactory agreement between the theory and experiments is obtained.

Microscale evaporative heat transfer has grown to an important research field in last decade, it can be considered as important area of microscale thermophysical sciences. The reasons for this trend are the miniaturization of systems and the recognition that microscale phenomena can be important for the understanding and performance prediction of microscopic and macroscopic devices. Micromechanical and microthermal systems are widely used now in computer technology, biological, medical and process engineering. The evaporative heat transfer often plays an important role in performance of miniaturized systems, such as micro heat exchangers or micro reactors as far as macroscopic devices such as heat pipes, falling film evaporators, boilers etc., which used mini structures on the surface to enhanced heat transfer. At last case the mini structures on heat transfer surface deform liquid interface to produce a very thin liquid layer or contact line and microscale evaporative heat transfer peculiarities becomes important to predict the overall system performance.

There have been many analytical and experimental investigations on microscale evaporation heat transfer as can be found in the text books and review papers [1], [2], [3], [4] and [5]. A microscale modelling of the evaporating extended meniscus focused on the role of the intermolecular forces on the phase change processes in thin liquid films and experimental study of thickness profiles of the evaporating meniscus has been developed in [5], [6], [7], [8], [9] and [10]. In the modelling, the wall temperature, the conductive resistance of the liquid film, the surface tension, the disjoining pressure, and the interfacial heat resistance are taken into account to describe the interface shape of the evaporating meniscus and defined the apparent contact angle. Using similar approach in [10], [11] was shown that the apparent contact angle is determined mostly by the capillary number based on the liquid surface tension, liquid viscosity, and the velocity scale set by the evaporation kinetics, so that the molecular forces have a minor effect on the apparent contact angle by evaporation. The study of microscale heat transfer and dynamic of evaporating thin films spreading on solid surfaces was done at [12], [13]. The effect of the local surface curvature and the disjoining pressure on the evaporation rate had been taken into account. In [14] the formation of a

S. Kakaç et al. (eds.), Microscale Heat Transfer, 303–320.

locally thinned liquid film near the corners of a vertical open channel has been demonstrated experimentally. They also solved numerically the governing equations for the thickness of the liquid film to predict a locally thin film produced by the capillary action in the trough. The distribution of liquid film around perimeter of vertical rectangular minichannel has been studied in [15]. The non-uniform nature of the local heat transfer around the perimeter has been discussed. It was shown also that dry spots are characteristic of liquid evaporation in non-cercular channel for small liquid flow rate. The dry spots formation supplied by contact line on heat transfer surface were observed in [16] just before boiling crises in narrow annular channels.

In the reviewed paper, it was shown that microscale transport phenomena at liquid-vapor interface can be stronger influenced by external and internal forces, such as capillary forces, intermolecular forces, or by temperature and concentration gradients. These phenomena are still not fully understood and are the subject of future study. The objective of this work is to study theoretically and experimentally the interface shape pattern and heat transfer characteristic of curved evaporating liquid film. We will discuss non-uniform nature of the evaporating heat transfer around the perimeter of minichannel, theory of fluid flow and evaporation heat transfer in a small rectangular channel, experiments on evaporating flow of refrigerant R21 in a small rectangular channel, peculiarities of evaporation heat transfer for liquid rivulet spreading and the nature of microscale heat transfer augmentation.

2. Mathematical Modeling of the Interface Shape for Gravity Induced Film Flow in Rectangular Minichannel

2.1. MODEL DEVELOPMENT

When a liquid flows down in a vertical rectangular channel under the action of gravity the flow characterizes by the formation of two zones: thin film flow along the sides of the channel and meniscus flow of a constant curvature in the corners of the channel [15], see Figure 1. Direct numerical flow modeling based on CFD codes cannot be applied for this case. The reason is the existing of the area where film thickness trends to zero in the solution. To avoid this singularity the flow model based on distinguishing of two specific areas of flow, which are the corner flow and film flow, was proposed in [15]. At corner flow, the interface has constant in cross section and variable along channel length the radius of curvature. At film area, the interface has variable along transversal direction the radius of curvature. The solutions for these areas are matched to each other near the channel corner to produce the total solution. For this case the flow problem has a small parameter which is $\varepsilon = \delta_0/a \ll 1$ (δ_0 is initial liquid layer thickness and a is half width of the channel wall).

To modeling the film flow, let us introduce the Cartesian coordinates shown in Figure 1 as follows: x-axis along the liquid flow direction, y-axis crosswise on the channel's surface and z-axis

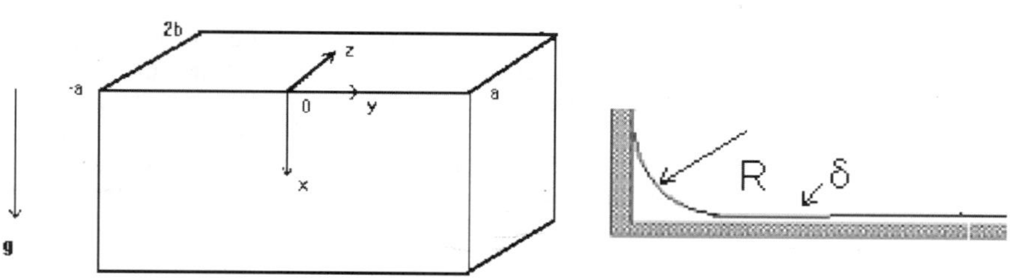

Figure 1. Coordinate system and scheme of the liquid flow.

perpendicular to the channel surface. Dropping the convective terms we can write the Navier-Stokes equations for steady film flow as follows:

$$-\frac{1}{\rho_L}\nabla p_L + \nu\nabla^2 U + g = 0 \qquad (2.1)$$

$$divU = 0, \quad U = (u, v, w), \qquad (2.2)$$

Here p_L is the pressure in the liquid and δ is the film thickness. The following conditions are valid on the film surface specified by equation $z=\delta(x,y)$:

$$w = u\frac{\partial \delta}{\partial x} + v\frac{\partial \delta}{\partial y} \qquad (2.3)$$

$$p_L - p_V + \sigma k = 0 \qquad (2.4)$$

where σ is the surface tension coefficient and k is the curvature of the liquid surface. The intermolecular forces are not accounted here.

Scaling the initial equations by the typical velocity δ_o^2/ν for u and v, $\varepsilon\delta_o^2/\nu$ for w, $\rho_L ga$ for pressure and typical size in the directions, Ox -a/ε, Oy - a and Oz - δ_0. Dropping terms that are $0(\varepsilon^2)$ and smaller we can obtain from (2.1)-(2.4) the velocity profiles as:

$$u = (1 - \varepsilon p_x)(mz - z^2/2)$$
$$v = -p_y(mz - z^2/2) \qquad (2.5)$$

Here $m=\delta/\delta_o$, the same symbols for dimensionless variables been used, and the subscript L for non dimensional pressure in liquid is dropped. With account for (2.5) and (2.3) it follows:

$$w_{z=m} = 0.5m^2(\varepsilon m_x - m_y p_y) \qquad (2.6)$$

After integration of (2.2) on z-coordinate and using (2.2), (2.6) we obtain the equation for dimensionless film thickness as:

$$(m^3)_x + (m^3 m_{yyy})_y = 0 \qquad (2.7)$$

here x is scaled by aBo/ε, and $Bo = \rho_L ga^2/\sigma$.

The boundary conditions for film flow are the symmetry condition in the center of the long side of the channel with y=l and m_y=O, $m_{yy}=1/(r\varepsilon)$ at the point where the solutions for film and meniscus are matched to each other. Last condition follows from equality of pressures in film and in meniscus at that point. Here $r=R/a$ where R is the radius of curvature of the meniscus.

For the solution to the total flow problem it is necessary to obtain the relationships between the flow rate Q_m in meniscus and values R, g, ν. They were obtained by numerical calculation of the Poisson equations for 'x' component of the velocity at the meniscus area and may be presented as:

$$q_m = \frac{Q_m \nu}{g a \delta_0^3} = \varepsilon^{-3} f(r, m_1, m_2, \Theta), \qquad \Theta = \arctan\left(\frac{r + \min(m_1, m_2)}{r + \max(m_1, m_2)}\right), \qquad m_2 > 0$$

$$q_m = \frac{Q_m \nu}{g a \delta_0^3} = \varepsilon^{-3} f(r, m_1, \varphi, \Theta), \qquad \Theta = \arctan\left(\frac{r \cos\varphi}{r + m_1}\right), \qquad m_2 = 0 \tag{2.8}$$

Here Θ is contact angle; m_1 and m_2 are layer thickness in the points were the solutions are matched to each other. The total flow rate conservation equation completes the systems (2.7) and (2.8):

$$\frac{d(Q_m + Q_f)}{dx} = 0 \tag{2.9}$$

where Q_f is quarter flow rate in film.

2.2. NUMERICAL SOLUTION FOR INTERFACE SHAPE DEVELOPMENT

The resulting equations were solved numerically to compute the film thickness distribution or the interface shapes and the radius of meniscus at the channel's corner. The implicit conservative scheme was used to obtain the solution. We considered the cases of uniform initial liquid distribution along the channel's wall as an initial condition.

The example of interface shape evolution is shown in Fig. 2 as a function of the distance from the point where the liquid was brought into the channel. The pictures on the left side are the experimental data on refrigerant R113 interface shape development obtained for rectangular 2.6x7.1 mm channel in [15]. The laser knife technique was used to record the interface shape.

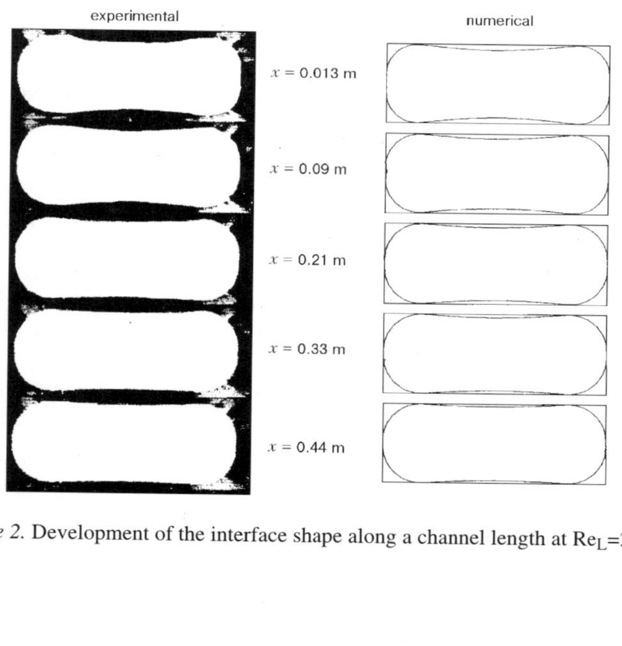

Figure 2. Development of the interface shape along a channel length at Re_L=200.

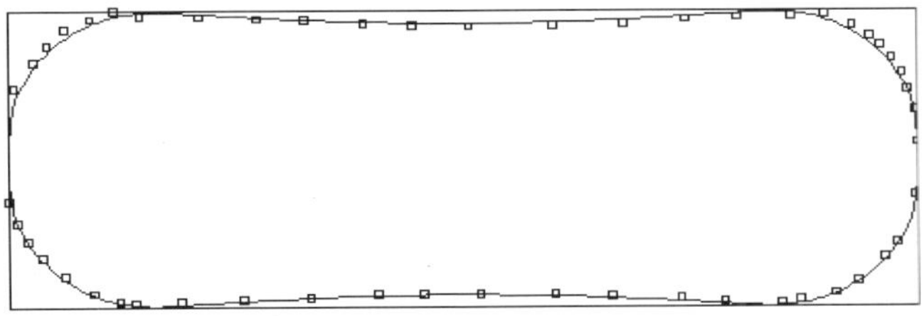

Figure 3. Interface shape of kerosene flow in 2.6x7.1 mm rectangular channel.

The corresponding figures on the right side are the result of numerical modeling. They are in agreement with experimental data. The main flow of liquid occurs in the channel corners and the film thickness on the long side of the rectangle is much less than the initial film thickness. It becomes thicker along the channel length showing the liquid accumulation in channel's corner area. Direct comparison of the experimental observation of interface shape for kerosene down flow in 2.6x7.1 mm channel obtained in [15] with data of calculation is shown on Fig. 3. Here the line shows experimental interface shape and points are the data of calculation at x=0.09 m and Re_L=14.

3. Film Flow Modeling for Co-current Vapor Flow with Evaporation

3.1. MODEL DEVELOPMENT

When a co-current vapor flow is present, the basic nature of this flow does not change, but the details differ because of the thinning of the liquid film by interfacial shear stress. Dropping the convective terms we can write the Navier-Stokes equations for steady film flow as follows:

$$-\frac{1}{\rho_L}\nabla p_L + \nu\nabla^2 U + g_e = 0, \qquad U = (u, v, w)$$
$$g_e = \left(g - \frac{1}{\rho_L}\frac{dp_V}{dx}, 0, 0\right) \tag{3.1}$$

Here p_V is the pressure in the gas or vapor. Mass balance equation for flow with evaporation is:

$$\int_0^\delta \left(\frac{\partial u}{\partial x} + \frac{\partial v}{\partial y}\right)dz + w\big|_{z=\delta} = -j/\rho_L \tag{3.2}$$

The following conditions are valid on the interface:

$$w = u\frac{\partial \delta}{\partial x} + v\frac{\partial \delta}{\partial y} \tag{3.3}$$
$$\mu e_{ij}\tau_i n_j = \chi \qquad e_{ij}t_i n_j = 0$$

Here e_{ij} are the components of the tensor of deformation rates, n_i, τ_i, t_i are the components of the normal, binormal and tangential vectors, μ is the liquid viscosity. The intermolecular forces are account as disjoining pressure component $p_d = A_0/(6\pi\delta^3)$, where A_0 is Hamaker constant, and χ is the shear stress produced by gas flow.

Scaling the equations (3.1) through (3.3) as in chapter 2.2 and dropping the terms that are $0(\varepsilon^2)$ and smaller we obtain:

$$u = (\gamma - \varepsilon p_x)(mz - z^2/2) + \kappa z/\varepsilon$$
$$v = -p_y(mz - z^2/2) \tag{3.4}$$

Here $\kappa=\chi/(\rho_L ga)$, $\gamma=1-(dp_V/dx)/(\rho_L g)=1-4a\kappa/d_h$ and d_h is the hydraulic diameter of the part of channel cross-section occupied by gas phase. The same symbols for dimensionless variables been used again and p is dimensionless pressure in the liquid. With account for (3.4) the kinematic condition on the interface (3.3) and (3.2) gives:

$$w \,\|_{z=m} = 0.5m^2(\gamma\varepsilon m_x - m_y p_y) + \kappa m m_x \tag{3.5}$$

$$w \,|_{z=m} = \frac{j\nu}{\varepsilon\rho_L g\delta_0^2} - 0.5m^2(\gamma\varepsilon m_x - m_y p_y) + \frac{m^3}{3}p \tag{3.6}$$

Finally from (3.5), (3.6) and (3.3) it follows:

$$(\gamma m^3 + 1.5\frac{\kappa}{\varepsilon}m^2)_x + (m^3 m_{yyy})_y = \frac{3}{\varepsilon^4}(Ga(\frac{m_y}{m})_y - \frac{G_0\Theta_{w,i}}{m\varepsilon}) \tag{3.7}$$

Here x is scaled by aBo/ε, $G_0 = \lambda_L T_* \nu/(i_{LV}\sigma a)$, i_{LV} is latent heat of evaporation, $Bo = \rho_L g a^2/\sigma$ $Ga = A_0/(6\pi a^2\sigma)$. For the electrically heated wall with constant volume heat production H the temperature scale is $T_* = Ha^2/\lambda_f$. When external wall temperature is given as $T_{w,e}$ the temperature scale becomes $T_* = T_{w,e} - T_v$. The dimensionless internal wall temperature is defined as $\Theta_{w,i}=(T_{w,i}-T_v)/T_*$ and mass flow in interface is $j=\lambda_L(T_{w,i}-T_v)/(\delta i_{LV})$.

If the heat produced in channel's walls is consumed for evaporation, then the total flow conservation equation is:

$$\frac{d(Q_m + Q_f)}{dx} = -\frac{1}{\rho_L r_0}\oint q_h dl, \qquad Q_m + Q_f \,|_{x=0} = Q_0 \tag{3.8}$$

The real surfaces are characterized by a certain extent of roughness. It is assumed that film thickness cannot be lower than the wall roughness value. When film thickness achieves this value, the film ruptured may occur depending on rivulet width, which corresponds to film flow rate at that time. With this statement the system of equations proves to contain the relationship describing the flow rate in a rivulet in the dependence on its length and contact angle. This correlation was obtained by the same way as that for the meniscus flow rate.

3.2. NUMERICAL MODELING OF INTERFACE SHAPE AND MICROSCALE HEAT TRANSFER

The resulting equations are solved numerically to compute the film thickness distribution or the interface shapes and the radius of the meniscus at the channel's corner. The implicit conservative scheme was used to obtain the solution. We used non-uniform grid pattern near moving film-meniscus contact point. Simultaneously with calculation of the interface shape, the heat transfer problem has been solved in the symmetry element of the channel. For the case of uniform heat production in the wall we have solved conservation energy equations in channel wall area:

$$\lambda_w \left(\frac{\partial^2 T}{\partial y^2} + \frac{\partial^2 T}{\partial z^2} \right) + H = 0 \tag{3.9}$$

and in liquid flow area:

$$\frac{\partial^2 T}{\partial y^2} + \frac{\partial^2 T}{\partial z^2} = 0 \tag{3.10}$$

Boundary conditions are: $\partial T/\partial n = 0$ (n is the normal to interface surface) on the gas-solid interface and on the symmetry lines as far as on the external wall surface (for case of channel with heat production in the wall); continuity both of heat flux and temperature on solid-liquid interface and the continuity of temperature on gas-liquid interface. Gauss-Zeidel iterative procedure has been used to solve the heat problem numerically. We used a non-uniform grid pattern near the vapor-liquid interface for higher computational accuracy.

The interface boundary inside the channel is shown on figures 4 and 5 as a function of the liquid Reynolds number. At large liquid flow rate the considerable part of the liquid flows in the corners and the film is thinned both on the long and short sides of the channel. This enhances the heat transfer in comparison with uniform film. At small liquid Reynolds numbers the minimum film thickness becomes the same as the wall roughness and film rupture occurs leading to stable rivulet flow. Dry areas exist on the wall, which are not wetted by liquid. This reduces the heat transfer. The

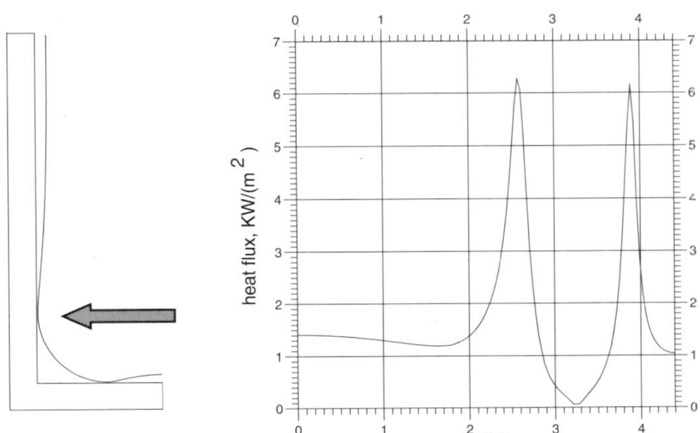

Figure 4. Liquid interface and distribution of the local heat flux at internal wall for Re=300.

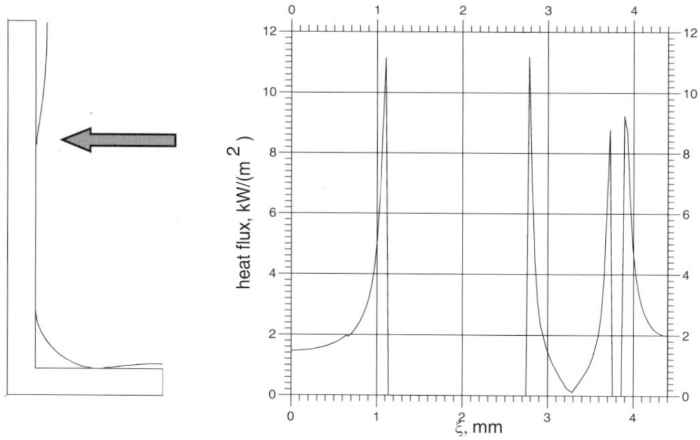

Figure 5. Liquid interface and distribution of the local heat flux at internal wall for Re=100.

contact angle used to model this behavior in calculations was set as ten degrees.

The local heat flux distribution along the perimeter of internal wall of the channel is shown on Figs. 4, 5 for non-uniform film thickness and for a film with a dry spots. Coordinate ξ =0 corresponds to the center of the long side of the channel, ξ =3.25 mm corresponds to the internal channel's corner and ξ =4.4 mm corresponds to the center of the narrow side. The calculations were performed with the ratio of wall and liquid heat conductivity equal to 114 and a volumetric heat production equal to 6 Mw/m^3. Figure 4 shows the consequence of the film being non-uniform in thickness. Here pointer shows the area of microscale heat transfer. The local heat flux around the internal wall surface in the normal direction to each side of the channel is presented. A large variation around the perimeter is seen with high local values near the edges of the long and short sides. Figure 5 shows the effect of the formation of dry regions and rivulet flow both near the center of narrow side and near the center of wide side. In case of flow with dry spots the rate of heat transfer achieves highest value in vicinity of liquid-vapor-solid contact line where the film thickness trends to zero. The calculations shows that for heat transfer surface with applied technical roughness, the intermolecular forces become not important in determine the aspect of flow such as microscale heat transfer.

3.3. COMPARISON OF THE SIMULATION DATA WITH EXPERIMENTAL ONES

Figure 6 shows the experimental apparatus used in [17, 18] to investigate the local evaporation heat transfer in the rectangular channel. Experiments have been performed in a closed loop with refrigerant R21, which operates in the regime of co-current down flow. The test section has size of 2.3x6.5 mm for internal cross section and length of 0.25 m. Liquid refrigerant is cooled after the pump and the flow rate was measured by turbine flow meter. The vapor flow rate at the inlet of the test section was calculated by noting the power consumed by the electrical coil, which was heated with AC. Immediately after electrical coil the flow is mixed to minimize the liquid super heat. To achieve thermal and hydrodynamic stabilization of the flow, an unheated pre-section with a length of 1 m was placed after the vapor generator. In order to observe and control the flow pattern a transparent section made from quartz glass was installed both at the inlet and outlet of the test section. At the outlet an additional evaporator was placed to avoid flooding in regimes of low vapor quality. The test section was manufactured from a stainless steel tube with a 0.25 mm wall thickness

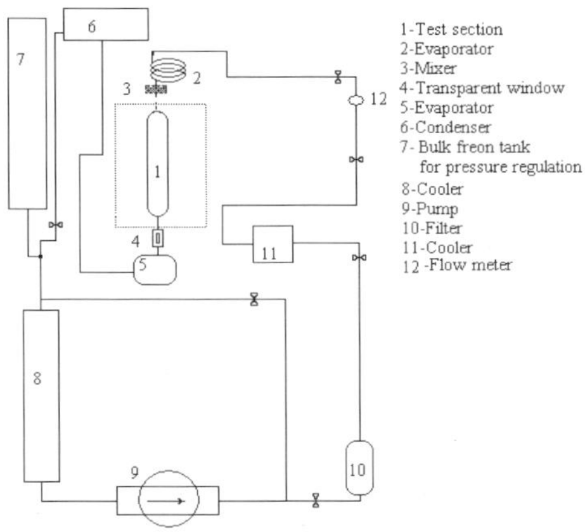

Figure 6. Experimental apparatuses.

by tube pressing The surface of the inner wall had a technical roughness equal to 10 microns. To measure the local heat transfer coefficient 16 thermocouples are installed along the perimeter at distances of 0.08 in and 0.22 m each from the channel inlet. Thermocouples made by contact welding are used for these experiments. The local heat transfer coefficient was determined as heat flux divided by temperature difference between the outer wall of the test section and the saturation temperature inside of the test section. Further discussion and details of the experimental procedure can be found in [17, 18].

Figure 7 presents the heat transfer coefficient data, which correspond to the center of the long side of the channel. The distance from the test section inlet equals 0.22 m. In these experiments the

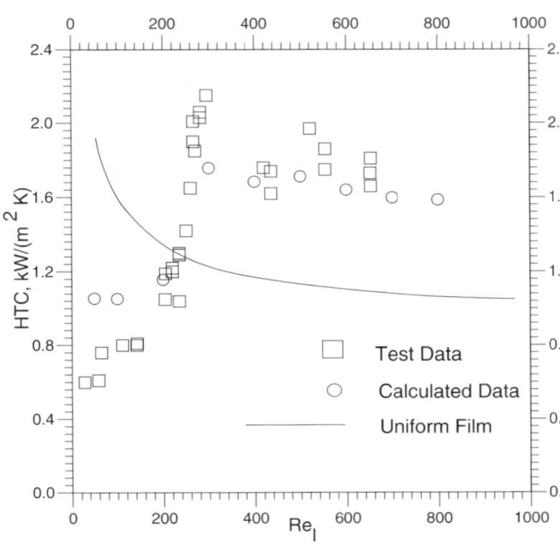

Figure 7. Heat transfer coefficient at the center of a long side at Re_V =4000 and q_h=1.5 kW/m^2.

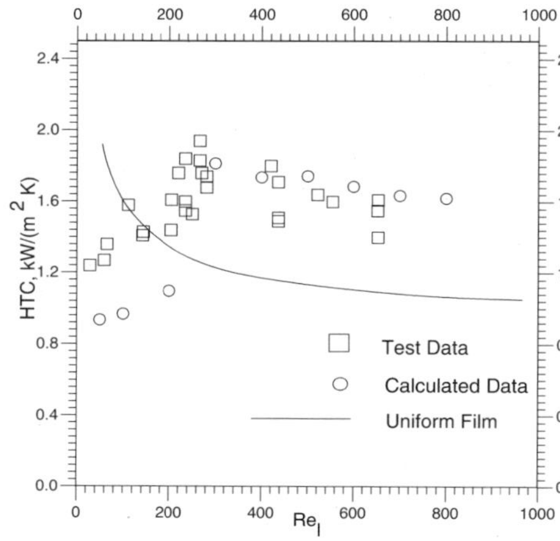

Figure 8. Heat transfer coefficient at corner of the channel at Re_V =4000 and q_h=1.5 kW/m^2.

vapor Reynolds number and heat flux density have been held constant. The liquid Reynolds number was defined as $\mathrm{Re_L} = 4AV_L/P\nu_L$. Here V_L is the superficial liquid velocity, A is the channel cross sectional area, P is the channel perimeter and ν_L is the kinematic liquid viscosity. Similarly the vapor Reynolds number is defined as $\mathrm{Re_V} = 4AV_V/P\nu_V$. Two typical heat transfer areas occur in the channel depending on flow regime. High heat transfer takes place at large liquid Reynolds numbers while the rate is reduced below liquid Reynolds numbers of 260. The line in Fig. 7 shows the predicted heat transfer coefficient for the evaporation of a liquid film that is uniformly distributed around the perimeter of the channel. To calculate the film thickness in co-current vapor flow, the correlation [19] for interface shear stress was used. In calculations the heat transfer enhancement by the surface waves was taken into account according to [20].

Figure 8 presents the heat transfer coefficient data for the corner of the test section. The circles in Figs. 7, 8 correspond to the calculated heat transfer for non-uniform liquid film thickness, which agree well with the data both for large and small liquid Reynolds numbers. In these calculations the temperature of the outer wall and the heat transfer coefficients were defined in accordance with the experimental procedure. The calculations show that liquid suction occurs toward to the channel's corner and liquid is therefore non-uniformly distributed along the perimeter.

4. Mathematical Modeling of Microscale Heat Transfer in Vicinity of Contact Line

4.1. MODELING OF EVAPORATING LIQUID SPREADING ALONG THE SURFACE

Let us consider a liquid rivulet flow over a vertical plate with given temperature T_w under the action of gravity and in surrounding stagnant vapor, which is saturated. The term ‘rivulet’ means that the size of flow region in gravity vector direction is much larger than in the perpendicular direction (the rivulet’s width). We suppose also that the liquid layer thickness in flow region is much less than the rivulet width. To model this phenomenon, let us introduce Cartesian coordinates as follows: x-axis along the gravity vector direction, y-axis is perpendicular to the gravity vector direction and z-axis is perpendicular to the plate surface.

The governing equations are Navier-Stokes equations for non-steady liquid flow with boundary conditions on a free liquid surface and mass conservation equation, which takes into account the mass flux from surface due to evaporation:

$$U_t + U\nabla U = -\frac{1}{\rho}\nabla p + \nu\nabla^2 U + G$$
$$U = (u, v, w) \qquad G = (g, 0, 0) \tag{4.1}$$

$$\int_0^\delta (u_x + v_y)dz + w|_{z=\delta} = -j/\rho \tag{4.2}$$

$$w = \delta_t + u\delta_x + v\delta_y \qquad z = \delta \tag{4.3}$$

$$p + \sigma k = \sum e_{ij} n_i n_j \tag{4.4}$$

$$\sum e_{ij}\tau_i n_j = 0 \qquad \sum e_{ij} t_i n_j = 0 \tag{4.5}$$

Here p is the pressure in liquid, ν is the kinematic viscosity, ρ is the liquid density, j is the mass flux on the liquid surface, σ is the surface tension, g is gravity acceleration, k is the curvature of the liquid surface, e_{ij} are the components of the tensor of deformation rates, n_i, τ_i, t_i are the components of the normal, binormal and tangential vectors on liquid surface. For a constant wall temperature we can write $j=\lambda\Delta T/(r_0\delta)$, where λ is the thermal conductivity, i_{LV} is the latent heat of evaporation, T_S is the vapor temperature and $\Delta T=T_W-T_S$.

Let us scale the initial equations by the typical velocity $V=Q_l/\delta_0$, pressure ρgb, time b/V and typical size in the directions x and y as b and for z as δ_0, where $\delta_0=(3Q_l\nu/g)^{1/3}$ is the initial liquid layer thickness at inlet section, $Q_l=Q/2b$, Q is the volumetric flow rate, b is the half-width of the open cross-section of nozzle. Let us denote $\varepsilon=\delta_0/b$, Reynolds number Re= Q_l/ν and Bond number $Bo=\rho g b^2/\sigma$. Let us suppose that $\varepsilon^2 \ll 1$. Dropping the members of the same order of magnitude as $0(\varepsilon^2)$ and smaller and supposing that velocity vector components in x and y directions u and v have a semi-parabolic profile in z-direction, we can integrate the momentum equations and mass balance equation by z from z=0 up to z=h(x,y) and using kinematic condition on the liquid surface can obtain the following equations of known the so-called 'integral flow model' [21] (mass flux from liquid surface due to evaporation is taking into account also):

$$\left(\frac{2}{3}(uh)_t + \frac{8}{15}\left((u^2h)_x + (uvh)_y\right)\right)\varepsilon\,\mathrm{Re} + 3hp_x = -\frac{2u}{h} + 3h$$
$$\left(\frac{2}{3}(vh)_t + \frac{8}{15}\left((uvh)_x + (v^2h)_y\right)\right)\varepsilon\,\mathrm{Re} + 3hp_y = -\frac{2v}{h} \tag{4.6}$$
$$p + \frac{\varepsilon}{Bo}(h_{xx} + h_{yy}) = 0$$
$$1.5h_t + (uh)_x + (vh)_y = -\frac{4.5E}{h\varepsilon^4} \qquad E = \frac{\lambda\Delta T\nu}{b^3\rho r_0 g}$$

Here $h=\delta/\delta_0$ and the same symbols for dimensionless variables are saved.

For creeping flow case ($\varepsilon Re<<1$) we can neglect the convective terms and obtain the single equation for dimensionless liquid layer thickness h of elliptic type:

$$(h^3)_x + \frac{\varepsilon}{Bo}\left((h^3 \Delta h)_x + (h^3 \Delta h)_y \right) = -\frac{3E}{h\varepsilon^4} \qquad (4.7)$$

$$\Delta h = h_{xx} + h_{yy}$$

The typical size of rivulet flow in x-direction is much higher than the one in y-direction, so we can obtain simplified equation for dimensionless liquid layer thickness h of parabolic type, where $x=x\varepsilon/Bo$:

$$(h^3)_x + (h^3 h_{yyy})_y = -3\frac{EBo}{h\varepsilon^5} \qquad EBo = \frac{\lambda \Delta T \nu}{r_0 b \sigma} \qquad (4.8)$$

The boundary conditions for equations (4.6) are the following: h(t,0,y)=1, u(t,0,y)=1.5, v(t,0,y)=0 for $|y| \leq 1$; h(t,0,y)=0 u(t,0,y)=0, v(t,0,y)=0 for $|y| > 1$; h=0, u=0, v=0 far from flow region. The boundary conditions for equation (4.8) are the following: condition of symmetry at line y=0; h(0,y)=1 for $y \leq 1$; h(0,y)=0 for y>1; h=0 far from flow region.

Using the implicit difference scheme with respect to time in equations for u and v, and non-implicit difference scheme with respect to time in equation for h the system (4.6) was solved. The equation (4.8) was solved by implicit difference scheme with respect to x. To solve the system of non-linear equations appearing after discretization of governing equations, the iteration procedures were implemented. To avoid the singularity caused by condition h=0 far from the flow region, this condition was replaced by condition $h=h^*>0$. The value of h^* was chosen in a way that the solution in the rivulet area was independent from this parameter. In calculations the value of h^* was equal to 10^{-5}.

4.2. NUMERICAL RESULTS ON MODELING OF FLOW AND MICROSCALE HEAT TRANSFER

The flow of rivulet of R11 was simulated at the temperature of 30° C at initial rivulet width of 4 mm for Reynolds numbers Re=2.5, 5, 25, and 75. The corresponding dimensionless parameters ε are ε=0.0194, 0.0244, 0.0419, and 0.0603. The values of the product εRe (characteristics of inertia impact on the flow) equal to 0.049, 0.122, 1.048, and 4.523. The first two cases refer to a creeping flow, and the other two correspond to the inertia mode of flow.

For creeping mode, the solution of (4.8) for interface shape was found numerically. The mesh size on the coordinate x equals to 1 micron (dimensionless is 0.005), and along the coordinate y equals to 2 micron (dimensionless is 0.01). As for simulation of rivulet in inertia mode, it was based on solution of the system of non-stationary equations of the integral model (4.6). The mesh size along the coordinates x and y was 10 micron (dimensionless 0.05), the time step was 1.43 microsecond (dimensionless 0.0001). A higher mesh size for spatial variables was caused by a significant growth in calculation domain because of elliptic type of the problem. We considered the rivulet flow caused by gravity forces on a cold wall; the wall superheat was ΔT=0.1K, 0.5K, 1K for creeping flow mode and ΔT=1K for inertial flow mode. The influence of the wall superheat on the apparent contact angle θ was studied; it was determined as the tangential angle in the inflection point of the liquid profile.

In our simulation, the initial liquid shape was set as a step-like profile; so the liquid flows down along non-heated wall, but starting from the level $x=c$ we assigned the heat flux. To estimate the

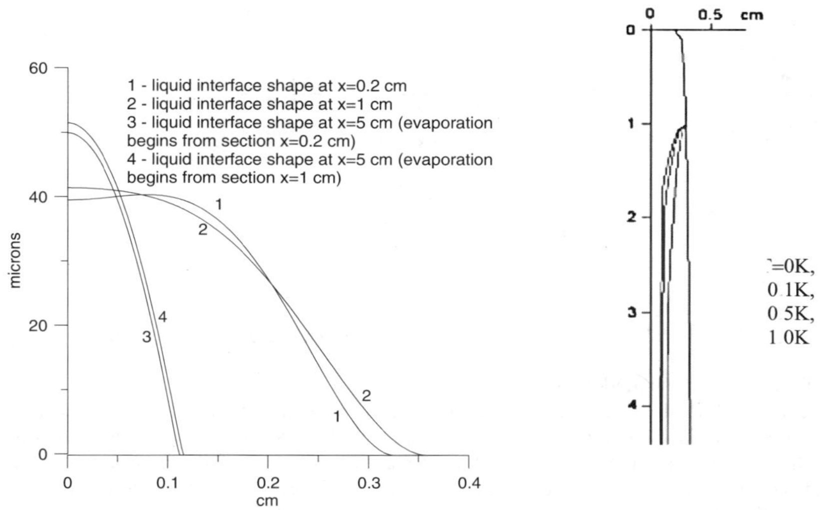

Figure 9. Interface shape before the evaporation start and in the area where evaporation occurs (wall superheat is 0.5K, Re=2.5) and rivulet boundary depending on wall superheat (right picture).

influence of parameter c on angle θ in our simulations at Re=2.5, we took two values of it: c=0.2 cm. The contact angle at the wall superheat of 0.1K is approximately 3.3 degrees, at the wall superheat of 0.5K it makes up 4.5 degrees, and at the wall superheat of 1K, this value equals to 5.3 degrees. Rivulet counters are presented in Fig. 9 (the isoclines for thickness level of 10 micron are presented) at Reynolds number Re=2.5 and wall superheat equals to 0K, 0.1K, 0.5K and 1K for two cases when evaporation begin from section x=0.2 cm and x=1 cm. The data of calculations according to inertial flow model (4.6) show that for given wall superheat the apparent (thermal) contact angle has the same

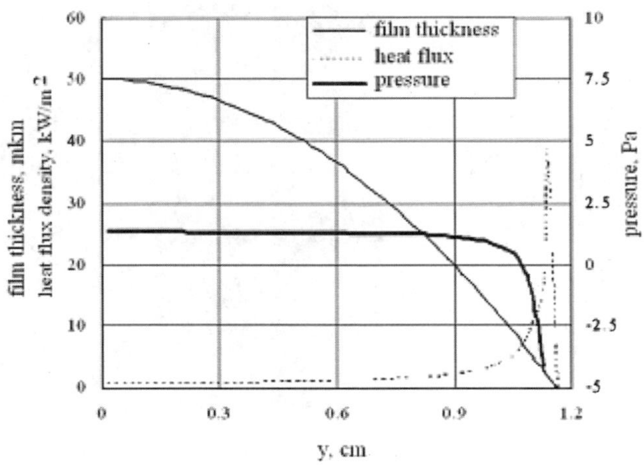

Figure 10. Interface shape, distribution of local heat flux and pressure in rivulet cross-section. Distance from liquid inlet equals to 5 cm, Re=2.5 and ΔT=0.5K.

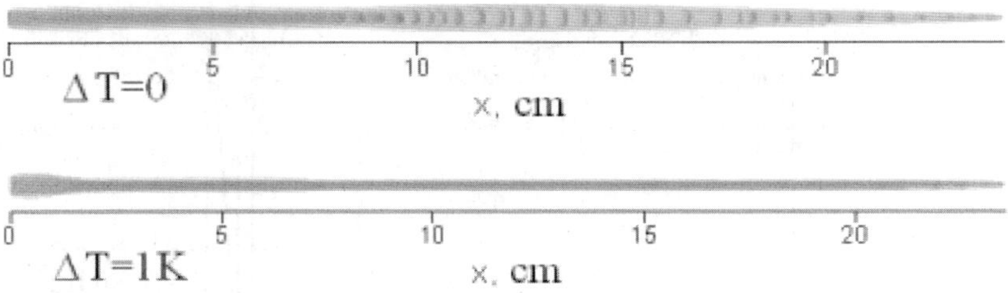

Figure 11. Wave patterns on rivulet interface depending on wall superheat.

value as for creeping flow model.

Figure 10 presents the interface shape of the rivulet for wall superheat as 0.5 K and Re = 2.5. Here also presented the data on pressure in liquid and heat flux density in rivulet cross-section. The intensive liquid evaporation in near contact line region causes the interface deformation. As a result the transversal pressure gradient creates the capillarity induced liquid cross flow in direction to contact line. Finally the balance of evaporated liquid and been bring by capillarity is established. This balance defines the interface shape and apparent contact angle value.For the inertia flow model, the solution is obtained from a non-stationary system of equations, i.e., it is time-dependable. In this case the disturbances in flow interface can create the wave flow patterns. The solutions of unsteady state liquid spreading on heat transfer surface without and with evaporation are presented on Fig. 11. When the evaporation is not included (for zero wall superheat) the wave pattern appears on the interface. When the evaporation includes, the apparent contact angle increase immediately and deform the interface. It causes the wave suppression due to increasing of the film curvature.

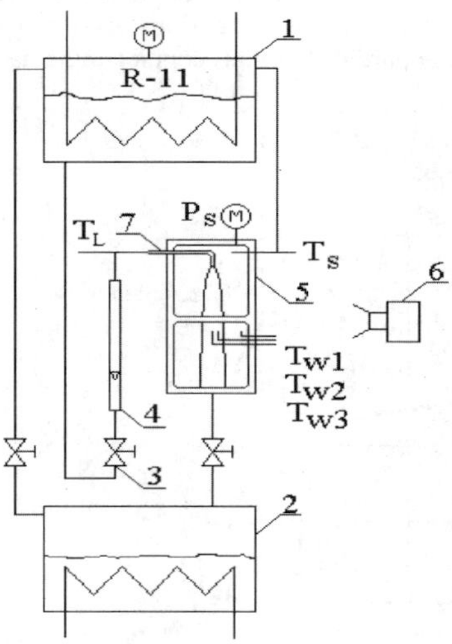

Figure 12. Experimental setup.

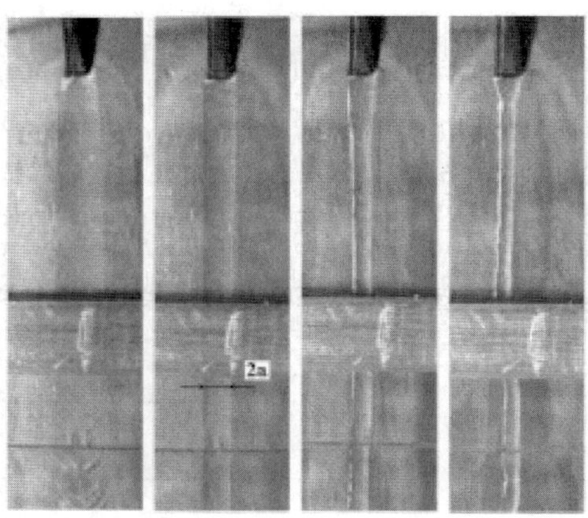

0.17 K 0.39 K 0.5 K 0.6 K

Figure 13. Rivulet flow patterns depending on wall superheat at Re=406.

4.3. COMPARISON OF SIMULATION DATA WITH EXPERIMENTAL ONES

The experiments on rivulet spreading on a vertical heated surface were carried out in [22] using the apparatus depicted in Fig. 12. The working liquid was refrigerant R11. Liquid refrigerant was fed from the top temperature-controlled tank 1 to the accumulating bottom tank 2 through the test section 5. The air was evacuated from the test section and it was filled by stagnant saturated vapor. The liquid flow rate was regulated by a valve 3 and controlled by flow meter 4. The process of liquid flow keeps for several hours, so the system may reach the thermodynamic equilibrium. The polished thick copper plate with the sizes of 285x120x10 mm was the wall of the test section. It was heated by outside cartridge electric heater with an external cooling to achieve the condition of given wall temperature at the experiments. The opposite wall of the test section was made of glass for recording the process with a video camera 6. There was a rivulet former 7 inside of the test section (the nozzle open cross-section is 4x0.5 mm and the former wall thickness is 0.75 mm), which touched the polished copper wall. Three copper-constantine thermocouples T_{W1}, T_{W2} and T_{W3} were fixed on the wall at the distance of 0.1 m from the rivulet former. The thermocouple T_{W1} was set on the axis line of the rivulet, the thermocouple T_{W2} was set on the distance of 3 mm from T_{W1} and the thermocouple T_{W3} was set on the distance of 25 mm from T_{W1}. The thermocouple T_{W2} measured the temperature of the wall near the boundary of the rivulet and T_{W3} measured the wall temperature far from the rivulet. All wall temperatures were the same within the accuracy of the measurements of 0.1K. In the tests the liquid temperature inside the nozzle T_L, the temperature of vapor T_S, and the pressure inside the test section P_S were measured.

Typical patterns of rivulet spreading on a vertical copper surface which was observed in [22] are shown in Fig. 13 for different overheating of the wall relative to saturation temperature $\Delta T=T_W-T_S$. The Reynolds number was defined as $Re=(Q/d\nu)$, where Q is the volumetric flow rate, and d is the width of the open cross-section of the nozzle. One can see that the higher the wall superheat, the narrower the spreading zone. At overheating of 0.39K the rivulet starts to contract on hot wall. The

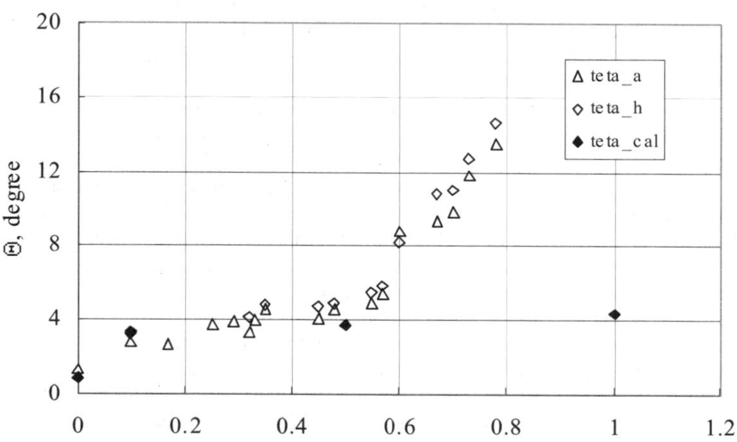

Figure 14. Apparent contact angle for rivulet evaporation on vertical heat transfer surface depending on wall superheat.

reason is intensive evaporation of liquid in the vicinity of gas-liquid-solid contact line where film thickness decreases up to zero. The thermal apparent contact angle is established, which is higher than the equilibrium contact angle. Capillary forces contract the film and it halts the spreading.

The apparent contact angle θ for a given rivulet flow rate was determined in [23], assuming semi-parabolic velocity profile in flow, through measurements of flow rate and width of the rivulet. Figure 14 shows the dependence of the magnitude of thermal apparent contact angle on wall superheat ΔT. For wall superheat below 0.17 K the value of the thermal contact angle is almost the same as the equilibrium contact angle value. This is in agreement with results of simulation presented before. As wall superheat increases higher than 0.5 K, the apparent contact angle increases up to calculated value and it tends to increase faster then in calculations. For this range of wall superheat the rivulet flow losses it's stability, it creates the longitudinal bending of the rivulet on heat transfer surface (see Fig. 13).

5. Conclusions

When evaporation occurs in mini and micro non-circular channels the microscale heat transfer to be characteristic of flow with non-uniform liquid layer thickness. For this case the capillary forces become important in determining aspects of the flow phenomena such as interface shape and local heat transfer.

- A model of gravity induced film flow is developed which can predict the shape of the interface, namely the film thickness distribution around the perimeter of rectangular minichannel. The flow is considered in terms of two components namely in the corners and on the sides of the channel. It gives the simplified equation that describes the flow interface, which is written and solved numerically. The effect of the strong capillary action that draws the liquid to the corner of the channel is noted. By comparing the model with the experimental data, it is shown that it predicts well the experimental results on interface shape and flow rate redistribution.
- A model of deformed evaporating liquid film, which is moved by co-current vapor flow and gravity, is developed. Shear stress from co-current vapor is included as a boundary condition on interface. Intermolecular forces are taking into account as disjoining pressure component and surface roughness is considered also.
- Numerical procedure to calculate the heat transfer during evaporation has been developed for the rectangular minichannel. Some conductive calculations are shown to compare with the experimental data. They illustrate the non-uniform nature of local heat transfer around the perimeter of the channel. Rivulet and dry spot formation are described. Numerical calculations show the enhancement of the heat transfer at large liquid Reynolds numbers caused at least in part by liquid suction to channel's corner where the area of microscale heat transfer arises. At small liquid Reynolds numbers the heat transfer reduction occurs due to dry spot formation. By comparing the model with the experimental data, it is shown that these conclusions are in consistent with the experimental results.
- Integral flow model and creeping flow model are developed for evaporating rivulet spreading on a vertical plate with given wall temperature. Numerical calculations show that rivulet is contracted on the heated wall and not spread. The reason is a very rapid evaporation of liquid in vicinity of liquid-vapor-solid contact line where the liquid film has a minimal thickness and microscale heat transfer arises. This rapid evaporation cannot be compensated by capillary inflow of liquid. Finally the thermal apparent contact angle establishes which is higher than the equilibrium contact angle; it is increased together with the wall superheat both for creeping and inertial flow models. By comparing the models with the experimental data, the range of wall superheat was defined where the simulation results are in consistent with the experimental results.

- The study of evaporating rivulet flow shows that macroscopic flow and heat transfer may be considerably changed itself by microscopic heat transfer, the example is establishing of thermal apparent contact angle in case of rivulet evaporation. Extremely high evaporation rate in thin film area changes the slope of interface as far as macro flow curvature and wave patterns.

NOMENCLATURE

a, Half width of the channel, [m];
A_0, Hamaker constant, [Nm];
b, Half width of rivulet, [m];
Bo, Bond number, [-];
d_h Hydraulic diameter, [m];
G, Mass velocity, [kg/(m^{2}s)];
$\mathbf{g}_e$, Effective gravity vector, [m/s^2];
g, Gravity vector x-component, [m/s^2];
G_0, Non-dimensional criteria, [-];
Ga, Non-dimensional criteria, [-];
h, Local heat transfer coefficient [W/m^2 K];
H, Volumetric heat generation rate, [W/m^3];
e_{ij}, Tensor of deformation rates, [N/m^2];
i_{LV}, Latent heat of evaporation, [J/kg];
j, Mass flow rate, [kg/m^{2}s];
k, Liquid surface curvature, [1/m];
m,h, Non-dimensional film thickness, [-];
n_i, Normal vector on liquid surface, [m];
p, Pressure [Pa];
Q, Flow rate [m^3/s]
Q_l, Flow rate per unit length, [m^2/s]
q_h, Heat flux density, [W/m^2];
R, Radius of curvature, [m];
r, Non-dimensional radius of curvature, [m];
Re, Reynolds number [-];
T, Temperature, [K];
t, Time [s];
t_i, Components of tangential vector [m];
U, Velocity vector;
u,v,w, Velocity vector components [m/s]
x,y,z, Cartesian coordinates, [m];

Greek symbols

χ, Shear stress on interface, [N/m^2];
δ, Film thickness, [m];
δ_0, Characteristic film thickness, [m];
ε, Non-dimensional criteria, [-];
γ, Non-dimensional criteria, [-];
κ, Non dimensional shear stress, [-];
λ, Heat conductivity, [W/(mK)];
μ, Dynamic viscosity, [kg/(ms)];
ν, Kinematic viscosity, [m^2/s];
ρ, Density, [kg/m^3];
σ, Surface tension, [N/m];
θ, Contact angle, [degree];
τ_i, Components of binormal vector [m]
ΔT, Wall superheat, [K];

Subscripts

e, External, effective;
f, Film;
d, Disjoining;
L, Liquid;
m, Meniscus;
V, Vapor;
s, Saturated;
x,y,z, Derivatives with respect to x, y and z;
w, Wall

REFERENCES

1. Abramson, A.R. and Tien, C.-L., (1999) Recent Developments in Microscale Thermophysical Engineering, *J. Microscale Thermophysical Engineering*, vol. 3, pp. 229-244.
2. Tien, C.-L., Majumdar, A. and Gerner, F.M., (ed.) (1998) Microscale Energy Transport, Taylor & Francis, Washington.
3. Wayner, P.C., (1998) Interfacial Forces and Phase Change in Thin Liquid Films, *Microscale Energy Transport,* Taylor & Francis, Washington. pp. 187-226.
4. Palm, B, (2001) Heat Transfer in Microchannels, *J. Microscale Thermophysical Engineering*, vol. 5, pp. 155-175.
5. Stephan, P. (2002) Microscale Evaporative Heat Transfer: Modelling and Experimental Validation, *Proceedings of 12th International Heat Transfer Conference*, Paper No. 08-KNL-02, Vol. 1, pp. 315-327.
6. Wayner, P.C., (1999) Intermolecular Forces in Phase – Change Heat Transfer, 1998 Kern Award Review, *AIChE Journal*, vol. 45, no. 10, pp. 2055-2068.

7. Pratt, D.M., Brown, J.R. and Hallinan, K.P. (1998) Thermocapillary Effects on the Stability of a Heated, Curved Meniscus, *ASME J. Heat Transfer*, vol. 120, pp. 220-226.
8. DasGupta, S., Schonberg, J.A., Kim, I.Y. and Wayner, P.C, (1993) Use of the Augmented Young-Laplace Equation to Model Equilibrium and Evaporating Extended Menisci, *J. Colloid Interface Science*, vol. 157, pp. 332-342.
9. Kim, I.Y. and Wayner, P.C. (1996) Shape of an Evaporating Completely Wetting Extended Meniscus, *J. of Thermophysics and Heat Transfer*, vol. 10, no. 2, pp. 320-325.
10. Morris, S.J.S. (2000) A Phenomenological Model for the Contact Region of an Evaporating Meniscus on a Superheated Slab, *J. Fluid Mechanics*, vol. 411, pp. 59-89.
11. Morris, S.J.S. (2001) Contact Angles for Evaporating Liquids Predicted and Compared with Existing Experiments, *J. Fluid Mechanics*, vol. 432, pp. 1-30.
12. Sharma, B (1998) Equilibrium and Dynamics of Evaporating or Condensing Thin Fluid Domains: Thin Film Stability and Heterogeneous Nucleation, ACS J. Langmuir, vol. 14, pp. 4915-4928.
13. Padmakar, S., Kargupta, K. and Sharma, A. (1999) Instability and Dewetting of Evaporating Thin Water Films on Partially and Completely Wetting Substrates, *J. Chemical Physics*, issue 110, pp. 1735-1744.
14. Hirasawa, S., Huikata, K., Mori, Y. and Nakayama, W. (1980) Effect of Surface Tension on Condensate Motion in Laminar Film Condensation (Study of Liquid Film in a Small Through), *Int. J. Heat Mass Transfer*, vol. 23, pp.1471-1478.
15. Kuznetsov, V.V., Safonov, S.A., Sunder, S. and Vitovsky, O.V. (1997), Capillary Controlled Two-Phase Flow in Rectangular Channel, *Proc. Int. Conf. on Compact Heat Exchangers for Process Industries*, Snowbird, pp.291-304.
16. Kozelupenko, Yu.D., Smirnov, G.F. and Koba, A.L. (1985) Heat transfer crisis in subcooled liquid in narrow annular channels at low velocities of motion, *Promushlennaya Teploenergetica*, 7, 1, pp 30-32.
17. Kuznetsov, V.V., Safonov, S.A., Shamirzaev, A.S., Houghton, P.A. and Sunder, S.D. (1999) Two-Phase Flow Patterns and Local Evaporating Heat Transfer in a Vertical Rectangular Channel, *Proc. Int. Conf. Compact Heat Exch. and Enh. Tech. Proc. Ind.*, Banff, Canada, pp. 311-320.
18. Kuznetsov, V.V., Safonov, S.A., Shamirzaev, A.S. and Sunder, S.D. (1998) Two-Phase Flow Pattern and Local Boiling Heat Transfer in Vertical Rectangular Channel, *Proc. 11th Int. Heat Transfer Conf.*, Kyongju, Korea, pp. 33-38.
19. Asali, J.C., Hanratty, T.J. and Andreussi, P. (1985) Interfacial Drag and Film Height for Annular Flow, *AIChE J.*, vol. 31, pp. 886-902.
20. Gimbutis, G. (1988) Heat Transfer of a Falling Fluid Film, Mokslas, Vilnius.
21. Nakoryakov, V.E., Pokusaev, B.G., and Shreiber, I.R. (1983) Wave Propagation Processes in Gas- and Vapor-Liquid Media, Institute of Thermophysics, Novosibirsk.
22. Kuznetsov, V.V., Safonov S.A. and Vitovsky O.V. (2002) Heat Transfer in the Contact Line Region of Evaporating Curved Microfilm, *Compact Heat Exchangers, Proc. Intern. Symp. on Compact Heat Exchangers*, Edizioni ETS, Piza, pp. 413-418.
23. Kuznetsov, V.V., Vitovsky, O.V., and Krasovsky, V.A. (2004) Experimental Study of Flow Regimes of the Evaporative Liquid on Vertical Heated Surface, *Proc. XXVII Siberian Thermophysical Seminar,* Institute of Thermophysics, Moscow-Novosibirsk, paper No. 077.

ULTRA – THIN FILM EVAPORATION (UTF) – APPLICATION TO EMERGING TECHNOLOGIES IN COOLING OF MICROELECTRONICS

M. OHADI and J. QI
Department of Mechanical Engineering Department – University of Maryland
College Park, Maryland, USA

J. Lawler
Advanced Thermal and Environmental Concepts, Inc.
387 Technology Drive
College Park, Maryland, USA

1. Introduction

The steady increase in microprocessor performance over the past three decades in accordance with "Moore's Law" [1] has introduced new opportunities and interesting challenges in thermal management of electronics (Figure 1). Although exponential growth may slow down at some point, the industry expects to continue to sustain Moore's law beyond this decade.

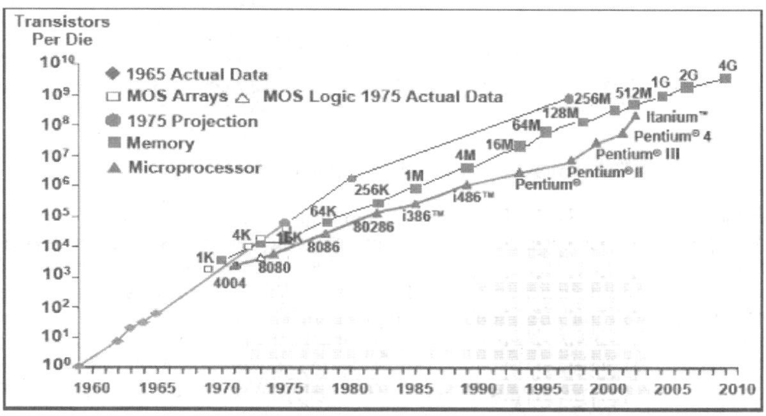

Figure 1: History of growth projection of transistors per die following "Moore's Law" [2]

S. Kakaç et al. (eds.), Microscale Heat Transfer, 321–338.

Transistor gate dimensions have been reduced 200x during the past 30 years (from 10mm in the 1970s to a present-day size of 60nm). The current state-of-the-art semiconductor chips feature size scaling dimensions of 0.13 micron (130 nanometers) entered the marketplace in 2001 and is expected to reduce to 0.045 microns (45 nanometers) by year 2007 (Figure 2). Transistor and feature size scaling have enabled microprocessor performance to increase exponentially with transistor density and microprocessor clock frequency doubling every two years. The 10-micron circuits of the first commercial microprocessor in 1971 are enormous in comparison to the latest generation of 0.13-micron chips. The number of transistors has grown from less than 3,000 in 1971 to more than 42 million on today's most advanced microprocessors and is expected to reach one billion transistors on the chip by year 2007, or four years ahead of the Moore's chart.

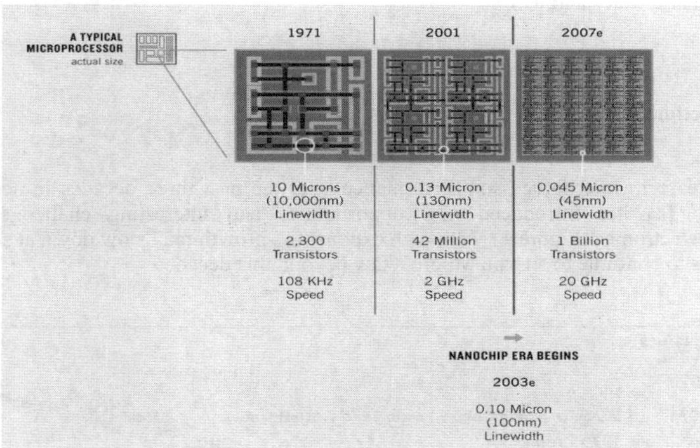

Figure 2: The number of transistors on a chip between 1971 and 2002 [3]

Over the early twenty years (1960s-1980s), chip heat generation increased by just one order of magnitude to approximately 1-5W in the 10mm, large-scale integration (LSI) bipolar devices and very large scale integration (VLSI) devices of the mid-1980s. But, by the early 1990s, larger, faster complementary metal-oxide-semiconductor (CMOS) chips pushed power dissipation to the range of 15-30W and set the stage for the prodigious thermal management needs ahead. In 1992, a 486 processor consumed only 5 Watts. Today's Pentium 4 consumes more than 80 Watts, about a sixteen fold increase. The recently released Itanium II has a thermal design power of 130 Watts. Intel's next 64-bit processor is expected to consume between 140 and 180 Watts. Projections of the 2002 Update Edition of the International Technology Roadmap for Semiconductors (ITRS) state that by the year 2010, on-chip clock frequencies and power for high performance processors will reach approximately 12 GHz and 120 Watts [4]. The increases in power and heat flux are driven by reduced feature sizes (small die) and higher frequencies. Additionally, lower operating temperatures are required to improve device reliability and performance. Each 10ºC drop in transistor temperature results in a 1-3% increase in semiconductor switching time [5]. The maximum junction temperature at the chip was reduced in 1999 from 100ºC to 85ºC, thus shrinking temperature budgets for thermal design. In the absence of cooling, the temperature of electronic components would rise until reaching a value at which the electronic operation of the device ceases or the component losses its physical integrity, resulting in device failure. Heat is one of the primary sources of electronic hardware failure.

The trend in microelectronic and optoelectronic devices is to increase the level of integration by minimizing the device size (higher density packaging) and rapidly increasing the performance of the device (higher frequency). This results in an increase in both power dissipation and power density on the device. Some of these next-generation devices are projected to dissipate local heat flux over $1000 W/cm^2$. Such high heat fluxes present a serious challenge to existing thermal management techniques to ensure device performance and reliability while maintaining acceptable temperatures. The critical need for advanced thermal management technologies capable of dissipating such high heat fluxes with low input power is thus well recognized in the electronics industry [6-7].

In this paper, we discuss several thermal management techniques that could serve as an effective means to cool the next generation high flux electronics. These include: immersion cooling, jet impingement, spray cooling, and ultra thin film evaporation (UTF), particularly to serve at the chip level for hot spot cooling of the chip. We will discuss the potential of these techniques for thermal management of current high flux electronics as well as the promise they hold for spot cooling of the next generation high flux electronics for commercial and military sectors.

2. Overview of Thermal Management Techniques

Thermal management techniques can be classified as passive, active, or mixed. Passive methods don't need any input power, and tend to be very reliable and relatively easy and low cost to implement. For low heat fluxes, passive thermal management techniques can be used that do not require expending external energy for the heat removal. Interest in such techniques continues very strongly, due to their simplicity, low cost, and high reliability. However, with reasonable pressure drop and noise levels and for cost effective utilization, only up to a few W/cm^2 can be removed by passive cooling techniques, thus inadequate for many high power applications. Active thermal management techniques can provide increased performance and thermal capacity, however they require external power and often present lower reliability and added cost and complexity.

In the 1960s, the power consumption/dissipation of early electronics was high, notably in response to the development of the solid-state transistor. Since there was no obvious spatial constraint, providing high capacity cooling in large rooms environment was the norm. Much of the technical effort during this decade was devoted to applying, documenting, and standardizing conventional air- and liquid-cooling techniques [8]. In 1990s the aggressive introduction of CMOS technology that continued to replace Bipolar technology amplified the false sense of security from heat issues. However, the CMOS technologies only bought the industry 10 years (Figure 3), before heat flux levels comparable to Bipolar technologies evolved.

In the early 1990s, with the popularity of and easy access to personal computers and the Internet, a paradigm shift occurred in the thermal management of electronics. The issue was speed. Mass access to electronics and higher consumer expectations for speed and miniaturization forced many companies to provide higher speed systems and services. Speed requirements necessitated smaller devices and systems. Consequently, power dissipation became a major challenge and its management is a critical technology in the electronics industry. Moreover, nothing indicates that circuit speed or space requirements will decrease. To address current and future issues, technological advancement is the only savior; that is the introduction of new interfacial and heat sink materials, new thermal management technologies, and a paradigm shift in thermal management technology concepts and modes of implementation.

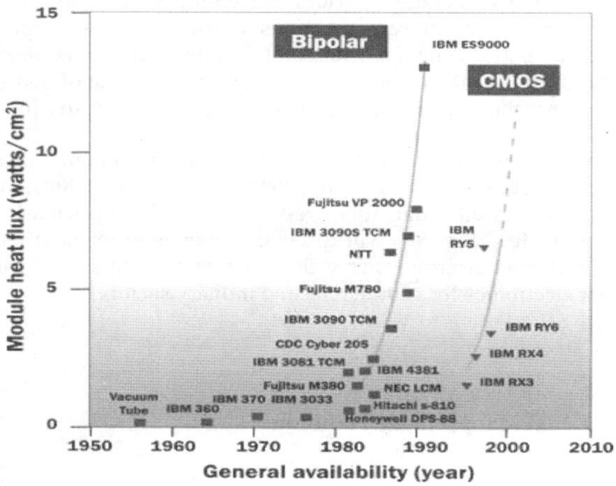

Figure 3: Trend of power dissipation [9]

In recent years, the successful application of thermal science and insightful thermal optimization have stabilized component temperatures at values below 100°C, despite this rapid increase in chip power dissipation, and even in the presence of a hostile external environment. Thermal management is gradually regarded as a key enabling technology in the development of advanced microelectronic systems and has facilitated many of the "Moore's Law" advances in consumer products and high-performance electronic devices that have defined the latter part of the 20th century. The recent SIA's "International Technology Roadmap for Semiconductors-2002 Edition" [4] and NEMI's 2000 edition of the "National Electronic Manufacturing Technology Roadmap" [10] have continued to affirm the expectation that "Moore's Law" improvements in CMOS semiconductor technology will occur up to 2006, at least.

Among various challenges, air-cooling of high flux electronics equipment continues at the forefront of electronic thermal management, partly due to its low cost and simplicity, as well as the fact that for most cases air is the ultimate heat sink. Air cooling is commonly used in commercial electronics such as desktop and notebook computers. State-of-the-art air cooling includes the use of air in a natural or forced convection over a finned surface, with a pin-fin heat sink being particularly common. Additional heat removal capacity is often achieved with increased fan speed, to achieve higher turbulence. However, with the increased power dissipation and the trend towards miniaturization from handheld electronics to high power electronics and mainframes, conventional finned-heat sink air cooling schemes are no longer capable of meeting the cooling demands of high performance electronics. The ever increasing packaging density in electronic systems requires a substantial enhancement of air-side heat transfer and, in some cases, its substitution with more complicated liquid or refrigeration cooling systems to achieve higher heat transfer coefficients.

Liquid cooling and refrigeration cooling techniques have been successfully introduced in high performance computers such as the Cray-2 supercomputer using immersion cooling [11], CDC's ETA-10 supercomputer using liquid nitrogen [12], and the IBM S/390-G5 sever using refrigeration [5]. Current microprocessors have an average heat flux of 10-50W/cm^2; future high performance microprocessors will have a heat flux above 100W/cm^2 by the end of this decade [10]. However, high-end military and aerospace wide band-gap amplifier, electromagnetic weapons (EMWs), and high power radar electronics will produce hot spots with heat fluxes on the order of 1000W/cm^2 [6-7]. Under such ultra high heat fluxes, existing thermal management can't meet the associated cooling requirement. Table 1 shows heat transfer coefficients for various processes [14]. From the data in Table 1, it is rather clear that for reasonable pressure drops and noise levels, traditional air cooling will not be able to address the cooling needs of high flux electronics. Single-phase liquid cooling is limited to a few hundred W/cm^2 for practical systems. Two-phase liquid systems, particularly like ultra thin-film (UTF) cooling, have the potential to provide cooling heat fluxes of several thousand W/cm^2 [6-7 and 15].

Table 1: Typical values of heat transfer coefficients

Description	Heat Transfer Coefficient (W/m^{2}K)
Natural convection, air	3-25
Natural convection, water	15-1000
Forced convection, air	10-200
Forced convection, water	50-10,000
Condensing steam	5000-50,000
Boiling water	3000-100,000
Ultra thin film evaporation	10,000-500,000
Microchannel Cooling	10,000-1,000,000

Spot cooling involves development of single phase or phase change heat transfer systems that rely on highly effective thermal management techniques. Additionally, the device must operate in near room temperature (300K), posing more engineering challenges. The high thermal load and exposure to harsh environmental temperatures complicate the development of advanced thermal management systems. Temperature stabilization is another very important issue that requires careful distribution of cooling medium over the device. Among other techniques, the next generation advanced thermal management schemes such as immersion cooling, jet-impingement, spray, and ultra thin film evaporation hold a promise to address the future needs of high flux electronics and are briefly discussed.

3. Emerging Thermal Management Techniques

3.1 IMMERSION COOLING

Immersion cooling, where the electronics are directly immersed in a pool of inert liquid, has been studied since the 1970s [16]. The pool boiling critical heat flux, or CHF, places an upper limit on this cooling mode and has been received considerable attention in the past decade. Although the effects of fluid properties, pressure, and subcooling, as well as heater geometry, on CHF, are relatively well established, explanations for the surface property effects remained controversial. Recently, a composite correlation for pool boiling CHF, embodying a dependence on the thermal effusivity of the surface material, and accounting for the hydrodynamic limits, as well as the effects of pressure, subcooling, and length, shown in Equation (1) below, was proposed [17]:

$$
\begin{aligned}
q_{CHF} &= \left\{ \frac{\pi}{24} h_{fv} \sqrt{\rho_v} \left[\sigma_f g (\rho_f - \rho_v) \right]^{1/4} \right\} \times \\
&\left\{ \frac{S}{S+0.1} \right\} \times \left\{ 1 + < \mathrm{a} - bL'(P) > \right\} \times \\
&\left\{ 1 + B \left[\left(\frac{\rho_f}{\rho_v} \right)^{0.75} \frac{c_f}{h_{fv}} \right] \Delta T_{sub} \right\}
\end{aligned}
\qquad (1)
$$

This effusivity-enhanced correlation was shown to predict a broad range of smooth surface dielectric liquid pool boiling CHF data, ranging from 6W/cm^2 to values as high as 60W/cm^2. It was found that the proposed correlation could predict the pool boiling critical heat flux with a standard deviation of 12.5% for horizontal heaters of various materials and geometries in a large range of subcooling and pressure (0-75 K and 1-4.5 Bar) conditions within a 95 percent confidence level.

A great variety of treated surfaces have been tested in the laboratory to investigate potential application in boiling heat transfer enhancement and boiling microporous coating surface [18-19], 3-D-porous heat sink studs [20], microfins, and micro-configured surfaces [13, 21-22]. More recently, Rainey and You [23] proposed a "double enhancement" technique using conventional extended surface techniques (pin fins) with the microporous coating surface. Critical heat flux (CHF) could be increased from about 16W/cm^2 for a flat, plain surface heater to 130W/cm^2 for a finned surface with microporous coating. The considerable heat transfer enhancement was achieved by using micro-pin-fins with submicron-scale roughness as opposed to a smooth surface in the nucleate boiling region [24]. The boiling enhancement is mainly attributed to increasing the heat transfer surface, bubble departure frequency, and number of nucleation sites per site. Moreover, electrohydrodynamic (EHD) enhanced boiling was reported under both normal gravity [25-26] and microgravity [27]. Since the power consumption of EHD devices is very low and no moving parts are required, this technique appears promising for enhancement of pool boiling over and above the base case (no EHD) for both smooth and enhanced surfaces [28]. Watwe demonstrated significant CHF enhancement by adding to the commonly used FC-72 dielectric liquid low concentrations of FC-40, which had a higher saturation temperature, higher molecular weight, higher viscosity, and higher surface tension than FC-72 liquid [29]. As seen in Figure 4 [30], increasing concentrations of the higher boiling point liquid shift the boiling curve to the right and also lead to higher CHF. The observed shift in the wall superheat, for the higher concentrations of FC-40, was used to relate the CHF enhancement to localized depletion of the lower boiling point liquid in the near heater regions.

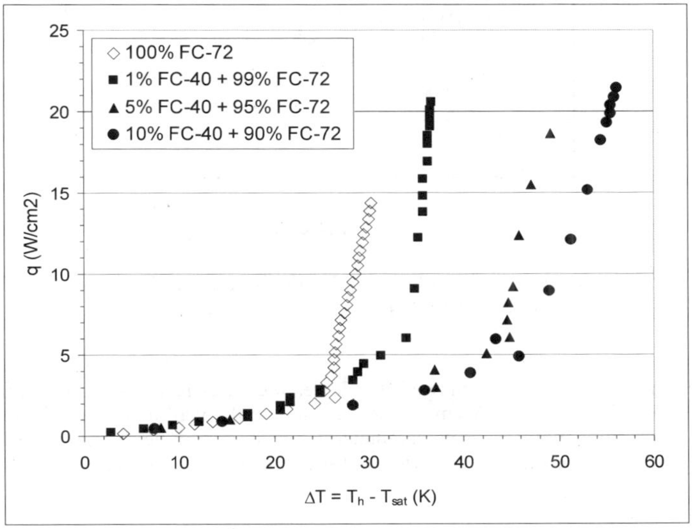

Figure 4: Pool boiling curves for FC-40/FC-72 mixtures at 101.3 kPa

Much research has been done on enhanced evaporators in thermosyphons for the cooling of electronics. Copper enhanced microstructure thermosyphon evaporators consisting of a stack of six copper plates, which are grooved with rectangular microchannels, have been investigated by Ramaswamy et al. [31] and Pal et al. [32]. They reported heat transfer rates up to 100 W/cm^2. The use of graphite foam as the evaporator in a thermosyphon was investigated by Coursey et al. [33]. These foams consist of a network of interconnected graphite ligaments whose thermal conductivities are up to five times higher than copper. Performance of the system with both PF-5060 and PF-5050 were examined as well as the effects of liquid level and chamber pressure (Figure 5).

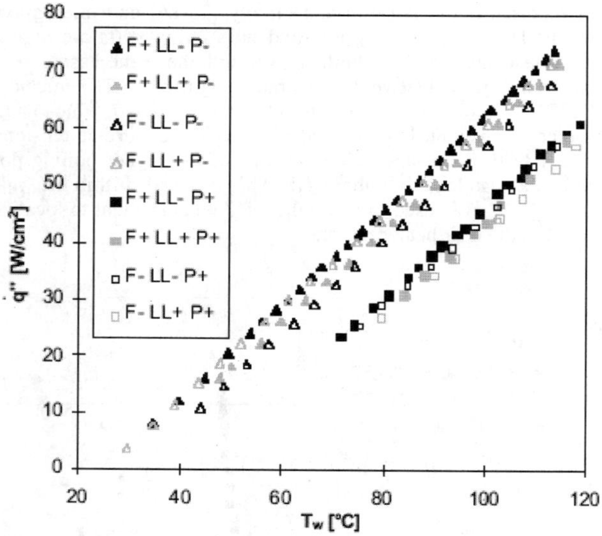

Figure 5: Nucleate boiling curves. F + (PF-5060), F- (PF-5050), LL+ (high liquid level), LL- (low liquid level), P+ (high pressure), and P- (low pressure)

The preliminary investigation showed that lower pressures result in significantly higher heat fluxes. It's possible to approach 50 W/cm^2 while maintaining the wall temperature below 85 °C with little optimization. A more comprehensive parameterization study will be addressed such as pore size, geometry, and other effects as the limits of graphite foam evaporator performance [34-35].

3.2 JET IMPINGEMENT AND SPRAY COOLING

One effective way of cooling high power electronics is through impinging fluid jets or spray cooling because of the high heat transfer coefficients achieved at the surface (Figure 6). Both cooling methods are initiated by forcing liquid through a small nozzle and can involve single phase and/or phase-change cooling. The impinging jet can be either submerged or free. Free jet impingement leads directly to significant evaporative cooling, while submerged jets can combine forced convection with flow boiling at the wall.

One of the earliest studies was investigating boiling jet impingement of R-113 on a simulated 5mm x 5mm microelectronic chip [36]. Heat fluxes up to 100W/cm^2 were reported. In the early of 1990s, IBM conducted experiments using single phase, boiling jet impingement and subcooled liquid nitrogen jet cooling on silicon chips [37-38]. One prototype of the SS-1 supercomputer using FC-72 submerged jet impingement was reported to remove 95W/cm^2 from chips in a multichip module [39].

(a) Jet Impingement (b) Spray Cooling

Figure 6: Cooling Schemes: (a) Jet Impingement and (b) Spray Cooling

In spray cooling, liquid is shattered into a dispersion of fine droplets that impinge individually upon the heated surface. The drop impingement enhances the spatial uniformity of heat removal and prevents liquid separation from the surface during vigorous boiling [15]. A spray can be formed either by high pressure (as a plain orifice spray) or atomized pressured gas (an atomized spray). Spray cooling performs well in applications with low to moderate heat removal requirements and is incorporated into Cray's SV modular supercomputer [40]. In this application heat fluxes as high as 60W/cm^2, on a downward facing chip, were achieved at near atmospheric pressure and the performance was projected to reach values in excess of 80W/cvm2 in an optimized configuration of the Cray module.

Spray cooling is also under evaluation for cooling commercial-off-the-shelf (COTS) electronics in harsh military environments like the Marine Corps Advanced Amphibious Assault Vehicle (AAAV) [41]. Jet impingement offers better cooling performance in low to moderately high heat flux applications. One disadvantage of jet-impingement cooling is that the concentration of heat removal within the impingement zone causes relatively large temperature gradients within the cooled device. Moreover, dryout occurs on the chip surface when the chip power is increased too quickly. Spray cooling provides some advantages over jet impingement such as smaller pressure, lower fluid volume, more uniform surface temperature, very small temperature overshot, and no surface erosion [42]. Practical application of jet impingement and spray cooling is determined by the reliability of fluid pump and the variation in nozzle performance due to contamination, corrosion, and clogging.

In both spray and jet impingement cooling technologies, the geometry and the placement of liquid nozzles and the flow rate of liquid into the nozzle must be designed very carefully to deliver liquid droplets of a certain size and quantity to generate and maintain the desired thin film of liquid. Insufficient delivery of the droplets will lead to uncovered surfaces, and excessive delivery will generate liquid films exceeding optimal thin film thickness. Producing extremely small droplets and achieving complete surface wetting with mechanical sprays is extremely difficult. The existing literature does not seem to have any published data on thermal mapping of a typical sample being cooled by spray cooling. The issues of space for nozzles and other accessories, such as fluid pump and filters, are other concerns for application.

3.3 ULTRA THIN FILM EVAPORATION COOLING

Thin film evaporation is one of several types of phase-change cooling processes. The phase change of a working fluid is recognized as an effective method to remove a large quantity of heat from a hot surface and with careful design relatively small temperature gradients across the surface can be achieved. The heat is absorbed by the working fluid as heat of vaporization or latent heat. The two main regimes of phase change for liquid converted to vapor cooling are: boiling and thin film evaporation. Boiling is a well-known heat transfer mechanism in which a superheat layer of liquid forms next to the heat transfer surface and helps the growth of nucleating bubbles on the surface cavities. Although pool boiling has been shown to yield cooling rates well above $50W/cm^2$ it use is limited by several concerns, among them a minimum superheat required for boiling inception, a relatively thick thermal boundary layer, and an inherently low critical heat flux. These limitations can be overcome if a thin film (several microns) of the working liquid continuously covers the heated surface. In operation, a temperature gradient forms across the film via heat conduction and the liquid simply vaporizes at the liquid-vapor interface. This process removes a very large amount of heat (Table 1) because the amount of heat removed is inversely proportional to the thickness of this thin liquid layer. Ultra Thin Film (UTF) evaporation is perhaps one of the most effective methods of heat removal from a high heat flux surface for several reasons:

- a small quantity of fluid is required to remove the heat by evaporation at the surface of the thin layer of fluid,
- a very high heat transfer coefficient results from the minimized thermal resistance across the thin liquid layer,
- the surface experiences a very small temperature rise above the saturation temperature of the working fluid, as long as a sufficient quantity of fluid is provided to wet the surface,
- a minimum amount of energy is required to circulate the working fluid due to no pressure drop across the thin film evaporator,
- unlike nucleation boiling where a complex set of parameters determine the stability of the system, UTF is virtually conduction across a very thin film, thus it can easily be modeled for performance characterization, and
- the upper limit on cooling performance would be limited by the homogeneous nucleation temperature and/or kinetics of vapor formation at a free interface, rather than the relatively low CHF.

Delivery of a continuous sub-micron (ideally) thin film and an in-depth understanding of the liquid-vapor interfacial forces and the resulting liquid-to-vapor phase transformation is critical to further development of UTF devices to meet high heat flux demands of the next generation electronics. This work is currently under progress through several on going projects in this field, through joint collaboration of Advanced Thermal Environmental Concepts Inc. (ATEC) at College Park, Maryland and the authors' thermal management group at the university of Maryland.

An alternative new technology, an electrohydrodynamic (EHD)-enhanced polarization pump, is being developed to deliver and maintain an ultra thin film of the fluid on a surface (Figure 7), which will enable high efficiency spot cooling of high heat flux electronics [6]. An array of EHD electrodes acts to draw an ultra thin film of fluid over the hot surface. The working fluid evaporates and removes its heat of vaporization from the surface [43]. EHD pumping is a promising option that offers the unique advantage of no moving parts and high reliability. The use of the EHD technique for microscale fluid pumping has been investigated by a number of researchers over the past decade [28]. In these studies, pump designs were based on either ion-drag [44] or induction pumping [45]. In ion-drag and induction pumping, the free charges and dipoles within the liquid are the driving forces that pump the liquid. This results in a high power consumption and a low pumping efficiency.

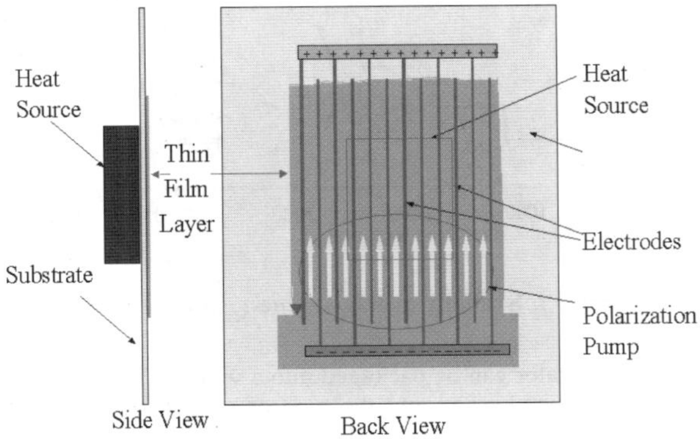

Figure 7: Conceptual Drawing of EHD Pumped Thin Film Evaporator

In a polarization pump, the average polarization force over all dipoles induces a body force on the dielectric liquid. Thus, the power required to achieve pumping is very low, leading to higher pumping efficiencies. Other benefits of the EHD pumping technique include low cost, very low power consumption, and minimal maintenance.

The thin film evaporator operates based on action of three forces applied by an electric field to polar molecules of a liquid. The first force pumps the liquid between the two electrodes that are generating the electric field. A second force generated due to the difference between the dielectric constant of the vapor and the liquid phase forces the liquid layer to the surface. A third force pushes down the liquid column raised between the two electrodes. A balance between the first and the third forces, gravity, and viscous forces at the fluid-solid interface determines the height of the liquid between the electrodes.

In the first generation prototype UTF system a pumping head of 250Pa and a maximum cooling capacity of 65W/cm^2 were reported [26] at an applied voltage of 150V for a 50 micron electrode gap when using R-134a as the working fluid (Figure 8). The total EHD power consumption is less than 0.02% of the total power input to the device, translating into a few milli Watts for the application at hand. In future designs, the voltage is being further reduced to the low DC voltage by scaling down the devices and reducing the electrode gap sizes. For example, 9-15V can be achieved by reducing the electrode gap to 3-5 microns.

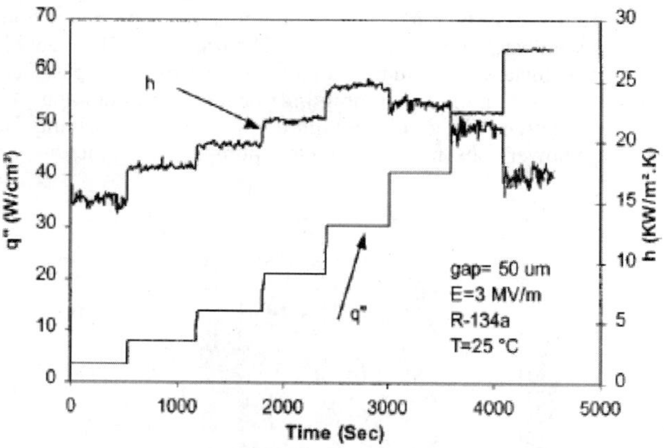

Figure 8: Spatially averaged time-resolved results

The ultra thin film evaporator can be packaged into a wide variety of configurations depending on the electronic components to be cooled and the type of secondary cooling options. A prototype based on a thermosyphon-type of cooling device (Figure 9) has been fabricated and tested [46]. The vapor leaving the thin film evaporator flows to the water cooled condenser wherein the vapor is condensed back to a liquid that then flows via gravity to a fluid reservoir at the base of the thin film evaporator.

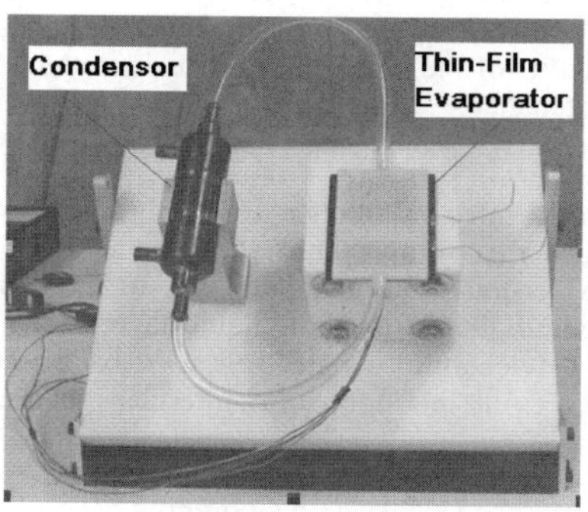

Figure 9: A photograph of the thermosyphon test system

A comparison of cooling performances of thin film evaporator with the performances of pool boiling and spray cooling techniques is shown in Figure 10. The data for pool boiling and spray cooling are reported by Bar-Cohen et al. [30] and Mudawar [15] for the 3M thermal fluid FC-72, which is quite similar to the 3M thermal fluid HFE-7100 used in the thin film evaporator test data as shown in Figure 10.

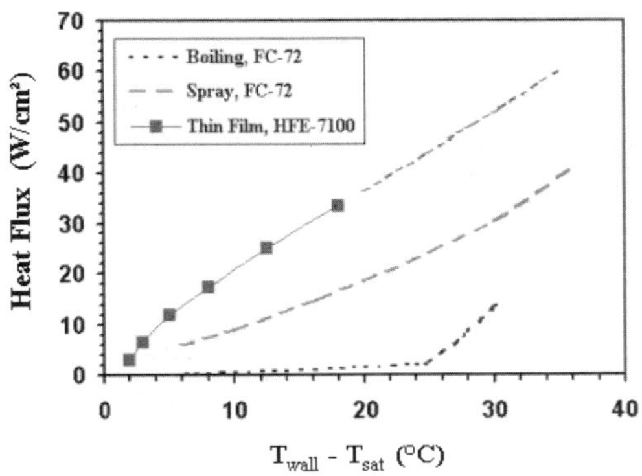

Figure 10: Comparison of cooling performances of thin film evaporator with the performances of pool boiling and spray cooling techniques

The thin film evaporator can remove heat fluxes of 20-40W/cm^2 with a temperature difference that is about 10-15°C less than spray cooling and about 30°C less than pool boiling. This reduced temperature difference is a big advantage, since almost all thermal management systems are required to operate within a very narrow temperature range to allow the ultimate transfer of this heat into warm ambient air or heat exchanger, often with an intervening secondary loop. Further optimizing the electrode pattern on the thin-film evaporator will generate ultra thin (micron-size) films on evaporator surfaces, higher cooling rates, and a more robust operation.

Cooling systems based on the ultra thin film evaporator will be quiet, compact, light-weight, and energy efficient. Their high rate of cooling will minimize the operating temperature of the electronic components being cooled. The MEMS/IC-compatible electrode fabrication process enables a wide latitude in the configuration of the cooling device architecture. One such architecture, a circuit board with embedded thin film evaporators, is shown in Figure 11.

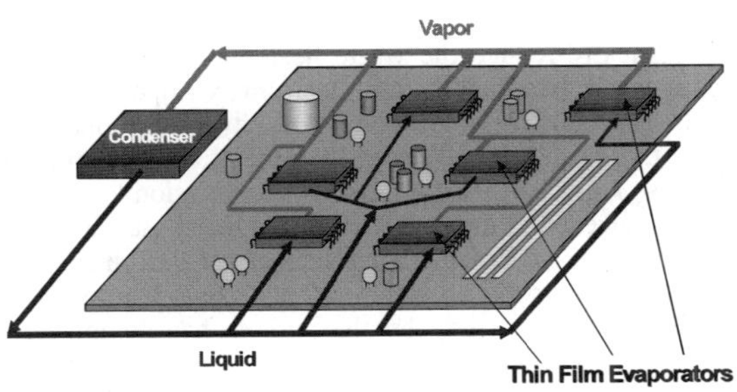

Figure 11: One example of thin film cooling architecture [Courtesy of ATEC]

3.4 SPOT COOLING WITH TE-POWER THIN FILM EVAPORATOR

An emerging solution for localized heat removal from high flux electronics is through "spot cooling" devices that incorporate both thin film evaporators and thermo-electric power generators (Figure 12). Waste heat from high flux electronics is used to generate power by the thermoelectric device, which in turn will locally power the thin film evaporator and the micro pump that may be needed to circulate the condensate back to the evaporator.

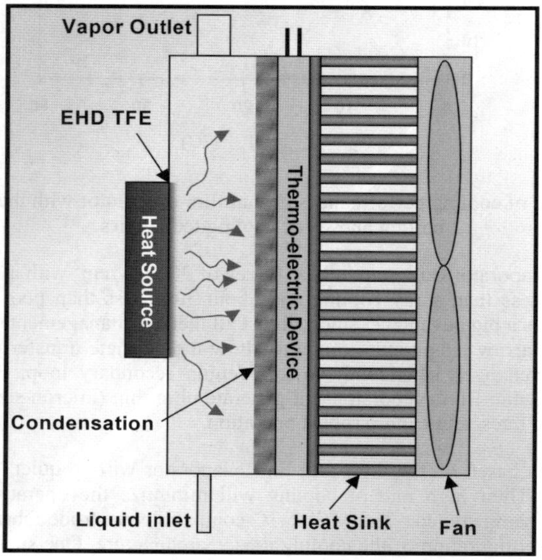

Figure 12: Schematic of a thermoelectric-powered thin film evaporator [Courtesy of ATEC]

This system will drastically reduce the waste heat that is entering the air space of systems surrounding the electronics, while maintaining the electronic components at a reduced temperature despite their high heat fluxes, yielding improved reliability. Additionally, this approach greatly increases the impact of the thermo-electric devices by generating and utilizing the electrical power locally within cooling devices. This configuration avoids transmission losses since the power does not have to be wired to cooling devices, and the power from the thermo-electric devices does not have to be wired into the power system of the cabinet. Rather extensive research is already in place to develop the future TE power generators, applicable in electronic packages, inherently requiring very thin and highly conductive materials. An earlier study by the authors has demonstrated the thermoelectric-driven cooling concept via TE-driven fans that cool the electronics. Details of this study can be found in Yazawa et al [47].

4. Conclusions

In this paper we briefly reviewed the existing and future thermal management technologies for thermal management of high heat flux electronics. Among other next generation technologies, the EHD-enhanced ultra thin film evaporator has a very high potential for spot cooling of high flux electronics and may provide a way for practical utilization of ultra thin-film evaporation in many electronic cooling applications. One unique feature of this novel cooling technology is that it integrates an active evaporative surface and an EHD micropump into a single chip, greatly facilitating both the manufacturing process and spot cooling at the chip level, enabling further system-level miniaturization. This novel technology can substantially increase cooling capacity and lead to substantial weight/volume reductions in the next generation thermal management systems. Among other benefits, only a fraction of the amount of fluid used in spray or jet cooling systems is necessary for UTF and with few ancillary components. Its power consumptions are extremely low and thus it is highly energy efficient for cooling high heat-flux devices. Advances in thermoelectric (TE) power generation may lead to incorporating TE power generators into EHD-pumped thin film evaporators to convert waste heat from high power electronics into electrical power that can be used for the EHD-micropumping due to the low voltages and low power consumption in EHD polarization mechanism.

NOMENCLATURE

AAAV	Advanced Amphibious Assault Vehicle
CHF	Critical Heat Flux
CMOS	Complementary Metal-Oxide-Semiconductor
COTS	Commercial Off-The-Shelf
EHD	Electrohydrodynamics
EMWs	Electric Magnetic Weapons
HFE	Hydro-Flouro-Ether
FC	Fluorochemical
IC	Integrated Circuit
ITRS	International Technology Roadmap for Semiconductors
LSI	Large-Scale Integration
MEMS	Micro-Electro-Mechanical System
NEMI	National Electronics Manufacturing Initiative
SIA	Semiconductor Industry Association
TE	Thermoelectric
TFE	Thin-Film Evaporator
UTF	Ultra Thin Film
VLSI	Very Large Scale Integration

REFERENCES

1. Moore, G.E., 1975, "Progress in Digital Integrated Electronics," Proceedings of the IEEE International electron devices Meeting, December, 1975, pp 11-1
2. Moore, G. E., 2003, "No Exponential is Forever, but We Can Delay 'Forever'", International Solid State Circuits Conference (ISSCC), February 10, 2003
3. Applied Materials Quarterly Report, April 2002
4. ITRS-2002 Update Edition, "International Technology Roadmap for Semiconductors – Assembly and Packaging"
5. Schmidt, R., 2000, "Low Temperature Electronic Cooling", Electronics Cooling, Vol.6 (3), pp.18-24
6. Ohadi, M.M., Bar-Cohen, A.V., and Cerza, M., 2002, "Thermal Packaging of High Flux Military and Commercial Electronics," Proceedings of the 1st annual workshop held on October 7, College Park, MD.
7. Ohadi, M.M., Bar-Cohen, A.V., and Cerza, M., 2003, "Thermal Packaging of High Flux Military and Commercial Electronics," Proceedings of the 2nd annual workshop held on October 7, College Park, MD.
8. Bergles, A.E. and Bar-Cohen, A, 1990, "Direct Liquid cooling of Microelectronic Components," in Chapter 5, in Advances in Thermal Modeling of Electronic Components and Systems, Vol 2, edited by A. Bar-Cohen and A.D. Kraus, ASME Press, pp 233-342
9. Chu, R.C., Simons, R.E., and Chrysler, G.M., 1999, "Experimental Investigation of an Enhanced Thermosyphon Heat Loop for Cooling of a High Performance Electronics Module," Proceedings of the 15th IEEE Semiconductor Thermal Measurement and Management Symposium, pp.1-9
10. NEMI-2000 Edition, "National Electronics Manufacturing Technology Roadmaps", National Electronics Manufacturing Initiative (NEMI), Inc, Virginia, US
11. Danielson, R., Krajewski, N. and Brost, J., 1986, "Cooling Superfast Computer", Electronic Packaging and Production, Vol.26 (7), pp.44-45
12. Vacca, A., Resnick, D., Frankel, D., Back, R., Kreilich, J. and Carlson, D., 1987, "A Cryogenically Cooled CMOS VLSI Supercomputer", VLSI Systems Design, Vol.8(7), pp.80-88
13. Miller, W.J., Gebhart, B., and Wright, N.T., 1990, "Effects of Boiling History on a Micro-Configured Surface in a Dielectric Liquid", International Communications in Heat and Mass Transfer, Vol. 17, pp.389-398
14. Mills, A.F., 1999, Basic Heat and Mass Transfer, 2nd, Prentice-Hall, New Jersey, p 22
15. Mudawar, I., 2001, "Assessment of High-Heat-Flux Thermal Management Schemes", IEEE Transactions on Components and Packaging Technologies, Part A, Vol. 24 (2), pp.122-141
16. Bar-Cohen, A., 1993, "Thermal Management of Electronic Components With Dielectric Liquids", JSME Int. J., Ser. B, Vol.36, pp.1-25
17. Arik, M. and Bar-Cohen, A., 2003, "Effusivity-Based Correlation of Surface Property Effects in Pool Boiling CHF of Dielectric Liquids," International Journal of Heat and Mass Transfer, 46, pp. 3755-3764
18. Yang, G.W., Liang, H.S., Yang, W.J. and Vrable, D., 1996, "Nucleate Pool Boiling on Micro Graphite-Copper Composite Surface", ASME J. Heat Transfer, Vol.118, pp.792-796
19. Chang, J.Y., and You, S.M., 1997, "Enhanced Boiling Heat Transfer from Micro-Porous Surfaces: Effects of Coating Composition and Methods", Int. J. Heat and Mass Transfer, Vol.40, pp.4449-4460

20. Nakayama, W., Nakajima, T., and Hirasawa, S., 1984, "Heta Sink Studs Having Enhanced Boiling Surfaces for Cooling Microelectronic Components", ASME Paper No. 84-WA/HT-89
21. Mudawar, I., and Anderson, T.M., 1993, "Optimization of Enhanced Surfaces for High Flux Chip Cooling by Pool Boiling", Trans. ASME J. Electronic Packaging, Vol.115, pp.89-100
22. Kubo, H., Takamatsu, H., and Honda, H., 1999, "Effects of Size and Number Density of Micro-Reentrant Cavities on Boiling Heat Transfer from a Silicion Chip Immersed in Degassed and Gas-Dissolved FC-72", J. Enhanced Heat Transfer, Vol.6, pp.151-160
23. Rainey, K.N. and You, S.M., 2000, "Pool Boiling Heat Transfer from Plain and Microporous, Square Pin Finned Surfaces in Saturated FC-72," ASME Journal of Heat Transfer, Vol. 122, pp. 509-516.
24. Honda, H., Takamastu, and Wei, J.J., 2002, "Enhanced Boiling of FC-72 on Silicon Chips with Micro-Pin-Fins and Submicron-Scale Roughness", J. Heat Transfer, Vol.124, pp.383-390
25. Singh, A., Ohadi, M.M., Dessiatour, S. and Chu, W., 1995, "In-Tube Boiling Heat Transfer Coefficients of R-123 and Their Enhancement Using the EHD Technique", J. Enhanced heat Transfer, Vol.2, pp.209-217
26. Darabi, J., Ohadi, M.M. and Dessiatoun, 1999, "EHD Enhancement of Thin Falling Film Evaporation", ASME/JSME Joint Thermal Engineering Series, Vol.5, pp.547-553
27. Shaw, B.D. and Stahl, S.L., "EHD Pool Boiling in a Reduced Gravity", Proc. of the 3rd Microgravity Fluid Physics Conference, NASA Conference Publication, Vol.3338, pp.213-218, 1996
28. Ohadi, M.M., Darabi, J., and Roget, B., 2001, "Electrode Design, Fabrication, and Materials Science for EHD-Enhanced Heat and Mass Transport," Annual Review of Heat Transfer, Edited by C. L. Tien, Vol. 11, pp. 563-623, Begell House, Inc, New York
29. Watwe, A. A., 1996, "Measurement and Prediction of the Pool Boiling Critical Heat Flux in Highly-Wetting Liquids," Ph. D. Thesis, Department of Mechanical Engineering, University of Minnesota, Minneapolis
30. Bar-Cohen, A., Watwe, A., and Arik, M., 2003, "Pool Boiling Critical Heat Flux in Dielectric Liquids," Keynote Lecture, 5th International Conference on Boiling Heat Transfer, Montego Bay, Jamaica, May 2003
31. Ramaswamy, C., Joshi, Y. and Nakayama, W., 2000, "Combined Effects of Sub-Cooling and Operating Pressure on the Performance of a Two-Chamber Thermosyphon," *IEEE* Transactions on Components and Packaging Technologies, pp. 61-69.
32. Pal, A., Joshi, Y., Beitelmal, M. H., Patel, C. D., and Wenger, T., 2002, "Design and Performace Evaluation of a Compact Thermosyphon," THERMES 2002, January 13-16, Santa Fe, USA.
33. Coursey, J. S., Roh, H., Kim, J., Boudreaux, P. J., 2002, "Performance of a Graphite Foam Thermosyphon Evaporator: Parametric Investigation of the Effects of Working Fluid, Liquid Level, and Chamber Pressure," Proc. of IMECE2002 ASME International Mechanical Engineering Congress & Exposition, New Orleans, LA November 17-22, 2002. IMECE2002-33733.
34. Coursey, J. S., 2003, "Performance and Parametric Investigation of a Graphite Foam Thermosyphon Evaporator", MSME Thesis, University of Maryland, College Park
35. Roh, H., Kim, J. and Boudreaux, P.J., 2003, "Thermal Characteristic of Graphite Foam Thermosyphon for Electronics Cooling", Proceedings of InterPack03, International Electronic Packaging Technical Conference and Exhibition, July 6-11, 2003, Hawaii, USA (InterPack2003-35002)
36. Ma, C.-F. and Bergles, A.E., 1983, "Boiling Jet Impingement Cooling of Simulated Microelectronic Chip", Heat Transfer in Electronic Equipment, ASME, NY, pp.5-12

37. Chrysler, G.M., Chu, R.C. and Simons, R.E., 1995, "Jet Impingement Boiling of a Dielectric Coolant in Narrow Gaps", IEEE Transactions on Components and Packaging Technologies, Part A, Vol.18 (3), pp.527-533
38. Vader, D.T., Chrysler, G.M., Chu, R.C. and Simons, R.E., 1995, "Experimental Investigation of Subcooled Liquid Nitrogen Impingement Cooling", IEEE Transactions on Components and Packaging Technologies, Part A, Vol. 18(4), pp.788-794
39. Ing, P., Sperry, R., Philstorm, R., Claybacker, P., Webster, J. and Cress, S., 1993, "SS-1 Supercomputer Cooling", Proceedings of the 1993 IEEE ECTC Conference, pp.219-237
40. Pautsch, G. and Bar-Cohen, A., 1999, "Thermal Management of Multichip Modules with Evaporative Spray Cooling," ASME Publication, EEP-Vol 26-1, Advances in Electronic Packaging -1999, Vol 1, Proceedings, InterPack'99, Maui, Hawaii
41. Tilton, C., 2000, "Applying Mist to COTS Convection Modules Is Really Cool", COTS Journal, pp.43-53
42. Yao, S.C., Deb, S. and Hammouda, N., 1989, "Impacting Spray Boiling for thermal Control of Electronic System", Heat Transfer in Electronics, ASME, Vol. HTD-111, pp.129-134
43. Darabi, J., Ohadi, M.M. and DeVoe, D., 2001, "An Electrohydrodynamic Polarization Micropump for Electronic Cooling", Journal of Microelectromechanical Systems, Vol.10 (1), pp.98-106
44. Ahn, S.H. and Kim, Y.K., 1997, "Fabrication and experiment of planar micro ion-drag pump," in Proc. Transducers'97, pp. 373-376
45. Fuhr, G., Schnelle, T. and Wagner, B., 1994, "Traveling Wave-driven micro- fabricated electrohydrodynamic pumps for liquids," J. Micromech. Microeng., Vol. 4, no. 4, pp. 217-226.
46. Moghaddam, S. and Ohadi, M., 2002, "Thin Film Evaporator with Splayed Electrodes", U.S. patent pending, Application number 60/419,690, October
47. Yazawa, K., Solbrekken, G.L., and Bar-Cohen, A., 2003, "Thermoelectric-Powered Convective Cooling of Microprocessors," IEEE Transactions on Advanced Packaging, accepted for publication.

BINARY-FLUID HEAT AND MASS TRANSFER IN MICROCHANNEL GEOMETRIES FOR MINIATURIZED THERMALLY ACTIVATED ABSORPTION HEAT PUMPS

Srinivas Garimella
George W. Woodruff School of Mechanical Engineering
Georgia Institute of Technology
Atlanta, GA 30332-0405
srinivas.garimella@me.gatech.edu

1. Introduction

International interest in the global climate change problem has focused renewed attention on the development of novel thermal systems that reduce the environmental impact of energy consumption in the space-conditioning, chemical processing and other energy-intensive industries. One response to this problem is the use of absorption heat pumps, which are environmentally sound and energy-efficient alternatives to CFC-based, ozone-depleting space-conditioning systems. These thermally activated systems are powered by recuperated waste heat or are gas-fired, as opposed to the electrical energy required for vapor compression systems, thus resulting in high energy efficiencies. They also have fewer moving parts. The principle of operation, illustrated in Figure 1, is as follows: thermal energy is used to boil a refrigerant from a concentrated refrigerant-absorbent solution in a generator at high pressure. The refrigerant is condensed using ambient air as the heat sink, and expanded to a low pressure across a valve. At this low pressure, the refrigerant is cold enough to effect space-conditioning as it evaporates in the evaporator, thus cooling room air. The evaporated refrigerant is combined with the dilute solution in an absorber releasing the heat of absorption, from where it is pumped back in liquid form to the generator, which requires orders of magnitude less electrical energy than the compression of the refrigerant vapor in conventional systems. This thermodynamic cycle can also be run in the heating mode in winter, with the evaporator coupled to the outdoor air to withdraw heat from the ambient, and the condenser and absorber coupled to the indoor air to provide space heating. Thus, these thermodynamically attractive absorption systems have been implemented in large commercial applications.

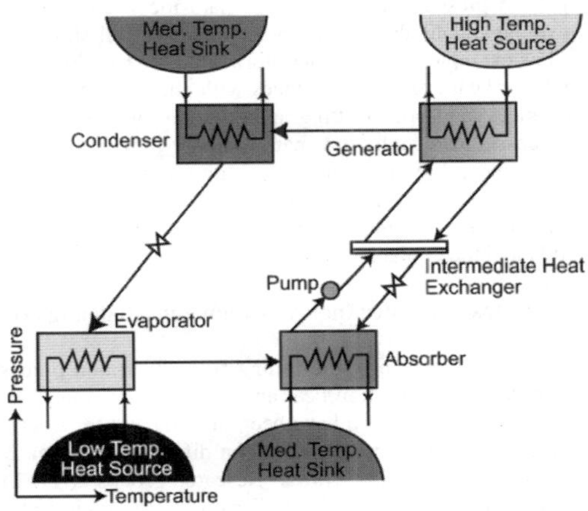

Figure 1: Schematic of an Absorption Heat Pump

1.1 ABSORPTION CYCLES

Various investigators have focused on increasingly complex thermodynamic cycles to obtain incremental improvements in the theoretical coefficients of performance (COPs). For example, Garimella *et al.* [1] and Engler *et al.* [2] have reported high cooling and heating mode COPs for the Generator Absorber Heat Exchange (GAX) Heat

S. Kakaç et al. (eds.), Microscale Heat Transfer, 339–368.

Pump cycle over a wide range of ambient conditions. Double-effect [3-5] and triple-effect cycles [6-9] were also investigated and showed the potential for high COPs. Even 4-effect absorption chillers [10] and other multiple-effect absorption cycles [11] have been investigated to achieve high COPs. However, the increased cycle complexity also results in the need for numerous heat exchangers and control systems. The performance potential of these advanced cycles cannot be realized without practically feasible and economically viable compact heat exchangers, which represents one of the biggest hurdles to commercialization. The development of such heat and mass exchangers has proved to be challenging, and has in fact hindered the adoption of absorption technology as a viable space-conditioning option. This is particularly true for the small-capacity residential market, where the lack of compact, inexpensive component designs for binary-fluid (refrigerant-absorbent) heat and mass transfer represents a crucial hurdle.

1.2 THE ABSORPTION PROCESS: SIGNIFICANCE AND CHALLENGES

In thermally activated absorption heat pumps, the absorber, in which refrigerant vapor is absorbed into the dilute solution with the release of the heat of absorption, governs the viability of the entire cycle and has been referred to as the "bottleneck" in the heat pump [12, 13]. Absorption systems commonly utilize two different kinds of fluid pairs. The ammonia-water fluid pair has a volatile absorbent, presenting heat and mass transfer resistances across the respective temperature and concentration gradients in both the liquid and vapor phases. The highly non-ideal ammonia-water fluid pair releases a considerable amount of heat of absorption at the vapor-liquid interface that must be transferred across a liquid film into the coolant. Some of this heat released at the interface is also transferred to the vapor, depending on the local temperature differences. In Lithium Bromide/Water systems, water is the refrigerant and evolves from the binary-fluid solution as a pure refrigerant vapor. Thus, there is inherently no mass transfer resistance in the vapor phase; however, the fluid properties of the $LiBr/H_2O$ pair typically present a considerable mass transfer resistance in the liquid phase.

Based on the above discussion, successful designs for such binary fluid heat and mass exchangers must address the often contradictory requirements of ensuring low heat and mass transfer resistances for the absorption/desorption side, adequate transfer surface area on both sides, low resistance of the coupling fluid, low coupling fluid pressure drop to reduce parasitic power consumption, and low absorption-side pressure drop. Most of the available absorber/desorber concepts fall short in one or more of these criteria essential for good design. For example, conventional horizontal tube falling-film absorbers have very low solution-side pressure drops and low heat transfer resistances with ammonia-water as the working fluid, but suffer from high coolant-side resistances, and poor wetting and solution distribution. The coolant-side resistance for falling films over vertical tubes is also high, and many parallel tubes must be used (with flow distribution and film splashing problems) to meet duty requirements within acceptable lengths. Falling films on the inside of vertical tubes are limited to very low vapor fluxes due to the flooding limitation. Forced-convective absorption inside tubes leads to high pressure drops, thus reducing the saturation temperature and the driving temperature difference between the solution and the coolant.

2. Heat and Mass Transfer Intensification and Modeling Overview

In view of the importance of the falling-film absorption process, several investigators have developed designs aimed at increasing absorption heat and mass transfer rates. Enhanced tubular surfaces (pin fins, integral fins; and fluted and grooved tubes) have been investigated for external falling-film absorption of $LiBr-H_2O$ by Miller and Perez-Blanco [14], who reasoned that diffusion of water absorbed at the surface will be augmented by the mixing induced by secondary flows. Research groups at TU Munchen, Germany and Technion, Israel [12, 13] [15] have conducted collaborative studies on absorption in NaOH/KOH mixtures, $LiBr/H_2O$, and NH_3/H_2O films over horizontal tube banks and vertical tubes. They reported that the solution did not fully cover the surface of the tubes and flowed as rivulets that correspond to the distributor locations. They also suggested that knurled tubes have higher heat transfer due to the additional surface area and better wetting, and higher mass transfer due to turbulence induced in the film. Spray absorbers, in which the absorption process is separated into a phase change process and a subcooling process [16] need solution recirculation pumps and subcoolers. Surfactants have also been investigated to enhance wetting of tubes and also to enhance mass-transfer coefficients [17-20] with mixed results, and the reliability and durability of these agents has not been well demonstrated.

For compact bubble absorbers in ammonia-water GAX systems, Merrill *et al.* [21, 22] used numerous passive enhancement techniques such as repeated roughness elements, internal spacers, and increased thermal conductivity metal to improve heat transfer. Mass transfer improvement was achieved through the use of static mixers, variable cross-section flow areas, and numerous vapor injector designs. Merrill and Perez-Blanco [23] investigated ammonia-water bubble absorption in a compact absorber in which the interfacial area per unit volume of vapor and the liquid mixing at the vapor-liquid interface were increased by breaking the vapor up into small bubbles and injecting them into the liquid. The increase in interfacial area per unit volume resulted in more compact absorbers. Herbine and Perez-Blanco [24] modeled the absorption process in an ammonia-water vertical-tube bubble absorber. Unlike previous models, their analysis was able to account for water desorption from the solution in some portions of the absorber due to the prevailing concentration gradients. Counter-current vertical fluted-tube absorbers for ammonia-water have been investigated by Kang and Christensen [25]. Kang *et al.* [26] also evaluated the heat and mass transfer resistances in both the liquid and vapor regions in a counter-current ammonia-water bubble absorber composed of a plate heat exchanger with an offset strip fin used to enhance heat transfer performance. Christensen *et al.* [27, 28] developed an absorption/desorption device that uses corrugated and perforated fins located between rectangular coolant channels. The vapor flows upward through perforations in the corrugated fins, while absorbent solution flows downward over the corrugated fins and through the perforations. For horizontal-tube, falling-film ammonia-water absorbers, Perez-Blanco [29] found that at typical operating conditions, the absorption rate is controlled by the mass transfer process in the falling film, with all other factors having negligible effects. An experimental study of a falling film over a coiled tube ammonia-water absorber was conducted by Jeong *et al.* [30], who found that the heat transfer coefficient of the falling film increased linearly with the solution flow rate both with and without absorption. Potnis *et al.* [31] developed a generalized approach for GAX component and system simulation that is useful for sizing equipment, estimating the cycle COP, and determining individual phase flow rates. They found that although the mass transfer resistance resides primarily in the vapor phase, the liquid-phase mass transfer resistance should not be considered negligible for an ammonia-water system.

It is clear from the above discussion that some of the binary-fluid devices investigated by the various researchers have yielded high heat and mass transfer rates, especially in commercial applications However, designs for small-capacity applications must of course yield these high transfer rates, but necessarily with simple and compact geometries. In addition, the coupled heat and mass transfer process with a large amount of heat released at the interface due to vapor absorption has not been understood well thus far. In the following sections, techniques for the miniaturization of binary-fluid heat and mass exchangers, and experimental and analytical studies that yield a comprehensive understanding of the relevant transport mechanisms are presented.

3. Miniaturization of Absorption Heat and Mass Transfer Devices – Volatile Absorbents

3.1 CONCEPT DESCRIPTION

To address the challenges in accomplishing binary-fluid absorption in a compact geometry, a novel miniaturization technology for binary-fluid heat and mass exchange was developed by Garimella [32], who first presented a semi-empirical model for ammonia-water absorption in this compact geometry [33, 34]. A schematic of this microchannel heat and mass transfer device is shown in Figure 2. Short lengths of microchannel tubes (similar to hypodermic needles) are placed in a square array. This array forms level 1, depicted by A1-B1-C1-D1. The second array (level 2) of thin tubes is placed above level 1, in a transverse orientation perpendicular to the tubes in level 1, depicted by A2-B2-C2-D2. A lattice of successive levels is formed, with the number of levels determined by the design requirements. Hydronic fluid (coolant) is manifolded through these tubes as shown in Figure 2. Thus the fluid enters level 1 at A1 and flows in the header in direction A1-B1. As it flows through the header, the flow is distributed in parallel through all the tubes in level 1. The number of parallel passes is determined by tube-side heat transfer coefficient requirements and pressure drop restrictions. The fluid flows through the tubes from A1-B1 to C1-D1. The fluid collected in the outlet header C1-D1 flows through the outlet connector tube D1-D2 to the upper level. The inlet and outlet headers are appropriately tapered to effect uniform hydronic flow distribution between the tubes. In level 2, the fluid flows in parallel through the second row of tubes from D2-B2 to C2-A2. This flow pattern is continued, maintaining a globally rotating coolant flow path until the fluid exits at the outlet of the upper-most header.

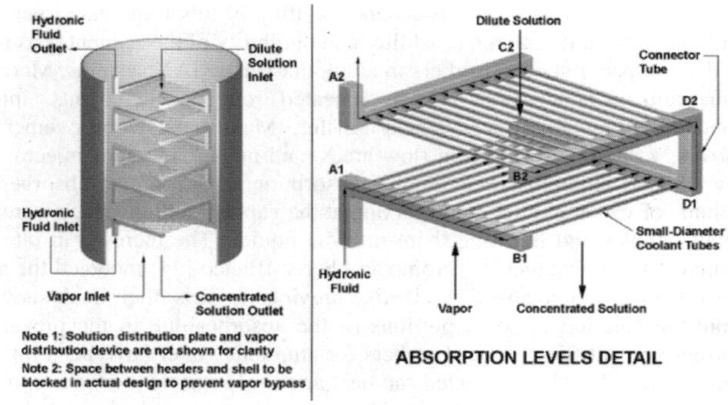

Figure 2: Microchannel Heat and Mass Transfer Device Concept

This configuration offers significant advances over the state of the art. The coolant-side heat transfer coefficients are extremely high even though the flow is laminar, due to the small diameter of the microchannel tubes ($h = Nu \bullet k / D_h ; D_h \to 0$). The high values are achieved with widely available smooth circular tubes without any passive or active heat transfer enhancement, which typically increases cost and complexity. In addition, coolant pressure drop can be minimized simply by modifying the pass arrangement (parallel flow within one level and/or across multiple levels), ensuring minimal parasitic power requirements. In an absorber application, a distribution device above the first row of tubes distributes solution so that it flows in the falling-film mode counter-current to the coolant through the heat exchanger rows. Vapor is introduced at the bottom, and flows upward through the lattice formed by the tube banks, counter-current to the gravity-driven falling solution. The gravity driven flows lead to negligible absorption-side pressure drops. The spacing (vertical and transverse) between the tubes is adjustable to obtain the desired vapor velocities while avoiding vapor hold-up, and also providing adequate adiabatic absorption of vapor between levels. Such an arrangement also minimizes the problem of inadequate wetting of the transfer surface. The effective vapor-solution contact coupled with the large surface area/volume ratio minimizes heat and mass transfer resistances. The heat of absorption is conveyed to the coolant with minimal tube-side resistance due to the high heat transfer coefficients described above. This concept, therefore, addresses all the requirements for absorber design cited above in an extremely compact and simple geometry. Furthermore, the concept is modular, with any heat duty being handled by changing the surface area per row (length and number of tubes) or the number of tube rows. The tube/header assembly can be brazed in place, even on a continuous, mass-production basis (similar to automotive heat exchangers). Finally, the geometry is suitable for almost all absorption heat pump components (absorbers, desorbers, condensers, rectifiers, and evaporators); the resulting uniformity of surface type and configuration throughout the system reduces capital costs and ensures commercial viability.

3.2 PROTOTYPE DESIGN AND FABRICATION

The semi-empirical model for ammonia-water absorption developed by Garimella [34] was used to design a hydronically cooled absorber for a single-effect heat pump with representative design conditions selected as inputs to the model. This absorber was designed for a 10.55 kW cooling load residential heat pump, i.e., an absorption load of about 19.3 kW. The resulting design in that analysis consisted of 5 passes of 15 tube rows each with 40, 1.587 mm OD × 127 mm long tubes per row. A tube center-to-center spacing of 3.175 mm was used in the horizontal direction, while the pitch in the vertical direction was 6.35 mm. The resulting absorber was 0.476 m high with a total surface area of 1.9 m^2. The 5-pass arrangement was selected to keep the tube-side pressure drop below 23.5 kPa. The Kutateladze number criterion provided by Richter [35] was used to ensure that the horizontal pitch chosen for the prototype did not cause flooding due to the countercurrent flow of the vapor. In addition, flow visualization studies were conducted to ensure optimum tube spacing. Seven test sections of varying horizontal and vertical tube pitches were fabricated to simulate the absorber tube array and placed in a transparent enclosure for the visualization tests. These tests indicated that a pitch of 4.76 mm in both directions minimizes bridge and column formation between adjacent tubes, while providing adequate flow area for the rising vapor.

Figure 3: Microchannel Absorber Prototype

A prototype absorber (Figure 3) was fabricated based on these analyses, with experimental results reported by Meacham and Garimella [36]. In Figure 3, the photograph on the left shows the absorber tube array before the end plates for the headers were brazed. The photograph in the center shows the crisscross structure of the tube array. The photograph on the right shows the baffle covers with thermocouple ports for each coolant-side pass welded to the outside of the absorber. This photograph shows the completed absorber with the drip tray and the dilute solution inlet tube attached to the top of the assembly. During testing, the entire assembly was enclosed in a flanged stainless steel shell with a domed top. This shell was bolted to a base flange, with all of the fluid inlets and outlets, thermocouple probes, and pressure transducer taps entering the absorber through this base flange. Details of the absorber geometry are shown in Table 1. It should be noted that the tube dimensions and spacing in both the vertical and horizontal directions are slightly different for this absorber tested by Meacham and Garimella [36] than for the preliminary configuration analyzed by Garimella [34]. (The total surface area of the prototype absorber is 79% of the area of the absorber modeled by Garimella [34]).

Table 1: Prototype Absorber Dimensions

Tube outer diameter, mm	1.575
Tube inner diameter, mm	1.067
Tube wall material and thermal conductivity	Stainless Steel, 15.4 W/m-K
Tube length, m	0.140
Number of tubes per row	27
Number of rows per pass	16
Number of passes	5
Tube transverse pitch, mm	4.76
Row vertical pitch, mm	4.76
Absorber height, m	0.508
Total surface area, m^2	1.50

3.3 EXPERIMENTS

The performance of the microchannel absorber described above was experimentally investigated in the test facility shown in Figure 4. A variable-speed pump delivered the desired concentrated solution flow rate to the steam-heated desorber where vapor generation was controlled by varying the inlet pressure of the steam. The vapor and dilute solution were separated downstream of the desorber outlet. The separator consists of concentric expansion chambers with wire-mesh screens that reduce the velocity of the incoming two-phase mixture. The vapor and dilute solution were then separated by buoyancy and gravity. Separation of the vapor and the dilute solution enabled these two streams to enter the absorber at different locations. The dilute solution was cooled as necessary to ensure that a single-phase liquid flowed through the dilute solution mass flow meter. This stream entered the absorber and was distributed over the coolant tubes through the drip tray. The vapor stream entered the absorber at the bottom of the tube array and flowed upward counter-current to the dilute solution. The heat of absorption was removed by the coolant flowing upward within the tube array. The exiting concentrated solution was further subcooled as necessary in a small shell-and-tube heat exchanger before returning to the solution pump. The coolant loop consisted of a circulation pump and a plate heat exchanger, where the heat of

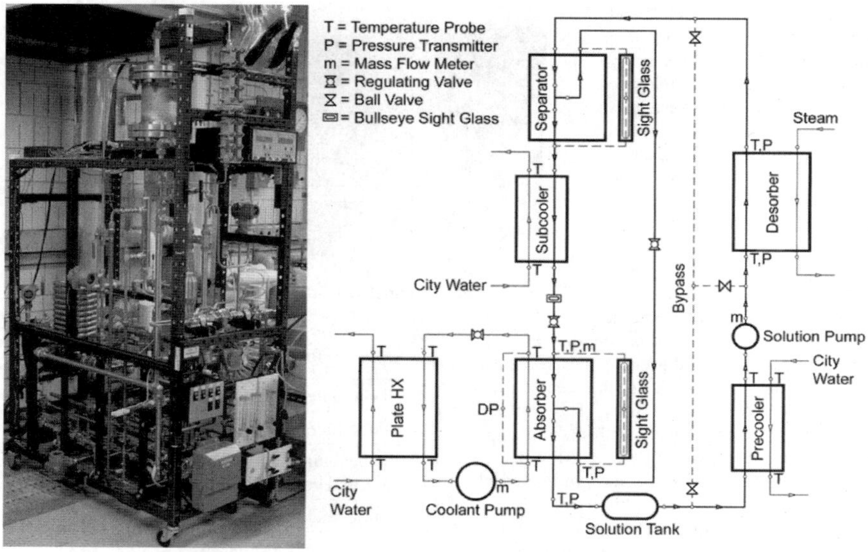

Figure 4: Absorption Heat and Mass Transfer Test Facility

absorption removed at the absorber was rejected to a city water stream. City water and cooling water flow rate variations provided the requisite control of the absorber heat sink.

Temperature, pressure, and mass flow rate measurement locations are shown in Figure 4. Two Coriolis mass flow meters measured the concentrated and dilute solution flow rates, respectively. The concentrated solution flow rate was controlled by a variable-speed solution pump, while the vapor generation rate was varied by increasing or decreasing the saturation pressure of the steam in the desorber. This vapor generation rate was calculated by taking the difference between the measured concentrated and dilute solution flow rates. Temperature and pressure measurements at the desorber outlet were used to determine the concentrations of the saturated vapor and dilute solution. This information, along with the measured mass flow rates of each of the solution streams, was used to calculate the concentration of the solution entering the desorber. Once the concentrations were determined, temperatures and pressures at the absorber inlet and outlet were used to determine the solution-side duty in the absorber. The temperatures and flow rate of the closed-loop cooling water were also used to calculate the heat duty, to help establish an energy balance between the two sides of the absorber. Other energy balances provided additional validation of the measured heat duties. Similarly, coolant temperature measurements at the inlet and outlet of each pass enabled the calculation of absorption rates on a *pass-by-pass basis*, and the modeling of the variation of transfer coefficients on a segmental basis. Additional details of the test facility and the measurements are available in Meacham and Garimella [36]. The experimental solution-side absorption heat transfer coefficient for each segment was deduced from the overall segment *UA* using a resistance network approach.

3.4 COUPLED BINARY-FLUID ABSORPTION HEAT AND MASS TRANSFER MODEL

Results from the experiments described above were analyzed by Meacham and Garimella [37] using a model that accounts for the incremental variations in the heat and mass transfer resistances through the length of the absorber as reported by Garimella [34]. The basic approach is well documented by Price and Bell [38], who in

Table 2: Falling-Film Absorption Heat and Mass Transfer Model Overview

Mass Transfer Rates	
Rate Equation: Molar flux of vapor absorbed into film: $\dot{n}_T = \beta_v C_T \ln\left[\dfrac{\tilde{z} - \tilde{x}_{v,int}}{\tilde{z} - \tilde{x}_{v,bulk}}\right]$ C_T = total molar concentration β_v = mass transfer coefficient [ ] = governing concentration gradient $\tilde{z}$ = molar concentration of the condensing flux	***Rates:*** Ammonia: $m_{abs,a} = M_a \dot{n}_a A_{SP}$ Water: $m_{abs} = m_{abs,a} + m_{abs,w}$ Total: $m_{abs,w} = M_w \dot{n}_w A_{SP}$ A_{SP} = Surface area per pass
$\tilde{z}$ based on the heat transfer rate that can be supported by the overall heat transfer coefficient	
Mass and Species Concentration Balances	
Vapor outlet:	$m_{vo} = m_{vi} - m_{abs}$; $m_{vo}x_{vo} = m_{vi}x_{vi} - m_{abs,a}$
Vapor-Liquid mass balance:	$m_{li} + m_{vi} = m_{lo} + m_{vo}$; $m_{li}x_{li} + m_{vi}x_{vi} = m_{lo}x_{lo} + m_{vo}x_{vo}$
Heat Transfer Rates	
Heat duty to coolant from enthalpy balance (latent heat of absorption and sensible cooling of liquid film)	$Q_c = m_{li}h_{li} + m_{vi}h_{vi} - m_{lo}h_{lo} - m_{vo}h_{vo}$
Vapor phase heat transfer: $Q_v = \alpha_v \left[\dfrac{\phi_T}{1 - e^{-\phi_T}}\right] A_{SP} LMTD$; $\phi_T = \dfrac{n_a \tilde{c}_{p,a} + \dot{n}_w \tilde{c}_{p,w}}{\alpha_v}$ α_v = vapor-phase heat transfer coefficient [ ] = correction for heat transfer coefficient due to mass transfer	Vapor phase temperature change: $Q_v = m_v c_{p,v}(T_{v,i} - T_{v,o})$ + or -, depending on interface and bulk vapor temperatures
Overall heat transfer resistance: $\dfrac{1}{U} = \dfrac{D_o}{D_i \alpha_c} + R_w + \dfrac{1}{\alpha_f}$; $Q_c = UA_{SP} LMTD$	Individual transfer coefficients determined by geometry, mode (falling film, droplet formation) etc.
Coolant heat duty and outlet temperature:	$Q_c = m_c c_{p,c}(T_{c,o} - T_{c,i})$
Note: Model closure obtained by using fluid properties, equilibrium relationships and transfer rate equations iteratively to convergence.	

turn used the technique developed by Colburn and Drew [39] for the condensation of a binary vapor with a miscible condensate. Other investigators [25, 26, 28, 40, 41] have also adapted this technique to the design of ammonia-water absorbers. Essentially, the absorption of the vapor into the falling liquid film is governed by the mass transfer coefficient in the vapor phase and the concentration gradient between the bulk vapor and that at the vapor-liquid interface. The geometry under consideration was assumed to induce a well-mixed liquid film. The model is summarized in Table 2; a detailed description is available in the papers by Garimella [34] and Meacham and Garimella [37]. Mass, species and energy conservation in the ammonia-water solution and vapor phases, and the corresponding heat transfer between the solution, vapor and coolant were computed simultaneously for the five segments representing the five coolant-side passes.

3.5 EXPERIMENTAL RESULTS AND MODEL VALIDATION

The data from the experiments showed that the overall heat transfer coefficient (for the whole absorber) varied from a minimum of 133 W/m^2-K at a concentrated solution flow rate of 0.0106 kg/s to a maximum of 403 W/m^2-K at 0.0361 kg/s. A relatively constant, high value of the coolant-side heat transfer coefficient, 2600 W/m^2-K, was seen for all the cases tested, in which the flow remains laminar. The absorber was found to be solution-side limited. Heat transfer rates as high as 16.2 kW, representative of residential heat pumps, were achieved in this small envelope; however, the performance was somewhat lower than that predicted by the preliminary model of Garimella [34]. The 21% smaller surface area of the prototype [37] compared to the modeled design [34] accounted for some of the discrepancy. Also, the single-pressure absorber test facility used in the experiments supplied vapor to the absorber at conditions representative of desorber outlet conditions (≈ 86.7% NH_3) rather than evaporator outlet conditions (> 99.5% NH_3) which also caused some of the difference in heat transfer rates. Finally, it was found that inadequate wetting of the tube surface and non-uniform distribution of the solution film due to poor dripper tray design contributed to the decreased performance. To account for these solution distribution and inadequate wetting problems, an area effectiveness ratio, $r = A_{eff}/A_{actual}$ was required to achieve a match between the predicted and measured absorber outlet conditions given the inlet conditions as inputs to the model described in Table 2. The area effectiveness ratio required to achieve this correspondence between model and experiment varied from 0.21 to 0.31 over the range of conditions tested.

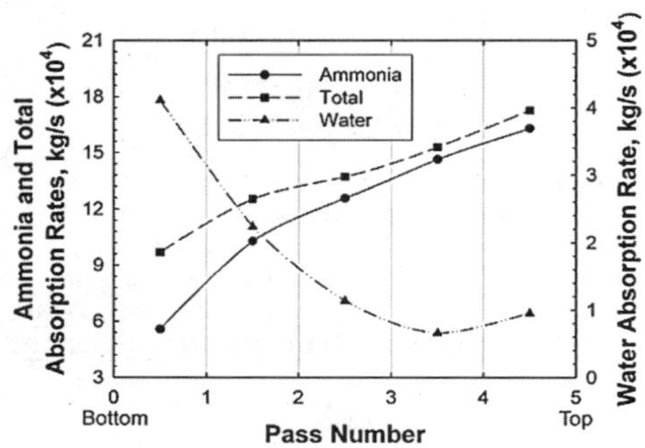

Figure 5: Total and Species Absorption Rates as Functions of Absorber Pass

The predictive capabilities of this refined model are illustrated in Figures 5-7 for a data point near the design condition. The total absorption rates and the absorption rates of the ammonia and water are shown in Figure 5. As expected, all three absorption rates increase steadily from the bottom to the top of the absorber, where the driving temperature difference is higher. (Note that the scale for the water absorption rate is different as this represents a small fraction of the overall absorption rate.) Predicted temperature profiles of the solution, coolant, and vapor, with the corresponding measured coolant temperatures, are shown in Figure 6. The vapor enters the absorber at a very high temperature and is initially cooled due to the sensible heat transfer from the bulk vapor to the interface. Midway through the absorber, the vapor temperature decreases to a temperature below that of the interface temperature (which is primarily governed by the liquid concentration), and the vapor temperature begins to increase steadily due to heat gain from the interface. The predicted coolant temperature increases steadily as it flows upward through the absorber. The measured coolant temperatures are in excellent agreement with the predicted values.

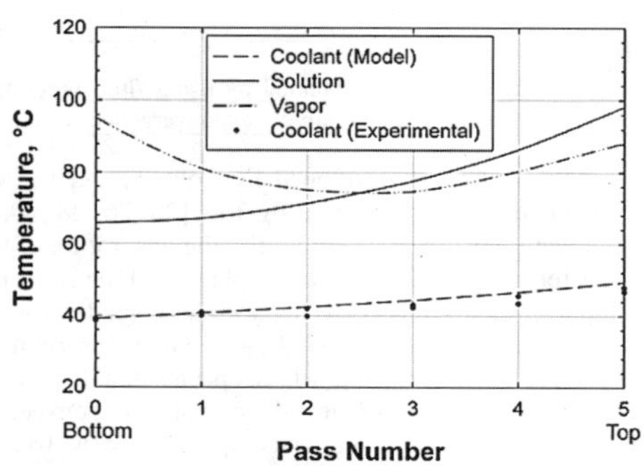

Figure 6: Measured and Predicted Temperature Profiles

Figure 7 shows the variation of the coolant heat duty and the sensible heat duty of the vapor with absorber pass. As the coolant flows upward through the absorber, the heat duty increases due to an increase in the *LMTD* between the coolant and the solution. This figure also indicates that the sensible heat transfer between the vapor and the liquid film is a small fraction of the coolant load. As explained above in the discussion of the vapor temperature profile, the sensible heat transfer between the vapor and the liquid film is positive as the high temperature vapor enters the absorber and becomes negative as the vapor temperature decreases below that of the liquid film.

3.6 PERFORMANCE IMPROVEMENT THROUGH HIGH-SPEED FLOW VISUALIZATION

The modeling of the local measured heat and mass transfer variations within the prototype absorber described above indicated that only 20 to 30% of the absorber surface area might have been effectively utilized. From these results, it was inferred that solution distribution problems at the top of the array might be responsible for the potential solution mal-distribution. It was also clear that the absorber performance was solution-side limited, due to the relatively small influence of the coolant flow rate on absorber performance. An improved configuration of the absorber was developed based on these inferences, primarily with the goal of addressing solution-side flow distribution and surface wetting issues. In addition, optical access to the absorption process was provided to enable confirmation of the improvement in flow distribution. Testing and analysis of the performance of the revised absorber, shown in Figure 8, was conducted by Meacham and Garimella [42] to document and interpret this improvement. The absorber microchannel tube array was housed in a flanged stainless steel 0.264 m ID schedule 10 shell with a large sight port for flow visualization as shown in Figure 8a. Three additional smaller sight ports at 90° intervals were also provided to enable viewing from different orientations (axial and transverse) as well as for the illumination of the absorber. The array (Figure 8b) consists of 1.575 mm OD, 0.2 mm wall tubes, sandblasted to facilitate surface wetting, with the straight section of each tube segment being 0.137 m long. There were 20 rows of tubes, with each row consisting of 33 tubes. Other details about the absorber geometry are provided in Table 3. It should be noted that in the first prototype investigated by Meacham and Garimella [36, 37], successive tube arrays were oriented transversely perpendicular to the adjacent arrays. However, in this absorber, that arrangement would prevent optical access due to the headers required for coolant inlet and outlet in both perpendicular directions. Therefore, the tubes in adjacent arrays were oriented in the same direction, as shown in Figure 8b. With such an arrangement, the solution flow and fluid distribution along the tube length could be viewed through the large sight port without obstruction by headers. Series flow between passes of the absorber was achieved through U-bends instead of headers, allowing viewing of the solution flow in the axial direction of the tubes. The coolant entered an inlet manifold, split into two rows of tubes in parallel, flowed through

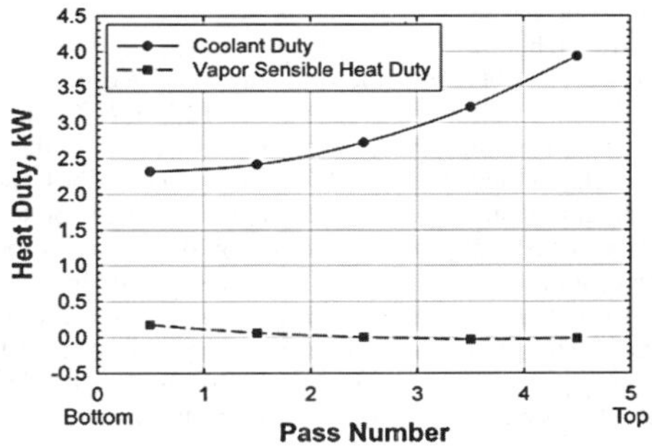

Figure 7: Coolant and Vapor Heat Duties as Functions of Absorber Pass

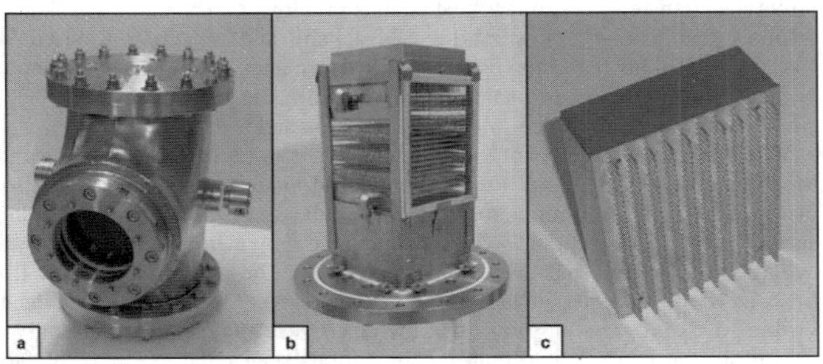

Figure 8: Absorber with Optical Access, a) External View, b) Microchannel Tube Array, c) Dripper Tray

the sets of tube rows (10 passes) in series and exited at the top of the array, providing an almost pure counterflow orientation with the solution flowing downward due to gravity. The horizontal pitch (tube centerline-to-centerline distance) of 4.76 mm was identical to that of the previous prototype, but due to the minimum bend radius of the tubes, the vertical pitch was increased to 7.94 mm. The overall dimensions of the absorber array were approximately 0.162 × 0.157 × 0.150 m, with a surface area of 0.456 m^2, about 30% of the surface area of the previous prototype.

Table 3: Geometry of Absorber with Optical Access

Tube outer diameter, mm	1.575
Tube inner diameter, mm	1.067
Tube wall material	Stainless Steel
Tube length, m	0.137
Number of tubes per row	33
Number of rows per pass	2
Number of passes	10
Tube transverse pitch, mm	4.76
Row vertical pitch, mm	7.94
Absorber height, m	0.150
Total surface area, m^2	0.456

The drip tray of the original absorber [36, 37] was thought to be the source of much of the distribution problems. That drip tray was essentially a perforated tray with minor warping of the underside of the tray causing the solution to collect at preferential locations before falling unevenly over the tube array. In the revised absorber [42], the drip tray consisted of a stainless steel tray with perforations at a pitch of 12.7 × 4.76 mm, with small lengths of tubing placed in each of the holes of the drip tray (Figure 8c). These tubes ensured that solution droplets formed directly above the coolant tubes and did not coalesce before detaching as large streams at preferential locations. In addition, the height of the tubes in the tray provided an even gravity head in the inlet solution pool above the tubes allowing for uniform flow through the dripper tubes to the tube array. The entire absorber tube array and drip tray assembly were brazed in place before being installed into the same test facility used for testing the first prototype.

Figure 9 shows the substantial improvement in the performance of this revised prototype over the performance of the absorber shown in Figure 3. As the solution flow rate was increased, the heat duty transferred by this absorber ranged from 4.51 to 15.1 kW, with 545 W/m^2-K < UA < 940 W/m^2-K, and 638 < $h_{solution}$ < 1648 W/m^2-K. With only 30% (0.162 × 0.157 × 0.150 m; surface area = 0.456 m^2) of the surface area of the previous absorber, this revised prototype transferred similar heat duties at lower solution and coolant flow rates. Meacham and Garimella [42] attributed this increase in performance primarily to the increase in solution-side heat transfer coefficients due to improved solution distribution and wetting.

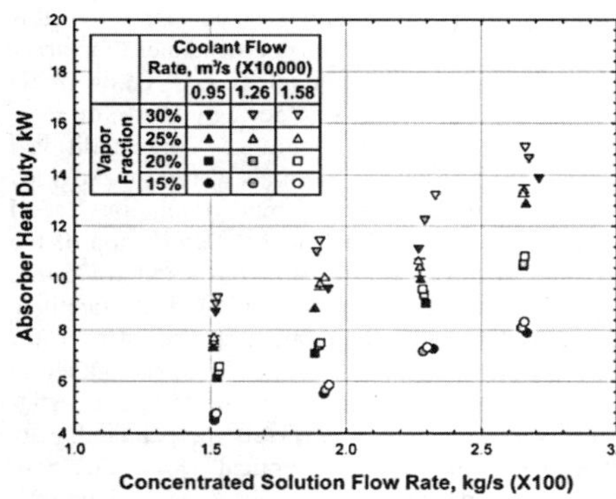

Figure 9: Revised Absorber Heat Duty as a Function of Concentrated Solution Flow Rate

Visualization of the flow of the ammonia-water solution during the absorption process was also conducted by Meacham and Garimella [42] using high-speed video equipment. This visual information confirmed uniform distribution of the solution across the entire tube array, and significantly improved utilization of the available surface area. Representative video frames are shown in Figure 10, and are helpful for the interpretation of evolution of the heat and mass transfer process. The progression of the flow of the solution over successive tubes at 10 ms intervals, including falling film, droplet formation, detachment, fall, and redistribution is clearly visible. This progression shows that the analysis of the data must not only consider the process on a pass-by-pass basis, but must also account for the various stages and modes of the absorption process around each tube and between tubes. This is addressed in greater detail in the following sections of this paper.

3.7 TRANSFORMATION TO FUNCTIONING AS A DESORBER

After demonstrating the performance of the microchannel heat and mass transfer device as an extremely compact absorber, the same geometry was experimentally demonstrated as an ammonia-water desorber by Determan *et al.* [43]. In this case, instead of a vapor stream supplied at the bottom of the assembly being absorbed into a falling dilute solution, concentrated solution was introduced at the top, with hot hydronic fluid flowing inside the tubes. Ammonia-water vapor was therefore desorbed in the process, and rose to the top of the desorber due to buoyancy, while the dilute solution flowed to the bottom due to gravity. The original absorber first fabricated and tested by Meacham and Garimella [36] was modified slightly (inlets and outlet locations and distribution device) to allow operation as a desorber; although the heat and mass transfer surface was not modified. The only modification made to that absorber was a modified distributor tray, as shown in Figure 11 (top and bottom view). The tray has numerous microchannel tubes of 15.9 mm length brazed into the bottom to ensure that solution supply droplets form directly above the tubes across the entire array. In addition, the height of the tubes in the tray provides an even gravity head for the inlet solution pool before flowing uniformly through the dripper tubes to the tube array. The tray also provides vapor bypass tubes (12.7 mm OD) that extend above the top edge of the tray to enable the vapor generated within the tube array to exit the desorber.

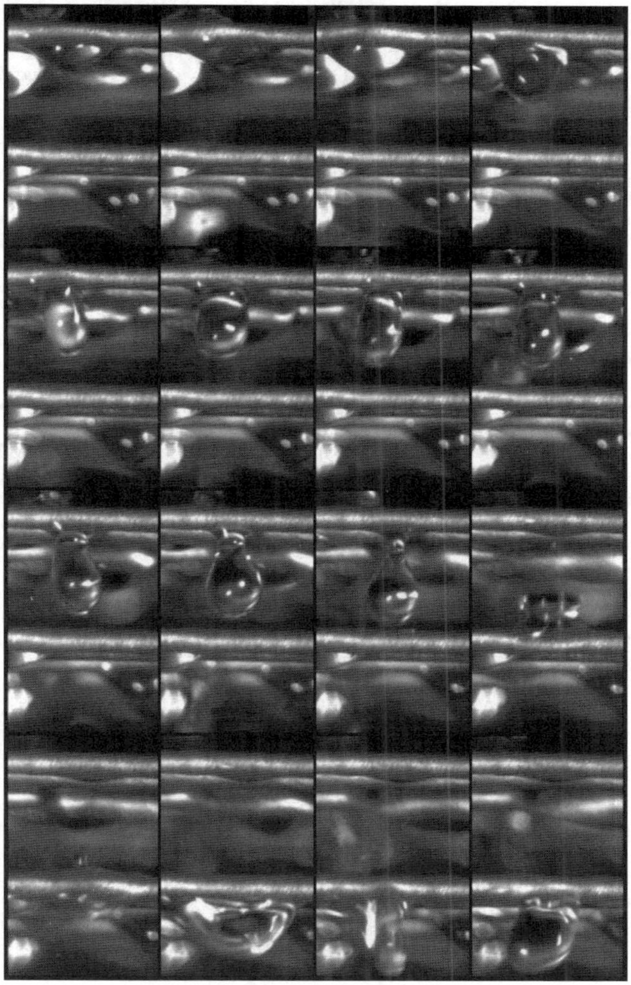

Figure 10: NH_3/H_2O Solution Flow Progression at 10 ms Intervals

The desorber was tested in the same facility that was used by Meacham and Garimella [36, 42] with some minor modifications, essentially switching the absorber and desorber functions from the facility shown in Figure 4. In this case, concentrated solution was supplied to the top of the desorber, and glycol solution heated by steam flowed through the microchannel tubes to supply the heat of desorption instead of removing the heat of absorption. These experiments by Determan *et al.* [43] indicated that the desorber transferred heat duties as high as

17.5 kW, with overall *UAs* of up to 617 W/m^2-K. The desorption heat transfer coefficient varied from 659 to 1795 W/m^2-K over the range of conditions tested. These heat transfer coefficients are much higher than those obtained in the first absorber prototype by Meacham and Garimella [36] ($145 < h_{absorption} < 510$ W/m^2-K) when they encountered solution distribution problems. It should be noted that these *desorption* heat transfer coefficients are almost the same as the *absorption* heat transfer coefficients measured by them in their second generation absorber ($638 < h_{absorption} < 1648$ W/m^2-K).

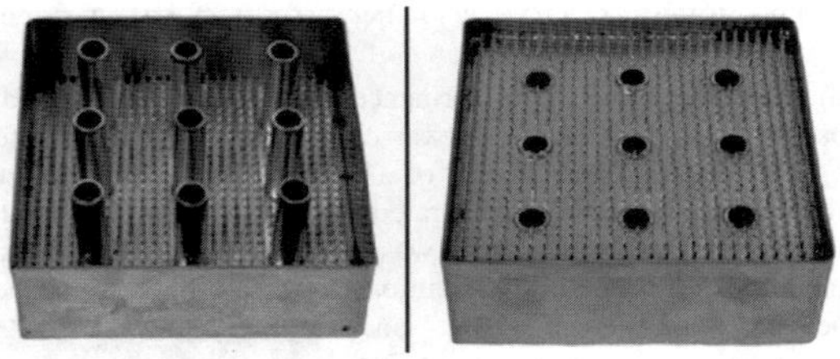

Figure 11: Distribution Device Modifications to Enable Desorber Mode

It is clear that the work of Garimella *et al.* [32-34, 36, 37, 42, 43] described above has successfully demonstrated the application of microchannel technology to the binary-fluid heat and mass transfer process encountered in thermally activated absorption heat pumps for fluid pairs with volatile absorbents. Heat and mass transfer experiments and multi-component heat and mass transfer analyses, augmented by high-speed flow visualization studies accomplished absorption and desorption at high transfer rates using essentially the same geometry, demonstrating the versatility of the concept. This microchannel technology can be used for absorption, desorption, condensation, evaporation and rectification to package an absorption heat pump system for small capacity applications in an extremely compact envelope. It can also be extended to other applications such as waste heat recovery, chemical and food processing, and others, i.e., applications that involve multi-component heat and mass transfer, particularly those with large enthalpies of phase change.

4. Miniaturization of Absorption Heat and Mass Transfer Devices - Non-Volatile Absorbents

Two key observations can be made about the miniaturization work reported above. First, the use of microchannel tubes leads to considerable reductions in component size for absorption systems with volatile absorbents such as ammonia water. It remains to be seen whether similar attempts can be made to reduce the size of absorption system components that utilize fluid pairs such as Lithium Bromide/Water, which have non-volatile absorbents. LiBr/H_2O systems are indeed used widely in large tonnage commercial systems. Since water is the refrigerant in this fluid pair, its large vapor-phase specific volume is a dominant factor in determining the compactness that can be achieved. Thus, if the vapor space (flow area) is constrained too much in the absorber, the resulting extremely large vapor velocities will result in appreciable pressure drops, which will in turn increase the pressure difference between the evaporator and the absorber in the cycle, consequently reducing the cooling that can be achieved in the evaporator. On the other hand, the mass transfer resistance of the liquid-phase LiBr/H_2O solution is considerable, and any measures that increase mixing in the liquid phase and increase the vapor-liquid interface area will result in more compact geometries regardless of the vapor velocities. This leads to the second observation from the above work, specifically, the insights from the flow visualization results. As seen in Figure 10, the liquid phase flow occurs in three modes: falling-film, droplet formation, and droplet fall/impact. Designs that recognize these different flow modes and tailor component geometries to judiciously modify the relative importance of these three modes are likely to yield compact geometries. These issues are addressed in this section on LiBr/H_2O absorption.

Killion and Garimella conducted comprehensive critical reviews of analytical and numerical models [44] and experimental investigations [45] of absorption heat and mass transfer. Killion and Garimella [44] found that most of the literature on absorption heat and mass transfer work has focused on the particularly simplified case of absorption in laminar vertical films of LiBr/H_2O. Fewer researchers have considered the important situations of wavy films, turbulent films, and films on horizontal tubes. They pointed out that attention must be paid to droplets and waves on horizontal tubes, and to the potential interaction of the heat and mass transfer process on the film hydrodynamics, surface wetting, and heat transfer in the vapor phase. In their review of experimental investigations [45] of $LiBr/H_2O$ absorption, Killion and Garimella found that although the literature contains a significant amount of data, much of the modeling of the data is empirical in nature. There are conflicting reports of the effects of absorber geometry including advanced tube surfaces, tube diameter and spacing, operating conditions such as liquid film flow rate and inlet conditions, and the influence of surfactants on absorption heat and mass transfer rates.

4.1 MULTIPLE FLOW REGIME HEAT AND MASS TRANSFER MODEL

Guided by these reviews of the literature, Jeong and Garimella [46] developed a realistic model for the absorption of water vapor into Lithium Bromide solutions flowing over horizontal tube banks. They noted that, in most practical cases, the solution mass flow rate in the absorber is maintained such that the solution falls as droplets between the consecutive tubes in the bank. Therefore, their model defined three different flow regimes: a falling-film region on the coolant tube, a region of drop formation on the underside of the tube, and a region of drop free fall between tubes. The solution passes through each of the three flow regimes in succession as it flows from one tube to the next. It was assumed that complete mixing occurs at the top of a tube due to droplet impingement. In the falling-film region, a fully developed laminar film was assumed and its flow characteristics were obtained using the well-known Nusselt equation. Instead of assuming a linear temperature profile in the film, the steep temperature increase near the interface due to the heat of absorption was taken into account. The set of differential algebraic equations, obtained from energy and concentration balance considerations, and mass transfer and equilibrium relations were solved using the DAE solver LSODI [47]. For these computations, the film region around each tube was divided into 100 segments. For the droplet formation regime, the shape of the droplet was idealized as a half-sphere. When the diameter of a droplet reaches the critical value, it detaches from the tube. Droplet formation was assumed to be adiabatic due to the negligible heat transfer between the drop and the coolant during this phase. The amount of vapor absorbed was calculated using empirical relations for the critical diameter and the mass transfer coefficient. The vapor absorption rate during the droplet formation phase was linearly dependent on the mass transfer coefficient, the surface area of the droplet, and the droplet formation time. The droplet formation time was calculated from the diameter of the droplet at the moment of the departure. Experimental relations from the literature were used for the mass transfer coefficient (see Jeong and Garimella [46] for additional details), which is inversely proportional to the square root of the droplet formation time. The droplet fall time was calculated from a free-fall expression. The equations for this regime were developed from a mass and energy balance on the falling droplet, applying equations similar to those that were used for the other regimes. For each calculation time step (one-hundredth of the time of droplet formation), these equations were solved numerically using LSODI [47]. It was found from these analyses that vapor absorbed in the droplet-fall regime is negligible.

This detailed, spatially and temporally resolved model provided valuable insights into the relative magnitudes of film-mode and droplet-mode absorption in $LiBr\text{-}H_2O$ absorbers over a wide range of conditions. Although much of the literature on falling-film heat and mass transfer has ignored the droplet formation mode, Jeong and Garimella [46] showed that the fraction of the total absorption that occurs in this mode is not insignificant, and in fact, for relatively higher mass flux cases, this represents the dominant mode for absorption (Figure 12). They found that at low solution flow rates, the tubes near the top of the bank absorb more vapor than the tubes at the bottom, and more vapor is absorbed in the falling-film region than in the droplet-formation regime. As solution flow rate increases, the droplet-formation regime plays a more important role. In addition, these models correctly predicted the rise in temperature of the solution between adjacent tubes in the bank due to quasi-adiabatic absorption, which is in excellent agreement with the independently obtained data of Nomura *et al.* [48] also shown in Figure 12. Jeong and Garimella also accounted for film distribution and surface wetting characteristics on tube banks – while it is obvious that inadequately wet surfaces will adversely affect absorption, this was quantified by them using a *wetting ratio* (wetted tube surface area/total tube surface area). As shown in Figure 13, as the wetting ratio decreases from 1 to 0.2, absorption rates can decrease by a factor of almost 3, with the absorption in the falling-film mode understandably being affected the most, due to its dependence on the wetted surface area. These findings are similar to those of Meacham and Garimella [37] for ammonia-water absorption that were discussed above for the microchannel absorbers.

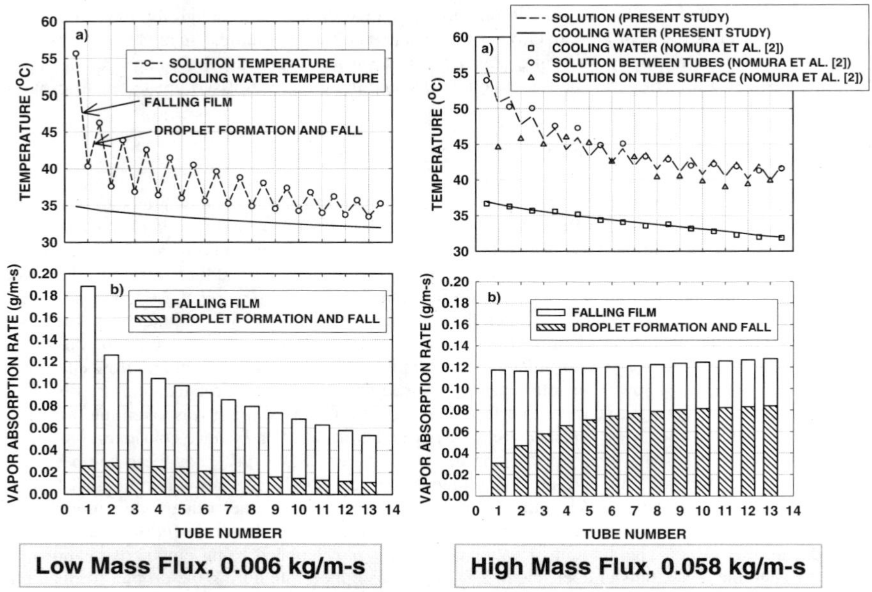

Figure 12: Relative Magnitudes of Falling-Film and Droplet Absorption

4.2 DEVELOPMENT OF COMPACT LITHIUM BROMIDE-WATER ABSORBERS

Jeong and Garimella [49] used the understanding obtained from the three-regime model described above to investigate the possibility of decreasing $LiBr/H_2O$ absorber sizes. They reasoned that if smaller tubes are used to build an absorber of a given surface area, the number of tubes required increases. Because the solution forms droplets at the bottom of each tube as it flows over successive tubes in the bank, and the time required for droplet formation is not greatly dependent on the tube diameter, the total residence time of the solution increases with the number of tubes per column of the tube bank. The increased number of tubes has another beneficial effect on mass transfer. As solution impinges on the tubes below, the solution concentration is redistributed to present a fresh surface for absorption. With more tubes, such mixing of the solution occurs more frequently as it flows down the absorber. The extended residence time and better mixing, in turn, result in the improvement of heat and mass transfer. Another advantage of smaller diameter tubes is that tube thickness can be reduced while maintaining the required mechanical strength of the tube, thus reducing the material requirements. Based on these considerations, they conducted a systematic investigation of tube diameters and pass

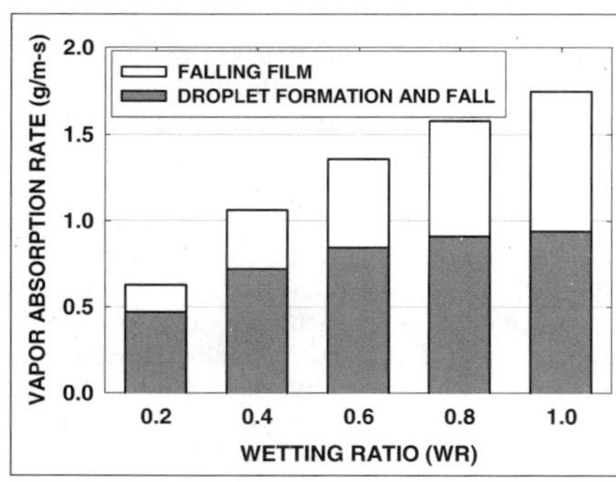

Figure 13: Effect of Wetting Ratio on Absorption Rates

arrangements and showed that for the same heat transfer area, tubes with a diameter of 6.35 mm deliver about 30% more cooling capacity than conventional tubes (OD = 15.88 mm). An even higher increase in capacity (about 55%) was found when 3.175 mm diameter tubes were used. This significant increase was primarily attributed to increased vapor absorption in the droplet-formation regime, which plays a progressively larger role in vapor absorption as the tube diameter decreases. However, they cautioned that because it is expected that wetting of tubes becomes poorer as the tube diameter decreases, wetting characteristics should be considered in conjunction with these findings to select the optimal tube diameter.

5. Understanding Flows over Tube Banks as Wavy Films and Droplets

The work of Jeong and Garimella [46, 49] discussed above made the advance of explicitly incorporating film and droplet flow modes separately into the calculation of absorption rates. While this pointed out the potential magnitude and importance of the droplet modes, the analyses had to rely on assumptions such as a hemisphere shaped drop on the underside of the tube, empirical mass transfer coefficients, and others. In cases where the liquid phase presents a significant resistance to either heat or mass transfer or both (low thermal or mass diffusivity), for example, in the case of the LiBr/H_2O fluid pair, the progression of the shape and motion of the film and droplet (e.g., Figure 10) profoundly influence the transfer processes. Assumptions such as smooth laminar films and hemispherical drop shapes represent a good first approximation, but do not capture the flow phenomena accurately enough to enable heat and mass transfer modeling that is representative of the actual flow mechanism. Droplet impact and detachment causes circumferential and axial waves and oscillations of the liquid film, which must be taken into account. The assumption that a hemispherical/spherical drop is initiated, and grows and falls as a sphere, does not account for the significant variations in the shape of the droplet (which only briefly resembles a sphere) during formation, detachment and freefall. This is particularly true when the focus shifts to miniaturization of absorption system components through the use of small diameter and microchannel tubes. In view of this, Killion and Garimella [50] undertook a study of the deviations of the flow over tube banks from idealized falling films. In fact, it should be noted that it is this "non-ideal" behavior of the falling film that directly leads to the high transfer rates achieved in practice. The results of their work are presented below, with a discussion of the implications of the flow mechanisms on the transfer processes.

5.1 HIGH-SPEED VISUALIZATION OF ABSORPTION MODES

Killion and Garimella [50] first investigated flow of water over a single column of aligned horizontal tubes under isothermal conditions. Typically the film Reynolds number was somewhat less than 100, and the water flowed in the droplet mode between successive tubes in the bank. (According to the work of Hu and Jacobi [51] the Reynolds number where the droplets in this system would begin to flow as continuous columns, which is not desirable for absorbers, is about 425.) High-speed digital video cameras were used to capture the images of the fluid flow at 500 frames per second and resolutions of 1024×1024 or 512×240 pixels, as necessary. Other details of the tube bank and experimental facilities and techniques are available in Killion and Garimella [50]. Significant features of the flow over tube banks that were observed by them are shown in Figure 14.

5.1.1 Droplet Formation

As the film thickens at the bottom of the tube (Figure 14a), it becomes unstable and disturbances spaced around the so-called most dangerous wavelength are amplified to become the sites where droplets develop. (Droplets impacting from a tube above can also lead to the inception of a drop formation site.) Axial motion of the liquid on the under-side of a tube due to propagating waves may generate a local excess of liquid that can also develop into a formation site. In the initial stages, a droplet forming pendant from a horizontal tube is not axisymmetric but is elongated in the axial direction of the tube due to surface tension. As liquid accumulates at the site, a wave of increasing amplitude forms and progresses to the shape of a droplet. Droplet formation also induces a transverse velocity component within the film surrounding the tube. As more liquid enters the forming droplet, the distance from the tube to the tip of the droplet increases and the tip of the droplet tends to a spherical cap, a more axisymmetric shape. The tip of the droplet has a nearly spherical cap and somewhat cylindrical sides. The portion of the droplet attached to the tube continues to contract in the plane of the tube axis during this transition, causing mass and species redistribution within the droplet, which affect its contribution to the absorption process. It has been shown [46, 48, 52] that the liquid film entering the droplets may be sufficiently cooled during its flow as a film around the tube that the absorption process continues on the forming droplet even though the heat transfer from the droplet to the tube is minimal due to the large thermal resistance between the liquid interface and the tube wall at this location. Since the droplet will be cool enough to continue absorbing vapor, the temperature and concentration of the liquid arriving at the new surface of the droplet and the circulation within the droplet itself will directly affect this absorption rate. Absorption models must also account for the continuously varying droplet shape and surface area that rarely, if ever, resembles a perfect sphere; and the bulk flows within the forming droplet.

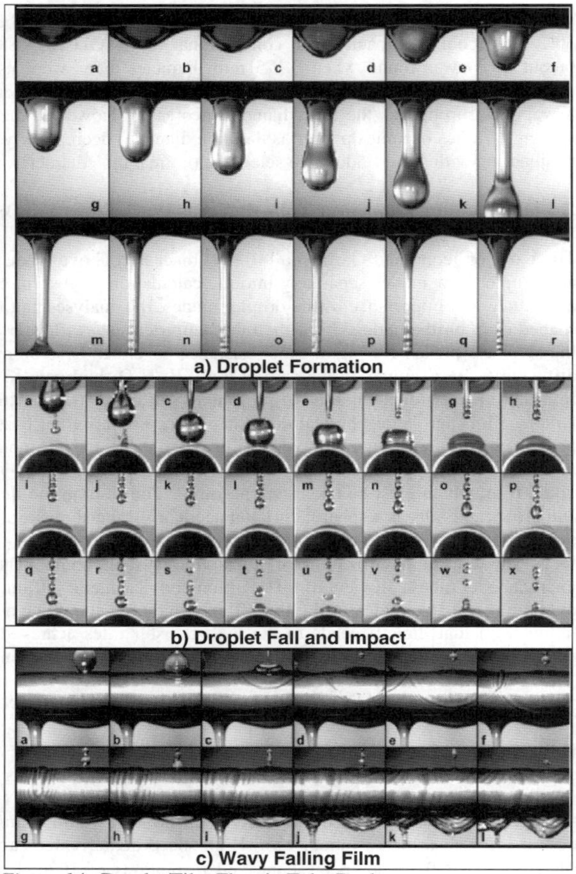

Figure 14: Droplet/Film Flow in Tube Banks

5.1.2 Detachment and Fall

As the droplet grows and extends away from the tube, its downward velocity increases quickly and the droplet pulls away from the tube faster than new liquid enters from the film. This causes necking in the liquid bridge between the droplet and the film on the tube, appearing as a large primary droplet attached to the bottom of a cylindrical filament. This liquid filament jet is subject to instabilities that are manifested as corrugations at the bottom of the filament. As the bridge continues to thin, the liquid bridge breaks at a point close to the droplet, which is now almost spherical because the surface tension forces from the liquid bridge are confined to a small area on the surface of the drop. The almost-point-contact between the filament and the detaching drop can be seen in Figure 14b. The shape transitions in the droplet impart enough momentum to generate oscillations in the shape of the falling droplet after detachment. The velocity when the liquid bridge breaks is extremely high [53, 54] and the recoil of the bridge is very fast and causes harmonic distortion of the bridge. This propagating waviness typically breaks the bridge again at the end nearest the tube, followed by breakup into a number of smaller satellite droplets. These shape oscillations, agglomeration, and the path of the satellites have received some attention [54-57] and will affect absorption rates during droplet fall. Although much smaller in volume and surface area than the primary droplet, satellite droplets contribute to the absorption process due to the additional surface area and the high-velocity shape transitions that mix the fluid within the satellites. The formation of these satellite drops has been largely ignored in studies of absorption.

5.1.3. Impact and Detachment Related Waves

Films on horizontal tubes typically do not have the required development length for roll or capillary waves to develop from the inherent film instability. In the present situation, however, film waviness is induced due to droplets impacting and detaching from the film. These waves mix the film and increase heat and mass transfer. However, in absorption models, it is commonly assumed that the film is completely mixed upon impact at the top of the tube and flows smoothly and without axial variations around the tube. Figure 14b,c shows that these assumptions are not really representative of the actual flow phenomenon. When the droplet impacts the film, it immediately begins to deform around the tube and forms a saddle-shaped wave. Ripples from the impact also travel upward through the droplet, which can affect the bridge (if still attached) to the tube above. The saddle-shaped wave generates smaller ripples in the film substrate ahead of it. Wave propagation slows somewhat as it spreads over an increasing portion of the tube surface, and also due to viscous effects. Other researchers have shown that the effect of a wave traveling over a substrate is to partially mix the film left behind [58-62]. The propagating saddle wave generates a fresh surface and partially mixed film behind its front. Figure 14c also shows the axial flow on the tube. When two droplets impact at nearly the same time, the saddle waves interact and give rise to a circumferential ring of liquid that is nearly stationary in the axial direction.

It should be noted that the detachment of a pendant drop causes a disturbance at the bottom of the tube, which generates waves on the film above and to the side. At the moment the bridge breaks, the liquid remaining attached to the tube is typically shaped like a stretched triangle. The surface tension forces at the tip of this shape furthest from the tube are very high and cause fast recoil of the liquid, which in turn leads to ripples that propagate up the tube. These waves can disturb the formation of neighboring droplets and cause some side-to-side motion of droplet formation sites.

5.1.4 Interacting Phenomena

While the above discussion focused on the behavior of individual droplets, this solitary behavior is infrequent, and neighboring droplet sites, interactions of forming and impacting droplets, and their effects on wave flow must all be considered to obtain a truly representative picture. The formation of circumferential rings due to collision of two adjacent advancing saddle waves was discussed above. Similarly, droplet impact and the resulting waves from one event on an upper tube in the bank can often lead to early droplet departure from a neighboring site on the lower tube. Bridge breakup and satellite drop formation are similarly affected by neighboring events. The overall effect of each of these interacting phenomena is to redistribute the species and thermal energy within the liquid. Other interacting phenomena include satellite droplets with upward velocity, agglomeration of neighboring droplet formation sites, droplet spinning during freefall, droplet slinging away from vertical, and others. The nature and frequency of these events depend on design parameters such as tube spacing and film flow rate as well as the thermophysical properties of the liquid, especially the fluid viscosity, surface tension and density (compared to the vapor). Additional details of these interacting phenomena are available in Killion and Garimella [50].

5.2 EXPERIMENTAL QUANTIFICATION OF FLOW MECHANISMS

5.2.1 Related Literature

With the qualitative understanding of the progression of the falling film and droplet behavior described above, Killion and Garimella [63] next addressed quantitative measurement of the evolution of flow around tube banks. In this study, they used aqueous Lithium-Bromide (53.44% LiBr by weight) solution. Many empirical/modeling techniques have been developed as a result of experimental investigations into the problem of droplet formation as summarized by Kumar and Kuloor [64], Clift *et al.* [65] and Frohn and Roth [66]. Eggers [67] provides an excellent summary of many modern mathematical, experimental, and computational methods utilized to understand the details of the droplet behavior. Other reviews covering similar topics can be found in [68-70]. Studies in the literature have primarily investigated axisymmetric cases. The case of formation under horizontal tubes, however, leads to significant differences from most of the literature in terms of the overall shape, size, and internal velocity fields of the droplets, especially in the early stages of formation where the extent of the droplet in the lengthwise direction of the tube may be several times its extent in the circumferential direction. In addition, the likelihood of the behavior of one droplet influencing a nearby droplet via impact and film waviness is high in the case of tube banks, especially at high flow rates. (Interactions between droplets in sprays, which have received considerable attention, are governed by completely different mechanisms than those in tube banks, which was the focus of attention for Killion and Garimella [63].) The impact of droplets in the case of horizontal tubes and falling films has not received a great deal of attention, although droplet impact in general is widely studied. Rein [71] and Tropea and Marengo [72] have reviewed the impact of droplets on walls, films and pools. When droplet impact leads to spreading without splashing, the thin film formed that propagates away from the point of impact with a decreasing velocity is called a lamella [72]. In the case of impact on a dry surface, the lamellae often expand to a maximum diameter and then retract, owing to surface tension, to an equilibrium shape. In contrast, with a wet surface, the lamellae expand until they are no longer distinguishable features. But in the case of impact onto a horizontal tube covered with a thin film, the geometry is not axisymmetric due to the preferential downward flow resulting from gravity, and the spreading lamella is no longer circular. This leads to the characteristic "saddle wave" observed by them [50] and discussed above, which must be studied and analyzed differently from axisymmetric cases.

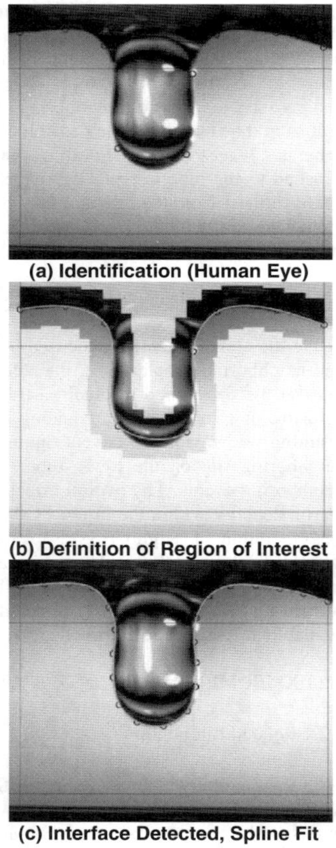

Figure 15: Interface Analysis Steps

5.2.2 Analysis of Interface Progression

Video frames for the overall progression of the liquid film and the typical evolution of a droplet from the early formation through detachment and subsequent impact on the tube below, similar to those reported above with water as the working fluid [50] were also recorded for LiBr/H_2O [63]. Details of this progression were discussed above in connection with similar water films and droplets [50]. To develop a quantitative understanding of the behavior of the droplets, a method for analyzing the video was developed. The method consists of the following steps: (a) identification of the location of the droplet interface in each frame, (b) generation of a mathematical description of the interface, and (c) utilization of the mathematical description to assess quantities of interest such as the surface area and volume of the droplets versus time.

In the semi-automated algorithm developed by Killion and Garimella [63], the approximate location and shape of the interface (edge of interest) is first identified by eye (Figure 15). This information is then used to guide the interface identification program on the initial and each subsequent frame. Edge detection algorithms and associated threshold parameters were adjusted as necessary for each frame to address the changing glares and shadows. A graphical user interface (GUI) was developed in the commercially available software program Matlab [73] to implement this algorithm. The GUI facilitates the limited human interaction required to guide the program through the evolution of a droplet. On the first frame of the sequence to be analyzed, a region of interest (ROI) that includes the droplet interface is defined by the user (see Figure 15a,b). An edge detection algorithm such as Canny [74] or Sobel [75], is selected along with the various edge intensity thresholds. After the ROI for the first frame is thus highlighted (Figure 15b), the ROI for each subsequent frame is defined based on the identified edge in the previous frame, which allows many consecutive frames to be analyzed with little human intervention. Once the edge is suitably identified, the mathematical description is accomplished using the image coordinates of a selected number of the identified edge pixels and cubic spline approximation. Detailed expositions of the practical application of splines are available in [76, 77]. The algorithms presented by de Boor [76] have been extended to handle vector-valued splines in the Matlab Spline Toolbox [78], which was used here. It should be noted that since systems having free surfaces with interfacial tension tend toward equilibrium states that minimize the surface energy, i.e. surface area [79] (p. 29), splines are uniquely suited for describing the shape of the interfaces of such systems. The two-valued (x and y coordinates) piece-wise-smooth cubic smoothing splines thus used are fit through a number of the identified edge pixels and parameterized by the distance between the selected edge points, completing the second step of the analysis.

For the final step, calculation of surface area and volume, one of the constraints due to the 2D images (only one camera used) was that information about the thickness dimension was unavailable. However, in the regions between the tubes, the droplet shape becomes axisymmetric and so the 2-D interface was treated as a surface of revolution. The vertical extent of the regions of such axial symmetry was determined visually and by interpreting the results of the analysis. The values are converted from image coordinates to SI units by calibrating the image using a dimension of known size within the image, such as the tube diameter. Further details of the analyses of the splines to obtain the areas and volumes are available in [63]. This procedure therefore provides the temporal development of droplet surface area and volume during droplet formation and detachment, which can be used for the analyses of the respective transport processes.

Figure 16 shows the analyzed interfaces as a function of time, with the spline fits shown as light solid curves. The horizontal lines show various limits used to analyze the droplet surface area and volume. Figure 17 shows surface area, internal volume, and surface area-to-volume ratio for these interfaces. The three curves on each plot in Figure 17 illustrate the effect of the upper and lower limits on the analysis of surface area and volume (i.e., limits of the assumed axisymmetric region.) The solid line illustrates an analysis where the chosen limits in fact represent axisymmetric regions; the two dashed lines represent limits that are too high or too low in the image. These latter limits lead to physically unrealistic variations of the area and volume and are discussed further in [63]. It can be seen that during droplet formation and the stages where the droplet pulls away from the tube, the volume and surface area increase to a maximum value just before impact. In reality, the volume does not truly decrease after the impact, as suggested by Figure 17; it just leaves the region of interest and is redistributed on the tube surface. Shortly after impact, the volume and surface area of the primary droplet is lost to the surface of the tube resulting in a steep decline in both curves. After this, the remaining liquid bridge continues to thin and drain until it breaks and forms satellites. During this process, the volume and surface area of liquid between the tubes

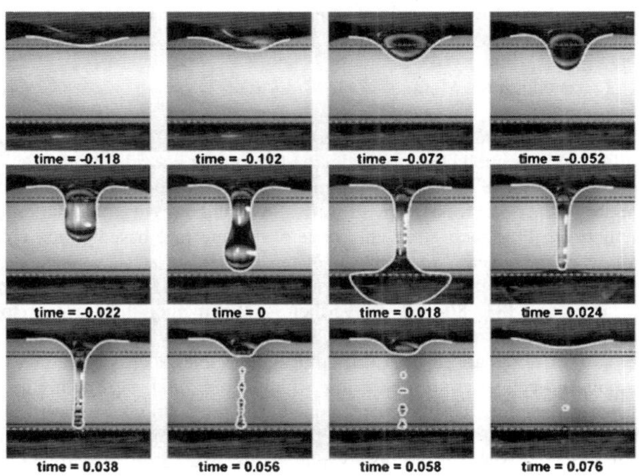

Figure 16: LiBr/H_2O Video Frames Superimposed with Analyzed Interfaces

continues to decrease, although at a slower rate than was associated with the impact of the primary drop. It is interesting to note however, that the surface area-to-volume ratio increases significantly and reaches a maximum with the development of satellite droplets. The unsteadiness in the surface/volume ratio in this region is due to shape oscillations, sequential impact of the satellite drops, and the increased uncertainty in the calculation of droplet sizes when the whole drop occupies just a small number of pixels. The chief implication of the rising area to volume ratio is that, although the total surface area and volume is small, satellite drops may absorb significant amount of vapor due to the reduced resistance to heat and mass transfer. These findings can now be incorporated into the multi-region absorption heat and mass transfer models of Jeong and Garimella [46, 49], to yield transfer rates and optimal sizes of miniaturized absorbers based on a realistic progression of the droplet formation phase of the process.

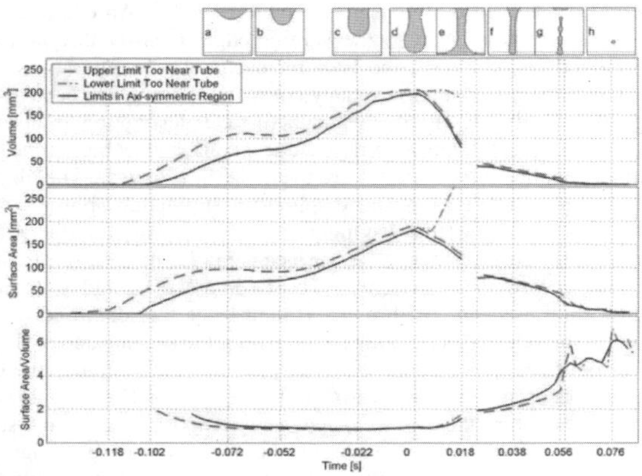

Figure 17: Temporal Variation of Surface Area, Volume and Area/Volume Ratio

5.3 COMPUTATIONAL MODELING OF ABSORPTION MODES

In parallel with the experimental investigation of flow of LiBr/H_2O solutions over horizontal tube banks, Killion and Garimella [80] also conducted 3-D computational simulations of the same problem, where the experimental results served to validate the results of these computational analyses. The results from this 3-D model were compared with 2-D, axisymmetric models, and it was shown that to adequately address all the features of these flows, 3-D models are indeed required.

5.3.1 Computational Method

The VOF technique, originally proposed in the mid 1970s by several different investigators [81], was chosen by Killion and Garimella [80] to model the flow under consideration. The VOF method has been advanced by many contributors [82-87]; it has been successfully extended to three dimensions [86, 87], and can appropriately handle surface-tension-driven flow [83, 84] and moving boundaries [85]. The two main challenges in this method are: 1) to reconstruct the shape of the interface based on the volume of fluid in neighboring cells, and 2) to advect fluid near the interface in such a way that the interface remains sharp (does not diffuse) while rigorously satisfying mass conservation. Methods for handling these challenges are still areas of current research, e.g. [81, 86-89]. Nevertheless, the method has proven to provide reasonable accuracy, is quite economical in terms of memory usage (requiring only one scalar value to be stored for each cell in the solution domain), and has no inherent difficulties when interfaces rupture or coalesce, unlike other methods such as the marker-and-cell or conformal grids, which can exhibit gaps or highly distorted cells when the interface shape becomes complex. It also provides one of the more straightforward ways to add interface tracking to an existing single-phase CFD solver, it works for three-dimensional non-axisymmetric problems, and is computationally economical. Some of its disadvantages are that, until recently [87], the advection methods could lead to "flotsam" or numerical debris, the interface reconstruction and advection methods could produce slightly jagged interface profiles when the velocity of the interface was not normal to the grid faces, and for an interface reversibly sheared forward and backward the same amount (typical of advection tests for different VOF implementations) the final interface location will show some disagreement or noise when compared with the initial position (see [81, 86-89]). Even with these potential pitfalls, the VOF method has seen widespread implementation. Applications include the prediction of droplet formation of a viscous fluid in air from the end of a capillary tube [53, 86], jet breakup in liquid-liquid flows [56], bubble formation and oscillation during the impact of raindrops into deep pools of liquid [90], the formation and break-up of crown splashes in droplet impacts on thin liquid layers [91], the formation of droplets from a capillary tube surrounded by a viscous medium [92], the formation [93], rise [86] and burst-through [94] of a bubble in a liquid medium.

A commercially available CFD package [95] was used to develop and implement the model. The flow was assumed to be incompressible, Newtonian and laminar throughout the solution domain due to the low film Reynolds numbers (< 500); even though bifurcation and impact events can lead to locally high velocities, they do not generate turbulence in a manner that requires special modeling. Killion and Garimella [80] considered both an approximate cylindrical (2-D case, column of spheres) formulation as well as a Cartesian (3-D case, horizontal tube bank). A finite-volume method was employed to translate these coupled, partial differential equations into algebraic expressions. The equations of motion and continuity were integrated over each computational cell and discretized using a second-order upwind scheme. The discretized equations were linearized and solved in a segregated, 1st-order implicit manner. The resulting linear system of simultaneous equations (one for each cell in the domain) was then solved using a Gauss-Siedel equation solver in combination with an algebraic multi-grid (AMG) method (which generates coarse level equations using a Galerkin method without actually performing any re-discretization). A pressure correction was used iteratively if necessary until convergence was achieved. To relate the solution of the continuity equation to the pressure correction, the PISO (Pressure-Implicit with Splitting of Operators) method [96], a part of the SIMPLE family of algorithms, was used. A piecewise-linear-scheme adapted for unstructured grids from [97] was used for interface reconstruction. Surface tension was handled using the so-called "continuum surface force (CSF)" method of Brackbill *et al.* [83, 84]. Thus, the pressure rise across the interface due to surface tension is converted to a source term dependent on the local curvature in the momentum equations. This source term is a body force similar to the way gravity appears in the equations. In addition, a wall contact angle was used to adjust the surface normal in the cells near the wall. In most cases, walls were specified to have a 0° contact angle to ensure complete wetting.

The computations were conducted on parallel processor machines with shared or distributed memory. When the fully three dimensional simulation was run in parallel on four processors, the speed-up was nearly fourfold, indicating that the overhead of parallel communication was quite low. The fixed grids used in the study were made up of quadrilateral (2-D) or hexahedral (3-D) volumes. Quadrilateral and hexahedral cells are known to be slightly more accurate when surface tension is an important force, as is the case here, than triangles/tetrahedrals. The grids were generated using the program GAMBIT [95].

5.3.2 Geometry and Model Initialization

The geometry under consideration (Figure 18) offers two vertical planes of symmetry that allow the model to be reduced to ¼ of a full droplet model. It was assumed that the vertical plane intersecting the axis of the tube that cuts the tube in half along its length and the vertical plane perpendicular to this intersecting the center of the droplet (vertical lines in box shown on Figure 18) were both planes of mirror symmetry. The model included one tube and half of the tube-to-tube spacing above and beneath the tube; the top and bottom boundaries (horizontal lines in box shown on Figure 18) were assumed to be periodic. Thus the droplet that fell from the tube would exit the bottom of the solution domain and reenter at the top to impact on the top of the tube making a continuous bank of tubes out of the solution domain. Finally, the tube length was taken to be half of the distance between droplet formation sites (as observed in the experimental work described in [63]). This was equivalent to assuming that droplets form and fall equally spaced and in perfect unison. This is a simplification compared to a real system where droplet spacing and sequencing would be variable. However, this assumption does lead to realistic interaction between the spreading lamellae of droplets impacting adjacent to one another. In the case where neighboring droplets are more out-of-phase, the spreading lamellae would continue to expand along the length of the tube until they become indistinguishable or interact with other out-of-phase droplet impacts creating more complicated wave patterns as reported by Killion and Garimella [50]. The initial conditions were taken to be quiescent with a smooth film surrounding the tube. The thickness of the film was varied in a sinusoidal fashion from 0.20 mm at one end to 0.53 mm at the other end, which ensured that the droplet formed at one of the planes of symmetry. Figure 18 shows the solution domain with an exaggerated initial film thickness profile. The solution domain was divided into quadrilateral elements that were finer near the tube walls and in regions where the two-phase interface was expected to pass, and coarser in regions where only the gas-phase was expected. The edge length varied from 1×10^{-4} m near the tube surface to a maximum of 2.5×10^{-4} m. The face mesh, shown in Figure 19, contained 16,941 quadrilaterals; this pattern was repeated along the length of the tube to form 1,185,870 hexahedral elements. The shading in Figure 19 corresponds to the partitioning of the grid to the four processors used for the parallel computation.

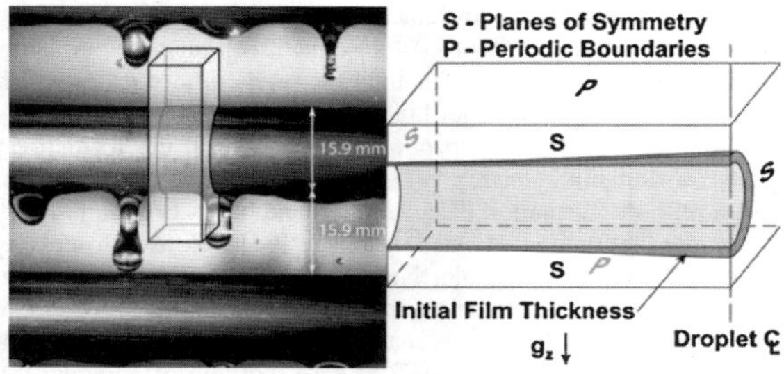

Figure 18: Computational Model Schematic (Not to Scale)

5.3.3 Computational Model Results

The model was solved using time steps of 0.2 ms. Larger time steps would lead to numerical instabilities at the liquid vapor interface; smaller time steps had no effect on the results. A typical droplet formation cycle required around 2000 time steps when beginning with a stagnant film. The interface was taken to be the surface where the volume of fluid was 0.5 (interpolated from the VOF values at each grid location). Defining this surface allowed rendered images of the droplet and film to be generated at regular time intervals. To aid visualization, the images were reflected about the planes of symmetry (to show a complete droplet) and duplicated about the periodic boundaries (to show multiple tubes and droplets). The results are shown in Figure 20. Because of the reflections, four identical droplets are shown, which aids in the visual interpretation of the interference patterns of the interacting lamellae. These computational results were validated using Killion and Garimella's earlier experimental flow visualization results [50, 63], as shown in Figure 21. The simulation exhibits with remarkable fidelity, all of the characteristics observed in the experiments including the stretching of the drop along the tube early in the formation process, the formation of a primary droplet and trailing liquid thread, the saddle-shaped spreading lamella, and the thinning and breakup of the liquid bridge into a number of satellite droplets. The droplet and thread volume in the simulation appear to be slightly smaller than in the experiment which leads to an earlier breakup of the liquid thread into satellites. In addition, the early droplet formation process took longer in the simulation than was observed experimentally. This is most probably due to the quiescent initial conditions used in the simulation. Early formation of the droplets is often accelerated by the arrival of a quantity of liquid from a previous droplet impact [50, 63], which might be captured by the simulation if continued for more droplet cycles.

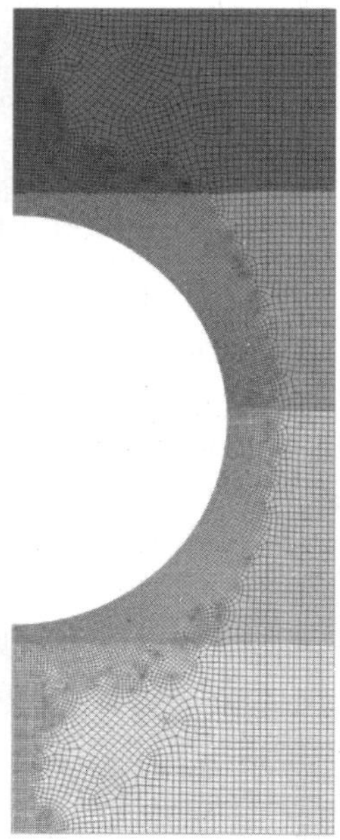

Figure 19: Mesh Used for 3-D Model

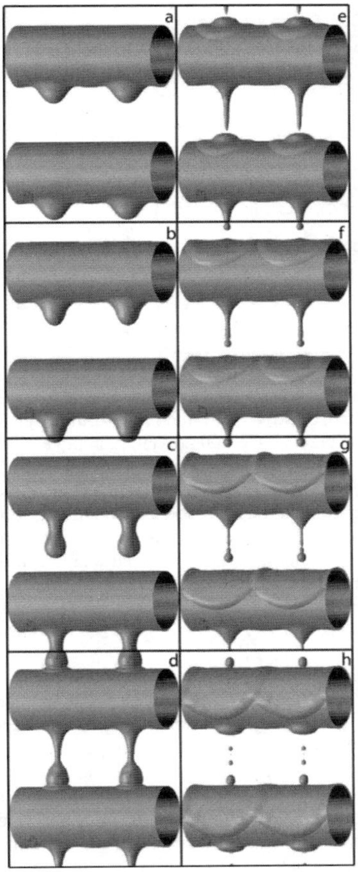

Figure 20: Computational Model Results

The digital image analysis routine developed by Killion and Garimella [63] was also applied to the rendered images produced from the CFD results to estimate droplet surface area and volume in the inter-tube region where the droplet is nearly axisymmetric over a sequence of images. A comparison is shown in Figure 22. The solid line shows the analysis of experimental results and the dashed line the simulation. Also shown on this plot is an axisymmetric model (dotted line) discussed later. The largest discrepancy between the model and experimental results appears early in the droplet formation. As mentioned above, the simulation begins slowly because of the assumed initial conditions, whereas the experimental results are typically accelerated early by the arrival of liquid from droplet impacts above. It is clear though, that after the droplet volume reaches about 75 mm^3, the results track very well, although the ultimate volume of the droplet is slightly smaller in the simulation, which results in the deviation seen near the point of impact (the peak in the curves at time 0). As the droplet spreads onto the tube and the thread breaks up, the volume and surface area between the tubes drop sharply in both the simulation and experiments. The liquid thread breaks up slightly earlier in the simulation. Thus, the increase in surface area-to-volume ratio associated with the creation of satellite droplets also occurs somewhat earlier in the simulation, although throughout the evolution the agreement in this parameter is quite good. The agreement between the simulation and experiment suggests that the model provides a sound basis for investigating the details of absorption heat and mass transfer in falling films on horizontal tubes.

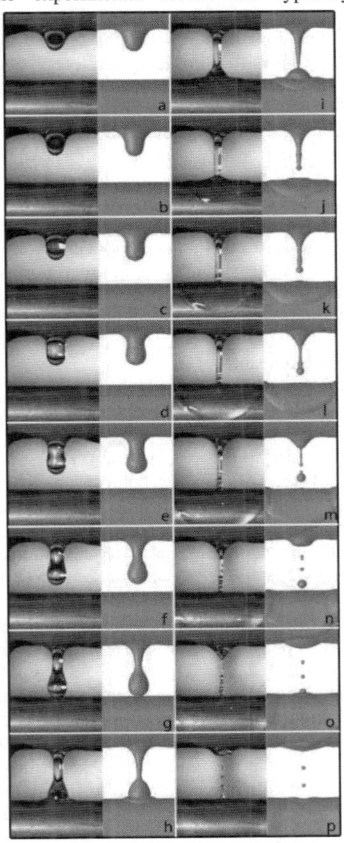

Figure 21: Comparison of Computational and Experimental Results

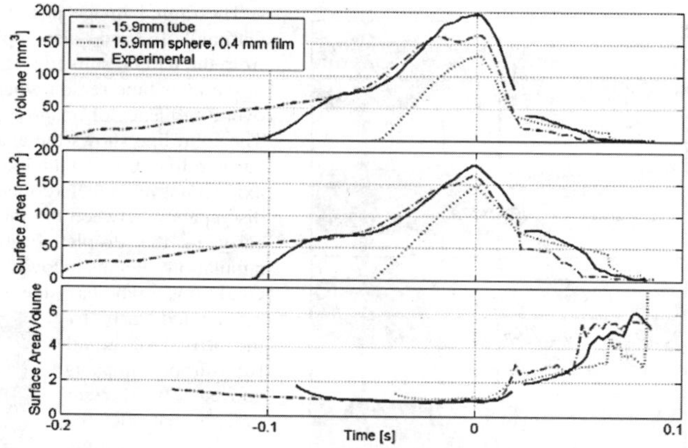

Figure 22: Comparison of Volume, Surface Area, and Area/Volume Ratio Using Experimental and Computational Techniques

Because the computational cost of performing a 3-D transient simulation is quite high, an axisymmetric 2-D approximation of the tube bank was also considered. A column of spheres of the same diameter and spacing as the tubes was modeled using a mesh similar to the 3-D case. It was found that several 2-D models with varying initial film thickness inputs all significantly under-predicted the droplet size and the time for droplet evolution. Both the initial droplet formation and the thread break-up and formation of satellite droplets occurred much quicker than observed for the horizontal tube case. In the case of horizontal tubes, surface tension in the direction of the axis of the tube tends to elongate the droplet and cause it to be larger. In addition, the flow of the film toward the droplet is more direct in the case of a sphere; in the horizontal tube case, some of the falling film must move axially along the tube to arrive at the forming droplet, which explained the longer evolution time observed in the horizontal tube case. These findings suggest that a 3-D model is essential to accurately capture all the flow phenomena observed in the flow of liquid films and droplets on horizontal tube banks.

6. Conclusions

A comprehensive treatment of the development of heat and mass transfer devices used in binary-fluid phase change processes was presented. Binary-fluid absorption, desorption, condensation and other phase change processes are prevalent in a variety of industries such as chemical processing, food processing, waste heat recovery, thermally activated space-conditioning systems, and others. In this paper, components required for the development of miniaturized thermally activated heat pumps that use working fluid pairs with volatile and non-volatile absorbents were addressed. A recently patented innovative geometry that uses microchannel tube arrays was shown to result in extremely compact components for ammonia-water heat pumps due to high heat and mass transfer coefficients in the binary fluid as well as the coupling fluid. Preliminary designs developed using empirical coupled heat and mass transfer considerations yielded prototype designs that were investigated experimentally to validate the miniaturization technology. High-speed flow visualization experiments conducted during the actual heat and mass transfer process clearly illustrated the details of the solution flows in these components. The importance of accomplishing uniform flow distribution in the tube banks was also illustrated. Initially encountered distribution problems were resolved by using improved solution distribution devices, which component heat duties considerably. The absorption heat load for a 10.55 kW cooling system was transferred in an extremely compact 0.127×0.127×0.476 m envelope. It was also demonstrated through experiments and analyses that the same microchannel tube array could function equally well as an absorber and a desorber. In fact, the utilization of this uniform microchannel geometry for all the major components of an absorption system (absorber, desorber, condenser, rectifier, and evaporator) will result in substantial reductions in the capital costs of the overall system, facilitating the widespread implementation of thermally activated heat pumps.

The initial high-speed video experiments on the miniaturized absorbers also demonstrated that the solution does not flow as an axially and circumferentially uniform, smooth film around the tubes, and simply appear on the next row in the tube bank as a well distributed film, as is commonly assumed in the literature. It was shown that the solution actually flows as non-uniform wavy films around the tubes, followed by droplet formation, elongation, detachment and impact phases that must be explicitly addressed to achieve success in modeling the coupled heat and mass transfer processes in these absorbers. These observations on the working fluid (NH_3/H_2O) with a volatile absorbent also guided the development of three-regime (falling film, droplet formation, droplet fall/impact) models for conventional Lithium Bromide absorbers. It was shown that not only does the droplet formation phase contribute significantly to the total absorption of vapor in the tube bank, but also the relative fractions of absorption in the droplet formation and falling-film modes change considerably from low solution flow rates to high solution flow rates. It is important to acknowledge this change in the relative importance of the different mechanisms when extrapolating component designs from small heat duties to large heat duties. These flow-regime-specific models also included a means to account for the incomplete wetting of tube surfaces that is encountered in actual absorbers.

The flow-regime specific $LiBr/H_2O$ absorption model guided the development of very compact absorbers for commercial applications. By using smaller diameter tubes, the number of tubes in the bank for any given surface area was increased, which proportionally increased the number of times droplet formation and impact took place. This presented added opportunities for absorption in the tube bank, and also increased the number of times the solution concentration is redistributed upon impact to provide a fresh surface with favorable solution concentrations for absorption. These considerations led to designs that transferred 55% additional heat duty when 3.175 mm diameter tubes were used in place of the conventional 15.88 mm tubes.

While the development of the three-regime absorption models represented an important advance in the modeling of absorption heat and mass transfer, a more accurate and realistic quantification of the wavy film and the stages in droplet evolution were felt necessary. This was addressed by first conducting visualization experiments on a column of tubes using high-speed video equipment. The resulting video frames documented in extensive detail the progression of the waves in the solution flowing around the tube, which rendered the film axially and circumferentially non-uniform. In addition, the growth of the droplet from the underside of the tube, including the formation of slender filaments of liquid with spherical caps at the bottom, the extension and thinning of the filament, and instabilities in the filament leading to breakup and formation of satellite droplets was demonstrated. It was also shown that the impact of the liquid from these inter-tube spaces on to succeeding tubes governed the growth and propagation of the wavy films around the tube. Interactions of impact and detachment events at neighboring sites led to even more complex flow phenomena. The ever changing shape and size of the waves and droplets clearly cause considerable internal circulation and mixing of the species that have a direct bearing on the absorption heat and mass transfer.

The progression of the droplets through the stages enumerated above was also quantitatively analyzed using image analysis techniques on successive frames, including edge detection, cubic spline fitting to define the evolving surface, and computation of the fluid volumes and surface areas in the inter-tube region. One of the more significant findings was that although the satellite droplets have small volumes and surface areas, the corresponding area/volume ratio is quite high, indicating that they contribute appreciably to the absorption process. This is particularly significant for the $LiBr/H_2O$ fluid pair, in which the properties of the liquid phase lead to a considerable mass transfer resistance. A 3-D computational model of these same phenomena using the volume of fluid method was also developed to complement the experimental work. Excellent agreement was achieved between the experimentally observed phenomena and the computations. The computations replicated accurately virtually all of the phenomena observed in the experiments. In addition, the droplet volume and surface area between the tubes obtained through image analysis of the experimentally obtained video frames and the computational results were in excellent agreement.

This multi-faceted approach toward understanding the fundamentals of coupled heat and mass transfer in binary-fluid absorption and the simultaneous application of this detailed understanding to the development of miniaturized components will assist the development of thermally activated systems for a wide range of small-capacity (residential) and large-capacity (commercial and industrial) space-conditioning systems. These insights can also be used in other industries that rely on multi-component phase-change heat and mass transfer. Ongoing research by the author's group is attempting to obtain, using experiments and numerical simulations, predictions of the heat and mass transfer rates in each phase with the same spatial and temporal resolution achieved on the fluid flow mechanisms reported here. In addition, the study of fluid flow and absorption in the presence of surfactants used in actual systems, and the study of desorption and other processes in such cycles with this level of detail will yield a comprehensive understanding of this field of binary-fluid heat and mass transfer.

References

1. Garimella, S., Christensen, R. N., and Lacy, D. (1996) Performance Evaluation of a Generator-Absorber Heat-Exchange Heat Pump, *Applied Thermal Engineering*, Vol. 16(7), pp. 591-604.
2. Engler, M., Grossman, G., and Hellmann, H.-M. (1997) Comparative Simulation and Investigation of Ammonia-Water: Absorption Cycles for Heat Pump Applications, *International Journal of Refrigeration*, Vol. 20(7), pp. 504-516.
3. Gommed, K., and Grossman, G. (1990) Performance Analysis of Staged Absorption Heat Pumps Water-Lithium Bromide Systems, *ASHRAE Transactions*, Vol. 96(1), pp. 1590 - 1598.
4. Garimella, S., Christensen, R. N., and Petty, S. E. (1992) Cycle Description and Performance Simulation of a Gas-Fired Hydronically Coupled Double-Effect Absorption Heat Pump System, *Winter Annual Meeting of the American Society of Mechanical Engineers*, Anaheim, CA, USA, Vol. 28, ASME, New York, NY, USA, pp. 7-14.
5. McGahey, K. R., Garimella, S., Cook, F. B., and Christensen, R. N. (1994) Enhancement of the ORNL Absorption Model and Simulation of a Double-Effect Absorption Heat Pump, *Proceedings of the International Absorption Heat Pump Conference*, New Orleans, LA, USA, ASME, New York, NY, USA, pp. 141-148.
6. Garimella, S., Lacy, D., and Stout, R. E. (1997) Space-Conditioning Using Triple-Effect Absorption Heat Pumps, *Applied Thermal Engineering*, Vol. 17(12), pp. 1183-1197.
7. Grossman, G., Wilk, M., and DeVault, R. C. (1994) Simulation and Performance Analysis of Triple-Effect Absorption Cycles, *Proceedings of the ASHRAE Winter Meeting*, New Orleans, LA, USA, Vol. 100, ASHRAE, Atlanta, GA, USA, pp. 452-462.
8. Ivester, D. N., and Shelton, S. V. (1994) Varying Heat Exchanger Parameters in the Triple-Effect Absorption Cycle, *Proceedings of the International Absorption Heat Pump Conference*, New Orleans, LA, USA, ASME, New York, NY, USA, pp. 243-250.
9. DeVault, R. C., and Marsala, J. (1990) Ammonia-Water Triple-Effect Absorption Cycle, *ASHRAE Transactions*, Vol. 96(1).
10. Grossman, G., Zaltash, A., and DeVault, R. C. (1995) Simulation and Performance Analysis of a Four-Effect Lithium Bromide-Water Absorption Chiller, *Proceedings of the 1995 ASHRAE Annual Meeting*, Chicago, IL, USA, Vol. 101, ASHRAE, Atlanta, GA, USA, pp. 1302-1312.
11. Ziegler, F., and Alefeld, G. (1994) Comparison of Multi-Effect Absorption Cycles, *Proceedings of the International Absorption Heat Pump Conference*, New Orleans, LA, USA, ASME, New York, NY, USA, pp. 257-264.
12. Beutler, A., Hoffmann, L., Ziegler, F., Alefeld, G., Gommed, K., Grossman, G., and Shavit, A. (1996) Experimental Investigation of Heat and Mass Transfer on Horizontal and Vertical Tubes, *Proceedings of the International Ab-Sorption Heat Pump Conference*, Montreal, Canada, pp. 409-419.
13. Beutler, A., Ziegler, F., and Alefeld, G. (1996) Falling Film Absorption with Solutions of a Hydroxide Mixture, *International Ab-Sorption Heat Pump Conference*, Montreal, Canada, pp. 303-309.
14. Miller, W. A., and Perez-Blanco, H. (1994) Vertical-Tube Aqueous LiBr Falling Film Absorption Using Advanced Surfaces, *Proceedings of the International Absorption Heat Pump Conference, Jan 19-21 1994*, New Orleans, LA, USA, ASME, New York, NY, USA, pp. 185-202.
15. Hoffmann, L., and Ziegler, F. (1996) Experimental Investigation of Heat and Mass Transfer with Aqueous Ammonia, *International Ab-sorption Heat Pump Conference*, Montreal, Canada, pp. 383-392.
16. Ryan, W. A. (1994) Water Absorption in an Adiabatic Spray of Aqueous Lithium Bromide Solution, *Proceedings of the International Absorption Heat Pump Conference, Jan 19-21 1994*, New Orleans, LA, USA, ASME, New York, NY, USA, pp. 155-162.
17. Michel, J. W., and Perez-Blanco, H. (1985) Influence of Surfactant Additive on Absorption Heat Pump Performance, *ASHRAE Transactions*, Vol. 91, Part 2B, p. 1847.

18. Kim, K. J., Berman, N. S., and Wood, B. D. (1994) Experimental Investigation of Enhanced Heat and Mass Transfer Mechanisms Using Additives for Vertical Falling Film Absorber, *Proceedings of the International Absorption Heat Pump Conference, Jan 19-21 1994*, New Orleans, LA, USA, ASME, New York, NY, USA, pp. 41-47.
19. Jung, S.-H., Sgamboti, C., and Perez-Blanco, H. (1994) Experimental Study of the Effect of Some Additives on Falling Film Absorption, *Proceedings of the International Absorption Heat Pump Conference, Jan 19-21 1994*, New Orleans, LA, USA, ASME, New York, NY, USA, pp. 49-55.
20. Ziegler, F., and Grossman, G. (1996) Heat-Transfer Enhancement by Additives, *International Journal of Refrigeration*, Vol. 19(5), pp. 301-309.
21. Merrill, T., Setoguchi, T., and Perez-Blanco, H. (1994) Compact Bubble Absorber Design and Analysis, *Proceedings of the International Absorption Heat Pump Conference, Jan 19-21 1994*, New Orleans, LA, USA, ASME, New York, NY, USA, pp. 217-223.
22. Merrill, T. L., Setoguchi, T., and Perez-Blanco, H. (1995) Passive Heat Transfer Enhancement Techniques Applied to Compact Bubble Absorber Design, *Journal of Enhanced Heat Transfer*, Vol. 2(3), pp. 199-208.
23. Merrill, T. L., and Perez-Blanco, H. (1997) Combined Heat and Mass Transfer During Bubble Absorption in Binary Solutions, *International Journal of Heat and Mass Transfer*, Vol. 40(3), pp. 589-603.
24. Herbine, G. S., and Perez-Blanco, H. (1995) Model of an Ammonia-Water Bubble Absorber, *Proceedings of the 1995 ASHRAE Annual Meeting, Jan 29-Feb 1 1995*, Chicago, IL, USA, Vol. 101, ASHRAE, Atlanta, GA, USA, pp. 1324-1332.
25. Kang, Y. T., and Christensen, R. N. (1994) Development of a Counter-Current Model for a Vertical Fluted Tube GAX Absorber, *Proceedings of the International Absorption Heat Pump Conference, Jan 19-21 1994*, New Orleans, LA, USA, ASME, New York, NY, USA, pp. 7-16.
26. Kang, Y. T., Kashiwagi, T., and Christensen, R. N. (1998) Ammonia-Water Bubble Absorber with a Plate Heat Exchanger, *Proceedings of the 1998 ASHRAE Winter Meeting. Part 2 (of 2), Jan 18-21 1998*, San Francisco, CA, USA, Vol. 104, ASHRAE, Atlanta, GA, USA, pp. 1565-1575.
27. Christensen, R. N., Garimella, S., Kang, Y. T., and Garrabrant, M. A. (1998) *Perforated Fin Heat and Mass Transfer Device*, Patent 5,704,417.
28. Garrabrant, M. A., and Christensen, R. N. (1997) Modeling and Experimental Verification of a Perforated Plate-Fin Absorber for Aqua-Ammonia Absorption Systems, *Proceedings of the 1997 ASME International Mechanical Engineering Congress and Exposition, Nov 16-21 1997*, Dallas, TX, USA, Vol. 37, ASME, Fairfield, NJ, USA, pp. 337-347.
29. Perez-Blanco, H. (1988) A Model of an Ammonia-Water Falling Film Absorber, *ASHRAE Transactions*, Vol. 94(1), pp. 467-483.
30. Jeong, S., Koo, K.-K., and Lee, S. K. (1998) Heat Transfer Performance of a Coiled Tube Absorber with Working Fluid of Ammonia/Water, *Proceedings of the 1998 ASHRAE Winter Meeting. Part 2 (of 2), Jan 18-21 1998*, San Francisco, CA, USA, Vol. 104, ASHRAE, Atlanta, GA, USA, pp. 1577-1583.
31. Potnis, S. V., Anand, G., Gomezplata, A., Erickson, D. C., and Papar, R. A. (1997) GAX Component Simulation and Validation, *Proceedings of the 1997 ASHRAE Winter Meeting, Jan 26-29 1997*, Philadelphia, PA, USA, Vol. 103, ASHRAE, Atlanta, GA, USA, pp. 454-459.
32. Garimella, S. (2004) *Method and Means for Miniaturization of Binary-Fluid Heat and Mass Exchangers*, Patent 6,802,364.
33. Garimella, S. (1999) Miniaturized Heat and Mass Transfer Technology for Absorption Heat Pumps, *Proceedings of the International Sorption Heat Pump Conference*, Munich, Germany, pp. 661-670.
34. Garimella, S. (2000) Microchannel Components for Absorption Space-Conditioning Systems, *2000 ASHRAE Winter Meeting, Feb 5-Feb 9 2000*, Dallas, TX, USA, Vol. 106 (PA, Amer. Soc. Heating, Ref. Air-Conditoning Eng. Inc., Atlanta, GA, USA, pp. 453-462.
35. Richter, H. J. (1981) Flooding in Tubes and Annuli, *International Journal of Multiphase Flow*, Vol. 7(6), pp. 647-658.

36. Meacham, J. M., and Garimella, S. (2002) Experimental Demonstration of a Prototype Microchannel Absorber for Space-Conditioning Systems, *International Sorption Heat Pump Conference*, Shanghai, China, pp. 270-276.
37. Meacham, J. M., and Garimella, S. (2003) Modeling of Local Measured Heat and Mass Transfer Variations in a Microchannel Ammonia-Water Absorber, *Technical and Symposium Papers Presented At the 2003 Winter Meeting of The ASHRAE, Jan 26-29 2003*, Chicago, IL, United States, Vol. 109 PART 1, Amer. Soc. Heating, Ref. Air-Conditoning Eng. Inc., pp. 412-422.
38. Price, B. C., and Bell, K. J. (1974) Design of Binary Vapor Condensers Using the Colburn-Drew Equations, *AIChE Symposium Series - Heat Transfer - Research and Design*, Vol. 70(138), pp. 163-171.
39. Colburn, A. P., and Drew, T. B. (1937) The Condensation of Mixed Vapours, *AIChE Transactions*, Vol. 33, pp. 197-212.
40. Kang, Y. T., and Christensen, R. N. (1995) Combined Heat and Mass Transfer Analysis for Absorption in a Fluted Tube with a Porous Medium in Confined Cross Flow, *Proceedings of the 1995 ASME/JSME Thermal Engineering Joint Conference. Part 1 (of 4), Mar 19-24 1995*, Maui, HI, USA, Vol. 1, ASME, New York, NY, USA, pp. 251-260.
41. Kang, Y. T., Chen, W., and Christensen, R. N. (1997) Generalized Component Design Model by Combined Heat and Mass Transfer Analysis in NH_3/H_2O Absorption Heat Pump Systems, *Proceedings of the 1997 ASHRAE Winter Meeting, Jan 26-29 1997*, Philadelphia, PA, USA, Vol. 103, ASHRAE, Atlanta, GA, USA, pp. 444-453.
42. Meacham, J. M., and Garimella, S. (2004) Ammonia-Water Absorption Heat and Mass Transfer in Microchannel Absorbers with Visual Confirmation, *ASHRAE Transactions*, Vol. 110(1), pp. 525-532.
43. Determan, M. D., Garimella, S., and Lee, S. (2004) Experimental Demonstration of a Microchannel Desorber for Ammonia-Water Heat Pumps, *Seventeenth National Heat and Mass Transfer Conference and Sixth ISHMT/ASME Heat and Mass Transfer Conference*, Kalpakkam, India, pp. 453-458.
44. Killion, J. D., and Garimella, S. (2001) A Critical Review of Models of Coupled Heat and Mass Transfer in Falling-Film Absorption, *International Journal of Refrigeration*, Vol. 24(8), pp. 755-797.
45. Killion, J. D., and Garimella, S. (2003) A Review of Experimental Investigations of Absorption of Water Vapor in Liquid Films Falling over Horizontal Tubes, *HVAC and R Research*, Vol. 9(2), pp. 111-136.
46. Jeong, S., and Garimella, S. (2002) Falling-Film and Droplet Mode Heat and Mass Transfer in a Horizontal Tube LiBr/Water Absorber, *International Journal of Heat and Mass Transfer*, Vol. 45(7), pp. 1445-1458.
47. Painter, J. F., and Hindmarsh, A. C. (1987) *Livermore Solver for Ordinary Differential Equations (LSODI)*, Lawrence Livermore National Laboratory.
48. Nomura, T., Nishimura, N., Wei, S., Yamaguchi, S., and Kawakami, R. (1994) Heat and Mass Transfer Mechanism in the Absorber of Water/LiBr Convectional Absorption Refrigerator: Experimental Examination by Visualized Model, *Proceedings of the International Absorption Heat Pump Conference, Jan 19-21 1994*, New Orleans, LA, USA, ASME, New York, NY, USA, pp. 203-208.
49. Jeong, S., and Garimella, S. (2005) Optimal Design of Compact Horizontal Tube LiBr/Water Absorbers, *HVAC and R Research*, Vol. 11(in press).
50. Killion, J. D., and Garimella, S. (2003) Gravity-Driven Flow of Liquid Films and Droplets in Horizontal Tube Banks, *International Journal of Refrigeration*, Vol. 26(5), pp. 516-526.
51. Hu, X., and Jacobi, A. M. (1996) The Inter Tube Falling Film: Part 1- Flow Characteristics, Mode Transition, and Hysteresis, *Journal of Heat Transfer*, Vol. 118, pp. 616-625.
52. Kirby, M. J., and Perez-Blanco, H. (1994) Design Model for Horizontal Tube Water/Lithium Bromide Absorbers, *Proceedings of the 1994 International Mechanical Engineering Congress and Exposition, Nov 6-11 1994*, Chicago, IL, USA, Vol. 32, ASME, New York, NY, USA, pp. 1-10.
53. Zhang, X. (1999) Dynamics of Growth and Breakup of Viscous Pendant Drops into Air, *Journal of Colloid and Interface Science*, Vol. 212(1), pp. 107-122.
54. Peregrine, D. H., Shoker, G., and Symon, A. (1990) Bifurcation of Liquid Bridges, *Journal of Fluid Mechanics*, Vol. 212, pp. 25-39.

55. Zhang, X., and Basaran, O. A. (1995) An Experimental Study of Dynamics of Drop Formation, *Physics of Fluids*, Vol. 7(6), pp. 1184-1203.
56. Richards, J. R., Beris, A. N., and Lenhoff, A. M. (1995) Drop Formation in Liquid--Liquid Systems before and after Jetting, *Physics of Fluids*, Vol. 7(11), pp. 2617-2630.
57. Zhang, D. F., and Stone, H. A. (1997) Drop Formation in Viscous Flows at a Vertical Capillary Tube, *Physics of Fluids*, Vol. 9(8), pp. 2234-2242.
58. Brauner, N., and Maron, D. M. (1982) Characteristics of Inclined Thin Films, Waviness and the Associated Mass Transfer, *International Journal of Heat and Mass Transfer*, Vol. 25(1), pp. 99-110.
59. Burdukov, A. P., Bufetov, N. S., Deriy, N. P., Dorokhov, A. R., and Kazakov, V. I. (1980) Experimental Study of the Absorption of Water Vapor by Thin Films of Aqueous Lithium Bromide., *Heat Transfer - Soviet Research*, Vol. 12(3), pp. 118-123.
60. Nakoryakov, V. Y., Burdukov, A. P., Bufetov, N. S., Grigor'eva, N. I., and Dorokhov, A. R. (1982) Coefficient of Heat and Mass Transfer in Falling Wavy Liquid Films, *Heat Transfer - Soviet Research*, Vol. 14(3), pp. 6-11.
61. Uddholm, H., and Setterwall, F. (1988) Model for Dimensioning a Falling Film Absorber in an Absorption Heat Pump, *International Journal of Refrigeration*, Vol. 11(1), pp. 41-45.
62. Mudawar, I., and Houpt, R. A. (1993) Measurement of Mass and Momentum Transport in Wavy-Laminar Falling Liquid Films, *International Journal of Heat and Mass Transfer*, Vol. 36(17), pp. 4151-4162.
63. Killion, J. D., and Garimella, S. (2004) Pendant Droplet Motion for Absorption on Horizontal Tube Banks, *International Journal of Heat and Mass Transfer*, Vol. 47(19-20), pp. 4403-4414.
64. Kumar, R., and Kuloor, N. R. (1970) The Formation of Bubbles and Drops, *Advances in Chemical Engineering*, Vol. 8, pp. 255-368.
65. Clift, R., Grace, J. R., and Weber, M. E. (1978) *Bubbles, Drops, and Particles*, Academic Press, New York, NY, USA.
66. Frohn, A., and Roth, N. (2000) *Dynamics of Droplets*, in *Experimental Fluid Mechanics*, Springer, Berlin, New York, p. 292.
67. Eggers, J. (1997) Nonlinear Dynamics and Breakup of Free-Surface Flows, *Reviews of Modern Physics*, Vol. 69(3), p. 865.
68. Middleman, S. (1995) *Modeling Axisymmetric Flows: Dynamics of Films, Jets, and Drops*, Academic Press, San Diego, p. 299.
69. Yarin, A. L. (1993) *Free Liquid Jets and Films: Hydrodynamics and Rheology*, Longman Publishing Group.
70. Bogy, D. B. (1979) Drop Formation in a Circular Liquid Jet, Vol. 11, pp. 207-228.
71. Rein, M. (1993) Phenomena of Liquid Drop Impact on Solid and Liquid Surfaces, *Fluid Dynamics Research*, Vol. 12(2), pp. 61-93.
72. Tropea, C., and Marengo, M. (1999) Impact of Drops on Walls and Films, *Multiphase Science and Technology*, Vol. 11(1), pp. 19-36.
73. The Mathworks Inc. (2002) *Matlab Version 6.5.0.180913a Release 13*, http://www.mathworks.com, Natick, MA.
74. Canny, J. (1986) A Computational Approach to Edge Detection, *IEEE Transactions on Pattern Analysis and Machine Intelligence*, Vol. PAMI-8(6), pp. 679-698.
75. Sobel, I. E. (1970) *Doctoral Thesis: Camera Models and Machine Perception*, Doctoral Thesis, *Dept. of Electrical Engineering*, Stanford University, Palo Alto, p. 89 (also in Stanford Computer Science Dept. Artificial Intelligence Laboratory. AIM-121 technical report).
76. de Boor, C. (1978) *A Practical Guide to Splines*, in *Applied Mathematical Sciences*, F. John, et al., Editors, Springer-Verlag, New York, p. xxiv, 392.
77. Cohen, E., Riesenfeld, R. F., and Elber, G. (2001) *Geometric Modeling with Splines: An Introduction*, AK Peters, Natick, Mass., p. xxii, 616.
78. The Mathworks Inc. (2002) *Matlab Spline Toolbox, Ver. 3.1.1*, Matlab Version 6.5.0.180913a Release 13, Natick, MA.

79. Carey, V. P. (1992) *Liquid-Vapor Phase-Change Phenomena : An Introduction to the Thermophysics of Vaporization and Condensation Processes in Heat Transfer Equipment*, in *Series in Chemical and Mechanical Engineering*, G.F. Hewitt and C.L. Tien, Editors, Taylor & Francis Series, Hemisphere Pub. Corp., Washington, D.C., p. xvii, 645.
80. Killion, J. D., and Garimella, S. (2005) Simulation of Pendant Droplets and Falling Films in Horizontal Tube Absorbers, *Journal of Heat Transfer*, Vol. 127 (in press).
81. Rider, W. J., and Kothe, D. B. (1998) Reconstructing Volume Tracking, *Journal of Computational Physics*, Vol. 141(2), pp. 112-152.
82. Kothe, D. B., and Mjolsness, R. C. (1992) Ripple: A New Model for Incompressible Flows with Free Surfaces, *AIAA Journal*, Vol. 30(11), pp. 2694-2700.
83. Brackbill, J. U., and Kothe, D. B. (1996) Dynamical Modeling of Surface Tension, *Proceedings of the 1996 3rd Microgravity Fluid Physics Conference, Jul 13-15 1996*, Cleveland, OH, USA, NASA, Cleveland, OH, USA, pp. 693-698.
84. Brackbill, J. U., Kothe, D. B., and Zemach, C. (1992) A Continuum Method for Modeling Surface Tension, *Journal of Computational Physics*, Vol. 100(2), pp. 335-354.
85. Kelkar, K. M., and Patankar, S. V. (1994) Numerical Method for the Prediction of Two-Fluid Flows in Domains with Moving Boundaries, *Proceedings of the 1994 ASME Fluids Engineering Division Summer Meeting. Part 7 (of 18), Jun 19-23 1994*, Lake Tahoe, NV, USA, Vol. 185, ASME, New York, NY, USA, pp. 169-176.
86. Gueyffier, D., Li, J., Nadim, A., Scardovelli, R., and Zaleski, S. (1999) Volume-of-Fluid Interface Tracking with Smoothed Surface Stress Methods for Three-Dimensional Flows, *Journal of Computational Physics*, Vol. 152(2), pp. 423-456.
87. Harvie, D. J. E., and Fletcher, D. F. (2001) A New Volume of Fluid Advection Algorithm: The Defined Donating Region Scheme, *International Journal for Numerical Methods in Fluids*, Vol. 35(2), pp. 151-172.
88. Rudman, M. (1997) Volume-Tracking Methods for Interfacial Flow Calculations, *International Journal for Numerical Methods in Fluids*, Vol. 24(7), pp. 671-691.
89. Rudman, M. (1998) Volume-Tracking Method for Incompressible Multifluid Flows with Large Density Variations, *International Journal for Numerical Methods in Fluids*, Vol. 28(2), pp. 357-378.
90. Prosperetti, A., and Oguz, H. N. (1993) Impact of Drops on Liquid Surfaces and the Underwater Noise of Rain, *Annual Review of Fluid Mechanics*, Vol. 25, p. 577.
91. Rieber, M., and Frohn, A. (1999) A Numerical Study on the Mechanism of Splashing, *International Journal of Heat and Fluid Flow*, Vol. 20(5), pp. 455-461.
92. Zhang, X. (1999) Dynamics of Drop Formation in Viscous Flows, *Chemical Engineering Science*, Vol. 54(12), pp. 1759-1774.
93. Oguz, H. N., and Prosperetti, A. (1993) Dynamics of Bubble Growth and Detachment from a Needle, *Journal of Fluid Mechanics*, Vol. 257, pp. 111-145.
94. Chen, L., Garimella, S. V., Reizes, J. A., and Leonardi, E. (1997) Motion of Interacting Gas Bubbles in a Viscous Liquid Including Wall Effects and Evaporation, *Numerical Heat Transfer; Part A: Applications*, Vol. 31(6), pp. 629-654.
95. Fluent Inc. (2003) *Fluent*, Lebanon, NH.
96. Issa, R. I. (1986) Solution of the Implicitly Discretised Fluid Flow Equations by Operator-Splitting, *Journal of Computational Physics*, Vol. 62(1), pp. 40-65.
97. Youngs, D. L. (1982) *Time-Dependent Multi-Material Flow with Large Fluid Distortion*, in *Numerical Methods for Fluid Dynamics*, K.W. Morton and M.J. Baines, Editors, Academic Press.

HETEROGENEOUS CRYSTALLIZATION OF AMORPHOUS SILICON ACCELERATED BY EXTERNAL FORCE FIELD: MOLECULAR DYNAMICS STUDY

J. S. LEE* and S. PARK**
*School of Mechanical & Aerospace Engineering-Seoul National University
**Department of Mechanical & System Design Engineering-Hongik University
Seoul, Korea

1. Introduction

Crystallization of amorphous silicon is a crucial technology in semiconductor industry. The key objectives in the development of crystallization process are to obtain high quality uniform crystalline structure, and to reduce processing time and temperature because the defect formation in the crystallization process is very sensitive to the growth conditions such as crystal growth rates and temperature gradient at the crystalline-silicon (*c*-Si) and liquid-silicon (*l*-Si) interface [1]. Over the last decade, quite a few techniques have been developed including solid phase crystallization, excimer laser crystallization, and metal induced crystallization. However, these techniques have shortcomings such as poor process uniformity, long process time, high temperature operation, narrow process window, metal contamination and so forth.

Recently low-temperature crystallization processes based on an alternating magnetic induction or microwave heating have drawn much attention because the problems normally associated with conventional fabrication methods can be avoided. This process could bring about a more effective crystallization of the amorphous-silicon (*a*-Si) on a glass substrate at temperatures below 600°C. Figure 1 compares the process time and temperature between the conventional solid phase crystallization (SPC) and the alternating magnetic field crystallization (AMFC) [2]. By the AMFC the process time is remarkably reduced to only one hour at a low process temperature of 430°C, whereas it takes more than ten hours at a much higher temperature of 600°C in the SPC. However, mechanisms for the AMFC or the microwave assisted crystallization (MAC) as well as those for metal induced crystallization (MIC) processes are not yet clearly understood [2, 3].

It can be conjectured that the AMFC or MAC is either due to heating of *a*-Si by eddy current losses in the alternating magnetic field or microwave heating, or due to field-enhanced oscillation or movement of defects or ions. In the SPC, the process temperature should be above 600°C to raise the temperature of *a*-Si film of about 50 nm thick up to an effective annealing temperature. Thus, heating due to eddy current loss in the AMFC is considered to be irrelevant. The heating contribution by microwave in the MAC is also negligible, since the specimen temperatures have been carefully controlled by adjusting the microwave power [3].

From the report by Yoon et al. [2] that the metal concentration in the material and external electric field strength are two most important factors in the MIC, it can be presumed that the field-enhanced molecular movements can be utilized in global crystallization below 600°C. Thus, the molecular dynamics (MD) simulation technique can be a most appropriate approach to determine whether the externally excited molecules will enhance crystallization or not, because there are not any other theoretical and microscopic tools better than the MD simulation for investigating the effect of the field-enhanced molecular movements during crystallization. This technique is known to be capable of providing a detailed information on the crystallization and nucleation processes and of providing critical tests of analytical theories or unexplained phenomena [4~10]. Recently, Caturla et al. [11], Weber et al. [12], and Motooka [13] showed that the MD simulation was a useful tool for investigating the microscopic process of the Si crystallization and amorphization induced by ions and defects. However, few reports are available to date that investigate the issue of the crystallization enhanced by externally imposed force fields using the MD simulation techniques.

In the investigation of feasibility associated with the constant-temperature external field-induced crystallization of

S. Kakaç et al. (eds.), Microscale Heat Transfer, 369–378.

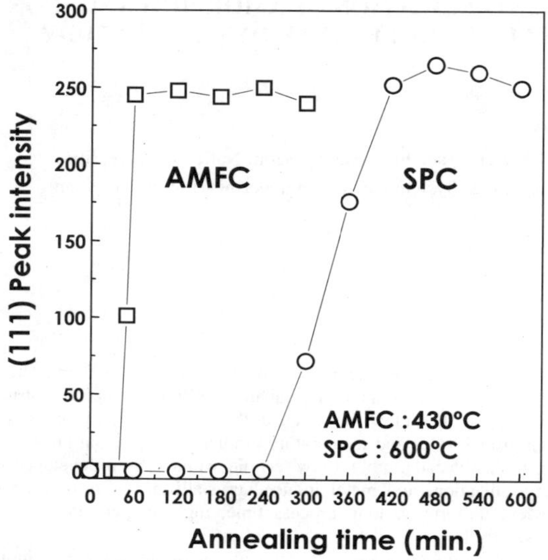

Figure 1 Comparison of process time and temperature between the conventional solid phase crystallization and the alternating magnetic field crystallization

a-Si at low temperatures, the MD simulation has recently been conducted [16] because the isothermal crystallization process will make significant contribution to advances in fabrication of the polycrystalline silicon (poly-Si) from *a*-Si at low temperatures by avoiding the problems present in high-temperature processes [2, 3, 14, 15]. In this study, the external field-expedited crystallization of the amorphous silicon is investigated using the Tersoff potential model. To obtain *a*-Si, liquid-phase silicon is quenched at a cooling rate of 10^{12} K/s, which results in physically and dynamically realistic structure [17]. In addition, the structural variations during the cooling process are observed by the Voronoi analysis.

2. Molecular Dynamics Method

The MD method is a computational simulation technique where the time evolution of interacting atoms is followed by integrating their equations of motion. The dynamic modeling problem is divided into two great tasks: developing a suitable model for the problem at hand and applying molecular dynamics to that model [18]. The task of model development is concerned with modeling of molecular interactions, system/environment interactions, and equation of motion. Having developed a model, the simulation itself divides into two jobs: solving the equations of motion to generate the molecular trajectories, and analyzing those trajectories for the properties of interest such as thermophysical and dynamic properties. To do this, the simulation procedure usually follows three steps: initialization, equilibration and production. In the initialization, the initial structure of the system and velocity should be defined, which is followed by the equilibration based on thermodynamic relations. From the obtained statistical data, temperature, internal energy, pressure, and transport properties can be evaluated based on the statistical mechanics. Once the potential of a system is given, it is straight forward to numerically integrate the equation of motion, however, simulations are usually confined to studies of relatively short-lived phenomena on the order of nanoseconds in timescale and hundred nanometers in length scale since MD simulations are limited largely by the speed and storage constraints of the computer. The limitations and current simulation records in time and length are shown in Fig. 2.

In the classical MD method, Newton's equations of motion are solved for atoms and molecules, which is given by

$$\vec{F}_i(t) = m_i \frac{d^2\vec{r}_i}{dt^2} = -\frac{\partial \Phi(\vec{r}^N)}{\partial \vec{r}_i} \tag{1}$$

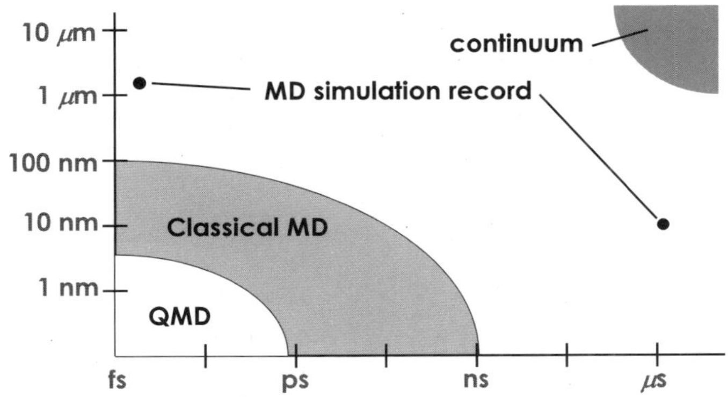

Figure 2 Limitations of time- and length-scales in MD simulations

where t denotes time, $\vec{F}_i$, m_i, $\vec{r}_i$ are force vector, mass, position vector of molecule i, respectively, Φ is the potential of the system, and N is the number of molecules. The above equation of motion is known to be a good approximation when the de Broglie thermal wavelength is much shorter than the mean nearest neighbor separation between molecules. Thus, it poorly predicts for very light systems, for example, H_2, He and Ne, and for very low temperature systems. For the simulation of these systems, quantum effects must be included. The potential of a system Φ can often be reasonably assumed as the sum of the effective pair potential $\phi(r_{ij})$ which can be expressed as

$$\Phi(\vec{r}^N) = \sum_i \sum_{j>1} \phi(r_{ij}) \tag{2}$$

where r_{ij} is the distance between molecules i and j. The pair-potential assumption is often used for simplicity even in the cases that the validity is questionable, however, this should not be accepted for the covalent system such as carbon and silicon. The detail descriptions about the available potential forms can be found in the literature [19].

3. Simulation of Crystallization of Heterogeneous Amorphous Silicon

An isothermal crystallization process of the amorphous silicon is investigated by applying a molecular dynamics simulation technique with the Tersoff potential [16]. In this study, the liquid-phase silicon is quenched at a cooling rate of 10^{12} K/s to obtain a realistic amorphous structure. During the cooling process the Voronoi analysis is conducted to visualize atomic structural changes. While the homogeneous crystallization of the amorphous silicon could not readily be achieved, the heterogeneous crystallization can significantly be accelerated by external force fields. This enhancement is owing to the increase in molecular jumping frequencies associated with the molecular potential energy. The potential energy is increased by external excitations, rather than due to thermal mechanisms normally found in conventional SPC processes.

The MD calculations using the Tersoff model are carried out under constant-volume and -temperature conditions. The melting temperatures predicted by the Tersoff model are much higher than the real phase-change temperatures. The model is, however, widely applied in the recent simulations, because it can reproduce well the structural and dynamical properties of elemental semiconductors such as silicon and carbon ranging from crystalline to amorphous structures

In the Tersoff model, the potential energy between two neighboring atoms i and j can be expressed as

$$\Phi_{ij} = f_c(r_{ij})[f_R(r_{ij}) + b_{ij} f_A(r_{ij})] \tag{3}$$

where

$$f_R(r_{ij}) = A_{ij} \exp(-\lambda_{ij} r_{ij}), \quad f_A(r_{ij}) = -B_{ij} \exp(-\mu_{ij} r_{ij})$$

TABLE 1. The Tersoff parameters for silicon [16]

Parameters	Tersoff (Si)
A_{ij} [eV]	1830.8
B_{ij} [eV]	471.18
λ_{ij} [1/ Å]	2.4799
μ_{ij} [1/ Å]	1.7322
χ_{ij}	1.0
β_i	1.0999×10^{-6}
n	0.78734
c_i	100390
d_i	16.218
h_i	-0.59826
R_{ij} [Å]	2.7
S_{ij} [Å]	3.0

$$f_c(r_{ij}) = \begin{cases} 1 & r_{ij} \le R_{ij} \\ \dfrac{1}{2} + \dfrac{1}{2}\cos\left[\dfrac{\pi(r_{ij} - R_{ij})}{(S_{ij} - R_{ij})}\right] & R_{ij} \le r_{ij} \le S_{ij} \\ 0 & r_{ij} \ge S_{ij} \end{cases}$$

$$b_{ij} = \chi_{ij}(1 + \beta_i^n \zeta_{ij}^n)^{-1/2n}, \quad \zeta_{ij} = \sum_{k \ne i,j} f_c(r_{ik})\ g(\theta_{ijk})$$

$$g(\theta_{ijk}) = 1 + \frac{c_i^2}{d_i^2} + \frac{c_i^2}{d_i^2 + (h_i - \cos\theta_{ijk})^2}$$

Here, b_{ij} is the multi-body parameter for bond-formation energy affected by local atomic arrangement, especially, by the presence of other neighboring atoms (atom k), ζ is the effective coordination number which is a function of the angle between r_{ij} and r_{ik}, and $g(\theta_{ijk})$ is fitted to stabilize the tetrahedral structure [20, 21]. The values of parameters are listed in Table 1.

In the simulation, the density of the silicon structure is chosen as that of *c*-Si (2.33 g/cm^3), since the density of *a*-Si without voids is close to that of *c*-Si [17]. The atoms are initially arranged as the diamond structure with periodic boundary conditions. They move according to the intermolecular forces based on the potential function, Eq. (3), and these movements can be described by the classical momentum equations. The momentum equations are integrated by the Gear algorithm with a time step of 0.002 ps and the average temperature of the structure is kept constant by the momentum scaling method.

In the case of the recrystallization progress of the amorphous structure, the determination of the *a*-Si microstructure is of great importance as an initial phase in the simulations. Many researchers have proposed the preparation methods of *a*-Si based on MD simulations and their computer-generated properties were in good agreement with those obtained by experiments. However, details of the microstructures have not yet fully examined [22, 23] because the calculation domain sizes in the *ab initio* MD simulations were quite limited [24, 25], and the preparation methods of *a*-Si using various simple empirical inter-atomic potentials involved unphysical treatments.

Recently, Ishimaru et al. [1, 17, 26] showed that the *l*-Si and *a*-Si networks generated using the Tersoff potential could reproduce the features of the structural properties reported by *ab initio* calculations. The procedure applied by Ishimaru et al. [17], therefore, is adopted to prepare a realistic amorphous structure of silicon. The main features are: 1) arrangement of atoms in the diamond structure; 2) equilibration of *l*-Si at 3500 K; and 3) quenching and solidification of the *l*-Si to 500 K at a cooling rate of 10^{12} K/s. They showed that the structural and dynamical properties of *a*-Si generated by a cooling rate of 10^{12} K/s were in excellent agreement with those of *a*-Si obtained experimentally, compared with those at other cooling rates. This cooling rate is estimated very close to that achieved in laboratories to prepare *a*-Si by the laser annealing.

The potential energy decreases during the cooling process as shown in Fig. 3 [16], and the kink around 1700 ~ 2000 K presumably corresponds to the enthalpy change upon melting of *a*-Si [17]. To understand the kink phenomenon more clearly, however, variation of structural characteristics such as the number of faces and asphericities are investigated by the Voronoi analysis. Here, the asphericity η is defined as

$$\eta = \frac{S^3}{36\pi V^2} \qquad (4)$$

where S denotes the surface area of the polyhedron and V, its volume. The topological properties obtained from the Voronoi analysis [27] are the number of faces, edges, and vertices of the polyhedron and the metric properties are its volume and area. These properties are related closely to the atomic structure of the material under investigation.

During the cooling process sudden increases in the face number and asphericity around 1700 ~ 2000 K coincide with a sudden decrease in the potential energy, as shown in Fig. 3. In addition, the comparison of the potential energy variations during the cooling and heating processes shows that the hysteresis of the potential variations during the cooling and heating cycle is very weak. It implies that the kink is due to the structural changes in solid phases, not due to the abrupt phase-change of the metastable liquid-phase. Here, the number of face and asphericity are normalized by the values of the diamond lattice of 16 and 2.16, respectively.

Figure 4 compares the unit cells of *l*-Si at 3500 K, and *a*-Si at 2000 and 500 K to *c*-Si (diamond structure) [16]. With decrease of temperature its amorphous structure becomes similar to that of the crystalline structure of large hexagonal and small triangular faces. This implies that the average coordination number becomes 4 and tetrahedral bonds predominated in *a*-Si, although its phase is still amorphous. Simulations for the homogeneous crystallization of the *a*-Si structure obtained by the above procedure are tried at various temperatures below the melting temperature. With and without the external field excitations, the crystallization of *a*-Si has not been observed within the simulation period (10 ns), while the crystallization is readily achieved for rapidly quenched amorphous argon [28]. Therefore, the next step considered is on the heterogeneous crystallization of *a*-Si.

For simulating the heterogeneous crystallization process, a test specimen with a dimension of $3a \times 3a \times 12a$ including two amorphous-crystalline interfaces is prepared by attaching 12 layers of *c*-Si (216 atoms), as a crystalline seed, to a bulk *a*-Si (648 atoms). The bulk *a*-Si is obtained by the above cooling process and pre-annealed at 500 K for 200 ps. The constant NVT MD simulations are carried out using the Newton equation with a time step for the integration set at 2 fs.

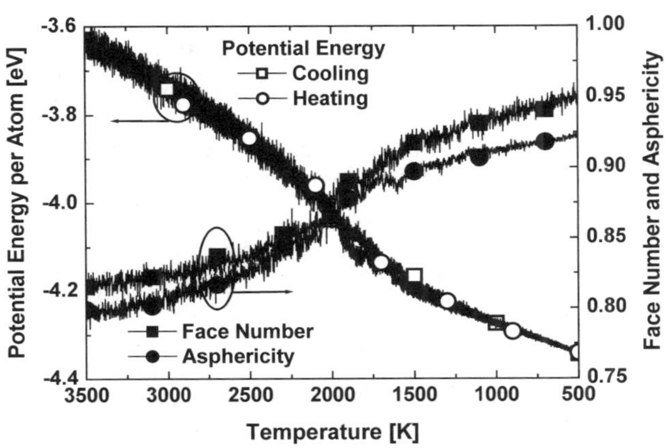

Figure 3 Potential energy during cooling/heating processes of *a*-Si, and face number and asphericity during cooling process at a rate of 10^{12} K/s [16]

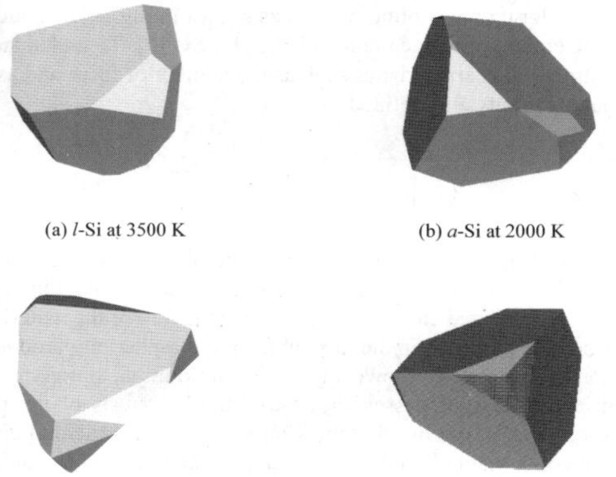

(a) l-Si at 3500 K (b) a-Si at 2000 K

(c) a-Si at 500 K (d) c-Si (diamond structure)

Figure 4 Voronoi polyhedra of silicons at various phases [16]

To investigate the crystallization phenomena induced by the external field, a cyclic force is imposed on a small portion of molecules in the amorphous structure, in addition to the inherent intermolecular forces. As a practical matter, this force may be applied to susceptor molecules (molecules with dangling or floating bonds, or other defects) through direct electric fields, electric fields induced by the external magnetic field, or other excitation sources. In this study, it is assumed that some molecules can act as susceptors of the external field, for the sole purpose of inducing artificial molecular movements selectively and of observing the subsequent crystallization behavior. As discussed previously, the results of this study could provide a clue to mechanisms involved in the athermal crystallization of a-Si to poly-Si, which is induced by the field-enhanced molecular movements.

The artificial force is estimated from the intermolecular force field experienced by each molecule. Since the molecules are not in a perfect crystalline position, there always exists a force imbalance, which may be induced by thermal excitation, defects, non-crystalline structures, or other disturbing sources. At each time step the forces acting on each molecule by neighboring molecules are averaged, and the averaged force is used as the amplitude of the external force. Another parameter to be determined is the period of the external cyclic field, which is estimated by the oscillating frequency of the lattice constant divided by the atomic velocity. Moreover, to speed up numerical calculations the external cyclic field frequency of 500 GHz are imposed on some molecules selected from the amorphous-phase, although 2.45 GHz is applied for the experimental microwave assisted crystallization [3], and much lower frequencies are currently tried for the AMFC.

Since the melting temperature of Si estimated in the simulation using the Tersoff potential is known to be around 3000 K [1], the crystallization temperatures in the current simulation range between 1300 and 2200 K. Each temperature the amorphous/crystalline sample has been attained by applying the heating rate of 10^{13} K/s from 500 K. Figure 5 shows the snapshots of the molecular distribution during the external field-induced crystallization at 1500 K where five molecules are assumed to be susceptors [16]. The initial distribution (time 0 ns) is very similar to that at 500 K. As time goes by, it can be seen that the a/c interfaces move towards the bulk amorphous region and full crystallization is achieved at around 1.2 ns. Here, open circles denote the c-Si seed atoms, while gray spheres do the atoms originally in the amorphous state.

During the annealing time from 0 to 1.6 ns, the RDF shown in Fig. 6 describes the crystallization progress quantitatively and especially the abrupt increases of the third peaks [16]. Figure 6(a) shows that the initial RDF is essentially similar to that of the Tersoff liquid with the first peak at 2.35 Å and the second one at 3.9 Å. As the crystallization proceeds, the third-nearest-neighbor peak of c-Si that is not observed in the amorphous phase appears clearly at 4.5 Å. In Fig. 6(b), the magnitudes of the third peaks during the crystallization process accelerated by the external field at annealing temperatures of 1400, 1500, and 1600 K are compared with those at 1500 K without imposing external field. The crystallization progress is very slow at that temperature in the case without external fields, while the external excitation can be an effective expediting mechanism for crystallization.

Based on the molecular distributions and radial distribution functions, crystallinities for the current heterogeneous condition are estimated in Fig. 7 [16]. Comparisons are also made for cases with and without the external fields at

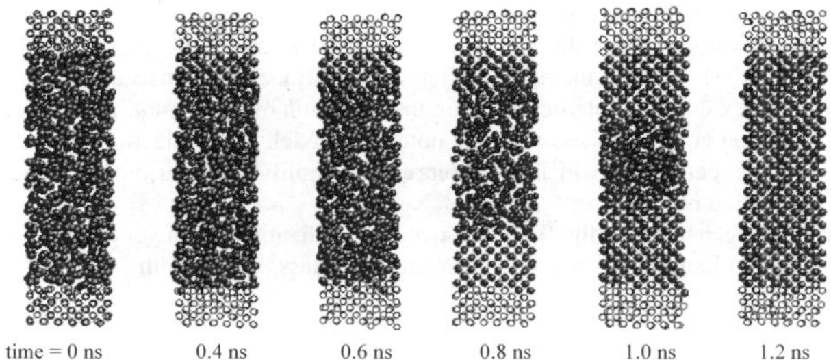

Figure 5 In the [100] direction, snapshots of molecular motion during heterogeneous crystallization process at 1500 K under external force fields [16]

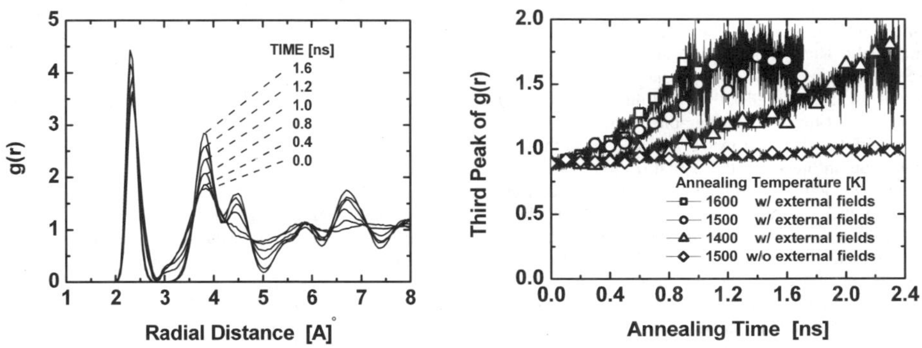

(a) (b)

Figure 6 (a) Radial distribution functions for various time steps at 1500 K, (b) Development of the third peaks of radial distribution functions during crystallization process under various conditions [16]

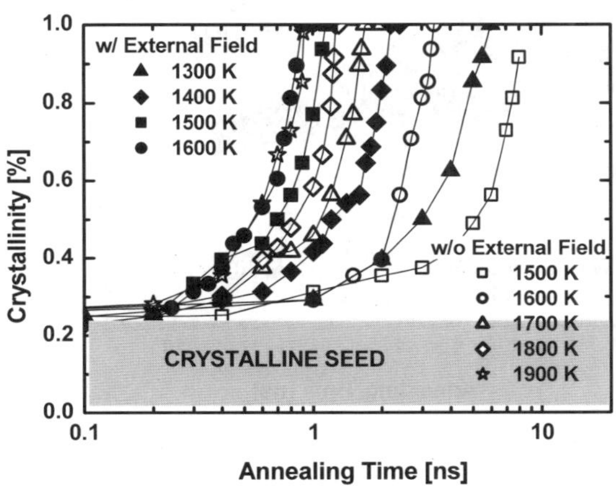

Figure 7 Crystallinities during heterogeneous crystallization process with and without external fields [16]

various process temperatures. It can clearly be seen that the crystallization growth rate increases with process temperature as expected. The development of the crystallinity at 1600 K under the external field (marked by solid circles) is very close to that at 1900 K without the external field (conventional SPC, marked by star symbols). It can be noted that the external fields exert a considerable influence on the reduction of the crystallization temperature by almost 300 K under the current simulation conditions and with the potential model. This reduction can be translated into as a reduction by 100 ~ 150 K in real experiments, which could represent a significant contribution to the development of an efficient low-temperature fabrication methodology.

As discussed previously, the mechanism of the field enhanced crystallization is not yet clearly understood. A simple thermal analysis on the silicon film in the form of a long strip with thickness d_f and width L_f in AMFC process provides the film temperature given by [16]

$$T_f = \left[T_\infty^4 + \frac{\left(\mu_0 H_0 \omega L_f\right)^2 d_f}{2\rho_f \varepsilon_f \sigma} \right]^{1/4} \tag{5}$$

where it is assumed that one side of the silicon film is exposed to the surroundings at T_∞ in vacuum, and the other side of the film is insulated. The imposed magnetic field and alternating angular frequency are denoted by H_0 and ω, respectively. The symbols μ_0, ρ_f, and ε_f denote the magnetic permeability, resistivity, and hemispherical emissivity of the Si film, respectively, and σ is the Stefan-Boltzmann constant. With properties of intrinsic *c*-Si at 500-600°C and for films of thickness smaller than 100 nm, it can easily be deduced from Eq. (4) that $T_f \approx T_\infty$. This implies that Joule heating cannot make a considerable contribution to the AMFC process at this low temperature. A similar analysis can be applied for the microwave assisted crystallization process. Thus, it is found that thermal contribution exerts insignificant influence on the crystallization process at low temperatures.

Figure 8 compares the potential energy distributions during the crystallization process [16]. Owing to the external excitation, the potential energy of the amorphous phase at 1500 K is considerably higher than that without the external field at the same temperature during the incubation period, which is close to that at 1600 K without external fields. After the incubation period the crystallization phase develops quickly according to the rapid drop in the potential energy. It can be concluded that the enhanced crystallization speed is due to the increased jump frequency from *a*-Si to *c*-Si, since the jump frequency has a functional dependency of $f_{a\rightarrow c} \propto \exp(-\Delta G / RT)$ and the increased potential energy reduces the potential barrier between the amorphous and crystalline phases.

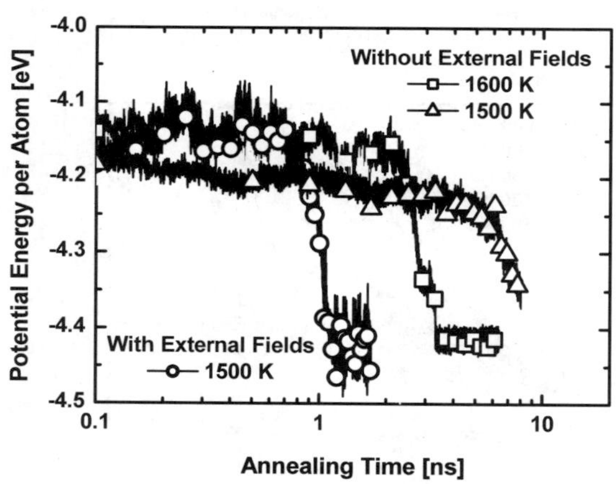

Figure 8 Potential energy variations during heterogeneous crystallization process with and without external force fields [16]

4. Summary

The MD simulations on isothermal crystallization processes of *a*-Si are introduced. To obtain a realistic *a*-Si structure, the *l*-Si prepared at 3500 K is rapidly quenched to 500 K at a cooling rate of 10^{12} K/s. During the cooling process the structural change is observed by the Voronoi polyhedron analysis. It is shown that the unit cell of the amorphous structure becomes similar to that of crystalline structure with decrease of temperature, although its phase is still amorphous.

Even though the homogeneous crystallization could not be observed, heterogeneous crystallizations under external fields as well as without the excitation field are simulated and compared. The external excitation can expedite the heterogeneous crystallization process significantly, since the field enhanced molecular movement increases the potential energy and increased potential energy augments the jump frequency from *a*-Si to *c*-Si.

NOMENCLATURE

A, Parameter in potential function
B, Parameter in potential function
b_{ij}, Multi-body parameter
d_f, Thickness of silicon film
$\vec{F}$, Force vector
f, Jump frequency
f_c, Cutoff function
f_A, Attractive potential function
f_R, Repulsive potential function
G, Gibbs energy
g, Radial distribution function
H_0, Magnetic field
L_f, Width of silicon film
m, Mass of a molecule
N, Number of molecules
$\vec{r}$, Position vector
R, Gas constant
R_{ij}, Cutoff radius
r_{ij}, distance between molecules *i* and *j*
S, Surface area of the polyhedron
T, Temperature
T_f, Temperature of thin film
T_∞, Ambient temperature
t, Time, s
V, Volume of the polyhedron

Greek symbols

ε_f, Hemispherical emissivity
ζ Effective coordination number
η Asphericity
θ_{ijk} Angle between r_{ij} and r_{ik}
λ, Parameter in potential function
μ, Parameter in potential function
μ_0, Magnetic permeability
ρ_f, Electric resistivity
σ, Stefan-Boltzmann constant
Φ, System potential
ϕ, Pair potential
ω, Alternating angular frequency

Subscripts

i, *i*th molecule
j, *j*th molecule
k, *k*th molecule

Acknowledgements: The authors gratefully acknowledge the financial support of the Micro Thermal System Research Center through the Korea Science and Engineering Foundation.

REFERENCES

1. Ishimaru, M., Yoshida, K., and Motooka, T. (1996) Application of Empirical Interatomic Potentials to Liquid Si, *Phys. Rev. B*, Vol. 53, pp.7176-7181.
2. Yoon, S.Y., Park, S.J., Kim , K.H., and Jang, J. (2001) Metal Induced Crystallization of Amorphous Silicon, *Thin Solid Films*, Vol. 383, pp.34-38.
3. Ahn, J.H., Lee, J.N., Kim, Y.C., and B.T. Ahn (2002) Microwave-Induced Low-Temperature Crystallization of Amorphous Si Thin Films, *Current Applied Physics*, Vol. 2, pp.135-139.
4. Mandell, M.J., McTague, J.P., and Rahman, A. (1976) Crystal Nucleation in a Three-Dimensional Lennard-Jones System: A Molecular Dynamics Study, *J. Chem. Phys.*, Vol. 64, pp.3699-3702.
5. Mandell, M.J., McTague, J.P. and Rahman, A. (1977) Crystal Nucleation in a Three-Dimensional Lennard-Jones System. II. Nucleation Kinetics for 256 and 500 Particles, *J. Chem. Phys.*, Vol. 66, pp.3070-3075.
6. Tanemura, M., Hiwatari, Y., Matsuda, H., Ogawa, T., Ogita, N., and Ueda, A. (1977) Geometrical Analysis of Crystallization of the Soft-Core Model, *Prog. Theor. Phys.*, Vol. 58, pp.1079-1095.
7. Alen, M.P., and Tildesley, D.J. (1987) *Computer Simulation of Liquids*, Oxford.
8. Yang, J., Gould, X. H., and Klein, W. (1998) Molecular-Dynamics Investigation of Deeply Quenched Liquids, *Phys. Rev. Lett.*, Vol. 60, pp.2665-2668.
9. Swope, W.C., and Anderson, H.C. (1990) 106-Particle Molecular-Dynamics Study of Homogeneous Nucleation of Crystals in a Supercooled Atomic Liquid, *Phys. Rev. B*, vol. 41, pp.7042-7054.
10. Pickering, S., and Snook, I. (1997) Molecular Dynamics Study of the Crystallisation of Metastable Fluids, *Physica A*, Vol. 240, pp.297-304.
11. Caturla, M.J., Diaz de la Rubia, T., and Gilmor, G.H. (1995) Recrystallization of a Planar Amorphous-Crystalline Interface in Silicon by Low Energy Recoils: A Molecular Dynamics Study, *J. Appl. Phys.*, Vol. 77, pp.3121-3125.

12. Weber, B., Gartner, K., and Stock, D.M. (1997) MD-Simulation of Ion Induced Crystallization and Amorphization Processes in Silicon, *Nuclear Instruments and Methods in Physics Research B*, Vol. 127/128, pp.239-243.
13. Motooka, T. (1997) Molecular Dynamics Simulations for Amorphous/Crystalline Si Interface: Amorphization and Crystallization Induced by Simple Defects, *Nuclear Instruments and Methods in Physics Research B*, Vol. 127/128, pp.244-247.
14. Sameshima, T. (1998) Status of Si Thin Film Transistors, *J. Non-Crystalline Solids*, Vol. 227-230, Part 2, pp.1196-1201.
15. Zhao, Y., Wang, W., Yun, F., Xu, Liao, X., Ma, Z., Yue, G. and Kong, G. (2000) Experimental Insight into the Performance Characteristics of Ni-mesh Semiconductor Photo-Electrochemical Cells , *Solar Energy Materials & Solar Cells*, Vol. 62, pp.43-49.
16. Park, S.H., Kim, H.J., Lee, D.B., Lee, J.S., Choi, Y.K., and Kwon, O.M. (2004) Heterogeneous Crystallization of Amorphous Silicon Expedited by External Force Fields: A Molecular Dynamics Study, *Superlattices and Microstructures*, Vol. 35, pp.205-215.
17. Ishimaru, M., Munetoh, S., and Motooka, T. (1997) Generation of Amorphous Silicon Structures by Rapid Quenching: A Molecular-Dynamics Study, *Phys. Rev. B*, Vol. 56, pp.15133-15138.
18. Haile, J.M. (1997) *Molecular Dynamics Simulation: Elementary Methods*, John Wiley & Sons, New York.
19. Maruyama, S. (2000) Molecular Dynamics Method for Microscale Heat Transfer, *Advances in Numerical Heat Transfer*, Minkowycz and Sparrow (Eds.), Vol. 2, Chap. 6, pp. 189-226, Taylor & Francis, New York.
20. Tersoff, J. (1988) New Empirical Approach for the Structure and Energy of Covalent Systems, *Phys. Rev. B*, Vol. 37, pp.6991-7000.
21. Tersoff, J. (1989) Modeling solid-state chemistry: Interatomic Potentials for Multicomponent Systems, *Phys. Rev. B*, Vol. 39, pp.5566-5568.
22. Luedtke, W.D., and Landman, U. (1988) Preparation and Melting of Amorphous Silicon by Molecular-Dynamics Simulations, *Phys. Rev. B*, Vol. 37, pp.4656-4663..
23. Luedtke, W.D., and Landman, U. (1989) Preparation, Structure, Dynamics, and Energetics of Amorphous Silicon: A Molecular-Dynamics Study, *Phys. Rev. B*, Vol. 40, pp.1164-1174.
24. Car, R., and Parrinello, M. (1988) Structural, Dymanical, and Electronic Properties of Amorphous Silicon: An ab initio Molecular-Dynamics Study, *Phys. Rev. Lett*. Vol. 60, pp.204-207.
25. Stich, I., Car, R., and Parrinello, M. (1991) Amorphous Silicon Studied by ab Initio Molecular Dynamics: Preparation, Structure, and Properties, *Phys. Rev. B*, Vol. 44, pp.11092-11104.
26. Ishimaru, M., Yoshida, K., Kumamoto, T., and Motooka, T. (1996) Molecular-Dynamics Study on Atomistic Structures of Liquid Silicon, *Phys. Rev. B*, Vol. 54, pp.4638-4641.
27. Jedlovszky, P. (1999) Voronoi Polyhedra Analysis of the Local Structure of Water from Ambient to Supercritical Conditions, *J. of Chem. Phys.*, Vol. 111, pp.5975-5985.
28. Park, S.H., Lee, J.S., Choi, Y.K., and Kim, H.J. (2002) Vibration Induced Crystallization of Amorphous Materials: Molecular Dynamics Study, *Proc. Int. Symp. On Micro/Nanoscale Energy Conversion and Transport*, Antalya, Turkey, pp.75-77.

HIERARCHICAL MODELING OF THERMAL TRANSPORT FROM NANO-TO-MACROSCALES

C. H. AMON[1,2], S.V.J. NARUMANCHI[1], M. MADRID[1,3], C. GOMES[1,2] AND J. GOICOCHEA[1,2]
[1]Institute for Complex Engineered Systems, [2]Department of Mechanical Engineering
Carnegie Mellon University, Pittsburgh, PA 15213, USA.
[3]Pittsburgh Supercomputing Center, Pittsburgh, PA 15213, USA.

1. Introduction

Understanding and predicting thermal transport at extremely short time and length scales is essential to further advance a variety of emerging technologies. The trend towards miniaturization of electronic devices has lead to device features in the sub-micron and nanometer range. In fact, transistors with gate lengths on the order of 65 nm are already in production, and silicon-on-insulator (SOI) transistors are predicted to reach a gate length of 45 nm by the year 2005 [1]. The thermal conductivity of silicon at these scales is smaller than that of bulk silicon due to the scattering of the energy carriers with the boundaries of the device [2, 3]. Heat dissipation becomes an important issue, since overheating can negatively affect the reliability of the transistor. In addition to applications in microelectronics, micro/nanostructures such as quantum wells, nanowires, nanotubes and superlattices (periodic arrays of thin films), are appearing in optoelectronic devices [4], semiconducting lasers [5] and thermoelectric applications [6]. The thermal properties of nanowires [7-9], superlattices [10-12] and nanotubes [13] have recently attracted attention because of their special characteristics [8, 9, 14-16]. Carbon nanotubes, for example, have the potential for very high thermal conductivity [13, 17], while the thermal properties of superlattices can be controlled by adjusting the thickness and composition of its components.

Theoretical modeling can be an important tool to further the understanding of thermal transport. Different approaches are needed to study the thermal properties of devices whose components span length scales from the nanoscale to the macroscopic [18]. Phonons, which are quantized lattice vibrations [19, 20], are the predominant energy carriers in dielectrics and undoped semiconductors. Fourier diffusion yields erroneous predictions under the following two conditions: (a) when the mean free path of the energy carriers (Λ) becomes comparable to or larger than the characteristic length scale of the device/system under consideration (L); and/or (b) when the time scale of the processes under consideration becomes comparable to or smaller than the relaxation time of the energy carriers [15, 21, 22]. Under these conditions, if the wave nature of the energy carriers can be neglected [23], the semi-classical Boltzmann transport equation (BTE) can be employed to solve for the distribution function of the energy carriers. Molecular dynamics (MD) involves solving the atomistic equations of motion, from which phonon and thermal properties can be obtained. Being an entirely classical method, molecular dynamics is strictly applicable to solids only at temperatures above Debye's temperature (for silicon, $\theta_D \sim 625$K [19]), where the system behaves classically. Several corrections have been proposed in the literature to partly account for quantum effects in molecular dynamics simulations of thermal transport when these effects become relevant, such as when the temperature of the system is smaller than its Debye temperature, as reviewed in [24]. However, when the temperature and dimensions of the system are such that the characteristic

S. Kakaç et al. (eds.), Microscale Heat Transfer, 379–400.

wavelength (which for silicon is on the order of a few nanometers at room temperature) is comparable to the system dimensions, quantum effects such as wave interference and quantum size effects [25, 26] become important and need to be incorporated into the formalisms [26]. The applicability of BTE and MD in this limit has not been tested. Ultimately, theories that take into account quantum mechanics, such as quantum BTE (QBTE) [27], although computationally far more demanding, need to be applied. Figure 1 shows the approximate regimes of applicability of Fourier diffusion, BTE-based approaches, MD and QBTE for the simulation of thermal transport in silicon at ambient temperature. The sizes of current devices are shown for comparison, as well as the dominant wavelength λ and the mean free path Λ of silicon at 300K.

In addition to the theoretical applicability of the different approaches, there are practical limitations to the size of the system that can be realistically simulated, due to the computational requirements. With the current computational power, for example, molecular dynamics is a viable approach when the system consists of several thousands of atoms. Bulk or infinite systems can be modeled using periodic boundary conditions. However, molecular dynamics simulations of larger systems with boundaries that cannot be approximated by periodic boundary conditions can be computationally prohibitive. Under these conditions, the Boltzmann equation is the ideal tool to study carrier transport. If the system under consideration involves multiple length scales, there is a need to apply different modeling approaches simultaneously to the same system. The strategy is to judiciously select the modeling approach based on the length and time scale of the device and processes to be simulated. In addition, as will be discussed subsequently in this section, some of these approaches can be coupled, such as using MD simulations to obtain phonon properties that can then be input into BTE.

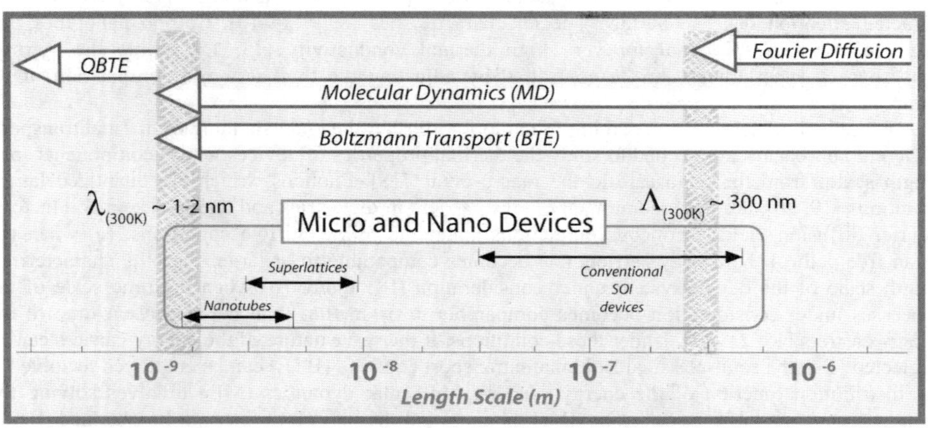

Figure 1. Range of applicability of different modeling approaches as a function of the system length. Current device sizes and the dominant wavelength λ and mean free path Λ for silicon at 300K are also shown.

Simulations of thermal conduction by means of molecular dynamics have the advantage that no prior assumptions are needed about the physical properties of phonons [15]. Instead, given an accurate interatomic potential and a starting structure, the equations of motion are integrated according to the laws of classical mechanics. As mentioned earlier, a drawback of molecular dynamics is the long computational time required to simulate relatively large systems. However, molecular dynamics can be an ideal complement to Boltzmann transport equation protocols, by providing needed phonon parameters, such as frequency-dependent relaxation times and dispersion curves. Currently, relaxation times are typically obtained from perturbation theory [28, 29] and most of the dispersion curves available in the literature are for bulk systems. Perturbation theory typically involves at least one empirical parameter which is determined by matching thermal conductivity predictions to experimental data [30]. Also, for low-dimensional systems, both the relaxation times and dispersion relations can undergo modifications due to size effects [31]. Relaxation times, as well as dispersion curves, can be computed from MD. These quantities can then be utilized in the BTE approach and applied to larger systems for which it may not be feasible to use MD. Recently, the Boltzmann transport equation was solved with parameters determined from molecular dynamics simulations of argon crystals with the Lennard-Jones potential [32]. It was shown that it is essential to take into account the temperature and frequency dependence of phonon relaxation times as well as phonon dispersion in order to obtain accurate predictions of thermal conductivity. Hierarchical modeling from atomistic molecular dynamics to phonon Boltzmann transport equation has the potential to substantially improve thermal predictions in nanoelectronics.

In this manuscript, we briefly describe the BTE and MD approaches and show how the two methods can be coupled. We validate both the BTE and MD approaches separately by comparing with experimental thermal conductivity data for bulk and thin-film silicon. Finally, we look at the problem of self-heating in silicon-on-insulator transistors. Differences in the predictions from the different BTE-based modeling approaches and Fourier diffusion are pointed out. The impact of boundary conditions on the thermal predictions of silicon-on-insulator transistors is also discussed.

2. BTE-based models

In the relaxation time approximation, the Boltzmann transport equation (BTE) takes the form [22, 33]:

$$\frac{\partial f_w}{\partial t} + \nabla \cdot (v_w \hat{s} f_w) = \frac{f_w^0 - f_w}{\tau_w} \tag{2.1}$$

where f_w is the frequency-dependent phonon distribution function, f_w^0 is the equilibrium Bose-Einstein distribution function, τ_w is the frequency-dependent phonon relaxation time, v_w is the frequency-dependent phonon group velocity and $\hat{s}$ is the unit direction vector. Next, we discuss several approaches to solve Eq. (2.1).

2.1 GRAY BTE MODEL

A large number of studies have appeared in the literature based on the "gray BTE" model (e.g., [15, 22, 33-39]). In this approach, the phonons are grouped together in one mode and characterized by a single group velocity and relaxation time. No distinction is made between the different phonon modes. This approach has been used to study a wide array of problems in sub-micron heat conduction.

In the energy formulation, the BTE can be expressed as:

$$\frac{\partial e''}{\partial t} + \nabla \cdot (v \hat{s} e'') = \frac{e^0 - e''}{\tau} + q_{vol} \tag{2.2}$$

$$e'' = \int_{\Delta w_i} \hbar w f_w D(w) dw; \; e^0 = \frac{1}{4\pi} \int_{4\pi} e'' d\Omega$$
$$e^0 = \frac{1}{4\pi} C(T_L - T_{ref}) \quad (2.3)$$

where e'' is the energy density per unit solid angle, e^o is the equilibrium energy density, f_w is the frequency-dependent phonon distribution function, $D(w)$ is the phonon density of states, C is the total specific heat, T_L is the lattice temperature, T_{ref} is an arbitrary reference temperature, q_{vol} is the volumetric heat generation and τ is the relaxation time. For the silicon-on-insulator transistor simulations with the gray BTE model presented later, v for silicon is chosen as 6400 m/s [40], while C is $1.66\text{x}10^6$ J/m^{3}K at 300K. The relaxation time of $\tau = 6.28$ ps is obtained from the kinetic relation for the thermal conductivity $k = \frac{1}{3} C v^2 \tau$, where $k = 142.3$ W/m-K for silicon. Hence, in the acoustically-thick limit (Knudsen number $K_n << 1$) this model yields results similar to the solutions from the Fourier diffusion equation.

2.2 SEMI-GRAY BTE MODEL

Another modeling approach proposed in the literature is the semi-gray model [41-43]. In this approach, the phonons are divided into propagating and reservoir modes. Propagating mode phonons are responsible for transporting energy, while the reservoir mode phonons are purely capacitative. For unsteady problems, the choice of which phonons fall in the reservoir and propagating modes is critical because it determines the effective capacitance of the system and, consequently, its time response. In references [42, 43], longitudinal acoustic phonons are considered to be the propagating mode, while the transverse acoustic and optical phonons are lumped together in the reservoir mode. Heat generation is incorporated via a source term for the reservoir mode BTE. The main equations are:

$$\frac{\partial e_P''}{\partial t} + \nabla \cdot (v_p \hat{s} e_P'') = \frac{\frac{1}{4\pi} C_P (T_L - T_{ref}) - e_P''}{\tau} \quad (2.4)$$

$$C_R \frac{\partial (T_R - T_{ref})}{\partial t} = \frac{C_R (T_L - T_{ref}) - C_R (T_R - T_{ref})}{\tau} + q_{vol} \quad (2.5)$$

$$C_P (T_P - T_{ref}) = \int_{4\pi} e_P'' d\Omega \quad (2.6)$$

$$T_L = \frac{C_R T_R + C_P T_P}{(C_R + C_P)} \quad (2.7)$$

where T_P is the propagating mode phonon temperature, T_R is the reservoir mode phonon temperature, C_P and C_R are the propagating and reservoir mode specific heats, respectively, τ is the relaxation time, v_P is the propagating mode group velocity, e_P'' is the propagating mode energy density per unit solid angle (J/m^3-sr), and, T_L is the overall lattice temperature interpreted as an average of the propagating and reservoir mode temperatures as expressed in Eq. (2.7).

For the modeling of silicon-on-insulator transistors using the semi-gray BTE presented in later sections, the value of C_P is $0.32\text{x}10^6$ J/m^{3}K, C_R is $1.34\text{x}10^6$ J/m^{3}K, τ is 74.2 ps, v_P is 4240 m/s (Eqs. 2.4-2.7) [42]. The rationale for choosing these values is as follows. The propagating and reservoir mode specific heats are obtained by evaluating their contributions based on the phonon dispersion curve [44]. The phonon velocity of 4240 m/s corresponds approximately to the group velocity of the higher frequency longitudinal acoustic phonons, which are assumed to be the main energy carriers [42]. The relaxation time is chosen so as to recover the bulk thermal conductivity of silicon (142.3 W/m-K) in the acoustically-thick limit ($K_n << 1$), using $k = \frac{1}{3}Cv^2\tau$. Hence, in the acoustically-thick limit for steady-state heat transfer, this model also yields results similar to those from the diffusion equation.

2.3 FULL PHONON DISPERSION BTE MODEL

Phonon dispersion has been incorporated in a more accurate way by considering the transverse and longitudinal acoustic phonons and the interactions between them in a Monte Carlo framework, while neglecting optical phonons [45]. However, optical phonons contribute significantly to the specific heat of silicon (e.g. about 35-40 % in bulk silicon at 300K). This aspect becomes very important for transient problems. Incorrect accounting for capacitance leads to erroneous transient predictions. It is also worth noting that in the semi-gray modeling approach described previously [42, 43], choosing only the longitudinal acoustic mode as the propagating mode and lumping the transverse acoustic and optical phonons together as the reservoir mode have a significant impact on the resulting specific heats for both the propagating and reservoir modes. The specific heats of the propagating and reservoir modes, in turn, affect the predictions of transient thermal phenomena. A more detailed model, based on the solution to the BTE in the relaxation time approximation, which incorporates the details of the phonon dispersion curves, has recently been proposed [44] and has been used in sub-micron transistor thermal simulations [46, 47]. This full phonon dispersion model accounts for optical phonons, and longitudinal and transverse acoustic phonons, and the interactions among them via frequency-dependent relaxation times. Frequency-dependent phonon group velocities are determined from bulk phonon dispersion curves for silicon [40]. The relaxation times are obtained by perturbation theory techniques [28, 31, 48-51]. The essential features of the model are as follows. Starting from Eq. (2.1), we can define the following quantities:

$$\begin{gathered} e_w'''(\vec{r},\hat{s},w) = \hbar w f_w D(w) \quad ; \quad e_w'' = \int_{\Delta w_i} e_w''' dw \\ e_w = \int_{4\pi} e_w'' d\Omega \quad ; \quad e_w^0 = \frac{1}{4\pi}\int_{4\pi} e_w'' d\Omega = \frac{1}{4\pi} e_w \end{gathered} \tag{2.8}$$

where e_w''' is the volumetric energy density per unit frequency per unit solid angle (Js/m^3-sr-rad), e_w'' is the volumetric energy density per unit solid angle (J/m^3-sr) for a given frequency band, $\vec{r}$ is the position vector, and $\hat{s}$ is the unit direction vector. The quantity e_w is the volumetric energy density (J/m^3) in the band and e_w^0 is the associated equilibrium volumetric energy density, $\hbar$ is the reduced Planck's constant, and $D(w)$ is the phonon density of states. The frequency integration is done over a discrete frequency band Δw_i. In a given symmetry direction, silicon has three acoustic branches and three optical branches. Of the three acoustic branches, one is longitudinal and two are degenerate transverse branches. Similarly, there is one longitudinal optical phonon branch and two degenerate transverse optical phonon branches. In our model, only one frequency band is used for the optical phonon branch, while there are N_{LA} and N_{TA} bands in the longitudinal acoustic (LA) and transverse acoustic (TA) branches, respectively [44, 52]. The experimental dispersion curves for the LA and TA branches [40] are fit by cubic splines (six each for LA and TA), and all relevant dispersion curve information (e.g. phonon group velocity, density of states) is extracted from these fits.

The optical mode for silicon has a negligible group velocity; therefore, the ballistic term in Eq. (2.1) is absent. The BTE for the optical mode can be written as:

$$\frac{\partial e_0}{\partial t} = \sum_{j=1}^{Nbands-1} \left(\int_{Tref}^{T_{oj}} C_0 dT - e_0 \right) \gamma_{oj} + q_{vol} \tag{2.9}$$

where $e_0 = \int_{Tref}^{T_0} C_0 dT$, $\gamma_{oj} = \overline{(1/\tau_{oj})} = \frac{1}{\Delta w_0} \int \frac{1}{\tau_{oj}} dw$ is the band averaged inverse relaxation time for the interaction between the optical phonons and the jth band of an acoustic branch and C_0 is the optical mode specific heat. In the development that follows, a number of "temperatures" are defined. Since thermodynamic temperatures cannot be defined for systems that are not in equilibrium, these temperatures must properly be interpreted as measures of the corresponding energy. In this spirit, T_0 is the temperature of the optical mode, T_{ref} is the reference temperature, T_{0j} is an *interaction* temperature specific to the optical mode and the jth band of the acoustic modes (either LA or TA), obtained from energy conservation requirements. The significance of T_{oj} is that if the optical branch and the jth band of the acoustic branches (LA or TA) were the only bands exchanging energy, and ballistic and storage effects were absent, T_{oj} would be the equilibrium "temperature" achieved by both bands. As an approximation, the term q_{vol} is the volumetric heat generation [43]. In microelectronics applications, it would represent the transfer of energy from the energetic electrons to the optical phonons (Joule heating).

The BTE for the ith frequency band of the acoustic branches in a direction $\hat{s}$ is written as:

$$\frac{\partial e_i''}{\partial t} + \nabla \cdot (v_i \hat{s} e_i'') = (e_i^0 - e_i'')\gamma_{ii} + \sum_{\substack{j=1 \\ j \neq i}}^{Nbands} \left\{ \left(\frac{1}{4\pi} \int_{Tref}^{T_{ij}} C_i dT - e_i'' \right) \gamma_{ij} \right\} \tag{2.10}$$

$$v_i = \frac{1}{\Delta w_i} \int_{\Delta w_i} v_w dw;\ C_i = \int_{\Delta w_i} C_{wi} dw;\ e_i'' = \int_{\Delta w_i} \hbar w f_w D(w) dw;$$
$$\gamma_{ii} = \overline{(1/\tau_{ii})} = \frac{1}{\Delta w_i} \int \frac{1}{\tau_{ii}} dw; \gamma_{ij} = \overline{(1/\tau_{ij})} = \frac{1}{\Delta w_i} \int \frac{1}{\tau_{ij}} dw \tag{2.11}$$

where v_i is the band averaged group velocity, C_{wi} is the specific heat per unit frequency in band *i*, C_i is the band integrated specific heat, e_i'' is the band integrated energy density per unit solid angle, γ_{ii} is the band averaged inverse relaxation time for interaction of band *i* with itself, and γ_{ij} is the band averaged inverse relaxation time for interaction of band *i* with band *j*. The first term on the right hand side (RHS) depicts scattering within a given frequency band, but across directions (elastic scattering). Physically, processes such as impurity scattering may be described by such a term. The second term on the RHS depicts the scattering from the ith band of the acoustic band to all other bands in all branches except to itself. T_{ij} again signifies an *interaction* temperature that the two bands i and j would attain if they were the only two bands exchanging energy. The model satisfies energy conservation, and is validated against known solutions from the radiative transport literature. It is also validated in the acoustically-thick limit ($K_n << 1$) by comparisons to solutions of the Fourier diffusion equation. The bulk thermal conductivity of silicon at different temperatures [40] is also recovered. In addition, the experimental in-plane thermal conductivities of silicon thin films of different thicknesses over a range of temperatures [3, 53, 54] are matched satisfactorily [30].

3. Description of the equilibrium MD approach

Molecular dynamics simulations entail integrating Newton's second law of motion for an ensemble of atoms in order to derive the thermodynamic and transport properties of the ensemble. The two most common approaches to predict thermal conductivities by means of molecular dynamics include the direct and the Green-Kubo methods. The direct method is a non-equilibrium molecular dynamics approach that simulates the experimental setup by imposing a temperature gradient across the simulation cell. The Green-Kubo method is an equilibrium molecular dynamics approach, in which the thermal conductivity is obtained from the heat current fluctuations by means of the fluctuation-dissipation theorem. Comparisons of both methods show that results obtained by either method are consistent with each other [55]. Studies have shown that molecular dynamics can predict the thermal conductivity of crystalline materials [24, 55-60], superlattices [10-12], silicon nanowires [7] and amorphous materials [61, 62]. Recently, non-equilibrium molecular dynamics was used to study the thermal conductivity of argon thin films, using a pair-wise Lennard-Jones interatomic potential [56].

Given an interatomic potential and the initial positions of the atoms, molecular dynamics simulates the time evolution of the atoms by integrating Newton's second law of motion:

$$m_i \frac{d^2\vec{r}_i}{dt^2} = \vec{F}_i \tag{3.1}$$

where m_i and $\vec{r}_i$ are the atom's mass and position, respectively; the total force $\vec{F}_i$ on atom i is calculated as the gradient of the interatomic potential.

The heat current is given by [63-65]:

$$\vec{J} = \frac{\partial}{\partial t}\left(\sum_i \vec{r}_i.E_i\right) \tag{3.2}$$

where the total energy E_i stored in atom i is the sum of its kinetic and potential energies.

Equilibrium molecular dynamics uses the Green-Kubo relationship between the heat current autocorrelation function and the thermal conductivity [66] to obtain the thermal conductivity as:

$$k = \frac{1}{3Vk_BT^2}\int_0^{\infty}\left\langle \vec{J}(\tau)\cdot\vec{J}(0)\right\rangle d\tau \tag{3.3}$$

where k is the thermal conductivity, V is the ensemble volume, T is the temperature, k_B is Boltzmann's constant, J is the heat current. The integrand is the ensemble average of the heat current autocorrelation function, which can be replaced by its time average in accordance with the ergodic hypothesis [67]:

$$k\left(\tau_M\right) = \frac{\Delta t}{3Vk_BT^2}\sum_{m=1}^{M}(N_s\text{-}m)^{-1}\sum_{n=1}^{N_s-m}\vec{J}(m+n)\cdot\vec{J}(n) \tag{3.4}$$

where k is the thermal conductivity, T is the temperature, Δt is the simulation time step, N_s is the number of heat current autocorrelation function averages, and M is the number of time steps required for the heat current autocorrelation function to decay to zero.

The temperature is defined according to the equipartition theorem [19] as:

$$\frac{3}{2}Nk_BT_{MD} = \frac{1}{2}\sum_{i=1}^{N} m_i \vec{v}_i \cdot \vec{v}_i \tag{3.5}$$

where N is the number of atoms in the ensemble and $\vec{v}_i$ is the velocity of atom i.

For temperatures below the Debye temperature (θ_D), quantum corrections must be applied to the temperature and thermal conductivity obtained from molecular dynamics. These quantum corrections are negligible for $T >> \theta_D$, where the system behaves classically. A quantum correction for the temperature can be estimated by equating the ensemble's total energy to the phonons' total energy [10, 57, 61] as:

$$3Nk_BT_{MD} = \sum_i \sum_j \left(f_{w_i}^0 + \frac{1}{2} \right) \hbar w_i \tag{3.6}$$

$$f_{w_i}^0 = \frac{1}{e^{\frac{\hbar w_i}{k_B T}} - 1} \tag{3.7}$$

where $f_{w_i}^0$ is the equilibrium phonon distribution function, given by the Bose-Einstein distribution function, Eq. (3.7), w is the phonon's frequency, $\hbar$ is Planck's constant divided by 2π, T_{MD} is the temperature of the molecular dynamics simulation, and T is the corrected temperature. The summation is over the phonons' frequencies, index i, and the phonons' branches, index j (longitudinal acoustic, transverse acoustic, longitudinal optical and transverse optical). The correction factor for quantum effects for the thermal conductivity [10] is given by:

$$\frac{k}{k_{MD}} = \frac{\partial T_{MD}}{\partial T} \tag{3.8}$$

where k is the corrected thermal conductivity and k_{MD} is the molecular dynamics predicted thermal conductivity. The quantum energy of a system includes a contribution given by the zero-point energy, the factor ½ $\hbar w$ in Eq. (3.6). Controversy surrounds the correction factors that should be used to account for quantum effects in a molecular dynamics simulation; in particular, whether the zero-point energy should or should not be included in Eq. (3.6). In this study we have included this factor, because our objective is to correct the classical energy and obtain the quantum one. The correction factors are taken from the reported experimental specific heat for silicon [68-70] (Fig. 2). It is seen that the quantum corrections for both T and k become significant only for $T < \theta_D$, with $\theta_D \sim 625$K for silicon [19].

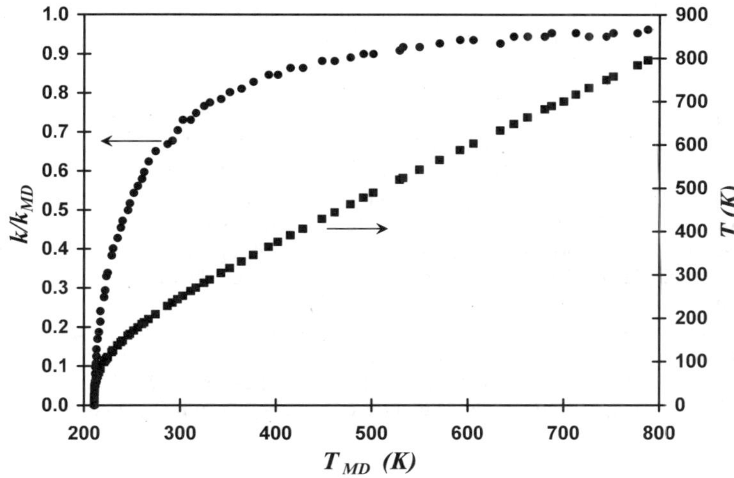

Figure 2. Quantum corrected temperature (right axis), and ratio of quantum corrected to molecular dynamics predicted thermal conductivity (k/k_{MD}) (left axis), as a function of the temperature of the MD simulations for silicon.

4. Coupled BTE-MD approach

As mentioned in section 1, the MD approach can be utilized to obtain parameters such as frequency-dependent relaxation times and phonon dispersion relations (the dependence of the frequency w on the wave vector κ). From the phonon dispersion curves, we can compute parameters such as the phonon group velocity v_w, density of states $D(w)$, and specific heat C_w. The phonon group velocities and relaxation times can then be input into the BTE (Eq. 2.1).

The relaxation time of a phonon mode is related to the temporal decay of the autocorrelation function of its energy components (potential and kinetic) [71, 72]. The total energy of each mode i for a classical system, under the harmonic approximation, is given by [73]:

$$E_{m,t} = \frac{w^2 Q_m^*(\vec{\kappa},\nu) Q_m(\vec{\kappa},\nu)}{2} + \frac{\dot{Q}_m^*(\vec{\kappa},\nu) \dot{Q}_m(\vec{\kappa},\nu)}{2} \tag{4.1}$$

The first and second terms on the right hand side of the equation represent the potential energy and kinetic energy of the mode m, respectively. $Q_m(\vec{\kappa},\mathrm{v})$ is the normal mode of the system or normal mode coordinate, defined mathematically as the inverse Fourier transform of the displacement of an atom in terms of waves as:

$$Q_m(\vec{\kappa},\nu) = \frac{1}{N_u^{1/2}} \sum_i m_i^{1/2} \exp(-\vec{\kappa} \cdot \vec{r}_{i,0}) \vec{e}_m^{\,*}(\vec{\kappa},\nu) \cdot \vec{u}_i(t) \tag{4.2}$$

where N_u is the number of unit cells, m_i is the mass of atom i, ν corresponds to the mode polarization (longitudinal or transverse) described by the polarization vector $\vec{e}_m^{\,*}(\vec{\kappa},\nu)$ - i.e. its complex conjugate, $\vec{\kappa}$ is the wave vector, $\vec{r}_{i,0}$ is the equilibrium position of atom i, $\vec{u}_i(t)$ is the relative displacement of atom i at time t from its equilibrium position. In Eq. (4.2), the different polarization vectors are calculated by solving the eigenvalue-eigenvector problem that arises from the Lagrangian description of the harmonic system.

The normal mode coordinates can be obtained from an MD simulation, where the motion of a set of atoms in an atomic structure is tracked over a period of time. The advantage of the use of MD is that it allows for an accounting of anharmonic effects, as well as the temperature dependence of physical properties [72]. Once the normal modes are computed, the autocorrelation function of the mean energy fluctuation (total or potential) for each vibration mode can be obtained.

The relaxation time for a given polarization mode (v) and wave vector ($\vec{\kappa}$) is calculated by: i) integrating the energy autocorrelation function [71], or ii) fitting the energy correlation function with an exponential [72] or damped oscillatory function [73], as shown in Fig. 3. For each wave vector, the anharmonic frequency can also be extracted.

After calculating discrete values of frequency and relaxation time for the polarization mode and different wave vectors, a continuous function for the relaxation time and dispersion relations can be established for each mode. The results shown in Fig. 3 are obtained with the Lennard-Jones potential for Argon. An approach similar to that described above can be applied with the Stillinger-Weber potential for silicon.

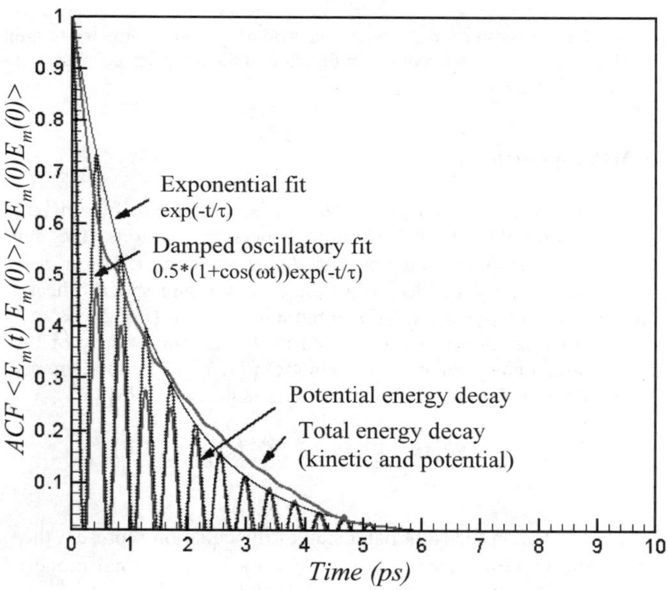

Figure 3. Autocorrelation curves (bold continuous lines) of the mean energy fluctuation of total and potential energy for mode m, and their corresponding exponential and damped oscillatory fits (thin dotted lines).

5. Thermal conductivity of bulk silicon

In this section, we compare BTE and MD predictions of bulk thermal conductivity of silicon to the available experimental data [74] over a temperature range of 500K to 1000K (Fig. 4). The BTE predictions are obtained from the full phonon dispersion model described in section 2.3. This full phonon dispersion model involves an adjustable parameter, the Gruneisen constant (γ), which is set to $\gamma = 0.59$ [44]. This value of γ gives a good match with experimental data [40] over a temperature range from 1K to 1000K [30]. Between 500K and 1000K, an even better match with experimental data [74] can be obtained by using a slightly different value of γ.

MD-predicted thermal conductivities are obtained at 700K and 1000K using equilibrium molecular dynamics and the Stillinger-Weber interatomic potential [75]. The simulation domain consists of 4x4x4, 5x5x5, 7x7x7 and 10x10x10 lattice constants. Periodic boundary conditions are used to simulate an infinite system. In molecular dynamics predictions of thermal conductivities performed under periodic boundary conditions, artifacts can arise as a consequence of the finite size of the simulation domain. The predicted thermal conductivity can increase with the size of the simulation domain, until a large enough simulation domain is used, beyond which the thermal conductivity is independent of the size of the simulation domain. This size effect has been attributed to the fact that periodic boundary conditions exclude from the simulation wavelengths larger than the simulation domain length [57]. Our results show no size effects for silicon for the simulation domains and temperatures reported in this paper (4x4x4, 5x5x5, 7x7x7 and 10x10x10 at 700K and 1000K), within the statistical error (Fig. 4) [60]. The MD results are in good agreement with the experimental values, considering that the experimental values could be reduced due to the presence of vacancies, isotopes and lattice dislocations.

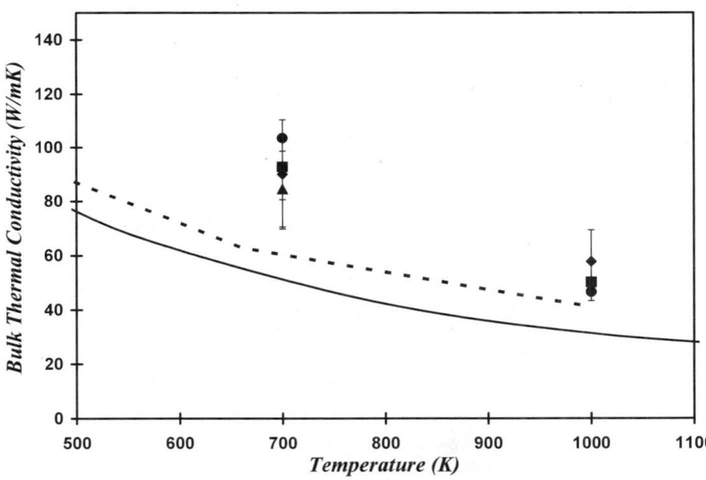

Figure 4. Thermal conductivity of single crystal bulk silicon predicted from MD for simulation domains of 4x4x4 (■), 5x5x5 (♦), 7x7x7 (●) and 10x10x10 (▲) lattice constants, predicted by BTE (dashed line) and compared to the experimentally determined thermal conductivity [74] (solid line).

6. Thermal conductivity of silicon thin films

We next present the BTE and MD predictions of thermal conductivities of silicon thin films. The thermal properties of silicon thin films are of paramount importance to the transistor industry. Silicon-on-insulator (SOI) and strained silicon transistors are composed of silicon thin films. In both cases, the thin silicon film is deposited on top of poor thermally conducting materials, and the thermal energy generated by the Joule effect has to be removed along the silicon film plane. A thorough understanding of the thermal properties of thin silicon films is essential for the accurate prediction of the thermal response of these transistors. The dimensions of the silicon thin film in

SOI and strained silicon transistors are comparable to or smaller than the phonon's mean free path (which, for silicon, has been estimated as 300 nm at 300K) [53]. In this limit, the film surfaces alter the phonon dispersion relations [76], and the phonon-surface scattering may become the predominant scattering mechanism [3, 53]. Since phonons are the main carriers of thermal energy in silicon, these effects alter the thermal conductivity, which differs from that of bulk silicon [10, 36, 77]. Measurements of the thermal conductivities of silicon films of thicknesses down to 74 nm found a reduction of 50% with respect to the bulk value at 300K [53]. This reduction depends on the temperature and the thickness of the film [3, 53].

Figure 5 shows a molecular dynamics simulation domain for a thin film of thickness d_s. Periodic boundary conditions are used in the directions parallel to the film surfaces. The atoms on the surfaces can be treated as free, delimited with a few layers of atoms kept frozen at their equilibrium positions or restrained by a weak harmonic potential [7, 56, 78]. Molecular dynamics simulations using free atoms on the surfaces may not successfully complete, because the atoms can escape the simulation domain. Delimiting the system with atoms kept fixed or subject to a harmonic potential does not allow the atoms on the surfaces to adopt configurations that minimize the number of dangling bonds. We developed a potential to treat the surface atoms in a realistic and computationally efficient manner [79]. Such a potential is needed for the accurate prediction of thin silicon film thermal conductivities by means of equilibrium molecular dynamics, because it allows for surface reconstruction while using a reasonable integration step. The effect of a surrounding gas on the surfaces of the film is replaced by a short-ranged, one dimensional potential whose force acts in the direction perpendicular to the film surface and towards the film. Molecular dynamics simulations of thin films using this repulsive potential conserve the number of atoms, volume and total energy of the ensemble, and therefore fulfill the requirements to apply the Green-Kubo formalism [79].

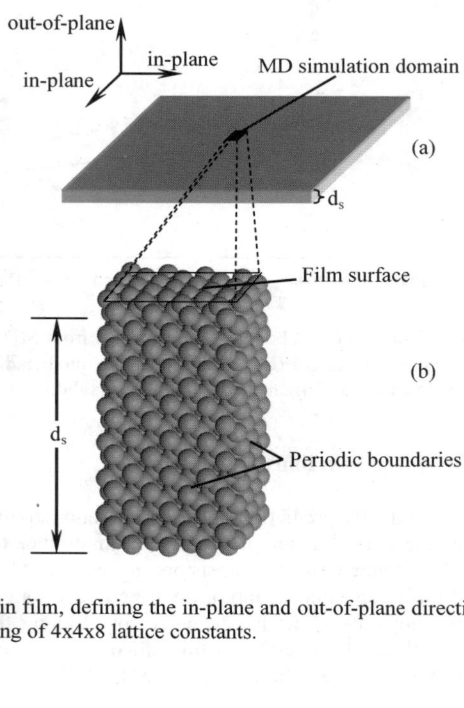

Figure 5. (a) Sketch of a thin film, defining the in-plane and out-of-plane directions and the thickness d_s; (b) Simulation domain consisting of 4x4x8 lattice constants.

Silicon thin film thermal conductivities are predicted using equilibrium molecular dynamics and the Green-Kubo relation. Periodic boundary conditions are applied in the directions parallel to the thin film surfaces (Fig. 5). Atoms near the surfaces of the thin film are subjected to the above-described repulsive potential in addition to the Stillinger-Weber potential [75]. Simulations were also performed adding to each surface four layers of atoms kept frozen at their crystallographic positions, in order to compare the dependence of the predicted thermal conductivities on the surface boundary conditions. We found that the thermal conductivities obtained using frozen atoms or the repulsive potential are identical within the statistical deviations, except for the in-plane thermal conductivity of films with thickness less than 10 nm [79]. Therefore, in the present study, we present only the predictions obtained with the repulsive potential.

The thermal conductivities of silicon thin films are predicted at 376K and 1000K. The simulation cross section is 2x2 lattice constants, with periodic boundary conditions. The choice of this small cross section is motivated by the observation that at 1000K, the thermal conductivity predicted by molecular dynamics appears to be independent of the number of lattice constants included in the simulation domain. In fact, our previous molecular dynamics simulations of bulk silicon using periodic boundary conditions have found no appreciable differences in the predicted thermal conductivity for simulations ranging from 2x2x2 to 7x7x7 lattice constants at 1000K [60]. Furthermore, three different simulations of a thin film of 4.344 nm with cross sections of 2x2, 4x4 and 7x7 lattice constants also showed no appreciable differences in the predicted thermal conductivity at 1000K. Simulations with larger cross sections at 376K are underway.

Figure 6 shows the MD predicted in-plane and out-of-plane thermal conductivities at 376K (Fig. 6a) and 1000K (Fig. 6b) as a function of film thickness. It is seen that both the in-plane and out-of-plane thermal conductivities are affected by the thickness of the film. For thickness smaller than the phonon mean free path (approximately 300 nm and 30 nm at 300K and 1000K, respectively), both the in-plane and out-of-plane thermal conductivities decrease with decreasing thickness, an effect attributed to the scattering of phonons with the boundaries of the thin film. This effect is more pronounced in the out-of-plane direction, where the dimensions of the thin film make the phonon transport ballistic. At large thicknesses, the thermal conductivities approach the bulk value (shown as dashed lines in Fig. 6). The bulk value is reached at smaller thicknesses at 1000K due to the smaller phonon mean free path at this temperature.

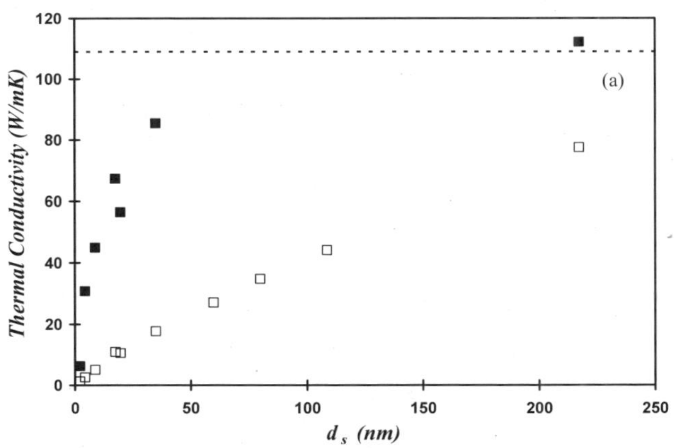

Figure 6 (a). MD predicted in-plane (■) and out-of-plane (□) silicon thermal conductivities at 376K as a function of film thickness d_s. Bulk silicon thermal conductivity at 376K is shown as a dashed line.

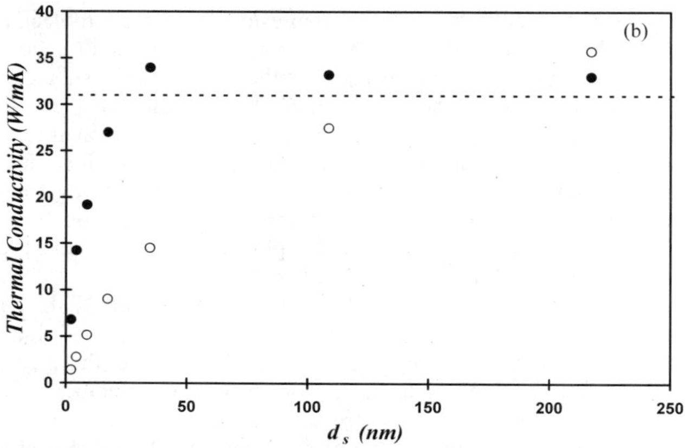

Figure 6 (b). MD predicted in-plane (•) and out-of-plane (○) silicon thermal conductivities at 1000K as a function of film thickness d_s. Bulk silicon thermal conductivity at 1000K is shown as a dashed line.

Figure 7 shows the MD predicted in-plane thermal conductivities at 376K (Fig. 7a) and 1000K (Fig. 7b) as a function of film thickness. Three different MD simulations were performed at 1000K for each thickness starting with different atomic velocities, from which the average thermal conductivity and the statistical deviations are calculated. The MD results are compared with predictions from BTE and to the available experimental data at 300K [53, 80]. BTE calculations are performed with specular factors of p=0.0 and 0.95, corresponding to diffusive and specular boundary conditions, respectively. The effect of isotopes is simulated in BTE by considering pure silicon (100% Si^{28}, dashed lines) and natural silicon (92.23% Si^{28}, 4.67% Si^{29} and 3.10% Si^{30}, solid lines). At 300K, the scattering with isotopes reduces the thermal conductivity (Fig. 7a). At 1000K, identical results are obtained for both pure and natural silicon. It is seen that the experimentally determined thermal conductivities are between the BTE predictions for the two boundary conditions considered. MD predictions agree with the BTE specular calculations and capture the expected trend of decreasing thermal conductivity with decreasing film thickness.

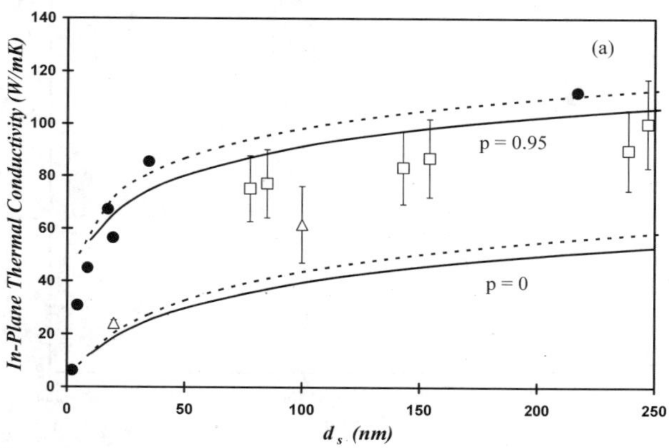

Figure 7 (a). In-plane silicon thermal conductivity predicted by molecular dynamics at 376K (•), predicted from BTE for pure (dashed lines) and natural (solid lines) silicon, and available experimental data (□) [53] and (Δ) [80] at 300K.

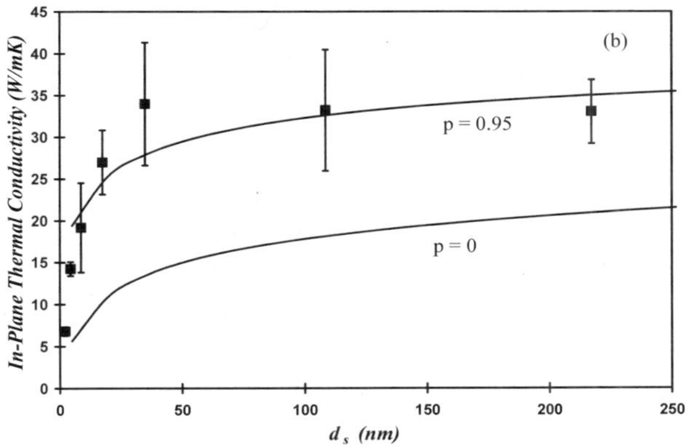

Figure 7 (b). In-plane silicon thermal conductivity at 1000K predicted by molecular dynamics (■), and from BTE (solid lines).

7. Modeling self-heating in silicon-on-insulator transistors

In this section, we focus on the problem of self-heating in silicon-on-insulator transistors. Fig. 8 shows a two-dimensional domain representing the silicon-on-insulator transistor, similar to that described in [43]. A thin silicon layer resides on top of a thicker insulating silicon dioxide (SiO_2) layer. The left, right, and bottom boundaries of the domain are maintained at 300K and serve as heat sinks, while the top silicon boundary is diffusely reflecting (adiabatic), as shown in Fig. 8.

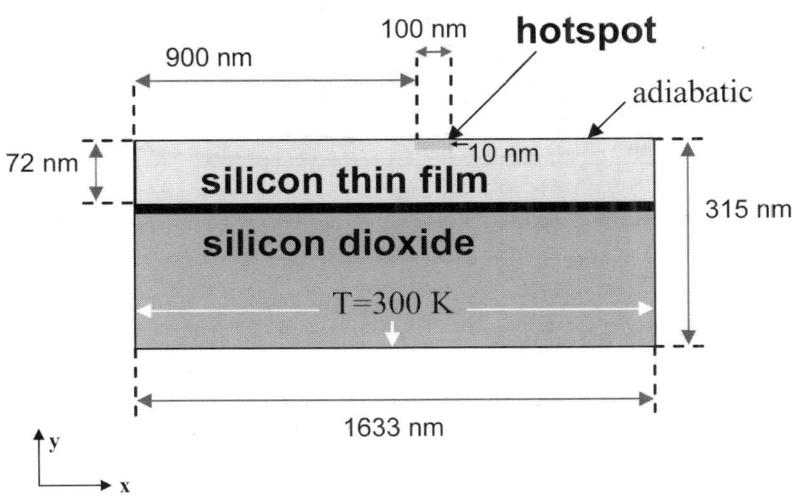

Figure 8. Two-dimensional silicon-on-insulator geometry and simulation domain.

Most of the electron-phonon scattering, and consequently the heat generation, occurs in the channel region of the transistor (labeled "hotspot" in Fig. 8). The dimensions of the heated region are 100 nm x 10 nm, with a volumetric heat generation of 6.0×10^{17} W/m^3, for a total heat generation per unit depth of 600 W/m. The entire heat generation is assumed to occur via electron-optical phonon interaction and is incorporated via the term q_{vol} in Eq. (2.9).

Since the mean free path of energy carriers in silicon dioxide is small compared to the dimensions of the domain, Fourier diffusion is applied in the silicon dioxide layer. The BTE-based approach is applied to the silicon layer. This is one example of hierarchical modeling in which we apply a micro/nanoscale modeling strategy to one part of the domain, while applying conventional Fourier diffusion approach to another part of the domain. One can easily visualize much more complicated geometries and systems, spanning several lengths and timescales, which would require different modeling strategies to be concurrently applied in different regimes. In the thin silicon layer, MD can be utilized to obtain phonon relaxation times and dispersion curves. In addition, the frequency dependent phonon transmission rates at the interface between silicon/silicon dioxide can also be obtained from MD. These phonon parameters can then be utilized in the BTE framework.

In Fig. 9, we show the temperature contours in the domain represented by Fig. 8, obtained by using the full phonon dispersion BTE model discussed in section 2.3. The maximum temperature occurs in the hotspot. The silicon layer is isothermal in the y-direction as a consequence of the ballistic phonon transport in the silicon thin film layer. Qualitatively, the results look similar when Fourier diffusion or other BTE models are applied in the silicon layer. However, quantitatively there are significant differences in the hotspot temperature obtained from the different models. Table 1 shows the maximum temperature in the hotspot, obtained by applying different BTE models in the silicon layer. There is a large difference between the results from Fourier diffusion and the BTE-based models [46]. Fourier diffusion underpredicts the temperature rise in the hotspot since it cannot capture the non-equilibrium effects at these small scales. This is the reason why sub-continuum modeling approaches are essential.

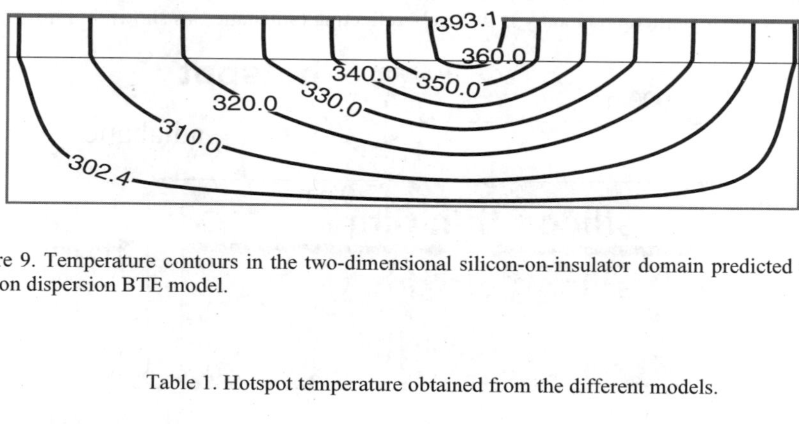

Figure 9. Temperature contours in the two-dimensional silicon-on-insulator domain predicted with the full phonon dispersion BTE model.

Table 1. Hotspot temperature obtained from the different models.

Model	**Hotspot temperature (K)**
Fourier Diffusion	320.7
Gray BTE	326.4
Semi-gray BTE	504.9
Full phonon dispersion BTE	393.1

7.1 IMPACT OF BOUNDARY CONDITIONS

We now examine the impact of changing the boundary conditions of the domain shown in Fig. 8. Typically, in practical device geometries, several transistors are placed side by side and there is a heat sink at the bottom. Most of the heat dissipation occurs from the bottom boundary. To capture this, we perform a study in which the side boundaries of the two-dimensional domain are adiabatic and there is a heat transfer coefficient (h) at the bottom boundary. Results obtained from applying Fourier diffusion and the full phonon dispersion BTE models in the silicon layer, with different values of the heat transfer coefficient, are shown in Table 2. T_{max} is the maximum temperature in the hotspot. $T_{b,avg}$ is the average temperature of the bottom boundary of the silicon dioxide layer. ΔT is the temperature difference between T_{max} and $T_{b,avg}$. For both the Fourier and full phonon dispersion BTE approaches, as the h value increases, the temperatures in the domain decrease. The value of ΔT remains nearly constant at approximately 72K and 123K for the Fourier and full phonon dispersion BTE models, respectively. It is expected that ΔT will remain constant for other values of h as well. Also, for the same h value, $T_{b,avg}$ from the two different models is the same. This is because all models have the same average heat flux entering the Si/SiO_2 boundary into the SiO_2 layer. Since the effective resistance of the SiO_2 layer is the same for all models, the bottom boundary temperature is the same.

Table 2. Temperatures in Fig. 8 domain with adiabatic side boundaries and a heat transfer coefficient at the bottom boundary.

	Fourier diffusion			**Full phonon dispersion BTE**		
h (W/m^{2}K)	T_{max} (K)	$T_{b,avg}$ (K)	ΔT (K)	T_{max} (K)	$T_{b,avg}$ (K)	ΔT (K)
1.0x10^7	409.9	337.8	72.1	460.5	337.2	123.4
1.5x10^8	375.6	302.5	73.0	425.4	302.5	123.0

The results in Table 2 imply that there is a significant difference in the predictions of the maximum hotspot temperature from the two different models. Fourier diffusion underpredicts the maximum temperature rise in the hotspot. Basically, the h boundary condition merely changes the reference temperature at the boundaries. Since the model is linear, temperature differences with respect to this reference boundary temperature are problem-invariant. That is, the dimensionless temperature:

$$\theta = \frac{T - T_{boundary}}{q_{vol} d^2 / k} \tag{7.1}$$

is invariant. Here, q_{vol} is the heat generation rate, d is a length scale on the device and k is the silicon bulk thermal conductivity. By using a constant reference temperature at the boundary, we produce results that are not dependent on the details of the package structure or external flow conditions. These details only change the boundary temperature, while the θ field remains similar.

It can therefore be concluded that irrespective of the precise boundary conditions, it is important to consider sub-continuum modeling approaches in the silicon layer. Notice that applying Fourier diffusion in the silicon channel layer leads to a substantial underprediction of the maximum hotspot temperature.

8. Summary and Conclusions

This paper discusses the Boltzmann transport equation and molecular dynamics approaches for modeling thermal transport in silicon devices of dimensions typically encountered in modern microelectronics and next generation devices. Both modeling approaches are validated against experimental thermal conductivity data of bulk and thin-film silicon. One particular focus is to establish a strategy for coupling the two approaches. Relaxation times and phonon dispersion relations can be obtained from molecular dynamics, and these parameters can then be utilized in the Boltzmann transport equation framework. Computing relaxation times from molecular dynamics provides an alternative to deriving them from perturbation theory.

It is important to judiciously apply different modeling approaches based on the length and time scales under consideration. Different modeling approaches need to be concurrently applied to systems that span different time and length scales. The results also reveal that Fourier diffusion is inadequate for making thermal predictions in the silicon layer of a sub-micron silicon-on-insulator transistor. Under such conditions, it becomes important to employ sub-continuum modeling approaches. Overall, hierarchical and concurrent modeling approaches are becoming very important in thermal modeling of microelectronics.

9. Nomenclature

a	Lattice parameter, [nm].
C	Volumetric heat capacity, [J/m^3-K].
C_P	Propagating mode specific heat, [J/m^3-K].
C_R	Reservoir mode specific heat, [J/m^3-K].
D	Phonon density of states, [m^{-3}].
d	Length scale of the device, [m].
d_s	Thin film thickness, [m].
e'''	Volumetric energy density per unit frequency per unit solid angle, [Js/m^3-sr-rad].
e''	Volumetric energy density per unit solid angle, [J/m^3-sr].
e^0	Equilibrium volumetric energy density, [J/m^3].
e_0	Volumetric energy density of the optical mode, [J/m^3].
$\vec{e}_m$	Polarization vector of mode m, [nondimensional].
$\vec{e}_m^{\,*}$	Complex conjugate of polarization vector of mode m, [nondimensional].
E	Energy stored in an atom, [J].
$E_{m,t}$	Total energy of mode m, [J].
f_w	Phonon distribution function, [nondimensional].
f_w^0	Equilibrium phonon distribution function, [nondimensional].
$\vec{F}$	Force, [N].
h	Heat transfer coefficient [W/m^{2}K].
$\hbar$	Planck's constant divided by 2π, [1.05456x10^{-34}Js].
k	Thermal conductivity, [W/mK].
k_{MD}	Molecular dynamics' thermal conductivity, [W/mK].
k_B	Boltzmann constant, [1.3807x10^{-23} J/K].
K_n	Knudsen parameter = Λ/L [nondimensional].
L	Simulation length, [nm].
$\vec{J}$	Heat current, [J.m/s].
m	Mass of the atom, [kg].
M	Number of time steps.
N	Number of atoms in the ensemble.
N_{LA}	Number of frequency bands in LA branch.
N_{TA}	Number of frequency bands in TA branch.
N_{bands}	Total number of frequency bands (N_{LA}+N_{TA}+1).
N_u	Number of unit cells.
N_s	Number of heat current autocorrelation function averages.
q_{vol}	Volumetric heat generation, [W/m^3].
Q_m	Normal mode coordinate m, [kg$^{1/2}$-m].
$\dot{Q}_m$	Time derivate of the normal mode coordinate m, [kg$^{1/2}$-m/s].
$\vec{r}$	Position vector, [m].
$\vec{r}_{i,0}$	Equilibrium position of atom i, [m].
$\hat{s}$	Unit vector, [nondimensional].
t	Time, [s].
T	Temperature, [K].
T_L	Lattice temperature, [K].
T_{MD}	Temperature of the molecular

	dynamics simulation, [K].
T_{max}	Maximum temperature in the hotspot, [K].
T_{ref}	Arbitrary reference temperature, [K].
$\vec{u}_i(t)$	Relative displacement of atom i at time t from its equilibrium position, [m].
V	Ensemble volume, [m^3].
$\vec{v}_i$	Velocity of atom, [m/s].

Greek:

Δt	Simulation time step, [s].
ΔT	Temperature variation, [K].
γ	Band averaged inverse relaxation time, [1/s].
$\vec{\kappa}$	Wave vector, [nondimensional].
λ	Phonon wave length [m].
Λ	Phonon mean free path [m].
τ	Relaxation time, [s].
τ_M	Integration time, [s].
θ	Dimensionless temperature, [K].
θ_D	Debye temperature, [K].
v	Phonon group velocity, [m/s].
v	Mode polarization in the MD context, [nondimensional].
w	Phonon's frequency, [rad/s or Hz].

Superscripts:

0	Equilibrium condition.

Subscripts:

w	Frequency dependence.
P	Propagating mode.
R	Reservoir mode.
L	Lattice.
LA	Longitudinal acoustic.
oj	Describes the interaction between the optical and the jth band of the acoustic mode.
O	Optical mode.
ii	Describes the interaction of band i with itself.
ij	Describes the interaction of band i with band j.
i	i^{th} band in the BTE context.
j	j^{th} band in the BTE context.
i	Atom i in the MD context.
TA	Transversal acoustic.

Acknowledgments. The authors gratefully acknowledge the funding of the National Science Foundation grant CTS-0103082 and the Pennsylvania Infrastructure Technology Alliance (PITA), a partnership of Carnegie Mellon, Lehigh University and the Commonwealth of Pennsylvania's Department of Community and Economic Development (DCED). Most of the MD computations were performed on the National Science Foundation Terascale Computing System at the Pittsburgh Supercomputing Center.

10. References

1. International Technology Roadmap for Semiconductors, http://public.itrs.net/.
2. Asheghi, M., Y.K. Leung, S.S. Wong, and K.E. Goodson, *Phonon-Boundary Scattering in Thin Silicon Layers.* Applied Physics Letters, 1997. **71**(13): p. 1798-1800.
3. Asheghi, M., M.N. Touzelbaev, K.E. Goodson, Y.K. Leung, and S.S. Wong, *Temperature-Dependent Thermal Conductivity of Single-Crystal Silicon Layers in SOI Substrates.* Journal of Heat Transfer, 1998. **120**: p. 30-36.
4. Chen, G., *Micro and Nanoscale Thermal Phenomena in Photonic Devices*, in *Annual Review of Heat Transfer*. 1996. p. 1-57.
5. Chen, G., *Nanoscale Heat Transfer and Information Technology.* Proceedings of 2003 Rohsenow Symposium on Future Trends of Heat Transfer, MIT, Cambridge, MA, 2003.
6. Dresselhaus, M.S., *Nanostructures and Energy Conversion.* Proceedings of 2003 Rohsenow Symposium on Future Trends of Heat Transfer, MIT, Cambridge, MA, 2003.
7. Volz, S.G. and G. Chen, *Molecular Dynamics Simulation of Thermal Conductivity of Silicon Nanowires.* Applied Physics Letters, 1999. **75**(14): p. 2056-2058.
8. Li, D., Y. Wu, P. Kim, P. Yang, and A. Majumdar, *Thermal Conductivity of Individual Silicon Nanowires.* Applied Physics Letters, 2003. **83**(14): p. 2394-2396.
9. Li, D., Y. Wu, F. Rong, P. Yang, and A. Majumdar, *Thermal Conductivity of Si/SiGe Superlattice Nanowires.* Applied Physics Letters, 2003. **83**(15): p. 3186-3188.

10. Volz, S., J.B. Saulnier, G. Chen, and P. Beauchamp, *Computation of Thermal Conductivity of Si/Ge Superlattice by Molecular Dynamics Technique.* Microelectronics Journal, 2000. **31**(9-10): p. 815-819.
11. Daly, B.C. and H.J. Maris, *Calculation of the Thermal Conductivity of Superlattices by Molecular Dynamics Simulation.* Physica B, 2002. **316-317**: p. 247-249.
12. Abramsom, A.R., C.-L. Tien, and A. Majumdar, *Interface and Strain Effects on the Thermal Conductivity of Heterostructures: A Molecular Dynamics Study.* Journal of Heat Transfer, 2002. **124**: p. 963-970.
13. Hone, J., M.C. Llaguno, M.J. Biercuk, A.T. Johnson, B. Batlogg, Z. Benes, and J.E. Fischer, *Thermal Properties of Carbon Nanotubes and Nanotube-based Materials.* Applied Physics A, 2002. **A74**(3): p. 339-43.
14. Cahill, D.G., K.E. Goodson, and A. Majumdar, *Thermometry and Thermal Transport in Micro/Nanoscale Solid-State Devices and Structures.* ASME Journal of Heat Transfer, 2002. **124**: p. 223-241.
15. Cahill, D.G., W.K. Ford, K.E. Goodson, G.D. Mahan, A. Majumdar, H.J. Maris, R. Merlin, and S.R. Phillpot, *Nanoscale Thermal Transport.* Journal of Applied Physics, 2003. **93**(2): p. 793-818.
16. Shi, L., D. Li, C. Yu, W. Jang, D. Kim, Z. Yao, P. Kim, and A. Majumdar, *Measuring Thermal and Thermoelectric Properties of One-dimensional Nanostructures Using a Microfabricated Device.* ASME Journal of Heat Transfer, 2003. **125**: p. 881-888.
17. Berber, S., Y.K. Kwon, and D. Tornanek, *Unusually High Thermal Conductivity of Carbon Nanotubes.* Physical Review Letters, 2000. **84**: p. 4613.
18. Amon, C.H. *Advances in Computational Modeling of Nano-scale Heat Transfer.* in *12th International Heat Transfer Conference*. 2002. Grenoble, France.
19. Ashcroft, N.W. and N.D. Mermin, *Solid State Physics*. 1976: Saunders College Publishers.
20. Kittel, C., *Introduction to Solid State Physics*. 1996, New York, NY: John Wiley.
21. Flik, M.I., B.I. Choi, and K.E. Goodson, *Heat Transfer Regimes in Microstructures.* ASME Journal of Heat Transfer, 1992. **114**: p. 666-674.
22. Majumdar, A., ed. *Microscale Energy Transport in Solids*. Microscale Energy Transport, ed. C.-L. Tien, A. Majumdar, and F.M. Gerner. 1998, Taylor and Francis: Washington DC. 1-94.
23. Chen, G., *Phonon Wave Heat Conduction in Thin Films and Superlattices.* Journal of Heat Transfer, 1999. **121**: p. 945-953.
24. Che, J., T. Çagin, W. Deng, and W.A. Goddard, *Thermal Conductivity of Diamond and Related Materials from Molecular Dynamics Simulations.* Journal of Chemical Physics, 2000. **113**(16): p. 6888-6900.
25. Ciraci, S., A. Buldum, and I.P. Batra, *Quantum Effects in Electrical and Thermal Transport Through Nanowires.* Journal of Physics: Condensed Matter, 2001. **13**: p. R537-R568.
26. Tien, C.L. and G. Chen, *Challenges in Microscale Conductive and Radiative Heat Transfer.* Journal of Heat Transfer, 1994. **116**: p. 799-807.
27. Mahan, G.D., *Many-Particle Physics*. 2000, Dordrecht: Kluwer-Plenum.
28. Klemens, P.G., *Thermal Conductivity and Lattice Vibrational Modes*, in *Solid State Physics*, F. Seitz and D. Thurnbull, Editors. 1958, Academic Press, New York. p. 1-98.
29. Klemens, P.G., *Theory of Thermal Conductivity of Solids*, in *Thermal Conductivity*, R.P. Tye, Editor. 1969, Academic Press: London. p. 1-68.
30. Narumanchi, S.V.J., J.Y. Murthy, and C.H. Amon. *Boltzmann Transport Equation-based Thermal Modeling Approaches for Microelectronics*. in *2nd International Thermal Sciences Seminar*. 2004. Bled, Slovenia.
31. Balandin, A., and Wang, K.L., *Significant Decrease of the Lattice Thermal Conductivity due to Phonon Confinement in a Free-standing Semiconductor Quantum Well.* Physical Review B, 1998. **58**(3): p. 1544-1549.
32. McGaughey, A.J.H., M. Kaviany, and J.D. Chung. *Integration of Molecular Dynamics and Boltzmann Transport Equation in Phonon Thermal Conductivity Analysis*. in *ASME International Mechanical Engineering Congress and Exposition IMECE2003-41899*. 2003. Washington, DC.
33. Majumdar, A., *Microscale Heat Conduction in Dielectric Thin Films.* ASME Journal of Heat Transfer, 1993. **115**: p. 7-16.

34. Joshi, A.A. and A. Majumdar, *Transient Ballistic and Diffusive Phonon Transport in Thin Films.* Journal of Applied Physics, 1993. **74**(1): p. 31-39.
35. Chen, G., *Size and Interface Effects on Thermal Conductivity of Superlattices and Periodic Thin-Film Structures.* Journal of Heat Transfer, 1997. **119**: p. 220-229.
36. Chen, G., *Thermal Conductivity and Ballistic-phonon Transport in the Cross-plane Direction of Superlattices.* Physical Review B, 1998. **57**(23): p. 14958-14973.
37. Goodson, K.E., *Thermal Conduction in Nonhomogeneous CVD Diamond Layers in Electronic Microstructures.* ASME Journal of Heat Transfer, 1996. **118**: p. 279-286.
38. Goodson, K.E., Y.S. Ju, and M. Asheghi, *Thermal Phenomena in Semiconductor Devices and Interconnects, in Microscale Energy Transport*, ed. C.L. Tien, A. Majumdar, and F.M. Gerner. 1998: Taylor & Francis.
39. Narumanchi, S.V.J., J.Y. Murthy, and C.H. Amon, *Simulation of Unsteady Small Heat Source Effects in Sub-micron Heat Conduction.* ASME Journal of Heat Transfer, 2003. **125**(5): p. 896-903.
40. Holland, M.G., *Analysis of Lattice Thermal Conductivity.* Physical Review, 1963. **132**(6): p. 2461-2471.
41. Ju, Y.S., *Microscale Heat Conduction in Integrated Circuits and their Constituent Films.* Ph.D. thesis, Stanford University, 1999.
42. Sverdrup, P.G., *Simulation and Thermometry of Sub-Continuum Heat Transport in Semiconductor Devices, Ph.D. Thesis.* 2000, Stanford University.
43. Sverdrup, P.G., Y.S. Ju, and K.E. Goodson, *Sub-continuum Simulations of Heat Conduction in Silicon-on-insulator Transistors.* ASME Journal of Heat Transfer, 2001. **123**: p. 130-137.
44. Narumanchi, S.V.J., J.Y. Murthy, and C.H. Amon. *Computations of Sub-micron Heat Transport in Silicon Accounting for Phonon Dispersion.* in *ASME Summer Transfer Conference, HT2003-47490.* 2003. Las Vegas, NV.
45. Mazumder, S., and Majumdar, A., *Monte Carlo Study of Phonon Transport in Solid Thin Films Including Dispersion and Polarization.* ASME Journal of Heat Transfer, 2001. **123**: p. 749-759.
46. Narumanchi, S.V.J., J.Y. Murthy, and C.H. Amon. *Simulations of Heat Conduction in Sub-micron Silicon-on-insulator Transistors Accounting for Phonon Dispersion and Polarization.* in *ASME International Mechanical Engineering Congress and Exposition, IMECE 2003-42447.* 2003. Washington, DC.
47. Narumanchi, S.V.J., J.Y. Murthy, and C.H. Amon, *Heat Transport During Transient Electrostatic Discharge Events in a Sub-micron Transistor.* Paper No. HT-FED2004-56252, Proceedings - ASME Heat Transfer/Fluids Engineering Summer Conference, Charlotte, North Carolina, 2004.
48. Klemens, P.G., *The Thermal Conductivity of Dielectric Solids at Low Temperatures.* Proceedings of the Royal Society of London, Series A, 1951. **208**(1092): p. 108-133.
49. Klemens, P.G., *The Scattering of Low Frequency Lattice Waves by Lattice Imperfections.* Proceedings of the Physical Society of London, 1955. **A68**: p. 1113-1128.
50. Han, Y.-J. and P.G. Klemens, *Anharmonic Thermal Resistivity of Dielectric Crystals at Low Temperatures.* Physical Review B, 1993. **48**: p. 6033-6042.
51. Zou, J., and Balandin, A., *Phonon Heat Conduction in a Semiconductor Nanowire.* Journal of Applied Physics, 2001. **89**(5): p. 2932-2938.
52. Narumanchi, S.V.J., J.Y. Murthy, and C.H. Amon, *Sub-micron Heat Transport Model in Silicon Accounting for Phonon Dispersion and Polarization.* ASME J. Heat Transfer,, 2004(in press).
53. Ju, Y.S. and K.E. Goodson, *Phonon Scattering in Silicon Thin Films with Thickness of Order 100 nm.* Applied Physics Letters, 1999. **74**(20): p. 3305-3307.
54. Asheghi, M., Kurabayashi, K., Kasnavi, R., and Goodson, K.E., *Thermal Conduction in Doped Single-crystal Silicon Films.* Journal of Applied Physics, 2002. **91**(8): p. 5079-5088.
55. Schelling, P.K., S.R. Phillpot, and P. Keblinski, *Comparison of Atomic-Level Simulation Methods for Computing Thermal Conductivity.* Physical Review B, 2002. **65**: p. 144306/1-12.
56. Lukes, J.R., D.Y. Li, X.-G. Liang, and C.-L. Tien, *Molecular Dynamics Study of Solid Thin-Film Thermal Conductivity.* Journal of Heat Transfer, 2000. **122**: p. 536-543.
57. Volz, S.G. and G. Chen, *Lattice Dynamic Simulation of Silicon Thermal Conductivity.* Physica B, 1999. **263-264**: p. 709-712.
58. Volz, S.G. and G. Chen, *Molecular-Dynamics Simulation of Thermal Conductivity of Silicon Crystals.* Physical Review B, 2000. **61**(4): p. 2651-2656.

59. Li, J., L.J. Porter, and S. Yip, *Atomistic Modeling of Finite-temperature Properties of Crystalline B-SiC. II.Thermal Conductivity and Effects of Point Defects.* Journal of Nuclear Materials, 1998. **255**: p. 139-152.
60. Gomes, C.J., M. Madrid, and C.H. Amon. *Parallel Molecular Dynamics Code Validation Through Bulk Silicon Thermal Conductivity Calculations.* in *Proceedings of the 2003 ASME International Mechanical Engineering Congress and Exposition, IMECE 2003-42352.* 2003. Washington, DC.
61. Lee, Y.H., R. Biswas, C.M. Soukoulis, C.Z. Wang, C.T. Chan, and K.M. Ho, *Molecular-Dynamics Simulation of Thermal Conductivity in Amorphous Silicon.* Physical Review B, 1991. **43**(8): p. 6573-6580.
62. Ding, K. and H.C. Andersen, *Molecular-Dynamics Simulation of Amorphous Germanium.* Physical Review B, 1986. **34**(10): p. 6987-6991.
63. McQuarrie, D.A., *Statistical Mechanics.* 1976, New York: Harper Collins. 641.
64. Allen, M.P. and D.J. Tildesley, *Computer Simulation of Liquids.* 1987, Oxford, UK: Clarendon Press. 385.
65. Hardy, R.J., *Energy-Flux Operator for a Lattice.* Physical Review, 1963. **132**(1): p. 168-177.
66. Kubo, R., *The Fluctuation-Dissipation Theorem.* Report of Progress in Physics, 1966. **29**: p. 255-284.
67. Frenkel, D. and B. Smit, *Understanding Molecular Simulation From Algorithms to Applications.* 2nd ed. 2001: Academic Press. p. 638.
68. Flubacher, P., A.J. Leadbetter, and J.A. Morrison, *Philosophical Magazine.* 1959. **4**: p. 273.
69. Hultgren, R., *Selected Values of Thermodynamic Properties of the Elements.* 1973, Metals Park: ASM.
70. Desai, P.D., *Thermodynamic Properties of Iron and Silicon.* Journal of Physical and Chemical Reference Data, 1986. **15**(3): p. 967-983.
71. Ladd, A., B. Moran, and W.G. Hoover, *Lattice Thermal Conductivity: A Comparison of Molecular Dynamics and Anharmonic Lattice Dynamics.* Physical Review B, 1986. **34**: p. 5058-5064.
72. McGaughey, A.J. and M. Kaviany, *Quantitative Validation of the Boltzmann Transport Equation Phonon Thermal Conductivity Model Under the Single-Mode Relaxation Time Approximation.* Physical Review B, 2004. **69**(9): p. 094303(1)-094303(11).
73. Dove, M.T., *Introduction to Lattice Dynamics.* Cambridge Topics in Mineral Physics and Chemistry. 1993: Cambridge University Press.
74. Ho, C.Y., R.W. Powell, and P.E. Liley, *Thermal Conductivity of the Elements.* Journal of Physical and Chemical Reference Data, 1972. **1**(2): p. 279-421.
75. Stillinger, F.H. and T.A. Weber, *Computer Simulation of Local Order in Condensed Phases of Silicon.* Physical Review B, 1985. **31**(8): p. 5262-5271.
76. Srivastava, G.P., *The Physics of Phonons*, ed. A. Hilger. 1990, New York.
77. Cahill, D.G., *Heat Transport in Dielectric Thin Films and at Solid-Solid Interfaces in: Microscale Energy Transport.* Series in Chemical and Mechanical Engineering, ed. C.L. Tien, A. Majumdar, and F.M. Gerner. 1998, Washington, DC: Taylor & Francis.
78. Chantrenne, P. and J.-L. Barrat, *Finite Size Effects in Determination of Thermal Conductivities: Comparing Molecular Dynamics Results with Simple Models.* Journal of Heat Transfer, 2004. **126**: p. 577-585.
79. Gomes, C.J., M. Madrid, and C.H. Amon. *Thin Film In-Plane Thermal Conductivity Dependence on Molecular Dynamics Surface Boundary Conditions.* in *Proceedings of the 2004 ASME International Mechanical Engineering Congress and Exposition, IMECE 2004-62264.* 2004. Anaheim, CA.
80. Liu, W. and M. Asheghi, *Phonon-Boundary Scattering in Ultrathin Single-Crystal Silicon Layers.* Applied Physics Letters, 2004. **84**(19): p. 3819-3821.

EVAPORATIVE HEAT TRANSFER ON HORIZONTAL POROUS TUBE

L. VASILIEV, A. ZHURAVLYOV, A. SHAPOVALOV
Porous Media Laboratory – Luikov Heat & Mass Transfer Institute
Minsk, Belarus

1. Introduction

Mini loop thermosyphons and heat pipes now are used for electronic components cooling, space systems of thermal control, transport, avionic and refrigeration. In some cases the temperature drop between the evaporator and condenser of a cooling device is a critical for temperature sensitive semiconductors and refrigerators and it is important to reduce its thermal resistance. Modern mini heat exchangers should be not only high effective but compact – such is the trend of late. Reduction of equipment size and mass leads to decrease of cost. Problem of miniaturization is important for the application in power engineering, refrigeration and cryogenic industries, electronics. Due to coupling phenomena between momentum and energy transport theoretical analysis of the loop thermosyphon is very complicate, therefore it is necessary that these problems be solved by experimental investigation before applying the loop thermosyphon or heat pipe as a heat exchanger. The problem of a new generation electronic chips cooling with its size reduction and increasing functionality is related with the use of confined and unconfined boiling and evaporation of some hydrocarbons ,for example R290 (propane) at low and moderate heat fluxes < 100 kW/m^2. Such type of cooling ought to be one of the most efficient ways to provide the necessary thermal control of chips. The experimental investigation of propane pool boiling and evaporation heat transfer on single horizontal tube (smooth and with porous coating) was performed for the reasons of analysis of its cooling efficiency. Many analytical and experimental investigations have been performed to provide a better understanding of liquid and two-phase heat transfer at microscale, which is very important in microdevice development and design. Evaporative and boiling heat transfer has been widely applied in industry. Different capillary structures were applied (mini-fins, sprayed and sintered metal porous structures) on a heat loaded element (evaporator) of thermosyphon and heat pipe. However, these studies haven't yet to lead to a general conclusion. Recently some experiments were carried out in Belarus (Luikov heat and mass transfer institute) at heat flux range from 10^2 to 10^5 W/m^2. The level of working liquid was varied to study the heat transfer between heat dissipating element (tube) and liquid. A set of experiments on two-phase heat transfer on tubes with porous layers obtained by spraying (stainless steel) or sintering of metal powder particles (copper) shows promising results with the point of view of heat transfer enhancement. The porosity of sprayed stainless steel coatings was varied from 4 to 20 %, coating thickness 0.1-0.3 mm. The open type porosity of the copper sintered powder wick was 55 %; the thickness of the coating was 0.1-0.5 mm. The data obtained on a flooded and partially flooded horizontal tube in the pool, and confined space with a sintered copper powder capillary-porous coating are testify the phenomena of micro heat pipe inside a porous structure. In series of experiments a horizontal tube with porous heat

S. Kakaç et al. (eds.), Microscale Heat Transfer, 401–412.

exchange surface was placed between two vertical restrictive plates or inside coaxial tube. A microscale effect took place inside porous body and a mini-scale effect is ensured due to flank mini-gaps or circular mini-gap in liquid pool.

Characteristics of heat exchangers depend on the design of evaporator essentially, that is why the microscale boiling and evaporating processes study is necessary. The complete definition of the channel size dimension in terms of hydraulic diameter D_h is proposed by Mehendale with co-authors [1]:

D_h=1-100 μm – micro heat exchanger
D_h=100-1 mm – meso heat exchanger
D_h=1-6 mm – compact heat exchanger
D_h>6 mm – conventional heat exchanger

Heat transfer at evaporation and boiling on surfaces covered by various kinds of porous layers is the subject of interest in other research centers of New Independent States (Moscow, Novosibirsk, Kiev, and Odessa). Comparison of published data show that the sintered powder porous coating is efficient, ensuring high heat characteristics, being firm and having a sufficiently good connection with a heat exchange surface.

2. Experiment

2.1 WORKING FLUID

Propane is beneficial working fluid due to its good thermodynamic properties, low cost, availability, compatibility with constructional materials, environmental friendliness. The latest circumstance is very important because of the ozone depleting by chlorfluorocarbons CFC and hydrochlorfluorocarbons HCFC. So, propane as other hydrocarbons is welcomed and used as working fluid in heat pumps for heating application in Europe [2, 3].

Some physical properties of propane at T_s=20°C are given in Table 1 [4].

Table 1: Propane properties at T_s=20°C

ρ_l (kg/m^3)	ρ_g (kg/m^3)	$\sigma \cdot 10^3$ (N/m^2)	p_s (Bar)	$p_r=p_s/p_{cr}$
502	18.2	7.35	8.4	0.197

Experimental results confirmed that propane (R290) is beneficial as a working fluid for low temperature heat pipes, thermosyphons, sorbers of chemical heat pumps, chillers of electronic components and for other evaporative heat transfer devices. The experimental data at the saturated condition T_s=20°C (p_s=8.4 bar, $p_r=p/p_{cr}$=0.197) are obtained in this series of experiments.

2.2 TEST APPARATUS AND EXPERIMENTAL PROCEDURE

The experimental set-up is shown on Figure 1. The main parts of the rig are test vessel, insulated chamber with thermo controlled liquid circuit, cooling machine (refrigerator), thermostats, condenser liquid loop, temperature control system, vacuum pump, liquid feed system. A heat to the heat exchange surface was supplied by the cartridge electric heater. The test sample 1 was placed horizontally in the boiling vessel 2. This chamber was made from stainless steel and has three windows for visual observing of the studied process. The test vessel is placed within the insulated thermal controlled box, 6 supplied by the controlled temperature liquid loop 7 to prevent heat exchange between vessel and environment. The experiments were performed in wide temperature range: from −10 to +60 °C. Temperature measurements were carried out by copper-constantan thermocouples in a steady-state regime, sensor signals were transferred to the measuring complex 11, and computer 3. The saturation temperature was measured by four thermocouples, the temperature difference between the heat exchange surface and liquid was measured by four differential thermocouples.

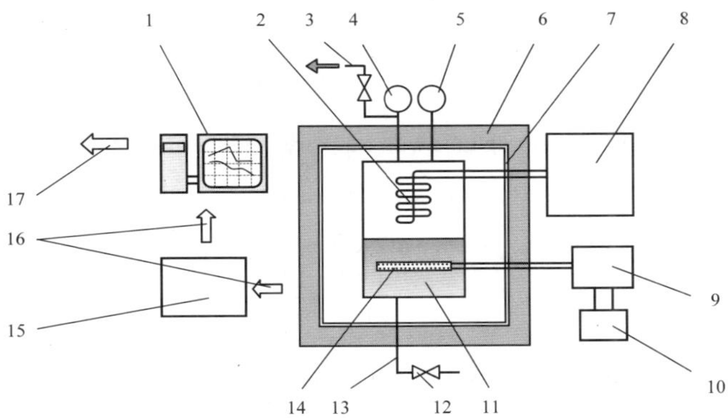

1 – computer, 2 – condenser, 3 – exhaust system, 4 – vacuometer, 5 – manometer, 6 – thermal controlled box, 7 – thermostated liquid loop, 8 – cooling machine, 9 – power supply, 10 – wattmeter, 11 – boiling vessel, 12 – valve, 13 – liquid feeding system, 14 – test sample, 15 – measuring complex, 16 – initial information, 17 – experimental results

Figure 1. Schematic diagram of the experimental set-up.

Evaporation heat transfer analysis was performed in a liquid pool or on a partly flooded horizontal tube (D=20 mm, L=100 mm). The level of liquid h was varied from 70 to 0 mm (Figure 2). The series of experiments was carried out with tested sample disposed between two vertical glass plates with side gaps about 2 mm (Figure 2, a), or inside of a glass tube so that the circular gap of the same size took place (Figure 2, b). Transparent material of the plates and glass cylinder allowed observing the two-phase heat transfer in confined space.

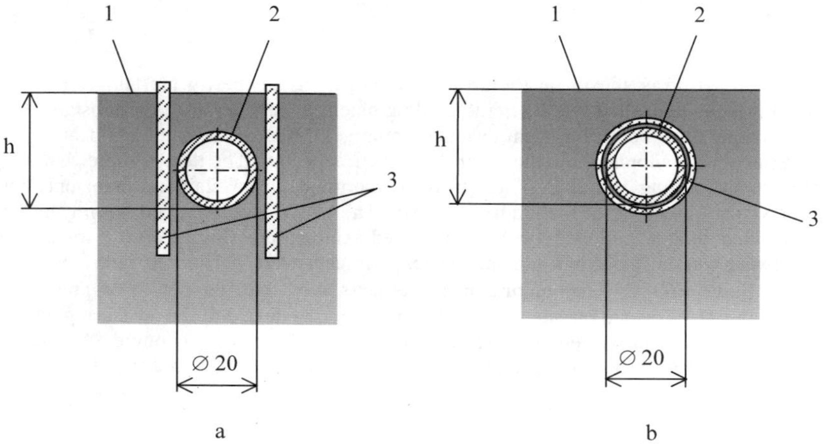

1 – liquid-vapor interface, 2 – tested sample, 3 – transparent vertical plates, 4 – glass coaxial cylinder

Figure 2. Cross-section of the tested sample.

The characteristics of a heat releasing tube with capillary – porous coating are given in Table 2.

Table 2: Characteristics of tested sample

Porosity (%)	Particle diameter (μm)	Mean particle dimension (μm)	Coating thickness (mm)
50-55	63-100	81.5	0.3

The sizes of flank and circular gaps were close to the capillary constant

$$\kappa = \left[2\sigma/(\rho_l - \rho_v)g^{1/2}\right] \tag{1}$$

In this case an influence of capillary forces on process is important. For propane at T=20 °C the capillary constant $\kappa \sim 1.8$ mm and the tested tube was installed in the working section with flank or circular mini-gaps equal to this dimension. The condenser above the liquid pool ensured the saturation temperature inside the chamber. To prevent heat transfer between test vessel and ambient medium the temperature inside a temperature-controlled chamber maintained equal to T_s. The electric heater inside the tube provided the constant heat input. Some auxiliary electric heaters guaranteed the constant saturation pressure, and liquid circuits ensured the level of saturation temperature by regulation of a cooling liquid flow rate through the condenser. A steady state heat transfer regime was necessary to obtain for all the measurements performed during the experiment. In order to verify the reliability of the method chosen or measuring, analysis of results and to compare the heat transfer intensity, a series of experiments on plain stainless steel and copper tubes [5-7] were carried out. Comparison of obtained data showed a good correlation with published results of the other authors.

3. Experimental Results

There were some stages of the boiling and evaporation heat transfer process investigation made in [7-10]. The first step was a study of pool boiling heat transfer on smooth surface. Then the process on a surface with metal gas sprayed porous coating was considered. The next step was a study of heat transfer at evaporation in sintered metal powder porous bodies in the liquid pool and on a partially flooded horizontal cylindrical tube. At last, the experiments with horizontal porous tube placed in confined space were realized. The results of experiments [7-10] testified, that the structure of gas-plasma sprayed coating due to closed pores inside is similar to results of heat transfer with boiling on a tube with microfins, so the heat transfer characteristics of the sprayed surface are similar to the data of P. Sokol, D. Gorenflo et al., obtained at propane boiling on tubes with the regular microgeometry of a Gewa-T-x type surface, [11-12] (T-shaped microfins with small depressions between fins to produce additional nucleation centres). The heat transfer intensity at boiling on surfaces with gas-plasma sprayed coating was 2.5-5 as higher as on a smooth tube [7, 8].

3.1 EVAPORATION HEAT TRANSFER ON PARTIALLY FLOODED CYLINDER

Heat transfer coefficients at vaporization in sintered powder porous bodies were up 6-8 times as much than at boiling on smooth surface [8-10] in the heat flux range up to q=100 kW/(m^2 K). This kind of porous coating is more efficient as compared with gas-sprayed one because of structure differences of these coatings. For better understanding of heat transfer mechanism of heat transfer at vaporization in sintered powder open type media the experiments at various levels of liquid were carried out, so that test sample was in liquid pool or partially flooded (Figure 3).

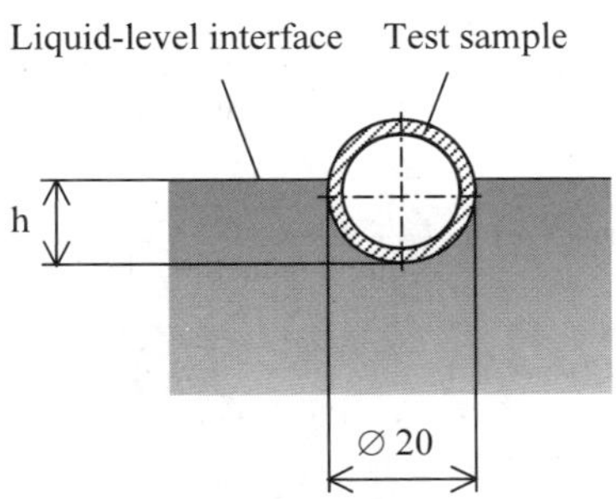

Figure 3. Cross-section of the partially flooded heat loaded tube.

The influence of the liquid level above the porous coating is important at h less than approximately 2 mm at low heat fluxes. In this case the heat transfer coefficients were higher than at a high liquid layer above the tested sample. The data obtained on flooded (h=70, 20 mm) and partially flooded (h=15, 10, 5 mm) porous cylinder are presented in Figure 4.

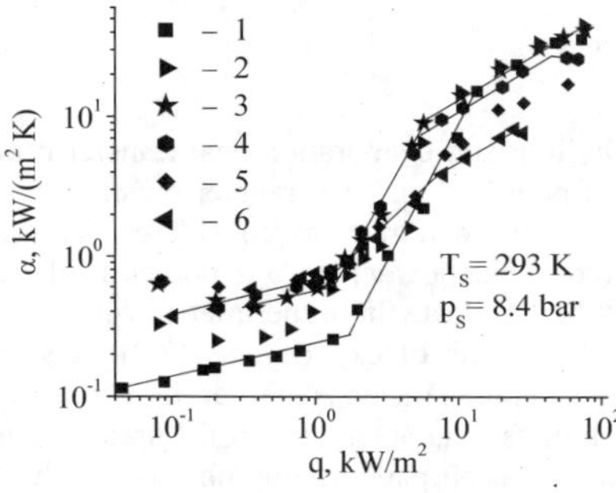

1 – h = 70 mm, 2 – h = 20 mm, 3 – h = 15 mm, 4 – h = 10 mm, 5 – h = 5 mm, 6 – h = 0 mm

Figure 4. Heat transfer intensity versus heat flow for propane (flooded and partially flooded tube).

The decrease of the liquid level above the tube by a quarter of diameter promotes the increase of the average heat transfer coefficient at low and moderate heat fluxes. The measuring of the temperature heads showed that it goes on due to surface superheat decrease in the unflooded part of a tube. It can be explained with help of following point model. There is not boiling but evaporative mechanism of heat transfer in a sintered powder porous media with the capillary transfer of the liquid from the pool to the zone of heat release.

3.2 MICRO HEAT PIPE EFFECT IN POROUS MEDIA

Sintered powder porous body can be considered as a system with micro- and macro-pores, (Figure 5). Micro-pores fulfill the function of capillary channels for transport of liquid to zones of vaporization. The macro pores represent the channels for vapor transfer. The vapor is generated on surfaces of meniscuses in orifices of micro-pores (Figure 6).

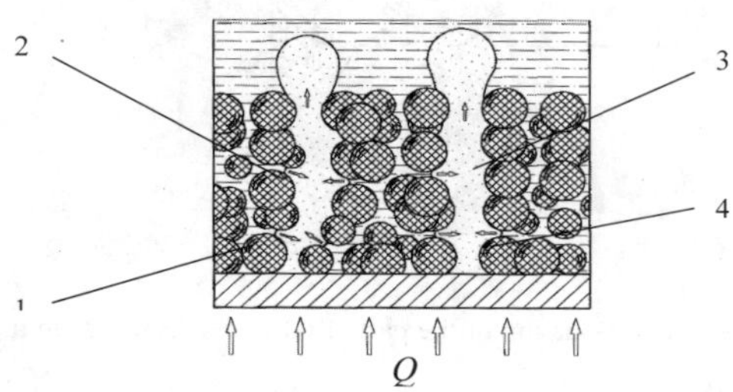

1 – micro-pore, 2 – meniscus, 3 – macro-pore, 4 – vapor

Figure 5. Cross-section of a powder porous coating

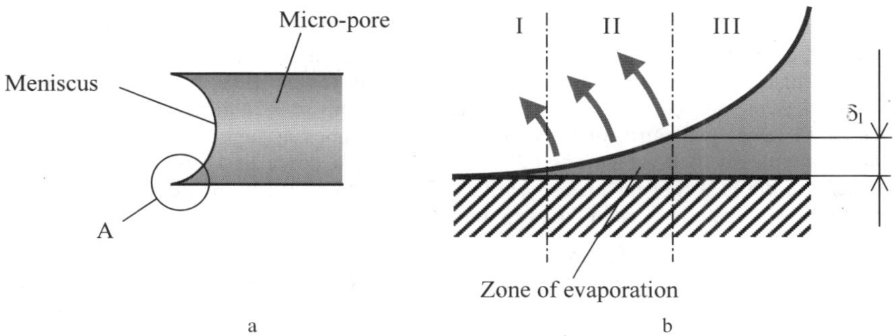

Figure 6. Zone of evaporation in orifice of micro-pore

The highest intensity of evaporation occurs in the zone II. The thickness of liquid film in the zone I is close to the size of molecular sorption film and there are no the favorable conditions for process of vaporization. In the zone III the liquid film is thick, so a thermal resistance is higher than in zone II. Thus the optimal conditions for intensive evaporation are in the zone II. There is great number of such zones over all porous surfaces, so the total area of evaporation is very large. Thus we have excellent condition for the intense heat transfer.

The process inside a sintered powder porous body is similar to the processes inside micro heat pipes (Figure 7) with zone of evaporation and condensation. If the tested sample is placed in liquid pool, we have micro heat pipe closed type (Figure 7, a), for partially flooded porous cylinder we have the surface with micro heat pipe open type (Figure 7, b).

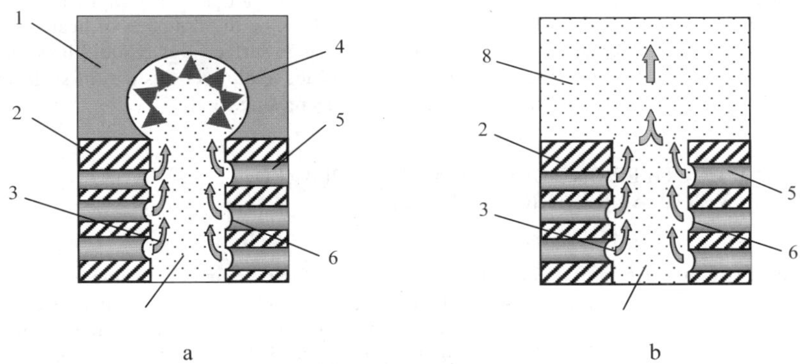

1 – liquid pool, 2 – porous body, 3 – rising vapor, 4 – zone of condensation,
5 – liquid filled micro-pore, 6 – meniscus, 7 – macro-pore, 8 – vapor space

Figure 7. Micro heat pipe effect in porous coating: a) closed type, b) open type

So, we have two different types of micro heat pipe systems inside the heat loaded porous coating. The first one is typical for the closed type micro heat pipe. The second is similar to the open type micro heat pipe system, depending on the heat flax density. A number of active centers of vaporization (meniscus of the evaporation) rise proportionally to heat flux. At the heat flux interval from 0.1 to 1.5 kW/m^2 the increase of heat transfer intensity up to 1.5 times was noticed, when the liquid covered an upper generatrix of a sample, and 2.5-3 times as high at h=15 mm (Fig. 3) to compare with completely flooded porous tube. Lowering of h down to 10 mm (a middle of tube diameter) decrease the heat transfer intensity at heat flux $q>(1.5\text{-}2)$ kW/m^2 , due to the insufficient liquid capillary flow to the meniscus of the evaporation.

There are two limitations for increasing of heat transfer intensity: hydrodynamic ability of porous coating to transport the liquid and finite number of potential centres of vaporization (micro-pores). On reaching the certain quantity of heat flux a heat transfer surface (meniscus) above liquid-level doesn't supplied a sufficient amount of liquid, a "dry spot" appears and then "dry spots" are spreading to all the surface. At some liquid-levels and heat fluxes heat transfer intensity for the opened tube surface of the evaporation is higher than for the tube immersed in the pool.

The increase of the porous layer thickness from 0.2 to 0.5 mm favoures the heat transfer intensification. It is true for all liquid levels and in the heat flux range (from 0.1 kW/m^2 up to 100 kW/m^2). But there is an optimum for the porous coating thickness. As a heat flux increases the tube with coating thickness 0.5 mm has better characteristics to compare with the coating thickness 0.8 mm.

The heat transfer intensity depends on the curvature of meniscuses in orifices of micro-pores

$$K = \frac{1}{\sigma}\left[p_v - p_l + \rho_v^2 v^2\left(\frac{1}{\rho_v} - \frac{1}{\rho_l}\right)\right] \tag{2}$$

While the curvature K doesn't exceed some value K_{max} it is enough of the capillary ability of porous coating to transport of liquid, when curvature rises to $K>K_{max}$ the drainage of heated surface begins. On reaching the certain quantity of heat flux a heat exchange surface above liquid-level doesn't get sufficient amount of liquid, "dry spots" appear and then spreads to all over this part of surface. The liquid level is lower; the maximum heat transfer coefficient is decreasing.

3.3 THE HEAT TRANSFER ON HORIZONTAL TUBE PLACED BETWEEN VERTICAL RESTRICTIVE PLATES

To study the influence of space limit on heat transfer the series of experiments with side vertical plates were carried out , Figure 2, a. The gaps between tube flank generatrixes and vertical walls were close to capillary constant ~1.8 mm. The Figure 8 shows the comparison of heat transfer characteristics in a presence and absence of vertical plates. A presence of vertical plates is limiting the space near the flank generatrixes and is changing the heat transfer characteristics of flooded sample up to heat load ~20 kW/m^2. An increase of heat transfer coefficients was registered at lowering of a liquid layer height down to h=20 mm (the interface "liquid-vapor" touched on the upper generatrix). Under this condition the temperature heads $\Delta T=T_w-T_l$ corresponding to first vapor bubbles arrival was less than for the case without vertical plates. The experimental data testifies that for the case with horizontal tube between the vertical plates the average superheats are decreasing with the temperature T_w decrease on the flank generatrixes.

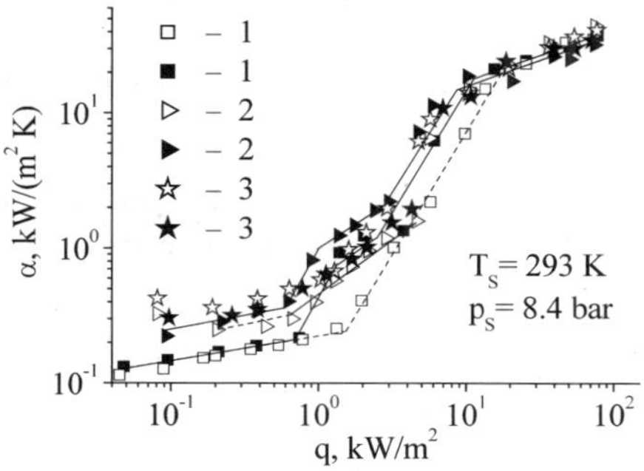

1 – H=70 mm, 2 – h=20 mm, 3 – h=15 mm
Dark symbols – the sample is placed between two vertical plates

Figure 8. Heat transfer coefficient versus heat flux q.

The possible reason of heat transfer intensification is assumed to be a liquid turbulence caused by vapor bubbles rise through the flank gaps. As a result the liquid at the saturation temperature occupies the place near the sample surface.

3.4 EVAPORATION HEAT TRANSFER IN COAXIAL MINIGAP

It is possible to combine microscale and mini-scale effects if to dispose the heat exchange cylindrical porous surface inside the tube coaxially with circular mini-gap (Figure 2, b). Experimental results (Figure 9) show, that such combination is favorable for the evaporation heat transfer.

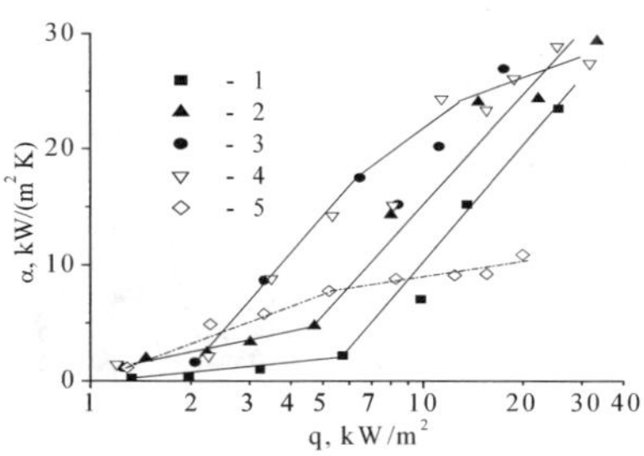

1 – pool liquid, 2-5 –circle minigap: 2 – h=75 mm, 3 – h=20 mm, 4 – h=15 mm, 5 – h=10 mm

Figure 9. Heat transfer intensification at evaporation in circular mini-gap

The availability of circular mini-gap significantly promotes the increase of heat transfer coefficient (up to 2.5-3 times as high) at low and moderate heat fluxes as compared with process in liquid pool. Figure 10 shows the comparison of heat transfer coefficients at height of liquid level h=75 mm for various conditions: 1) porous sample is in liquid pool, 2) porous sample is placed between vertical plates with flank mini-gaps, 3) the same sample is inside coaxial tube with circular mini-gap and 4) the plain tube is in liquid pool.

The sample with porous coating has advantage to compare with plain tube. The best results were obtained with heat transfer in circular mini-channel. To understand these phenomena the visual observing and photographing of various stage of process were curried out. The photos are presented in Figure 11.

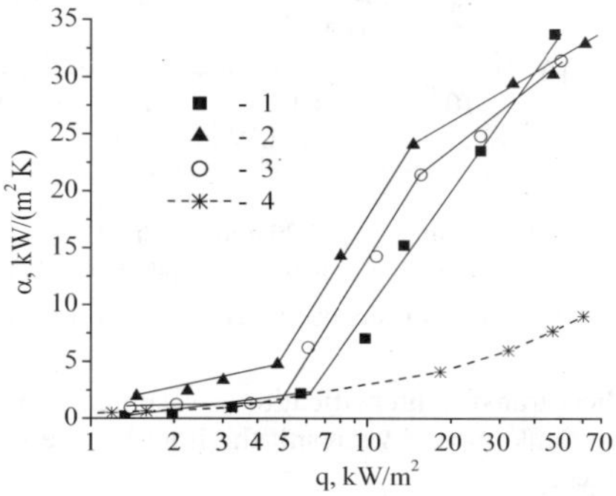

Porous tube: 1 – liquid pool, 2 – circular mini-gap, 3 – flank mini-gaps; plain tube – 4

Figure 10. Comparison of heat transfer coefficients (h=75 mm)

$q=0.25$ kW/m^2 $\qquad$ $q=34$ kW/m^2

Figure 11. Visualization of process

Visual observing of vaporization process testify that vapour bubbles movement in the circular mini-channel has a complicated character: bubbles move not only perpendicularly to heat exchange surface but also along the axe of tube toward the outlets to liquid pool. The saturated heat transfer coefficient was found to be dependent on the interface liquid/vapour position in the pool, the most high, when there was a partial flooding of the tube, Figure 7. The average heat transfer coefficient for the sample inserted inside the glass tube was higher to compare with pool and confined space ensured by vertical plates. Thus the two-phase flow through the mini-channel takes place. This is the reason of heat transfer intensification.

4. Conclusions

Porous covering with micro heat pipe phenomena stimulates the evaporative heat transfer near 8-10 times to compare with pool boiling heat transfer on the smooth horizontal tube.

Reducing the size of cooling system we increase its efficiency, improve system performance by adding micro scale function (microporous heat pipe effect) to macro scale engineering application.

At vaporization occurring in minichannels (coaxial gap near 1.8 mm) two-phase flow takes place without additional power supplying (no mechanical pumping).

Micro heat pipe effect + two-phase forced convection in the coaxial gap (porous tube inside the glass tube) increase the heat transfer 2-2.5 times as higher as in liquid pool at low and moderate heat fluxes.

NOMENCLATURE

D,d Diameter
g, Gravitation constant
h, Height of liquid level
K, Curvature
L, Length
p, Pressure
q, Heat flux
T, Temperature

Greek symbols
α, Heat transfer coefficient
δ, Thickness
κ, Capillary constant
ρ, Density
σ, Surface tension coefficient

Subscripts
cr, Critical
g, Gas
h, Hydraulic
l, Liquid
r, Reduced
s, Saturation

REFERENCES

1. Mehendale, S.S., Jacobi, A.M., and Shan, P.K., (2000) Fluid Flow and Heat Transfer at Micro- and Meso-Scales With Application to Heat Exchanger Design, *Applied Mechanics Reviews*, Vol. 7 (53), pp.175-193.
2. Informative Fact Sheet: Hydrocarbons as Refrigerants in Residential Heat Pumps and Air-Conditioners, (2002) Int. Energy Agency's Heat Pump Center.
3. Jung, D., Lee, H., Bae, D., and Oho, S., (2004) Nucleate Boiling Heat Transfer Coefficients of Flammable Refrigerants, *International Journal of Refrigeration*, Vol. 27 (4), pp. 409-414.
4. Heat Exchanger Design Handbook, (1983) Vol. 2, Part 5. Hemisphere Publishing Corporation.
5. Klimenko, A.P., and Kozitskey, V.I., (1967) Experimental Heat Transfer Investigation at Propane Boiling, *Oil and Gas Industry (Neftyanaia i Gazovaia Promishlennost)*, No. 1, pp. 40-43. – In Russian.
6. Gorenflo, D., Sokol, P., and Caplanis, S., (1990) Pool Boiling Heat Transfer from Plain Tubes to Various Hydrocarbons, *International Journal of Refrigeration,* Vol. 13, pp. 286-292.
7. Vasiliev, L.L., Khrolenok, V.V., and Zhuravlyov, A.S., (1998) Intensification of Heat Transfer at Propane Pool Boiling on Single Horizontal Tubes, *Revue Générale de Thermique,* Vol. 37, (11), pp. 962-967.
8. Vasiliev, L.L., Zhuravlyov, A.S., Novikov, M.N., and Vasiliev, L.L. Jr., (2001) Heat Transfer with Propane Evaporation from a Porous Wick of Heat Pipe, *Journal of Porous Media*, Vol. 4 (2), pp. 103-111.
9. Vasiliev, L.L., Zhuravlyov, A.S., Novikov, M.N., and Vasiliev L.L. Jr., (2000) Experimental Investigation of Propane Boiling in Porous Structures, *Proceedings of The IV Minsk International Seminar "Heat Pipes, Heat Pumps, Refrigerators", Minsk, Belarus,* pp. 245-255.
10. Vasiliev, L.L., Zhuravlyov, A.S., Novikov, M.N., Shapovalov, A.V., and Litvinenko, V.V. (2003) Heat Transfer at Propane Pool Boiling and Evaporation in Capillary-Porous Evaporators *Proceedings of The 4th Baltic Heat Transfer Conference "Advances in Heat Transfer Engineering", Kaunas, Lithuania,* pp. 739-746.
11. Sokol, P., Blein, P., Gorenflo, D., Rott, W., and Schömann, H., (1990) Pool Boiling Heat Transfer from Plain and Finned Tubes to Propane and Propilene, *Proceedings of 9th International Heat Transfer Conference, Jerusalem, Israel,* Vol. 2, pp. 75-80.
12. Gorenflo, D., Blein, P., Rott, W., et al., (1989) Pool Boiling Heat Transfer from GEWA-T-x Finned Tube to Propane and Propylene, *Eurotherm № 8. Advanced in Pool Boiling Heat Transfer: Proceedings of International Seminar, Paderborn, F.R.G.,* pp. 116-126.

MICRO AND MINIATURE HEAT PIPES

L.L. VASILIEV
Luikov Heat and Mass Transfer Institute, 220072,
P. Brovka 15, Minsk, Belarus

1.Introduction

Micro (MHP) and miniature heat pipes (mHP) are small scale devices that are used to cool microelectronic chips. The term "micro heat pipes" was suggested for the first time in 1984 by Cotter [1]. The hydraulic diameter of MHPs is on the order of 10 -500 μm, the hydraulic diameter of mHPs is on the order 2 - 4 mm. Actually new cooling techniques are being attempted to dissipate fluxes in electronic components in order of 100 up to 1000 W/cm^2. Besides electronic cooling, there are many other applications where MHPs may be useful. For example, MHPs are interesting to be used in implanted neural stimulators, sensors and pumps, electronic wrist watches, active transponders, self – powered temperature displays, temperature warning systems. MHPs are promising to cool and heat some biological micro objects. MHPs are used to eliminate hot spots and to reduce temperature gradients of electronic components. In this review paper as an example of this process some micro and miniature heat pipes (mHP) R&D, both for passive systems of electronics cooling and for use in refrigerating machines is presented with stringent cooling requirements posed by the electronics industry. The mHP is one of the promising technologies for the achievement of high local heat removal rate and uniform temperatures in computer chips. Flat heat pipes are successfully applied for the cooling of semiconductor elements. The flat heat pipe can be installed on a semiconductor element or a semiconductor substrate with the flat surface of the heat pipe parallel to the substrate. The heat sink can be the existing components of the notebook, from Electro-Magnetic Interference shielding under the key pad to metal structural components. High power mainframe, mini-mainframe, server and workstation chips may also employ heat pipe heat sinks. For example, high end chips dissipating up to 100 W are outside the capabilities of conventional sinks. Heat pipes used to transfer heat from the chip to fin stack enough to convert the heat to supplied air stream. A series of dual-in-line packages are in direct contact with heat pipes aligned in parallel.

High-performance mHPs were designed and manufactured in the Luikov Institute, Belarus, Figure 1 [1]. These mHPs have a sintered porous wafer with arteries. This wafer is sintered between two metal walls with triangular mini-grooves on its inner surface working as micro-heat pipes. Working fluid is water, methanol, and propane. Modern flat and cylindrical miniature heat pipes are used now to cool millions of notebook computers and processors in desktop PCs. Efficiency of processors and power dissipation of CPU and other electronic components increase, so we need to improve heat pipe parameters, particularly wick structure. A lot of mHPs with different wick structures (sintered powder, mesh structure, wire bundle, axial and radial micro grooves) is fabricated now by some leading companies. The mHPs are used often flat, or flattened (1-3 mm thick), 4-6 mm in width and 150-300 mm in length. Such mHPs are bonded to PCBs with epoxy and dual-in-line packages are soldered in place over heat pipes and thermally connected to heat pipes with thermal grease, or thermal epoxy. The mHPs require an effective wick design because they are made with 90 ^{0} bends to improve heat transfer from circuit boards that do not plug in. When cold plate side walls are used in a chassis, the 90 ^{0} bend permits the heat pipe transport the heat flow directly from the PCB to the side wall of the chassis. Some mHPs ensure nearly isothermal temperature field along its entire length with a temperature rise above the heat sink only

S. Kakaç et al. (eds.), Microscale Heat Transfer, 413–428.

$5 - 7\,^0$ C, when the copper and aluminum strips exhibit a temperature rise above the heat sink near 40 ^{0}C to 95 ^{0}C respectively.
The time of beginning of heat pipe science was near 40 years ago with first heat pipe definition and prediction of most simple cases. Now at the computer age some changes of basic equations are performed, more powerful predicting methods are available with increasing awareness of the complexity of heat pipes and new heat pipe generations. But even today heat pipes are still not completely understood and solution strategies still contain significant simplifications.
Standard heat pipe is shown on Figure 1. Basic phenomena and equations are related with liquid-vapour interface, heat transport between the outside and the interface ("radial" heat transfer), vapour flow and liquid flow.

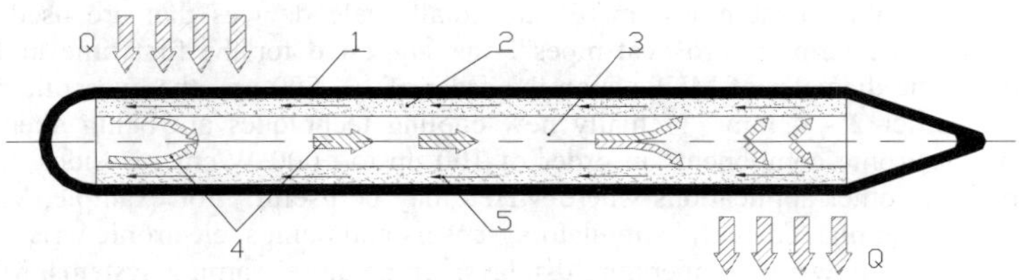

Figure 1. 1 Heat pipe schematic
1 – envelope, 2 – porous wick, 3 – vapour channel, 4 – vapour, 5 – liquid

There is a strong interaction between basic phenomena in heat pipes, Figure 2. Feedbacks may cause instabilities, such as waves, flooding, performance jumps. Basic equations are related to vapour flow in the MHP channel, liquid flow in the capillary structure, interface position between the vapour and liquid (mechanical equilibrium yields interface curvature K), radial heat transfer, vapour flow limit, capillary limit [2]. MHPs and mHPs are sensitive to the surplus liquid inside. Surplus liquid tends to be accumulated at the wet point defined by $K = K_{min}$. Sometimes the wet point is not at an end of the heat pipe and there could be deterioration of the radial heat transfer coefficient. Interface instability is the reason of the liquid accumulation in the condenser and leads to dry-out in the evaporator. While traditionally condenser resistance is deemed small and often neglected in MHP/mHP the detailed tests revealed substantial temperature drop across the length of the condenser in some devices. Potential sources of this temperature drop may be non-condensable gases, the surplus liquid and constrained vapour space. A deformation of the interface changes the vapour flow. The resulting change of the vapour stress on the interface tends to increase the deformation of the interface.

2. Micro heat pipes

Micro heat pipes phenomena are available in nature. For example there is an analogy between micro heat pipe operation and functioning of a sweat gland [3]. Open – type mini/micro heat pipes are suggested in [4 -5], as a system of thermal control of biological objects and drying technology. Some new possibilities are available to enhance heat and mass transfer in evaporators of heat pumps and refrigerators covered by capillary – porous coatings of heat releasing components applying the micro heat pipe phenomena [6]. For example for copper sintered powder structure disposed on the surface of horizontal copper heat releasing tube and propane as a working fluid evaporative heat transfer coefficient can be 8 times as high as boiling heat transfer coefficient on smooth tube at heat flux up to $q = 10^4$ W/m^2, and 6 times at $q > 10^4$ W/m^2. The liquid evaporation mostly is realized near the inter line and intrinsic meniscus region on the micropore outlets. A liquid is supplied to zones of vaporization by capillary force; a vapor is generated on the annular surfaces of meniscuses in orifices of micro-pores and goes out through macropores (Figure 3).

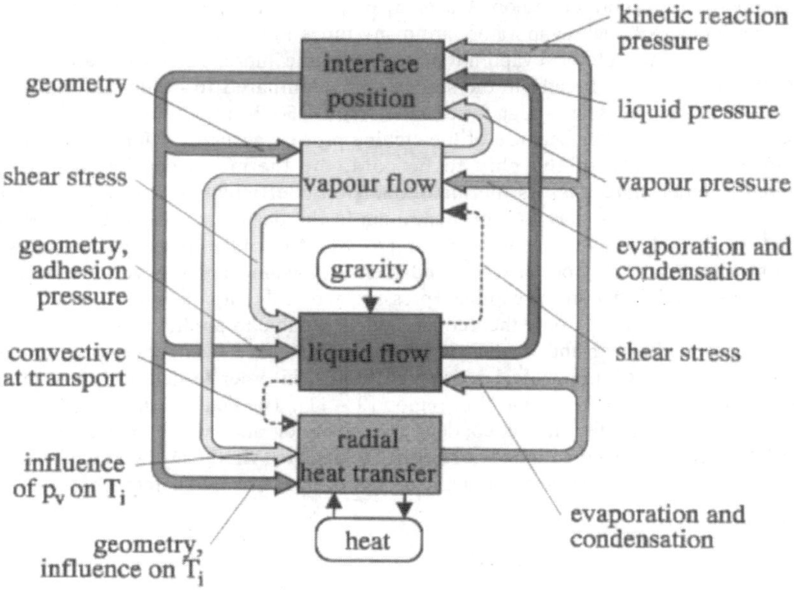

Figure 2. Interaction between basic phenomena in heat pipes

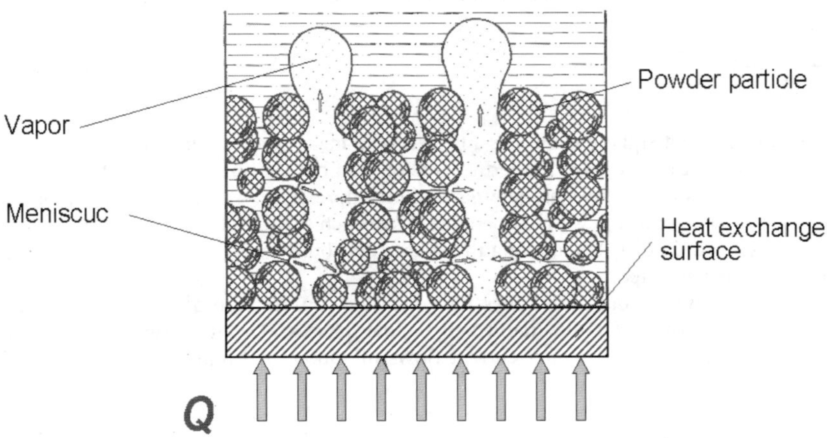

Figure 3. Micro heat pipe phenomena in the capillary - porous structure, [4-6].

So, the heat transfer with evaporation is similar to the heat transfer in the evaporator of micro heat pipe. In contrast to the conventional MHPs with polygon, triangular, trapezoidal-grooved capillary system, MHPs available inside the sintered powder structure with micro and macro pores have a complex shape and the evaporator region is near cylindrical with diameter of the order of some microns, but the number of such evaporators are many times more to compare with conventional MHPs, so the total surface of the evaporation is many time more. MHPs actually are mostly interesting to be implemented directly in the silicon chip. Compared to other materials, silicon provides several advantages. It has a high heat conductivity (150 W/mK). It permits to obtain much smaller devices than other metals because of the etching process accuracy. Moreover, as the MHP can be machined in the core of the chip to be cold, the thermo-mechanical constraints are lower compared to other materials. Its thermal expansion coefficient is 7times lower than that of copper and 10 times lower than that of aluminium.In the design of MHPs a number of heat transfer limitations should be taken in consideration [7 – 8]. Most of the investigations focus on the capillary heat transport capability because the fundamental phenomena that govern the operation of MHPs, arise from the difference in the capillary pressure across the liquid – vapour interface in the evaporator and condenser zones. Some theoretical models capable to predict the effects of the thin film region on the evaporating and condensing heat transfer have been developed in the past, particularly for triangular and trapezoidal – grooved MHPs, in order to determine the maximum evaporation heat transfer through the thin film region [9 – 11]. The detailed theoretical analysis of capillary flow, the heat transfer in the condenser, evaporator and macro region, Figure 4, is presented in [12]. In all abovementioned references related to MHP 1D theoretical analysis is available with emphasizes on one microchannel hydrodynamic and heat transfer:

2.1 CAPILLARY FLOW ALONG THE MHP

The triangle cross section of the microchannel is considered with its apex angle is made in the silicon plate. The working fluid during the MHP operation recedes into the triangle corners, generating capillary forces to move the liquid from condenser to the evaporator. Following the authors [7 -8] the liquid flow in the triangular channel is considered as:

$$\frac{dP_l}{dz} = \frac{dP_v}{dz} - \frac{d}{dz}\left(\frac{\sigma}{r}\right) \tag{2.1}$$

In order to analyze the different equations there is a necessity to make some assumptions:
1) steady – state conditions; 2) the vapour properties are constant but variable along the MHP axial direction; 3) both liquid and vapour flows are laminar and incompressible; 4) the interface curvature radius is supposed equal to zero; 5) the wall temperature T_w in each section of MHP is constant;
6) the heat flux Q_E is uniformly distributed along the evaporator, but varies on the MHP perimeter;
7) axial heat conduction is neglected.
The MHP is divided into several control volumes for which the conservation of mass, momentum and energy equations are written for the liquid and vapour phases. The numerical model was developed in [24-25] considering the counter current flows of vapour and liquid in microchannel. These conservation equations can be written as

$$\rho_v \frac{d(A_v u_v)}{dz} = -A_i \rho_l u_i \tag{2.2}$$

where ρ_l, ρ_v and u_l, u_v are the density and the velocity of the liquid and the vapour, repectively

$$\frac{d(A_l u_l)}{dz} = -A_i u_i \tag{2.3}$$

$$u_i = -\frac{Q_E}{L_E \rho_l A_i h_{lv}}$$ - radial phase change velocity

$$\rho_v \frac{d(A_v u_v^2)}{dz} = -\frac{d(A_v P_v)}{dz} - |\tau_i| p_i - |\tau_{vw} p_{vw} - \rho_v A_v g \sin\theta| \tag{2.4}$$

$$\rho_l \frac{d(A_l u_l^2)}{dz} = -\frac{d(A_l P_l)}{dz} - |\tau_i| p_i - |\tau_{lw} p_{lw} - \rho_l A_l g \sin\theta| \tag{2.5}$$

h_{vl} is the latent heat of vaporization, g – gravitational constant, τ - the shear stress , and Θ – the MHP tilt angle.

The equations (2.1 – 2.5) are the set of five first-order, nonlinear, coupled ordinary differential equations with five unknown variables: r ,u_l , u_v, P_l and P_v. This system need to be solved numerically with the following boundary conditions:

$$r\Big|_{z=L_T} = r_{max}, \quad u_l\Big|_{z=L_T} = u_v\Big|_{z=L_T} = 0, \quad P_v\Big|_{z=L_T} = P_{sat}, \quad P_l\Big|_{z=L_T} = P_{sat} - \frac{\sigma}{r_{max}}$$

The maximum curvature radius r_{max} is the radius of the inscribed circle in the inner triangular cross section, L_T – total length of MHP.

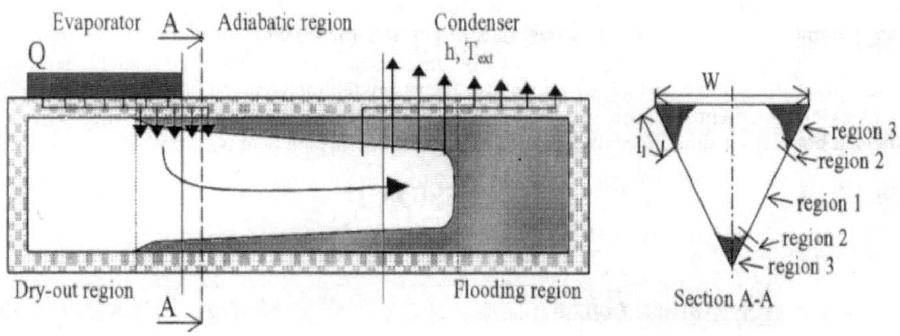

Figure 4. Schematic of triangular grooves (evaporator, adiabatic zone and condenser) in MHP, [25].

The numerical solution of governing equations of heat transfer for the evaporator and condenser regions allows determining the temperature distribution along the MHP axial direction.

$$\frac{1}{R_{th}} = \sum_{j=1}^{n} \frac{1}{R_{th,j}} = \sum_{j=1}^{n} \frac{\lambda_l}{\delta_{\perp j}} dz d\xi \tag{2.6}$$

The experimental data on silicon micro heat pipe arrays filled with methanol or water were published in [13] .Two arrays were fabricated with silicon wafers of thickness equal to 279 μm. A Pyrex glass wafer with a thickness of 0.760 mm was used to close the MHP array. The results show a maximum improvement of 11% in effective thermal conductivity at higher power levels for water and 6% for methanol. These results are disappointing compared to the improvement calculated from theoretical models. Simple two-phase MHPs were fabricated by sandwiching parallel metal wires between two square metal sheets of length 152.4 mm and thickness 0.406 mm [14 – 15]. The wire-bonded MHPs are made of aluminium and filled with acetone. Several MHPs were tested with different wire diameters. A maximum heat flux of 31 W (near 0.4 W/cm^2) is obtained for a wire diameter of approximately 1 mm and a saturation temperature of 55°C. An innovative wick design for flat plate heat pipes (copper – water device) used for the cooling of multiple sources was described in [16]. The capillary limitation was observed at 64 W/cm^2 with a maximum HP effective thermal conductivity of 700 W/mK. The radial MHPs with a three layers structure made of copper and filled with methanol,[17], were designed to allow separation of the liquid and vapour flows in order to reduce the liquid-vapour interfacial shear stress. This structure consists of three copper wafers which are bonded together: one for the upper vapour phase structure, one for the middle interface and one for the lower liquid structure. Two wick designs, one using 0.2 mm wide etched radial grooves and the other with 100 mesh copper screen, were investigated. The external dimensions of the MHP are 31 x 31 x 2.7 mm^3. The maximum heat fluxes are 45 W (23 W/cm^2) and 35 W (17.8 W/cm^2) and the thermal resistance is approximately 1.6 K /W. Recently a review paper on MHP/mHP for the cooling of electronic devices was published in [18]. Some new MHP designs are presented in the literature mostly related with an increasing of the surface of the evaporation and condensation and vapour pressure drops decreasing in the vapour channels (heat pipe spreaders, flat plate micro heat pipes, ets.), [19 – 20]. Analysis of the applicability of different grooved MHPs show, that there are some advantages of this heat pipe design (simple geometry of the micro channel, low cost of fabrication, using etching technology in silicon chip) and drawbacks such as sensitivity to the presence of non-condensable gases in the vapour channel, the strong liquid - vapour interfacial shear stress, dry-out effects with liquid accumulation in the vapour channel and hot spots arrival, low heat transfer output due to the low surface of the evaporation. The most efficient improvement of the MHP/mHP parameters could be obtained, if the surface of the evaporation and condensation zones would be dramatically increased. An example of such heat transfer enhancement due to the surface of two-phase heat transfer increasing is the design of capillary – porous MHP element (Figure 3), made from the sintered powder, when the heat transfer enhancement is 3 - 4 time more to compare with grooved surface heat transfer.

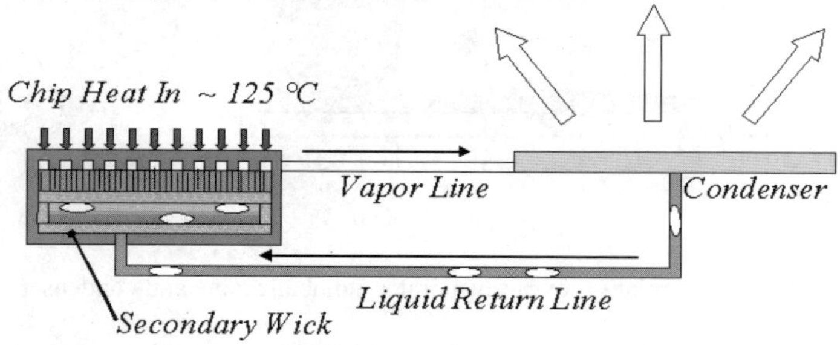

Figure 5. Micro - loop heat pipe - the electronic microchip cooler, [21].

Such types of MHP evaporators are also convenient for micro loop heat pipes, Figure 5, when it is necessary to ensure the heat dissipation outside the chip.

3. Miniature heat pipes

Actually a lot of miniature heat pipes with different wick structures (sintered powder, mesh structure, grooves, fiber bundle) are fabricated by some leading companies. The typical mHP application for the electronic components cooling is shown on Figure 6, [22]. Let us consider the typical mHP with diameter 4 mm and the length 200 mm.The maximum heat transport capability of the map is governed by several limiting factors which ought to be considered when designing a heat pipe. There are five primary heat pipe heat transport limitations: viscous, sonic, capillary pumping, entrainment or flooding and boiling. For the low temperature heat pipes (for example miniature copper/water heat pipe) the most important are capillary pumping and boiling limits. In some cases the flooding limit (in condenser zone) is also important.

Two most important properties of the wick are the pore size and the permeability. The pore size (radius) determines the fluid pumping pressure (capillary head) of the wick. The permeability determines the frictional losses of the fluid as it flows through the wick. Actually there are several types of the wick structures available including: metal sintered powder; fine fiber bundle, axially grooves, screen mesh.

1. Metal sintered powder wicks have a high fluid pumping pressure (can work against gravity field), low effective thermal resistance (high effective thermal conductivity), can be partially dried (still working efficiently), boiling crisis is smooth between Q_{max1} and Q_{max2} , but have low fluid permeability (the pressure losses are relatively high). Have a good reliability (wettability) after the crisis phenomena

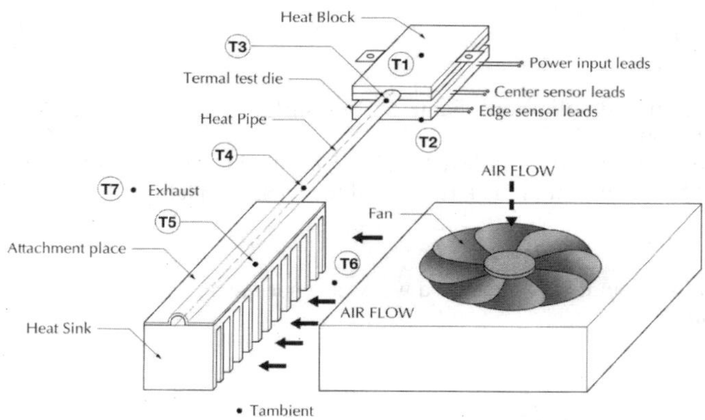

Figure 6. mHP for the electronic components cooling, (Fujikura Ltd).

2. Grooves as a wicks have a large pore radius and high permeability (the pressure losses are low), but it pumping head (fluid pumping pressure) is low (can't be used against the gravity field), it can't be functioning with partially dried evaporator zone. Boiling crisis is sharp at Q_{max}., structures have a bed reliability (wettability) after the crisis phenomena (dry-out).
3. Fine fiber bundle wicks have a good capillary pumping head, but have low permeability and high effective thermal resistance (low thermal conductivity) across the wick. They have the

low heat transfer coefficient between the envelope and the wick. Can't be used with partial wick drying, the boiling crisis (dry-out) is sharp. Have low wettability after the dry-out.

4. Screen mesh wicks have a moderate capillary pumping head, but have low permeability and high effective thermal resistance. Have low wettability after the dry-out.

The effective thermal resistance (or thermal conductivity) of heat pipe is one of the important parameters and is not constant but a function of a large number of variables, such as heat pipe geometry, evaporator length, condenser length, wick structure and working fluid. The total thermal resistance of a heat pipe is the sum of resistances due to conduction through the wall (heat pipe envelope), conduction through the wick, evaporation or boiling, axial vapor flow, condensation, and conduction losses back through the condenser section wick and heat pipe wall. The detailed thermal analysis of different heat pipes is rather complicated, but now is it clear, that a heat pipe with a metal sintered powder wick is the most efficient in its function in any position of heat pipe in the gravity field with good heat input capability. Sintered powder wick because of the close particle to particle spacing, generate very high pumping capabilities as compared to more conventional grooves or mesh screen wicks. Sintered powder wick has an additional advantage over screen wicks, it has relatively high thermal conductivity, and it means a higher heat flux performance capability due to the enhanced fin effect, which the sintered powder provides in the evaporator region of the heat pipe. But an efficient sintered powder wick needs to be optimized to ensure the high heat flux performance capability. The problem of the wick structure optimization is related with structural porous wick parameters: the particle size and its form, wick porosity, specific surface of porous wick, pore diameter.

Basically some of these parameters are related with each other, the hydraulic pore passages are one of the main parameters of a wick. For example the capillary pressure p_c is determined as:

$$p_c = \frac{4\sigma\cos\theta}{d_0}, \quad (3.1)$$

where d_0 – the mean hydraulic pore diameter; σ – surface tension coefficient, θ – the angle of wetting of the wick pore.

The coefficient of permeability of the wick "k" is related with porosity of the wick Π, diameter of the particles and the coefficient of the tortuosity of the wick channels β

$$k = \frac{f(\Pi)}{\beta} D^2, \quad (3.2)$$

where $f(\Pi)$ – is a function of Π, which has a week dependence on D, β – coefficient of the tortuosity of the wick channels

$$k = \xi d_0^{\nu}, \quad (3.3)$$

where ξ, ν – are the constants, $1 < \nu < 2$

The experimentally determined dependence between k and $d_0{}^{\nu}$ (Fig.7) for the copper sintered powder wick is:

$$k = 0{,}00144 d_0^{1{,}79}. \quad (3.4)$$

The heat flux transferred through mHP depends on the distance between the condenser and evaporator zone, the wall superheat and the liquid subcooling, the thermal contact between the heater and wick and the superficial boundary conditions of the wick.

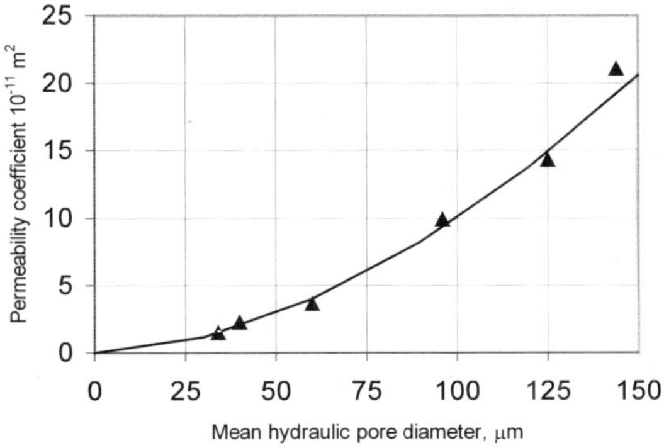

Figure 7. The coefficient of the wick permeability as function of the mean hydraulic pore diameter

The thermal hydraulic parameter of the wick is determined through the experimental measurement of:

- capillary height (through which the equivalent porous radius can be evaluated);
- liquid hydraulic head (through which the liquid pressure drop in the wick is determined);
- wick permeability is found from the hydraulic head (Darcy's law);
- heat flux;
- wick mass flow rate during the evaporation (through which, from the knowledge of other measured wick parameters, the wick two-phase pressure drop is calculated);
- wick porosity (through which the thermal conductivity of the wick saturated with liquid can be determined).

Let us consider the miniature cylindrical heat pipe with the length l, the condenser length l_c, evaporator length l_e, and the transport zone length l_t (effective transport zone length l_{ef}). Heat pipe is inclined to the horizon on the angle $\varphi > 0$ (evaporator is disposed above to the condenser), the wick cross-section square is S (heat pipe outer diameter D_p and inner diameter D_{ch}). The sintered powder wick is saturated with liquid. The liquid at the heat pipe working temperature T has a density ρ_l, surface tension σ, dynamic viscosity μ_l and the latent heat of evaporation L. The vapour has the density ρ_v , viscosity μ_v. The angle of the wick wetting is θ.

The following assumptions are adopted:

1) Wick parameters are constant along the heat pipe;
2) Evaporation of the liquid is on the surface of the evaporator;
3) The heat flux in the evaporator and in the condenser is constant;
4) There is a saturated vapor in the transport zone, and its temperature is T_s;
5) The liquid and the vapor motion is described by the Navier-Stocks set of equations, valid for the non-compressible fluid;
6) No heat sources and heat sinks are in the vapor media;
7) The liquid movement inside the porous wick is followed by the Darcy law;
8) The friction forces on the vapor-liquid interface in negligibly small to compare with the friction forces inside the wick;

9) The hydrodynamic and heat transfer in the heat pipe are considered as 1D model.
The capillary pressure, which we need to calculate Q_{max} is equal:

$$p_c = \Delta p_v + \Delta p_l + \Delta p_g, \tag{3.5}$$

where Δp_c ,Δp_l ,Δp_v and Δp_g – pressure drop due to the capillary, liquid, vapor, and gravity forces.
The capillary drop is described by the equation (3.5). The liquid pressure drop (Darcy law) is described as:

$$\Delta p_l = \frac{Q\mu_l l_{ef}}{\rho_l LS\xi d_0^v}; \tag{3.6}$$

The vapor pressure drop is described by the Poiselle equation:

$$\Delta p_v = \frac{128 Q\mu_v l_{ef}}{\pi D_{ch}^4 \rho_v L}; \tag{3.7}$$

The gravity pressure drop is equal to:

$$\Delta p_g = \rho_l gl \sin\varphi, \tag{3.8}$$

where g – is the gravity constant.
The heat flow is determined as:

$$Q = \frac{\pi L}{4 l_{ef}} \frac{\dfrac{4\sigma\cos\vartheta}{d_0} - \rho gl\sin\varphi}{\dfrac{\mu_l}{\rho_l (D_p^2 - D_{ch}^2)\xi d_0^v} + \dfrac{32\mu_v}{D^4 \rho_v}}. \tag{3.9}$$

Following this analysis Q depends on two capillary structure parameters – the mean hydraulic pore diameter and the inner diameter of the porous wick. To find the Q_{max} we need equation (3.9) analyze for the extreme function finding. Due to the temperature dependence of the thermo-physical properties of the working fluid the maximum heat flow Q_{max} will be different for different saturated vapor temperatures T_{sat} in the heat pipe transport zone, Figure 8. For different angles of heat pipe inclination to the horizon we need to determine Q_{max} at the worst situation with the point of view of the heat transfer, when the heat pipe evaporator is above the heat pipe condenser, vertical (inverted) heat pipe disposition.
So, the analysis show the possibility to transport along the copper heat pipe with water and copper sintered powder wick a heat flow Q = 50 – 60 W at the saturated temperature neat 100 ^{0}C.
The experimental set-up designed and made to determine mHPs (flat and cylindrical) parameters and mHPs predicting software was done in [23]. The general goal of this set-up was to determine the temperature distribution along the heat pipe for different heat loads, estimate the heat pipe maximum capacity in any position, and evaluate the dependency between a heat pipe thermal resistance and heat dissipation. The cylindrical mHP has dimensions: L = 200 mm, L_e = 70 mm, L_c = 85 mm, L_a = 45 mm. Outer diameter D_p = 4mm, mHP wall thickness – 0.2 mm, diameter of the vapor channel D_{ch} = 2 mm, size of the copper powder particles - <100 μm.

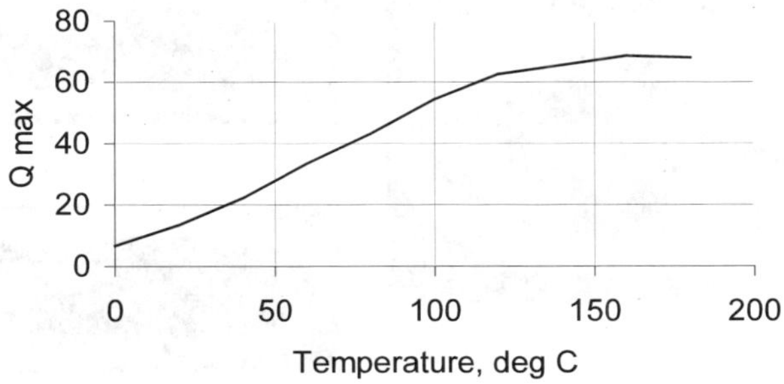

Figure 8. Q max as a function on T

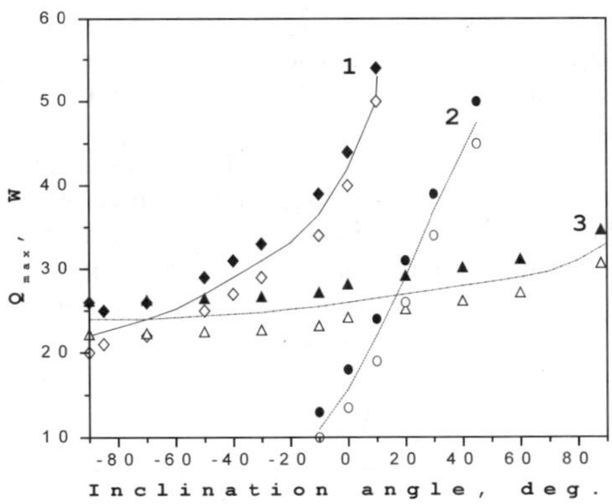

Figure 9. Q_{max} dependence for cylindrical water mHP inclination angle to the horizon: 1 – advanced mHP with sintered copper powder; 2 – mHP with longitudinal grooves; 3 - conventional mHP with sintered copper powder

Following the analysis of Q_{max} for different cylindrical mHPs , Figure 9 it is clear, that copper sintered powder as a wick has some advantages to compare with another wicks in the field of gravity, when the evaporator is disposed above the condenser. Advanced copper sintered powder wick (curve 1, Figure 9) has some advantages to compare with such wicks as mesh structure, grooves, fiber bundle.

Experimental samples of the flat miniature heat pipe were chosen as copper /water heat pipe with three types of a wick: sintered powder wick, wire mesh wick and wire bundle wick (Figure 10). The experimental data of q_{max}, Q_{max}, R_{HP}, h_e and h_c as a function of heat pipe tilt were obtained for different heat load and heat pipe orientation in space and temperature level.

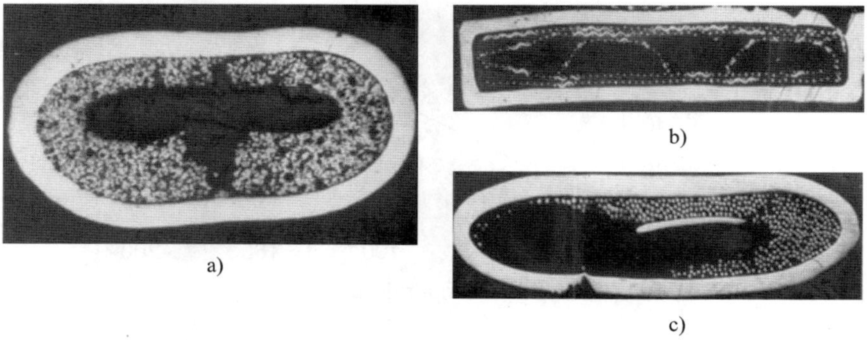

Figure 10. Cross section of flat copper mHP with a copper sintered powder (obtained by the cylindrical heat pipe flattening) (a), mesh structure (b) and wire bundle (c)

We can conclude that the heat transfer limit for two different orientations of heat pipe in space (vertical and horizontal) is essentially different, Figure 11. Temperature field along the heat pipe for flat miniature heat pipes is represented on Figure 12. It is interesting to note that the temperature field along the mHP in conditions, when Q_{max} is achieved, is different for different mHP wicks. The main difference of these data is related with different wick parameters (effective thermal conductivity, pore size, pore distribution, thermal contact of the wick and mHP envelope, and heat transfer in the evaporator and condenser zones). Analyzing the examples of numerical modeling and experimental data comparison which were realized for three different wicks of flat miniature heat pipes, we can conclude that the theoretical and experimental data are coinciding with small limit for large specter of heat flow value transferred by a heat pipe. 2D steady-state mathematical model for predicting thermal performances (maximum transport capacity, thermal resistance, heat pipe's temperature axial profile and temperature drop between a heat source and heat sink) was developed [23].

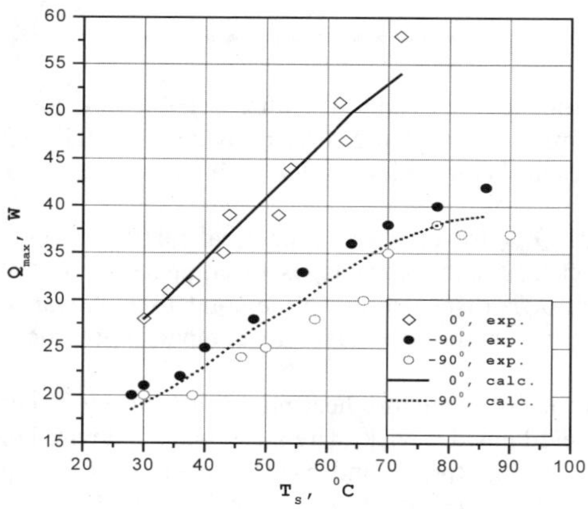

Figure 11. Capillary limit for the flat mHP with a copper sintered powder as a wick

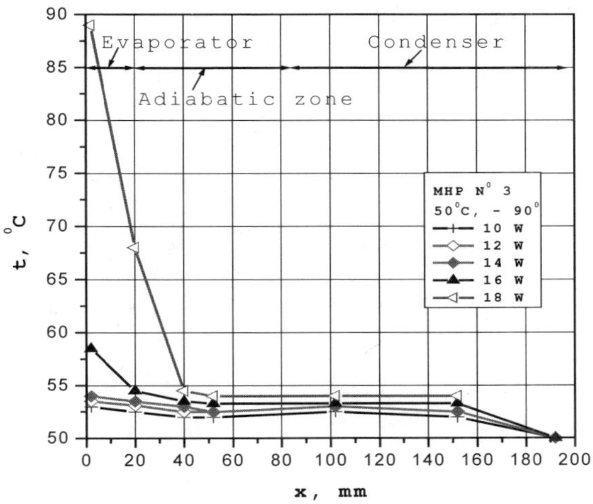

Figure 12. Temperature distribution along the flat mHP with a copper sintered powder as a wick

Heat pipe thermal resistance (and the heat transfer coefficient in the evaporator and condenser zones) was found using the data of the vapour temperature in the adiabatic zone and the mean temperature in the evaporator and in the condenser. The heat transfer coefficients in the evaporator and condenser of the flat mHPs depend on two- dimensional hydraulic (pore saturation, capillary permeability, capillary pressure) and thermal (temperature distribution along the heat pipe envelope) parameters of device. The temperature in the middle of the heated side (heat load input) of the evaporator can exceed the symmetric point temperature on the opposite (non-heated) surface of the envelope by nearly 10 ^{0}C.

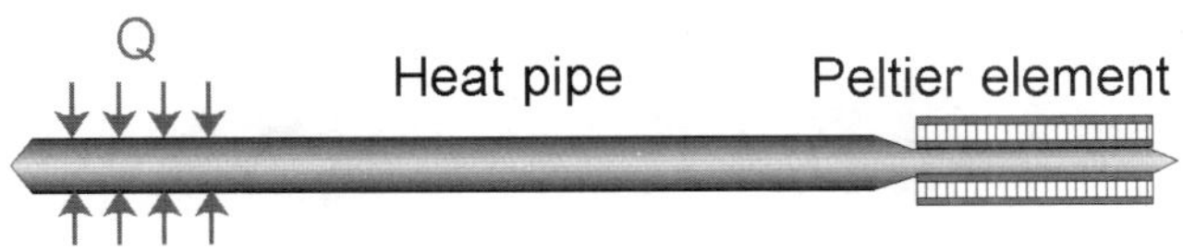

Figure 13 mHP for electronic component cooling (using Peltier element)

It is necessary to note that mHPs can be applied with success in other kinds of refrigerating machines, for example in thermoelectric refrigerators. Heat pipe here is used to transfer and to transform heat flow. That is sometimes that the surface of heat input and heat output can be not identical. The Peltier element usually is small, the surface of heat output sometimes should have the large surface or certain form (Figure 13).The combination "heat pipe - Peltier element " for such cases is convenient and can be used for processors cooling and for cooling/heating in medical devices, for example in cryogenic surgery. The thermal control system for local hypothermia with heat pipe instrument was developed in the Luikov Institute. It successfully tested and now is recommended for medical practice [26].

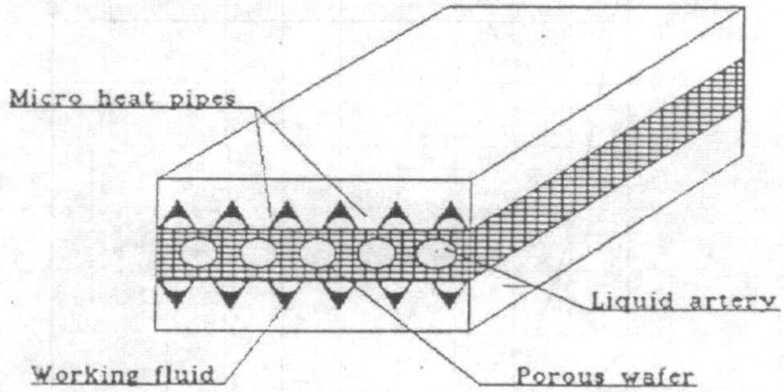

Figure 14 .Flat mHP with longitudinal grooves, porous wafer and arteries. [27]

In 1991 year a flat mHP was suggested by Luikov Institute [27], which has a capillary - porous wafer with arteries. The triangular minichannels for vapour transfer from the evaporator to condenser are made on the inner surface of the mHP envelope. When there is a low density heat transfer the triangular minichannels are used as in MHPs to ensure the evaporation and condensation in the interface liquid/vapour. When the heat load is high the evaporator of this heat pipe is cooling by the "inverted meniscus" of evaporation on the interface capillary – porous wafer and heat loaded wall. The vapour is condensing in the condenser, the liquid is sucking by the wafer and transported to the evaporator by capillary forces. Such heat pipe in insensitive to the gravity and has low thermal resistance 0.1 K/W, Q_{max} = 100 W.

4. Conclusion

- The existing technologies of MHP/ mHP production ought to be significantly improved in order to face the new challenges in electronic cooling
- The heat transfer limit of MHP/ mHP should be increased by optimizing the geometric and operating parameters.
- Thermal modelling is a powerful way to predict the performance and the temperature response of MHP/ mHP. Unfortunately most of developed thermal models are 1-D models and empirical correlations are employed to determine fraction factor of vapour flow.
- In order to predict the heat transfer limit and temperature distribution of MHP/ mHP, a comprehensive 3-D model that includes heat transfer in liquid and vapour must be developed.
- In order for MHP/ mHP to find commercial application in microelectronic cooling it must compete with other cooling methods, such as forced convection, impingement and two phase direct cooling in areas such as manufacturing cost and reliability.

  Optimisation of copper sintered powder wick in miniature copper/water heat pipes with outer diameter 4 mm and length 200 mm is a good challenge to improve the mHP parameters.

  Analysis of the experimental data for such a new optimised miniature heat pipe with copper sintered powder wick proves the possibility to use mHPs independently of its orientation with the maximum heat transport capability near 50 W.

Theoretical simulation of mHPs with different wick structures (sintered powder, mesh structure, wire bundle) is an efficient tool to perform the comparisons of mHP efficiency. Experimental verification of mHP parameters proves validity of the simulation software.

REFERENCES

1. Cotter T. M., (1984), "Principles and Prospects of Micro Heat Pipes", *Proceedings of the 5th International Heat Pipe Conference, Tsukuba, Japan,* 328 - 335
2. Busse Claus A., (1992), Heat Pipe Science, *Proceedings of the 8th International Heat Pipe Conference*, September 14 – 18, Beijing, China, 3 – 8.
3. Dunn P.D., Reay D.A., (1976), Heat Pipes, Pergamon Press, First edition, 258 - 260
4. Vasiliev L.L., (1993), Open – type miniature heat pipes, *Journal of Engineering Physics and Thermophysics,* Vol.65, No.1, 625 - 631
5. Reutskii V.G., Vasiliev L.L., (1981*), Dokl. Akad. Nauk BSSR*, 24, No.11, 1033 – 1036
6. Vasiliev L., Zhuravlyov A., Shapovalov A., (2004), Comparative analysis of heat transfer efficiency in evaporators of loop thermosyphons and heat pipes, *Preprints of the 13th International Heat Pipe Conference*, September 21 – 25, Shanghai, China, 52 – 59
7. Ma H., Peterson G.P., (1996), Experimental investigation of the maximum heat transport in triangular grooves, *Journal of heat transfer,* Vol.118,740 -745
8. Hopkins R., Faghri A., Khrustalev D., (1999), Flat miniature heat pipes with micro capillary grooves, *Journal of heat transfer,* Vol.121,102 – 109
9. Lefevre E., Revelin R.,. Lallemand M, (2003), Theoretical Analysis of Two – phase Heat Spreaders with Different Cross – section Micro Grooves, *The Preprints of the 7th International Heat Pipe Symposium,* October 12 – 16, Jeju, Korea, 97 – 102
10. Kim S.J., Seo J.K., Do K.H., (2003), Analytical and experimental investigation on the operational characteristics and the thermal optimization ob a miniature heat pipe with a grooved wick structure, *International Journal of heat and Mass Transfer,* Vol.46, 2051 – 2063
11. Ma H., Jiao A.J., (2004), Thin film evaporation on heat transport capability in a grooved heat pipe, *Proc. of the Sixth International Symposium on Heat Transfer,* Beijing, 340 – 345
12. Jiao Anjun, Riegler Rob, Ma Hongbin, (2004), Groove geometry effects on thin film evaporation and heat transport capability in grooved heat pipe, *Preprints of the 13th International Heat Pipe Conference,* Shanghai, China, September 21 -25, 44 -51
13. Badran B., Gerner F., Ramada P., Henderson T., Baker K., (1997), Experimental results for low – temperature silicon micromashined micro heat pipe arrays using water and methanol as working fluids, *Experimental Heat Transfer,* 10, 253 – 272
14. Wang Y., Ma H., Peterson G., (2001), Investigation of the temperature distribution on the radiator fins with micro heat pipes, Journal of Thermophysics and Heat Transfer, 15 42 – 49
15. Wang Y., Peterson G., (2002), Analysis of wire bonded micro heat pipe arrays, *Journal of Thermophysics and Heat Transfer,* 16, 346 – 355
16. Rightley M.J., Tigges C.P., Civler R.C., Robino C.V., Mulhall J.J., Smith P.M., (2003), Innovative wick design for multi – source, flat plate heat pipes, *Microelectronic Journal,* 34, 187 – 194
17. Kang S., Tsai S., Ko M., (2004), Metallic micro heat pipe heat spreader fabrication, *Applied Thermal Engineering*, 24, 299 – 309.
18. Lallemand M., Lefevre F., (2004), Micro/Mini heat pipes for the cooling of electronic devices, *Preprints of the 13th International Heat Pipe Conference*, Shanghai, China, September 21 -25, 12 - 23
19. Katsuta M., Homma Y., Hosova N., Shino T., Sotani J., Rimura Y., Nakamura Y., (2003), Heat

transfer characteristics in Flat Plate Micro heat pipe, *Proceedings of the 7th International Heat Pipes Symposium, October 12- -16, Jeju, Korea. 103 - 108*

20. Lefevre F., Revelin R., Lallemand M.,(2003), Theoretical analysis of the two – phase heat spreader with different cross – section micro grooves, *Proceedings of the 7th International Heat Pipes Symposium,* October 12- -16,Jeju, Korea, 97 – 102
21. Golliher Eric, Mellott Ken, NASA GRC (2002) Hamdan Mohammed, Gerner Frank, Henderson Thurman, University of Cincinnati Jake Kim, Jinha Sciences, Inc., while at TTH, Inc. Oinuma Ryoji, Kurwitz Cable, Best Fred, Texas A & M
22. Ektummakij Prasong, Kumthonkittikun Vichan, Kuriyama Hiroyuki, Mashiko Koichi, Mochizuki Masataka, Saito Yuji, Nguyen Thang, (2004) New Composite wick heat pipe for cooling personal computers, *Preprints of the 13th International Heat Pipe Conference,* Shanghai, China, September 21 – 25, 263 – 268
23. Vasiliev L.L., Antukh A.A., Maziuk V.V., Kulakov A.G., Rabetsky M.I., Vasiliev L.L. Jr., Oh Se Min, (2002), Miniature Heat Pipes Experimental Analysis and Software Development, *Proceedings of the 12th International Heat Pipe Conference,* Moscow, 329 – 335.
24. Launay S., Sartre V., Lallemand M., (2002), Thermal study of water-filled micro heat pipe including heat transfer in evaporating and condensing microfilms, *Proceedings of the 12th International Heat Transfer Conference*, Grenoble, France, August 18 – 23, 6 p.
25. Zaghdoudi M.C., Sartre V., Lallemand M., (1997),Theoretical *investigation of micro heat pipes performance, 10th* IHPC, Stuttgart, Germany, 21 -25 Sept., 6 p.,
26. Vasiliev L.L. ,. Zhuravlyov A.S , Molodkin F.F. , Khrolenok V.V., Adamov S.L., Turin A.A. , (1995) , Medical heat pipe instrument for local cavitary hypothermia, *Preprints of 1st Minsk International Seminar "Heat Pipes, Heat Pumps, Refrigerators",* September 12-15, Minsk, Belarus, pp. 104 -114
27. Vasiliev, L.L., Maziuk, V.V., Kulakov, A.G., Rabetsky, M.I., (2000), Miniature heat pipes software development and experimental verification, *Proceedings of the IV Minsk International Seminar "Heat pipes, Heat pumps, Refrigerators*", September 4-7, Minsk, Belarus, pp. 270-278.

ROLE OF MICROSCALE HEAT TRANSFER IN UNDERSTANDING FLOW BOILING HEAT TRANSFER AND ITS ENHANCEMENT

K. SEFIANE
School of Engineering and Electronics – University of Edinburgh
Edinburgh, UK

V. V. Wadekar
HTFS – Aspen Technology
Harwell, UK

1. Introduction

Due to the trend towards miniaturization and intensification of processes, a large demand is placed on associated cooling/heating systems, for example, the intensification of chip power, heat exchange to micro-reactor, or reduced weight/volume in off shore applications and avionics. To meet this large demand, two-phase heating/cooling systems become increasingly necessary. Boiling heat transfer has long been recognized as one of the most efficient means for transferring very high heat fluxes at relatively low wall superheats. This has made boiling a very attractive technique to be used in a wide range of cooling technologies. In addition to this specific interest in boiling in microchannels, the general area of microscale heat transfer itself is receiving increasing attention. For example, environmental concerns in an increasing number of applications have necessitated the use of new working fluids (pure and blends) and natural fluids such as CO_2 in refrigeration and in some process duties, where micro-technology is increasingly being considered for heat transfer components. Thermal control in fuel cells, now proposed for small devices such as mobile phones, is an additional challenge, with implications extending to other micro-reactor uses and process intensification applications. Other related applications range from biological systems to industrial applications like: transportation, electronics, material processing, energy and environment.

The number of studies performed in the field of microscale heat transfer is rapidly increasing. The research in microscale heat transfer contributes to the development of design techniques and design methods in terms of performance data for microchannel heat exchangers. Another aspect of microscale heat transfer research is the contribution in the understanding of underlying physical mechanisms for macroscale phenomena. The contribution of microscale investigations can be extended to be used in improving existing enhancement techniques used at the macroscale or even develop new techniques for heat transfer augmentation.

In this paper we attempt to review some of the recent studies on microscale heat transfer involving vapour and liquid two-phase flow covering both experimental and analytical aspects. Furthermore we link these studies to interpreting and understanding boiling heat transfer, and in particular augmentation of flow boiling heat transfer.

S. Kakaç et al. (eds.), Microscale Heat Transfer, 429–444.

2. Boiling Heat Transfer Enhancement

Enhancement of nucleate boiling heat transfer in 'large' diameter channels is often used to augment flow boiling heat transfer. Enhancement techniques can be classified either as passive (without external power) or as active (with external power or external additives). Table 1 gives a short list of the various techniques used in the area of boiling heat transfer. From process industry perspective the passive techniques are more important than active techniques because of the benefit of simplicity.

Table 1: Enhancement techniques in boiling heat transfer

Passive techniques	**Active techniques**
Treated surfaces	Mechanical aids
Rough surfaces	Surface vibration
Extended surfaces	Fluid vibration
Displaced enhancement devices	Electro-static fields
Swirl flow devices	Suction or injection
Coiled tubes	Additives for fluids
Surface tension devices	

A more detailed description of the above techniques can be found in many reviews in the literature. The progress on the microscale heat transfer can contribute indeed to the understanding of the fundamental mechanisms in the various enhancement techniques used. For example the use of microstructures and treated surfaces has been widely used to enhance boiling heat transfer [1], Figure 1. The mechanisms by which the enhancement is achieved are still debatable and in some cases not yet fully understood. The use of porous surfaces and some specific structures is found to increase the number of nucleation sites in boiling heat transfer. The effect of well wetting fluids where enhancement is observed is still open to question and the role of wetting contact angle in the enhancement of boiling heat transfer in the various regimes (nucleate boiling, critical heat flux) is not elucidated yet.

In the past the role of microscale heat transfer in enhancement of nucleate boiling heat transfer was not fully appreciated. However it is now recognised that as size reduces so the influence of the surface tension and wetting phenomena play a greater role in the dynamics of the process. More recently, Stephan [2] has reported a model, which incorporates a combination of micro and macroscale heat transfer for boiling enhancement situations, where evaporation takes place from thin liquid films. Using this model Stephan was able to predict augmentation of nucleate boiling heat transfer by some structured boiling surfaces. In this approach heat transfer is assumed to be governed by one dimensional heat conduction normal to the wall, the molecular interfacial phase change resistance and intermolecular forces of adsorption. It is shown that mass transfer is influenced by capillary forces and an evaporation induced flow.

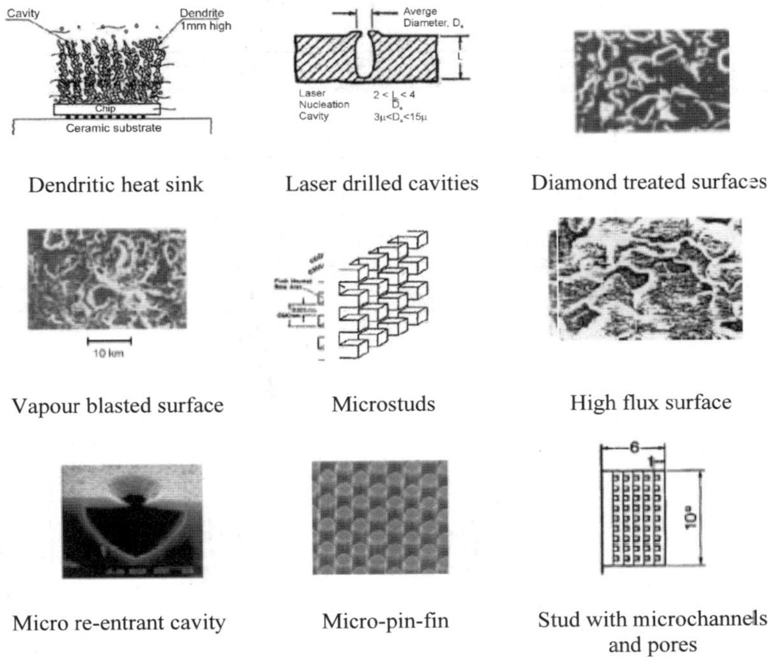

Figure 1. Surface microstructures [1]

Another interesting area is the enhancement achieved by adding surfactants to boiling liquids. Figure 2 shows an example of considerable enhancement obtained in nucleate boiling heat transfer by adding very small amounts of surfactant to boiling fluid.

It is well known that surfactants generally lower the surface tension but why should that lead to the enhancement of boiling heat transfer is not fully understood. A partial explanation of this could be attempted as follows in terms of the *Marangoni effect*. Concentration gradients may exist at the interface which can result in surface tension gradients at the interface. The gradients in the surface pressure contribute to tangential stresses. It has the same role in the surface momentum balance as pressure gradient in the momentum balance. As liquid drains from the film, the drag of the liquid on the interface stretches the film. The resulting expansion of the interface reduces the surface concentration on the interfaces. This establishes a surface tension or surface pressure gradient between the interface in the film and the interface in the meniscus. This gradient tends to oppose the motion of the interface and thus tends to maintain the interface immobile as the liquid drains from between the two near-immobile interfaces. At the same time this surface tension gradient tends to pull the interface from the meniscus back into the film. This leads to a turbulent motion of the interface at the boundary between the meniscus and the film. This effect is the *Marangoni effect* caused by the surface tension gradient. If the effect of Marangoni forces on boiling heat transfer is precisely understood, this would lead to effective use of these surfactants in the enhancement techniques

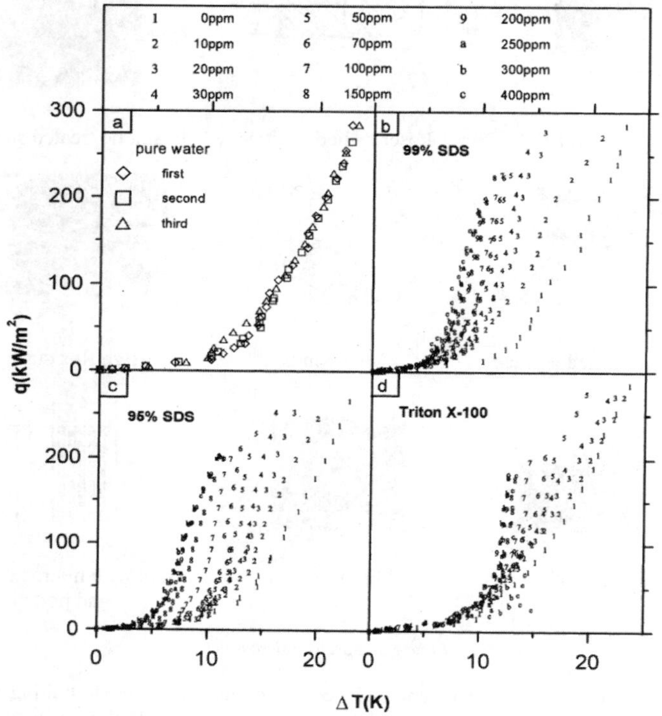

Figure 2. Enhancement observed with surfactant additives [3]

3. Flow Boiling Heat Transfer in Large Diameter Channels

The conventional picture of flow boiling heat transfer is of the presence of two-phase convective heat transfer and nucleate boiling heat transfer as two competing mechanisms of heat transfer. This gives rise to nucleate boiling dominated region, occurring at high wall superheats or high heat fluxes, where the flow boiling heat transfer coefficient increases with heat flux, pressure and is independent of mass flux and vapour quality. The convective heat transfer region dominates at low wall superheats or low heat flux. In this region the flow boiling heat transfer coefficient increases with mass flux, vapour quality and decreasing pressure, and it is independent of heat flux. However, some of the recent data [4] obtained with a large diameter tube shows that, under certain conditions, the experimental data may present a different picture. Figure 3 for example, shows a region of decreasing heat transfer coefficient with respect to increasing vapour quality for higher heat fluxes in 5 to 15% quality range. Furthermore, at the highest heat flux of 60 kW/m^2, there is a peak in the heat transfer coefficient profile just before the fall in the coefficient due to dryout. These trends cannot be explained on the basis of existing models for flow boiling heat transfer. It is worth examining a possibility of whether these anomalous trends could be explained on the basis of microscale heat transfer. For example, regarding the peak in the coefficient profile, it is possible that this enhancement in heat transfer is due to two microscale features. Firstly, before the dryout the liquid film is likely to be very thin, with the thickness being of the order of few hundred microns at the most. This in itself would give very high rates of two-phase convective heat transfer. Secondly, it is possible that such thin liquid film contains nucleation activity where the bubble sizes could be as small as at least one tenth of the liquid film thickness. Such bubble activity on microscale would increase the heat transfer coefficient even further, explaining the peak in the heat transfer coefficient profile.

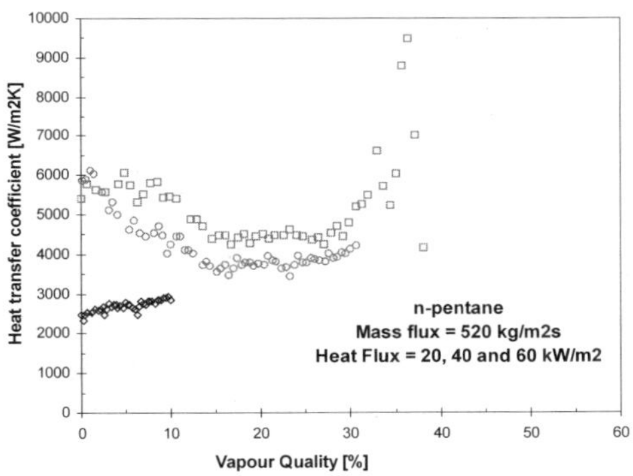

Figure 3. Flow boiling heat transfer in a 25.4 mm i.d. tube at different heat fluxes

4. Two Phase Flow Heat Transfer in Micro- and Mini-Channels

A huge interest has been shown in heat transfer in general and boiling in mini- and micro-channels in particular. Many international meeting and conferences were devoted to the subject. Recently Kandlikar [5] published a literature survey on boiling in micro and mini channels summarizing the main investigations in the area. An adapted version of the summary presented by Kandlikar [5] is shown in Appendix 1.

Similar to the large diameter channels, channels with relatively small diameters, of about few millimetres also exhibit some anomalous trends. This is illustrated in Figure 4, where some typical data reported by Huo et al [6], are shown. It can be seen that at high heat fluxes the flow boiling heat transfer coefficient decreases with increasing vapour quality. At the high vapour quality, where the coefficient is independent of heat flux, the mechanism of heat transfer is likely to be two-phase convective heat transfer. Comparison of Figure 3 and 4 shows that there is an important difference between these two cases, regarding how rapidly the heat transfer coefficient falls with respect to vapour quality. The rapid fall seen in Figure 3 is characteristic of dryout of liquid film in annular two-phase region. The gradual fall in Figure 4 cannot therefore be attributed to a similar phenomenon. However, it is possible that the dryout phenomenon in mini and micro size channels occurs in a different manner, resulting in the gradual fall of the coefficient. Jacobi and Thome [7] have developed a model for boiling in small diameter channels, which predicts that the convective heat transfer coefficients increase with the heat flux but are nearly independent of the vapour quality; a trend that is normally associated with nucleate boiling. Such model would explain the trend obtained by Huo et al [6] at the two low heat fluxes of 27 and 40 kW/m^2 in Figure 4, but it would not explain the trend, discussed above, at higher heat fluxes.

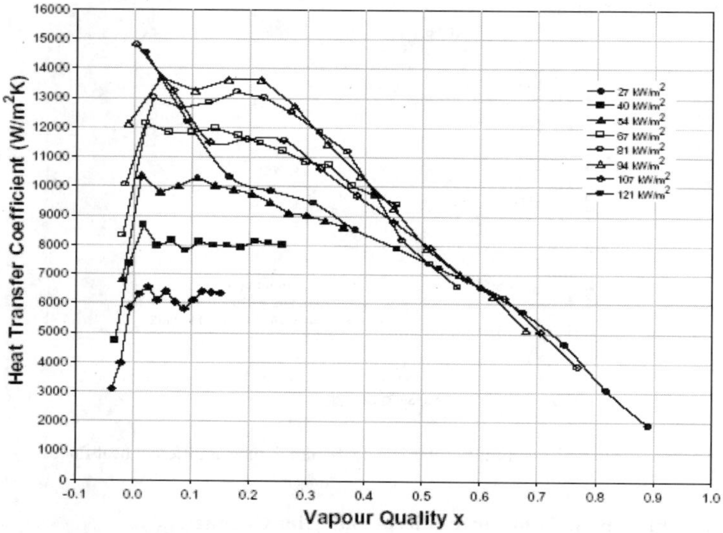

Figure 4. Flow boiling heat transfer in a 2.01 mm i.d. tube at different heat fluxes (mass flux =300 kg/m^{2}s, pressure = 8 bar) [6]

The fact that boiling heat transfer at the microscale is governed by the evaporation, hydrodynamics, capillary effects of the thin liquid film, has led to the introduction of some relevant dimensionless numbers. These numbers are used by many authors to analyse the behaviour of boiling heat transfer at the microscale (see Table 2).

As described by many authors, for example Jacobi and Thome [7], the primary flow regimes observed are elongated bubble flow (also referred to as slug flow) and annular flow. Other authors report various flow regimes observed under different experimental conditions, for example Figure 5. For Figure 5 the internal diameter of tube was 4.26 mm and operating pressure of 10 bar with R134a as the test fluid. Despite the differences in flow regimes observed, it is recognized that at these scales, the evaporation of the thin liquid layer between the bubbles and the wall plays a major role in the heat transfer process. Also capillary effects and the hydrodynamics of this film could be determining the flow boiling behaviour.

Table 2: Relevant dimensionless groups used in boiling heat transfer in microscale configurations

Dimensionless number	Expression	Relevance
Bond number	$Bo = \frac{(\sigma/(g(\rho_l - \rho_g)))^{0.5}}{D_h}$	$\frac{\text{Bubble dep. dia}}{\text{Channel dia}}$
Confinement number	$Co^2 = \frac{1}{Bo}$	$\frac{\text{Channel dia}}{\text{Bubble dep. dia}}$
Eötvös number	$Eo = \frac{g(\rho_l - \rho_g)L^2}{\sigma}$	$\frac{\text{gravitational forces}}{\text{capillary forces}}$
Capillary number	$Ca = \frac{\mu u}{\sigma}$	$\frac{\text{viscous forces}}{\text{capillary forces}}$
Weber number	$We = \frac{L\dot{m}^2}{\rho\sigma}$	$\frac{\text{inertial forces}}{\text{capillary forces}}$
Jacob number	$Ja = \frac{\rho_l}{\rho_g}\frac{c_{pl}\Delta T}{h_{lg}}$	$\frac{\textit{specific heat}}{\text{latent heat}}$

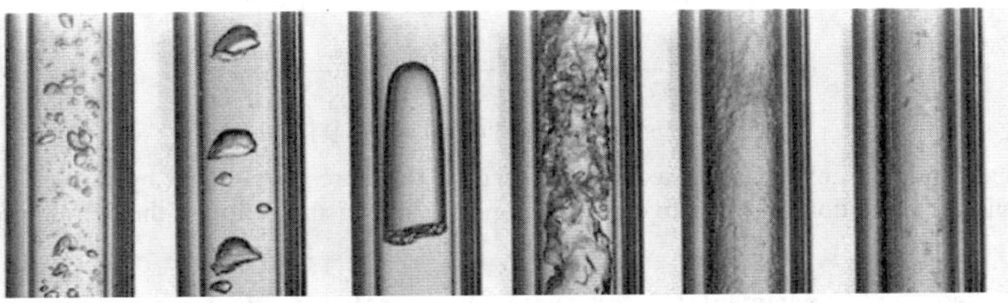

Dispersed Bubble   Bubbly   Slug   Churn   Annular   Mist

Figure 5. Flow patterns observed by Huo et al [6]

Many investigations have shown that during boiling in micro and mini channels the transition between the various flow regimes is different from the one observed for conventional channels. It is demonstrated for instance that the transition to annular flow takes place at lower superficial velocities compared to conventional channels. The following observations have been reported by many researchers:

1. Nucleation takes place in microchannels followed by the formation of vapour mass.
2. Dry out of the wall and its rewetting by the liquid is observed. This seems to be an unstable phenomena occurring intermittently.
3. Reverse flow has also been observed in many studies.

Brutin et al. [8] have investigated flow boiling in small minichannels; steady and unsteady behaviours were observed and found to be a function of the upstream boundary conditions. A critical Reynolds number is found to delimit steady and unsteady state behaviours. The same authors investigated two-phase flow oscillation phenomena occurring in flow boiling in narrow channels (Figure 6). An analysis of the steady and unsteady behaviour is made through a flow pattern analysis and a stability delimitation of the unsteady zone. Video analysis of a reverse flow for an unsteady period is also proposed. This phenomenon shows the vapour slug formation which blocks the two phase flow and pushes the two-phase flow back to the entrance.

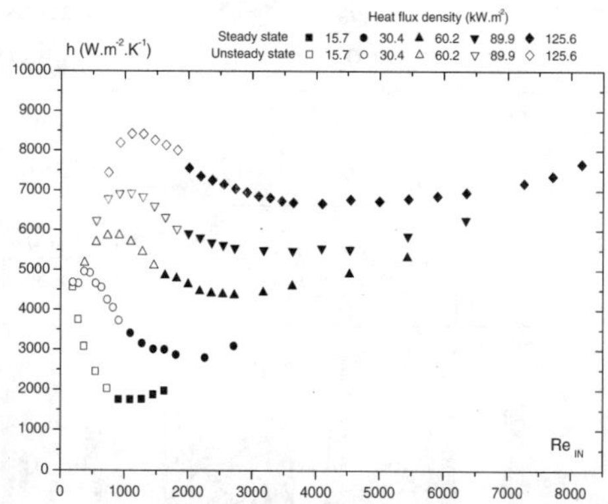

Figure 6. Instabilities observed [8]

In general, the investigation techniques used so far were constrained by the use of intrusive methods and the limitation in access to clear and non-intrusive visualization of the boiling process.

5. Challenges of Experimental Techniques in Microscale Investigations

The investigation of microscale heat transfer involves the implementation of experimental techniques at the microscale level. Few decades ago this was almost impossible for many areas to access to measurements at the microscale level. The parameters governing the evaporation/condensation process have a localised effect necessitating measurement of velocity, temperature, heat flux on the microscale.

Due to difficulties of making microscale measurements, the mechanisms governing the phase change have not yet been fully elucidated. The availability of high speed, high resolution digital photography, high resolution IR imaging, thermochromic liquid crystals and the development of novel microsensors has opened up new opportunities for measurement of the influences characterising the processes. Kenning and Yan [9] pioneered the use of thermographic liquid crystals (TLC's) to map local temperatures of the substrate during nucleate boiling on a thin plate (Figure 7). The results obtained by Kenning et al. reinforced the conviction of the need of local and microscale measurements in order to better understand boiling heat transfer.

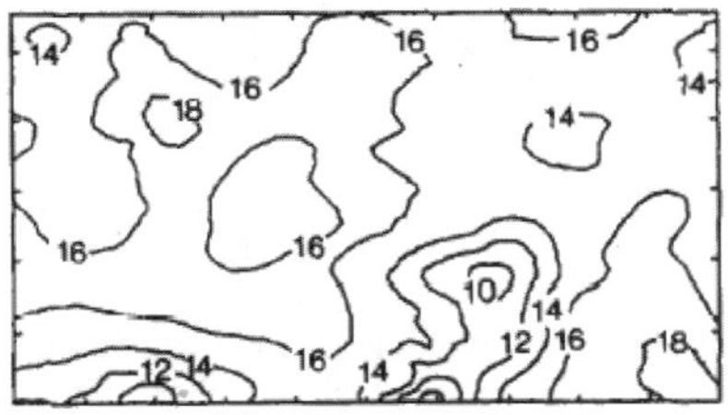

Figure 7. Temperature map during nucleate boiling on a thin plate [9]

Bae et al. [10] at Maryland University (USA) have developed a local heat transfer rates measurement device. The device was built for measuring heat-transfer rates at many points underneath areas of interest in heat transfer systems, in order to determine the heat-transfer coefficient as a function of position and time (Figure 8). The device is a planar array of small heaters that also serve as heat-flux sensors. The device enables measurement of the local heat transfer from the surface, with very high temporal and spatial resolution. The information provided by this, and other similar devices, could be used to validate or improve analytical and numerical models used in computational simulations.

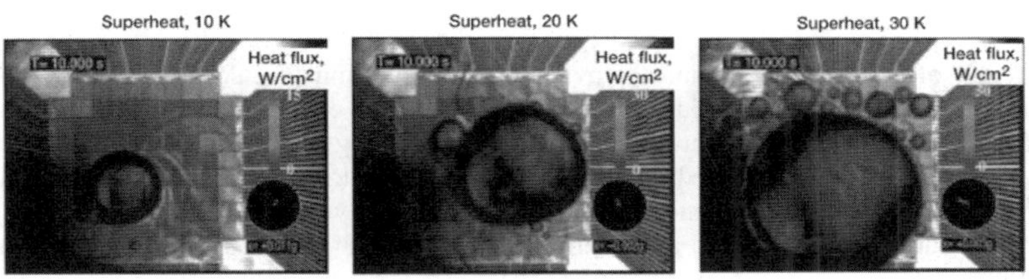

Figure 8. An array of micro-sensors for local heat transfer rates measurements [10]

The use of microscale techniques has been demonstrated to be useful in understanding the fundamental mechanisms of simultaneous heat and mass transfer. Buffone et al. [11] have used micro-particle image velocimetry (μ-PIV) to map the velocity field of evaporating thin films in small capillaries (Figure 9).

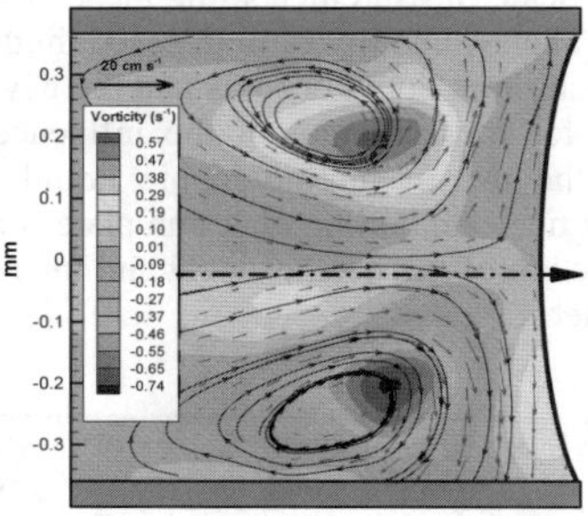

Figure 9. Particle Image Velocimetry (PIV) measurement during the evaporation of a liquid in a capillary tube, ID=200 μm [11]

The heat transfer and fluid flow processes associated with liquid-vapour phase change phenomena are typically among the most complex transport circumstances encountered in engineering applications. These processes may have all the complexity of single phase convective transport (nonlinearities, transition to turbulence, three-dimensional or time varying behaviour) plus additional elements resulting from motion of the interface, non-equilibrium effects or other complex dynamic interactions between the phases. Due to highly complex nature of these processes, development of methods to predict the associated heat and mass transfer has often proven to be a formidable task. Although empirical correlations describing heat transfer mechanism have been around for some time, reliable microscale devices measuring local parameters to confirm their validity are still lacking, primarily due to the complexity of these processes and difficulties in setting up and conducting experiments. A wide range of applications are dependent on the understanding of the local parameters of heat and mass transfer at a microscale level.

6. Further discussion

The behaviour of a single bubble and the evaporation of the thin film separating the vapour and the solid wall are crucial in understanding flow boiling in confined environments. It is well accepted that surface tension and interface curvature considerations affecting condensation, bubble formation and liquid transport at these scales need comprehensive investigation. Moreover, the reduction of bulk volume means that transport phenomena are taking place very close to interfaces. Interfaces themselves are not well understood and even less known are the mass or heat transport mechanisms through them.

The nature of the triple interline region is dependent on the shape of the meniscus, and more importantly on the meniscus contact angle with the wall. Based on motivations other than boiling, a lot of work has been carried out to try to model the contact angle and how it is affected by different parameters. Wayner has made major contribution in unravelling some aspects of this subject. Potash and Wayner [12] investigated the transport processes that occurred in a two-dimensional evaporating meniscus and adsorbed thin film on a superheated flat glass plate immersed in a liquid. The work was able to conclude that a change in the profile of the meniscus brings about a pressure drop that is sufficient to give the fluid flow required for evaporation. In a further work, Renk and Wayner [13] reaffirmed the fact that the meniscus profile was a function of the rate of evaporation and demonstrated that the fluid flow resulting from a change in the meniscus profile is able to replenish the liquid evaporated from a stationary meniscus. Given the size of the local evaporative heat fluxes from the meniscus in these cases they can act as heat sinks. Morris [14] in a recent paper presented an exhaustive analysis for contact angles for evaporating liquids. He established that the apparent contact angle of an evaporating meniscus is fundamental for computing heat flow. He derived an expression for the contact angle which he compared to the results of Kim et al. [15]. Höhmann et al. [16] have experimentally investigated the temperature profile underneath an evaporating meniscus. They have used liquid crystals to map the temperature at a very small scale. They have demonstrated a dip in the temperature near the triple line related to the high evaporative flux.

Recently, Gokhale et al. [17] used experimental and analytical techniques to measure the liquid pressure field in a meniscus formed by a constrained vapour bubble. They have developed an analytical expression for the curvature as a function of the film thickness profile and the apparent contact angle was deduced. The dynamics of bubbles in a confined environment is still the subject of wide investigations. With the interferometry technique to measure profiles of evaporating menisci they were able to measure thicknesses of films having very small contact angle (Figure 10).

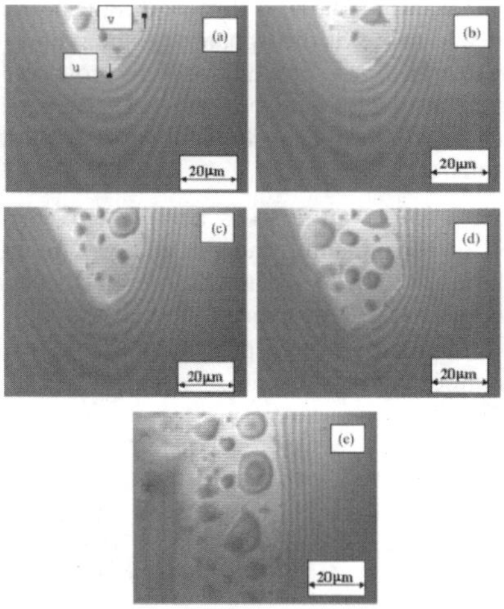

Figure 10. Snapshots of the interference fringes of the corner meniscus during break-up from drop on slightly hydrophobic quartz surface, Gokhale et al. (2003)

The effect of geometry size reduction on the dynamics of bubbles is crucial. It is important indeed to be able to predict the scale below which the confinement starts to influent the bubbles growth and dynamics. In order to be able to answer these and many other questions, precise experimental measurements are required. At these small scales developing non-intrusive measurement techniques remains a big challenge. The growth of bubbles in microchannels influences and interacts with the fluid flow. Many forces are considered to be determining in these process: inertial, capillary and viscous forces are thought to be the dominant ones at these small scales. Another important phenomenon taking place is the interaction between vapour and the solid wall. The liquid film separating the vapour and the wall can become very thin, very high evaporation rates can be achieved from these thin films. The stability of these thin evaporating films and their rupture can lead to dry out. Dry out leads to the formation of hot spots and triple lines. When investigating the fundamentals of the meniscus region close to the tube wall it is usually [18, 19] divided in three different sub-regions. In the adsorbed layer of constant thickness the Van-der-Walls forces dominate and no evaporation takes place. In a very tiny region (usually few percents of the entire meniscus length) most of the evaporation takes place because the adhesion forces are balanced by the capillary ones; this sub-region is called micro-region. As the meniscus thickness increases in the macro-region, the thermal resistance of the layer increases accordingly and the heat transfer coefficient decreases. In the macro-region the capillary forces dominate and set about a flow to re-supply the liquid evaporated in the micro-region. In these situations many factors play major role: wetability, solid surface energy and disjoining pressure are amongst these. The interaction between the thin film and the solid is determining of the behaviour of the dry spot and consequently the dry out.

7. Concluding Remarks

Role of microscale heat transfer is examined here in flow boiling heat transfer in channels of large and small diameters. It is concluded that some of the apparently anomalous trends can only be explained on the basis of microscale heat transfer. In addition, the role of microscale heat transfer in enhancement of nucleate boiling heat transfer is also discussed. The paper also highlights the role that can be played by fundamental microscale heat transfer investigation in the understanding and improving of flow boiling heat transfer.

As discussed in the paper a lot of progress is being made in developing new non-intrusive measurement techniques as well in fundamental understanding of microscale heat transfer. It does appear that many practical hurdles need to be overcome before microchannel heat exchanger is adopted for the main stream industrial applications. However in view of the rapid developments in this area, it is best to keep a watching brief on this area from a practicing thermal research engineer's perspective and the present paper serves this important purpose.

NOMENCLATURE

Bo	Bond number
Ca	Capillary number
Co	Confinement number
c_p	Specific heat capacity; (J/kgK)
D_H	Hydraulic diameter
Eo	Eötvös number
g	Gravitational acceleration; (m/s²)

Ja	Jacob number
L	Length; (m)
$\dot{m}$	Mass flux; (kg/m2s
T	Temperature; (K)
We	Weber number

Greek symbols

ρ	Density; (kg/m3)
σ	Surface tension; (N/m)

Subscripts

g	Gas
l	Liquid

REFERENCES

1. Honda, H., Wei J.J., (2003) Enhanced boiling heat transfer from electronic components by use of surface microstructures, *Experimental Thermal and Fluid Science*, Vol.28, , pp.159-169.
2. Stephan, P., (2002) Microscale Evaporation Heat Transfer: Modelling and Experimental Validation, Heat Transfer 2002, *Proceedings of 12th International Heat Transfer Conference*, Paper No. 08-KNL-02, Vol.1, pp.315-327.
3. Wu. W.-T. W.,Yang. Y.-M and Maa. J.-R, (1998) Nucleate pool boiling enhancement by means of surfactant additives, *Experimental Thermal and Fluid Science*, Vol.18, November, pp.195-209.
4. Urso, M. E. D., Wadekar V. V. and Hewitt, G. F., (2002) Heat Transfer at the Dryout and Near-Dryout Regions in Flow Boiling. Transfer, *Proc.12th International Heat Transfer Conference*, pp. 701-706.
5. Kandlikar ,S.G., (2002) Fundamental issues related to flow boiling in minichannels and microchannels, *Experimental Thermal and Fluid Science*, Vol.26, pp.389-407.
6. Huo, X., Chen, L., Tian, W. and Karayiannis, T., (2003) Flow boiling and flow regimes in small diameter tubes, *8th UK National Heat Transfer Conference*, Oxford, September 2003, Paper No. FB 3.
7. Jacobi, A. M. and. Thome J. R., (2002) Heat Transfer Model for Evaporation of Elongated Bubble Flows in Microchannels, *ASME Journal of Heat Transfer*, Vol.124, pp.1131-1136.
8. Brutin, D., Topin, F. and Tadrist, L., (2003) Experimental study of unsteady convective boiling

in heated minichannels, *International Journal of Heat and Mass Transfer*, Vol.46, pp.2957-2965.

9. Kenning, D. B. R. and Yan, Y., (1996) Pool boiling heat transfer on a thin plate: features revealed by liquid crystal thermography, *International Journal of Heat and Mass Transfer*, Vol.39, pp.3117-3137.

10. Bae, S., Kim, M.H., and Kim, J., (1999) Improved technique to measure time and space resolved heat transfer under single bubbles during saturated pool boiling of FC-72, *Exp. Heat Transfer*, Vol.12(3), pp.265–278.

11. Buffone, C., and Sefiane, K., (2004) Investigation of thermocapillary convective patterns and their role in the enhancement of evaporation from pores, *International Journal of Multiphase Flow*, Vol.30, pp. 1071-1091.

12. Potash, J., Wayner, P., (1972) Evaporation from a two-dimensional extended meniscus, *International Journal of Heat and Mass Transfer*, Vol.15, No.3, pp. 1851-1863.

13. Renk, F. and Wayner, J.P., (1979) An Evaporating Ethanol Meniscus. Part I: Experimental Studies, *Journal of Heat Transfer*, Vol.101, pp.55-58

14. Morris, S., (2001) Contact angles for evaporating liquids predicted and compared with existing results, *Journal of Fluid Mechanics*, Vol.432: pp.1-30

15. Kim, I. and Wayner, P., (1996) Shape of an evaporating completely wetting extended meniscus, *Journal of Thermophysics and Heat Transfer*, Vol.10, pp.320-325

16. Höhmann, C., Stephan, P., (2002) Microscale temperature measurement at an evaporating liquid meniscus, *Experimental Thermal and Fluid Science*, Vol. 26, pp.157-162.

17. Gokhale, S. J., Plawsky. J. L., and Wayner Jr., P. C., (2003) Effect of interfacial phenomena on dewetting in dropwise condensation, *Advances in Colloid and Interface Science*, Vol.104, pp.175-190,.

18. Swanson, L.W., Herdt, G.C. (1992) Model of evaporating meniscus in a capillary tube, *Journal of Heat Transfer* ,Vol.114, pp.434-440

19. Schonberg, J.A., Dagupta, S., Wayner, P.C., (1995) An augmented Young-Laplace model of an evaporating meniscus in a microchannel with high heat flux, *Experimental Thermal and Fluid Science*, Vol.10, pp163-170.

Appendix 1 Summary of Investigation on Evaporation in Mini and Microchannels (Adapted from Kandikar [5])

Author/Year	Fluid and ranges of parameters, G(kg/m²s), q(kW/m²)	Channel size/D_a(mm)	Heat transfer	Pressure drop	Flow patterns	Remarks
Cornwell and Kew, 1992	R-113, G=124-627 q=3-33	Parallel rectangular, 75 channels-1.2×0.9 deep 36 channels-3.25×1.1deep	Heat transfer coefficient	Not reported	Isolated bubble, Confined bubble, annular-slug	The heat transfer coefficient was found to be dependent on the existing flow pattern. In the isolated bubble region, $h \sim q^{0.7}$, lower q effect in confined bubble region, convection dominant in annular-slug region.
Wambsganss et al., 1993	R-113, G=50-100, q=8.8-90.7	Circular, D=2.92 mm	h as a function of x, G and q	Not reported	Not reported	Except at the lowest heat and mass fluxes, both nucleate boiling and convective boiling components were present.
Bowers and Mudawar, 1994	R-113 q=1000-2000, 0.28-1.1ml/s	Mini and microchannels, D=2.54 and 0.51	Heat transfer rate	Pressure drop components	Not studied	Analytical (1993) and experimental (1994) studies comparing the performance of mini and micro channels. Mini channels are preferable unless liquid inventory or weight constraints are severe.
Peng et al., 1998	Theoretical	Theoretical	Bubble nucleation model	Not studied	Not studied	Bubble nucleation model uses a vapour bubble growing on a heater surface with heat diffusion in the vapour phase. This corresponds to post-CHF heat transfer and is an inaccurate model of heat transfer during nucleation and bubble growth.
Kuznetsov and Shamirzaev	R=318C, G=200 900	Annulus, 0.9 gap×500	h~1-20 kW/m²C	Not studied	Confined bubble	Capillary forces important in flow

Shamirzaev, 1999	G=200-900, q=2-110	gap×500	kW/m^{2}C		bubble, Cell, Annular	important in flow patterns, thin film suppresses nucleation, leads to convective boiling.
Hestroni et al., 2000	Water, Rc=20-70, q=80-360	Triangular parallel channels, θ=55°, n=21.26, Dh=0.129-0.103, L=15	Measured, but data not reported	Not reported	Periodic annular	Periodic annular flow observed in microchannels. There is a significant enhancement of heat transfer during flow boiling in microchannels.
Serizawa and Feng, 2001	Air=Water, j_L=0.003-17.52 m/s, j_G=0.0012-295.3 m/s, Steam water, ranges not given	Circular tubes, diameters of 50 μm for air-water, and 25 μm for steam-water	Air-water two-phase flow, steam-water details not given	Not studied	Flow patterns identified over the ranges of flow rates studied	Two new flow patterns identified: liquid ring flow and liquid lump flow.

HEAT TRANSFER ISSUES IN CRYOGENIC CATHETERS*

RAY RADEBAUGH
Cryogenic Technologies Group
National Institute of Standards and Technology
Boulder, Colorado, USA

1. Introduction

Catheters with diameters of 3 mm or smaller are capable of accessing many internal organs through arteries and veins. They provide the medical community with tools to operate on internal organs without the need to cut through the body to gain access. As a result the use of catheters greatly speeds recovery times, reduces costs, and often reduces risks. The removal of unwanted tissue, such as cancer tumors or malfunctioning tissue can be carried out through cryoablation with cryogenic catheters. The effectiveness of these catheters depends very much on enhanced heat transfer at the very small tip. Issues to be discussed in this paper include the heat transfer at the boundaries between the working fluid and the catheter tip and between the tip and the tissue. Also of importance is the transient heat transfer within the surrounding tissue. In some cases a miniature heat exchanger between high and low pressure streams of a Joule-Thomson cycle may be required at the cold tip. Hydraulic diameters less than 100 μm are required in such miniature heat exchangers to provide high effectiveness and high heat flux densities. Steep temperature gradients are required for the destruction of cancer tumors or other unwanted tissue without excessive damage to surrounding normal tissue. Such a requirement can be met by making the catheter tip as cold as possible. However, achieving lower temperatures with a Joule-Thomson cycle requires higher refrigerant flow rates because the heat of vaporization of lower temperature refrigerants is less than that of refrigerants with higher boiling points. The refrigerant flow rate is limited by the catheter diameter. Thus, there is some optimum temperature and associated refrigerant that provides the maximum temperature gradient in the tissue being destroyed. In this paper we examine the use of mixed refrigerants that can be tailored for any temperature. Factors that influence the lower size limit of cryogenic catheters, such as choice of working fluid, flow rates, tip temperature, and pressures will be covered. An interesting example for cryogenic catheters, which will be discussed here, is the application of freezing locations on the heart that are triggering cardiac arrhythmias or irregular heart beats.

The recent development of small diameter and flexible cryogenic catheters has opened up the possibility of treating many abnormalities, including cancer, in internal organs without major surgery to gain access to the organ or to remove the organ. Instead of a several-day hospital stay and about a 6-week recovery period, the cryosurgical procedure can usually be performed as an outpatient service with local anesthesia that has a one-day recovery period.

S. Kakaç et al. (eds.), Microscale Heat Transfer, 445–464.

Successful application of cryosurgical probes is very much dependent on providing a high cooling rate at the tip with little or no cooling along the length of the catheter. To reach further into the body, catheters need to be made smaller and smaller. Large amounts of heat (often 10s of watts) need to be removed at the catheter tip in order to freeze a large enough region at a fast enough rate to provide well-defined regions of tissue destruction. These requirements lead to high heat flux densities and microscale geometries inside the catheter tip to provide high surface areas for heat transfer. Rapid cooling and freezing of the surrounding tissue requires the internal working fluid at the tip of the catheter to be at a significantly lower temperature than the tissue in contact with the tip. In general, refrigerants with lower boiling points have lower heats of vaporization. Catheter operation at lower temperatures then requires higher flow rates, but the diameter and length of the catheter provides an upper limit to this flow rate. Thus, it is important to optimize the refrigerant operating temperature. In this paper we discuss the competing catheter requirements, such as the need for small size and high heat transfer densities in the presence of high refrigerant flow rates. An approach to the optimization of the various competing catheter requirements is explained in this paper.

2. History of Cryosurgery

Cryosurgery had its beginnings in 1851 when Dr. James Arnott used iced saline solutions to treat carcinomas of the breast and cervix [1,2]. The minimum temperature of –21 °C achievable with this technique is just at the upper limit for cell destruction [3], so Arnott had little success in curing the cancers, but achieved some beneficial effects, such as pain relief and reduction of bleeding. The use of solid carbon dioxide and liquid nitrogen around 1900 for cryosurgery [4] permitted much lower temperatures and more complete tissue destruction. These lower temperatures were commonly used in dermatological applications after 1900, but were not used much for deep-seated malignancies. One exception was the work of the neurosurgeon Temple Fay between 1936 and 1940 [5]. He used implanted metal capsules connected to an external refrigeration system to treat brain tumors. He also used refrigerated liquids to treat large inoperable cancers of the cervix and breast. Few advances in cryosurgery followed until in 1961 the New York neurosurgeon Dr. Irving Cooper designed and used a cryosurgical probe cooled with liquid nitrogen to treat Parkinson's disease [6]. This event is often taken as the beginning of modern cryosurgery, although subsequent developments in medication have gone on to replace the cryosurgical approach for this disease.

In the mid 1960s Gonder and colleagues developed a modified liquid nitrogen cryosurgical systems for prostate cryosurgery [7]. Extensive animal experiments were carried out that lead to a broader clinical use of cryosurgery. So far liquid nitrogen had become the most common cryogen for use in cryosurgery, but it had certain disadvantages. It could not be stored for long periods of time and it required bulky equipment for its use in cryosurgery. Thus, the equipment could not be easily moved from room to room. It was not easy to use and control. The freezing process could not be stopped quickly because of the time for the liquid to drain from the probe. Vacuum insulated probes were required to prevent freezing along the length of the probe. Many advances in cryosurgical probes have occurred since 1961, and a much better understanding of tissue destruction at cold temperatures has developed. Advances in cryobiology have shown that cell temperatures of –40 °C or lower result in complete destruction of cancerous tissue, though somewhat higher temperatures up to about –20 °C may lead to destruction of some healthy tissue [8,9]. It has also been found that the lethal cell temperature can be increased by repeated freeze/thaw cycles and by rapid cooling and slow warming

[3]. In the last 40 years several cooling methods have been developed and used for cryogenic catheters and probes. It was during this period that Joule-Thomson (JT) systems with high-pressure argon and nitrous oxide (N_2O) were introduced. Expansion of these fluids from high to low pressure results in cooling and liquefaction of the fluid. Liquid argon (-186 °C), liquid nitrous oxide (N_2O) (-88 °C), and various hydroflourocarbons (HFCs) and fluorocarbons (FCs) (-30 °C to –80 °C) are now used as refrigerants in addition to liquid nitrogen (-196 °C). Mixtures of pure fluids have recently been used [3,10] that allow for a wide range of boiling temperatures between –30 °C and –196 °C.

These fluids can either be introduced to the cryogenic probe at near-atmospheric pressure and already cold, for example, liquid nitrogen, in which case the entire length of the probe will be cold unless it is insulated, or they can be introduced at high pressure and allowed to undergo a Joule-Thomson expansion and cooling at the tip to near ambient pressure. Pure fluids like nitrogen and argon require pressures of at least 15 MPa (2200 psia) and an efficient heat exchanger at the tip to cool from ambient to their normal boiling points. Nitrous oxide requires a pressure of about 5.9 MPa and no heat exchanger to reach its normal boiling point of 185 K (–88 °C) if the condenser temperature is at 300 K (27 °C). If the high pressure is reduced to 2.5 MPa, a heat exchanger is required to reach 185 K for the aftercooler (condenser) at 300 K. However, if the aftercooler is held at 263 K (-10 °C) condensation of the N_2O will occur at the pressure of 2.5 MPa, and the low temperature of 185 K can be reached after expansion without the use of a heat exchanger. For this reason some two-stage systems are being used for cryogenic catheters. For a pressure of 2.5 MPa with mixed gases and with a small heat exchanger at the tip, temperatures down to about 85 K (-188 °C) are possible [10]. Thus, there are now available a wide range of refrigerant options for use in cryosurgical probes that span the temperature range from about –196 °C to about –30 °C.

The 1980s saw the development of a competing technology, that of electrosurgical or radio frequency (rf) catheters that destroy tissue by heating them to temperatures above about 42 °C. These rf catheters are simpler and easy to control. However, they are limited to the amount of tissue they can destroy at one location because the temperature difference between the 42 °C destruction temperature and the catheter tip is limited by the 100 °C boiling point of the water in the tissue. The rf catheters have become quite widely accepted by the medical community at this time. The later development of cryogenic catheters has hindered their use in place of the rf catheters

Beginning in the 1990s the development of improved cryosurgical probes along with advances in ultrasound and MRI imaging to locate the ice front have resulted in considerable interest within the medical community in the use of cryosurgery for a variety of applications [11]. Early applications were mostly for treatment of abnormalities near the body surface. However, the recent development of small diameter and flexible cryogenic catheters has opened up the possibility of treating many abnormalities, including cancer, in internal organs without major surgery to gain access to the organ or to remove the organ. Instead of a several-day hospital stay and about a 6-week recovery period, the cryosurgical procedure can usually be performed as an outpatient service with local anesthesia that has a one-day recovery period. For example a new cryogenic catheter [12] received the approval by the U.S. Food and Drug Administration (FDA) in 2001 to treat women with abnormal menstrual bleeding by freezing the uterine lining instead of surgically removing the uterus in a hysterectomy [13]. Another new cryogenic catheter has been developed for the treatment of certain types of cardiac arrhythmia or irregular heart beating [14]. Over 5 million people in the world's developed countries suffer from some form of heart arrhythmia. Current treatments involve medication or the use of rf catheters. Heart arrhythmia occurs when a portion of the heart distorts the heart's electrical signals. Ablating the tissue that distorts these electrical signals with either rf or cryogenic catheters eliminates the arrhythmia in almost all cases. Though newer than rf catheters, the cryogenic cardiac catheters offer the following advantages: (1) cryomapping capability, (2) not susceptible to forming blood clots, and (3) will stick to the tissue once freezing begins and not be moved out of place by heart movement. The cryogenic cardiac catheters are one of the most challenging cryogenic catheters in terms of refrigeration and heat transfer issues. Figure 1 shows the procedure used with the cardiac catheters for treating heart arrhythmia. The catheter is about 3 mm in diameter and about 1 m long. It is inserted through a small incision into a large vein in the leg in the groin area as shown in Figure 1(a). It enters the right atrium of the heart via this vein and continues into the left atrium through a hole punctured in the wall between the two chambers as shown in Figure 1(b). A needle-tipped catheter was used previously to make the puncture and then withdrawn and leaving a tube through which the cryogenic catheter is inserted. With the catheter tip in the proper location at the junction to the pulmonary vein coming from the lungs, a small balloon is inflated to temporarily block the blood flow, as shown in Figure 1(c). Finally, refrigerant flow to the catheter tip is initiated to cause freezing of the necessary tissue.

Further improvements in cryosurgical probes could open up many more medical applications, including the treatment of cancers of the liver, breast, and lung. Successful application of cryosurgical probes is very much dependent on providing a well controlled cooling rate at the tip with little or no cooling along the length of the probe. To reach further into the body, catheters need to be made smaller and smaller. Large amounts of heat (often 10s of watts) need to be removed at the catheter tip in order to freeze a large enough region at a fast enough rate to provide well-defined regions of tissue destruction. Cryosurgery applications have the potential to expand considerably and become a major medical procedure if the following developments occur: (a) the medical device community has the information needed to help them select the optimum cooling method in a cryosurgical probe for a particular application, (b) a given heat removal rate can be achieved in smaller and smaller cryogenic catheters, and (c) compact, reliable, and-easy-to-use catheter systems are readily available that can be applied to various procedures. Recent reviews of cryosurgery have been given by Dobak [3], Rubinsky [11], and Theodorescu [15].

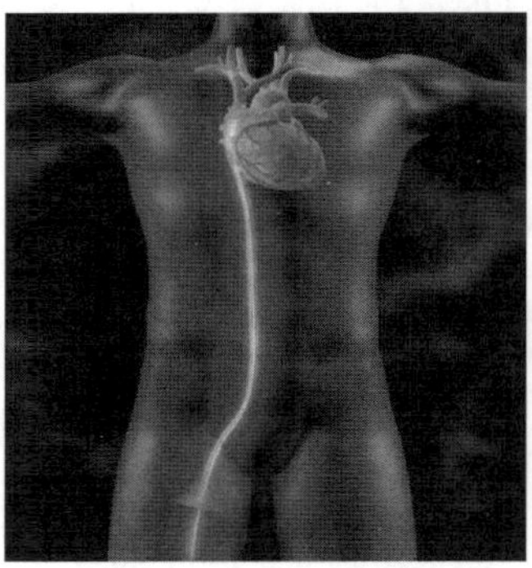

Figure 1a. Cryogenic heart catheter inserted into vein at the groin and advanced to the right atrium.

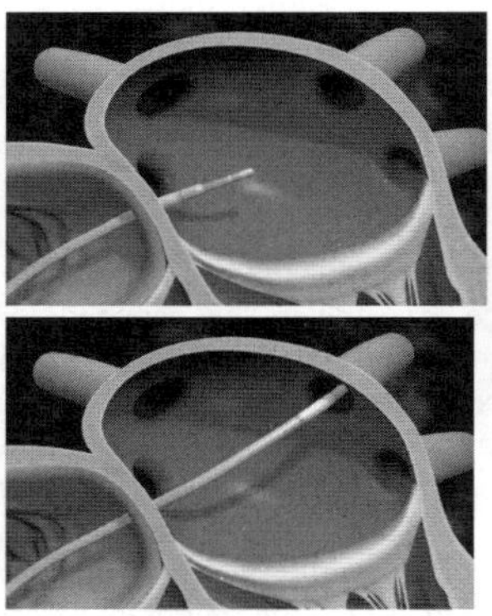

Figure 1b. After the septum is punctured the catheter is passed into the left atrium.

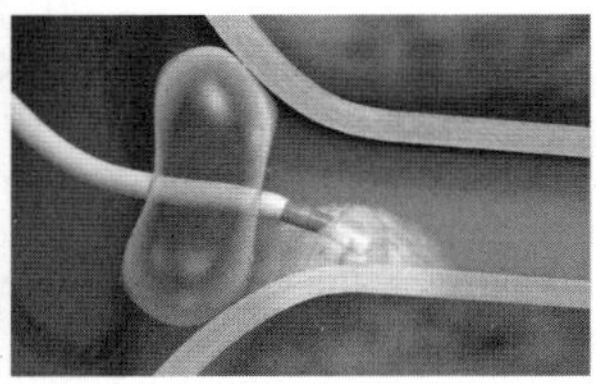

Figure 1c. With catheter in place balloon is inflated to block blood flow and catheter tip freezes damaged tissue.

Figures courtesy of CryoCor.

3. Advantages and Disadvantages of Cryosurgery

Table 1 shows the advantages and disadvantages of cryosurgery compared with conventional surgery or hyperthermia (heating). One of the major advantages compared with conventional surgery where tissue is cut is the rapid recovery times of about one day. The use of catheters permits access to internal organs without cutting through the external part of the body. The cryogenic catheters have two significant advantages compared with rf or laser catheters. There is some tendency for blood clots to form during the burning process with rf or laser catheters, whereas such clots do not tend to form during freezing. Another significant advantage is the cryomapping ability of most cryogenic catheters. Cryomapping is the process of cooling tissue to a low enough temperature that its normal activity ceases, but the activity resumes with no side effects after the tissue is warmed back to normal temperature. This capability is particularly useful in the treatment of cardiac arrhythmias. In this case a local disturbance of the heart's electrical signal causes the arrhythmia, but the surgeon may not know the precise location. The cryogenic catheter can be positioned at the suspected location and the refrigerant flow in the catheter is turned on at a low rate. The suspected tissue is cooled enough that the arrhythmia would temporarily stop if the catheter were in the proper location, but the tissue would not be destroyed at that temperature. If the arrhythmia did not stop with the preliminary cooling, the surgeon would move the catheter tip to another location and try the procedure again. Once the correct location is found the refrigerant flow is increased and the tissue cooled to a low enough temperature to bring about tissue destruction. The arrhythmia is then permanently stopped. The major disadvantages to the use of cryogenic catheters are that at present physicians have limited experience with the actual use of the devices, and the long-term effects of cryoablation are not well characterized yet. The Joule-Thomson catheters utilize high-pressure fluids to provide the cooling at the tip. A leak in the catheter that allows the fluid to escape into the body could be dangerous even though nontoxic fluids are used. Imaging of the ice ball is often done with MRI, which adds to the cost of performing cryosurgery. With a single catheter it is difficult to achieve the desired temperature profiles over a given volume.

Table 1. Advantages and disadvantages of cryogenic catheters compared with conventional surgery and with rf heating catheters.

Advantages	Disadvantages
• Minimally invasive • Few side effects • Treatable as outpatient (less expensive) • Short recovery period (about 1 day) • Cryomapping capability • Not prone to blood clots • Sticks to tissue during freezing • Preserves structural integrity of tissue • Larger lesions than rf catheters • Repeatable	• Short history • Long-term effectiveness uncertain • Few physicians familiar with procedures • Bulky equipment in some cases • High pressure fluids present inside body • Expensive MRI for observing ice ball • Precise temperature profiles difficult

4. Heat Transfer in Tissue and Catheters

A typical cryogenic catheter tip with a surrounding ice ball is shown in Figure 2. The refrigerant is brought in to the tip through the smaller inside tube. The refrigerant absorbs heat from inside the tip and exits from the tip through the annular region formed by the coaxial tubes. In many cases the refrigerant may be vented to the atmosphere once it leaves the catheter. In other cases it will enter a compressor to be circulated back into the catheter through the small inner tube. A guide wire and an internal thermocouple are usually a part of the cryogenic catheter as shown in this cutaway view in Figure 2. The surgeon can steer the tip of the catheter by manipulating the guide wire back at the handle of the catheter. The temperature profile for a cross-section of a cryogenic catheter and the surrounding tissue is shown in Figure 3. The lowest temperature is that of the cryogen inside the tip of the catheter. It is absorbing heat from the surrounding tissue, and because of a finite heat transfer coefficient between the inside of the tip and the cryogen, there is a finite temperature difference at the interface, as shown in Figure 3. Because the diameter of the tip is quite small, typically less than 5 mm and often less than 3 mm, the surface area for heat transfer at the internal and external surfaces of the tip are quite limited. Boiling heat transfer usually occurs inside the tip and enhanced surface areas could be used to reduce the temperature difference ΔT at the inside surface. Contact with the tissue occurs at the outside interface, where another finite ΔT exists. Freezing usually causes a good bond between the tip and the tissue. Base on measurements of joint thermal conductance for grease joints we estimate that the thermal conductance at the interface with the tissue would be about 1-2 W/(cm^2·K). For a heat flow of 10 W into a 3 mm diameter tip, 10 mm long, the heat flux Q_A is 10.6 W/cm^2 and the ΔT is about 5-10 K at the outside interface. For boiling heat transfer inside the catheter tip the 10.6 W/cm^2 heat flux would be near the critical value for liquid nitrogen [16] and cause a ΔT of about 10 K. For refrigerants with higher boiling points the critical heat flux will generally be higher than for liquid nitrogen.

The temperature profile within the tissue can be determined from the Pennes bio-heat equation [17]

$$\rho_t c_t \frac{\partial T}{\partial t} = \nabla(k_t \nabla T) + q_b + q_m, \tag{1}$$

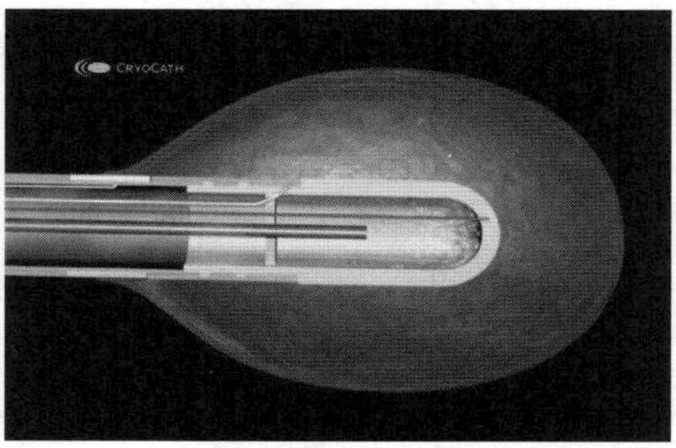

Figure 2. Cutaway view of a coaxial catheter surrounded by an ice ball. Steering wire and thermocouple are inside the catheter. Courtesy of CryoCath.

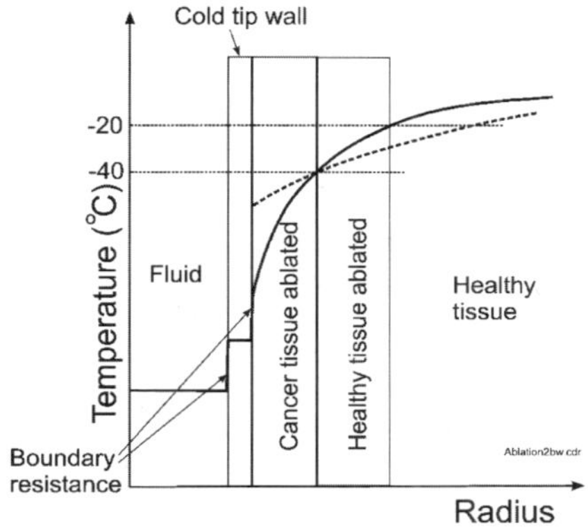

Figure 3. Cutaway schematic of catheter tip and ice ball showing two different temperature profiles and the location destroyed tissue. With the steeper gradient less healthy tissue is destroyed with temperatures below –20 °C.

where ρ_t is the density of the tissue, c_t is the specific heat of the tissue, T is the tissue temperature, t is time, k_t is the thermal conductivity of the tissue, q_b is the heat source density due to blood profusion, and q_m is the heat source density due to metabolism. Usually the metabolism heat source can be neglected in this application. For tissue freezing an additional term should be added to equation (1) to account for the latent heat of fusion during the freezing process. The blood profusion heat source density is given by

$$q_b = w_b c_b (T - T_b), \tag{2}$$

where w_b is the profusion rate (mass flow per unit volume) of blood flow [kg/(s·m^3)], c_b is the specific heat of blood, and T_b is the temperature of arterial blood (37 °C). Usually we can take $c_t = c_b = c$. Table 2 lists thermophysical properties of selected biological materials [18]. Numerical techniques are often used to solve equation (1) for a particular geometry. For a simple steady-state case of the catheter tip being approximated by an infinitely long cylinder, negligible blood profusion, and a fixed temperature T at a radius r, the heat flux into the catheter of radius r_1 is given by

$$Q_A = \frac{\dot{Q}}{2\pi r_1 L} = \frac{k_t (T - T_1)}{r_1 \ln(r / r_1)}. \tag{3}$$

For r_1 = 1.5 mm at a temperature of –70 °C and the –20 °C isotherm at a radius r = 15 mm the heat flux

Table 2. Thermophysical properties of human tissue and water [18].

Material	Density, ρ_t [kg/m^3]	Conductivity, k_t [W/(m·K)]	Specific Heat, c_t [J/(kg·K]	Blood Profusion, w_b [kg/(s·m^3)]
Muscle	1010-1050	0.38-0.54	3600-3800	0.0016
Fat	850-940	0.19-0.20	2200-2400	0.013
Kidney	1050	0.54	3900	0.18
Heart	1060	0.59	3700	0.18
Liver	1050	0.57	3600	
Brain	1040-1050	0.16-0.57	3600-3700	0.06
Water 37 °C	990	0.63	4300	

is $Q_A = 0.72$ W/cm^2 for a tissue thermal conductivity of $k_t = 0.50$ W/(m·K). With this low heat flux there is only about a 1 K temperature drop at the interface between the tissue and the catheter tip. The temperature gradient at the radius r becomes

$$\frac{dT}{dr} = \frac{Q_A(r_1/r)}{k_t}. \tag{4}$$

For this example the temperature gradient at the –20 °C isotherm ($r = 15$ mm) is 14.5 K/cm and at the interface with the catheter the gradient is 145 K/cm. For tumor destruction at –40 °C and undamaged healthy tissue at –20 °C a 15 K/cm gradient would cause about 13 mm of healthy tissue being destroyed by cooling below its lethal temperature. The temperature profile shown by the dashed curve in Figure 3 would be representative of a profile with a small temperature gradient. By reducing the temperature of the refrigerant and imposing a rapid temperature change at the surface of the catheter tip, a steeper temperature gradient as shown by the solid curve in Figure 3 can be obtained. The temperature profiles shown in Figure 3 are not for steady state but are for one instant of time. At a later time they will have shifted some to the right except at the interface to the catheter. Thus, the cooling of the tip must be stopped at just the right time. A steeper gradient would lessen the amount of healthy tissue being destroyed. The steeper gradient requires a lower temperature in the catheter and more refrigeration power. However, with the 10 W refrigeration power at the tip and the heat flux of 10.6 W/cm^2, equation (4) indicates a very large gradient of 2120 K/cm at the catheter interface.

5 Cryogenic Refrigeration Techniques

5.1 CONVENTIONAL REFRIGERATION

The refrigeration techniques required to reach cryogenic temperatures are different than those of conventional vapor-compression refrigeration, which is used for most cooling applications closer to ambient temperatures. Most domestic refrigerators and air conditioners use the vapor-compression cycle. Figure 4a shows a schematic of the vapor-compression cycle, and Figure 4b shows the path of

Figure 4. (a) Schematic of the vapor-compression cycle with an oil-lubricated compressor. (b) The vapor-compression cycle shown on a temperature entropy diagram operating between a low pressure P_L and a high pressure P_H and between a low temperature T_c and ambient temperature T_0.

the cycle in the temperature-entropy (*T-S*) diagram. In this cycle heat is absorbed at some low temperature during the boiling of the liquid at a pressure near 0.1 MPa (1 bar). Typically the temperatures may be about 250 K (-23 °C) for most domestic refrigerators. At this temperature oil can remain dissolved in the refrigerant and not freeze. The vapor being boiled off in the evaporator then passes to the oil-lubricated compressor where it is compressed to about 2.5 MPa (25 bar). As the compressed vapor travels through the condenser it cools to ambient temperature and condenses into the liquid phase. The oil used for lubrication of the compressor is soluble in the refrigerant and a small amount of oil then completes the entire refrigerant cycle dissolved in the refrigerant. The condensed liquid then passes to the expansion capillary where the pressure is reduced to about 0.1 MPa and the temperature drops from ambient to about 250 K during this isenthalpic process between c and d in Figure 4.

5.2 RECUPERATIVE CYCLES

To achieve lower temperatures, particularly cryogenic temperatures of 120 K or less, the process shown in Figure 4 must be modified in two ways. First, the solubility of lubricating oil in the working fluid at such low temperatures is extremely small and any excess will freeze and cause plugging of the expansion channels. Thus, either the compressor must be oil free, which introduces reliability issues, or the system must have oil removable equipment utilizing complex processes (cost issues) to remove the oil before it reaches such low temperatures. Second, no fluid exists which can be expanded in an isenthalpic process (no expansion work) from room temperature to cryogenic temperatures. Even with a work-recovery process the initial pressure would need to be impractically high to achieve such low temperatures after expansion. Thus, it is necessary to precool the high-pressure gas in a heat exchanger prior to the expansion, as is shown schematically in the Joule-Thomson cryocooler in Figure 5a. The path followed on the *T*-S diagram is shown in Figure 5b. Because the heat transferred in the heat exchanger to provide sufficient precooling is much larger than

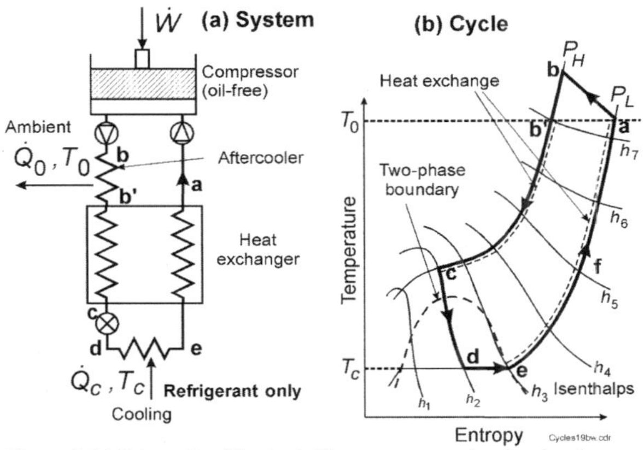

Figure 5. (a) Schematic of the Joule-Thomson cryocooler showing the use of an oil-free compressor and a high-effectiveness heat exchanger. (b) The Joule-Thomson cycle shown on a temperature-entropy diagram. Dashed lines indicate the heat exchange process in the heat exchanger.

the refrigeration power, the effectiveness of the heat exchanger must be very high, often higher than 95%. Small hydraulic diameters are needed in the heat exchanger to obtain such high effectiveness, especially for miniature cryocoolers. Hydraulic diameters of 50 to 100 μm may be required in some compact heat exchangers. When an expansion engine or turbine replaces the expansion orifice the cycle is called the Brayton cycle. Both it and the Joule-Thomson cryocooler are classified as recuperative types because of the use of recuperative heat exchangers throughout the cycle.

6 Refrigerants and Cryogens

6 1 PURE FLUIDS

Table 3 lists possible refrigerants for use in Joule-Thomson catheters. They are listed in order of decreasing normal boiling point temperatures T_{nbp}. The first fluid, R134a, is listed simply as a reference because it is the most common refrigerant now used in vapor-compression refrigerators, but its normal boiling point is too high for most any cryoablation application. The refrigerant R504 is a mixture of two components (R32 and R115), but it is listed here because it is an azeotrope. An azeotrope is a liquid mixture that has a unique constant boiling point and the liquid and vapor have the same composition. The next column gives the temperature $T_{2.5}$ of the vapor-liquid phase equilibrium at a pressure of 2.5 MPa. Such a pressure can be achieved with some commercial compressors for vapor-compression refrigerators. The latent heat of vaporization L_v is taken at atmospheric pressure, 1 bar. For a gross refrigeration power $\dot{Q}_r$ of a vapor-compression or Joule-Thomson (JT) refrigerator the gross refrigeration power per unit molar flow rate $\dot{n}$ is given by

$$q_r = \frac{\dot{Q}_r}{\dot{n}} = \Delta h_{\min}, \tag{5}$$

where $\Delta h_{\min}$ is the minimum enthalpy difference between the low (0.1 MPa) and high (2.5 MPa) pressure streams for temperatures between T_{nbp} and 300 K. For all fluids in Table 3 the minimum

Table 3. Thermodynamic and transport properties of various pure refrigerants. Conditions are for a high pressure of 2.5 MPa and a low pressure of 0.1 MPa with an ambient temperature of 300 K. Flow and ΔP are for 10 W of refrigeration at the cold tip in a 2.5 mm diameter circular tube.

Fluid	Mol. Wt. (kg/mol)	T_{nbp} (K)	$T_{2.5}$ (K)	L_v (J/mol)	$\Delta h_{\min}$ (J/mol)	q_r*	μ (μPa·s)	ρ (kg/m^3)	$\dot{n}$ (mmol/s)	$\Delta P/L$ (kPa/m)
R134a	0.10203	247.08	350.7	22138	19272	4.362	11.89	4.173	0.519	0.157
C_3H_8	0.04410	231.06	341.4	18777	16019	3.364	8.196	1.796	0.624	0.131
R32	0.05024	221.50	313.5	19865	16278	4.799	12.69	2.111	0.614	0.200
R504	0.07925	215.5	309.8	18684	14558	2.974	13.60	3.218	0.687	0.240
R41	0.03403	195.03	280.7	16635	2075	0.523	11.02	1.376	4.819	4.87
N_2O	0.04401	184.70	264.6	16484	1156	0.272	11.48	1.774	8.651	18.99
Xe	0.13130	165.03	251.3	12543	1029	0.355	23.11	5.291	9.718	60.28
R14	0.08801	145.10	214.1	11831	831	0.105	17.34	3.541	12.03	60.95
Kr	0.08380	119.78	183.2	8991	411	0.107	25.27	3.367	24.33	225.5
Ar	0.04000	87.302	134.6	6437	179.8	0.040	22.68	1.603	55.62	531.7
N_2	0.02801	77.355	119.9	5580	144.8	0.022	17.90	1.123	69.06	564.3

difference occurs at 300 K, which is represented by points **a** and **b'** in Figure 5. The relative refrigeration power q_r^* and the relative enthalpy difference Δh^* are given by

$$q_r^* = \frac{q_r}{[h(T_h) - h(T_c)]_{\min}} = \frac{\Delta h_{\min}}{[h(T_h) - h(T_c]_{\min}} = \Delta h^*, \tag{6}$$

where the enthalpy difference in the denominator is between the enthalpy at $T_h = 300$ K and $T_c = T_{nbp}$ and for the pressure that has the smaller difference (the low pressure in these cases). For refrigerators with isenthalpic expansions $q_r^* = \Delta h^*$, but they are different for other types of refrigerators. The heat exchanger ineffectiveness (1 - ε), where ε is the effectiveness, causes a heat load q_{hx} per unit flow to the cold end. The ineffectiveness is given by

$$(1-\varepsilon) = \frac{q_{hx}}{[h(T_h) - h(T_c)]_{\min}}. \tag{7}$$

The heat load q_{hx} must always be less than q_r if there is to be any net refrigeration power remaining. Thus, q_r^* (or Δh^* for the JT cycle) represents the maximum allowed value of the heat exchanger ineffectiveness. Whenever $q_r^* < 1$ a heat exchanger is required. For cases where $q_r^* > 1$ no heat exchanger is required and indicates the high pressure fluid has condensed at 300 K with a pressure of 2.5 MPa. That situation is characteristic of vapor-compression refrigerators. Larger values of q_r^* indicate less need for a heat exchanger. The ratio of net to gross refrigeration power (neglecting other losses) when there is no heat exchanger is given by

$$\frac{q_{net}}{q_r} = 1 - \frac{1}{q_r^*}. \tag{8}$$

Figure 6 shows how L_v, $\Delta h_{\min}$, and q_r^* vary with the normal boiling points of the fluids. Notice how $\Delta h_{\min}$ drops rapidly below about 200 K because the high-pressure fluid is not condensed at 300 K. At

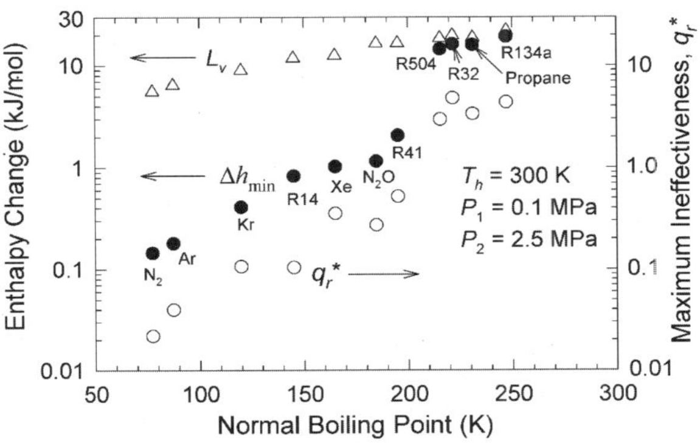

Figure 6. Latent heat of vaporization at 1 bar, gross refrigeration power, and maximum ineffectiveness for various refrigerants.

the same time the maximum allowed ineffectiveness drops below 1.0.

The last four columns in Table 3 relate to the pressure drop calculated in the low-pressure (0.1 MPa) stream of a hypothetical catheter. For most of the distance the low-pressure stream in the catheter will have warmed to 300 K, so the viscosity μ and density ρ are taken at 300 K. The molar flow rate given in Table 3 is the flow required from equation (5) to provide 10 W of cooling at the tip assuming a perfect heat exchanger has been used in the system. The last column in Table 2 gives the pressure drop per unit length in the low-pressure stream as it flows through a 2.5 mm diameter circular tube. In practice the flow is usually in an annular space outside the tube containing the high-pressure stream. For the heart catheter the outer tube is usually about 3 mm od with a 1 mm od inner tube for the high pressure stream. For this example the 2.5 mm diameter circular tube would be a good approximation for the flow pressure drop in the low-pressure stream. For a 1 m long catheter, as used for the heart catheter, the pressure drops for the last three fluids, and possibly the last five fluids, would be excessive. Those fluids, particularly argon, are generally used with much higher inlet pressures to increase Δh_{min}, which reduces the require flow rate. The high-pressure gas supply is from commercial cylinders and the expanded gas is vented to the atmosphere after leaving the catheter.

6.2 MIXED REFRIGERANTS

Mixed-refrigerant or mixed-gas Joule-Thomson (JT) refrigerators have been used since the mid-1930's when Podbielniak [19] received a U.S. patent on a system with mixed refrigerants in a single flow stream using a series of heat exchangers and phase separators. The phase separators allow only the liquid phase to be expanded at each stage, thus maintaining a high efficiency in the expansion process. Kleemenko [20], Missimer [21], and Little [22] have made subsequent improvements to this mixed refrigerant cascade MRC cycle that uses phase separators. In 1969, Fuderer and Andrija [23] first used mixed refrigerants in a Joule-Thomson cooler with no phase separators. In this simplified cooler, the entire fluid mixture flows through the heat exchanger and expands at the JT valve or orifice. Fuderer and Andrija pointed out that because the mixture is in the two-phase region in most of the heat exchanger, much greater heat transfer coefficients are possible compared with pure gas streams. The mixture is usually 100% liquid at the cold end just prior to expansion. Alfeev et al. [24], Little [25], and Longsworth [26,27] have described subsequent improvements to this type of mixed-refrigerant JT system. The mixed-refrigerant systems without phase separators are better suited to catheter applications because of the extreme space limitations. The advantage of phase separators is their ability to separate any lubricating oil in the refrigerant into the liquid phase, which returns to the compressor without going to the cold end. The use of mixed refrigerants in a system without phase separators that achieves temperatures below about 200 K may cause plugging of the JT orifice after some time, unless an oil-free compressor is used.

There are several recent reviews of the use of mixed refrigerants in JT cryocoolers [21,28-30]. The higher efficiency possible with mixed refrigerants yields significant refrigeration with a high pressure of only 2.5 MPa. Such pressures can be achieved in conventional compressors used for domestic or commercial refrigeration, thereby reducing costs. The higher boiling point components must remain a liquid and not freeze at the lowest temperature. In general, the freezing point of a mixture is less than that of the pure fluids, so temperatures of 77 K are possible with nitrogen-hydrocarbon mixtures even though the pure hydrocarbons freeze in the range of 85 to 91 K. The presence of propane also increases the solubility of oil in the mixture at 77 K so that less care is needed in removing oil from the mixture when using an oil-lubricated compressor. Much research is currently underway pertaining to the solubility of oil and the freezing point in various mixtures.

Other refrigerants that are not flammable, have low ozone depletion potentials, and low freezing points have been investigated in mixed-refrigerant JT systems. Practical mixed-refrigerant JT systems have achieved Carnot efficiencies near 10% for temperatures around 90 K, but their efficiencies drop off quickly for lower temperatures. Dobak *et al.* [31] describe a JT catheter system that used precooling of the mixed refrigerant with a conventional vapor compression system. This two-stage system provided 20 W at the tip of the catheter at 150 K compared with only 9 W without the precooling. Alexeev *et al.* [32] showed that precooling a mixed-refrigerant system increase the overall Carnot efficiency to 18% at 100 K. Marquardt *et al.* [10] discuss the procedure for optimizing gas mixtures for a given temperature range. They also discuss the development of a mixed-refrigerant JT cryocooler for use as a cardiac catheter only 3 mm in diameter and 1 m long. Some of those discussions are repeated here.

Mixed-gas refrigerants allow for higher cycle efficiency by maintaining a more uniform enthalpy difference between the high and low-pressure streams. Traditional JT systems have used a pure fluid for the refrigerant. This results in a very inefficient system. Joule-Thomson systems rely on the fact that there is a reduction in enthalpy with increasing pressure. An ideal gas has no change in enthalpy with pressure; therefore, a nonideal working fluid must be used. Fluids are most nonideal in the liquid or near-liquid state. The vapor-compression cycle works well because the fluids are nonideal over the limited temperature range of operation. Cryogenic systems must operate over a very large temperature range, making the fluids very close to ideal near the warm end of the temperature range, as can be seen in Figure 7 by the enthalpy of nitrogen. As indicated by equation (5) the refrigeration power of a JT crycooler is given by Δh_{min}, which for nitrogen occurs at the warm end of the cycle (300 K). Iso-butane has a normal boiling point of 261 K, so its Δh_{min} value of 19.6 kJ/mol is quite large. By adding it to nitrogen we hope to increase the Δh_{min} of the mixture. Figure 8 shows the enthalpy of the mixture with 20% (molar) of iso-butane. The Δh at 300 K has been greatly increased over that of pure nitrogen, but the Δh_{min} now occurs at about 170 K and is only slightly larger than Δh_{min} of pure nitrogen. We see from this exercise that some other component must be added to the mixture that has a normal boiling point near 170 K. Ethylene (T_{nbp} = 169.4 K) would be a good choice. Instead we choose to add

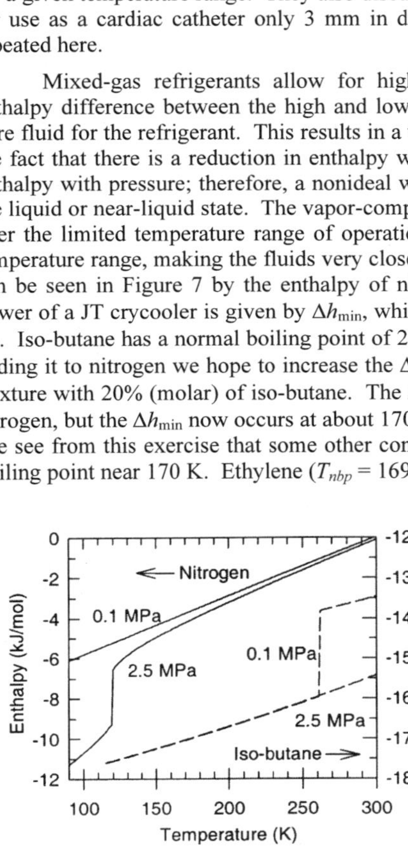

Figure 7. Molar enthalpy of nitrogen and iso-butane for different pressures.

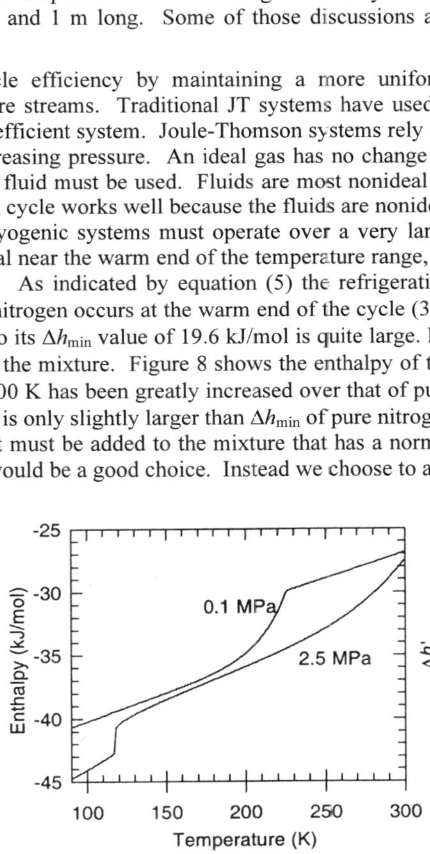

Figure 8. Molar enthalpy of a mixture of 80% nitrogen and 20% iso-butane (mol%) at different pressures.

both methane (T_{nbp} = 111.7 K) and ethane (T_{nbp} = 184.6 K) to try to smooth both enthalpy curves more and increase $\Delta h_{\min}$.

Because of severe size constraints on catheter heat exchangers we do not want to be required to use very high effectiveness heat exchangers, which increases their size [33]. According to equations (6) and (7) Δh^* equals the maximum ineffectiveness of the heat exchanger. Thus, maximizing Δh^*, which minimizes the total heat transfer in the heat exchanger, becomes our goal in optimizing the refrigerant mixture. The NIST computer code NIST14 for the thermodynamic properties of mixtures was used for the many trial and error calculations to find the maximum Δh^* in fluid mixtures for a cold temperature of 90 K and a warm temperature of 300 K and with 0.1 MPa and 2.5 MPa as the low and high pressure. With pure nitrogen the relative enthalpy difference is $\Delta h^* = 0.0236$, but with a mixture of 38.5% nitrogen, 27.5% methane, 10.5% ethane, and 23.5% isobutane (molar %) $\Delta h^* = 0.0958$ [10]. Figure 9 shows how the relative enthalpy difference for this mixture varies with temperature. Other mixtures have been found that do not contain flammable components and have $\Delta h^* = 0.16$ for the same conditions. For pure argon $\Delta h^* = 0.040$ at about the same temperature, which indicates that a factor of 4 better performance by the mixed refrigerant compared with pure argon. We expect that mixed refrigerants could be found to have $\Delta h^* > 1$ at temperatures of 150 K or higher. Lower temperatures would still require some heat exchanger with a 300 K upper temperature and a 2.5 MPa high pressure.

7. Experimental Cryogenic Catheter System

7.1 HEAT EXCHANGERS

If the length of the catheter is to be at body temperature and only the tip is to be cooled, a miniature heat exchanger must be incorporated in the catheter tip when temperatures below 150 K are required. The cardiac catheter discussed earlier had such requirements when development of the device began at NIST. The small diameter and the need for flexibility prevented the use of insulation around the catheter that would have allowed the heat exchanger to be place in the handle instead of the tip. Originally, a cold temperature below 150 K was the design goal. Figure 10 is a schematic of the heat exchanger and the tip of the catheter.

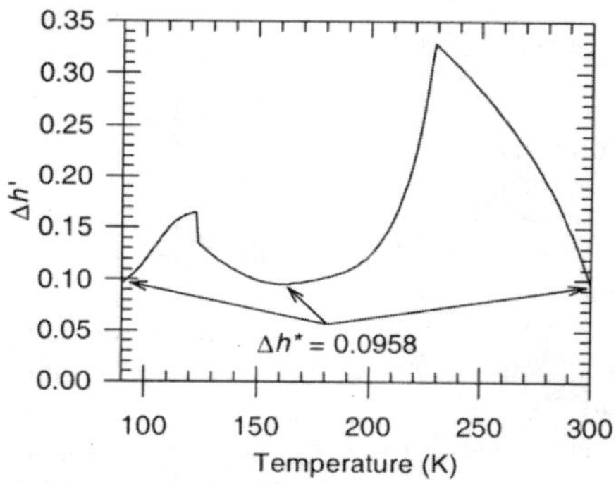

Figure 9. Relative enthalpy difference for 38.5% nitrogen 27.5% methane, 10.5% ethane, and 23.5% isobutane (mole %).

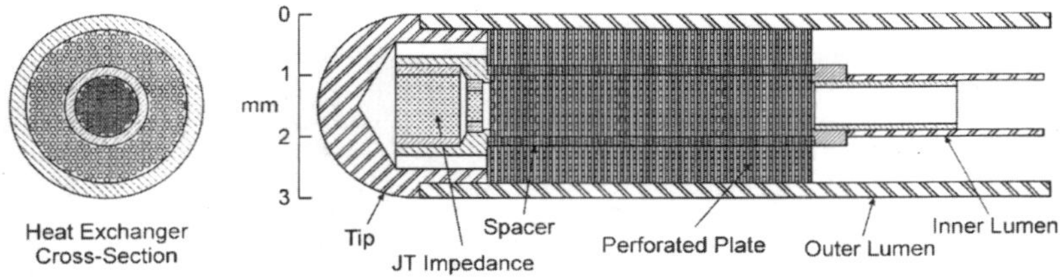

Figure 10. Cross-sectional drawing of the experimental catheter tip for a cryogenic cardiac catheter.

The micro heat exchanger at the cold end is fabricated by diffusion bonding perforated plates of copper alternated with stainless steel spacers. Photos of the heat exchanger are shown in Figures 11 and 12. By diffusion bonding large metal sheets containing many individual heat exchanger layers, we are able to make large numbers of cold ends at one time, significantly reducing the cost in order to make them disposable. Figure 13 shows the technique used to fabricate several hundred micro heat exchangers at one time. The hole diameters for the high pressure side are about 50 μm and are somewhat larger for the low pressure side. Prototype heat exchangers varied in length from 5 to 15 mm. The outer diameter of the heat exchangers was 2.5 mm. The enthalpy is more affected by changes in pressure at lower pressures, so the low-pressure side of the heat exchanger must have a much lower pressure drop than the high-pressure side.

7.2 EXPANSION ORIFICE

The J-T impedance shown in Figure 10 was fabricated of sintered copper powder. This permitted many flow channels, limited plugging problems , and provided a large area for heat transfer. Later catheters were fabricated using a single knife-edge orifice and provided similar results with much less effort.

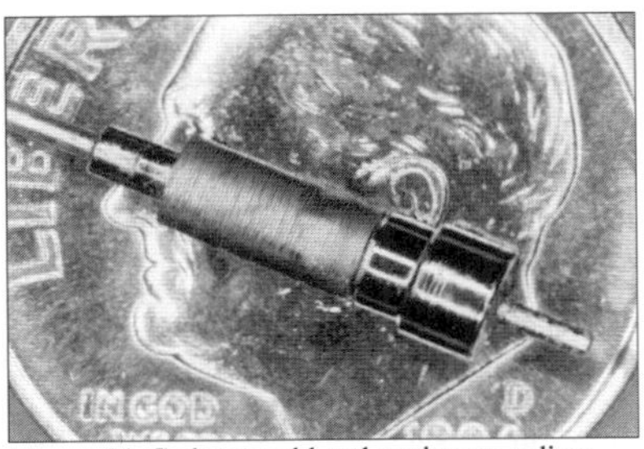

Figure 11. Catheter cold end resting on a dime. The micro heat exchanger is in the center. The tube to the right is for a pressure transducer.

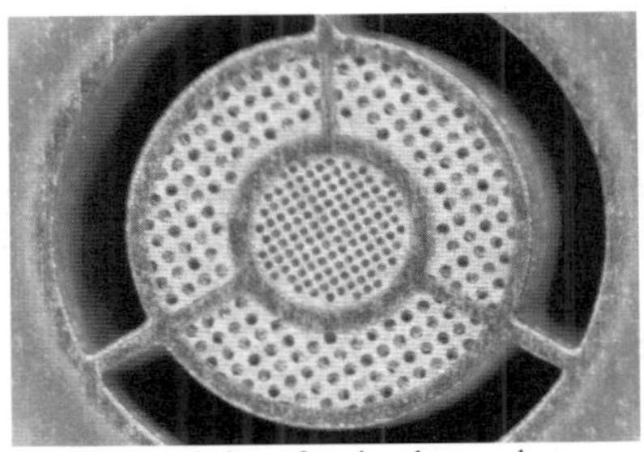

Figure 12. End view of a micro heat exchanger held in the plate with breakoff tabs. The top layer is a stainless steel spacer used between the copper layers. Diameter of heat exchanger is 2.5 mm.

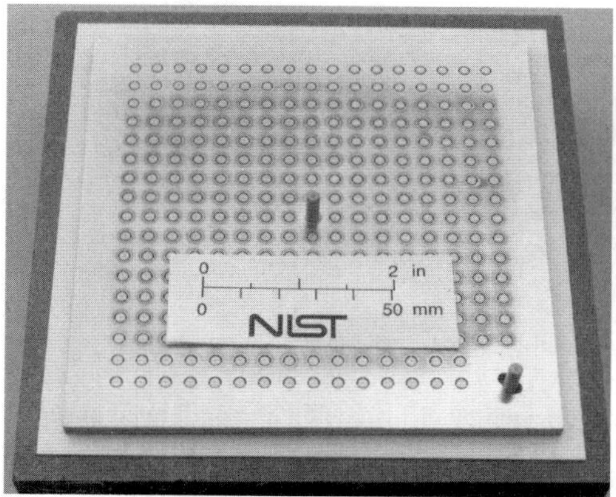

Figure 13. Assembly of photoetched stainless steel and copper foil positioned on a ceramic and graphite block with alignment pins for diffusion bonding. The finished plate contains 284 micro heat exchangers that are twisted out with a special tool.

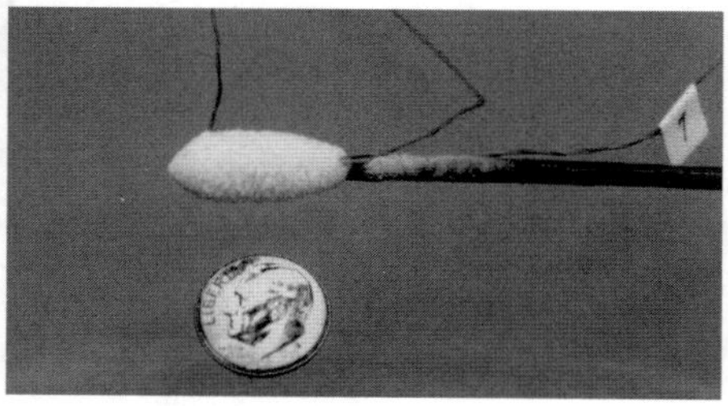

Figure 14. Cryogenic catheter with micro heat exchanger in tip during cooling to 140 K while exposed to air. The outer lumen was above the ice point everywhere along the I-m length except very near the tip.

7.3 LUMENS

A lumen is simply the tubing used for the flexible catheter pressure lines. The inner lumen was a 1 mm outer diameter polyimide tube with a stainless steel reinforcing braid. The reinforcing braid provided not only strength against the 2.5 MPa pressure but also helped to reduce the possibility of kinks in the tube. The outer lumen was braided nylon-derivative tube with a 0.25 mm wall. For safety, the outer lumen was designed to withstand the highest pressure in the system, although a ballast volume was added to the low-pressure side to reduce the average system pressure.

7.4 COMPRESSOR

The compressor was a commercial single-stage oil-lubricated compressor that required input powers from 300 to 500 W. Some experiments operated two compressors in parallel to increase the mass flow rate. Later experiments were performed using a custom built oil-free compressor. This simplified problems with the gas mixture since higher boiling-point components are more soluble in the oil than the lower boiling-point components.

7.5 RESULTS

Most experiments were performed with the catheter sitting on the bench top exposed to ambient air as shown in Figure 14. The catheter had fast cool-downs and warm-ups of about 2 minutes between room temperature and 150 K. The lowest temperature achieved was 85 K with no load on the cold end.

Typical operating conditions were 140 K with about 3 W of additional heat added to the cold end. A few experiments were performed with the cold end inserted into a room temperature gelatin to simulate biological heat loads. Catheter tip temperatures of 160 to 175 K were achieved and ice balls with diameters of about 26 mm and a mass of 11 g were created.

8. Comparison of Cryogenic Catheter Systems

The earliest cryogenic catheter systems used liquid nitrogen and many of them still do. The nitrogen is vented to atmosphere. Figure 15 is a schematic of such systems that utilize two dewars of liquid nitrogen. A vacuum pump reduces the pressure and therefore, the temperature of one dewar. It is used to subcool the liquid nitrogen being transferred at slightly elevated pressures from the other dewar and into the catheter. The subcooling prevents the liquid nitrogen from boiling until it reaches the tip of the catheter and permits high flow rates. This technique provides very high cooling powers because of the low temperature achieved (<77 K) and high flow rates. However, it is very bulky. Figure 16 shows a schematic of the high-pressure Joule-Thomson catheters. Argon or nitrous oxide at high pressure in a cylinder is expanded at the tip to provide cooling. Very low temperatures can be achieved with the argon, but the high inlet pressures (about 15 MPa) from the cylinder can be

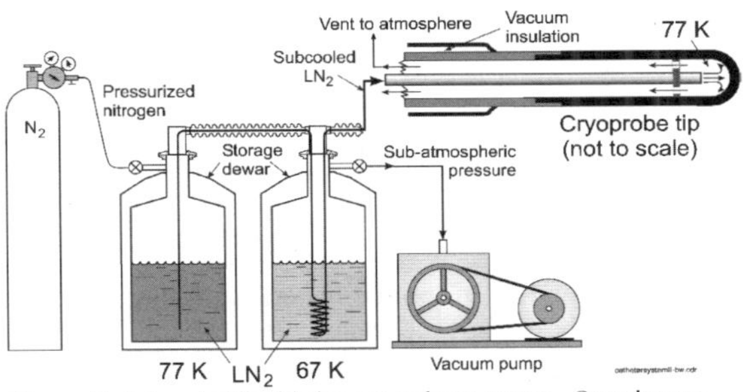

Figure 15. Subcooled liquid nitrogen catheter system. One nitrogen dewar is pumped on with a vacuum pump to cool it below 77 K.

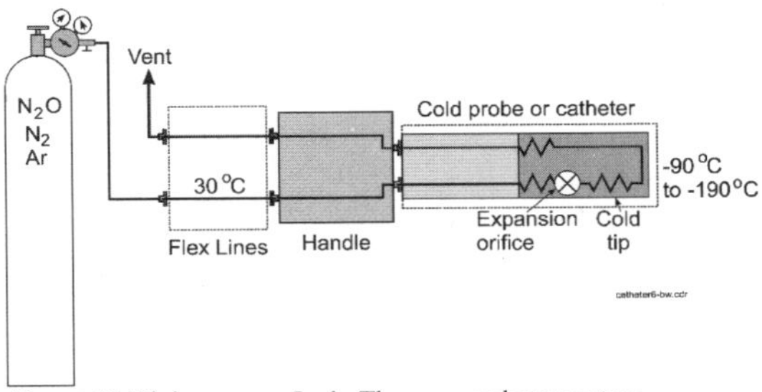

Figure 16. High pressure Joule-Thomson catheter system.

dangerous for use internal to the body if a leak were to develop in the catheter. The gas cylinders used in this approach also makes this system bulky.

Figure 17 shows the arrangement of components in a system that uses vapor-compression (fluid is in liquid state at the high pressure and at ambient temperature). Such a system is very simple and compact. It is a closed-loop system where the compressor circulates the refrigerant through the system. As discussed earlier this approach is limited to cold end temperatures of about 220 K or higher. Because no heat exchanger is required the catheter surface is maintained at ambient without the need for any insulation. To reach temperatures lower than about 220 K a precooling stage can be added, as shown in Figure 18, but now the catheter surface will be cold unless insulation is used. The other option for lower temperatures is to use a single stage with a mixed refrigerant. A heat exchanger is required. Ideally it should be placed in the cold tip to keep the catheter surface at near ambient temperature. However, that requires the development of micro heat exchangers as described in section 7. Alternatively the heat exchanger could be placed in the handle where there is much more room, but the catheter surface then will be cold unless insulated.

There are many options for cryogenic catheters, and the method used often depends on the particular application. In some case the catheter must be small, long, and flexible, such as the heart catheter. That is difficult to insulate. A uterine catheter can be nearly twice as large, much shorter, and be rigid. Insulation is easy to apply in that case along the entire length except at the tip. The ease of use for the closed-cycle systems makes them attractive to physicians because they can be ready to use at most anytime and are easy to move around.

9. Conclusions

Cryosurgery has a long but checkered history. As the effect of freezing on various tissue is better understood cryosurgery is growing in usefulness. The 1990s saw a rapid growth in the use of cryosergery for a wide variety of applications primarily because the ice front could be visualized better using ultrasound and MRI. At the same time many new approaches to cryogenic catheters were developed that simplified the hardware. Improved refrigerants, particularly mixed refrigerants, can

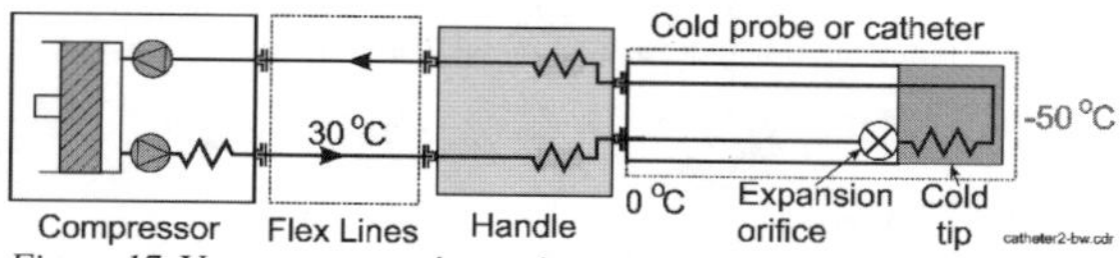

Figure 17. Vapor-compression catheter system. Heat exchanger in handle is optional, but will yield lower temperatures.

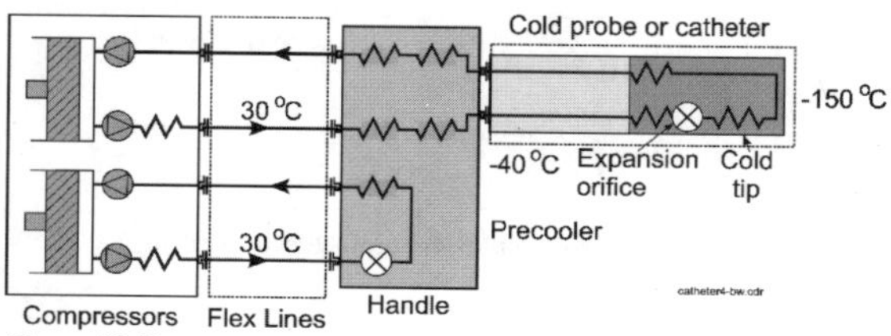

Figure 18. Two-stage vapor-compression catheter system.

simplify the hardware and/or increase the cooling power. Many heat transfer problems occur in the use of cryogenic catheters. Design data is not available regarding two-phase heat transfer in micro heat exchangers utilizing mixed refrigerants. At this time there still is some uncertainty regarding the best temperature that maximizes the temperature gradients in tissue. As cryogenic catheters are improved the applications for them keep increasing. If smaller catheters can be developed, then more remote parts of the body can be treated for abnormal behavior without having surgeons cut into the body to gain access. The short recovery times of about one day for cryosurgery are a strong driving force to use it more often.

REFERENCES

1. Arnott, J., (1851) On the Treatment of Cancer Through the Regulated Application of an Anaesthetic Temperature, J. & A. Churchill, London.
2. Fraser, J., (1979) Cryogenic Techniques in Surgery, *Cryogenics*, Vol. 19, pp. 375-381.
3. Dobak, J., (1998) A Review of Cryobiology and Cryosurgery, *Adv. Cryogenic Engineering*, Vol. 43, pp. 889-896.
4. Le Pivert, P., (1984) Cryosurgery: Current Issues and Future Trends, Proc. *10th International Cryogenic Engineering Conf.*, Butterworth, Surrey, pp. 551-557.
5. Fay, T., (1959) Early Experiences with Local and Generalized Refrigeration of the Human Brain, *J. Neurosurg.*, Vol. 16, pp. 239-259.
6. Cooper, I., S., and Lee, A., S., (1961) Cryostatic Congelation: A System for Producing a Limited Controlled Region of Cooling or Freezing of Biological Tissues, *J. Nerv. Ment. Dis.*, Vol. 133, pp. 259-269.
7. Gonder, M., Soannes, W., and Smith, V., (1964) Experimental Prostate Cryosurgery, *Invest. Urol.,* Vol. 1, pp. 610-619.
8. Gage, A. A., (1979) What Temperature is Lethal for Cells, *J, Derm. Surg. Oncol.* Vol. 464, pp.459-460.
9. Gage, A. A., (1992) Cryosurgery in the Treatment of Cancer, *Surgery, Gynecology, and Obstetrics* Vol.174, pp. 73-92.
10. Marquardt, E. D., Radebaugh, R., and Dobak, J., A (1998) Cryogenic Catheter for Treating Heart Arrhythmia, *Adv. Cryogenic Engineering,* Vol. 43, pp. 903-910.
11. Rubinsky, B., (2000) Cryosurgery, *Annual Review of Biomedical Engineering*, Vol. 2, pp. 157-187.
12. Dobak, J. D., Ryba, E., and Kovalcheck, S. (2000) A New Closed-Loop Cryosurgical Device for Endometrial Ablation, *J. Am. Assoc. Gynecol. Laparosc.,* Vol. 7(2), pp. 245-249.
13. Parker-Pope, T., (2001) FDA Clears Devices That Help to Treat Difficult Menstruation, *The Wall Street Journal, Health Journal*, May 4.
14. Ryba, E., (2001) CryoCor, Inc., San Diego, CA, personal communication.

15. Theodorescu, D., (2004) Cancer Cryotherapy: Evolution and Biology, *Reviews in Urology*, Vol. 6, suppl. 4, pp. S9-S19.

16. Brentari, E. G., Giarratano, P. J., and Smith, R. V., (1965) Boiling Heat Transfer for Oxygen, Nitrogen, Hydrogen, and Helium, *NBS Tech. Note 317*.

17. Pennes, H. H., (1948) Analysis of Tissue and Arterial Blood Temperatures in the Resting Human Forearm, *J. Appl. Physiol.*, Vol. 1, pp. 93-122.

18. Sekins, K. M., and Emery, A. F. (1982) Thermal Science for Physical Medicine, *Therapeutic Heat and Cold*, J. F. Lehman (ed.) Williams and Wilkins, Baltimore, pp. 70-132.

19. Podbielniak, W. J., (1936) Art of refrigeration, U.S. Patent 2,041,725.

20. Kleemenko, A. P., (1959) One Flow Cascade Cycle (in Schemes of Natural Gas Liquefaction and Separation), *International Institute of Refrigeration,* paper 1-a-6.

21. Missimer, D. J., (1994) Auto-Refrigerating Cascade (ARC) Systems – an Overview, *Tenth Intersociety Cryogenic Symposium*, (AIChE Spring National Meeting, March.

22. Little, W. A., and Sapozhnikov, I., (1997) Low Cost Cryocoolers for Cryoelectronics, *Cryocoolers 9*, Plenum Press, New York, pp. 509-513.

23. Fuderer, M., and Andrija, A., (1969) Verfahren zur Tiefkühling, German Patent 1426956

24. Alfeev, V. N., Brodyanski, V. M., Yagodin, V. M., Nikolsky, V. A., and Ivantsov, A. V., (1973) Refrigerant for a Cryogenic Throttling Unit, UK Patent 1,336,892.

25. Little, W. A., (1990) Advances in Joule-Thomson Cooling, *Adv. Cryogenic Engineering*, Vol. 35, Plenum Press, pp.1305-1314.

26. Longsworth, R. C., (1994) Cryogenic Refrigeration with a Single Stage Compressor, U. S. Patent 5,337,572.

27. Longsworth, R. C., Boiarski, M. J., and Klusmier, L. A., (1995) 80 K Closed-Cycle Throttle Refrigerator, *Cryocoolers 8*, Plenum Press, pp. 537-541.

28. Little, W. A., (1998) Kleemenko Cycle Coolers: Low Cost Refrigeration at Cryogenic Temperatures, *Proc. 17th International Cryogenic Engineering Conference*, pp. 1-9.

29. Boiarski, M. J., Brodianski, V. M., and Longsworth, R. C., (1998) Retrospective of Mixed-Refrigerant Technology and Modern Status of Cryocoolers Based on One-Stage, Oil-Lubricated Compressors, *Adv. in Cryogenic Engineering,* Vol. 43, Plenum Press, pp. 1701-1708.

30. Radebaugh, R. (1996) Recent Developments in Cryocoolers," *Proc. 19th International Congress of Refrigeration*, (The Hague, Netherlands), pp. 973-989.

31. Dobak, J. Xiaoyu, Y., and Ghaerzadeh, K, (1998) A Novel Closed loop Cryosurgical Device, *Adv. in Cryogenic Engineering,* Vol. 43, Plenum Press, pp. 897-902.

32. Alexeev, A., Haberstroh, C., and Quack, H., (1999) Mixed Gas J-T Cryocooler with Precooling Stage, *Cryocoolers 10*, Plenum Press, New York, pp. 475-479.

33. Radebaugh, R., (2005) Microscale Heat Transfer at Low Temperatures, *Microscale Heat Transfer – Fundamentals and Applications*, Springer, (S. Kakac, *et al.,* eds.), This publication.

SORPTION HEAT PIPE – A NEW DEVICE FOR THERMAL CONTROL AND ACTIVE COOLING

L. Vasiliev and L. Vasiliev Jr.
Luikov Heat and Mass Transfer Institute
P. Brovka 15, 220072, Minsk, Belarus
vll@ns1.hmti.ac.by

1. Introduction

Sorption heat pipe (SHP) is beneficial for the power electronic components cooling (IGBT, thyristors, et.) especially for transport application, high power electronic component cooling (laser diodes) and for the space two-phase thermal control systems. SHP is a combination of a heat pipe and solid sorption cooler with some specific interaction between these elements. This device is based on the enhanced heat and mass transfer in conventional heat pipes with sorption phenomena of sorbent bed inside it [1]. Now the electronic component cooling is a key problem in the industry and space application, Hoang, 2002. The major problems associated with cryo-coolers are reliability, efficiency, vibration, size and mass, electromagnetic interference, heat rejection, and cost. Besides classical techniques of cryogenic refrigeration by dumping in cryogenic fluid (nitrogen, hydrogen, or liquid helium), magnetic refrigeration, Vuilleumier machine, Stirling and pulsed tubes actually many attempts of adsorption systems applications have been carried out to reduce the constraints of less vibration, low power consumption, reliability and long-term life, Bard, 1986, Chan, 1986. This problem can be solved successfully with a coupled use of heat pipe and solid sorption cryo-coolers, realized in sorption heat pipe. A good example of its application is for lunar missions where during the day, conventional unshaded space radiators would have to look at either the sun, or the hot lunar surface, making difficult heat rejection directly to the environment. Another example is for cooling of electronic components during the lunar or Martian day, when the temperature of the surrounding could be higher than 50° C or even more. . Sorption Heat Pipe is very attractive for cooling IGBT modules widely used for railway train application. For such cases it is necessary to dissipate high level heat flux, releasing from small heat loaded surface and, of cause, we need to use ammonia or water as a working fluids. Ammonia is not comfortable as a working fluid for transport applications and has low latent heat of evaporation and liquid tension. Water is not convenient also, if the system is stored at the temperature below zero C^0. By using SHP we can keep adsorbed water at the temperature below $0\ ^0$ C. Sorption heat pipe includes the advantages of conventional heat pipes and sorption machines in one unit. The major it advantage to compare with conventional sorption machines is convective (two-phase flow) mode of the heating/cooling of sorbent structure inside the heat pipe. The same working fluid is used as a sorbate and heat transfer media. SHP has a sorbent bed (adsorber/desorber and evaporator) at one end and a condenser + evaporator at the other end of heat pipe (Fig.1-2). This device is working in such a way: Phase 1. At the beginning it is necessary to desorb a sorption structure (2), Fig.1 of heat pipe due to absorption of the heat of a heat source. During desorption of a sorbent bed the vapor (1) of a working fluid is transporting outside of a porous structure (2) and condensed in evaporator/condenser (3). The vapor is generating inside the sorbent bed, the vapor pressure is increasing, and the vapor flow enters the condenser and is condensing releasing the heat to

S. Kakaç et al. (eds.), Microscale Heat Transfer, 465–477.

surroundings. The part of fluid due to the pressure drop between the hot part of a heat pipe and the evaporator is filtrated through the porous valve (5) and enters the evaporator. The other part of the fluid is going back to the sorbent bed due to capillary forces of the wick (4) and heat the sorbent material due to the micro-heat pipe phenomena inside the sorbent bed. When desorption of the sorbent structure is accomplished, the heater is switched of, the pressure in the sorbent bed is decreasing and the working fluid is accumulated inside the evaporator.

Phase 2. After the Phase 1 accomplishment the porous valve (5) is opened and the vapor pressure inside the heat pipe is equalizing following the procedure of the liquid evaporation inside the porous structure of the evaporator (6). The schematic of sorption heat pipe evaporator is shown on Fig.1, Fig.3. During the liquid (7) evaporation, the air is cooling inside the cold box, (8). When the liquid evaporation is finished and the sorbent bed is saturated, a porous valve is closed and the sorbent bed is cooled down to the ambient temperature with the help of the heat pipe condenser (3). The phase 2 is finished.

In this paper we consider SHP as a combination of a loop heat pipe (LHP) and ammonia/ (active carbon fiber + chemicals) solid sorption cooler. Such system extends the limits of two-phase thermal control and ensures successful mode of electronic components cooling even in very harsh environmental conditions (ambient temperature 40 ^{0}C, or more) and ensures a deep cooling of space sensors down to the triple point of the hydrogen.

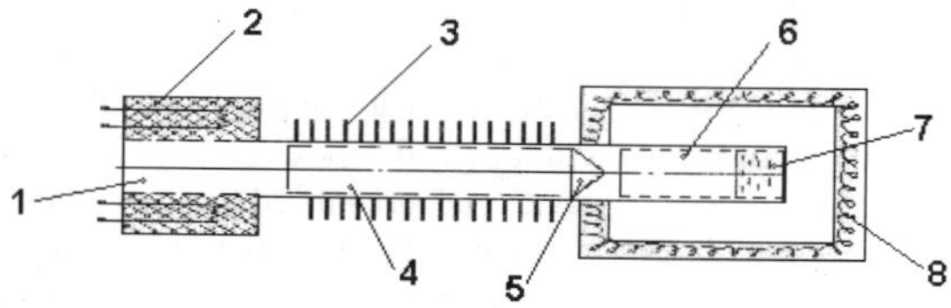

Figure 1: Sorption heat pipe. 1 – vapor channel; 2 – porous sorption structure; 3 – finned surface of heat pipe evaporator/condenser; 4 – porous wick inside heat pipe; 5- porous valve; 6 – heat pipe low temperature evaporator with porous wick; 7- working fluid accumulated inside the evaporator; 8 – cold box.

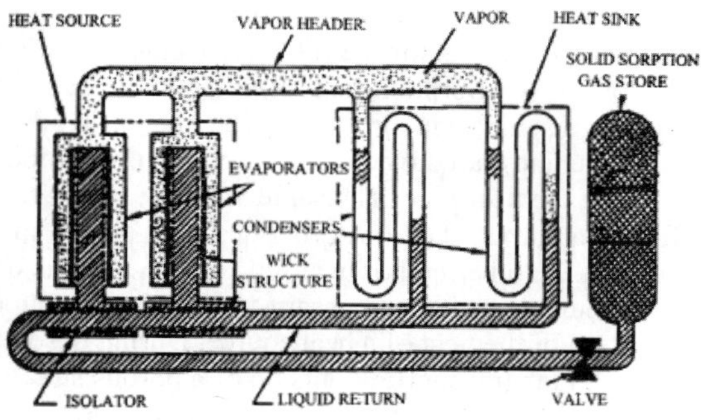

Figure 2: The schematic of sorption heat pipe.

The conventional two-phase thermal control system for space application is sensitive to the vehicle acceleration and vibration (spacecraft launching time, change the altitude of the orbit, est.). SHP avoids this inconvenience. The active mode of cooling/heating of the electronic components can be realised using SHP integrated directly to the two-phase heat transport system. The solid sorption cooler of SHP begins to function (to be switched on), when the cooling possibilities of an ordinary heat pipe are exhausted (for example, the condenser is damaged, or covered by some insulation media, the special situation with LHP during the space vehicle launch to the orbit). Actually a lot of low temperature heat pipes are used for space applications (as a cold plate for infrared observation of the Earth or Space), or effective electronic components cooling. It is known, that conventional LHP seems to be very promising heat transfer devices [2-4] for space applications in a large temperature range from 60 K up to 400 K. Cryogenic LHP (hydrogen, oxygen, and nitrogen) needs to be protected against super pressure influence at room temperatures. The combination LHP with solid sorption gas storage canister (SHP) can be useful to solve this problem. There are some problems with the start of LHP at cryogenic temperatures from supercritical state. Solid sorption gas storage device facilitates adsorption of cryogenic fluid at room temperature. When the LHP condenser is cooling and the solid sorption gas storage system is heating the working fluid, being desorbed from the sorbent bed, is condensing in condenser and is sucking by the porous wick of the LHP evaporator. Now SHP is ready to cool the electronic components (infrared sensor).

In SHP there are some basic phenomena interacting with each other:

1. A vapour flow (two phase flow) with kinetic reaction rate and pressure, vapour pressure, geometry, conductive and convective heat transport with radial heat transfer inside the sorbent material;

2.In the condenser and evaporator there is a vapour flow, liquid flow, interface position, radial heat transfer with kinetic reaction pressure, liquid pressure, vapour pressure, condensation and evaporation, shear stress, geometry, adhesion pressure, convective heat transport, radial heat transfer under the influence of the gravity field.

2. The experimental set-up

The experimental set-up is composed of two-phase loop, valves and compact solid sorption cooler with its system of thermal control, Fig.3. The core of this set-up is a capillary pumped evaporator with the inverted meniscus of the evaporation. Capillary pumped evaporator is a key element of LHP and SHP. The evaporator design, Fig. 4 (wick structure, geometry, and internal volumetric surface of pores, surface of the liquid meniscus and surface of liquid entrance, thermal conductivity, permeability, and est.) is the dominating factor in SHP operation. In our experiments the heat pipe evaporator was made from Ti sintered powder as a compact cylindrical heat transfer device. This evaporator has a liquid accumulator inside it and a set of longitudinal and circumferential vapor channels (grooves) on the outer surface of the wick contacting with SS tube (heating element). The Ti wick was force-fitted within the stainless steel tube. Working liquid (Fig.3) is entering from the heat pipe condenser to the liquid accumulator through the liquid subcooler. The liquid subcooler heat exchanger ensures the reliable operation of the LHP evaporator at the time of its transient heating. The radial liquid flow is going from the liquid accumulator through the wick to the hot wall, cools it and the vapor is moving along the grooves to the vapor outlet. The Ti wick structure has porosity 45%, the length - 280 mm, the outer diameter - 38 mm, maximum pore diameter - 10 microns, medium pore diameter - 3-5 microns, wick thickness - 4 mm. The heat pipe condenser is made as a tube in tube heat exchanger. The heat output of the system is up to 900 W, the mean thermal resistance of the evaporator – 0.03 K/W, pressure drop $\Delta P_c = 400$ mbar.

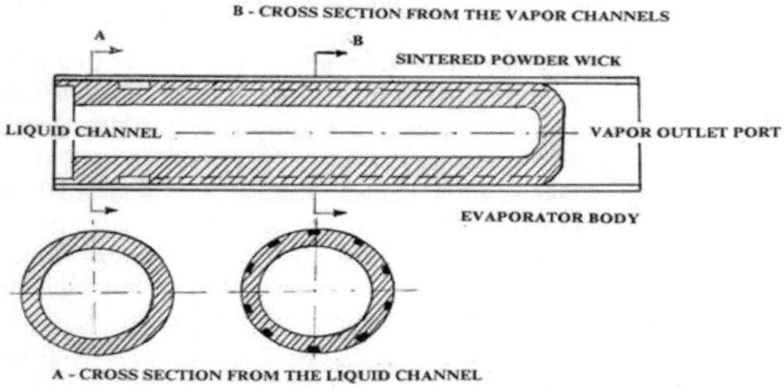

Figure 3: Porous evaporator, used for SHP and LHP.

This set-up is convenient to determine LHP and SHP parameters, due to its ability to joint during the experiment the evaporator to SHP, or LHP loop, Fig. 3. When the valve 10 is opened and the valve 11 is closed this set-up is functioning as LHP. When the valve 10 is closed and the valve 11 is opened a set-up is functioning as SHP.

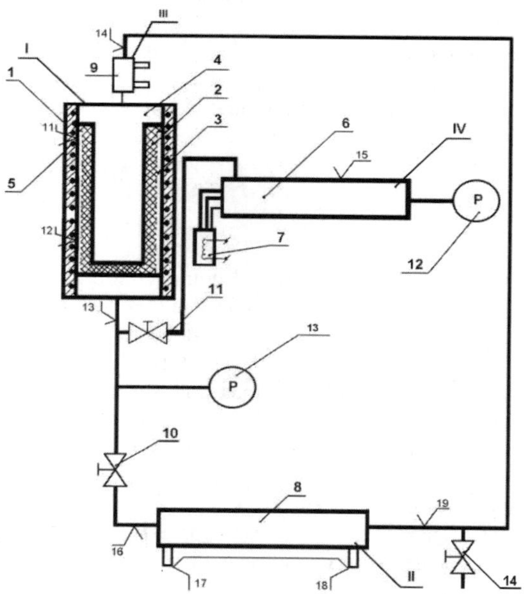

Figure 4: Schematic of the experimental set-up

I - heat pipe evaporator, II – he condenser, III - liquid subcooler, sorption canister.

1- heat pipe envelope, 2 – capillary-por wick, 3 – vapor channels disposed along the heat pipe envelope , 4 - liquid compensation chamber inside the porous wick, 5 -electrical heater, 6 - sorption canister, 7 - sorption bloc thermal control system, 8 - heat pipe condenser, 9 - liquid subcooler heat exchanger, 10 - valve , 11- regulated valve, 12 - pressure sensor , 13 - vacuum sensor , 14 - valve for the fluid charging, 11-19 thermocouples

3. Analysis

The considered experimented set - up can be applied in two different ways. The first one is to use is as a semiconductor sensor cooler with low heat dissipation to cool the sensor down to the ambient temperature. It is interesting to be applied in cryogenic range of temperatures. The second option is related with the cooler for high energy dissipation devices (for example laser diode cooler). The first set of experiments was performed with sorption heat pipe and ammonia as a working fluid to demonstrate the basic possibility to decrease the temperature of the heat loaded wall to compare with the temperature of this wall in the phase of loop heat pipe cooling mode.

On Fig. 5 - 6 heat transfer coefficient in LHP as a function of the heat load on the evaporator is shown. It is clear, that the most effective heat transfer coefficient (4500 W/m^2 K) is for the heat flow range 700 W – 900 W, when LHP thermal resistance R is minimal. For this heat load maximal surface of evaporation inside the pores is activated. For LHP the maximum pressure rise due in the wick can be evaluated by Laplace equation:

$$(p_c)_{max} = 2\sigma / r_c \quad , \tag{1}$$

where σ - is a surface tension and r_c is the effective capillary radius of the wick.

In the real LHP design capillary pressure drop ΔP_c depends on some LHP parameters:

$$\Delta P_c \geq \Delta P_v + \Delta P_l + \Delta P_w + \Delta P_g, \tag{2}$$

where ΔP_v and ΔP_l - are the pressure drop in the vapor and liquid lines, ΔP_w - is the pressure drop in the wick pores and ΔP_g is a pressure drop due to the gravity field.

In SHP the pressure rise is equal to the vapor pressure difference between the evaporator and adsorber (Clausius-Clapeyron equation), Fig.7:

$$dLnP/d\,(1/T) = -\,L/R, \text{ or } -\,\Delta H/R. \tag{3}$$

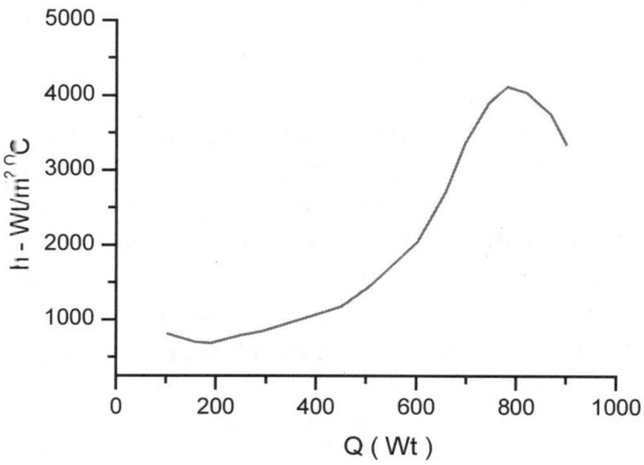

Figure 5: Heat transfer between the hot wall of the LHP and the vapor at the evaporator exit vs. a heat flow Q .

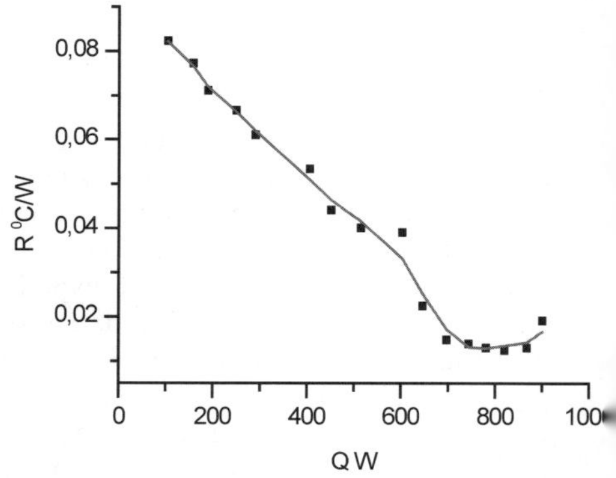

Figure 6: LHP thermal resistance as a funct of the heat flow Q.

For ammonia the pressure drop in SHP could be near 10 bars, it is 10 times more to compare with LHP. For LHP the external source of energy is the heat input to the evaporator and the heat sink is the condenser. For SHP the external source of energy is also the heat input to the evaporator, but the heat sink is the adsorber. The second external source of energy for SHP needs to be used periodically for the sorbent material regeneration. LHP can be used constantly, SHP needs to be used periodically (or at least two adsorbers ought to function, working out of phase). Naturally, such evaporator is compatible with these two cooling systems. SHP experimental set-up is composed of the evaporator, condenser and adsorber – solid sorption cooler SSC (sorbent bed - active carbon fiber "Busofit" saturated with salts). By choosing a working fluid having a large heat of evaporation (water, ammonia) one may expect to absorb significant heat flux in the evaporator. The adsorber is filled with an active carbon fiber "Busofit" (120 g) and saturated with $CaCl_2$ salt (in the mass ratio 70% of Busofit , 30% salt). SSC adsorber increases the heat transfer intensity inside the evaporator due to the volumetric evaporation of the liquid in the pores under the heat load and pressure drop influence (some bars). In order to calculate the total sorption capacity of the sorbent material we use the Dubinin-Radushkevich equation :

$$a = \left({W_o}/{v} \right) \exp\left(-BT^2 \left[\lg\left({P_s}/{P} \right) \right]^2 \right) \tag{4}$$

where: a- total sorption capacity , W_0- total volume of micro pores, B – structural constant , P_s, P – pressures of equilibrium and saturation. The experimental and calculated data of the sorption ammonia capacity for “Busofit” + $CaCl_2$ is a function of pressure and temperature, Fig. 8. There is a two - phase flow forced convection of ammonia through the porous structure following with an intense evaporator wall cooling.

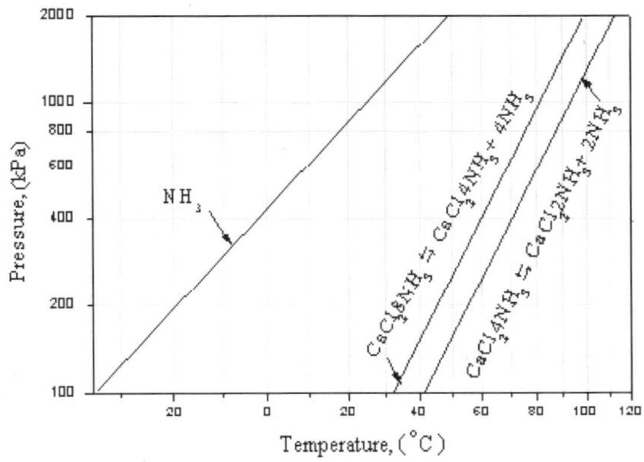

Figure 7: Pressure as a function of the temperature for the ammonia and $CaCl_2$ saturated with ammonia.

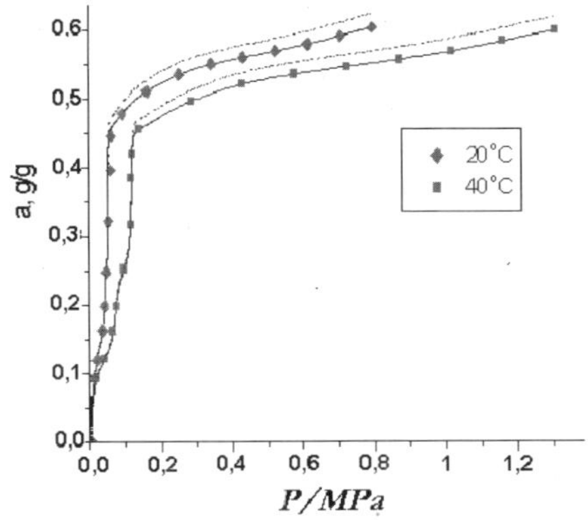

Figure 8: Experimental and calculated ammonia Isotherms of "Busofit+ $CaCl_2$". Dashed lines – calculated data at different temperature of equilibrium (1- T_{eqi} =20 ^{0}C, 2- T_{eqi} =40 ^{0}C).

To evaluate heat transfer intensity inside the porous wick with two-phase fluid filtration through the porous media a volumetric coefficient of the heat transfer h_v (W/m^3K) need to be determined. The energy absorbed by the cold fluid can be estimated by the number h_v (T – t) (W/m^3), which represents the heat energy dissipation in the unit of the volume of porous structure per unit of time. The temperature of the porous structure in this elementary volume is "T", and the temperature of the fluid is "t". The temperature field inside the porous wall is determined as:

$$\lambda \frac{d^2T}{dZ^2} = Gc\frac{dt}{dZ} \tag{5}$$

When the heat flow dissipation through the porous layer is increasing, the temperature difference between the pore wall and fluid in the pore is also increasing. The temperature field inside the porous wall needs to be determined as:

$$\lambda \frac{d^2T}{dZ^2} = h_v(T - t) \tag{6}$$

$$Gc\frac{dt}{dZ} = h_v(T - t) \tag{7}$$

The set of equations can be transformed into:

$$\frac{d^2T}{dz^3} + A\frac{d^2T}{dz^2} - AB\frac{dT}{dz} = 0 \tag{8}$$

$$\mathrm{T} = \mathrm{t} + \frac{1}{A}\frac{dt}{dz} \tag{9}$$

Here some non-dimensional terms are applied:

$$Z = Z/\delta;\ A = h_v\delta/Gc;\ B = G\delta c/\lambda;\quad D_{1,2} = \frac{A}{z}[-1 \pm (1+4B/A)^{1/2}]. \tag{10}$$

Non-dimensional complex B characterizes the ratio of the heat absorbed by the cold fluid and the heat transferred through the porous skeleton due to its thermal conductivity.
δ - is the wall thickness. The equation (5) now can be presented as:

$$\frac{d^2T}{dz^2} - B\frac{dt}{dz} = 0 \tag{11}$$

with the solution of the set of equations (6 – 7) as:

$$t = C_1 + C_2 \exp D_1 z + C_3 \exp D_2 z \tag{12}$$

$$T = C_1 + C_2 (1 - D_1/A) \exp D_1 z + C_3 (1 + D_2/A) \exp D_2 z. \tag{13}$$

The boundary condition for the case when the fluid flow with initial temperature t_0 is filtered through the porous wall is:

$$\text{At } Z = 0 \quad \lambda dT/dZ = Gc\,(t_1 - t_o) \tag{14}$$

On the other hand the porous coating on the inner surface of the evaporator can be considered as mini-fins set and we can determine the heat transfer efficiency between the fluid and porous structure as:

$$Q = \alpha S\,(T_{wl} - t_v), \tag{15}$$

where α is the effective heat transfer coefficient between the cooling fluid and the wall of the evaporator?
S is the heat loaded wall; Q - heat flow; T_{wl} - the temperature of the wall; t_v - the temperature of the vapor on the evaporator exit.

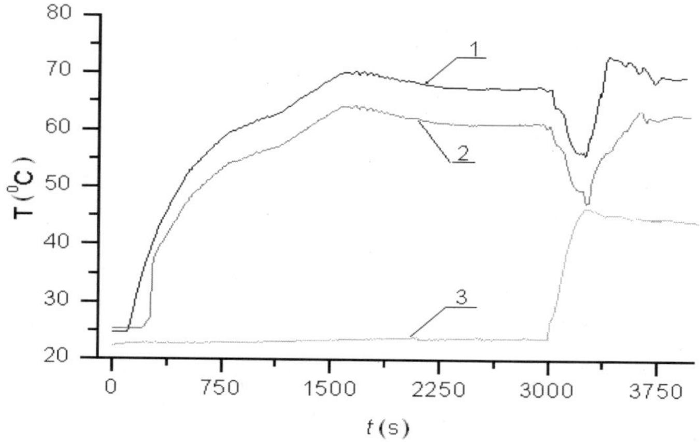

Figure 8: Temperature field evolution in SHP. 1 – surface of the evaporator, 2 – vapor in the transport zone, 3 – surface of the SSC (adsorber)

On Fig.8 the temperature evolution of the heat loaded wall of the evaporator (1), vapor on the evaporator exit (2), and the temperature of the sorbent bed canister surface (3) are demonstrated as a function of the time. When LHP is switched on, during the time of its preheating (0 – 1500 s) the temperature of the evaporator surface and the temperature of the vapor is steadily increasing till the steady state heat pipe operation (time interval 1500 – 3000 s) is fixed. After the time interval of 3000 s, when the steady state condition of heat pipe operation was determined and , for example, the Q max continue to increase , a canister with sorbent bed need to be switched on connecting with the evaporator through the valve 11 Fig.3. At the same time we need to close valve 10 (Fig.3), other ways the liquid will go directly to the sorbent bed. A sharp temperature decrease of the evaporator wall and vapor are immediately checked on during the time interval 3000 – 3500 s. When the sorbent bed saturation is finished and the temperature of the sorbent material begins to increase (curve 3) together with the increase of the temperature of the evaporator, the temperature of the vapor at the output starts to increase also (time interval 3500 – 3750 s). By opening the valve 10 (Fig.3) during the desorbtion we put the vapor flow in two directions: one from the evaporator to condenser, another from the sorbent canister to condenser. After the sorbent material dessorption is finished we close valve 10 and start the system operation as loop heat pipe. This enhanced mode of the heat pipe functioning can be quasi – stationary, if we use the system with at least two sorbent beds, swiching on and off alternatively. The time interval 0 - 3000 s. (Fig.8 - Fig.9) is typical for LHP, the time interval 3000 – 3500 s is typical for the SHP, operating during 500 s. The coefficient of heat transfer between the evaporator and the cooling fluid is shown on Fig.9. When the sorbent bed structure (SSC) is jointed to the evaporator, a sharp increase of the heat transfer coefficient (three times more to compare with conventional heat pipe function) is measured for the forced convection of two-phase flow inside the pores. The sorption block efficiency as a cooler depends on several parameters such as: total mass of working fluid in HP, sorption capacity of sorbent material, flow rate of ammonia from evaporator to a sorption block and heat transfer rate inside of the evaporator.

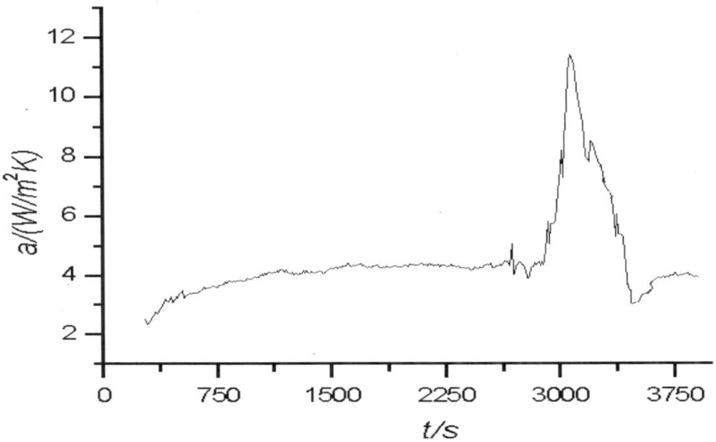

Figure 9: Heat transfer coefficient as a function of time for SHP evaporator.

There is another possibility to apply this experimental set – up to estimate SHP parameters as a high heat releasing components cooling system (Laser diodes stack, IGBT transistors). The goal of experiments is to estimate the possibilities of SHP cooler capable of removing high heat fluxes > 200 W/cm^2 with low surface superheat for ammonia as a working fluid. This new cooling device can be applicable to large area cooling at high heat fluxes, operating in microgravity space environment. Its relatively low pressure drop across the porous structure is an important advantage to compare with conventional systems of semiconductor components cooling. A sorption heat pipe as a cooler has a compact evaporator/micro multi nozzle spray chamber, where there is a possibility to subcool the heat loaded wall and enhance heat transfer coefficient.

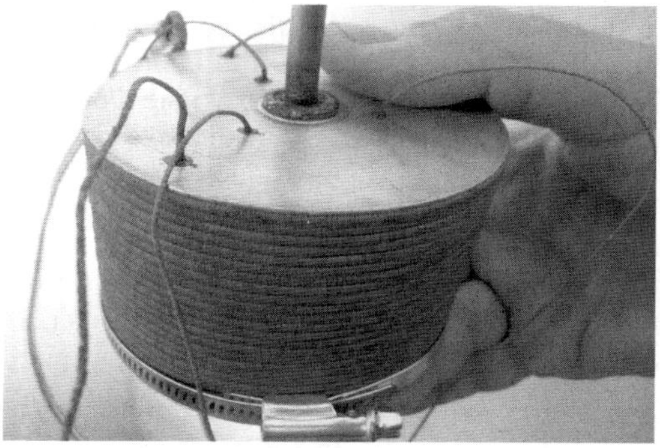

Figure 10: Photo of SHP evaporator with electrical heater and thermocouples.

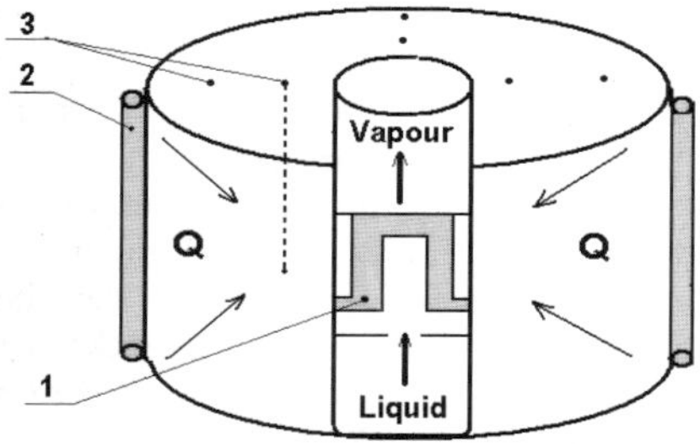

1. Porous media body
2. Electrical heater
3. Thermocouples

Figure 11: Schema of SHP evaporator with electrical heater and thermo-couples.

The experimental evaporator (micro two-phase jets generator) is shown on Fig.10 -11. The heater block of the evaporator includes a hole machined directly in the centre of the copper cylinder with thick walls and is used for the installation of a nickel sintered powder evaporator. Some thermocouples are disposed inside the copper block to control the heat flow to the evaporator from the electric heater disposed on its outer surface. To minimize heat losses, the heater block was insulated. Heat input to the evaporator was calculated by conduction analysis using thermocouples that were placed at a known distance apart in the copper heater block.

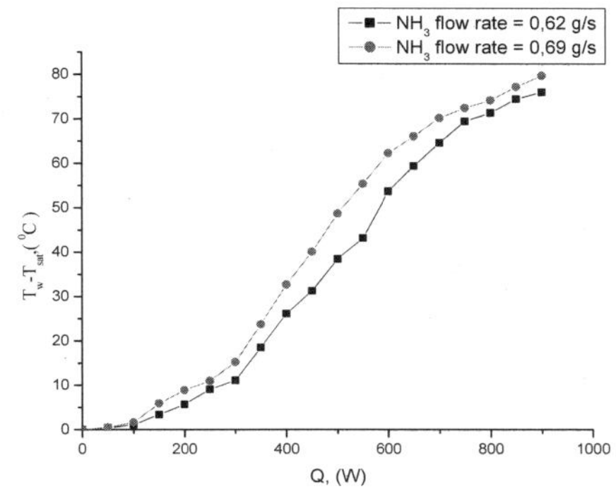

Figure 13: Temperature drop between the heat loaded wall and the vapour (evaporator outlet) ($\Delta T = T_W - T_V$) as a function of heat load Q

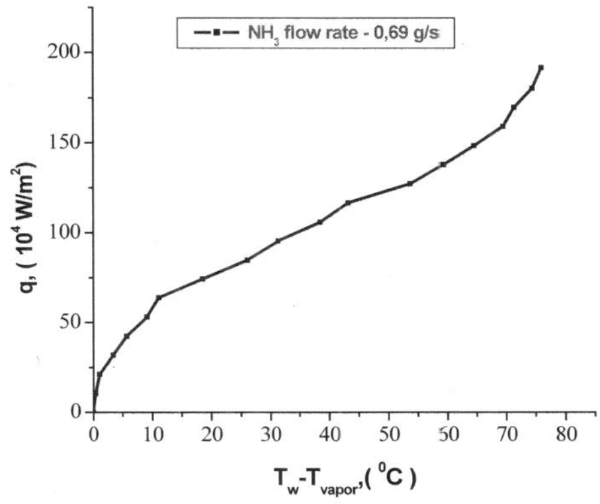

Figure 14: Input heat flux as a function of temperature difference T_W - T_V

The SHP evaporator, Fig 13 –Fig.14 represents the largest thermal resistance of the circuit. This resistance includes the temperature drop due to the thermal conduction through the heat pipe wall and the porous structure. Within the heat flux range of interest (100 – 200 W/cm^2 and more) the average evaporator resistance typically is between 0.07 – 0.08 K/W. So there is a possibility to cool successfully the electronic components, using new generation of SHP. The limited possibility to increase the heat flux value more than 200 W/cm^2 is the electric heater, which didn't allow increasing the heat input to the evaporator. This analysis testify the dominant mode of heat transfer as a volumetric evaporation of the liquid inside the evaporator porous structure, micro - jet spray cooling of the heat loaded wall by the ammonia with film evaporation of micro droplets on its surface. The efficiency of the cooling process is the best for the heat loads more 200 W/cm^2 with complete phase transfer liquid/vapour on the evaporator exit. Such type of cooling is subject to a large variation in the surface temperature depending on the heat load as the liquid moves through the porous structure with volumetric phase change in pores and the micro droplets in the jets guarantee the additional thin film evaporation on the hot wall whereas the surface temperature is quite uniform.

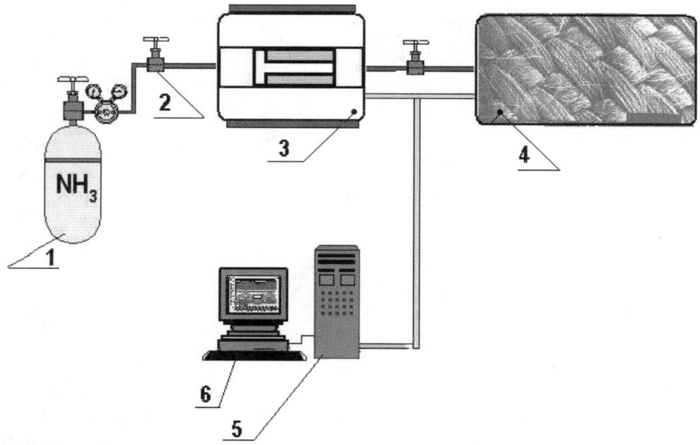

Figure 15: Experimental set-up. 1- Ammonia Tank, 2- Valve, 3- SHP Evaporator, 4- Sorbent Canister, 5- Data Acquisition System with Thermocouples Set, 6- Personal Computer

4. Conclusions

1. Sorption heat pipe is a novelty and combines the enhanced heat and mass transfer typical for conventional heat pipe with sorption phenomena in the sorbent structure. The sorption heat pipe parameters are insensitive to some "g" acceleration and such heat pipe can be suggested for space and ground application.
2. Significant heat transfer enhancement is obtained in the sorption heat pipe to compare with a conventional loop heat pipe for the same fluid (ammonia) and evaporator dimensions during operation of a sorption bloc.
3. Sorption heat pipe can be considered as a cryocooler, applied for deep cooling of sensors down to the triple point of the hydrogen (nitrogen) with cyclic or periodic operation in space. The working fluid can be stored in sorption material of SHP at low pressure.
4. Sorption heat pipe can be used as an efficient multi micro jet cooler of the laser diode with good thermal contact of the SPH porous wick and heat loaded wall, insensitive to some "g" acceleration and capable to dissipate high heat fluxes q > 200 W/cm^2 for ammonia with low surface superheat.

Nomenclature:

a- Sorption capacity (g/g)
c- Specific heat (J/kg K)
g- Gravitational acceleration (m^2/s)
G- Flow rate per unit area or mass flux (kg m^{-2}s^{-1})
h_v- Volumetric heat transfer coefficient (W/m^{3}K)
H- Enthalpy (J/kg)
L- Latent heat (J mol^{-1})
P –Pressure (Mpa)
r- Capillary radius of the wick (m^{-6})
R- Thermal resistance (K/W)
S- Surface (m^2)
T- Temperature (K)

Greek letters:

α- heat transfer coefficient (W/m^2K)
δ- thickness (m^{-6})
θ- angel of wetting (^{0})
λ- thermal conductivity (W/mK)
ρ- density of working fluid (kg/m^3)
σ- coefficient of surface tension (H/m)
τ- time (s)

Subscripts:

c - capillary
eff - effective
max - maximal
l - liquid
v - vapor
w - wick
wl - wall

References

1. L.L.Vasiliev, A.A. Antukh, L.L. Vasiliev Jr. " Electronic cooling system with a loop heat pipe and solid sorption cooler" in Proceedings of the 11 th IHPC, Tokyo, September 12-16, 1999, Japan, Preprint Vol. 1, pp. 54-60.
2. Yu.F. Gerasimov, Yu. F. Maidanik, Yu. E. Dolgirev, V. M. Kiseev "Antigravitational heat pipes - development, experimental and analytical investigation" in Proceedings of the 5th IHPC, Tsukuba, Japan, 14-17 May,1974.
3. Yu. Maidanik, Yu. Fershtater, V. Pastukhov " Some results of Development of Loop Heat Pipes" in Proceedings of the CPL-96 Workshop on Capillary pumped Two-phase Loops, Noordwijk, Netherlands, 1996 .
4. V. Kiseev, N. Pogorelov, "A Study of Loop Heat Pipes Thermal Resistance", in Proceedings of the 11th IHPC, September 21-25, 1997, Stuttgart, Germany, Sessions A1, A2, p. 45.
5. L. Vasiliev, V. Senin, "Device for Cooling Semiconductors", USSR Patent 306,320, Bulletin of Discoveries, Inventions, Industrial Models, Trade Marks, 19, 1971.
6. L.L. Vasiliev, A.S. Zhuravlyov, M. N. Novikov, L.L. Vasiliev Jr. , " Heat transfer with propane evaporation from porous wick of heat pipe", in Proceedings of the 11th IHPC , Tokyo, Japan, September 12-16, 1999, Vol.2,pp.1214-129.
7. D. Gorenflo, P. Sokol, "Pool Boiling Heat Transfer from Single Plain Tubes to Various Hydrocarbons", Int. J. Refrig., Vol.13, 1990, pp. 286-292.
8. R. Mertz, M. Groll, L. Vasiliev, V. Khrolenok, A. Khalatov, G. Kovalenko, G. Geletuha "Pool boiling from enhanced tubular heat transfer surfaces", Proceedings of the 11th Int. Heat Transfer Conference, Kyongju, Korea, 23 -28 August, 1998
9. J. Gottschlich, R. Richter, 1991 " Thermal Power Loops" SAE 91 - 1188, Proc. SAE Aerospace Atlantic, Dayton, OH.
10. L.L. Vasiliev, D. Nikanpour, A.A. Antukh, K. Snelson, L.L. Vasiliev Jr., A. Lebru "Multisalt-carbon chemical cooler for space applications", Journal of Engineering Physics and Thermophysics, Vol.72, No.3, 1999.

THERMAL MANAGEMENT OF HARSH-ENVIRONMENT ELECTRONICS

M. OHADI* and J. QI
Department of Mechanical Engineering Department – University of Maryland
College Park, Maryland, USA

1. Introduction

As the trend of highly integrated electronics and simultaneous miniaturization escalates to include faster processors, more functions, and higher bandwidths, electronics continue to become more compact in response to size limitations and strict reliability requirements. The result is an increasing heat flux at both the component and circuit board levels. In the last decade, average power densities and heat dissipation rates have increased nearly two-fold [1]. It is expected that heat flux levels in excess of $100W/cm^2$ for commercial electronics and over $1000W/cm^2$ for selected military high power electronics will soon become a realistic and immediate challenge to overcome. There is also a growing demand for more sophisticated and capable electronics used in harsh environment applications such as those found in defense, automotive and oil exploration systems. Thermal management of harsh environment electronics is vital to the successful design, manufacture, and tactical operation of a variety of electronic systems to meet the high temperature, environmental, reliability, and cost effectiveness requirements.

This paper will look into fundamental characteristics and thermal management challenges for practical harsh environment electronics and will overview the most widely known, as well as emerging technology solutions for such applications. Future thermal management of harsh environment electronics at the chip, board and system levels will be also discussed. The paper concludes with a ranking of the potential applicability of these techniques according to several criteria, including cost, ease of use, thermal performance, and reliability concludes the paper.

2. Problem Statement

The main factor that distinguishes harsh environment electronics from commercial electronics is the environment in which they perform. Harsh-environment microelectronics operate at temperatures well above the traditional maximum allowable operating temperatures of 70°C for consumer electronics, as in Table 1.

S. Kakaç et al. (eds.), Microscale Heat Transfer, 479–498.

Table 1: Thermal operating environments

Electronics	Operating Temperature
Consumer	0 °C to +70 °C
Industry	-40 °C to +85 °C
Automotive	-40 °C to +125 °C
Military	-55 °C to +125 °C

The three main harsh environment electronics categories are: military electronics including land-based, shipboard, airborne, missile-based, and space-based applications; automotive; and oil exploration systems. Some applications in each of these categories in actuality require performance at even more extreme temperatures, ranging down to -65°C and lower for avionics in cold climates, and up to 225°C for avionics distributed control systems. For the drilling of well holes of 2 km or higher in length, temperatures often reach 200°C, with pressures up to 20,000psi. Under-the-hood temperatures in a car can be as low as -40°C in some areas of the world such as Alaska and above 200°C in other areas of the world such as Death Valley in the United States. Figure 1 shows typical automotive under-hood temperatures [2].

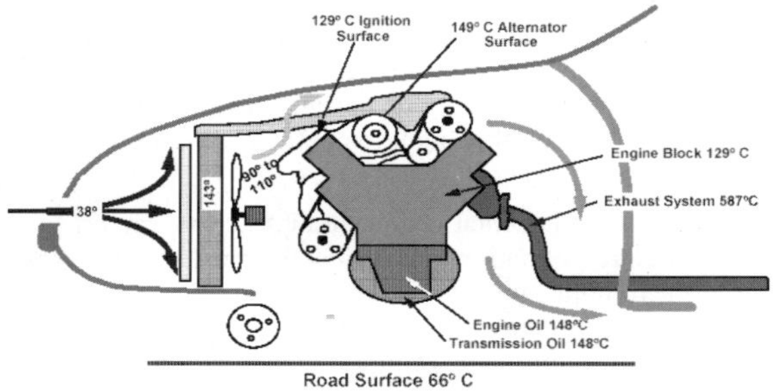

Figure 1: Typical automotive under-hood temperatures

The electronics products in vehicles, especially under-hood components, operate in a very harsh environment, including petroleum vapors, random vibration (up to 10G on engine), moisture, various fluids, dirt, and chemicals. High reliability and durability are simultaneously required, as the automotive manufacturers are offering extended warranties to consumers for 100,000 miles/10 years or even more. Added to these requirements is the fact that automotive applications are extremely cost sensitive and therefore require low cost targets referred to as the "convergence of automotive electronics and consumer electronics." In addition to the traditional engine management, comfort, and entertainment systems of the past, there is a growing market for office and entertainment systems that rival those in an office or home. Furthermore, more efficient thermal management is needed for increased power requirements that have made the automotive suppliers turn to 42 V sources.

In military environments, operating conditions are extremely harsh: the hardware must maintain reliability and ability in severe conditions, including thermal and mechanical shock vibration, cycling, humidity, corrosive, chemical/biological, and radiation degradation, and altitude changes [3]. Military electronics are required to have a functional lifetime of 20 to 30 years, more than that of rapidly obsolete consumer electronics, which have an average service life of less than 5 years. For example, missiles are required to be stored for 10 to 20 years in any location in the world and still meet safety and reliability requirements. Avionics must undergo severe changes in temperature, humidity, acceleration, and atmospheric pressure (altitude) within very short periods of time. Some military electronics require survivability when exposed to nuclear attack.

For more than three decades, military electronics have been governed by performance-based specifications. They are typically in the upper cost, high reliability, and high-density range, and only a small portion of the electronics industry in relationship to consumer electronics. Furthermore, the weight and volume occupied by cooling devices and hardware become even greater constraints in military electronics than in conventional applications. Since the electronics industry has been gradually shifting away from the production of military grade electronic parts, mainly due to high costs and very low volume production, the use of commercial off-the-shelf (COTS) electronic components, which have lower maximum temperature ratings than MIL-SPEC devices, has become more frequent, offering major benefits in the areas of supply and cost. The gradual shift in industry towards all COTS parts, the increasing chip-level temperatures, and the higher circuit density of next generation electronics, together with the need for high reliability and long life cycles of all the parts and materials, has introduced new challenges for harsh-environment applications. With miniaturization, the greatest dissipation requirement for high energy military lasers and MEMS devices is expected to be on the order of 100 W/cm^2 for high performance microprocessors and 1 kW/cm^2 for high power electronics components with a smaller allowable temperature difference (Figure 2). The heat flux associated with laser diodes is on the order of several kW/cm^2, comparable to the heat flux associated with ballistic missile entry.

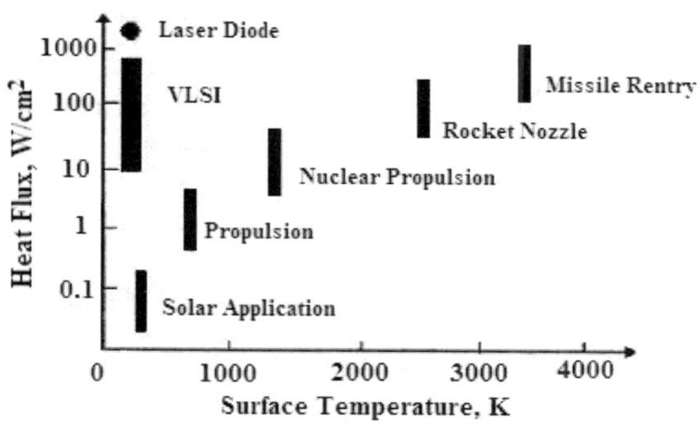

Figure 2: Heat dissipation for various events

Therefore, thermal management is now becoming increasingly critical to the design of harsh environment electronics to satisfy the increasing market demand for faster, smaller, lighter, cheaper and more reliable products. More aggressive thermal management techniques are required to handle the high heat loads on the all-electric military systems, which are expected to increase by several fold compared to existing electronic technologies [4]. In this paper an overview of the current thermal management technologies, as well as emerging solutions will be offered, with additional details available in Ohadi et al [4].

3. Overview of thermal management for harsh environment electronics

In a typical electronics system, heat removal from the chip may require the use of several heat transfer mechanisms to transport heat to the coolant or the surrounding environment. There are three basic heat transfer modes (including phase change): conduction, convection, and radiation.

Thermal management techniques can be characterized as passive, active, or a combination of the two (hybrid). The passive techniques, absence of external power, are relatively reliable and simple to implement. However, they are performance-limited for many high power applications. The main passive thermal management techniques are:

- *Conduction* (metal spreader, interface materials, adhesives, pads, pastes, epoxy bond)
- *Natural convection* (finned heat sinks, ventilation slots, liquid immersion cooling)
- *Radiation* (paints, coatings, mechanical surface treatments)
- *Phase change* (phase change materials, heat pipes, thermosyphons, vapor phase chambers)

Active thermal management techniques, requiring input power, provide increased performance/capacity, but also reduced reliability and added complexity. The essential active techniques include:

- *Forced convection* (fans, nozzles)
- *Pumped loops* (heat exchangers, cold plates, jet/spray)
- *Refrigerators & coolers* (vapor-compression, vortex, thermoacoustic, thermoelectric/Peltier)

Figure 3 shows the values of thermal resistance for a variety of coolants and heat transfer mechanisms with a typical component wetted area of 10 cm^2 [5]. Thermal resistances vary from 100K/W for natural air convection to 33K/W for forced air convection, to 1K/W in FC forced liquid convection, and to less than 0.5K/W for boiling in FC liquids.

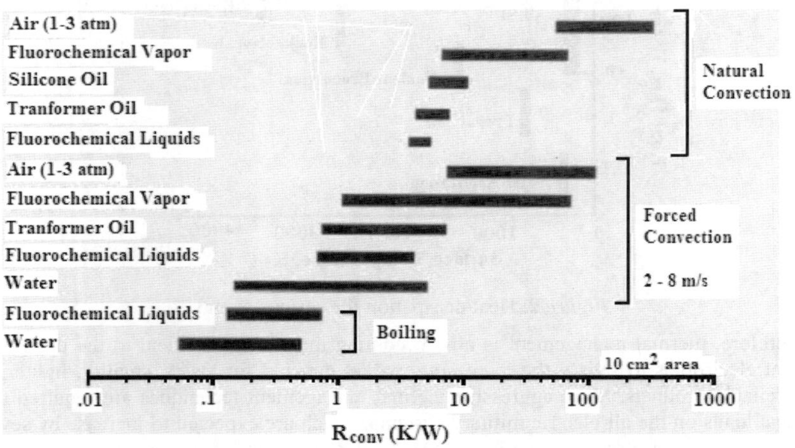

Figure 3: Typical thermal resistances for various thermal management modes.

Although the primary thermal transport mechanisms and the commonly used heat removal techniques vary substantially from one packaging level to the next, in general, heat removal can be addressed hierarchically. The first level of the hierarchy is at the chip package (IC) level where heat conducts from the chip or component to the package surfaces through interface materials and is then rejected from the outer surfaces (heat sink and the board) into ambient air (Figure 4).

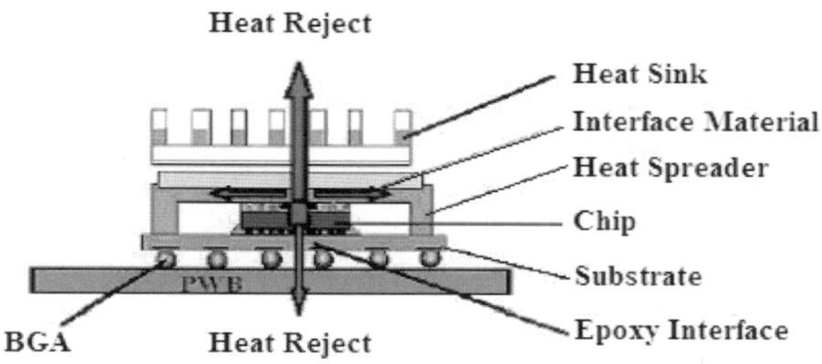

Figure 4: Schematic of thermal packaging architecture

Reduction of thermal resistance between the die and the outer surface is the most effective way to lower the chip temperature. A variety of passive techniques are available to reduce the interface thermal resistance, such as using die-attach adhesives with diamond, silver, or other highly conductive material, or thermal greases, thermal epoxies, and phase change materials (PCMs). Successful thermal management requires the development of a Thermal Interface Material (TIM) that connects the die and heat sink. Alternatively, attaching metal-plate heat spreaders to the chip while using thermally-enhanced molding compounds, embedded heat slugs or heat pipes for printing wiring board (PWB) and lead frame packages can improve heat spreading effectiveness by one order of magnitude. The Integrated Heat Transfer Spreader (IHS) is a good example of this option (Figure 4), which is used in the packaging for Pentium® 4 processors as integrated heat pipe lids in the Itanium™ processor.

Heat sinks are commonly attached to the surface of the spreader to provide additional surface area for heat removal by convection. The convection may be natural air convection or forced air convection via a fan or duct. For very high power applications, it may be necessary to cool the chip directly with a heat pipe attachment, high-speed air jets, a direct heat sink attachment (cold plate), or dielectric liquid immersion.

The second level is at the board, which provides the means for chip-to-chip communication with the backplane or motherboard to interconnect PWBs where heat removal typically occurs (through conduction in PWBs and convection to the ambient air). Use of PWBs with thick, high conductivity power and ground planes that include insulated metal substrates and/or embedded heat pipes provides improved thermal spreading at this level of packaging. Heat sinks are often attached to the back surface of PWBs. In systems designed for very harsh environments, e.g., avionics systems, the convective air-cooling is limited due to low gravity and air density at high altitudes. Instead, heat is transported to the edge of the PWB via conduction and removed by a cold plate or a heat exchanger attached at this edge (Figure 5), where wedge locks are used to ensure complete contact and efficient conducting. "System on a chip" or "computer on a chip" heat sinks, or finned surfaces protruding into the air, can be applied at the first and second levels to directly dissipate heat into ambient air.

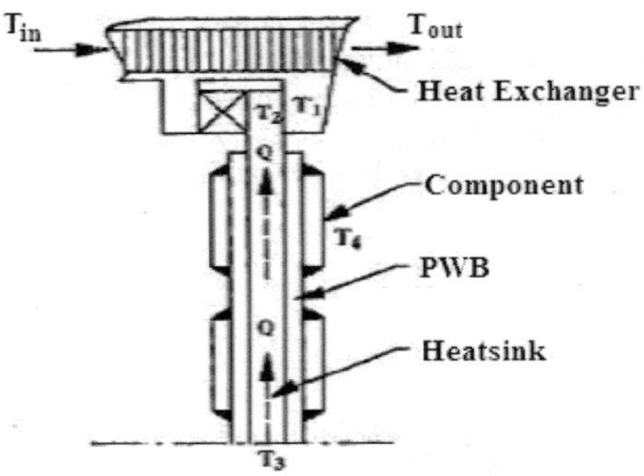

Figure 5: Heat transfer path to the cold wall of heat exchanger [6]

The third level is the system level, such as the box, rack, or cabinet, which provides the anticipated operating environment for the electronic device. An air-cooled chassis, shown in Figure 6, with modular integrated racks (MIR) to accommodate a complement of Line Replaceable Modules (LRMs), is suitable for shallow avionics. The thermal-mechanical interface to the LRMs is provided by removable, air/liquid cooled rack heat exchangers (top and bottom). Heat dissipated by the LRMs is conducted into each heat exchanger and is removed by coolant that is supplied from the aircraft's environmental control system (ECS). The coolant circulates through the MIR. The rack heat exchangers contain an internal serpentine channel that the coolant passes through [7]. Offset fins [8] or pin fins [9] are attached between the walls of the channel to provide structural integrity and enhance thermal performance.

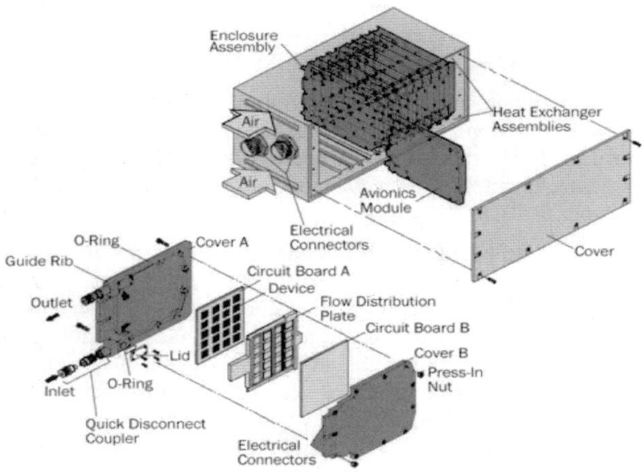

Figure 6: Air-cooled chassis system [6]

When the third level exists, thermal packaging generally involves the application of active thermal management techniques, such as air handling systems, refrigeration systems, or heat pipes, heat exchangers, and pumps. Routine fan cooling will still be maintained among the next generation electronic cooling solutions due to its simplicity and cost-effectiveness.

However, rapidly increasing heat flux, low junction temperatures and high ambient temperatures, and concerns over volume, weight, cost, and acoustic noise are limiting the successful application of fan cooling, particularly in harsh environment electronic systems. Therefore, the cooling resolutions will be focused on performance optimization for a particular application (system package level) and in some cases integrated into the electronics themselves (chip package level) to meet the junction temperature and power dissipation requirements. Enhanced air-cooled heat sinks, direct liquid cooling, phase change cooling, and refrigeration, along with design for manufacturability, sustainability and availability, can be expected to play pivotal roles in future electronic systems. In the following section more specific thermal management methodologies for selected harsh environment electronics are discussed.

3.1 AUTOMOTIVE ELECTRONICS

Currently, most of the thermal management solutions used in automotive applications rely on a combination of passive cooling such as conduction, and active air/liquid cooling. Most designs use conduction to transport heat from electronic components to the surface of the electronic enclosure and natural and/or existing convection to dissipate this heat into the ambient air. Figure 7 shows a typical automotive cooling system that works by moving coolant (water plus antifreeze) through the engine, and moving that heated coolant through the radiator, where its heat is transferred to the surrounding air.

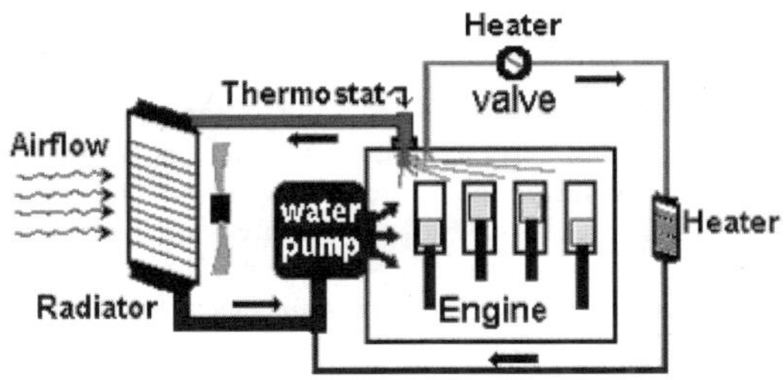

Figure 7: Typical automotive cooling system

The engine cooling system keeps the engine at its most efficient temperature at all speeds and operating conditions. It consists of a radiator, radiator pressure cap, coolant recovery tank, hoses, thermostat, water pump, fan and fan belt. The water pump sucks cooled coolant from the radiator and pushes it into the engine. The coolant flows through the engine, absorbing the engine's heat. The thermostat is an automatic valve to control the coolant circulation to keep the engine at a normal operating temperature. It closes when engine is cold and opens when engine is hot. If the thermostat is open, coolant flows into the radiator for cooling. At low speeds, airflow is maintained by the fan, and at high speeds, it is maintained by the relative velocity of the vehicle in relation to the outside air.

As more automotive mechanical functions are converted to electronic and electrical functions, and as recently introduced hybrid vehicles (which use internal combustion engines in conjunction with electric drive motors) and emerging fuel cell based electric vehicles (which use electric motors alone without internal combustion engines) become more common, these vehicles will use high power motor controls and drive electronics that will likely dissipate kilowatts of thermal energy. A thermal power dissipation summary for many current and future automotive electronic systems is shown in Figure 8.

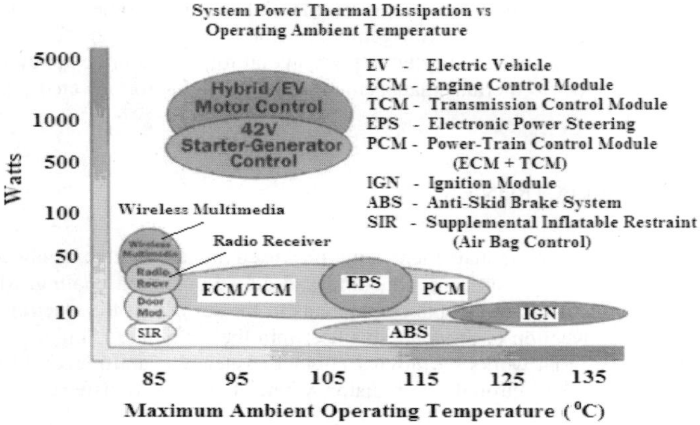

Figure 8: Current and future thermal power dissipation in automotive electronic systems [10]

The applications that operate in the highest ambient temperatures (i.e., ignition) and that have the highest power dissipation (i.e., hybrid and electric vehicle motor controllers) present the greatest challenge to current thermal management system design. For low heat fluxes, passive thermal management techniques can be used that do not require expending external energy for the heat removal. Interest in such techniques is currently very strong, due to their design simplicity, low cost, and high reliability. New thermal packaging materials for electronic components and system level that can reliably operate at junction temperatures of 175°C for digital and analog devices and 200°C for power drivers are needed to reduce the need for higher cost and more complex thermal cooling systems. In order to meet future higher power densities, the use of more efficient and feasible cooling technologies are anticipated to update those used in most automotive electronic applications today. Several emerging cooling technologies include advanced thermal packaging materials (i.e., PWBs with high-efficiency, copper, power and signal plane layers), reliable heat pipes and self-contained PCMs with solid-to-solid, solid-to-liquid, or liquid-to-gas, thermo-siphons, and liquid or refrigerant cooling systems. For example, Thermocore proposed one integrated cooling concept using heat pipe technology as a more efficient remote heat dissipating solution in automotive electronics systems (Figure 9). However, the greater challenge will lie in applying these available technologies in high volume and low cost, low weight and high reliability.

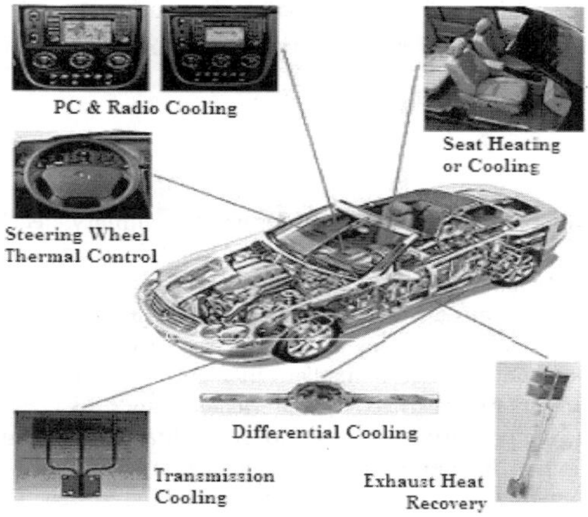

Figure 9: Application of heat pipe technology for automotive electronics [11]

3.2 MILITARY ELECTRONICS

Currently, heat flux at the chip level is in the range of 1-10W/cm^2 for avionics. Generally, conduction cooling has been widely used in military thermal management. For example, circuit boards in missiles are typically attached on one side to an aluminum structure resulting in a conduction path through the board. Many avionic-based electronics are conduction cooled using the military standard electronic module-format E (SEM-E) module that consists of conducting the heat away from the board through a thermal path parallel to the plane of the boards and rejecting the heat through air or liquid heat exchangers along two edges of the module (Figure 5). The most commonly known thermal management techniques for harsh environment electronics are summarized in Table 2.

Table 2: Overview of current thermal management techniques for harsh environment electronics

	Techniques	Comments
Passive Technology	High conductivity thin materials - thin diamond film - grease/adhesive with high k fill material (i.e., silver, graphite, diamond, MMC)	Simple and conventional High pressure between contact surfaces. Limited capability Inefficient for a non-uniform heat flux (i.e., hot spot)
	Heat sink	Sensitive to gravity and altitude Not feasible due to space limited
	Heat spreader	Effective and reliable Advanced MEMS heat spreader needed
	Phase change materials (PCMs)	Effective for intermittently operated avionics
	Heat pipe	Effective heat transport Miniature needed insensitive to gravity
Active Technology	Air jet impingement	Require cleaning and dehumidification Difficult integrated at chip level
	Indirect liquid cooling (cold plate)	More efficient than thermal conduction Require pump to overcome overall pressure drop in the loop Require low thermal resistance packaging at component level
	Immersion in dielectric liquid	Direct contact with chip surface Require high reliable liquid, complex hardware and high cost

Passive cooling is prevailing in current harsh-environment application because it is simple and reliable. However, as discussed in next section, passive cooling may not be sufficient to satisfy future generation high flux electronic cooling.

3.2.1 AIR COOLING

For decades, air cooling has been preferred for cooling military electronics ranging from PWBs to chips, multichip modules, and rack heat exchangers. Although air has less attractive thermophysical properties than most liquids, such as a low thermal conductivity (about 0.026W/mK) and a small Prandtl number (about 0.70), air has advantages over other coolants: it is available on most platforms, it is simple and inexpensive to implement, and it is easy to maintain and highly reliable without complex and expensive sealing devices. It appears that until two-phase and liquid cooling systems reach a stage where they can be inexpensively fabricated and packaged with high reliability, air-cooling will continue to be the primary choice for most of military thermal management systems where possible. As shown in Figure 10, four air-cooling schemes are used in today's military electronics, including (a) indirect air conducting cooling, (b) direct air cooling, (c) air flow through cooling and (d) air flow around cooling [12].

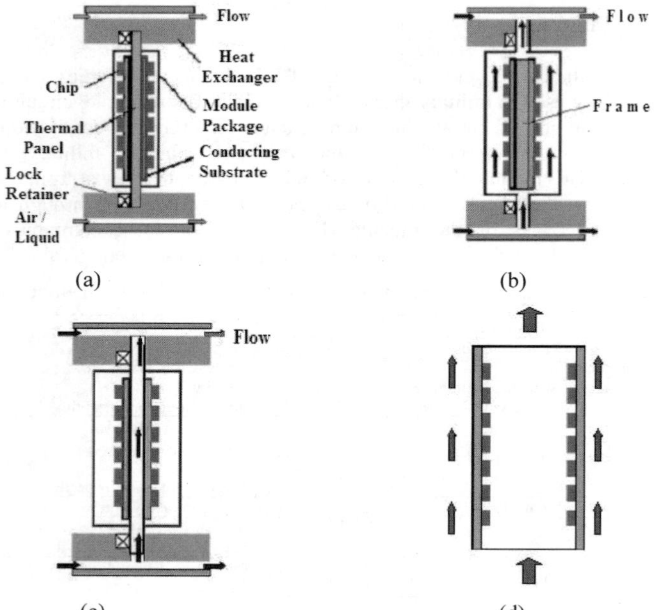

Figure 10: Main air cooling schemes for military electronics

Table 3 summarizes current air cooling techniques being used for military electronics modules. Due to increasing heat dissipation requirements, air cooling schemes are no longer capable of meeting demands of high performance military electronics modules, which require more than several hundred watts of cooling per module.

Table 3: Summary of available air cooling techniques

	Conducting	Air Flow Through (AFT)	Air Direction Flow (ADF)
Cooling Capacity (SEM module)	*< 50W*	*< 90W*	*< 100W*
Advantage	Low cost and high reliability Low mass and easy installation No pumps, ducting, filters, etc.	Abundant, free supply if taken from atmosphere No atmospheric altitude impact if using engine bleed air More efficient than conducting	Directly contact with components Eliminating thermal path resistance between air and components Some improvement in thermal performance than AFT
Disadvantage	Low cooling capacity Thermal contact resistance Substrate thermal conductivity dependant Need high pressure, intimate contact	Air supply need conditioning (throttling to low temperature) More complex cooling hardware Relatively large thermal resistance associated with edge heat exchanger	Atmospheric or cabin air need cleaning, filtering and dehumidification Component surfaces be free from corrosion
Comment	Air cooling schemes longer for demands of high performance avionics (up to 200 to 300W for SEM modules) Air cooling techniques are being replaced by liquid cooling systems. Single-phase fluid systems provide higher heat fluxes than conventional air cooling systems, phase change cooling system would provide even higher heat fluxes and allow compact packaging.		

One effort in air cooling seeks to improve air cooling's capability to reduce noise levels, pressure losses and heat sink volume, such as MEMS air cooling approaches, i.e., micromachined air jet arrays, or synthetic jets. A MEMS impinging-jet cooling device was developed for chip level cooling with a single-phase direct air cooling, or micro-jet array (MJA) (Figure 11a).

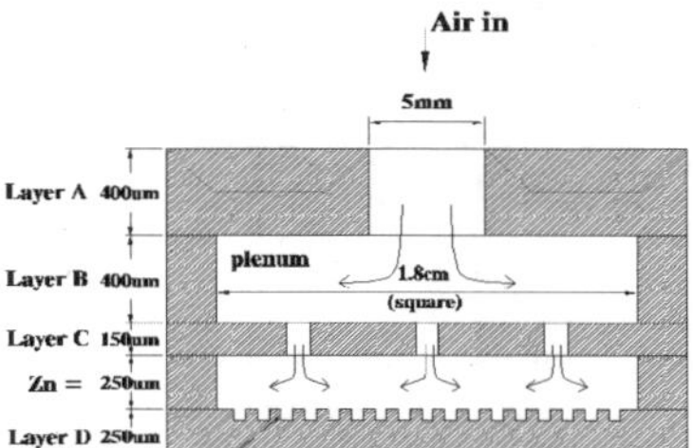

(a) Micro-jet air impingement

(b) Microchannel air cooler

Figure 11: MEMS based air cooling technologies

Single and multi-jet arrays with orifice diameters ranging from 50-800 micron were investigated including integration with actuation by magnetically driven membranes [14]. Heat transfer coefficients of 2500W/m^{2}K were reported with a 1.3×10^{7}ml/min air flow rate [15], which showed better than conventional air forced convection (around 50W/m^{2}K). An innovative cooling concept using a micro-channel air cooler was proposed for highly integrated avionics modules [12]. The micro cooler was fabricated by stacking precision etched thin copper foils that were fused together by direct bonded copper (DBC) technology. Several microcoolers can be integrated into one board to directly cool heat generating components (Figure 11b), which saves space and reduces system weight.

3.2.2 LIQUID COOLING

Due to the limits of air cooling, air cooling technology will be no longer capable of meeting the demands of future high performance military electronics in harsh environments. The rapidly increasing thermal management requirements of advanced electronics in harsh environments have lead to the use of liquid cooling techniques with the superior thermal transport properties of liquid coolants and the merits of phase change. For example, unlike other fighter aircraft, the next generation military aircraft, the F-22, successfully uses Polyalphaolefin (PAO) liquid flow through (LFT) cooling, rather than air cooling for its mission avionics. The LFT cooling configuration is similar to air flow-through (AFT) cooling scheme as shown in Figure 10c, except that liquid instead of air is circulated in the LRM core. The main difference with AFT is the need for quick disconnect couplings between the LRM and the rack distribution network. Currently, LFT using polyalphaolefin (PAO) liquid can cool several hundred Watts for today's military Standard Electronic Module-format E (SEM-E) modules. The use of the porous metal matrix as a heat exchanger core and as chassis racks and aircraft sandwich structures combines structural efficiency, heat transfer efficiency, and reduction in the volume of the heat exchanger core [16]. Mudawar (2001) investigated various liquid cooling schemes compatible with the existing SEM-E military enclosures. Figure 12 shows the removal capabilities for different liquid cooling arrangements.

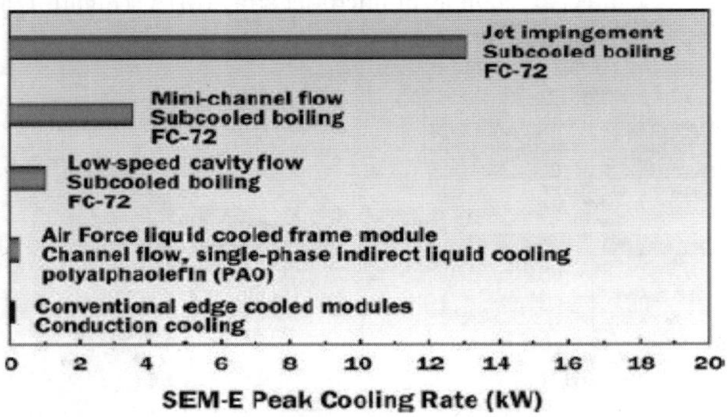

Figure 12: Heat removal capacity for different liquid cooling modules [6]

The single-phase PAO flow-through module demonstrated heat removal capability around 250W, and about 1000W heat dissipation was achieved with an FC-72 immersion cooling clamshell module. Mini- and micro-channel cooling techniques enabled heat removal capability over 3000W, one order of magnitude greater than the single-phase PAO flow-through module. The performance of the subcooled jet impingement showed great advantage over a conventional immersion cooling technique by approximately a factor of 13. Over the past decade, the Air Force Research Laboratory and Navy have invested in the development of a spray-cooled chassis integrated into the environmental control system on an advanced aircraft or other military electronic systems [17]. Isothermal Systems Research (ISR) proposed a spray-cooled chassis to the U.S. Marine Corps for the Advanced Amphibious Assault Vehicle (AAAV) [18]. A comparison of power consumed per 1kW of cooling between traditional air refrigeration and spray cool systems is shown in Table 4.

Table 4: Comparison of power consumption for air refrigeration cooling and spray cooling

Cooling System Power for 1 kW Elect. System	Air + Refrigeration	Spray Cooling
Fan Power	250/1000	30/1000
Pump Power		20/1000
ACU Power	412/1000	0
Total Power	662/1000	50/100

Using spray cooling, the total power consumption is much lower than a fan-cooled electronics system in an air-conditioned environment. Other candidates for spray cooling include the U.S. Navy's EA-6B carried-based jet and the U.S. Air Force's F-16 jet fighter. While jet impingement and spray cooling have been proven in the laboratory, they still require significant development before they are ready for application to a real, military, harsh environment system. Their practical application is dependent on the reliability of fluid pump and the variation in nozzle performance due to contamination, corrosion, and clogging.

Although pool boiling has been shown to yield cooling rates well above $50W/cm^2$, its use is limited by several concerns, such as a minimum superheat required for boiling inception, a relatively thick thermal boundary layer, and an inherently low critical heat flux. These limitations can be overcome if a thin film (several microns) of the working liquid continuously covers the heated surface. In operation, a temperature gradient forms across the film via heat conduction and the liquid simply vaporizes at the liquid-vapor interface. This process has the potential to remove a very large amount of heat because the amount of heat removed is inversely proportional to the thickness of this thin liquid layer. Emerging Ultra Thin Film (UTF) evaporation is perhaps one of the most effective methods of heat removal from a high heat flux surface. One prototype of an EHD pumped UTF evaporator was reported to achieve a maximum cooling capacity of $65W/cm^2$ at an applied voltage of 150V for a 50 micron electrode gap when using R-134a as the working fluid. The total EHD power consumption was less than 0.02% of the total power input to the device, translating into a few mille Watts for the application at hand [19]. A comparison of cooling performances of the thin film evaporator with the performances of pool boiling and spray cooling techniques is shown in Figure 13 [20].

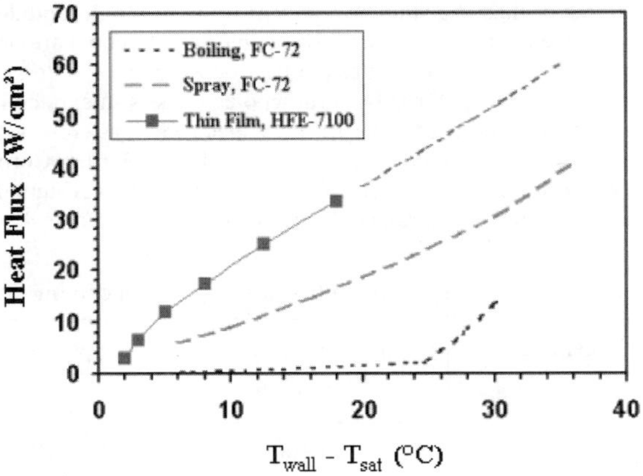

Figure 13: Comparison of cooling performances of thin film evaporator with the performances of pool boiling and spray cooling techniques

The data for pool boiling and spray cooling are reported by Bar-Cohen et al. [21] and Mudawar [6] for the 3M thermal fluid FC-72, which is quite similar to the 3M thermal fluid HFE-7100 used in the thin film evaporator test data shown in Figure 13. The thin film evaporator can remove heat fluxes of 20-40W/cm^2 with a temperature difference that is about 10-15°C less than spray cooling and about 30°C less than pool boiling. Further optimizing the electrode pattern and gap on the thin-film evaporator will generate ultra thin (micron-size) films on evaporator surfaces with higher cooling rates, low voltage and a more robust operation.

3.2.3 SPACE-BASED ELECTRONICS

On Earth, heat travels by conduction, convection, and radiation. However, conduction and natural convection are almost entirely nonexistent in the vacuum of space. Radiation is the primary method of heat transport in space. Space-based electronics that need to be kept cold are attached to radiators that face deep space and radiate excess heat into space. These electronics (i.e., space based phased-array-radar and laser systems) and radiators are thermally insulated from the rest of the spacecraft. Cooling is achieved through surface thermal radiation to deep space. Space-based electronics thermal management encompasses not only the removal of waste heat, but also the conservation of heat to provide a benign environment for the instruments and on-board electronic equipment.

The three types of cooling used for the extreme environmental requirements in space applications are compared in Table 5.

Table 5: Three types of space cooling technologies

Cooling Type	Advantage	Disadvantage
Passive cooling - *Blankets and barriers* - *Louvers* - *Heat pipes, pumped loops* - *Radiators*	- *simple* - *reliable* - *long lifetime (15yrs)* - *low cost and mass* - *proven technology*	- *limited cooling range (>80K)* - *depends on surroundings (orbit condition)*
Stored cryogen - *Superfluid helium* - *Solid nitrogen*	- *stable and passive* - *low temperature (1.3K using sup He)*	- *limited lifetime (1-2 yrs)* - *massive and bulky* - *complex interfacing (storage)*
Active cooling - *Mechanically pumped loops* - *Recuperative cryocooler (Joule-Thomson and Brayton)* - *Regenerative cryocooler (Stirling and Pulse-Tube)* - *Adiabatic demagnetization refrigerator (ADR)* - *Optical cryocooler*	- *low temperature (down to 10K now)* - *good lifetime (5-10yrs)* - *easily regulated, flexible, stable and uniform* - *enabling technology (Oxford 80K Stirling cooler succeeded in the late 1980s)*	- *power consuming* - *low efficiencies (1 to 3% of Carnot)* - *expensive/impossible to repair produce vibrations*

Currently, space-based radar (SBR) systems have a time-averaged power dissipation level in the 50 W/m^2 to 100W/m^2 range. The heat dissipation of future radar systems is anticipated to reach 1 kW/m^2 to 2 kW/m^2 in response to greater function and higher density packaging [22]. Because the dissipated power from the SBR modules is highly distributed, one cost- effective thermal solution is to use the large antenna area as a thermal radiator. The more concentrated heat dissipation from the other electronics (such as the power converters) is about 20% or less of the total dissipation and can be controlled with local heat spreaders. Moreover, low weight requirements will necessitate new materials of low weight and high thermal conductivity like graphite and/or phase change materials. Thermal performance of PCMs can be further improved by filling them with metallic foams [23].

Future high power SBR, space-based laser (SBL), Integrated Power Systems (IPSs), and electromagnetic weapons (EWs) will require the utilization of two-phase thermal management systems that employ capillary-pumped-loops or loop heat pipes with thin, flat evaporators and multiple parallel evaporators [24 and 25] or even more aggressive use of refrigeration (i.e., thermoelectronic devices, magnetic or thermo-acoustic refrigeration and cryogenic coolers) [26, 27 and 28] and emerging MEMS based cooling techniques (Microchannel cooling and ultra thin film cooling) [29, 30 and 31]. The potential application of such cooling techniques will depend largely on the development of low-cost, use-based technologies that can be proven suitable and reliable for long-life, harsh-environment applications [32].

4. Summary

Continued electronic circuit miniaturization, coupled with increased performance requirements and severe mechanical/environmental constraints in harsh environment systems, has resulted in ever increasing circuit power densities and dissipation levels. Furthermore, the future electrical generation for the all-electric ships, more-electric aircraft (MEA), and more-electric vehicles (EV) will require higher levels of power generation and increased thermal management challenges. Power conditioning circuits will switch from silicon substrates to silicon carbide (SiC) substrates with a five-times increase in heat flux for air-cooled applications and a more than three-times increase in heat flux for liquid cooled applications [NEMI 2002]. These increases result in the need for advanced and more efficient thermal management techniques. With the current emphasis on the use of COTS components in military electronics systems, there will be an increased emphasis on utilizing commercially available cooling technology to meet these demands and reduce costs.

The authors' qualitative ranking of the potential applicability of various thermal management technologies for harsh environment electronics is shown in Figure 14. The potential applicability of these techniques in terms of criteria such as cost, ease of use, thermal performance, and reliability is compared.

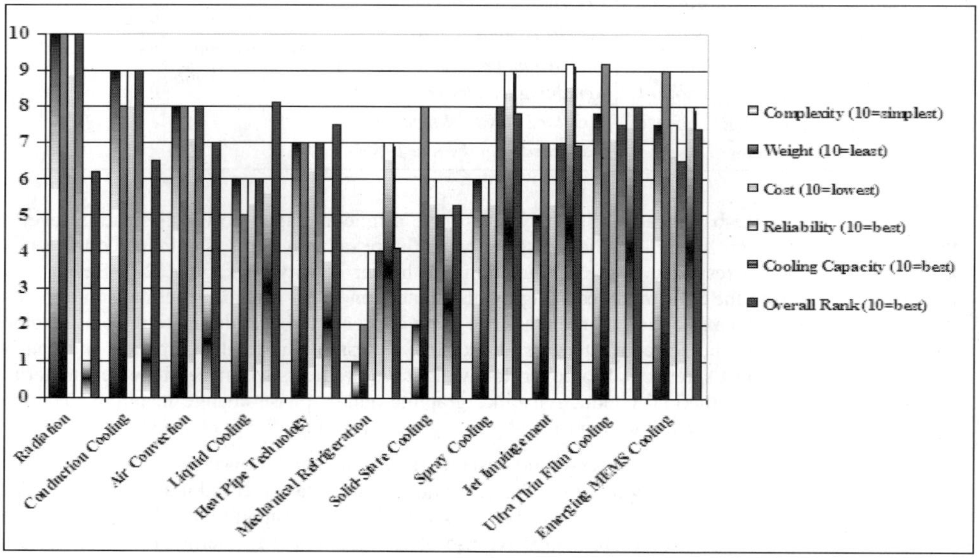

Figure 14: Summary of thermal management techniques for harsh environment electronics

For low heat flux systems, passive cooling (no machinery to move coolant) for new-generation electronics is as an attractive approach that can produce overall improvements in system reliability and reduced costs of operation. High conductivity materials and lightweight micro-heat pipes may possibly be used as thermal backplanes within circuit cards or enclosures. Since heat pipes do not rely on gravity, heat pipes can be used in any orientation. High thermal conductivities such as those found in graphite composites impregnated with unidirectional carbon fibers can be up to 800 W/mK, which can be used as heat sinks or for efficient heat removal.

For high heat flux high performance systems, a combination of active and passive techniques can be used. These may include using a high thermal conductivity substrate at the board level, along with forced air cooling or direct-contact liquid cooling at the board level. For this range of heat fluxes, the use of single phase cold plate technology and flow through cooling are also possible. For example, when current air cooling methods couldn't meet the future high performance military electronics system, single-phase liquid cooling methods (Liquid Flow-Through Cooling) became available in the next generation of military aircraft (the F-22). For even higher heat fluxes, a variety of cooling techniques are available that require the use of external power for cooling and can operate a coolant circulator (forced convection), direct liquid-immersion, spray/jet impingement, and ultra thin film evaporative cooling techniques which have recently been examined for removal of very high heat fluxes to meet future cooling requirements in harsh environment systems.

Future thermal management technology for harsh environment electronics will be focused on the following areas.

1. At the chip level, thermal management solutions need high performance heat spreaders to minimize thermal contact resistance:

- low CTE and low-coat, high thermal conductivity packaging materials, i.e., metal matrix composite and AlSiC
- embedded micro-heat pipes or vapor chamber cold plates
- micro- or mini-channel heat sinks
- micro-machined air-jet array impingement
- high performance thermal interfacial materials, such as thin film thermal grease, thermal pastes and PCM interfacial materials.

2. At the board level, compact single or phase change cooling methods have potential application in future high heat-flux avionics:

- single phase micro cooling, i.e., single-phase micro/mini-channel liquid cooling
- phase change cooling, i.e., spray or jet impingement cooling, and ultra thin-film evaporative cooling (UTF)
- refrigeration cooling, i.e., high COP micro-refrigerators and advanced thermoelectric cooling.

3. At the system level, several important criteria should be considered for selecting a suitable thermal management solution for a particular harsh environment system:

- heat dissipation potential
- cost and reliability
- packaging concerns, i.e., weight, volume and power
- complexity and flexibility, i.e., enhanced usability and minimal interface incompatibilities
- integration of thermal subsystem in the existing environmental control system
- minimal impact to the environment.

NOMENCLATURE

AAAV	Advanced Amphibious Assault Vehicle
ABS	Anti-Skid Brake System
ADR	Adiabatic Demagnetization Refrigerator
AFT	Air-Flow-Through
BGA	Ball Grid Array
COP	Coefficient of Performance
COTS	Commercial Off-The-Shelf
CTE	Coefficient of Thermal Expansion
DBC	Direct Bonded Copper
ECM	Engine Control Module
ECS	Environmental Control System
EHD	Electrohydrodynamics
EPS	Electronics Power Steering
EV	Electric Vehicle
EWs	Electromagnetic Weapons
FC	Fluorochemical
HEF	Hydro-Flouro-Ether
HEV	Hybrid Electric Vehicle
IC	Integrated Circuit
IGN	Ignition Module
IHS	Integrated Heat Spreader
IPS	Integrated Power System
LFT	Liquid-Flow-Through
LRMs	Line Replaceable Modules
MEA	More Electric Aircraft
MEMS	Micro-Electro-Mechanical System
MIR	Module Integrated Rack
MJA	Mciro-Jet Array
NEMI	National Electronics Manufacturing Initiative
PAO	Polyalphaolefin
PCM	Powertrain Control Module (ECM+TCM)
PCMs	Phase Change Materials
PWB	Printing Wiring Board
SBL	Space-Based Laser
SBR	Space-Based Radar
SEM	Standard Electronic Module
SiC	Silicon Carbide
SIR	Supplemental Inflatable Restraint (Air Bag Control)
TCM	Transmission Control Module
TIM	Thermal Interface Material
UTF	Ultra Thin Film
VLSI	Very Large Scale Integration

REFERENCES

1. Patterson, D., 2002, "Novel Approaches to Conduction Cooling," COTS Journal, pp. 45 –48, May, pp.51.
2. National Electronics Manufacturing Technology Roadmaps, NEMI 2002 Edition.
3. Sandeau, R., 1992, "Interconnection Trends: View of A Supplier to the Military," Proc. Int. Electron. Packag. Conf., Austin, TX, pp.23-24.
4. Ohadi, M., Cerza, M., and A. Bar-Cohen, eds., 2003, "Thermal Management of High Flux Military and Commercial Electronics," Workshop Proceedings, College Park, MD, Oct. 2003.
5. Bar-Cohen A., 1993, "Physical Design of Electronic Systems –Methodology, Technology Trends, and Future Challenges." In Advances in Thermal Modeling of Electronic Components and Systems, Vol. 3, ASME Press, pp. 45.
6. Mudawar, I. 2001, "Assessment of High-Heat-Flux Thermal Management Schemes," IEEE Transactions on Components and Packaging Technologies, Vol. 24 (2), pp.122-141.
7. LeVasseur, R., 1991, "Liquid Cooled Approaches for High Density Avionics," 1991 IEEE/AIAA 10th Digital Avionics Systems Conference, Los Angeles, CA.
8. Tang, L., Moores, K.A., Ramaswamy, C. and Joshi, Y., 1998, "Characterizing the Thermal Performance of a Flow Through Electronics Module (SEM-E Format) Using a Porous Media Model," IEEE Fourteenth IEEE SEMI- THERMTM Symposium.
9. Buechler, A., Sims, C.W., 1996, "Liquid-Flow-Through Module/ Cardcage Thermal and Environmental Test Report (NSWC-6043-AECS-5)," Naval Surface Warfare Center, Crane Division.
10. Myers, B. A., 2003, "Cooling Issues for Automotive Electronics," Electronics Cooling, Volume 9, No 3.
11. Thermocore Inc., Automotive Thermal Control (http://www.thermacore.com/automotive.htm)
12. Schulz-Harder, J., Exel, K., Hermann, L., 2001, "Innovative Cooling Concept for Highly Integrated avionics modules," 13th European Microelectronics and Packaging Conference, Strasbourg, France, May 31.
13. Ohadi, M., Cerza, M., and Bar-Cohen, A., 2004, "Thermal Management of High Flux Military and commercial electronics, Book manuscript in preparation.
14. Coe, D.J., Allen, M.G., Smith, B.L. and Glezer, A., 1995, "Addressable Micromachined Jet Arrays," International Conference on Solid-State Sensors and Actuators and Eurosensors IX, pp.329-332.
15. Lin, Q., Wu, S., Yuen, Y., Tai, Y.C. and Ho, C.M., 2000, "MEMS Impingement-Jet Cooling," Proceedings of 2000 ASME International Mechanical Engineering Congress and Exposition, Orlando, FL, pp.137-142.
16. Price, D.C., Weber, R. M., Schwartz, G. J., McDaniel, J. and Lage, J. L., 1999, "Cold plate design for thermal management of phased-array radar systems," U.S. Patent 5 960 861.
17. Bass, M. and Chow, L., 2000, "Spray Cooling: Enabling High Power Solid State and Diode Lasers," Air Force Research Laboratory.
18. Tilton, C., 2000, "Applying Mist to COTS Convection Modules Is Really Cool," COTS Journal, pp.43-53.
19. Darabi, J., Ohadi, M.M. and Dessiatoun, 1999, "EHD Enhancement of Thin Falling Film Evaporation," ASME/JSME Joint Thermal Engineering Series, Vol.5, pp.547-553.
20. Moghaddam, S. and Ohadi, M., 2002, "Thin Film Evaporator with Splayed Electrodes," U.S. patent pending, Application number 60/419,690, October.

21. Bar-Cohen, A., Watwe, A., and Arik, M. "Pool Boiling Critical Heat Flux in Dielectric Liquids," Keynote Lecture, 5th International Conference on Boiling Heat Transfer, Montego Bay, Jamaica, May 2003.
22. Bran, M, Thibodeaux, R, Fronista, G, Wilson, J, and Levin, S., 2000, "Dual Use Power Management System for Space Radar," 38th AIAA Aerospace Sciences Meeting & Exhibit, AIAA paper A00-37724.
23. Haws, J. and Short, E., 2002, "Method and Apparatus for Cooling with Phase Change Materials and Heat Pipes," European Paten No. 00965034.2-2220-US0025297.
24. Meyer, L., Dasgupta, S., Shaddock, D., Tucker, J. and Fillion, R., 1993, "A Silicon-Carbide Micro-Capillary Pumped Loop for Cooling High Power Devices," 9th Semiconductor Thermal Measurement and Management Symposium, pp. 364-368.
25. Butler, D., Ku, J. and Swanson, T., 2002, "Loop Heat Pipes and Capilary Pump Loops – an Application Perspective," Space Technology and Applications International Forum-STAIF, Albuquerque, NM, pp.49-56.
26. Fan, X., Zeng, G., LaBounty, C., Croke, E., Vashaee, D., Shakouri, A., Ahn, C. and Bowers, J.E., 2001, "High Cooling Power Density SiGe/Si Micro Coolers," Elec. Lett., Vol. 37(2).
27. Zimm, C. et al., 1998, "Description and performance of a near-room temperature magnetic refrigerator. Adv. Cryogen. Eng., Vol. 43, pp.1759-1766
28. Swfit, G.W., 2002, *Thermoacoustics: A unifying Perspective for Some Engines and Refrigerators*, Acoustical Society of America (ASA) Publications, NY
29. Golliher, E.L., 2002, "Microscale Technology Electronics Cooling Overview," Space Technology and Applications International Forum-STAIF, Albuquerque, NM, pp.250-257
30. Beach, R., Freitas, B. and Rotter, M., 2002, "Compact Microchannel-Cooled Laser-Diode Arrays Deliver over Tens of kW/cm^2 onto the Working Surface," Lawrence Livermore National Laboratory (LLNL), Laser Science and Technology Monthly Highlights Newsletter (March)
31. Heffington, S.N., W.Z. Black, and A. Glezer. 2002, "Vibration-Induced Droplet Atomization Heat Transfer Cell for High-Heat Flux Dissipation," Thermal Challenges in Next Generation Electronic Systems (THERMES-2002), Santa Fe, NM
32. Werner, M.R.; Fahrner, W.R., 2001, "Review on Materials, Microsensors, Systems and Devices for High-temperature and Harsh-environment Applications," IEEE Transactions on Industrial Electronics, Vol. 48, No.2, pp.249-257

Thermal Transport Phenomenon In Micro Film Heated By Laser Heat Source

S. TORII* and W.-J. YANG**
* Department of Mechanical Engineering and Materials Science
Kumamoto University, Kumamoto, JAPAN
** Mechanical Engineering and Applied Mechanics
University of Michigan, Ann Arbor, USA

1. Introduction

Several issues of basic scientific interest arise in cases such as laser penetration and welding, explosive bonding, electrical discharge machining, and heating and cooling of micro-electronic elements involving a duration time of nanosecond or even picosecond in which energy is absorbed within a distance of microns from the surface. For example, the issue of energy transfer into a lattice and resulting temperature in the lattices during such a short period of time and over such a tiny region is of fundamental importance but remains a matter of controversy [1]. This is because the wave nature of thermal propagation is dominant, that is, a thermal disturbance travels in the medium with a finite speed of propagation [2], [3], [4] and [5].

The above phenomena are physically anomalous and can be remedied through the introduction of a hyperbolic equation based on a relaxation model for heat conduction, which accounts for a finite thermal propagation speed. Recently, considerable interest has been generated toward the hyperbolic heat conduction (HHC) equation and its potential applications in engineering and technology. A comprehensive survey of the relevant literature is available in reference [6]. Some researchers dealt with wave characteristics and finite propagation speed in transient heat transfer conduction [3], [7], [8], [9] and [10]. Several analytical and numerical solutions of the HHC equation have been presented in the literature.

Several authors have studied analytically the parabolic and hyperbolic models of heat conduction with the laser heat source and with a convective boundary condition [11]. Using both models, Kar et al. [12] studied heat conduction due to shortpulse heating for various boundary conditions. They reported that the predicted temperature distribution is substantially affected by the temperature dependent thermal properties. Lewandowska [13] also dealt with the parabolic and hyperbolic heat conduction in one-dimensional, semi-infinite body with the insulated boundary and discussed different time characteristics of the heat source capacity. It is disclosed that (i) for small dimensionless Bouguer number the temperature distribution in the body results from the heat generation process, and (ii) the significant difference between the hyperbolic and parabolic solutions appears in only an edge of the body, where the hyperbolic temperature is higher than the parabolic one. Size effects on nonequilibrium laser heating of metal films were investigated by Qiu and Tien [14].

This paper deals with thermal wave behavior during transient heat conduction in a film (solid plate) subjected to a laser heat source with various time characteristics from both side surfaces. Emphasis is placed on the effect of the time characteristics of the laser heat source (constant, pulsed and periodic) on thermal wave propagation. Analytical solutions are obtained by means of a numerical technique based on MacCormack's predictor-corrector scheme to solve the non-Fourier, hyperbolic heat conduction equation.

2. Formulation and Numerical Method

The constitutive equation used in the linearized thermal wave theory is expressed as

$$\mathbf{q}(\mathbf{x},\mathbf{t}) + \tau \frac{\partial \mathbf{q}(\mathbf{x},\mathbf{t})}{\partial \mathbf{t}} = -\mathbf{k}\nabla \mathbf{T}(\mathbf{x},\mathbf{t}) \tag{2.1}$$

S. Kakaç et al. (eds.), Microscale Heat Transfer, 499–506.

This is referred to as the Maxwell-Cattanero equation, because it was postulated first by Maxwell [15] and later on by Cattaneo [16]. Note that the relaxation time τ defined as $\tau=\alpha/C^2$ is assumed to be constant, where C is the speed of "second sound" (thermal shock wave).

The energy equation including the heat generation reads

$$\rho \mathbf{c_p} \frac{\partial \mathbf{T}(\mathbf{x},\mathbf{t})}{\partial \mathbf{t}} = \mathbf{S}(\mathbf{x},\mathbf{t}) - \nabla \mathbf{q}(\mathbf{x},\mathbf{t}) \tag{2.2}$$

where **S** is the heat source term, i.e., the heat generation term.

One-dimensional thermal propagation in a film with thickness of x_0 is analyzed, as shown in Fig. 1. At t=0, a very thin film with thickness of x_0 is maintained at a uniform, initial temperature T_0. For t>0, the wall surfaces at x=0 and x_0 are suddenly heated due to the laser heat source. Nonequilibrium convection and radiation are assumed to be negligible. Under these conditions and assumptions, the modified Fourier equation including the relaxation time and the energy equation with internal heat sources can be represented as

$$\tau \frac{\partial q}{\partial t} + q + k \frac{\partial T}{\partial x} = 0 \tag{2.3}$$

$$\rho c_p \frac{\partial T}{\partial t} + \frac{\partial q}{\partial x} - g = 0 \ , \tag{2.4}$$

respectively. For a metal (absorption coefficient m of the order of 10^7-10^8 m^{-1}) which absorbs laser energy internally, many researchers (for examples, Vick and Ozisik [17]; Ozisik and Vick [18]; Tang and Araki [19]) reported that almost all energy is absorbed within a depth of the order of 0.1 mm which can be treated as a skin effect. Thus, the model considers the laser radiation as a heat source, which is x-independent and non-zero only within a layer of the body or even as a surface heat flux. Based on this idea, the energy sources term in Eq. (2.4), for a material that absorbs laser energy internally, is modeled by Blackwell [20] and Zubai and Aslam [21] as

$$g(x,t) = I(t)(1 - R)\mu \exp(-\mu x) \tag{2.5}$$

Here I(t) is the laser incident intensity and R is the surface reflectance of the body. Note that this model assumes no spatial variations of I(t) in the plane perpendicular to the laser beam and no heat transport in the direction perpendicular to the beam. α is thermal diffusivity, t is time, c_p is specific heat, k is heat conduction, ρ is density, and q is heat flux.

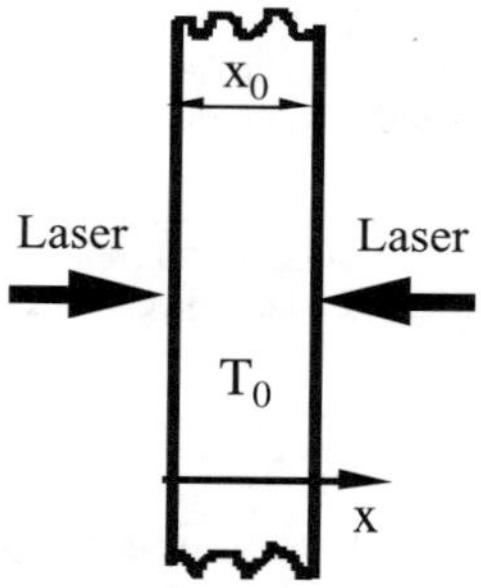

Figure 1. Physical configuration and coordinate system

The following dimensionless quantities, i.e., dimensionless temperature, dimensionless heat flux, and dimensionless time and space variables are introduced as

$$\theta(\xi,\eta) = \frac{T - T_0}{T_{w1} - T_0} \tag{2.6a}$$

$$Q(\xi,\eta) = \frac{\alpha_0 q}{(T_{w1} - T_0)k_0 c_0} \tag{2.6b}$$

$$\xi = \frac{c_0^2 t}{2\alpha_0} \tag{2.6c}$$

$$\eta = \frac{c_0 x}{2\alpha_0} = \frac{x}{2\tau c_0} \tag{2.6d}$$

Equations (2.3) and (2.4) are expressed in terms of the above dimensionless variables as

$$\frac{\partial Q}{\partial \xi} + \frac{\partial \theta}{\partial \eta} + 2Q = 0 \tag{2.7}$$

and

$$\frac{\partial Q}{\partial \eta} + \frac{\partial \theta}{\partial \xi} - 2\psi_o \phi(\xi)\exp(-\beta\eta) = 0 \tag{2.8}$$

Initial and boundary conditions are represented, as

$$\theta = 0,\ Q = 0 \qquad \text{at } \xi = 0,\ 0 < \eta < \frac{c_0 x_0}{2\alpha_0}$$

$$\frac{\partial \theta}{\partial \eta} = 0,\ Q=0 \qquad \text{at } \xi > 0,\ \eta = 0 \text{ and } \frac{c_0 x_0}{2\alpha_0}$$

Note that the boundary condition of Q at $\xi > 0$ is derived from Eqs.(2.7) and (2.8). For comparison, consider a very thin film with thickness of x_0 maintained at a uniform, initial temperature T_0. The walls at x=0 and x_0 are suddenly heated to a temperature T_w. Equations to be employed are Eq. (2.7) and the following dimensionless one as:

$$\frac{\partial Q}{\partial \eta} + \frac{\partial \theta}{\partial \xi} = 0 \tag{2.9}$$

Glass et al. [7] reported that MacCormack's method [22], which is a second-order accurate explicit scheme, can handle these moving discontinuities quite well and is valid for the HHC problems. Since the hyperbolic problems considered here have step discontinuities at the thermal wave front, MacCormack's prediction-correction scheme is used in the present study. When MacCormack's method is applied to Eqs. (2.7) and (2.8), the following finite difference formulation results:

Predictor:

$$\overline{\theta_i^{n+1}} = \theta_i^n - \frac{\Delta\xi}{\Delta\eta}\left(Q_{i+1}^n - Q_i^n\right) + \Delta\xi\left[-2\psi_0\phi(\xi)\exp(-2\beta\eta)\right] \tag{2.10}$$

$$\overline{Q_i^{n+1}} = Q_i^n - \frac{\Delta\xi}{\Delta\eta}\left(\theta_{i+1}^n - \theta_i^n\right) - 2\Delta\xi Q_i^n \tag{2.11}$$

Corrector:

$$\theta_i^{n+1} = \frac{1}{2}\left\{\theta_i^n + \overline{\theta_i^{n+1}} - \frac{\Delta\xi}{\Delta\eta}\left(\overline{Q_i^{n+1}} - \overline{Q_{i-1}^{n+1}}\right) + \Delta\xi\left[-2\psi_0\phi(\xi)\exp(-\beta\eta)\right]\right\} \tag{2.12}$$

$$Q_i^{n+1} = \frac{1}{2}\left[Q_i^n + \overline{Q_i^{n+1}} - \frac{\Delta\xi}{\Delta\eta}\left(\overline{\theta_i^{n+1}} - \overline{\theta_{i-1}^{n+1}}\right) - 2\Delta\xi\overline{Q_i^{n+1}}\right], \tag{2.13}$$

where the subscript i denotes the grid points in the space domain, superscript n denotes the time level, and $\Delta\eta$ and $\Delta\xi$ are the space and time steps, respectively. The circumflex terms, i.e., $\overline{Q_i^{n+1}}$, $\overline{\theta_i^{n+1}}$, etc. are a temporary predicted value at the time level n+1.

Computations are processed in the following order:

1. Specify the values of Q and θ at n=0, i.e., the initial values.
2. Solve Eqs. (2.10) and (2.11) for θ and Q to obtain temporary predicted value at the time level n+1.
3. Calculate new values of θ and Q at the time step n+1 using Eqs. (2.12) and (2.13).
4. Repeat steps 2-3 until n reaches a desired time level from the onset of calculation.

Throughout numerical calculations, the number of grids is properly selected between 1,000 and 5,000 to obtain a grid-independent solution, resulting in no appreciable difference between the numerical results with different grid spacing. The ranges of the parameters are nondimensional plate thickness $cx_0/2\alpha$=0.5 and 5.

3. Numerical Results and Discussion

Figures 2(a) and (b), for $\phi(\xi)=1$, $\psi_0=1$, and $c_0x_0/\alpha=1.0$, depicts the time-histories of the temperature distribution, θ, in a film for $\beta=1$ and 10, respectively. Since the space region of the heat source capacity increases for larger β, an increase in the film temperature yields over the whole film, as seen in Fig. 2(a). Note that at ξ fixed, the film temperature for $\beta=1$ is higher than that for $\beta=10$. It is observed in Fig. 2(b) that as time progresses, the film temperature for $\beta=10$ gradually increases because almost all energy is absorbed in the vicinity of both side walls and after $\xi=0.6$, the film temperature substantially induces in the centre region of the film, that is the temperature overshoot occurs. This trend becomes minor in a thick film, as seen in Fig. 3. Figure 3 illustrates the time-histories of the temperature distribution, θ, in a film for $c_0x_0/\alpha=10.0$. Notice that $\phi(x)$ and ψ_0 are the same as the corresponding values in Fig. 2. The film temperature gradually increases in the absence of temperature overshoot even for different β. The temperature profiles behave like diffusion domination and are in accordance with theoretical results predicted by the classical heat-conduction theory.

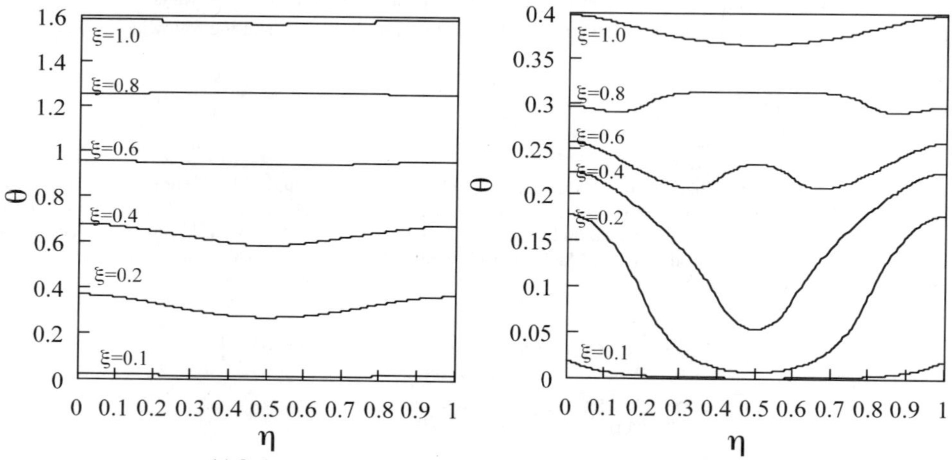

(a) $\beta=1$ (b) $\beta=10$

Figure2. Instantaneous dimensionless temperature distributions in the film for $\phi(\xi)=1$, $\psi_0=1$ and $c_0x_0/\alpha=1.0$

Next is to investigate the effect of time-dependence of laser heat source on the time history of the film temperature, for $\psi_0=1$ and $\beta=10.0$. Figure 4 illustrates the time-history of the temperature distribution in the film with $c_0x_0/a=1$, in which the periodic laser source is modeled as $\phi(\xi)=(1+\sin\omega\xi))/2$. This phenomenon implies that the heat source is periodically oscillated in the vicinity of both side-walls of the film. (a) and (b) of the figure depict numerical results for $\omega=1$ and 10, respectively. It is observed in Fig. 4(a) that although the propagation

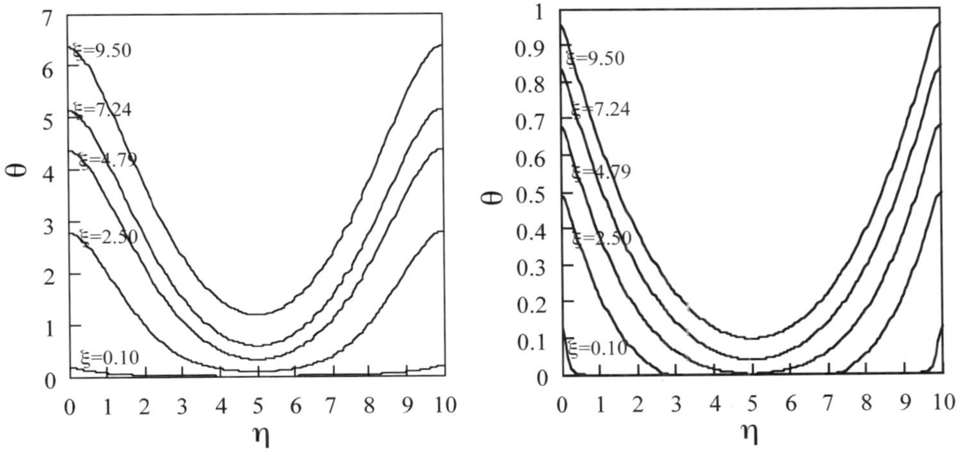

(a) β=1 (b) β=10

Figure 3. Instantaneous dimensionless temperature distributions in the film for $\phi(x)=1$, $\psi_0=1$ and $c_0x_0/\alpha=10.0$

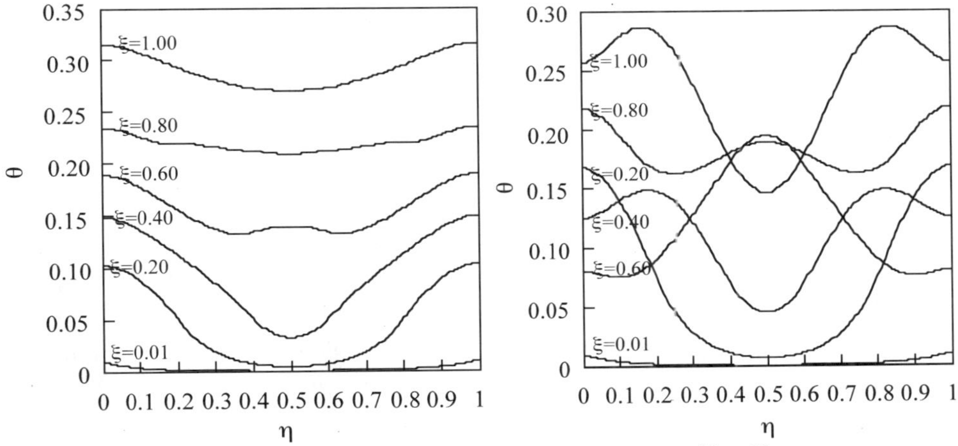

(a) ω=1 (b) ω=10

Figure 4. Instantaneous dimensionless temperature distributions in the film for $\phi(\xi)=1$, $\psi_0=(1+\sin(\omega\xi))/2$ and $c_0x_0/\alpha=1.0$

process of thermal waves in a film, for $\omega=1$, is similar to the numerical result for constant laser source source, as seen in Fig. 2(b), the temperature overshoot disappears. By contrast, Fig. 4(b) shows that when the frequency of the periodic laser heat source becomes larger, a substantial change in wave propagation is caused through time. One observes that (i) when wavefronts from both sides arrive at the center of the film, the temperature is substantially increased and; (ii) the temperature overshoot takes place; and (iii) thought the film temperatures at both-side walls are changed with time, the inner film temperature is increased over the whole region of the film as time progresses. The effect of the time-dependence of laser heat source becomes minor for the thick film, as seen in Fig. 5, which illustrates the timewise variation of the film temperature profile with $c_0x_0/\alpha=10.0$. Here the values of ψ_0, $\phi(\xi)$, β and ω are the same as these of Fig. 4(b). One observes that as time progresses, the film temperature increases gradually, whose behavior is similar to that shown in Fig. 3(b), thought the oscillate temperature profile yields near the sidewalls because of the periodic laser heat source. Note that for the small value of β, i.e., β=1, the calculated temperature distribution is similar to that for β=10 (not shown), but the absolute value of the film temperature is different for β=1 and 10. It is found that (i) the effect of the frequency of a periodic heat source w on the temperature distribution become considerably greater in the very thin film, while its oscillation is affected only near the wall of the thick film.

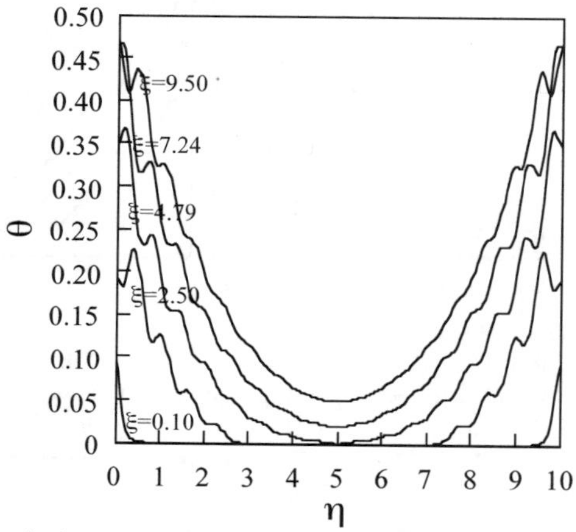

Figure 5. Instantaneous dimensionless temperature distributions in the film for $\phi(\xi)=(1+\sin(10\xi))/2$, $\psi_0=1$, $\beta=10$ and $c_0x_0/\alpha=10.0$

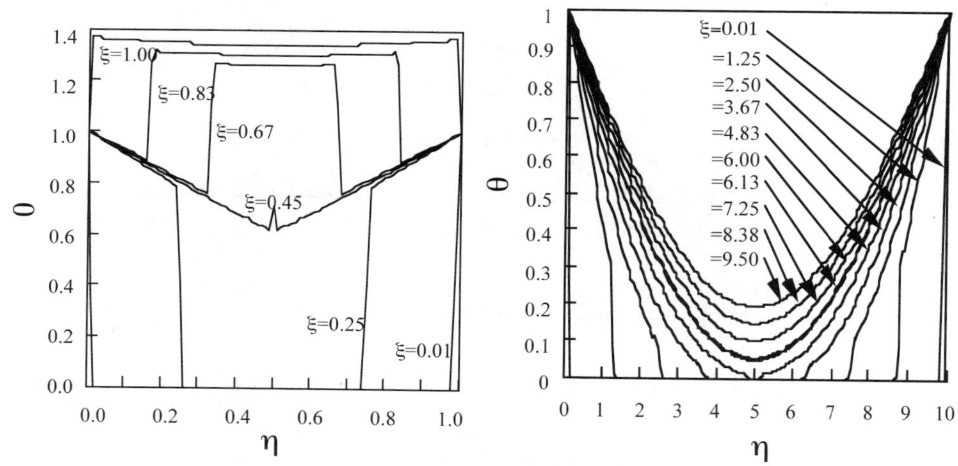

(a) $c_0x_0/\alpha=1$ (b) $c_0x_0/\alpha=10$

Figure 6. Instantaneous dimensionless temperature distribution in the film with an symmetrical temperature change

Figures 6(a) and (b) illustrate the timewise variations of the temperature distribution, θ, in films having $c_0x_0/\alpha=1.0$ and 10.0, respectively. Note that Fig. 6 reproduces precisely the theoretical results of Tan and Yang [23], which show in detail the propagation process of thermal waves in a film. Figure 6(a) depicts that (i) after the wall temperatures on two sides are suddenly raised, a set of sharp wavefronts exists in the thermal wave propagation and advances towards the center in the physical domain which separates the heat-affected zone from the thermally undisturbed zone; (ii) thermal wavefronts in the linear case meet and collide with each other at the center of the film; (iii) after first collision, the center temperature in a film causes a significant amplification resulting a much higher temperature in this region, (iv) after that, reverse thermal wavefronts take place and travel towards both side walls of the film, and (v) when thermal wavefronts reach at both side walls, the film temperatures at both sides of strongly heated walls exceed the imposed wall temperature, called temperature overshoot. The numerical solution predicts the existence of thermal waves, particularly in a very thin film and presents the propagation process of thermal waves, the magnitude and shape of thermal waves. On the contrary, Fig. 6(b) depicts that (i) after wavefronts arrive at the center of the film with $c_0x_0/\alpha=10$, they gradually disappear in the absence of reverse temperature waves and temperature overshoot, and (ii) similar temperature distribution in the film yields in the linear and nonlinear cases. It is found from these results that the thermal relaxation time τ plays a primary role in distinguishing a domain to be wave dominating or diffusion dominating and the effect of temperature-dependent thermophysical property on the thermal wave propagation becomes larger in the very thin film.

It is found from these results that the thermal relaxation time τ plays a primary role in distinguishing a domain to be wave dominating or diffusion dominating. Several investigators have estimated the magnitude of thermal relaxation time τ to range from 10^{-10} second for gases at standard conditions to 10^{-14} second for metals with that for liquids and insulators, falling within this range. If τ is known, one can obtain the range of film thickness within which heat propagates as a wave. The criterion for thermal wave dominating in the present study is $c_0x_0/\alpha<10$. For example, the value of silicon corresponds to the thickness of the film in the order of about 0.01 micron using 10^{-14} s and $93.4\text{x}10^{-6}$ m^2/s as the relaxation time and thermal difffusivity [24], respectively.

4. Summary

Heat waves have been theoretically studied in a very thin film subjected to a laser heat source and a sudden symmetric temperature change at two side walls. The non-Fourier, hyperbolic heat conduction equation is solved using a numerical technique based on MacCormak's predictor-corrector scheme. Results have been obtained for the propagation process, magnitude and shape of thermal waves and the range of film thickness and duration time within which heat propagates as wave.

If a film is heated by the continuous-operated or oscillated lasers, temperature overshoot takes place in the films of smaller values of c_0x_0/α within a very short period of time. The effect of the laser heat source on the temperature distribution in the film becomes larger in the thin film. In other words, if the absorption coefficient, b, of the laser increases, the temperature is more dependent on the laser heat source in a thin film than in a thick film. Overshoot and oscillation of thermal wave depend on the frequency ω of the heat source time characteristics.

NOMENCLATURE

c	speed of thermal wave, m/s
c_p	specific heat
g_0	reference capacity of internal heat source $Ir(1-R)\mu$ (W/m^3)
Ir	arbitrary reference laser intensity (W/m^2)
k	thermal conductivity, w/km
$Q(\eta, \xi)$	dimensionless heat flux
$q(x,t)$	heat flux, W/m^2
$T(x,t)$	temperature, K
T_0	reference temperature, K
t	time, sec.
x	space variable
x_0	film thickness, m

Greek Letters

α	thermal diffusivity, m^2/s
β	dimensionless absorption coefficient, $2\tau\mu c_0$
η	dimensionless space variable
$\theta(\eta, \xi)$	dimensionless temperature
ξ	dimensionless time variable
ρ	density, kg/m^3

τ relaxation time, α/C^2, sec.

ψ_0 constant coefficient related to the dimensionless strength of internal heat source, $\dfrac{k_0 g_0}{(T_{w1} - T_0)c_0^2 \rho^2 c_p^2}$

$\phi(\xi)$ dimensionless rate of energy absorbed in the medium, $I(2\tau\xi)/Ir$

REFERENCES

1. Bloembergen, N. Kurz, H. Kiu, J. M. and Yen, R. (1982) Fundamentals of Energy Transfer during Picosecond Irradiation of Silicon, in Laser and Electron-Beam Interactions with Solids, Elsevier Science Publishing Co., Inc., Neq York, pp. 3-20.
2. Chan, S. H. Low, J. D. and Mueller, W. K. (1971) Hyperbolic Heat Conduction in Catalytic Supported Crystallites, *AIChE Journal*, Vol. 17, pp. 1499-1507.
3. Kazimi, M. S. and Erdman, C. A. (1975) On the Interface Temperature of Two Suddenly Contacting Materials, *Journal of Heat Transfer*, Vol. 97, pp. 615-617.
4. Mourer, M. J. and Thompson, H. A. (1973) Non-Fourier Effects at High Heat Flux, *Journal of Heat Transfer*, Vol. 95, pp. 284-286.
5. Baumeister, K.J. and Hamill, T.D. (1969) Hyperbolic heat conduction equation - a solution for the semi-infinite body problem, *Journal of Heat Transfer*, Vol. 91, pp. 543-548.
6. Ozisik, M. N. and Tzou, D. Y. (1994) On the Wave Theory in Heat Conduction, *Journal of Heat Transfer*, Vol. 116, pp. 526-535.
7. Glass, D. E. Ozisik, M. N. and Vick. B. (1985) Hyperbolic Heat Conduction with Surface Radiation, *International Journal of Heat and Mass Transfer*, Vol. 28, pp. 1823-1830.
8. Wiggert, D. C. (1977) Analysis of Early-Time Transient Heat Conduction by Method of Characteristics, *Journal of Heat Transfer*, Vol. 99, pp. 35-40.
9. Yuen, W. W. and Lee, S. C. (1989) Non-Fourier Heat Conduction in a Semi-Infinite Solid Subjected to Oscillatory Surface Thermal Disturbances, *Journal of Heat Transfer*, Vol. 111, pp. 178-181.
10. Kao, T. T. (1977) Non-Fourier Heat Conduction in Thin Surface Layers, *Journal of Heat Transfer*, Vol. 99, pp. 343-345.
11. Glass, D. E., Ozisik, M. N., McRae, D. S. and Vick, B. (1985) On the numerical solution of hyperbolic heat conduction, *Numerical Heat Transfer*, Vol. 8, pp. 497-504.
12. Kar, A., Chan, C.L. and Mazumuder, J. (1992) Comparative Studies on Nonlinear Hyperbolic Heat Conduction for Various Boundary Conditions: Analytical and Numerical Solutions, *Journal of Heat Transfer*, Vol.114, pp.14-20.
13. Lewandowska, M. (2001) Hyperbolic Heat Conduction in the Semi-Infinite Body with a Time-Dependent Laser Heat Source, *Int. J. Heat Mass Transfer*, Vol. 37, pp.333-342.
14. Qin, T.Q. and Tien, C.L. (1993) Size Effect on Nonequilibrium Laser Heating of Metals, *Journal of Heat Transfer*, Vol.115, pp.842-847.
15. Maxwell, J. C. (1867) On the Dynamic Theory of Gases, *Philosophical Transactions*, London, Vol. 157, pp. 49-88.
16. Cattaneo, C. (1958) A Form of Heat Conduction Equation which Eliminates the Paradox of Instantaneous Propagation, *Compte Rendus*, Vol. 247, pp. 431-433.
17. Vick, B., and Ozisik, M.N. (1983) Growth and Decay of a Thermal Pulse Predicted by the Hyperbolic Heat Conduction Equation, *Journal of Heat Transfer*, Vol. 105, pp. 902-907.
18 Ozisik, M.N., and Vick, B. (1984) Propagation and Reflection of Thermal Waves in a Finite Medium, *Int. J. Heat Mass Transfer*, Vol. 27, pp. 1845-1854.
19 Tang, D.W., and Araki, N. (1997) On Non-Fourier Temperature Wave and Thermal Relaxation Time, *Int. J. Thermophys,* Vol. 18, pp. 493-504.
20. Blackwell, B. F. (1990) Temperature Profile in Semi-Infinite Body with Exponential Sources and Convective Boundary Condition, *Journal of Heat Transfer*, Vol. 112, pp. 567-571.
21. Zubair, A.S., and Aslam Chaudhry, M. (1996) Heat Conduction in a Semi-Infinite Solid due to Time-Dependent Laser Source, *Int. J. Heat Mass Transfer*, Vol. 39, pp. 3067-3074.
22. Anderson, D. A. Tannehill, J. C. and Pletcher, R. H. (1983) Computational Fluid Mechanics and Heat Transfer, Hemispherer, New York.
23. Tan, Z. M., and Yang, W.-J. (1997) Non-Fourier Heat Conduction in a Thin Film Subjected to a Sudden Temperature Change on Two Sides, *Journal of Non-Equilibrium Thermodynamics*, Vol. 22, pp. 75-87.
24. Kreith, F., and Bohn, M. S. (1986) Principles of Heat Transfer, Fourth Edition, Harper & Row, New York.

INDEX